# Ribozymes and siRNA Protocols
## Second Edition

# METHODS IN MOLECULAR BIOLOGY™

## John M. Walker, SERIES EDITOR

# Ribozymes and siRNA Protocols

*Second Edition*

Edited by

## Mouldy Sioud

*Institute of Cancer Research, The Norwegian Radium Hospital, Oslo, Norway*

**Humana Press** ✳ **Totowa, New Jersey**

This publication is printed on acid-free paper. ∞

ANSI Z39.48-1984 (American Standards Institute)

Permanence of Paper for Printed Library Materials.

Cover illustration: Inset: A newborn "green" mouse without an additional transgene (GM #1), a"green" mouse containing the pRed transgene (GM$_R$ #2), and a "silenced green" mouse with the pGtoR transgene (GM$_{GR}$ #4). *See* Fig. 3C on p. 505. Background: Micro- and macro-seeding to optimize crystal size and morphology. *See* Fig. 3B on p. 308.

Production Editor: Mark J. Breaugh.

Cover design by Patricia F. Cleary.

For additional copies, pricing for bulk purchases, and/or information about other Humana titles, contact Humana at the above address or at any of the following numbers: Tel.: 973-256-1699; Fax: 973-256-8341; E-mail: humana@humanapr.com; or visit our website: www.humanapress.com

Printed in the United States of America. 10 9 8 7 6 5 4 3 2 1

E-ISBN: 1-59259-746-7

Library of Congress Cataloging in Publication Data
Ribozymes and siRNA protocols / edited by Mouldy Sioud.-- 2nd ed.
    p. ; cm. -- (Methods in molecular biology ; v. 252)
Includes bibliographical references and index.
  ISBN 1-58829-226-6 (alk. paper)    ISSN 1064-3745
  1. Catalytic RNA--Laboratory manuals. 2. Small interfering
RNA--Laboratory manuals.
  [DNLM: 1. RNA, Catalytic--Laboratory Manuals. 2. Gene Expression
Profiling--Laboratory Manuals. 3. Gene Therapy--Laboratory Manuals. 4.
RNA, Small Interfering--Laboratory Manuals. QU 25 R486 2004] I. Sioud,
Mouldy. II. Series: Methods in molecular biology (Totowa, N.J.) ; v.
252.
  QP623.5.C36R535 2004
  572.8'8--dc22

                              2003016680

# Preface

During the last few years, new discoveries in the RNA field have opened up a wealth of opportunities to specifically target mRNA for functional analysis and target validation. Contained in this volume of *Methods in Molecular Biology* are methods useful for the design and application of ribozymes, DNAzymes, and small interfering RNAs (siRNAs). In addition, a number of illustrations aiming to facilitate the understanding and application of each method is included.

Like any new field, ribozymes and siRNA research is rapidly evolving, and so too are the methods used to study, select, express, and control their structure and in vivo biological activities. *Ribozymes and siRNA Protocols, Second Edition* focuses on the latest technical advances in interfering with gene expression. It contains a short introduction followed by general and specific protocols for using hammerhead ribozymes and derivates, DNAzymes, hairpin ribozymes, group I intron ribozymes, RNase P ribozymes, and siRNAs. Also in this volume are general methods for RNA structure analysis, delivery of oligonucleotides, and gene therapy protocols using ribozymes.

Multiple mechanisms exist by which short synthetic or expressed oligonucleotides can be used to modulate gene expression in mammalian cells. These include ribozymes, DNAzymes, and siRNAs. Although these are different strategies, all recognize their target mRNA via Watson-Crick base pairing and mediate the cleavage of cognate mRNAs. Hence, the successful use of these technologies requires good accessibility to cellular mRNA. Chapters 8, 9, 16, 19, 20, and 23 cover novel methods by which accessible sites are identified. Additionally, a PCR-based approach is used to select active siRNA in vivo (Chapter 39).

Combination of chemical and enzymatic synthesis of ribozymes and siRNAs is covered in Chapters 2 and 34, while Chapters 3–5 cover the analysis of ribozyme kinetic parameters and real-time monitoring of ribozymes and DNAzyme cleavage reactions by fluorescence detection. These chapters are followed by articles on recent technical advances in analyzing ribozyme structures and conformational transitions using nucleotide analog interference mapping and fluorescence resonance energy transfer (Chapters 6 and 7).

Using rational design and in vitro selection, a number of allosteric ribozymes that can be triggered by a variety of effectors have been engineered. Chapter 10 and 11 describe these novel molecular switches that undergo ribozyme-mediated cleavage when they bind to specific ligands. Additionally, the use of inducible promoters for switching the expression of ribozymes off or on is described in Chapter 12.

Chapters 13–18 cover the design, and in vivo and in vivo application of hammerhead ribozymes and derivates (e.g., maxizymes). Also included is functional gene discovery using hybrid ribozyme libraries that are expected to facilitate functional genomic studies.

A group of three chapters cover the DNAzyme technology. These include design rules, target site selection, and mutation analysis. The next four chapters are on hairpin ribozymes describing, first, methods for crystallization and, second, selection of effective target sites and hairpin ribozymes for therapeutic application (Chapters 22–25).

Group I intron and RNAse P ribozyme protocols are covered in Chapters 26–32. These seven chapters describe protocols for effective design, selection, and therapeutic applications. The use of ribozymes and DNAzymes in rodent models of human disease is described in Chapter 33.

A block of nine chapters (34–42) address the latest RNAi technology. This has emerged as a powerful technique for sequence-specific gene silencing in a wide variety of organisms. The chapters cover a wide range of methods and application of siRNAs. These include potential design rules, production, target site selection, plasmid/viral expression vectors, and knockout animals. These methods should benefit not only those experienced in siRNAs, but also those applying this technology for the first time.

Delivery agents from different companies and methods for the selection of cell binding peptides for specific delivery of oligonucleotides into cancer cells are described in Chapters 43 and 44. The book ends with clinical and gene therapy protocols using ribozymes (Chapters 45 and 46). Such protocols are also relevant for the future use of siRNA in humans.

It has been an honor to work with each of the authors in assembling this compilation of protocols and procedures. I congratulate them all on their achievement in making their methods available, and hope you will find *Ribozymes and siRNA Protocols, Second Edition* a useful reference easily used at the lab bench.

*Mouldy Sioud*

# Contents

# Contributors

ROGER ABOUNADER • *Department of Neurology, Johns Hopkins University School of Medicine and Kennedy Krieger Research Institute, Baltimore, MD*

HIDEO AKASHI • *JSPS Research Fellowships for Young Scientists; Gene Function Research Center, National Institute of Advanced Industrial Science and Technology (AIST), Tsukuba, Japan*

RAFAEL G. AMADO • *Department of Medicine and UCLA AIDS Institute, University of California, Los Angeles, CA*

AKHIL C. BANERJEA • *Laboratory of Virology, National Institute of Immunology, New Dehli, India*

ALICIA BARROSO-DELJESUS • *Instituto de Parasitología y Biomedicina "López-Neyra," CSIC, Granada, Spain*

SOUMITRA BASU • *Center for Pharmacogenetics, Department of Pharmaceutical Sciences, School of Pharmacy, University of Pittsburgh, Pittsburgh, PA*

ALFREDO BERZAL-HERRANZ • *Instituto de Parasitología y Biomedicina "López-Neyra," CSIC, Granada, Spain*

J. MICHAEL BISHOP • *G. W. Hooper Foundation, and the Department of Microbiology and Immunology, University of California, San Francisco, CA*

JACQUES BOLARD • *Laboratoire de Physicochimie Biomoléculaire et Cellulaire, Université Pierre & Marie Curie, Paris, France*

EMMA T. BOWDEN • *Department of Oncology and the Lombardi Comprehensive Cancer Center, Georgetown University Medical School, Washington DC*

MAUREEN P. BOYD • *Johnson & Johnson Research Pty Limited, Eveleigh Sydney, Australia*

RONALD R. BREAKER • *Department of Molecular, Cellular and Developmental Biology, Yale University, New Haven, CT*

MURRAY J. CAIRNS • *Johnson and Johnson Research Laboratories, Australian Technology Park, Eveleigh, Australia*

DANIELA CASTANOTTO • *Department of Biology and Molecular Biology, Beckman Research Institute of the City of Hope, Duarte, CA*

SAMITABH CHAKRABORTI • *Laboratory of Virology, National Institute of Immunology, New Dehli, India*

KAREN CHAN • *Division of Infectious Diseases, School of Public Health, University of California, Berkeley, CA*

HSIU-HUA CHEN • *Department of Biology and Molecular Biology, Beckman Research Institute of the City of Hope, Duarte, CA*

GOPESWAR CHOWDHURY • *Center for Pharmacogenetics, Department of Pharmaceutical Sciences, School of Pharmacy, University of Pittsburgh, Pittsburgh, PA*

GARY A. CLAWSON • *Departments of Pathology & Biochemistry and Molecular Biology, The Gittlen Cancer Research Institute, Pennsylvania State University, Hershey, PA*

*xi*

HARRY DIETZ • *Department of Pediatrics, Institute of Genetic Medicine, John Hopkins University School of Medicine and Howard Hughes Medical Institute, Baltimore, MD*

MICHELLE DOBSON • *Department of Biological Sciences, Northern Illinois University, DeKalb, IL*

WALTER DUNN • *Division of Infectious Diseases, School of Public Health, University of California, Berkeley, CA*

CHRISTER EINVIK • *Department of Molecular Biotechnology - RNA Research Group, Institute of Medical Biology, University of Tromsø, Tromsø, Norway*

MARTHA J. FEDOR • *Department of Molecular Biology and the Skaggs Institute for Chemical Biology, The Scripps Research Institute, San Diego, CA*

ADRIAN R. FERRÉ-D'AMARÉ • *Division of Basic Sciences, Fred Hutchinson Cancer Research Center, Seattle, WA*

TONJE FISKAA • *Department of Molecular Biotechnology - RNA Research Group, Institute of Medical Biology, University of Tromsø, Tromsø, Norway*

JASON J. FRITZ • *Department of Molecular Genetics and Microbiology, College of Medicine, University of Florida, Gainesville, FL*

RAJESH K. GAUR • *Graduate School of Biological Sciences, City of Hope National Medical Center and Beckman Research Institute, Duarte, CA*

ANDREI GOGA • *Division of Hematology/Oncology, Department of Medicine, University of California, San Francisco, CA*

MARINA GORBATYUK • *Department of Molecular Genetics and Microbiology, College of Medicine, University of Florida, Gainesville, FL*

ARNOLD HAMPEL • *Department of Biological Sciences, Northern Illinois University, DeKalb, IL*

ANDREAS HANNE • *artus GmbH, Hamburg, Germany*

HIDETOSHI HASUWA • *Genome Information Research Center, Osaka University, Osaka, Japan*

WILLIAM W. HAUSWIRTH • *Department of Molecular Genetics and Microbiology, Powell Gene Therapy Centre, Department of Ophthalmology, College of Medicine, University of Florida, Gainesville, FL, USA*

LINDA HOSTALEK • *Department of Biological Sciences, Northern Illinois University, DeKalb, IL*

JASON HU • *Sophie Davis School of Biomedical Education, City University of New York Medical School, New York, NY*

PER OLE IVERSEN • *Institute for Nutrition, University of Oslo, Oslo, Norway*

MAYU IYO • *Department of Chemistry and Biotechnology, School of Engineering, The University of Tokyo, Tokyo, Japan*

STEINAR JOHANSEN • *Department of Molecular Biotechnology - RNA Research Group, Institute of Medical Biology, University of Tromsø, Tromsø, Norway*

YOSHIO KATO • *Department of Chemistry and Biotechnology, School of Engineering, The University of Tokyo, Tokyo, Japan*

HIROAKI KAWASAKI • *Department of Chemistry and Biotechnology, School of Engineering, The University of Tokyo, Tokyo, Japan*

KAMEL KHALILI • *Center for Neurovirology and Cancer Biology, Temple University, Philadelphia, PA*

KIHOON KIM • *Division of Infectious Diseases, School of Public Health, University of California, Berkeley, CA*

DAGMAR KLEIN • *Diabetes Research Institute, University of Miami School of Medicine, Miami, FL*

PAUL E. KLOTMAN • *Department of Medicine, Mt. Sinai Medical Center, New York, NY*

YASUO KOMATSU • *DNA Chip Reseach Inc., Yokohama, Japan*

GUIDO KRUPP • *artus GmbH, Hamburg, Germany*

TOMOKO KUWABARA • *Gene Function Research Laboratory, National Institute of Advanced Industrial Science and Technology (AIST), Tsukuba, Japan*

JOHN LATERRA • *Departments of Neurology, Neuroscience and Oncology, Johns Hopkins University School of Medicine and Kennedy Krieger Research Institute, Baltimore, MD*

JEANNE LEBON • *Department of Biology and Molecular Biology, Beckman Research Institute of the City of Hope, Duarte, CA*

MARIANNE LEIRDAL • *Department of Immunology, Molecular Medicine Group, The Norwegian Radium Hospital, Oslo, Norway*

ALFRED S. LEWIN • *Department of Molecular Genetics and Microbiology, Powell Gene Therapy Centre, College of Medicine, University of Florida, Gainesville, FL*

DAVID M. J. LILLEY • *Cancer Research UK Nucleic Acid Structure Research Group, Department of Biochemistry, MSI/WTB Complex, The University of Dundee, Dundee, UK*

FENYONG LIU • *Division of Infectious Diseases, School of Public Health, University of California, Berkeley, CA*

EIRIK W. LUNDBLAD • *Department of Molecular Biotechnology - RNA Research Group, Institute of Medical Biology, University of Tromsø, Tromsø, Norway*

JANET L. MACPHERSON • *Johnson & Johnson Research Pty Limited, Eveleigh Sydney, Australia*

JONATHAN D. MARMUR • *Department of Medicine, Mt. Sinai Medical Center, New York, NY*

RONALD T. MITSUYASU • *Department of Medicine and UCLA AIDS Institute, University of California, Los Angeles, CA*

MAKOTO MIYAGISHI • *Department of Chemistry and Biotechnology, School of Engineering, The University of Tokyo, Tokyo, Japan*

ROBERT MONTGOMERY • *Department of Surgery, Johns Hopkins University School of Medicine, Baltimore, MD*

FRANCES K. NGOK • *Department of Medicine and UCLA AIDS Institute, University of California, Los Angeles, CA*

EIKO OHTSUKA • *National Institute of Advanced Industrial Science and Technology, Sapporo, Japan*

MASARU OKABE • *Genome Information Research Center, Osaka University, Osaka, Japan*

WEI-HUA PAN • *Departments of Pathology & Biochemistry and Molecular Biology, The Gittlen Cancer Research Institute, Pennsylvania State University, Hershey, PA*

RICARDO L. PASTORI • *Diabetes Research Institute, University of Miami School of Medicine, Miami, FL*

CATHERINE PAZSINT • *Department of Pharmaceutical Sciences, Center for Pharmacogenetics, School of Pharmacy, University of Pittsbursgh, Pittsburgh, PA*

LEONIDAS A. PHYLACTOU • *Laboratory of Molecular Function & Therapy, The Cyprus Institute of Neurology and Genetics, Nicosia, Cyprus*

ELENA PUERTA-FERNÁNDEZ• *Instituto de Parasitología y Biomedicina "López-Neyra," CSIC, Granada, Spain*

STEPHEN RAJ • *Division of Infectious Diseases, School of Public Health, University of California, Berkeley, CA*

JAY RAPPAPORT • *Center for Neurovirology and Cancer Biology, Temple University, Philadelphia, PA*

SVEN N. RESKE • *Department of Nuclear Medicine, University Hospital of Ulm, Ulm, Germany*

MAX W. RICHARDSON • *Center for Neurovirology and Cancer Biology, Temple University, Philadelphia, PA*

CAMILLO RICORDI • *Diabetes Research Institute, University of Miami School of Medicine, Miami, FL*

ANNA T. RIEGEL • *Department of Oncology and the Lombardi Comprehensive Cancer Center, Georgetown University Medical School, Washington DC*

ARTHUR D. RIGGS • *Biology Division, Beckman Research Institute of the City of Hope, Duarte, CA*

CRISTINA ROMERO-LÓPEZ • *Instituto de Parasitología y Biomedicina "López-Neyra," CSIC, Granada, Spain*

JOHN J. ROSSI • *Department of Biology and Molecular Biology, Beckman Research Institute of the City of Hope, Duarte, CA*

ADAM ROTH • *Department of Molecular, Cellular and Developmental Biology, Yale University, New Haven, CT*

PETER B. RUPERT • *Division of Basic Sciences, Fred Hutchinson Cancer Research Center, Seattle, WA*

MASAYUKI SANO • *Gene Function Research Center, National Institute of Advanced Industrial Science and Technology (AIST), Tsukuba, Japan*

OLIVIER SEKSEK • *Laboratoire de Physicochimie Biomoléculaire et Cellulaire, Université Pierre & Marie Curie, Paris, France*

MOHSEN SHADIDI • *Department of Immunology, Molecular Medicine Group, The Norwegian Radium Hospital, Oslo, Norway*

CHANGXIAN SHEN • *Department of Nuclear Medicine, University Hospital of Ulm, Ulm, Germany.*

RICHARD SHIPPY • *Department of Biological Sciences, Northern Illinois University, DeKalb, IL*

KUMUD K. SINGH • *Department of Pediatrics, Division of Infectious Diseases, University of California, San Diego, CA*

MOULDY SIOUD • *Department of Immunology, Molecular Medicine Group, The Norwegian Radium Hospital, Oslo, Norway*

ANDREW SIWKOWSKI • *Department of Biological Sciences, Northern Illinois University, DeKalb, IL*

DAG R. SØRENSEN • *Department of Comparative Medicine, Rikshospitalet, University of Oslo, Norway*

BANDI SRIRAM • *Laboratory of Virology, National Institute of Immunology, New Dehli, India*

LUN-QUAN SUN • *Johnson and Johnson Research Laboratories, Australian Technology Park, Eveleigh, Australia*

GEOFF P. SYMONDS • *Johnson & Johnson Research Pty Limited, Eveleigh Sydney, Australia*

KAZUNARI TAIRA • *Department of Chemistry and Biotechnology, School of Engineering, The University of Tokyo, Tokyo, Japan*

ALISON V. TODD • *Johnson & Johnson Research Pty Limited, Eveleigh Sydney, Australia*

PHONG TRANG • *Division of Infectious Diseases, School of Public Health, University of California, Berkeley, CA*

MASARU TSUNEMI • *Department of Chemistry and Biotechnology, School of Engineering, The University of Tokyo, Tokyo, Japan*

MASAKI WARASHINA • *Gene Function Research Laboratory, National Institute of Advanced Industrial Science and Technology (AIST), Tsukuba, Japan*

DUN YANG • *G. W. Hooper Foundation, Department of Microbiology and Immunology, University of California, San Francisco, CA*

HUA ZOU • *Division of Infectious Diseases, School of Public Health, University of California, Berkeley, CA*

# 1

## Ribozyme- and siRNA-Mediated mRNA Degradation

*A General Introduction*

**Mouldy Sioud**

## 1. Introduction

A number of recent discoveries in the RNA field have opened up a wealth of opportunities to specifically target mRNA for the development of therapeutics and/or the elucidation of gene function. Novel agents such as ribozymes, DNAzymes, and siRNAs are emerging as effective strategies that are antigene agents *(1)*.

Ribozymes are naturally occurring RNA sequences with catalytic activity *(2–4)*. For *trans*-cleaving RNAs such as the hammerhead and hairpin ribozymes, the cleaved RNA can dissociate from the ribozyme, and thereby allow turnover for signal amplification *(5)*. Using in vitro selection protocols, DNAzymes capable of cleaving mRNAs were selected from a random library of oligonucleotides, and shown to be a versatile tool for gene inactivation *(6)*. Recently, the well-preserved phenomenon known as RNA interference (RNAi) has become a powerful technique for sequence-specific gene silencing in a wide variety of cells and organisms *(7)*. This short introduction provides a brief description of ribozymes, DNAzymes, RNA interference, and delivery agents, which are described in subsequent chapters.

### 1.1. Hammerhead Ribozyme

The hammerhead-type ribozyme was originally discovered as a self-cleaving RNA molecule in certain plant viroids and satellite RNAs *(8)*. Naturally, this ribozyme is used during the rolling-circle replication, which involves a self-cleaving pathway also known as *cis*-reaction. Intermolecular cleavage in a *trans* reaction was achieved by dividing the domain into ribozyme and substrate fragments *(9,10)*. This novel *trans*-acting hammerhead ribozyme contains three

From: *Methods in Molecular Biology, vol. 252: Ribozymes and siRNA Protocols, Second Edition*
Edited by: M. Sioud © Humana Press Inc., Totowa, NJ

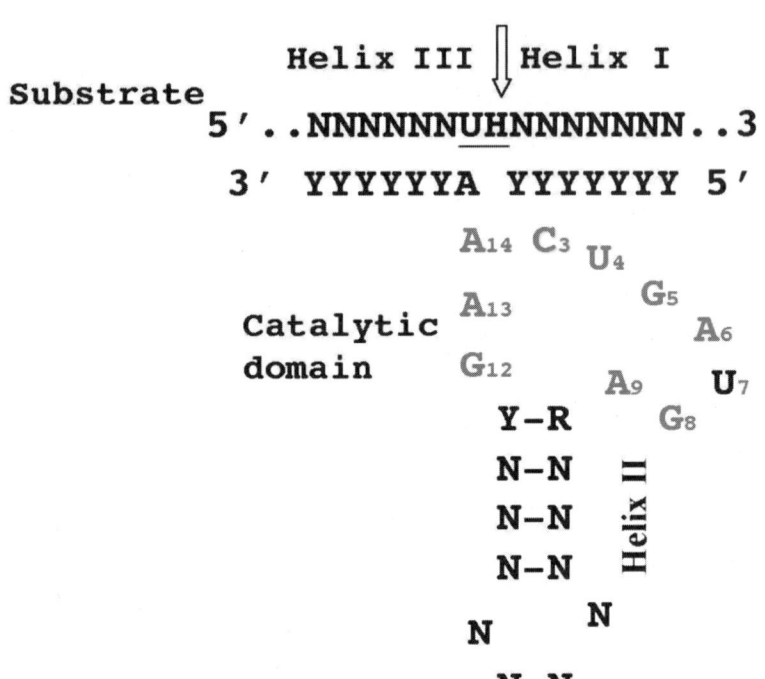

Fig. 1. Secondary structure of the hammerhead ribozyme/RNA target complex. Gray sequences are conserved. Nucleotides numbering is according to **ref. 24**. The arrow indicates the cleavage site.

helical stems—I, II, and III—which flank the nine conserved bases of the catalytic core (**Fig. 1**). The core sequence is believed to be involved in the formation of the tertiary structure necessary for cleavage, whereas the 5' and the 3' antisense arms that form stem I and III, respectively, define the ribozyme cleavage specificity. The cleavage site is a 5'-UH-3' sequence in which H is any nucleotide except G. However, the identification of active sites can be influenced by RNA structure and other factors that can be easily resolved (*see* Chapters 8, 9, and 16). The cleavage reaction proceeds through an in-line SN2 mechanism in which the 2'-hydroxyl group of the substrate cleavage site is the initiating nucleophile. The ribozymes can be chemically synthesized or intracellulary expressed from recombinant vectors. Aside from their general interest in structural and mechanistic studies (*see* Chapters 3–7), hammerhead ribozymes have been used for various biological applications such as the regulation of gene expression (*see* Chapters 12–15 and 17).

New versions of minimized hammerhead ribozymes—so-called maxizymes—were also engineered (*11*). They can form active conformation

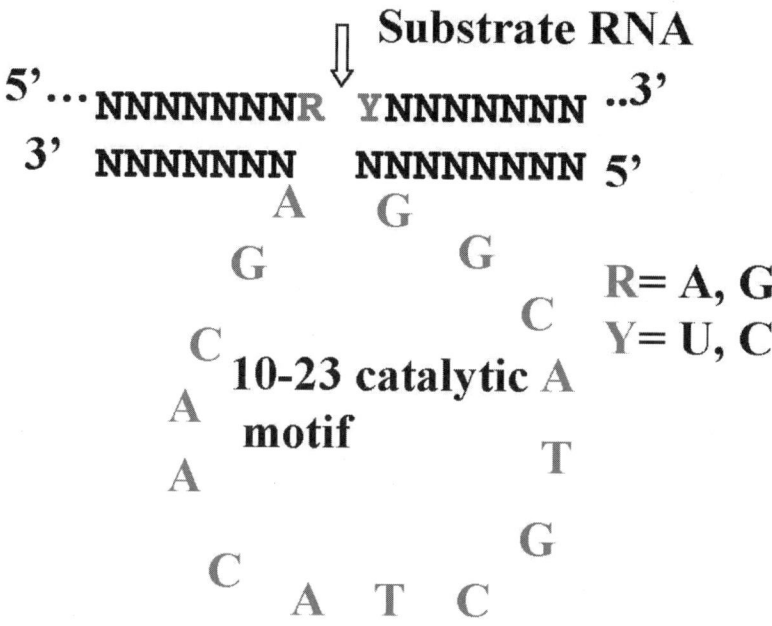

Fig. 2. Sequence and secondary structure of the 10–23 DNAzyme.

only when they specifically bind to two target sites (*see* Chapter 18). Recently, molecular engineering efforts have demonstrated that ligand-dependent ribozymes (allosteric ribozymes) that respond to the intended targets with high specificities can be designed *(12)*. Thus, in vitro and in vivo ribozyme function can now be controlled (*see* Chapters 10 and 11).

### 1.2. DNAzyme

Since the discovery of RNA catalysis, various combinatorial and rational design strategies have been used to expand the type of chemical reactions catalyzed by nucleic acids. As a result, a new generation of ribozymes known as artificial ribozymes have been discovered. Using the in vitro selection strategy, Sontoro and Joyce *(6)* have selected DNA sequences that are capable of sequence-specific cleavage of mRNA. The 10–23 DNAzyme can recognize RNA through Watson-Crick basepairing, and cleaves its target at a phosphodiester bond located between an unpaired purine and paired pyrimidine. This consensus sequence is frequently found in mRNAs (**Fig. 2**). The mechanism of cleavage is similar to that of the hammerhead ribozyme. DNAzymes have been also shown to be susceptible to engineered ligand sensitivity. Chapters 19–21 detail the design, target selection, and application of the DNAzymes.

### 1.3. Hairpin Ribozyme

The hairpin ribozyme is found in the negative strand of the satellite RNA of tobacco ringspot virus and chicory yellow mottle virus *(13)*. It has four helical domains and five loops. These ribozymes can be engineered to cleave *in trans* heterogonous RNAs *(14)*. The cleavage site has the sequence 5'–XN*GUC-3', in which X is any base except A, N is any base, and * denotes the site of cleavage. The GUC triplet is required. Recently, the development of optimized hairpin ribozymes for cleaving mRNA *in trans* has generated considerable interest. Recent methods for analyzing the hairpin ribozyme structure, target-site selection, and application as antigene agents are described in Chapters 22–25.

### 1.4. Group I Intron Ribozyme

The *Tetrahymena* group I intron is the best-characterized example of naturally occurring ribozymes. In the presence of guanosine cofactor, the intron is excised and two exons are ligated. In addition to many important RNA stem structures, an important RNA element is the internal guide sequence (IGS), which is located at the 5' end of the intron *(2)*. This sequence defines the specificity of the ribozyme. As for the naturally occurring hammerhead ribozyme, the group I intron ribozyme was modified to perform the reaction *in trans (15)*. In this respect, it *trans*-splices a part of an mRNA linked to its 3' end onto a separate 5' target RNA through a two-step *trans*-splicing reaction. Therefore, the ribozyme can be used as an RNA repair of somatic mutations on the mRNA level. Examples of such medical applications are described in Chapters 26 and 27.

### 1.5. RNase P Ribozyme

Ribonuclease P (RNase P) is a ubiquitous enzyme that cleaves the 5' leader sequences of pre-tRNA to generate mature tRNAs. RNase P contains two components: a RNA moiety and a protein moiety. The RNA moiety has been found to be a catalyst *(3)*. Notably, the enzyme can recognize and process all types of tRNA precursors, among which there is no sequence homology around the cleavage site. However, the cleavage reaction requires RNA-RNA basepairing interactions between nucleotides near the cleavage site and a guide sequence that can either be part of the substrate molecule (as in unprocessed tRNA) or be provided by an unattached, short ribonucleotide that is complementary to nucleotides adjacent to the cleavage site. Based upon this structure requirement, external guide sequences (EGSs) were designed *(16)*. When complexed with target RNA, they generate a structure RNA that is susceptible to cleavage by RNase P. The latest improvements of RNase P ribozyme design, EGS selection, and application are described in Chapters 28–32.

## 1.6. RNA Interference and siRNAs

RNA interference (RNAi) is a newly discovered cellular pathway in which double-stranded RNA (dsRNA) induces the degradation of its cognate mRNA in a wide variety of organisms (for review, *see 7*). In this process, the double-stranded RNA is recognized by an RNase III nuclease, which processes the dsRNA into small interfering (siRNAs) of 21–23 nt (*see* **Fig. 3**). siRNAs are incorporated into the RNA interfering silencing complex (RISC), which contains the proteins needed to unwind the double-stranded siRNA and cleave the target mRNAs at the site where the antisense RNA are bound *(7)*. However, in mammalian somatic cells, long dsRNAs (>30 nt) activate the interferon responses that are mainly mediated via the activation of a dsRNA-dependent protein kinase (PKR) and 2', 5'-oligoadenylate synthetase (**Fig. 3**).

Recently, it was demonstrated that small synthetic duplexes of 21–23-nt siRNAs have gene-specific silencing function in vitro and in vivo *(17,18)*. In contrast to long double-stranded RNA, in somatic mammalian siRNAs can bypass the activation of PKR. The technology has been rapidly adapted for silencing gene expression in vitro and in vivo, and new vectors for siRNAs expression have been designed *(19–21)*. Chapters 34–42 detail the design, production, and expression of siRNAs in mammalian cells. A protocol for the generation of transgenic mouse lines expressing active siRNAs is also included (*see* Chapter 38).

## 1.7. Delivery

The development of efficient methods for introducing ribozymes, DNAzymes, and siRNA into mammalian cells could be the key element in treating genetic and acquired disease. There are two types of nucleic acid delivery: endogenous and exogenous delivery. Both strategies have advantages and disadvantages.

### 1.7.1. Exogenous Delivery

This strategy involves the in vitro synthesis of the molecules and their delivery to cells. Since the cell membrane presents a substantial barrier to the entry of highly charged, high-mol-wt molecules, delivering these into the cytoplasm is a major challenge. To overcome this problem, many transfection techniques have been used, including electroporation, microinjection, and cationic liposome-mediated transfection. Notably, exogenous delivery offers the possibility to develop compounds with a therapeutic potential that can be applied locally or systemically (*see* Chapter 33). In addition, when nucleic acids are made synthetically, a variety of chemical modifications can be introduced to

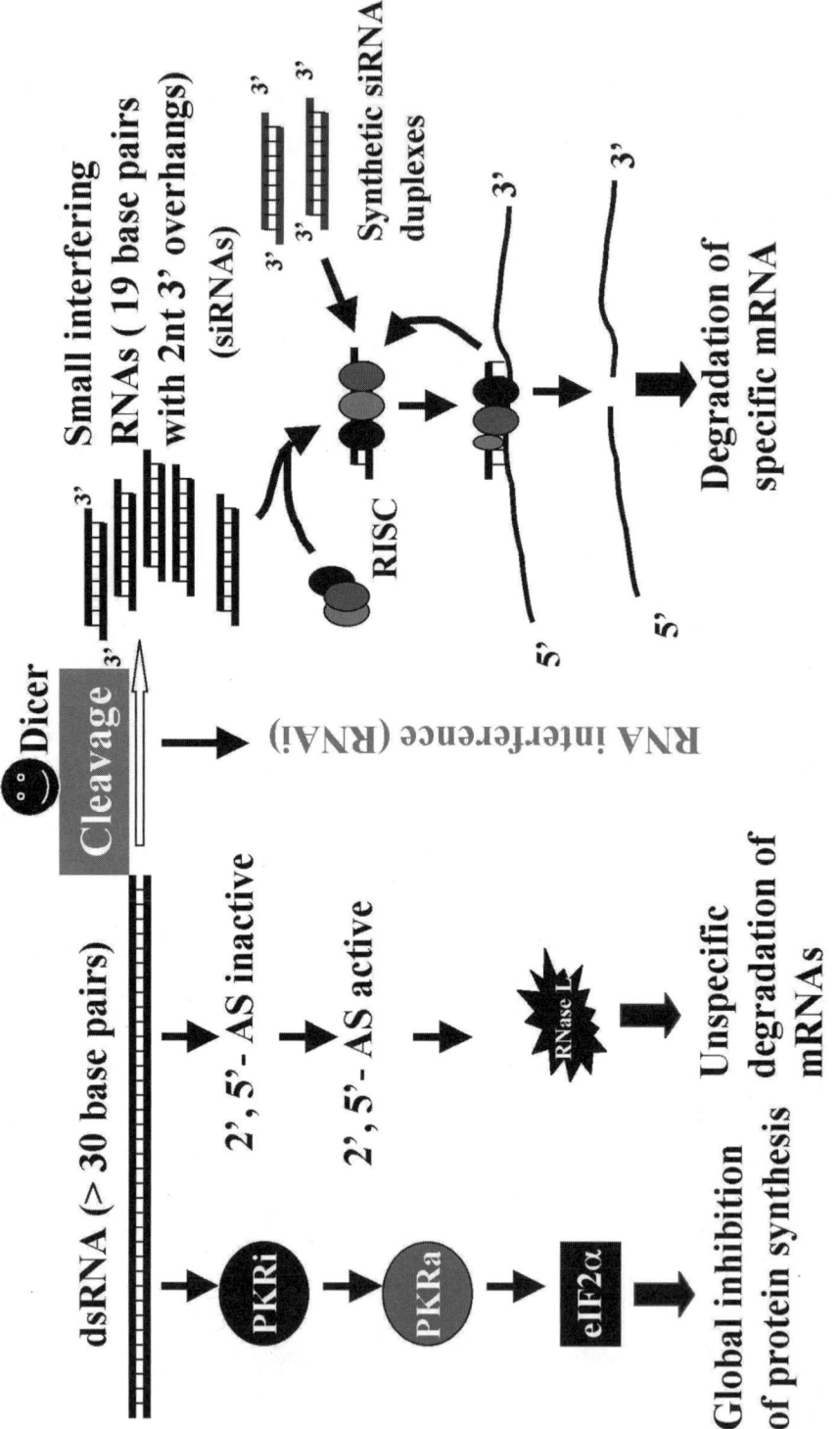

Fig. 3. Gene silencing by small interfering RNAs (siRNA).

increase their half-life in biological fluids. Chapter 43 describes a variety of delivery agents that are suitable for both DNA and RNA oligonucleotides.

### 1.7.2. Endogenous Delivery

Endogenous delivery of ribozymes and siRNA involves the cloning of these molecules into viral or non-viral vectors behind a suitable promoter. The major advantages of the endogenous application of ribozymes and siRNAs are related to their continual expression. In addition, the expression can be switched on and off when inducible promoters are used (*see* Chapter 12). However, when ribozymes are expressed intracellularly, the vector-derived transcribed sequences that usually flank the ribozyme sequence may interfere with the ribozyme structure and activity (*see* Chapters 14 and 15). Therefore, appropriate vectors should be used. In regard to siRNAs, U6, and H1 promoters were found to be the vector of choice (*see* Chapters 36, 38, and 41).

### 1.7.3. Specific Delivery

Gene therapy is currently limited by the difficulty of achieving efficient delivery into target cells. Thus, there is a need for developing cell- or tissue-specific delivery agents. Selective delivery of nucleic acids such as antisense, ribozymes, and siRNAs would improve their efficacy and minimize potential adverse side effects. Recently, cell surface-binding peptides were found to be useful agents for targeting cancer cells (*22,23*). The selection of such peptide is detailed in Chapter 44.

## References

1. Sioud, M. (2001) Nucleic acid enzymes as a novel generation of anti-gene agents. *Curr. Mol. Med.* **1,** 575–588.
2. Zaug, A. J., Been, M. D., and Cech, T. R. (1986) The *Tetrahymena* ribozyme acts like an RNA restriction endonuclease. *Nature* **324,** 429–433.
3. Guerrier-Takada, C., Gardiner, K., Marsh, T., Pace, N., and Altman, S. (1983) The RNA moiety of ribonuclease P is the catalytic subunit of the enzyme. *Cell* **35,** 849–857.
4. Forster, A. C. and Altman, S. (1990) External guide sequences for an RNA enzyme. *Science* **249,** 783–786.
5. Symons, R. H. (1994) Ribozymes. *Curr. Opin. Struct. Biol.* **4,** 322–330.
6. Santoro, S. W. and Joyce, G. F. (1996). A general purpose RNA-cleaving DNA enzyme. *Proc. Natl. Acad. Sci. USA* **94,** 4264–4266.
7. Hannon, G. J. (2002) RNA interference. *Nature* **418,** 244–251.
8. Forster, A. C. and Symons, R. H. (1987) Self-cleavage of plus and minus RNAs of a virusoid and a structural model for the active sites. *Cell* **49,** 211–220.
9. Uhlenbeck, O. C. (1987) A small catalytic oligoribonucleotide. *Nature* **328,** 596–600.

10. Haseloff, J. and Gerlach, W. L. (1988) Simple RNA enzymes with new and highly specific endoribonuclease activities. *Nature* **334,** 585–591.
11. Kuwabara, T., Warashina, M., Orita, M., Koseki, S., Ohkawa, J., and Taira, K. (1998) Formation *in vitro* and in cells of a catalytically active dimmer by tRNA^val^-driven short ribozymes. *Nat. Biotechnol.* **16,** 961–965.
12. Breaker, R.R. (2002) Engineered allosteric ribozymes as biosensor components. *Curr. Opin. Biotechnol.* **13,** 31–39.
13. Hampel, A. and Tritz, R. (1989) RNA catalytic properties of the minimum (-) sTRSV sequence. *Biochemistry* **28,** 4929–4933.
14. Berzal-Herranz, A., Joseph, S., Chowrira, B. M., Butcher, S. E., and Bruke, J. M. (1993) Essential nucleotide sequences and secondary structure elements of the hairpin ribozyme. *EMBO J.* **12,** 2567–2574.
15. Sullenger, B.A. and Cech, T.R. (1994) Ribozyme-mediated repair of defective mRNA by targeted, trans-splicing. *Nature* **371,** 619–622
16. Foster, A. C. and Altman, S. (1990) External guide sequences for an RNA enzyme. *Science* **249,** 783–786.
17. Elbashir, S. M., Harborth, J., Lendeckel, W., Yalcin, A., Weber, K., and Tuschl, T. (2001) Duplexes of 21-nucleotide RNAs mediate RNA interference in cultured mammalian cells. *Nature* **411,** 494–498.
18. Caplen, N. J., Parrish, S., Imani, F., Fire, A., and Morgan, R. A. (2001) Specific inhibition of gene expression by small double-stranded RNAs in invertebrate and vertebrate systems. *Proc. Natl. Acad. Sci. USA* **98,** 9742–9747.
19. Brummelkamp, T. R., Bernards, R., and Agami, R. (2002) A system for stable expression of short interfering RNAs in mammalian cells. *Science* **296,** 550–553.
20. Miyagishi, M. and Taira, K. (2002) U6 promoter-driven siRNAs with four uridine 3' overhangs efficiently suppress targeted gene expression in mammalian cells. *Nat. Biotechnol.* **20,** 497–501.
21. Lee, N. S., Dohjima, T., Bauer, G., Li, H., Li, M.-J., Ehsani, A., et al. (2002) Expression of small interfering RNAs targeted against HIV-1 *rev* transcripts in human cells. *Nat. Biotechnol.* **20,** 500–505.
22. Arap, W., Pasqualini, R., and Ruoslahti, E. (1988) Cancer treatment by targeted drug delivery to tumor vasculature in a mouse model. *Science* **279,** 377–380.
23. Shadidi, M. and Sioud, M. (2003) Identification of novel carrier peptides for the specific delivery of therapeutics into cancer cells. *FASEB J.* **17,** 256–258.
24. Hertel, K. J., Pardi, A., Uhlenbeck, O. C., Koizumi, M., Ohtsuka, E., Uesugi, S., et al. (1992) Numbering system for the hammerhead. *Nucleic Acids Res.* **20,** 3252.

# 2

## Combination of Chemical and Enzymatic RNA Synthesis

### Rajesh K. Gaur, Andreas Hanne, and Guido Krupp

#### Summary

The potential of standard in vitro transcription reactions can be dramatically expanded, if chemically synthesized low-mol-wt compounds are used as building blocks in combination with standard nucleotide 5' triphosphates (NTPs). Short oligonucleotides that terminate in guanosine effectively compete with guanosine 5' triphosphate (GTP) as starter building blocks, and they are incorporated at the 5'-end of transcripts. Applications include production of RNAs with "unfriendly 5'-ends" (they do not begin with G), variations of the 5'-sequence are possible with the same DNA template, site-specific insertion of nucleotide modifications, and addition of 5'-labels, such as fluorescein for detection or biotin for capture. Clearly, chemically synthesized, modified NTPs are inserted at internal sites. The combination with phosphorothioate linkages for detection has been developed into a powerful high-throughput method to study site-specific interference of modifications with RNA function.

**Key Words:** Biotin; digoxygenin; FAM; fluorescence; initiator oligonucleotide; 5'-label; modification; mutation; NAIM; nonradioactive; 5'-$^{32}$P-label; phosphorothioate.

## 1. Introduction

In vitro transcription reactions with bacteriophage RNA polymerases (SP6, T3, and now, most used T7) have been developed into a very powerful technique to produce large quantities of long RNA molecules. Although all effective DNA templates include the homologous double-stranded promoter, the template types vary from the standard transcription plasmid to specifically designed PCR products and to mostly single-stranded templates, containing only the promoter in double stranded form *(1)*. The power of this technology can be dramatically expanded by combining chemical synthesis of low-mol-wt compounds with standard NTPs as RNA building blocks.

From: *Methods in Molecular Biology, vol. 252: Ribozymes and siRNA Protocols, Second Edition*
Edited by: M. Sioud © Humana Press Inc., Totowa, NJ

The discovery that short, synthetic oligonucleotides, terminating with guanosine, effectively compete with GTP as starter building blocks enables the convenient and precise manipulation of the 5'-proximal section of RNA transcripts. Otherwise, this is only possible in the complete chemical RNA synthesis that is limited to short lengths. Applications of these so-called initiator oligonucleotides *(2)* include: i) overcoming the limitation that in vitro transcripts must begin with G, ii) variations of the 5'-sequence without the need for a series of different templates, iii) site-specific insertion of nucleotide modifications, iv) direct 5'-labeling during the transcription reaction with fluorescein for detection or with biotin for capture, and v) the direct production of transcripts with 5'-OH, for simplified and very effective 5'-$^{32}$P-labeling, avoiding removal of the recalcitrant 5'-triphosphate.

Chemically synthesized, modified NTPs offer a wide range, and are clearly inserted at many internal sites. The combination with phosphorothioate linkages for detection has been developed into a powerful high-throughput method to study site-specific interference of modifications with RNA function *(3,4)*.

## 2. Materials

1. Template DNA.
2. 10X transcription buffer: 400 m$M$ Tris-HCl, pH 8.0, 200 m$M$ MgCl$_2$, 20 m$M$ spermidine.
3. Ribonucleoside triphosphates (NTP): a solution containing each NTP (adenosine 5' triphosphate [ATP], cyndine 5' triphosphate [CTP], guanosine 5' triphosphate [GTP], uridine 5' triphosphate [UTP]) at 10 m$M$.
4. 100 m$M$ dithiothreitol (DTT); do not autoclave.
5. RNase-inhibitor RNasin from human placenta (e.g., Fermentas, Roche, Promega).
6. 50% (w/v) Polyethylene glycol (PEG), M$_r$ 6000; can be autoclaved.
7. 0.1% Triton X-100 (Roche); do not autoclave.
8. T7 RNA polymerase or other appropriate phage RNA polymerase (Fermentas, NE-Biolabs, Roche).
9. Optional: [α-$^{32}$P]-UTP (Amersham, ICN, Hartmann-Analytic).
10. 4 $M$ ammonium acetate, 20 m$M$ ehtylenediaminetetraacetic acid (EDTA). Adjust to pH 7.0, autoclave.
11. Cold ethanol, p.a. (stored at −20°C).
12. Equipment for polyacrylamide gel electrophoresis and elution.
13. If appropriate: Replace **items 2–8** by High-yield transcription kit, e.g., AmpliScribe (Epicentre), MEGAscript, or MEGAshortscript (Ambion).
14. Initiator oligonucleotides—a wide range is commercially available (e-mail: krupp@artus-biotech.com). Fluorescent-labeled materials should be stored in the dark (wrapped in aluminum foil). Purity is a very important issue, since these short oligos are difficult to separate from work-up products from the chemical synthesis.
15. Modified NTPαS: a wide range is commercially available (e-mail: krupp@artus-biotech.com).

## 3. Methods

### 3.1. Overcoming the Limitation of Standard Protocols: In Vitro Transcription of RNAs That Have No Guanosine as Their 5'-End

Commercially available transcription systems with bacteriophage RNA polymerases (T7, T3, or SP6) all require guanosine as the 5'-terminal first nucleotide in the transcript. Frequently, functional RNA molecules do not start with G—for example, many tRNAs.

One approach to overcome this limitation is the introduction of a ribozyme structure that cleaves the primary transcript and liberates the desired 5'-end *(5,6)*. Based on our previously published observations *(2)*, we present a simple protocol if the desired RNAs have a G at least near the required 5'-end, at the second, third, or fourth nucleotide.

For this purpose, the template DNA codes for a transcript beginning with the first G in your RNA. The in vitro transcription reactions are performed as usual, but in addition, a short "initiator oligonucleotide" is added. This oligonucleotide contains your desired 5'-sequence ending at the first G. If preferred, the oligonucleotide may already contain a 5'-phosphate, and a schematic example would be:

| | |
|---|---|
| 5'-terminal sequence of desired RNA | 5'-CAGGCCAGUAAA....... |
| template-encoded transcript | 5'-pppGGCCAGUAAA....... |
| in vitro transcript with the trinucleotide (p)CAG | 5'-(p) CAGGCCAGUAAA....... |

The incorporation efficiencies listed in **Table 1** were obtained using a twofold molar excess of the initiator oligonucleotide over GTP that competes as an initiator in the transcription reaction. Example results are shown in **Fig. 1**. Reactions can be performed with all four NTPs at the same concentration—e.g., all in the standard range of 0.5–2 m$M$. The "high-yield transcription kits," such as Ampliscribe (from Epicentre) or Megascript (from Ambion) contain much higher NTPs (4–7 m$M$), and in this case, a lower GTP concentration of approx 1 m$M$ can be used to reduce the required amount of the more expensive oligonucleotide.

### 3.1.1. Protocol With Standard Transcription Method

1. For a 100-µL reaction: Use approx 1–10 pmoles of DNA template (e.g., standard transcription plasmid, PCR product, or a combination of synthetic oligos (*7, see* **Note 1** and **2**).
2. Set up the reaction with final concentrations of 40 m$M$ Tris-HCl, pH 8.0, 20 m$M$ MgCl$_2$, 2 m$M$ spermidine, 10 m$M$ DTT, 1 m$M$ NTPs each (up to 2 m$M$). Optional additions: 50 U of RNasin; the enhancing additives 8% polyethylene glycol 6000 and 0.01% Triton X-100.

   If desired, a tracer amount of [α-$^{32}$P]-UTP can be added, for visualization by autoradiography and quantification by scintillation counter.

**Table 1**
**Incorporation Efficiency of Initiator Oligonucleotides**

| | |
|---|---|
| All sequences NxG are possible, but oligo(G) homopolymers should be avoided | |
| Dinucleotides (*one* extra nucleoside at 5'-end) unmodified, or with label—e.g., **biotin** or **fluorescein** | >95% |
| Trinucleotides (*two* extra nucleosides at 5'-end) | >85% |
| Tetranucleotides (*three* extra nucleosides at 5'-end) | >80% |
| Pentanucleotides (*four* extra nucleosides at 5'-end) | >60% |
| Hexanucleotides (*five* extra nucleosides at 5'-end) | about 40% |

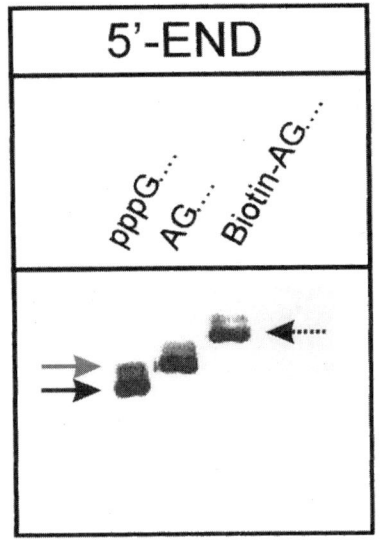

Fig. 1. Incorporation of initiator oligonucleotides in transcripts. Transcriptions were performed as described in **Subheading 3.1.1.**, including a tracer amount of [α-$^{32}$P]-UTP. The plasmid template encodes mature tRNA$^{Phe}$ from yeast *(2)*, T7 RNA polymerase was used. Analysis of transcripts was performed by 8% denaturing PAGE, followed by autoradiography. The 5'-terminal sequence is indicated above the lanes. pppG: normal triphosphate end in standard transcription reaction. ApG: addition of dinucleotide AG results in extra adenosine with 5'-OH end. Biotin-AG: addition of biotinylated dinucleotide results in extra adenosine with 5'-biotin end. As usual, transcripts terminate with the last template-encoded nucleotide (black arrow at left side), and about 30% are extended by one extra nucleotide (gray arrow at left side). The product with one extra 5'-terminal adenosine migrates slightly above the gray arrow, because of the missing negative charges (5'-OH instead of triphosphate). Addition of the bulky group biotin results in further shift (dotted arrow at right side). Please note: the initiator oligonucleotides (twofold molar excess over GTP) effectively outcompete formation of standard transcript; further, all products display a similar 3'-heterogeneity.

3. Add the appropriate initiator oligonucleotide at twofold excess—e.g., at 2 m*M* (or at 4 m*M*).

4. Add 100 U (or up to 10-fold higher amount, but not exceeding 10% of the total reaction volume to avoid excessive glycerol addition) of T7 RNA polymerase.

5. Incubate at 37°C for 1–4 h.

6. If desired, remove DNA template by adding 20 U of RNase-free DNase, incubate an additional 30 min at 37°C.

7. Transcripts are recovered by ethanol precipitation: add 100 µL of 4 *M* ammonium acetate/20 m*M* EDTA, mix, add 500 µL cold ethanol, and mix again. Chill for 15 min on dry ice (or 30 min at –70°C, or >60 min at –20°C), microfuge for 15 min, and discard supernatant. Dry briefly.

8. Dissolve pellet in 10–20 µL gel loading solution, denature by heating for 2 min at 96°C, and load on denaturing polyacrylamide gel.

9. After electrophoresis, RNA can be visualized by autoradiography or for unlabeled RNA, by UV-shadowing or by staining—e.g., with ethidium bromide.

### 3.1.2. Protocol for Using a High-Yield Transcription Kit

1. Kits are available—for example, from Epicentre (Ampliscribe) or from Ambion (MEGAscript or MEGAshortscript).

2. For a 20-µL reaction, 1–10 pmols of DNA template.

3. Set up the reaction as specified in the kit. The NTPs are used at high concentrations, about 5–7 m*M* each. Although this will compromise the transcript yields, reduce GTP concentration to 2 m*M*, thus reducing the required amount of initiator oligonucleotide. If desired, a tracer amount of [α-$^{32}$P]-UTP can be added, for visualization by autoradiography and quantification by scintillation counter.

4. Add the appropriate initiator oligonucleotide at twofold excess—e.g., at 4 m*M*.

5. Add the RNA polymerase from the kit.

6. Incubate at 37°C for 1–4 h.

7. If desired, remove DNA template by adding 20 U of RNase-free DNase, incubate an additional 30 min at 37°C.

8. Transcripts are recovered by ethanol-precipitation: add 20 µL of 4 *M* ammonium acetate/20 m*M* EDTA, mix, add 100 µL cold ethanol, and mix again. Chill for 15 min on dry ice (or 30 min at –70°C, or >60 min at –20°C), microfuge for 15 min, and discard supernatant. Dry briefly.

9. Dissolve pellet in 10–20 µL gel-loading solution, denature by heating for 2 min at 96°C, and load on denaturing polyacrylamide gel.

10. After electrophoresis, RNA can be visualized by autoradiography or for unlabeled RNA, by UV-shadowing or by staining—e.g., with ethidium bromide.

### 3.1.3. Introducing Defined Sequence Changes in the 5'-Terminal Sequence, Without Using Different Templates

An example of this approach is the generation of tRNAs with different extra 5'-terminal sequences as 5'-flanks, suitable for studies of pre-tRNA processing by RNase P *(3,8)*.

The approach is very similar. In this case, the provided template DNA codes for a transcript beginning with 5'-terminal G of the mature tRNA. The in vitro transcription reactions are performed as usual, but in addition, a short "initiator oligonucleotide" is added. This oligonucleotide contains the desired extra 5'-sequence, including the 5'-terminal G of the mature tRNA.

| | |
|---|---|
| 5'-terminal sequence mature $tRNA^{Phe}$ from yeast | 5'-GCGGAUUUAGC....... |
| template-encoded transcript | 5'-pppGCGGAUUUAGC....... |
| in vitro transcript with the trinucleotide AAG | 5'-AAGCGGAUUUAGC....... |

Protocols are exactly as described in **Subheading 3.1.**

### *3.1.4. Producing RNAs With Modified Nucleotides in the 5'-Terminal Sequence*

Another example is the generation of RNAs that contain 5'-proximal, well-defined nucleotide modifications, suitable for studies of RNA processing. Already, this 5'-modified RNA can be the desired final product ***(9)***, or the modifications can be internalized by combining two RNA molecules ***(10,11)***.

Again, the approach is very similar, and the provided template DNA codes for a transcript beginning with 5'-terminal G. The in vitro transcription reactions are performed as usual, but in addition, the "initiator oligonucleotide" contains a well-defined modification, and includes the 5'-terminal G of the normal transcript. An example is the site-specific introduction of a 2'-deoxyribose:

| | |
|---|---|
| 5'-terminal sequence of normal transcript | 5'-pppGCGGAUUUAGC....... |
| in vitro transcript with the trinucleotide ***dA***AG | 5'-***dA***AGCGGAUUUAGC....... |

Another example is the introduction of a fully characterized stereoisomer of a phosphorothioate (R or S isomer; as a reminder: at internal sites, only the R isomer can be introduced by in vitro transcription):

| | |
|---|---|
| 5'-terminal sequence of normal transcript | 5'-pppGCGGAUUUAGC....... |
| in vitro transcript with the dinucleotide ***A(pS)***G | 5'-***A(pS)***GCGGAUUUAGC....... |

A further example is the introduction of a modified base in long RNA transcripts, such as 7-deazaadenine ($c^7A$):

| | |
|---|---|
| 5'-terminal sequence of normal transcript | 5'-pppGCGGAUUUAGC....... |
| in vitro transcript with the dinucleotide ***c7A***G | 5'-***c7A***GCGGAUUUAGC....... |

Protocols are exactly as described in **Subheading 3.1.**, using the proper modified initiator oligonucleotide.

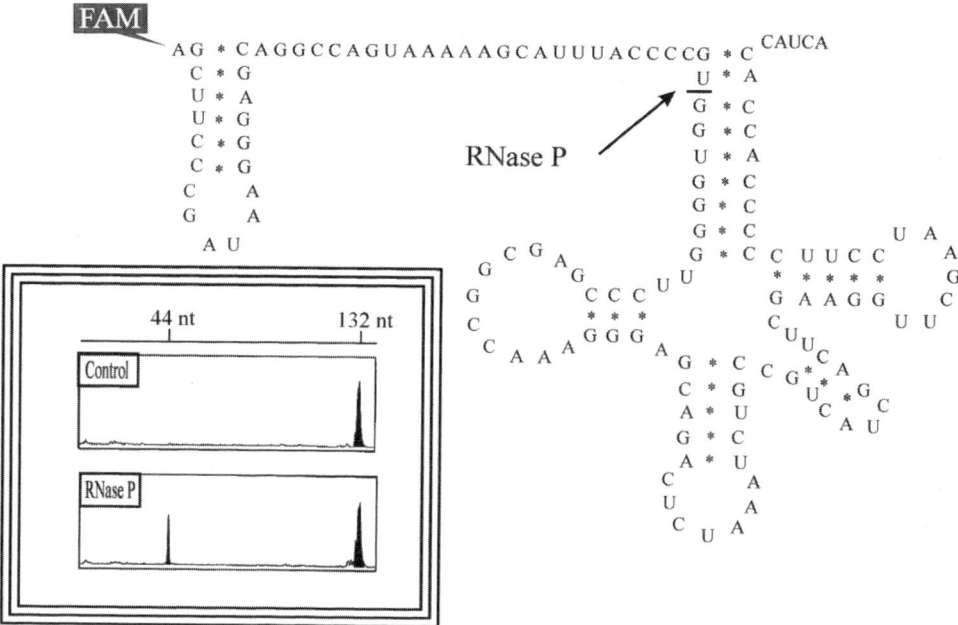

Fig. 2. Processing of flourescent-labeled pre-tRNA, monitored by automated sequencer. Transcriptions were performed as described in **Subheadings 3.1.1.** and **3.1.3.** The plasmid template encodes pre-tRNA[Tyr] from *E. coli (3)*, T7 RNA polymerase, and the initiator oligonucleotide FAM-AG was used. Transcripts were purified by 8% denaturing PAGE, transcript was directly visible in the gel as green band, or visualized by fluorescence scanning (Storm 860 from Amersham-Pharmacia). The transcript structure is shown, including the extra 5'-terminal A, linked to the fluorescent dye. The cleavage position of the pre-tRNA processing RNase P is indicated *(3)*. Insert: two runs on the ABI Prism 310 capillary sequencer. Control: incubation without enzyme, only peak for full-size 132-nucleotide pe-tRNA is visible. RNase P: treatment with RNase P from yeast *(8)* results in additional peak for liberated 44-nucleotide 5'-flank.

### 3.1.5. Direct Nonradioactive 5'-Labeling of RNAs During In Vitro Transcription (e.g., With fluorescein or With biotin)

The 5'-fluorescent-labeled RNAs are convenient for analysis with polyacrylamide gel electrophoresis combined with a fluorescence scanner or for use in standard automated DNA sequencers. An example is shown in **Fig. 2** with a 5'-FAM-labeled pre-tRNA, processed by RNase P and analyzed in an ABI 310 capillary sequencer.

Furthermore, even real-time analysis of ribozyme reactions is possible by observing changes in fluorescence polarization (*12*; *see* also Chapter 4).

5'-biotinylated RNAs were previously used for nonradioactive detection in polyacrylamide gels *(2)*, and an equivalent option would be digoxygenin. An attractive property of these site-specifically biotinylated RNAs is their highly efficient recovery with streptavidin-beads. Applications could be the isolation of high-affinity binding compounds after incubation with complex biological samples, or the immobilization of RNA aptamers without compromising their activity and without requiring a chemical synthesis of the full-size RNA.

Again, the template DNA codes for a transcript beginning with 5'-terminal G. The in vitro transcription reactions are performed as usual, but in addition, a biotin- or FAM-AG (indicated as X-AG in the following scheme) is used as "initiator oligonucleotide."

| | |
|---|---|
| 5'-terminal sequence of normal transcript | 5'-pppGCGGAUUUAGC....... |
| in vitro transcript with the dinucleotide *X*-AG | 5'-*X*-AGCGGAUUUAGC....... |

Protocols are exactly as described in **Subheading 3.1.** Illustrative results are shown in **Table 1**, and a biotinylated RNA is shown in **Fig. 1**.

### 3.2. Functional RNA Studies With Transcripts Containing Internal, "Partially Modified" Nucleotides

This approach is only briefly presented, to show another context in which chemically synthesized RNA building blocks are used (*see also* Chapter 6). Here, internal sites can be screened for functional importance of ribose or base moieties. The crucial step is a semi-quantitative, site-specific detection of modification levels in RNA transcripts. This can be achieved by combining a phosphorothioate linkage (specifically cleaved and thus semi-quantitatively detected by iodine/ethanol treatment) with the modification of interest (*see* **Note 3**). Initially, the only commercially available RNA modification type was deoxyribose, in the form of dNTPαS, and the technique was established in the identification of important ribose moieties in RNase P substrates *(3)*. Subsequently, it was used to define chemical groups in base moieties that were essential for the function of other ribozymes *(4,13)* and the technique was known as nucleotide analog interference mapping (NAIM). This technique awaits further use, since the number of commercially available, modified NTPαS building blocks has dramatically increased (e-mail: krupp@artus-biotech.com).

### 4. Notes

1. Avoid using plasmids linearized with a restriction enzyme such as P*st*I that generates 3'-protruding ends. If unavoidable, blunt ends can be generated by brief treatment with T4 DNA polymerase.

2. If synthetic oligos or PCR products are used as templates, DNA and transcript size are similar, and to ensure DNA removal, a DNase treatment is advisable.
3. Phosphorothioate and other modified RNAs are more sensitive to degradation, and elution buffers should be adjusted to pH 7.0 (measuring in the final mixture).

## References

1. Gaur, R. K. and Krupp, G. (1997) Preparation of templates for enzymatic RNA synthesis. *Methods Mol. Biol.* **74,** 69–78.
2. Pitulle, C., Kleineidam, R. G., Sproat, B., and Krupp, G. (1992) Initiator oligonucleotides for the combination of chemical and enzymatic RNA synthesis. *Gene* **112,** 101–105.
3. Conrad, F., Hanne, A., Gaur, R. K., and Krupp, G. (1995) Enzymatic synthesis of 2'-modified nucleic acids: identification of important phosphate and ribose moieties in RNase P substrates. *Nucleic Acids Res.* **23,** 1845–1853.
4. Strobel, S. A. and Shetty, K. (1997) Defining the chemical groups essential for *Tetrahymena* group I intron function by nucleotide analog interference mapping. *Proc. Natl. Acad. Sci. USA* **94,** 2903–2908.
5. Fechter, P., Rudinger, J., Giege, R., and Theobald-Dietrich, A. (1998) Ribozyme processed tRNA transcripts with unfriendly internal promoter for T7 RNA polymerase: production and activity. *FEBS Lett* **436,** 99–103.
6. Ferré-D'Amaré, A. R. and Doudna, J. A. (1996) Use of cis- and trans-ribozymes to remove 5' and 3' heterogeneities from milligrams of in vitro transcribed RNA. *Nucleic Acids Res.* **24,** 977–978.
7. Milligan, J. F. and Uhlenbeck, O. C. (1989) Synthesis of small RNAs using T7 RNA polymerase. *Methods Enzymol.* **180,** 51–62.
8. Krupp, G., Kahle, D., Vogt, T., and Char, S. (1991) Sequence changes in both flanking sequences of a pre-tRNA influence the cleavage specificity of RNase P. *J. Mol. Biol.* **217,** 637–648.
9. Kleineidam, R. G., Pitulle, C., Sproat, B., and Krupp, G. (1993) Efficient cleavage of pre-tRNAs by *E. coli* RNase P RNA requires the 2'-hydroxyl of the ribose at the cleavage site. *Nucleic Acids Res.* **21,** 1097–1101.
10. Moore, M. J. and Sharp, P. A. (1992) Site-specific modification of pre-mRNA: the 2'-hydroxyl groups at the splice sites. *Science* **256,** 992–997.
11. Gaur, R. K., Beigelman, L., Haeberli, P., and Maniatis, T. (2000) Role of adenine functional groups in the recognition of the 3'-splice-site AG during the second step of pre-mRNA splicing. *Proc. Natl. Acad. Sci. USA* **97,** 115–120.
12. Singh, K. K., Rücker, T., Hanne, A., Parwaresch, R., and Krupp, G. (2000) Fluorescence polarization for monitoring ribozyme reactions in real time. *BioTechniques* **29,** 344–351.
13. Oyelere, A. K., Kardon, J. R., and Strobel, S. A. (2002) pK(a) perturbation in genomic Hepatitis Delta Virus ribozyme catalysis evidenced by nucleotide analogue interference mapping. *Biochemistry* **41,** 3667–3675.

# 3

## Determination of Kinetic Parameters for Hammerhead and Hairpin Ribozymes

### Martha J. Fedor

#### Summary

The application of conventional enzymological methods to the study of hairpin and hammerhead ribozymes has led to valuable insights into the mechanisms by which these small RNAs catalyze phosphodiester cleavage and ligation reactions. Here, protocols are presented for measuring rate constants for simple cleavage and ligation reactions mediated by minimal hammerhead and hairpin ribozymes under standard experimental conditions. Information is also provided to help researchers recognize and interpret more complex reaction kinetics that can be observed for ribozyme-sequence variants under a variety of reaction conditions.

**Key Words:** Ribozyme; catalytic RNA; RNA; nucleic acid; kinetics; ribonuclease; RNA ligase; hairpin ribozyme; hammerhead ribozyme.

## 1. Introduction

Hammerhead and hairpin ribozymes belong to the family of small RNA enzymes that catalyze a reversible phosphodiester cleavage reaction that produces 2',3'-cyclic phosphate and 5' hydroxyl termini (**Fig. 1**). These catalytic RNA motifs were first discovered in plant satellite RNAs, where self-cleavage and ligation reactions participate in processing intermediates of rolling-circle transcription (*1–3*). Hammerhead and hairpin motifs catalyze the same chemical reactions, but they have different structures and appear to exploit distinct catalytic and kinetic mechanisms. Although hammerhead and hairpin motifs assemble from sequences within single-plant satellite RNAs in nature, they can be divided into separate ribozyme and substrate RNAs that assemble through formation of intermolecular basepaired helices (*4–6*). Dividing self-cleaving motifs into separate ribozymes and substrates allows the application

From: *Methods in Molecular Biology, vol. 252: Ribozymes and siRNA Protocols, Second Edition*
Edited by: M. Sioud © Humana Press Inc., Totowa, NJ

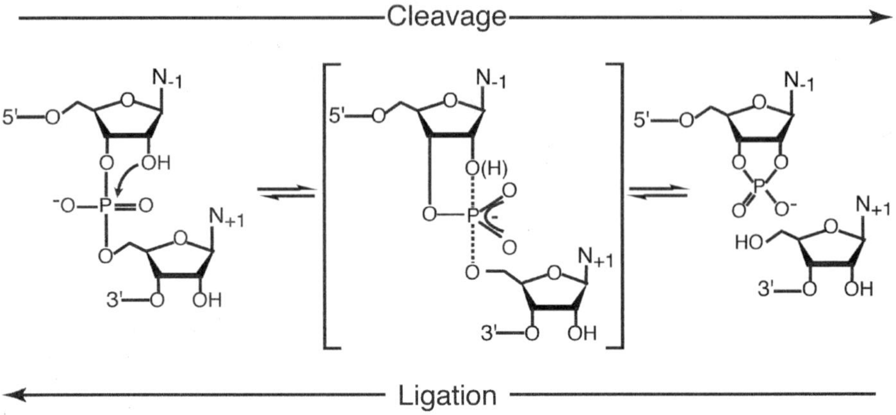

Fig. 1. Chemical mechanism of the reversible cleavage reaction mediated by hammerhead and hairpin ribozymes.

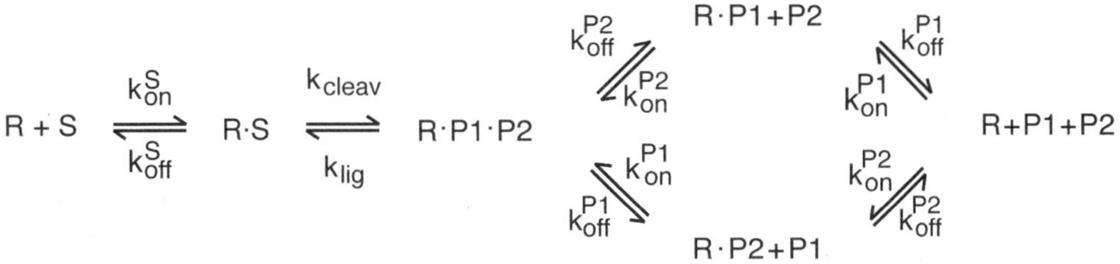

Fig. 2. Minimal kinetic mechanism for intermolecular reactions mediated by hairpin and hammerhead ribozymes.

of conventional enzymological methods to investigate the structure-function relationships that govern activity.

Reaction pathways for ribozyme-mediated cleavage and ligation include assembly and dissociation steps, as well as the transesterification steps that break and form phosphodiester bonds (**Fig. 2**). The cleavage and ligation rates that are observed in a particular experiment can reflect the kinetics of any of these steps, depending on which step is slowest for specific ribozyme and substrate sequences under a chosen set of reaction conditions. This chapter presents basic experiments that are useful for the initial characterization of cleavage and ligation activity for a new ribozyme sequence or new set of reaction conditions, along with additional experiments that can help to identify which step(s) in the reaction pathway are rate-determining. Determination of kinetic parameters using real-time PCR is described in Chapters 4 and 5.

## 2. Materials

1. Sterile, RNase-free, siliconized microfuge tubes.
2. 96-well microtiter plates with U-shaped wells and parafilm or tape for sealing wells.
3. Pipetmen and tips capable of accurately delivering volumes ranging from 1–200 μL.
4. Thermal cycler or dry block capable of maintaining temperatures between 25 and 95°C.
5. Stopwatch or digital timer.
6. Ribozyme and substrate RNA stocks, ≥20 μ$M$, prepared through chemical synthesis or T7 RNA polymerase transcription of DNA templates and purified through denaturing gel electrophoresis and ion-exchange chromatography as Na$^+$ salts.
7. 5'-$^{32}$P substrate RNA prepared through reaction with T4 polynucleotide kinase and [γ-$^{32}$P] adenosine 5' triphosphate (ATP) using conventional methods.
8. Stock of 3' cleavage product RNA (P2) with 5' hydroxyl termini, ≥5 μ$M$, prepared through chemical synthesis and purified through denaturing gel electrophoresis and ion-exchange chromatography as Na$^+$ salts.
9. $^{32}$P-5' end-labeled 5' cleavage product RNA with 2',3'-cyclic phosphate termini ([5'-$^{32}$P]P1), prepared through ribozyme-mediated cleavage of [5'-$^{32}$P] substrate RNA.
10. Stock solutions of 1 $M$ NaHEPES, pH 7.5; 100 m$M$ MgCl$_2$; and 250 m$M$ EDTA.
11. 5X reaction buffer: A "standard" 5X buffer includes 250 m$M$ buffer, 50 m$M$ MgCl$_2$, and 0.5 m$M$ ethylenediaminetetraacetic acid (EDTA) for final reaction concentrations of 50 m$M$ buffer, 10 m$M$ MgCl$_2$, 0.1 m$M$ EDTA (*see* **Note 1**).
12. Stop solution: 8 $M$ urea, 25 m$M$ EDTA, 0.005% bromophenol blue, and 0.005% xylene cyanol.
13. 19:1, acrylamide:*bis*acrylamide gels with 7 $M$ urea and 1X TBE buffer, which consists of 0.1 $M$ Tris-borate, pH 8.3, 1 m$M$ EDTA (*see* **Note 3**). A 120-mL gel with dimensions of 40 × 20 × 1.5 m$M$ (W × H × D) can be prepared with wells to accommodate 32 samples.
14. Gel electrophoresis apparatus and power supply.
15. Radioanalytic scanner or scintillation counter.
16. Appropriate safety equipment, including absorbent bench paper, radiation shield, gloves, lab coat, goggles, a Geiger counter suitable for monitoring $^{32}$P, and a radioactive-waste receptacle.

## 3. Methods

### 3.1. Measuring k$_{cleav}$ and K$_M$' in Reactions With Ribozyme in Excess of Substrate

In reactions with ribozyme in excess of substrate, substrate can cleave to completion in a single catalytic cycle so—at least in the simplest case—observed cleavage rates are not complicated by product dissociation steps in the reaction pathway. This method can be used to monitor the central conversion of the Michaelis complex, E·S, to E·P1·P2 only for ribozyme variants or reaction conditions in which dissociation one or both cleavage products is much faster than ligation—that is, when $k_{\text{off}}^{\text{P1}} \gg k_{\text{lig}}$ and/or $k_{\text{off}}^{\text{P2}} \gg k_{\text{lig}}$. This is likely to be the case for most minimal hammerhead ribozymes under standard conditions because $k_{\text{lig}}$ is likely to be slow *(7)*, and for most minimal hairpin ribozymes because 5' cleavage product dissociation is likely to be rapid *(8)*. With ribozyme variants or reaction conditions for which these assumptions are not correct, measurement of cleavage rates will be complicated by rapid re-ligation of bound products. By evaluating the ribozyme concentration dependence of observed cleavage rates, this experiment reveals the maximum rate of cleavage that can be achieved when all substrate is bound to ribozyme—that is, the cleavage-rate constant or $k_{\text{cleav}}$. $k_{\text{cleav}}$ is sometimes called $k_2$ to indicate that this rate constant does not necessarily reflect the rate of the chemical step of the reaction. It also reveals the concentration of ribozyme that is required to achieve half-maximal cleavage rates—that is, $K_{\text{M}}$ (sometimes called $K_{\text{M}}'$) to indicate a value obtained from reactions with ribozyme in excess of substrate.

1. Choose four ribozyme concentrations below $K_{\text{M}}$ and four concentrations above $K_{\text{M}}$ (*see* **Note 4**).
2. Choose a [5'-$^{32}$P] substrate concentration that is at least 10-fold lower than the lowest ribozyme concentration chosen in **step 1** (*see* **Note 5**).
3. Plan reaction time-courses so that one-half of the time-points fall in the first one-half of the reaction and one-half of the time-points fall in the second half of the reaction (*see* **Note 6**). Design a series of eight time-courses to stagger initiation and reaction time-points.
4. Label one tube for each ribozyme concentration, one tube for [5'-$^{32}$P]substrate, and place 35 µL of stop solution into each of 64 microtiter-plate wells (*see* **Note 7**). Seal microtiter-plate wells with parafilm or tape to prevent evaporation until needed.
5. Combine ribozyme stock solution with water to obtain a ribozyme concentration that is 2.5× the final desired concentration in a volume of 20 µL. (For a final ribozyme concentration of 20 n*M*, for example, prepare 20 µL of 50 n*M* ribozyme.) Combine [5'-$^{32}$P] substrate stock solution with water to obtain a substrate concentration that is 2.5× the final desired concentration in a volume of 200 µL. (For a final concentration of 0.1 n*M* [5'-$^{32}$P] substrate, for example, prepare 200 µL of 2.5 n*M* [5'-$^{32}$P]substrate.) Heat solutions to 95°C for 30 s and cool to the reaction temperature of 25°C. Add one-fourth vol of 5X reaction buffer to ribozyme (5 µL of 5X buffer) and substrate (50 µL of 5X buffer) solutions. Preincubate ribozyme and substrate solutions in 1X reaction buffer for 10 min or longer.
6. Mix 2.5 µL of [5'$^{32}$P] substrate with 20 µL of stop solution for a sample at a time-point of $t = 0$.
7. Mix 25 µL of [5'-$^{32}$P] substrate with 25 µL of ribozyme to start the reaction. Mix 5 µL of the reaction solution with 35 µL of the stop solution in a microtiter-plate well at each of the remaining seven time-points.

8. Load samples onto an acrylamide gel (19:1, acrylamide:*bis*acrylamide) and electrophorese long enough to separate substrates and products (*see* **Note 3**).
9. Quantify the amount of substrate and product at each time-point using a radioanalytic scanner, or scintillation counting of excised bands.
10. For time-courses at each ribozyme concentration, calculate $k_{obs, cleav}$ from the fraction of product formed as a function of time by computing the nonlinear, least-squares fit to $P/(P + S) = P/(P + S)_0 + P/(P + S)_\infty(1 - e^{-kobst})$ (*see* **Note 8**).
11. Plot $k_{obs, cleav}$ (y-axis) vs $k_{obs, cleav}/[R]$ (x-axis). The y intercept of this Eadie-Hofstee plot gives $k_{cleav}$, the cleavage-rate constant. The absolute value of the slope gives $K_M'$, the ribozyme concentration at which observed cleavage rates are half-maximal.

## 3.2. Measuring $k_{cat}$ and $K_M$ in Reactions With Substrate in Excess of Ribozyme

In reactions with substrate in excess of ribozyme, the first catalytic cycle resembles a ribozyme excess reaction, and subsequent catalytic cycles require product dissociation to regenerate free ribozyme. Comparison of kinetic parameters obtained from ribozyme-excess experiments and the single- and multiple-turnover phases of substrate-excess reactions can reveal important information about the relationship between cleavage and product dissociation rate constants and the fraction of functional ribozyme and substrate RNAs.

1. Estimate four substrate concentrations below $K_M$ and four substrate concentrations above $K_M$ (*see* **Note 4**).
2. Choose eight ribozyme concentration that are at least 20-fold lower than the substrate concentrations chosen in **step 1** (*see* **Note 10**).
3. Plan reaction time-courses so that all time-points fall in the first 10–15% of the reaction, before the initial concentration of substrate has been significantly reduced through cleavage (*see* **Note 10**). Design a series of time-courses to stagger initiation ($t = 0$) and reaction time-points.
4. Label one tube for each substrate concentration and one tube for each ribozyme concentration, and place 35 μL of stop solution into each of 64 microtiter-plate wells (*see* **Note 7**). Seal microtiter-plate wells with parafilm or tape to prevent the stop solution from evaporating until it is needed.
5. Combine [5-$^{32}$P]substrate stock solution with unlabeled substrate stock solution and water to obtain a substrate concentration that is 2.5× the final desired concentration in a volume of 20 μL. Combine ribozyme stock solution with water to obtain a ribozyme concentration that is 2.5× the final desired concentration in a volume of 20 μL. Heat the solutions to 95°C for 30 s, then cool them to the reaction temperature of 25°C. Add 5 μL of 5X reaction buffer to the ribozyme and substrate solutions. Pre-incubate the ribozyme and substrate solutions in 1X reaction buffer at 25°C for 10 min or longer.
6. Mix 2.5 μL of [5'-$^{32}$P] substrate with 20 μL of stop solution for a sample at time-point at $t = 0$.

7. Mix 25 μL of [5'-$^{32}$P] substrate with 25 μL of ribozyme to start the reaction. Mix 5 μL of the reaction solution with 35 μL of the stop solution in a microtiter-plate well at each of the remaining time-points.

8. Follow **steps 8** and **9** as described in **Subheading 3.1.**

9. For time-courses at each substrate concentration, calculate $k_{obs, cleav}$ from the fraction of product formed as a function of time by computing the fit to [P]/[R] vs time during the initial linear phase of the reaction when less than 15% of the substrate has been converted to product (*see* **Note 11**).

10. Plot $k_{obs, cleav}$ (y-axis) vs $k_{obs,cleav}$/[S] (x-axis). The y intercept of this Eadie-Hofstee plot gives $k_{cat}$, the cleavage-rate constant. The absolute value of the slope yields $K_M$, the substrate concentration at which observed cleavage rates are half-maximal (*see* **Note 12**).

## 3.3. Measuring k$_{lig}$ From the Internal Equilibrium Between Cleavage and Ligation and the Rate of Approach to Equilibrium in Single-Turnover Reactions With Small Amounts of [5'-$^{32}$P]P1 and Saturating Concentrations of R·P2

Rate constants for ligation can be calculated from the internal equilibrium between cleavage and ligation of bound products, $K_{eq}^{int} = k_{lig}/k_{cleav}$, and the rate of approach to equilibrium, $k_{\rightarrow\infty} = k_{cleav} + k_{lig}$ *(7,9)*. Single-turnover ligation reactions are carried out at a saturating concentration of a binary complex that contains the ribozyme in complex with the 3' cleavage product RNA, P2, and a small amount of [5'-$^{32}$P] 5' cleavage product RNA, [5'-$^{32}$P]P1. This approach is appropriate only for hammerhead ribozymes that form stable complexes with 5' and 3' cleavage products. Minimal hairpin ribozymes typically bind 5' cleavage products with affinities that are too low to allow saturating concentrations of the ribozyme-P2 complex to be experimentally accessible. It also is important that reactions contain RNA concentrations that are high enough to ensure that ligation kinetics are truly limited by the rate of approach to equilibrium, $k_{\rightarrow\infty}$, and not by slow 5' cleavage product binding. Hammerhead ribozyme ligation is much slower than hairpin ribozyme ligation, making it possible to prepare hammerhead ligation reactions with RNA concentrations that promote complex formation at rates that are faster than the sum of cleavage- and ligation-rate constants.

1. Plan reaction time-courses so that one-half of the time-points fall in the first half of the reaction and one-half of the time-points fall in the second half of the reaction (*see* **Note 13**). Include two additional time-points at $t = 2$ h and $t = 4$ h.

2. Prepare a microtiter plate with 20 μL of stop solution in each of ten wells. Seal wells with tape or parafilm until needed.

3. Prepare a 30-μL solution that contains 0.22 n$M$ [5'-$^{32}$P]P1, 550 n$M$ ribozyme, and 1100 n$M$ P2 in 55 m$M$ NaHEPES, pH 7.5 (*see* **Note 14**). Heat to 95°C for 1 min. Incubate at 25°C for 10 min.

4. Mix 3 μL of the RNA solution with 10 μL of stop solution for a time point at $t = 0$.
5. Add 3 μL 100 m$M$ $MgCl_2$ to the remaining RNA solution to start the reaction.
6. At each time-point, mix 3 μL of the reaction solution with 20 μL of stop solution in a microtiter-plate well.
7. Follow **steps 8** and **9** as described in **Subheading 3.1.**
8. Calculate $k_{obs, lig}$ from the fraction of substrate formed as a function of time by computing the nonlinear, least squares fit to $S/(P + S) = S/(P + S)_0 + S/(P + S)_\infty (1 - e^{-k_{obs}t})$.
9. Calculate $k_{lig}$ from $k_{\to\infty}$ according to $k_{\to\infty} = k_{lig} + k_{cleav}$ or $k_{lig} = k_{\to\infty} - k_{cleav}$.
10. Calculate $K_{eq}^{int}$ from the maximum extent of ligation observed when the reaction has reached equilibrium (*see* **Note 15**) according to $K_{eq}^{int} = [S]_\infty/[P1]_\infty$.
11. Calculate $k_{lig}$ from $K_{eq}^{int}$ and $k_{cleav}$ according to $K_{eq}^{int} = k_{lig}/k_{cleav}$ or $k_{lig} = K_{eq}^{int} \times k_{cleav}$ (*see* **Note 16**).

## 3.4. Measuring $k_{lig}$ From in Single-Turnover Reactions With Small Amounts of $[5'-^{32}P]P1$ From the Dependence of the Rate of Approach to Equilibrium on the Concentration of R·P2

The approach described in **Subheading 3.3.** is usually not appropriate for minimal hairpin ribozymes under standard conditions because the ribozyme complex with the 5' cleavage product RNA is usually not stable enough to allow saturation at experimentally accessible concentrations. Ligation-rate constants for hairpin ribozymes can be determined by extrapolating maximum ligation rates from the ligation rates observed in reactions with a small amount of $[5'-^{32}P]$ 5' cleavage product, $[5'-^{32}P]P1$, and increasing amounts of a binary complex that contains ribozyme and 3' cleavage product RNAs, R·P2 (*8*). Hairpin ribozyme variants used for this type of experiment must bind P2 RNA with high affinity to ensure that all ribozyme RNA in the reaction is present as an R·P2 complex.

1. Choose eight concentrations of the R·P2 binary complex that fall above and below an estimated $K_M^{P1}$ value (*see* **Note 17**).
2. Plan reaction time-courses so that one-half of the time-points fall in the first half of the reaction and one-half of the time-points fall in the second half of the reaction (*see* **Note 18**). Design a series of eight time-courses to stagger initiation and reaction time-points.
3. Prepare a microtiter plate with 20 μL of stop solution in each of 64 wells. Seal wells with tape or parafilm until needed.
4. Prepare a 240-μL solution that contains 0.5 n$M$ $[5'-^{32}P]P1$. Heat to 95°C for 1 min. Add 60 μL of 5X reaction buffer. Incubate at 25°C for 10 min.
5. Prepare eight 24-μL solutions that contain ribozyme and P2 RNAs at 2.5× and 2.8× the final desired concentrations, respectively. Heat to 95°C for 1 min. Add 6 μL of 5X reaction buffer. Incubate at 25°C for 10 min or longer.
6. Mix 3 μL of the $[5'-^{32}P]P1$ solution with 24 μL of stop solution for a time-point at $t = 0$.

7. Mix 30 μL of the [5'-$^{32}$P]P1 solution with 30 μL of the R·P2 solution to start the reaction.

8. At each time-point, mix 6 μL of the reaction solution with 24 μL of stop solution in a microtiter-plate well.

9. Follow **steps 8** and **9** as described in **Subheading 3.1.**

10. Calculate $k_{obs, lig}$ from the fraction of substrate formed as a function of time by computing the nonlinear, least-squares fit to $S/(P + S) = S/(P + S)_0 + S/(P + S)_\infty(1 - e^{-kobst})$ for reactions with each R·P2 concentration.

11. Plot $k_{obs, lig}$ (y-axis) vs [R·P2] (x-axis). Calculate $k_{lig}$ from the nonlinear, least-squares fit to $k_{obs, lig} = k_{lig}([R·P2]/(R·P2] + K_M^{P1})) + k_{cleav}$.

## 4. Notes

1. "Standard reaction conditions" include 50 m$M$ NaHEPES, pH 7.5, 10 m$M$ MgCl$_2$, and 0.1 m$M$ EDTA at 25°C. Use of standard conditions for initial experiments facilitates comparison among ribozyme variants.

2. NaHEPES, with a pK$_a$ value of 7.5 at 25°C, is an appropriate buffer for reactions to be carried out under standard conditions. For reactions carried out at higher or lower pH, other "Good" buffers should be chosen with pK$_a$ values in the appropriate range *(10)*. Buffer pH depends on temperature. Care should be taken to adjust the buffer pH at the temperature at which it is to be used, or to calculate the appropriate correction for differences between buffer preparation and reaction temperatures. This is particularly important for experiments with hammerhead ribozyme reactions, in which rates show a log-linear pH dependence.

3. The optimal concentration of acrylamide depends on the range of RNA fragment sizes to be separated. An acrylamide concentration of 20% is suitable for fractionation of substrate and product RNAs in the 5- to 20-nucleotide range, and acrylamide concentrations of 10–15% are suitable for fractionation of RNAs in the 25- to 50-nucleotide range.

4. $K_M$ values ranging from approx 1 n$M$ to 1 μ$M$ have been measured for hammerhead and hairpin ribozymes under standard conditions, with values typically falling in the low nanomolar range. A reasonable choice of ribozyme or substrate concentrations for initial ribozyme-excess or substrate-excess experiments, respectively, would be 5, 10, 20, 40, 80, 200, 400, and 800 n$M$.

5. In order to calculate accurate cleavage-rates using a simple exponential rate equation, the reaction must be pseudo first-order—that is, the reaction rate should be independent of substrate concentration and the concentration of free ribozyme must remain virtually unchanged through the course of the reaction. To avoid a significant change in the concentration of free ribozyme upon substrate binding and cleavage, the concentration of substrate must be no more than one-tenth the concentration of ribozyme. Reactions with 0.1–0.2 n$M$ [5'-$^{32}$P substrate RNA that has a specific activity of 6000 Ci/mmol will allow accurate quantification from a PhosphorImager screen that has been exposed for less than 12 h. Inaccuracies in RNA concentration determinations, preparations of ribozyme that are not fully functional or substrate that is not completely labeled can interfere with

pseudo first-order reaction kinetics even when ribozyme is believed to be in 10-fold excess of substrate. To ensure that reactions are truly pseudo first-order, control experiments can be carried out to measure cleavage rates in reactions with threefold higher or lower substrate concentrations. When observed reaction kinetics truly are pseudo first-order, control reactions with moderately different amounts of substrate will yield the same observed cleavage rates if ribozyme remains in sufficient excess.

6. Accurate determination of kinetic parameters usually occurs through successive approximations, beginning with good estimates of the expected values for $k_{cleav}$ and $K_M'$. Observed cleavage rates can be predicted from the Michaelis-Menten equation:

$$k_{obs,\,cleav} = (k_{cleav}\,[R])/[R] + K_M') \qquad (1)$$

in which $k_{cleav}$ is the cleavage-rate constant, [R] is the ribozyme concentration, and $K_M'$ is the ribozyme concentration at which observed cleavage rates are half-maximal. Inspection of this equation shows that $k_{obs,\,cleav}$ will increase linearly in reactions with ribozyme concentrations that are far below $K_M'$ and $k_{obs,\,cleav}$ will be virtually the same as $k_{cleav}$ in reactions with ribozyme concentrations that are far above $K_M'$. The half-time of the reaction is described by $t_{1/2} = \ln2/k_{obs}$—that is, $t_{1/2} = 0.693/k_{obs}$. Reasonable estimates of $k_{cleav}$ and $K_M'$ for hammerhead and hairpin reactions under standard conditions are $0.3–2.0$ min$^{-1}$ and $10–50$ n$M$, respectively. These approximate values can be inserted into **Eq. 1** to give an estimate of $k_{obs,\,cleav} \approx 0.2$ min$^{-1}$ for a reaction with [R] $\approx 40$ n$M$, for example. A reaction with $k_{obs,\,cleav} = 0.2$ min$^{-1}$ will have a $t_{1/2}$ of 3 min. In this example, time-points of 20 s, 45 s, and 1.5 min will fall in the first half of the reaction, and time-points of 6 min, 12 min, 24 min will fall in the second half of the reaction. After the first experiment with a new ribozyme variant or set of reaction conditions, more accurate estimates of $k_{cleav}$ and $K_M'$ that are calculated from the data can be used to further determine the choice of optimal ribozyme concentrations and the details of optimal time-courses.

7. The volume of stop solution that is needed to quench the reaction varies. Reactions with minimal hammerhead ribozymes can usually be quenched with an equal volume of stop solution because the EDTA in the stop solution chelates essential divalent cations. Hairpin ribozyme reactions can proceed without divalent cations, so these RNAs must be denatured by the urea in the stop solution to quench the reaction. As a control experiment to ensure that the volume of stop buffer is sufficient to quench the reaction, ribozyme and substrate solutions can be added separately to the stop solution and allowed to incubate for the duration of the time-course.

8. Considerable information can be obtained from examining the fit of the fraction of product vs time to a single exponential-rate equation. The fit to the simple exponential-rate equation can provide values for the observed cleavage rate, $k_{obs,\,cleav}$, and the extent of cleavage at the end point of the reaction, $P/(P + S)_\infty$. In the simplest reactions with ribozyme concentrations in large excess relative to substrate concentrations, $k_{obs,\,cleav}$ monitors steps in the reaction pathway that

precede product dissociation because substrate can cleave to completion in a single catalytic cycle (**Fig. 2**). Evaluating the quality of the fit and considering deviations from the fit can help in understanding more complicated reaction pathways and guide the design of follow-up experiments.

a. Inaccuracies in $k_{obs, cleav}$ values result when reactions do not proceed far enough toward completion to provide well-determined end points. A calculated end point that yeilds a fraction of cleavage products that is higher than the true end point leads to an underestimation of $k_{obs, cleav}$. Conversely, a calculated end point that is lower than the true end point leads to an overestimation of $k_{obs, cleav}$. Inaccuracies also can result if the reaction has progressed too far toward completion by the first time-point sample is collected, because a small range of values of the fraction of product is associated with larger experimental error. The solution to these problems is to repeat the experiment using a more appropriate time-course that captures the end point and the full range of change in the fraction of product from the beginning to the end of the reaction.

b. In most experiments, some fraction of the substrate will remain uncleaved, even after very long incubation times. The failure of cleavage reactions to proceed to completion can indicate that a fraction of the substrate RNA adopts an inactive conformation that fails to bind ribozyme, or binds ribozyme aberrantly to form a nonfunctional complex. If inactive RNA structures are stable throughout the time-course, the final reaction extent will be low, but reasonable kinetic parameters can be obtained for the fraction of the substrate that does react. However, active and inactive structures can sometimes exchange on the same time-scale as a cleavage reaction. Instead of yeilding a low reaction extent, cleavage can proceed to completion when reactions contain a mixture of inactive and active RNA structures that interconvert on a similar time-scale as cleavage. However, the data will fit poorly to a single exponential-rate equation because cleavage of the misfolded substrate RNA will be delayed by the time required for it to exchange into a reactive conformation. This type of reaction can display biphasic or multiphasic cleavage kinetics that do not allow a simple interpretation.

c. Hairpin and hammerhead ribozyme cleavage reactions are reversible. In fact, the hairpin ribozyme—but not the hammerhead ribozyme—is a better ligase than it is a nuclease when cleavage-product RNAs are bound in the active site *(7,8)*. This feature of the hairpin ribozyme kinetic mechanism also can lead to low cleavage extents at the reaction end point. Products that remain bound in a stable ribozyme-product complex can undergo re-ligation so that a low cleavage extent reflects the balance between cleavage and ligation of bound products at equilibrium. When product dissociation rates are similar to ligation rates, ribozyme-product complexes partition between dissociation and re-ligation. In this case, observed cleavage rates can reflect a complex interplay of cleavage, product dissociation, and ligation rates.

9. Hammerhead and hairpin ribozymes typically display cleavage-rate constants on the order of 1 min$^{-1}$, and $K_M'$ values typically fall in the low nanomolar range

*(11,12)*. Interpretation of structure-function studies relies on an understanding of how changes in RNA sequences and reaction conditions lead to changes in kinetic parameters. Kinetic parameters outside the typical range can be an indication of complications in the kinetic mechanism.

a. In simple reactions with ribozyme in excess of substrate, $K_M'$ is a second-order rate constant that is related to substrate association and dissociation and cleavage-rate constants according to $K_M' = (k_{off}^S + k_{cleav})/k_{on}^S$. Most hairpin ribozyme-substrate complexes and many hammerhead ribozyme-substrate complexes dissociate much more slowly than they cleave *(8,9,13)*. When $k_{off}^S$ values are much smaller than $k_{cleav}$ values, an analysis of this equation shows that $K_M'$ values will be virtually the same as the ratio of cleavage and binding-rate constants, or $K_M' \sim k_{cleav}/k_{on}^S$. Thus, an increase in $K_M'$ that is observed in a reaction with a typical cleavage-rate constant can signify a decrease in the apparent substrate binding-rate constant. For example, such a change in $K_M'$ can occur if a significant fraction of the ribozyme RNA misfolds into structure that is incompatible with substrate binding. In this case, the effective concentration of functional ribozyme is lower than the total RNA concentration in the ribozyme stock solution. Likewise, a decrease in the cleavage-rate constant will be accompanied by a decrease in $K_M'$ if the substrate binding-rate constant remains unchanged.

b. For some hammerhead-substrate complexes, substrate dissociation-rate constants can be much faster than the cleavage-rate constant. In this case, $K_M'$ will be virtually the same as the equilibrium dissociation constant—that is, $K_M' \sim k_{off}^S/k_{on}^S \sim K_d^S$. When $K_M' \sim K_d^S$, changes in $K_M'$ reflect changes in the relative stability of the ribozyme-substrate complex. In this case, changes in $K_M'$ values are associated with changes in $k_{on}^S$, $k_{off}^S$, or both.

10. Ribozyme concentrations must be at least 20-fold lower than substrate concentrations to ensure that initial rates for single- and multiple-turnover phases of the reaction can be measured before a significant amount of substrate has been depleted through cleavage. Ribozyme concentrations that are much lower than substrate concentrations can be chosen to collect data exclusively from the multiple-turnover phase of the reaction.

11. In the simplest case, a plot of [P]/[R] vs time will be linear until progress of the cleavage reaction leads to a significant reduction in the concentration of substrate. However, certain features of the kinetic mechanism can cause the rate of appearance of cleavage products to decline, even before much substrate has been depleted through cleavage.

a. A fraction of the substrate can fold into nonreactive structures. If only 50% of the substrate RNA is able to bind ribozyme and cleave, for example, 30% of the functional substrate will have cleaved when only 15% of the total substrate RNA has been converted to products. In a reaction with a large fraction of nonreactive substrate RNA, initial rates of cleavage of the active substrate fraction can be measured over a shorter time-course if active and inactive structures do not interconvert on the time-scale of the cleavage reaction.

     b. Nonlinear cleavage rates can be observed at early times if cleavage products remain bound to the ribozyme long enough to undergo ligation. An initial rapid rate of cleavage can be followed by a slower phase that reflects slow product dissociation.

12. In the simplest case, $k_{cat}$ and $K_M$ values measured in substrate-excess experiments will be the same as $k_{cleav}$ and $K_M'$ values measured in ribozyme-excess experiments. Deviations from the behavior expected in the simplest case can provide insight into the details of the reaction pathway.

     a. Ribozyme-excess and substrate-excess reactions can display different kinetic parameters if ribozyme and/or substrate RNAs tend to misfold into nonreactive conformations. If a significant fraction of ribozyme adopts a nonfunctional structure, $k_{cat}$ values determined in substrate-excess experiments can be lower than $k_{cleav}$ values measured in ribozyme-excess experiments by the same fraction. $K_M$ values measured in a substrate-excess experiment can also be lower than $K_M'$ values measured in ribozyme-excess experiments if the substrate RNA is fully functional while the ribozyme RNA is partially inactive (*see* **Note 9**). Conversely, $K_M$ values calculated from substrate-excess reactions can be higher than $K_M'$ values calculated from ribozyme-excess reactions when a significant fraction of the substrate adopts nonfunctional structure(s). In this case, a high $K_M$ value for substrate-excess reactions can be associated with low end points in ribozyme-excess reactions with the same substrate. In some cases, substrate RNAs can aggregate at the high concentrations used in substrate-excess experiments, leading to high $K_M$ values, although the same substrate is fully reactive at the low substrate concentrations used in ribozyme-excess experiments. Concentration-dependent substrate RNA aggregation can lead to nonlinear Eadie-Hofstee plots.

     b. If dissociation of P1 (the 5' cleavage product), P2 (the 3' cleavage product), or both cleavage products is slower than cleavage, the rate of the first turnover can match $k_{obs, cleav}$ in ribozyme-excess experiments at similar concentrations, yet cleavage rates observed during subsequent turnovers will be slower because they are limited by rates of product dissociation. When dissociation of one or both products is much slower than cleavage, $k_{cat}$ values determined from substrate-excess experiments will be much lower than $k_{cleav}$ values determined from ribozyme excess experiments. This feature of the kinetic mechanism can be used in an "active-site titration" experiment to determine the fraction of functional ribozyme from the size of the "burst" of products formed during the initial turnover *(13)*.

     c. The kinetics of substrate-excess reactions can be complicated by re-ligation of products. Re-binding of both cleavage products to the same ribozyme molecule virtually never occurs in reactions with a large excess of ribozyme (*see* **Note 11**). However, in a substrate-excess reaction, cleavage products can accumulate to concentrations high enough that re-binding and ligation occur even when product dissociation is rapid.

13. At saturating concentrations, ligation is expected to occur at a rate that is the sum of the cleavage- and ligation-rate constants. Hammerhead ribozyme ligation-rate

constants are 100-fold lower than cleavage-rate constants under standard conditions *(7)*. Therefore, the cleavage-rate constant will dominate the rate of approach to equilibrium, which can be estimated as approx 1 min$^{-1}$, a typical cleavage-rate constant for hammerhead ribozymes under standard conditions. With an estimated $t_{1/2}$ of ln2/1 = 0.693 min, a reasonable time-course could include time-points at 0, 8, 15, 30, 60, 120, 240, and 480 s.

14. Reactions with ribozyme and P2 RNAs at three or four times these concentrations should yield the same values for $k_{\to\infty}$ if RNA concentrations truly are saturating.

15. The fit to the simple exponential-rate equation gives values for the observed ligation rate and for the extent of cleavage at the end point. The fraction of ligated products calculated at the end point should agree with the fraction of ligated product measured after long incubation times if the reaction has truly reached equilibrium.

16. In the simplest case, measurements of the internal equilibrium from the fraction of substrate at equilibrium and from cleavage- and ligation-rate constants will yeild the same values—that is, $K_{eq}^{int} = [S]_\infty/[5'P]_\infty = k_{lig}/k_{cleav}$. When the small difference between $k_{\to\infty}$ and $k_{cleav}$ is close to the range of experimental error, $k_{lig}$ values obtained from $K_{eq}^{int}$ and $k_{cleav}$ can be more accurate than values obtained from the difference between $k_{\to\infty}$ and $k_{cleav}$. However, if some fraction of the RNA is trapped in a nonreactive conformations, values obtained from the relative concentrations of substrate and products at equilibrium can underestimate $k_{lig}$.

17. Under standard conditions, the most stable minimal hairpin ribozyme variants display $K_M^{P1}$ values in the low micromolar range *(8,14)*. Therefore, concentrations of R·P2 complex ranging from 0.1 μM–15 μM are suitable for an initial experiment. P2 RNA concentrations should be approx 10% higher than ribozyme RNA concentrations to ensure that all ribozyme forms a binary complex.

18. Ligation will occur at the rate of approach to equilibrium, which is equal to the sum of the forward and reverse rates. For design of an initial experiment, observed ligation rates can be estimated according to $k_{obs, \, lig} = k_{lig} ([R·P2]/[R·P2] + K_M^{P1})$ + $k_{cleav}$, using an estimate of 0.3 min$^{-1}$ for $k_{cleav}$, an estimate of 3 min$^{-1}$ for $k_{lig}$, and an estimate of 3 μM for $K_M^{P1}$. Note that observed ligation rates will be greater than $k_{cleav}$ for all R·P2 concentrations.

## Acknowledgments

Our work is supported by research grants GM46422 and GM62277 from the National Institutes of Health and by the Skaggs Institute for Chemical Biology. I thank members of the Fedor laboratory for their comments on the manuscript.

## References

1. Hutchins, C. J., Rathjen, P. D., Forster, A. C., and Symons, R. H. (1986) Self-cleavage of plus and minus transcripts of avocado sunblotch viroid. *Nucleic Acids Res.* **14**, 3627–3640.

2. Buzayan, J. M., Gerlach, W. L., and Bruening, G. (1986) Nonenzymatic cleavage and ligation of RNAs complementary to a plant virus satellite RNA. *Nature* **323**, 349–353.

3. Buzayan, J. M., Hampel, A., and Bruening, G. (1986) Nucleotide sequence and newly formed phosphodiester bond of spontaneously ligated satellite tobacco ringspot virus RNA. *Nucleic Acids Res.* **14,** 9729–9743.
4. Uhlenbeck, O. C. (1987) A small catalytic oligoribonucleotide. *Nature* **328,** 596–600.
5. Haseloff, J. and Gerlach, W. L. (1988) Simple RNA enzymes with new and highly specific endoribonuclease activities. *Nature* **334,** 585–591.
6. Hampel, A. and Tritz, R. (1989) RNA catalytic properties of the minimum (-)sTRSV sequence. *Biochemistry* **28,** 4929–4933.
7. Hertel, K. J. and Uhlenbeck, O. C. (1995) The internal equilibrium of the hammerhead ribozyme reaction. *Biochemistry* **34,** 1744–1749.
8. Hegg, L. A. and Fedor, M. J. (1995) Kinetics and thermodynamics of intermolecular catalysis by hairpin ribozymes. *Biochemistry* **34,** 15,813–15,828.
9. Hertel, K. J., Herschlag, D., and Uhlenbeck, O. C. (1994) A kinetic and thermodynamic framework for the hammerhead ribozyme reaction. *Biochemistry* **33,** 3374–3385.
10. Ferguson, W. J., Braunschweiger, K. I., Braunschweiger, W. R., Smith, J. R., McCormick, J. J., Wasmann, C. C., et al. (1980) Hydrogen ion buffers for biological research. *Anal. Biochem.* **104,** 300–310.
11. Stage-Zimmermann, T. K. and Uhlenbeck, O. C. (1998) Hammerhead ribozyme kinetics. *RNA* **4,** 875–889.
12. Fedor, M. J. (2000) Structure and function of the hairpin ribozyme. *J. Mol. Biol.* **297,** 269–291.
13. Fedor, M. J. and Uhlenbeck, O. C. (1992) Kinetics of intermolecular cleavage by hammerhead ribozymes. *Biochemistry* **31,** 12,042–12,054.
14. Nesbitt, S., Hegg, L. A., and Fedor, M. J. (1997) An unusual pH-independent and metal-ion-independent mechanism for hairpin ribozyme catalysis. *Chem. Biol.* **4,** 619–630.

# 4

# Real-Time Monitoring of RNA and DNA Reactions by Fluorescence Detection

## Kumud K. Singh, Andreas Hanne, and Guido Krupp

### Summary

Conventional analysis of nucleic acid reactions (cleavages, ligations) is performed with a (radioactive) labeled substrate, and reaction products (one aliqout for each time-point) are analyzed by gel electrophoresis followed by detection (X-ray film + densitometer or phosphoimager). This is a cumbersome approach, and involves frequent preparation of fresh labeled substrate, and rather limited time resolution. Real-time monitoring with non-decaying fluorescent labels overcomes these problems. Two analysis methods are presented here. Fluorescence polarization with a very broad range of applications, can be used for any type of RNA or DNA conversion with significant change of mol wt (including "gel shift"). FRET (fluorescence resonance energy transfer) depends on intramolecular interaction (or in a multimolecular complex) of a fluorescent dye ("reporter") with a quencher moiety (a guanine base or a second dye). Applications include detailed kinetic studies and optimizing reactions with a wide range of conditions and highly specific nucleic acid sequence detections, challenging other hybridization-based methods. Special applications are the introduction of catalytic nucleic acids for real-time monitoring of nucleic acid amplification reactions such as NASBA and PCR.

**Key Words:** Amplification; cleavage; Cy5; dark quencher; DNAzyme; FAM; fluorescence polarization; FRET; kinetics; ligation; NASBA; PCR; ribozyme; TAMRA.

## 1. Introduction

Real-time monitoring has revolutionized routine PCR application. In this chapter, we provide examples of how this convenient and high-throughput technological platform can be adapted for monitoring nucleic acid conversions with cleavages and ligations as a prototype for catalytic nucleic acids. The use of nucleic acids labeled with only a single fluorescent dye allows monitoring with fluorescence polarization; fluorescence resonance energy transfer (FRET) is also possible in special constructs. However, standard FRET analytes are

From: *Methods in Molecular Biology, vol. 252: Ribozymes and siRNA Protocols, Second Edition*
Edited by: M. Sioud © Humana Press Inc., Totowa, NJ

double fluorescent-labeled nucleic acids (in one molecule or in a multimolecular complex). It is important to note that the same reaction types are catalyzed by conventional protein enzymes, and they can be monitored with the same approach. A special application is real-time monitoring of enzyme-based nucleic acid amplification in (NASBA) and in polymerase chain reaction (PCR) by means of fitted ribozyme or DNAzyme cleavage reactions.

## 2. Materials

### 2.1. Special Equipment

1. For **Subheading 3.1.**: Fluorescence polarization can be measured with advanced flourimeters such as POLARstar from BMG LabTechnologies.
2. For **Subheading 3.2.**: The minimal instrument required is a standard fluorimeter, such as SFM 25 fluorimeter from Kontron Instruments. Ideally, equipped with a thermostated microcuvet holder. More advanced FRET analyses require expensive instruments developed for real-time PCR-monitoring, such as a "TaqMan" system (SDS 7700 and similar instruments) from Applied Biosystems, a LightCycler from Roche, or a RotorGene from Corbett Research (or a similar instrument).

Complementing quantitative analyses of radioactive or fluorescent nucleic acids by polyacrylamide gel electrophoresis (PAGE) require an imager instrument, such as Storm or Typhoon from Amersham-Pharmacia.

### 2.2. Consumables

1. Fluorescent-labeled substrate RNA (light-sensitive, stable at –20°C, may be wrapped in aluminium foil). Enzymatically or chemically synthesized RNA is preferably purified by polyacrylamide gel electrophoresis to insure homogeneity and to remove "tightly adsorbed" fluorescent dyes. Several companies provide this service, including iba GmbH (Göttingen, Germany; www.iba-go.de).
   Single, 5'-fluorescent-labeled RNAs can be prepared by one-step in vitro transcription with initiator oligos, as described in Chapter 2.
2. Matching catalytic nucleic acid.
3. Reaction buffer.
4. Reaction and measuring "tubes," depending on the type of fluorescence instrument. Standard flourimeters such as SFM 25 from Kontron Instruments: very small sample volumes can be analyzed in sub-microcuvets (as low as 15 µL), black with three clear windows (e.g., from Starna). POLARstar instrument: transparent, flat-bottom 96-well microtiter plates (e.g., from Costar). TaqMan system (SDS 7700): MicroAmp® 96-well Tubes/Tray/Retainer Assemblies (from Applied Biosystems) or similar. LightCycler: glass capillaries with stoppers (from Roche). RotorGene: special plastic tubes (from Corbett Research).
5. Equipment for PAGE.
6. Ideally, a dedicated software for kinetic analyses (e.g., EnzPack from Biosoft).

## 3. Methods

Although fluorescence polarization is less demanding in substrate preparation (requiring single labels only), the technical feasibility was demonstrated only recently *(1,2)*. More extensive data from quantitative analyses are available for FRET studies, which include determination of kinetic parameters *(3,4)*.

### 3.1. Single-Labeled Substrates for Monitoring With Fluorescence Polarization

#### 3.1.1. RNA Cleavage Reactions

1. For a 50-µL hammerhead ribozyme cleavage reaction (*see* **Figs. 1–3**): Prepare 40 µL containing 5 pmol substrate and varying amounts of ribozyme (0.3 pmol, 0.1 pmol, 0.03 pmol) in 50 m$M$ Tris-HCl, pH 7.5 and 10% ethanol (*see* **Notes 1** and **2**).
2. Transfer the reaction mix to the wells of a 96-well plate (Costar) and preincubate for 15 min at 37°C in the POLARstar instrument (BMG LabTechnologies). Choose appropriate filter settings for (FAM) fluorescence (excitation 485 nm; emission 535 nm) in the polarization mode.
3. Start the reaction by adding 10 µL of 100 m$M$ MgCl$_2$.
4. Measurements were taken every 3 min for 100 cycles. Cleavage reaction is evident in decreasing polarization values (*see* **Fig. 3** and **Note 3**).

#### 3.1.2. RNA Ligation Reactions

1. For 50-µL ligation reaction: Prepare 40 µL containing 5 µ$M$ unlabeled heptamer oligo and 2 µ$M$ FAM-labeled "sticky hairpin" oligo (*see* **Fig. 4**) and 1 µ$M$ of sunY ribozyme in 30 m$M$ Tris-HCl (pH 7.4), 400 m$M$ KCl, 10 m$M$ NH$_4$Cl, and 10% ethanol.
2. Transfer the reaction mix to the wells of a 96-well plate (Costar) and preincubate for 15 min at 25°C in the POLARstar instrument (BMG LabTechnologies). Choose appropriate filter settings for FAM fluorescence (excitation 485 nm; emission 535 nm) in the polarization mode.
3. Start the reaction by adding 10 µL of 750 m$M$ MgCl$_2$.
4. Measurements were taken every 3 min for 100 cycles.

Ligation is coupled with liberation of the small FAM-labeled trinucleotide. Thus, ligation is evident in decreasing polarization values (**Fig. 5**).

### 3.2. Single-Labeled Substrates for Monitoring With FRET

In brief, under special circumstances FRET is also possible with single-labeled oligos. In addition to dyes, also the natural base guanine is an efficient quencher for FAM *(5)*, and for (TAMRA) *(2,3)*. This property was used to follow association kinetics of FAM-labeled substrate with a hairpin ribozyme *(5)*, and tertiary structure formation of the hairpin ribozyme *(6)*. The approach is outlined in **Fig. 6**.

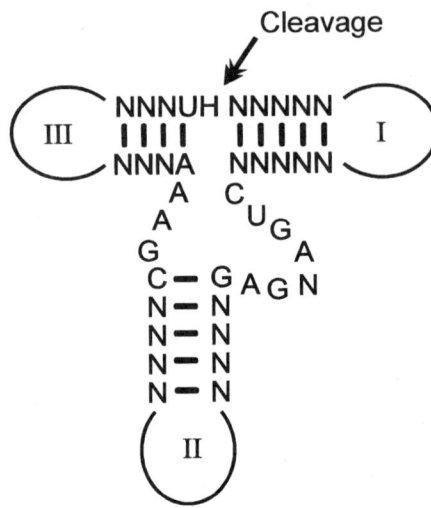

Fig. 1. Schematic hammerhead ribozyme structure with conserved sequence segments (H=A,C,U). Three alternative substrate/ribozyme formats can be derived by opening sets of two different loops. Format I/II after opening loops I and II results in short ribozyme with long substrate (*see* **Figs. 2** and **7**), other options are format I/III (large ribozyme with short substrate) and format II/III (short ribozyme with long substrate).

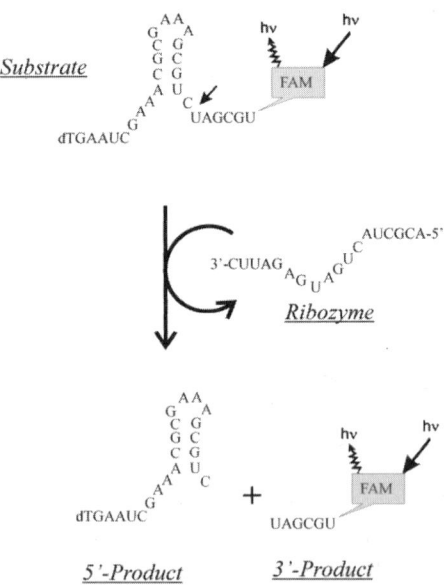

Fig. 2. Reaction scheme of format I/II hammerhead ribozyme. A single fluorescent label (FAM) is attached at the substrate, yielding a short FAM-labeled product.

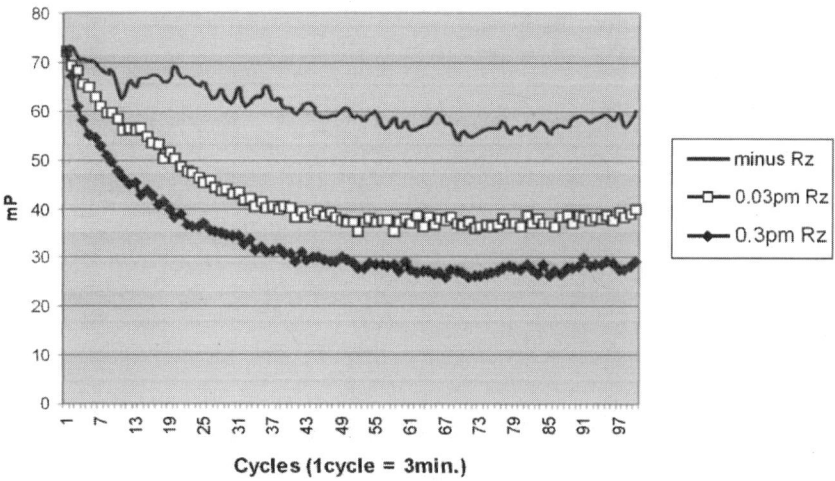

Fig. 3. Decreasing fluorescence polarization to follow hammerhead ribozyme cleavage reaction. The reaction depicted in **Fig. 2** was analyzed using 5 pmols substrate and varying amounts of ribozyme.

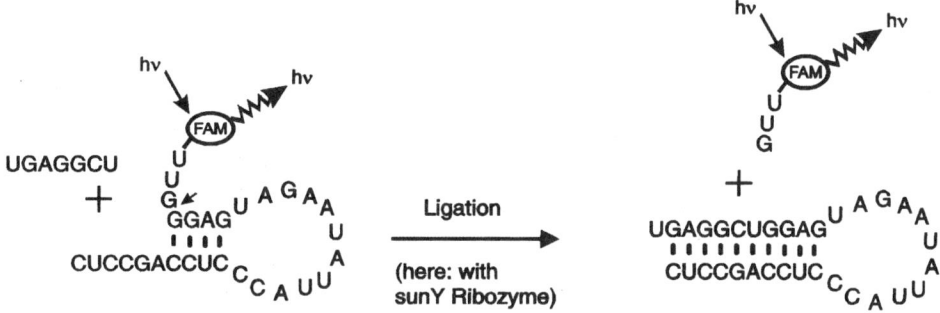

Fig. 4. Reaction scheme of RNA ligation catalyzed by sunY, a group I intron ribozyme. A single fluorescent label (FAM) is attached at the larger substrate hairpin, yielding a short FAM-labeled trinucleotide product.

## 3.3. Double-Labeled Substrates for Monitoring With FRET

### 3.3.1. FRET Pair With High Spectral Overlap: FAM and TAMRA

The high overlap of both fluorescent dyes requires: i) sophisticated software to calculate reliable FRET values, as provided by the SDS 7700 and similar instruments from Applied Biosystems; ii) accurate measurements require an additional means to calibrate the FRET data (**steps 6–8**). A I/II version of the hammerhead ribozyme was used (*see* **Figs. 1** and **7**). A chemically synthesized

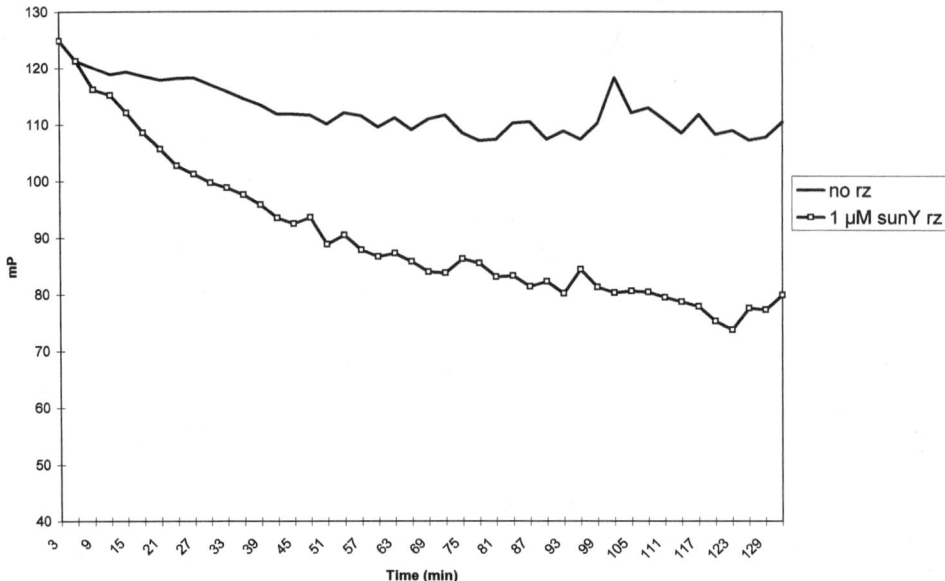

Fig. 5. Decreasing fluorescence polarization to follow sunY ribozyme ligation reaction. The reaction depicted in **Fig. 4** was analyzed using 5 μ*M* heptamer oligo, 2 μ*M* FAM-labeled hairpin and no or 1 μ*M* ribozyme.

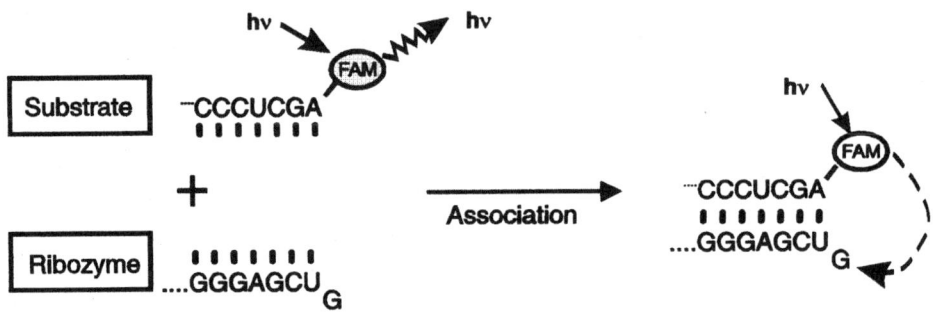

Fig. 6. Scheme for FRET with single fluorescent label. Approachment of FAM and guanine results in quenching and reduced fluorescence intensity. TAMRA may also be used in this fashion.

substrate was used. The fluorescent dyes TAMRA were attached to the 5'-terminal dT nucleotide (leaving a free 5'-hydroxyl for [32]P-labeling), and FAM to the 3'-end (FAM was provided by the CPG support). These constructs can be used for multiple- (excess substrate), and for single-turnover conditions (excess catalyst) *(3,4)*. Here, an example for multiple turnover is presented.

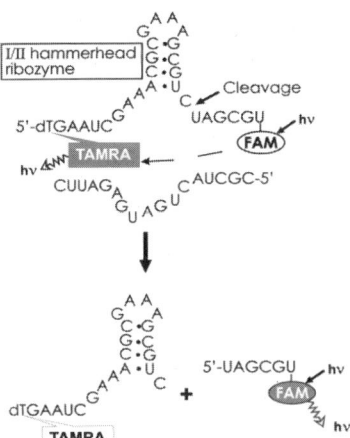

Fig. 7. Reaction scheme of format I/II hammerhead ribozyme. Two fluorescent labels with high spectral overlap (FAM and TAMRA) are attached at both ends of the substrate. Although the structure implies spatial separation of both dyes, FRET can occur as a result of the tight association of both hydrophobic dyes. Cleavage results in separation of both dyes, and FAM-fluorescence increases.

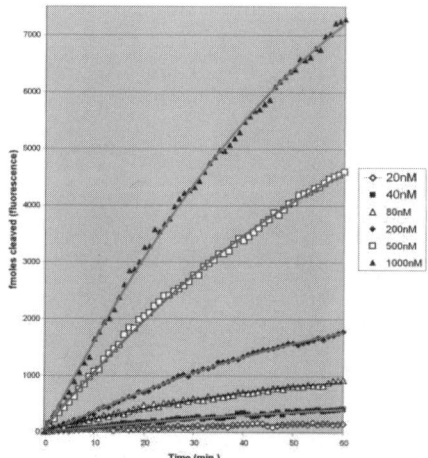

Fig. 8. Increasing fluorescence intensity to follow hammerhead ribozyme cleavage reaction. The reaction depicted in **Fig. 7** was analyzed under multiple-turnover conditions in the SDS 7700 instrument using 10 n*M* ribozyme and the indicated concentration range of substrate.

1. A 20-µL ribozyme reaction contained 10 n*M* ribozyme (200 fmols), and excess substrate (20–1000 n*M*; as specified in **Fig. 8**) in 50 m*M* Tris-HCl, pH 7.5, 20 m*M* MgCl$_2$ and 10% ethanol. For normalization of FRET data, a tracer amount of 5'-$^{32}$P-labeled substrate was included (*see* **steps 6–8**).

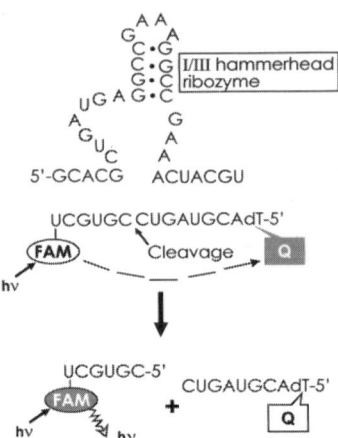

Fig. 9. Reaction scheme of format I/III hammerhead ribozyme. Both ends of the substrate carry dyes, either two fluorescent labels with low spectral overlap (FAM and Q = Cy5) or with one non-fluorescent dye (Q = dark quencher, e.g., Dabcyl). Although the structure implies spatial separation of both dyes, FRET can occur because of the tight association of both hydrophobic dyes. Cleavage results in separation of both dyes, and FAM-fluorescence increases. Substrate and product can be quantitatively analyzed in polyacrylamide gels with a fluorimager because unlike FAM, fluorescence of the quencher Cy5 (excited with red light) is not affected by variable quenching in substrate vs product.

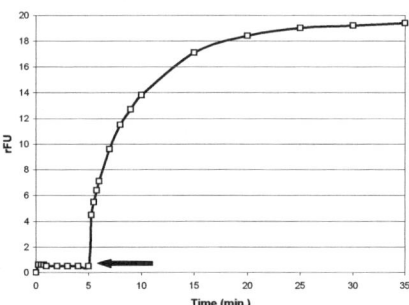

Fig. 10. Increasing fluorescence intensity to follow hammerhead ribozyme cleavage reaction. The reaction depicted in **Fig. 9** was analyzed under single-turnover conditions in the standard cuvet fluorimeter SFM 25, using 60 n$M$ substrate and 1 μ$M$ ribozyme.

2. For format I/III (the "normal substrate" type for hammerhead ribozymes, no multiple-turnover results are shown here. For single-turnover reactions (*see* **Subheading 3.3.2.** and **Figs. 9** and **10**), ribozyme and substrate were preincubated (95°C for 1 min, 5 min at 37°C) and the reaction was started by adding MgCl$_2$.

3. For format I/II (as shown in **Fig. 7**), start with MgCl$_2$ caused a time lag, therefore ribozyme and substrate were preincubated separately (95°C for 1 min, 5 min at 37°C), and the reaction was started by combining both components.

4. For following the reaction in real time: transfer reaction mix to a 96-well plate (Costar) and start the reaction by adding MgCl$_2$ (format I/III) or by adding ribozyme (format I/II).

5. Readings in the 'TaqMan system' (SDS 7700 from Applied Biosystems) were recorded (continuously for 15 s) in 1-min intervals at 37°C for 60 cycles (*see* **Fig. 8**).

6. For subsequent control of reaction products, stop reaction immediately by adding an equal volume of gel-loading buffer (containing 50 m$M$ ethylenediamine-tetraacetic acid [EDTA] to chelate Mg$^{2+}$), followed by gel electrophoresis and fluorescence imager analysis (e.g., Storm or Typhoon from Amersham-Pharmacia).

7. The fraction of cleaved substrate is determined with 5'-$^{32}$P-labeled RNA and phosphorimaging.

8. For calibration of FRET data, the raw multicomponent data are exported to Excel software, and FAM readings for every 1-min cycle were averaged. Data were normalized by setting the first readings of all wells as zero. Then, the determined cleavage fraction (**steps 6** and **7**) was equated to the final value for FAM fluorescence, and values for shorter times were calculated accordingly.

9. Determination of kinetic parameters for multiple-turnover reactions: initial reaction rates (fmols/min) were plotted against substrate concentrations. EnzPack software (Biosoft,) was used to obtain values for $K_m$ and $k_{cat}$ using the nonlinear Wilkinson method (*see* **Table 1**).

### 3.3.2. FRET With Minor Spectral Overlap

If the FRET pair has very little spectral overlap in the emission wavelengths (e.g., FAM at 520 nM and Cy5 at 660 nM), the approach is drastically simplified. The same also holds true if a "dark quencher" is used. This means that the dye absorbs light (quenches), but the dye is non-fluorescent and emits no fluorescent light of its own—e.g., Dabcyl can be used. i) No sophisticated software or instruments are required; a simple fluorimeter, ideally equipped with a thermostated microcuvet is sufficient. ii) No additional $^{32}$P-labeling is required for calibrating the FRET data (**steps 6–8**). **Please note:** This is not possible if a dark quencher is used (*see* **steps 7** and **8**).

A I/III version of the hammerhead ribozyme was used (**Fig. 9**). Instead of TAMRA, Cy5 was attached to the 5'-terminal dT nucleotide, and FAM (provided by the CPG support) was added at the 3'-end. These constructs can be used for multiple- (excess substrate), as well as for single-turnover conditions (excess catalyst). Here, an example with a FAM-Cy5-labeled substrate under single-turnover conditions is presented.

1. For a 50-μL ribozyme reaction: Mix 10 μL of 250 m$M$ Tris-HCl, pH 6.5 with 3 pmols substrate (60 n$M$ final concentration), and water to a total volume of 40 μL.

**Table 1**
**Example Results for Kinetic Parameters Under Multiple-Turnover Conditions**

Multiple turnover with hammerhead ribozyme in format I/II derived from **Figs. 7** and **8**.

| Kinetic method | 5'-$^{32}$P-RNA | 5'-$^{32}$P-RNA | FRET |
|---|---|---|---|
| Fluorescent label | None | 3'-FAM/5'-TAMRA | 3'-FAM/5'-TAMRA |
| $K_m$ [µ$M$] | $0.5 \pm 0.15$ | $1.0 \pm 0.3$ | $1.0 \pm 0.2$ |
| $k_{cat}$ [min$^{-1}$] | $0.5 \pm 0.1$ | $3.0 \pm 0.1$ | $3.0 \pm 0.2$ |
| $k_{cat}/K_m$ [min$^{-1}$ µ$M^{-1}$] | $1.0 \pm 0.4$ | $3.0 \pm 1$ | $3.0 \pm 1$ |

Data from direct measurements using the $^{32}$P-label are compared with FRET. Comparison with the non-fluorescent RNA substrate reveals a slight increase of $k_{cat}$.

**Table 2**
**Example Results for Kinetic Parameters Under Single-Turnover Conditions**

Single turnover with hammerhead ribozyme in format I/II derived from **Figs. 9** and **10**.

| Kinetic method | FRET |
|---|---|
| Fluorescent label | 3'-FAM/5'-Cy5 |
| $k_{obst}$ [min$^{-1}$] | $0.5 \pm 0.03^a$ |

$^a$The value for $k_{obs}$ is determined at pH 6.5 to slow down the fast reaction with high excess of ribozyme and make kinetic measurements more convenient. For comparison with data in **Table 1** (reactions with excess of substrate at pH 7.5) this value must be multiplied by a factor of 10 (*10*).

**Please note:** For better quantitative detections, the reaction with excess ribozyme is slowed down 10-fold by analysis at pH 6.5 (*see* **Table 2**).

2. Transfer to a preincubated cuvet (at 37°C or at RT) in SFM 25 Fluorimeter (Kontron Instruments), set to 488-nm excitation and 518-nm emission.
3. Add 10 µL of ribozyme (excess requires >50 pmols or >1 µ$M$ final concentration).
4. Take readings immediately, then after 15, 30, 45, and 60 s, and then every minute until a stable signal is reached (about 5 min; *see* **Fig. 10**).
5. Start reaction by adding 5 µL of 200 m$M$ MgCl$_2$.
6. Take further readings immediately, then after 15, 30, 45, and 60 s, and then every minute for 30 min (*see* **Fig. 10**).
7. For subsequent control of reaction products, stop reaction by adding an equal volume of gel loading buffer (containing 50 m$M$ EDTA to chelate Mg$^{2+}$), followed by gel electrophoresis and fluorescence imager analysis.
8. Using red light (>600 nm—e.g., in Storm or Typhoon from Amersham-Pharmacia) for excitation, only fluorescence from the quencher Cy5 is emitted, and this signal is not affected by variable amounts of FAM. This means that Cy5

signal per molecule is the same for substrate and for 5'-product. Thus, the residual amount (if any) of uncleavable substrate can be determined, requiring only the fluorescent-labeled RNA. **Please note:** If a non-fluorescent dark quencher such as Dabcyl is used, this option is obviously not possible.

9. The fraction of finally cleaved substrate is $F_\infty$ (*see* **step 10**) and it is used to calibrate the fluorescence readings from **step 6**.

10. Determination of kinetic parameters for single-turnover reactions: values for time-dependent cleavage fractions were used to calculate $k_{obs}$ by fitting experimental data to the equation $F_t = F_0 + F_\infty(1 - e^{-k_{obs}t})$, using $F_0 = 0$ and $F_\infty$ as determined in **steps 8** and **9** (*see* **Table 2**).

## 3.4. Double-Labeled Substrates for Simultaneous Monitoring With FRET and With Fluorescence Polarization

This example is similar to the ribozyme reaction in **Subheading 3.1.2.** (*see* **Figs. 4** and **5**), but here a somewhat modified substrate type was used, including a double fluorescent-labeled short RNA/DNA chimera. Aside from some sequence changes, the substrate structure was modified. The hairpin loop was deleted, the stem was replaced by a short unmodified RNA (bottom strand in **Fig. 11**), hybridized to six nucleotides of an 8-mer nucleic acid. In addition, ribo-Us in the 8-mer were replaced by dT as potential dye-coupling positions, here the 3' terminal dT was coupled to TAMRA, and FAM was attached at the 5'-end.

Reactions are performed as described in **Subheading 3.1.2.** Here, each oligo of the three-part substrate was used at 5 µ*M*, and incubated with 1 µ*M* sunY ribozyme. Results in **Fig. 12** display FRET data (increasing fluorescence intensity of FAM) and fluorescence polarization data (decreasing polarization values). Both data sets were obtained simultaneously with the POLARstar instrument.

## 3.5. Catalytic Nucleic Acids in Real-Time Detection of Nucleic Acid Amplification

The application of polymerase chain reaction (PCR) in routine diagnostics was boosted by the availability of real-time detection procedures to avoid post-PCR sample handling and potential laboratory contamination with the highly amplified nucleic acids from positive samples or positive control reactions. Several different technologies have been developed for this purpose. Artus GmbH, "the PCR reference," has developed numerous real-time PCR kits for convenient, reliable, high-throughput detection of pathogens such as viruses, bacteria, and protists (*see* www.artus-biotech.com). These kits are based on different formats of hybridizing DNA probes, and they are available for the major real-time PCR instruments, such as the various SDS or TaqMan systems from Applied Biosystems, the LightCycler from Roche and the RotorGene

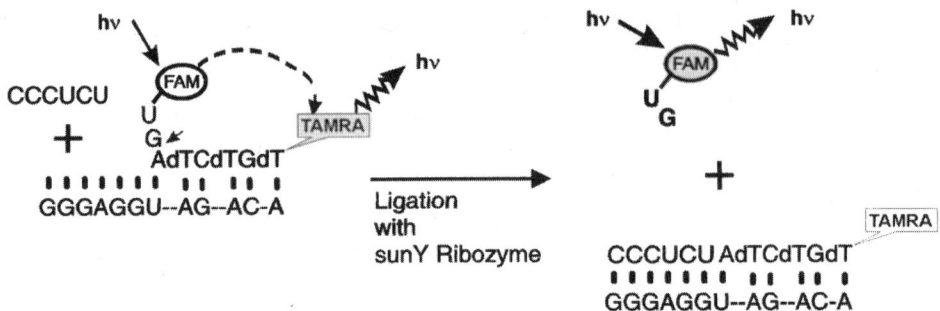

Fig. 11. Reaction scheme of RNA ligation catalyzed by sunY, a group I intron ribozyme. In contrast to **Fig. 4**, a three-composite substrate was used. The chimeric RNA/DNA oligo carries two fluorescent labels (FAM and TAMRA). Ligation disrupts the FRET pair and yields a short FAM-labeled dinucleotide product.

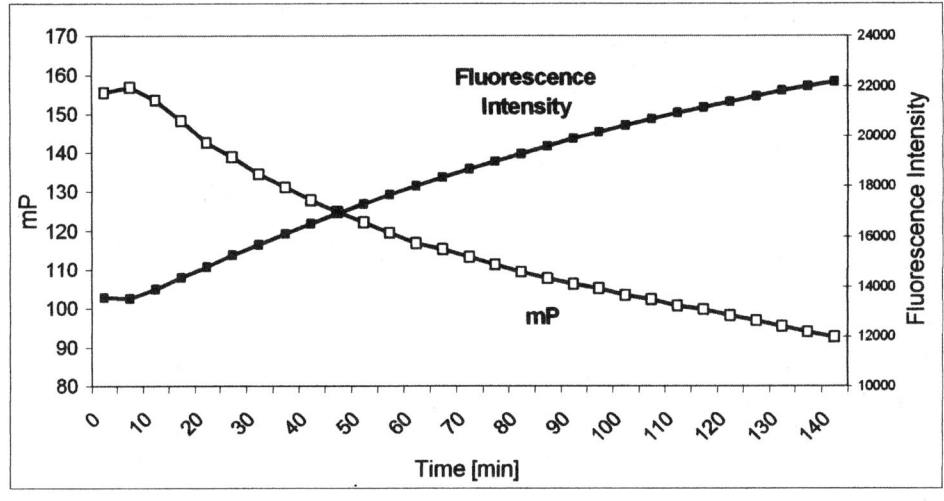

Fig. 12. Decreasing fluorescence polarization and simultaneously increasing fluorescence intensity to follow sunY ribozyme ligation reaction. The reaction depicted in **Fig. 11** was analyzed using 5 μM each of the three substrate oligos, and 1 μM sunY ribozyme. FRET data (increasing fluorescence intensity of FAM) and fluorescence polarization data (decreasing polarization values) were obtained simultaneously with the POLARstar instrument.

from Corbett Research (7). These kits use internal target-specific probes, which allow the detection of specific target-sequences in the amplified nucleic acid. However, the application of DNAzymes (catalytic DNA) in PCR-monitoring is presently limited to the primer-dependent introduction of catalytic nucleic-acid motifs. Therefore, they are similar to PCR-monitoring with intercalating dyes such as SYBR, and they are limited to the monitoring of *all and any* primer-derived PCR amplification reactions *(8)*. Unlike intercalating dyes, which detect the formation of double-stranded DNA, catalytic nucleic acids are not limited to PCR monitoring. Prior to the studies with DNAzymes in PCR, it was demonstrated that amplification of single-stranded RNA in NASBA can be monitored by coupled ribozyme reactions and specific detection of internal target sequences is possible *(9)*.

### 3.5.1. Ribozymes in Real-Time Detection of Amplification by In Vitro Transcription: NASBA (Nucleic Acid Sequence-Based Amplification) or 3SR (Self-Sustained Sequence Replication)

Similar to PCR, these systems apply two primers enclosing the nucleic acid segment to be amplified. As outlined in **Fig. 13**, a hammerhead ribozyme motif can be introduced by one of the primers, and the resulting amplified RNA segment will contain an active ribozyme *(9)*.

The short sequence motif GAAA in the core of the hammerhead ribozyme (*see* **Fig. 1**) can be exploited in yet another alternative, the format II/III, creating a short ribozyme and a rather long substrate. The tetranucleotide GAAA is very frequent in natural sequences and can be exploited for the detection of specific target-sequences in the amplified RNAs (*see* **Fig. 14**). Practical possibilities for species-specific NASBA detection assays have also been explored *(9)*.

The quantitative detection of the ribozyme activity thus created is possible with the same approaches as described in **Subheadings 3.1.** and **3.3.**

### 3.5.2. DNAzymes in Real-Time Detection of Amplification by PCR

As in conventional PCR, this system applies two primers enclosing the nucleic acid segment to be amplified. As indicated in **Fig. 15**, a DNAzyme motif can be introduced by one of the primers, and the resulting amplified DNA segment will contain an active DNAzyme *(8)*.

The quantitative detection of the this DNAzyme activity is possible with the same approaches described in **Subheadings 3.1.** and **3.3.**

## 4. Notes

1. Problems may occur with insufficient purity, resulting in high background fluorescence intensity or low fluorescence polarization value. Remove traces of free fluorescent dyes by PAGE purification. Some commercial suppliers of synthetic RNA provide this option when obtaining synthetic RNA, such as iba GmbH (Göttingen, Germany; www.iba-go.de).
2. Dilutions of fluorescent nucleic acids may not be linear. Due to the hydrophobic dyes, multi-molecular aggregates can form. This effect can be reduced by including 10% ethanol, 0.1% sodium dodecyl sulfate (SDS) (with the further advantage

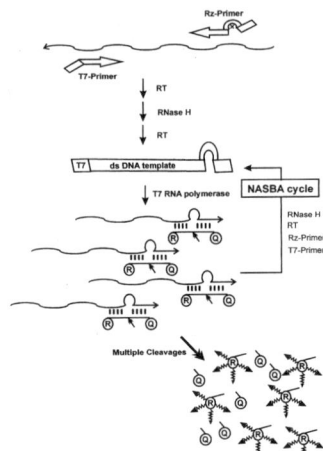

Fig. 13. Schematic NASBA reaction including a primer-derived ribozyme for real-time monitoring. Both primers contain a target-specific 3'-terminal segment. In addition, one primer includes a promotor for T7 RNA polymerase, and the second a ribozyme motif in antisense orientation (indicated by X). NASBA cycles generate RNA transcripts with a 3'-terminal, active ribozyme. Each RNA associates transiently with double fluorescent-labeled substrate (in high excess; multiple-turnover conditions) and yields multiple cleavage products with increasing fluorescence signal of the reporter dye.

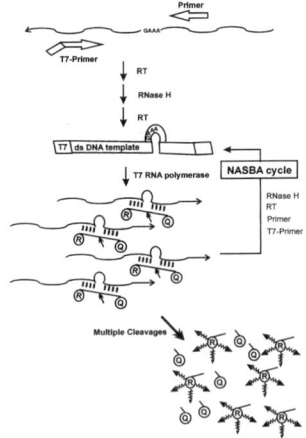

Fig. 14. Schematic NASBA reaction with a target-derived ribozyme for real-time monitoring. The target sequence contains the minimal GAAA hammerhead sequence (*see* **Fig. 1**) and NASBA cycles generate RNA transcripts with an internal, active ribozyme in format II/III. With high excess of double fluorescent-labeled substrate (multiple-turnover conditions) each RNA yields multiple cleavage products, with increasing fluorescence signal of the reporter dye.

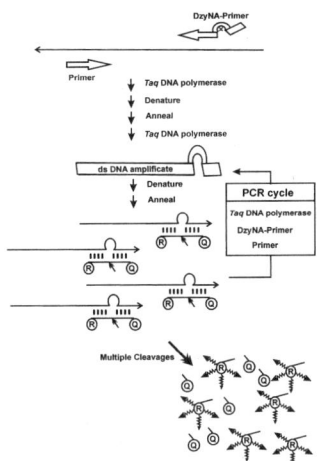

Fig. 15. Schematic PCR including a primer-derived DNAzyme for real-time monitoring. Both primers contain a target-specific sequence. In addition, one primer includes a DNAzyme motive in antisense orientation (indicated by X). PCR cycles generate DNA amplicons with a 3'-terminal, active DNAzyme. After denaturation, in the annealing step, DNAzyme and substrate can associate. Each single-stranded DNA associates transiently with double fluorescent-labeled substrate (in high excess; multiple-turnover conditions) and yields multiple cleavage products with increasing fluorescence signal of the reporter dye.

    of inactivating protein nucleases, without interference with ribozyme reactions) or a nonionic detergent.

3. In general, a fraction of the substrate RNA is not cleaved (can be more than 10%), even after prolonged incubation and with excess ribozyme, this is presumably caused by alternate, inactive conformations. The fraction of active substrate should be determined and used to quantitate substrate amounts and concentrations.

## References

1. Singh, K. K., Rücker, T., Hanne, A., Parwaresch, R., and Krupp, G. (2000) Fluorescence polarization for monitoring ribozyme reactions in real time. *BioTechniques* **29,** 344–351.
2. Singh, K. K., Rücker, T., Rörig, C., Hanne, A., Klapper, W., Paulsen, H., et al. (2000) Application of fluorescence for RNA studies. In: *Ribozyme Biochemistry and Biotechnology* (Krupp, G. and Gaur, R. K., eds.), Eaton, Natick, MA, pp. 373–399.
3. Singh, K. K., Parwaresch, R., and Krupp, G. (1999). Rapid kinetic characterization of hammerhead ribozymes by real-time monitoring of fluorescence resonance energy transfer (FRET). *RNA* **5,** 1348–1356.

4. Singh, K. K., Parwaresch, R., and Krupp, G. (2000) Real time kinetics of ribozyme reactions. In: *Ribozyme Biochemistry and Biotechnology* (Krupp, G. and Gaur, R. K., eds.), Eaton, Natick, MA, pp. 351–371.
5. Walter, N. G. and Burke, J. M. (1997) Real-time monitoring of hairpin ribozyme kinetics through base-specific quenching of fluorescein-labeled substrates. *RNA* **3(4),** 392–404.
6. Walter, N. G., Hampel, K. J., Brown, K. M., and Burke, J. M. (1998) Tertiary structure formation in the hairpin ribozyme monitored by fluorescence resonance energy transfer. *EMBO J.* **17,** 2378–2391.
7. Krupp, G., Laue, T., Söller, R., Grewing, T., Cramer, S., Heß, M., and Spengler, U. (2003) Nucleic acid preparations of pathogens from biological samples for real-time PCR analysis. In: *Nucleic Acids Isolation Methods* (Bowien, B. and Dürre, P., eds.), American Scientific Publishers, Stevenson Ranch, pp. 95–135.
8. Todd, A. V., Fuery, C. J., Impey, H. L., Applegate, T. L., and Haughton, M. A. (2000) DzyNA-PCR: use of DNAzymes to detect and quantify nucleic acid sequences in a real-time fluorescent format. *Clin. Chem.* **46,** 625–630.
9. Krupp, G. (2000) Detection of nucleic acid amplified products. Patent Numbers *WO005850; DE19915141.* artus GmbH.
10. Stage-Zimmermann, T. K. and Uhlenbeck, O. C. (1998) Hammerhead ribozyme kinetics. *RNA* **4,** 875–889.

# 5

## Quantification of Ribozyme Target RNA Using Real-Time PCR

### Dagmar Klein, Camillo Ricordi, and Ricardo L. Pastori

#### Summary

An important part of the ribozyme efficiency-screening process is to have a fast and accurate way to measure steady-state levels of the target RNA. Here, we describe the use of real-time polymerase chain reaction (PCR) for quantification of ribozyme target transcripts. In contrast to classical quantitative PCR, real-time PCR does not require extensive manipulation or generation of relatively complex reagents, thus reducing the risk of contamination. PCR products generated by Taq polymerase in the presence of SYBR Green dye I can be monitored each cycle by collecting fluorescence signals emitted only as the double-stranded DNA is formed. The temperature at which the fluorescent data used for quantification are collected is based on the melting-curve analysis of the amplified product. After constructing a standard curve by plotting the log of the standards' copy number vs their fractional cycle number, the copy number of the unknown samples is automatically determined by interpolation of this curve. However it is very important to validate the melting curve profile with standard gel electrophoresis, particularly while setting up the technique. Real-time PCR is fast and reproducible. Excluding the isolation of RNA and synthesis of cDNA, the results can be obtained in less than 1 h. The coefficient of variance is 15% in the range of 104–106 gene copies.

**Key Words:** RNA quantitation; real-time PCR; CD95 (Fas); ribozymes; insulinoma cell lines; βTC-3 cells.

## 1. Introduction

The use of ribozymes for human gene therapy requires proof of their cellular efficacy. Designing ribozymes against a specific RNA remains a semi-empirical task, in which site accessibility on the target RNA is a critical factor for ribozyme efficiency *(1)*. Many strategies had been developed to predict ribozyme/target RNA interactions. However, testing the effectiveness of a

From: *Methods in Molecular Biology, vol. 252: Ribozymes and siRNA Protocols, Second Edition*
Edited by: M. Sioud © Humana Press Inc., Totowa, NJ

relatively large number of ribozymes directed against different sites in the same RNA is essential.

Thus, an important part of the ribozyme screening process is to utilize a fast and accurate way to measure steady-state levels of the target RNA. This can be achieved by Northern analysis, RNase protection assays, or quantitative reverse transcriptase-polymerase chain reaction (RT-PCR). Although the first two methods are suitable for relative quantification of RNA, they are also time-consuming, the third method, quantitative RT-PCR, requires extensive post-PCR manipulations, thus increasing the possibilities of contamination *(2)*. Recently, real-time PCR—a new product detection and quantification technique—was described *(3,4)*. This method simultaneously carries out and evaluates reactions as they are progressing in the thermal cycler. Real-time PCR is much faster and far more accurate than "end-point" PCR in which amplified product is measured after the PCR is completed.

The quantification is most accurate when the log of template-amplified copies or amplicons increases linearly with the number of cycles. Utilizing other stages of a PCR is highly inaccurate and irreproducible. The identification of these PCR cycles with end point PCR is fairly difficult to achieve. However, real-time PCR has the capability to exactly identify these cycles, improving the quantification of nucleic acids considerably. It has been reported that real-time PCR can amplify as low as 103 copies *(5)*, in a reliable and reproducible fashion. Monitoring formation of the amplified product in each cycle can be performed either by detection of fluorescence generated by using hybridization probes or by SYBR Green I dye. In case of hybridization probes, the fluorescence is generated during the annealing phase by fluorescence resonance energy transfer (FRET) from two specific contiguous primers, each labeled with a different fluorophore. Hybridization probes provide the highest specificity to real-time PCR amplifications. SYBR Green I, is a dye that only becomes fluorescent when bound to double-stranded DNA, and thus, fluorescence is monitored after each elongation step. Fluorescence monitoring with SYBR Green I combined with melting-curve analysis is highly sensitive, specific, and cost effective.

Our laboratory is focusing on designing ribozymes that will inhibit the expression of pro-apoptotic genes, induced by cytokine-mediated nonspecific inflammatory events, in transplanted pancreatic islets of Langerhans *(6)*. Fas is one of the death receptors that mediates the detrimental effect of cytokines on the insulin-producing β cells. We have designed ribozymes that target Fas mRNA, and have expressed them in mouse insulinoma cell lines derived from RIP-Tag mice *(7)*. To illustrate the value of real-time PCR for quantification of ribozyme target RNA, we describe the method of a SYBR Green I fluorescence detection in RT-real-time PCR of Fas mRNA from insulinoma-cell line βTC-3.

## 2. Materials

All reagents used for RNA manipulation must be prepared using double-distilled, water purified with a Millipore system and treated with diethylpyrocarbonate (DEPC) to remove possible traces of nucleases. DNA oligonucleotides may be ordered from numerous companies. Purification by reverse-phase cartridge provides the required purity level for real-time PCR applications.

### 2.1. RNA Extraction and cDNA Synthesis

Total RNAs were extracted from control and ribozyme-transfected cells using RNeasy Total RNA kit (*see* **Note 1**). RNA concentrations were determined by spectrophotometry and its integrity was verified by electrophoretic analysis (*see* **Note 1**). The RT reaction was performed according to the manufacturer specifications (Invitrogen).

### 2.2. Real-Time PCR Amplification Procedure

Real-time PCR was performed using a ready-to-use reaction mix 10X DNA Master SYBR Green I (Roche Molecular Biochemicals, Indianapolis, IN) preincubated for 5 min at room temperature with 0.55 µg of TaqStart antibody (Clontech Laboratories, Palo Alto, CA).

## 3. Methods
### 3.1. Melting-Curve Analysis and Determination of the Temperature at Which to Acquire Fluorescence

To avoid the contribution of fluorescence generated by nonspecific PCR products (*see* **Notes 2** and **3**), the fluorescence data for quantification should be acquired at no more than 2°C below the melting temperature (Tm) of the amplified product. Therefore, the first step in real-time PCR amplification is to determine the Tm of the amplicon. **Figure 1** shows the melting-curve profile of the Fas amplicon, performed by the LightCycler software. The SYBR Green I dye is fluorescent only when bound to double-stranded DNA.

1. To determine the melting curve of an amplicon, increase the temperature at a slow rate of 0.2°C/s and monitor the fluorescence continuously. Sharp decrease in the fluorescence signal should be observed when the characteristic melting temperature of the amplified product is reached. The Tm of the Fas amplicon is displayed when the negative derivative of the fluorescence (-dF/dT) is plotted against the temperature. A sharp peak with a Tm of 87°C represents specific Fas amplification product, and the primer-dimer amplification is readily observed as a broad peak with melting temperatures ranging between 72 and 83°C. The size of the Fas amplicon was confirmed by agarose electrophoresis (**Fig. 1**).

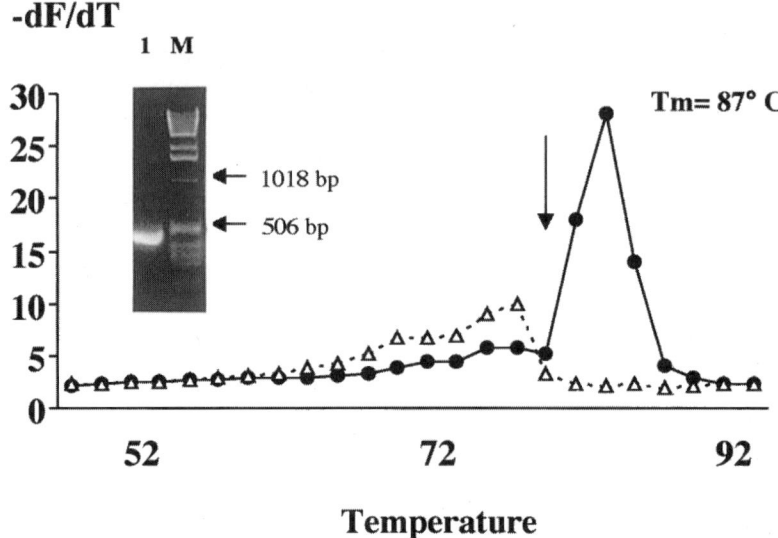

Fig. 1. Melting-curve profile of Fas amplicon. Melting-curve analysis carried out after Fas amplification of cDNA from βTC-3 cells induced for 7 h with IL-1 ($10^4$ U/L) and IFNγ ($10^5$ U/L) (●) or without DNA template (◊). The first negative derivative of fluorescence vs time (-dF/dT) is plotted against temperature. The Fas product has a Tm of 87°C. The arrow marks the temperature (85°C) at which fluorescent data will be acquired for quantification of Fas transcripts. Agarose gel electrophoresis of the PCR product is shown. M is a 1-kb ladder DNA marker.

2. To avoid fluorescence contribution from primer dimers, in all subsequent Fas transcript quantifications, the fluorescence data was collected at 85°C, two degrees below the Tm.

### 3.2. Description of Melting-Curve Analysis Procedure

After completion of PCR, a melting-curve analysis is performed by heating the PCR reaction to 95°C, and cooling it to 55°C at 20°C/s, followed by raising the temperature to 94°C at much slower rate (0.2°C/s) and acquiring fluorescence continuously.

### 3.3. Quantification of Fas Transcripts

To quantify ribozyme target transcripts, the following procedure was performed:

1. The cDNA samples were amplified simultaneously with serially diluted standards representing $8 \times 10^4$ to $16 \times 10^6$ copies of the Fas-amplified fragment.

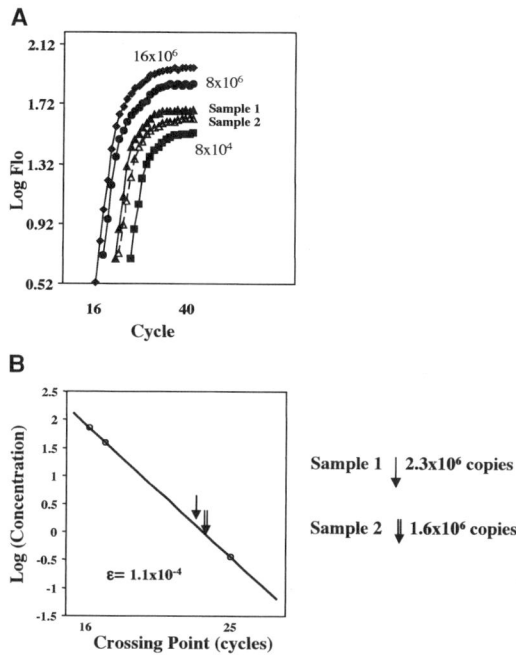

Fig. 2. Quantification of Fas transcripts. **(A)** The logarithmic plot Log Flo vs Cycle shows the log-linear phase of Fas amplification from reactions containing different standard Fas copy numbers after the background fluorescence was automatically substracted. Sample 1: cDNA from mock transfected βTC-3 cells and sample 2 cDNA from βTC-3 cells transfected with an anti-Fas ribozyme. **(B)** By plotting the log of standard concentrations vs the fractional cycle number "crossing points," a standard curve is constructed. The amount of unknown samples 1 and 2 is obtained by the best-fit interpolation of the standard curve carried out automatically by the "LightCycler" software. ($\varepsilon = 1.1 \times 10^{-4}$, the sum of squares of deviation/number of measuring points). Arrows mark the interpolations for sample 1 and 2. Each measurement was performed in duplicates. At this copy number, the coefficient of variation is 15%.

2. The fluorometric data was displayed and analyzed with the "LightCycler" software throughout the amplification. The program has the capability to extrapolate the log-linear values of the log fluorescence vs cycle graphic (**Fig. 2A**) to reach a fractional cycle number known as "crossing number" (the fractional cycle number at which the onset fluorescence was detected).
3. A standard curve was constructed by plotting the log concentration of the standards vs their respective "crossing points" (**Fig. 2B**).
4. The number of copies originally present in the cDNA samples was calculated and adjusted automatically from the best-fit interpolation of the standards curve, to

reach the lowest mean squared error (**Fig. 2B**). Samples were analyzed in duplicates to minimize the tube-to-tube reaction variance.

5. Because of the variability introduced by the RT reaction, the cDNA samples must be standardized to have the same copy number of a housekeeping gene to fully utilize all of the valuable features of the real-time PCR. In the experiments shown in the study, we have used the β-actin gene as an adequate normalizing factor. Therefore, a separate amplification of the same cDNA was performed to normalize samples by copy number ratios between the housekeeping gene β-actin and Fas. The same consideration described here also applies for the housekeeping gene amplification. In the experiment shown in **Fig. 2A** and **2B**, the level of Fas RNA in βTC-3 cells expressing one of our anti-Fas ribozymes was 32% lower than corresponding controls (coefficient of variance = 15%). Both analyses of the samples required less than 1 h.

## 3.4. Description of Real-Time PCR Amplification Procedure

The fluorescent thermal cycler utilized in our experiments ("LightCycler"/ Roche Molecular Biochemicals, Indianapolis, IN) is equipped with software with which we can follow the PCR reaction "on-line" step-by-step through all the phases. It also provides the melting-curve analysis and calculations of melting temperature of the PCR product (*see* **Notes 2** and **3**).

1. The primers used in the amplification of Fas and the housekeeping gene β-actin were:
   Fas forward: 5'-TGTTTTCCCTTGCTGCAGAC-3'
   Fas reverse: 5'-TGGCTCAAGGGTTCCATGTT-3'
   Actin forward: 5'-GATGACCCAGATCATGTTTG-3'
   Actin reverse: 5'-AGGCTGGAAGAGTGCCTCA-3'
   PCR amplifications were performed in a fluorescence temperature cycler (LightCycler/ Roche Molecular Biochemicals) in a final volume of 20 µL containing 2 µL of ready-to-use reaction mix 10X DNA Master SYBR Green I preincubated 5 min at room temperature with 0.55 µg of TaqStart antibody, 3 m$M$ MgCl$_2$, 0.5 µ$M$ of each primer and 2 µL of the cDNA as a template. The template was set-up in three different dilutions to fit the quantitative curve and minimize the error. The amplification program for Fas was designed as follows: Transition rate of 20°C/s for all steps involved. One cycle of denaturation 95°C 20s, followed by 40 cycles of:
   Denaturation 95°C for 2 s
   Annealing to 60°C for1 s
   Extension 72°C for 10 s

2. Acquisition of Fluorescence: heating to 85°C for 1 s. (The Tm of Fas is 87°C.) The β-actin gene transcript was amplified using the same cDNA with Actin-forward and Actin-reverse primers. The amplification program was similar to that of Fas, except that the annealing temperature was 58°C. Fluorescence was acquired at 88°C. PCR amplifications generated amplicons of 470 bp for Fas and 439 bp for β-actin.

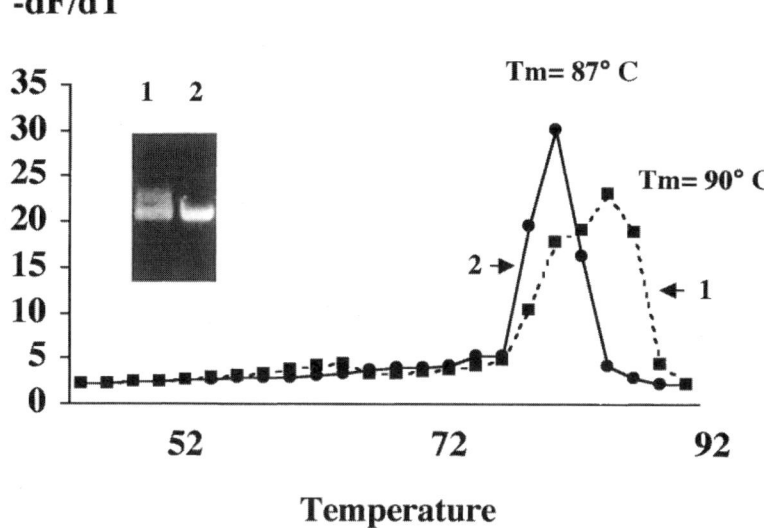

Fig. 3. Melting-curve profile of PCR products containing specific and nonspecific amplicons. Melting-curve analysis performed after PCR amplification of cDNA from βTC-3 cells induced for 7 h with various cytokine concentrations; $10^4$ U/L and $10^5$ U/L (●) or $0.3 \times 10^4$ U/L and $0.3 \times 10^5$ U/L of IL-1 and IFN-γ (■), respectively. An agarose gel electrophoresis of PCR products from both reactions is shown in the inset.

## 4. Notes

1. As in standard PCR amplification, in real-time PCR the good quality of RNA is essential in order to achieve reproducible results. Several commercially available RNA purification kits yield RNA of sufficient quality for real-time PCR (Invitrogen, Quigen, Ambion).
2. In general, designing primers for real-time PCR using SYBR green monitoring requires the same considerations as that for standard PCR. Several free programs are accessible from the Internet (e.g., Primer3 from MIT, www-genome.wi.mit.edu/cgi-bin/primer/primer3_www.cgi). Amplicon should be preferentially less than 500 bp, and when possible, the primers should span through different exons to avoid spurious fluorescence from contaminating genomic DNA. If that is not possible, the RNA should be treated with DNase I to eliminate genomic DNA contamination.
3. The melting-curve analysis of the amplified products is a valuable feature provided by this technology. Real-time PCR displays accumulated fluorescence regardless of the size of the amplified product; therefore it is important to verify that the peak of the melting curve corresponds to a fragment of the expected size. As we illustrate in **Fig. 3**, small variations in the Tm indicate a different pattern of amplification. The melting curve obtained by amplification of cDNA from β TC-3 cells induced with suboptimal concentration of cytokines yielding a mixture

of specific and nonspecific products. The melting-curve profile presented a specific peak of 87°C and an additional peak of 90°C. Gel electrophoresis confirmed the presence of two fragments: a specific product of 480 bp and a larger fragment of 500 bp, demonstrating that even a slight shift in Tm of the amplified product indicates the amplification of a nonspecific product.

If spurious amplified fragments have a Tm lower than the expected fragment, acquiring the fluorescent data at temperature between the nonspecific and specific products allows precise quantification of nucleic acids. If the nonspecific product has a Tm sufficiently higher than the specific product, acquiring fluorescence at temperatures flanking the specific peak could use a subtractive strategy. Alternatively different sets of primers can be used to generate only specific amplicons or the reaction can be monitored by FRET analysis of specific fluorescent primers *(9)*.

# References

1. Birikh, K. R., Heaton, P. A., and Eckstein, F. (1997) The structure, function and application of the hammerhead ribozyme. *Eur. J. Biochem.* **245,** 1–16.
2. Freeman, W. M., Walker, S. J., and Vrana, K. E. (1999) Quantitative RT-PCR: pitfalls and potential. *BioTechniques* **26,** 112–125.
3. Heid, C. A., Stevens, J., Livak, K. J., and Williams, P. M. (1996) Real time quantitative PCR. *Genome. Res.* **6,** 986–994.
4. Wittwer, C. T., Ririe, K. M., Andrew, R. V., David, D. A., Gundry, R. A., and Balis, U. J. (1997) The LightCycler: a microvolume multisample fluorimeter with rapid temperature control. *BioTechniques* **22,** 176–181.
5. Morrison, T. B., Weis, J. J., and Wittwer, C. T. (1998) Quantification of Low copy transcripts by continuous SYBR Green I monitoring during amplification. *BioTechniques* **24,** 954–962.
6. Berney, T. and Ricordi, C. (1999) Islet transplantation. *Cell Transplant.* **8,** 461–464.
7. Hanahan, D. (1985) Heritable formation of pancreatic beta-cell tumors in transgenic mice expressing recombinant insulin/simian virus 40 oncogenes. *Nature* **315,** 33–40.
8. Efrat, S., Linde S, Kofod, H., Spector, D., Delannoy, M., Grant, S., et al. (1988) Beta-cell lines derived from transgenic mice expressing a hybrid insulin gene-oncogene. *Proc. Natl. Acad. Sci. USA* **85,** 9037–9041.
9. De Silva, D., Reiser, A., Herrimann, M., Tabiti, K., and Wittwer, K. (1998) Rapid genotyping and quantification on the LightCycler with hybridization probes. *Biochemica* **2,** 11–14.

# 6

# Analysis of Ribozyme Structure and Function by Nucleotide Analog Interference Mapping

**Soumitra Basu, Catherine Pazsint, and Gopeswar Chowdhury**

## Summary

Nucleotide analog interference mapping (NAIM) is a quick and effective method to define concurrently, yet singly, the importance of specific functional groups at particular nucleotide residues in relation to the structure and function of an RNA. NAIM can be utilized on virtually any RNA with an assayable function, including catalytic RNAs. The method hinges on the ability to successfully incorporate, within an RNA transcript, various 5'-$O$-(1-thio)nucleoside analogs randomly via in vitro transcription. This can be achieved by using wild-type or Y639F mutant T7 RNA polymerase, thus creating a pool of analog-doped RNAs. When subjected to a selection step to separate the active transcripts from the inactive ones, the pool helps to identify functional groups that are crucial for RNA activity. The technique can be used to study ribozyme structure and function via monitoring of cleavage or ligation reactions, or define functional groups that are critical for RNA folding, RNA-RNA interactions, and RNA interactions with proteins, metals, or other small molecules. All major classes of catalytic RNAs have been examined by NAIM. This is a generalized approach that should provide the scientific community with the tools to better understand the RNA structure-activity relationship (SAR).

**Key Words:** Nucleotide-analog interference mapping; ribozyme; RNA folding.

## 1. Introduction

Biochemical studies that rely on chemical modifications using footprinting or interference mapping have been extensively used to derive information on RNA structure and function *(1,2)*. NAIM is a technique that blends the phosphorothioate modification interference with the inherent simplicity of RNA sequencing to generate information on critical functional groups within an RNA (**Fig. 1**; *3–5*). A major difference between NAIM and site-directed

From: *Methods in Molecular Biology, vol. 252: Ribozymes and siRNA Protocols, Second Edition*
Edited by: M. Sioud © Humana Press Inc., Totowa, NJ

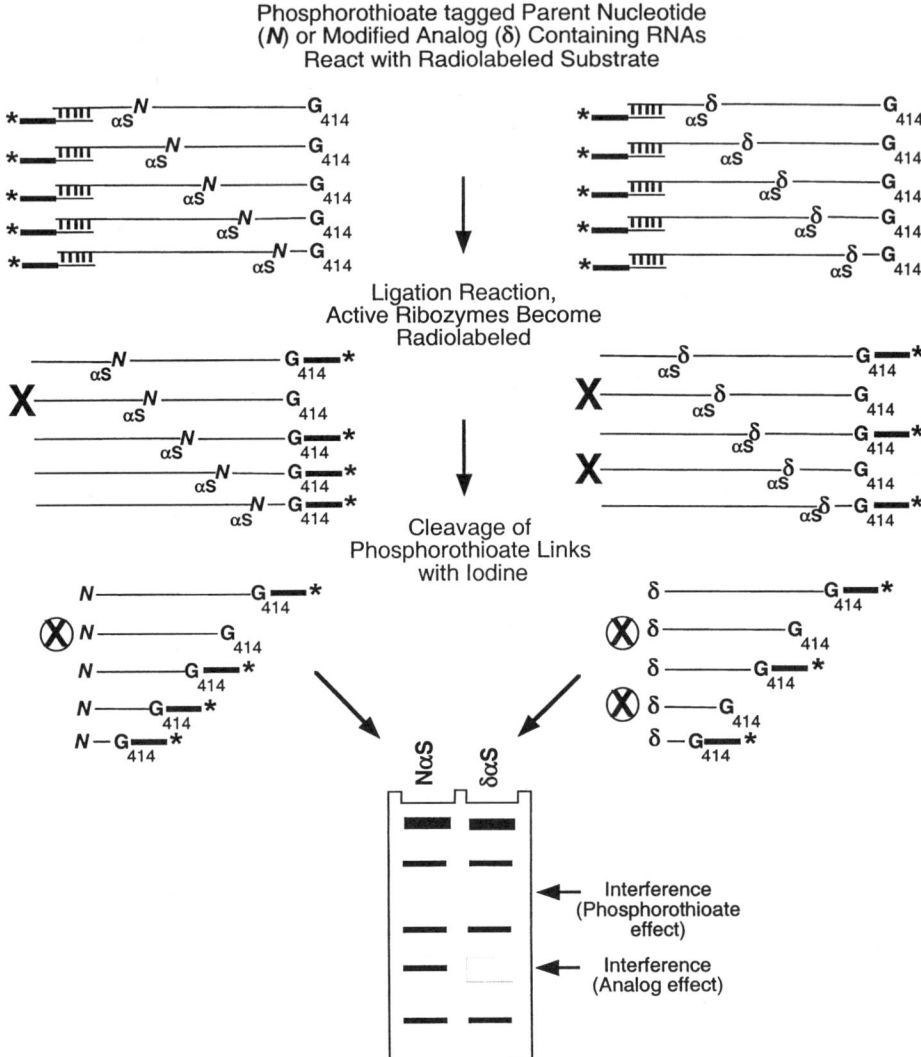

Fig. 1. Schematic of NAIM. In this example, the active ribozymes perform a ligation reaction and covalently transfer the radioactive to its 3' end from a radiolabeled substrate. N$\alpha$S indicates the parent nucleotide, and $\delta\alpha$S the modified nucleotide analog.

mutagenesis is that in NAIM, the smallest mutated unit is a single functional group, whereas in site-directed mutagenesis the entire nucleotide must be changed. This enables NAIM to generate data at the functional group level, often providing atomic level details such as contribution of a single hydrogen bond. NAIM has been used extensively to study almost all major classes of catalytic RNAs, including, group I introns (*6,7*), group II intron (*8*), RNase P

*(9)*, hairpin ribozyme *(10)*, hepatitis delta virus (HDV) ribozyme *(11)*, and varkud satellite (VS) ribozyme *(12)*.

The first step in performing NAIM is to randomly incorporate the α-phosphorothioate-tagged nucleotides with a functional group substitution or deletion into a RNA molecule by in vitro transcription. This produces a pool of transcripts with substitutions of analogs at different nucleotide positions along the RNA molecule (**Fig. 1**). The analog substitutions will then determine whether the particular ribozymes (or any other RNA) among the pool are still functional or are inactive depending on the importance of the functional group at that exact position. The ribozyme function is generally a cleavage or a ligation reaction; in either case, the active population must be radioactively labeled. Because the analogs are incorporated randomly, sites of incorporation are precisely mapped by cleaving the phosphorothioate tag with iodine followed by polyacrylamide gel electrophoresis (PAGE) to clarify the fragments *(13)*. The RNA is then visualized by autoradiography. Each band on the sequencing gel represents a particular nucleotide. If a band is missing from the sequence then the activity of the RNA was lost as a result of that specific analog substitution at that site (**Fig. 1**; *5*). The ability to determine the importance of each nucleotide and its functional groups in such an efficient and high-throughput manner explains why NAIM can be effectively applied to study the structure and function of ribozymes.

This chapter outlines the general methods for synthesis of the phosphorothioate-tagged, modified nucleotide analogs, incorporation of the analogs into the *Tetrahymena* group I intron ribozyme, the NAIM experiment performed on one of the engineered version of the catalytic RNA from *Tetrahymena*, and interpretation and quantitation of the data.

## 2. Materials

### 2.1. Reagent Preparation

#### 2.1.1. Synthesis of α-Phosphorothioate Tagged Nucleotide Triphosphates

The four parent 5'-*O*-(1-thio)nucleoside triphosphates are available from Ambion, Texas. These compounds and several of the nucleotide analogs described in this chapter are also available from Glen Research (Sterling, VA). Many of the modified nucleosides can be obtained from Sigma (St. Louis, MO).

1. 100 mg unprotected nucleoside.
2. 50 mL anhydrous pyridine.
3. 2 mL triethyl phosphate (TEP) (*see* **Note 1**).
4. $PSCl_3$.
5. Trioctylamine (TEA).
6. 0.5 *M* LiCl in $H_2O$.

7. Tributylammonium pyrophosphate (TBAP).
8. TEA.
9. 50 m*M* triethyl ammonium bicarbonate (TEAB).
10. *n*-propanol.
11. Ammonium hydroxide.
12. Argon.
13. Cellulose chromatographic plate (for thin-layer chromatography [TLC]).
14. DEAE Sephadex A-25.

## 2.1.2. In Vitro Transcription With α-Phosphorothioate Analogs

1. Linearized plasmid DNA template (1 mg/mL).
2. 10X transcription buffer: 400 m*M* Tris-HCl, pH 7.5, 40 m*M* spermidine, 100 m*M* dithiothreitol (DTT), 150 m*M* MgCl$_2$, 0.5% Triton X-100.
3. 10X nucleotide triphosphates (NTPs): 10 m*M* each of NTP, pH 7.5. Individual NTP is weighed to make approx 1 mL of 100 m*M* of stock solution in diethylpyrocarbonate (DEPC)-treated ddH$_2$O. The solution is adjusted to pH 7.5 and filtered through a 22-μ syringe filter. Absorbance is measured on an ultraviolet spectrophotometer, and the exact concentration is calculated using the appropriate extinction coefficient. 10X NTP mixture is prepared by diluting the stock to a solution containing 10 m*M* of each of the NTPs.
4. Nucleotide analogs: Often, a diastereomeric mixture is used, unless the S$_p$ isomer is purified from the R$_p$ by high-performance liquid chromatography (HPLC). The T7 RNA polymerase only recognizes the S$_p$ from a diastereomeric mixture, and during incorporation there is a reversal of configuration to become R$_p$.
5. T7 RNA polymerase and Y639F mutant version of the enzyme.
6. Urea, Premade polyacrylamide solution (National Diagnostics, Atlanta, GA), Ammonium persulfate, and *N*,*N*,*N'*,*N'*-tetramethyl-ethylenediamine (TEMED).
7. 5 *M* NaCl.
8. 100 m*M* Tris-EDTA, pH 7.0 (TE).
9. 100% and 70% ethanol.
10. Formamide loading buffer: 90% formamide, 100 m*M* Tris-HCl, pH 7.0, 22 m*M* EDTA, 0.001% bromophenol blue, and 0.001% xylene cyanol (0.22-μ, filter-sterilized).

## 2.1.3. Radiolabeling of RNA

1. Purified RNA.
2. Alkaline phosphatase, calf intestinal (CIP).
3. T4 polynucleotide kinase (PNK).
4. Yeast PolyA polymerase.
5. Phenol-chloroform (Tris-buffered).
6. 100% and 70% ethanol.
7. 100 m*M* TE.
8. [γ-$^{32}$P]-adenosine 5' triphosphate (ATP) (6000 Ci/mmol) and [α-$^{32}$P]cordycepin triphosphate (5000 Ci/mmol), NEN Life Sciences.

9. Kodak X-OMAT Film.
10. 5 *M* NaCl.

## 3. Methods

### 3.1. Synthesis of NTPαS

The 5'-*O*-(1-thio)nucleoside triphosphates are prepared from unprotected nucleosides on a scale of approx 50–200 mg of starting material, depending on the availability. The synthesis involves a one-pot, two-step reaction, adopted from the method developed by Arabashi and Frey *(14)*.

100 mg of nucleoside precursor is transferred to a 25-mL round-bottomed flask, and co-evaporated with 3 × 10 mL of anhydrous pyridine. The nucleoside is further co-evaporated with (1–2 mL) of toluene to remove any residual pyridine. After drying, the flask is capped with a septum, and the inside is purged with argon.

The dry nucleoside is dissolved in a minimal volume of TEP. The volume is maintained less than 1 mL. Nucleotides that have difficulty becoming soluble are heated with an air-gun while constantly stirring the mixture.

The synthesis of purine analogs are slightly different than pyrimidines. To the purine analog, 1.1 equivalent of trioctylamine (Aldrich) (MW 353.7, density 0.816 g/mL, approx 176 μL depending upon the mol wt of the analog) and 1.1 equivalent of $PSCl_3$ (Aldrich) (approx 45 μL, MW 169.4, density 1.67 g/mL) are added. The reaction is stirred for 30–60 min under argon. The progress of the reaction is monitored by analyzing a small portion of the reaction (100–200 μL) via TLC. The reaction is quenched with water, and the products are resolved by cellulose TLC (Aldrich) in a solvent system of 0.5 *M* LiCl in water. The products on the TLC plate are visualized by hand-held uv light. The monophosphate migrates more slowly than the unreacted starting material. To drive the reaction forward, additional amounts of trioctyl amine and $PSCl_3$ are added.

We took a slightly modified approach to synthesize the 5'-*O*-(1-thio-1, 1-dicloro)phosphoryl nucleoside intermediate on our way to prepare the triphosphates of the pyrimidine analogs *(15)*. The nucleoside is dried as previously described. The dry nucleoside (approx 100 mg) is dissolved in minimum amount of TEP. The solution is cooled to 0°C and trioctylamine (1.2 equivalents) and collidine (Sigma) (1.2 equivalents, MW 121.8, density 0.917 g/mL) are added under anhydrous conditions. $PSCl_3$ (about 1.5 equivalents) is added dropwise to the cooled solution, and allowed to react for 30 min before warming to room temperature and continuing the reaction at that condition for another 45 min. Monitoring the progress of the reaction and other steps are similar to the processes used for purine analogs.

To the 5'-*O*-(1-thio-1, 1-dicloro)phosphoryl nucleoside formed at the first step, the TBAP (Sigma) (MW 451.5, 4 equivalents of TBAP per mole of $PSCl_3$)

solution is added, and stirred for 30 min. The TBAP solution is prepared by dissolving in TEP (0.1 g/mL) by mild heating with an air gun. A small portion (100–200 µL) of the reaction is removed and quenched by adding a few drops of TEA (Aldrich). The TEA precipitates the phosphates, which is collected by centrifugation and dissolved in 50 m$M$ TEAB, and the products are analyzed by silica TLC using a solvent system of $n$-propanol:ammonium hydroxide:water (6:3:1) v/v. TLC analysis shows at least two products and both of these migrate slower than the unreacted nucleoside precursor and the nucleoside monophosphate. One of the products is the desired triphosphate and the other is the cyclic triphosphate, which hydrolyzes to the linear triphosphate after several hours at room temperature. After the reaction is over, the triphosphates are precipitated by adding about 50 equivalents of TEA (MW 101.19, density 0.726 g/mL, calculated per mol of the nucleoside precursor). The resulting precipitate is collected by centrifugation and dissolved in 5 mL of 50 m$M$ TEAB and kept overnight at room temperature. The cyclic intermediate of the nucleoside triphosphate is hydrolyzed under this condition. The nucleoside triphosphate is purified by DEAE Sephadex chromatography using a linear gradient from 50–800 m$M$ TEAB in a total volume of 1 L (16). For guanosine analogs using a gradient from 50 to 1 $M$, TEAB in a total volume of 1 L yields a better result. The nucleotide typically elutes at approx 0.6 $M$ TEAB. The fractions (approx 15 mL each) are checked for the presence of the nucleoside by UV measurements ($A_{260}$). Positive fractions are pooled as a batch of 2–3, and lyopholized. The reaction products are characterized by [31]P NMR and mass spectroscopy. The [31]P NMR spectra of the correct nucleotide shows three resonance peaks at 42–43 (a), –5–10 (g), and –20–25 (b) ppm relative to an 85% phosphoric acid standard (17). The nucleotide amount is quantitated by measuring the UV absorbance, and the identity is established by mass spectroscopy.

### 3.2. Incorporation of Phosphorothioate Nucleotide Analogs Via In Vitro Transcription (see Notes 2–4)

One of the basic requirements of NAIM is to incorporate the nucleotide analogs randomly within the RNA, which is usually accomplished enzymatically via T7 RNA polymerase (18,19). Certain analogs especially the minor groove modified analogs are incorporated better by the Y639F mutant version of T7 RNA polymerase (20,21). The incorporation level is chosen to generate sufficient signals within the detectable range, and also to limit cooperative interference due to too many substitutions. Usually, the incorporation level is kept at about 5% to fulfill the above criteria (7,22). These conditions are selected based upon a transcription condition where each of the NTP except the parent is present at 1 m$M$. The concentration of the parent nucleotide may need to be varied to create a suitable ratio with the modified analog to achieve 5% incorporation level (**Fig. 1**).

The reactions are performed according to standard procedures for in vitro transcription. The template may be derived from plasmid linearized with the suitable restriction enzyme, or may be a synthetic DNA template. The linearized plasmid is phenol-chloroform extracted, and is then ethanol-precipitated. The precipitated DNA is resuspended in a volume of TE to make the stock 1 mg/mL. If using a synthetic template (typically used for transcription of small RNAs), DNAs longer than 35–40 nucleotides must be purified by denaturing polyacrylamide gel electrophoresis (PAGE) before using them for transcription *(23)*. A typical transcription reaction will contain buffer, plasmid DNA template (0.05 µg/µL), T7 RNA polymerase (5 U/µL), inorganic pyrophosphatase (0.001 U/µL), nucleotide analog, and NTPs (**Fig. 1**). The reactions are incubated for 3 h at 37°C. The reaction is purified by 7 $M$ urea denaturing PAGE. The RNA transcripts are visualized by uv shadowing, excised from the gel with a clean razor blade, and eluted overnight in TE, 250 m$M$ NaCl at 4°C. The eluted RNA is ethanol-precipitated, and resuspended in 100 µL of TE. The RNA is quantitated by measuring the absorbance at 260 nm.

### 3.2.1. Optimization of Analog Incorporation

The incorporation of nucleotide analogs at particular sites within an RNA transcript or the overall incorporation level may vary, and thus it is necessary to normalize the incorporation efficiency. The incorporation efficiency is determined by resolving the cleavage products of $^{32}$P-5'-end-labeled RNA transcripts containing the parental nucleotide or the nucleotide analog by denaturing gel electrophoresis, and comparing individual band intensities of the parental nucleotide containing lane with the corresponding bands of the nucleotide analog lane. Twenty picomoles of RNA are treated with 200 U of CIP for 1 h at 37°C to remove the 5'-terminal phosphate group. The reaction is phenol-chloroform extracted and the RNA is ethanol-precipitated. The RNA is 5'-radiolabeled by incubating it with 10 U of T4 PNK and 50 mCi of [γ-$^{32}$P]ATP for 1 h at 37°C. The reaction is passed through a ProbeQuant G-50 spin-column (Amersham Biosciences, Piscataway, NJ) to remove unincorporated radionucleotides. The labeled RNA is purified by denaturing PAGE and visualized by autoradiography, the associated band is cut off the gel and eluted into 1% sodium dodecyl sulfate (SDS) in TE overnight at 4°C. To remove the SDS and precipitate the RNA the eluent is extracted with an equal volume of phenol-chloroform and is ethanol-precipitated. The labeled RNAs containing equal amount of counts are mixed with two volumes of loading buffer, reacted with 1/10th vol of $I_2$ in ethanol (50 m$M$), and heated at 90°C for 2 min *(13)*. The iodine selectively cleaves the phosphorothioate linkages, which are the sites of parent or analog incorporation at about 10–15% efficiency. The cleavage products are resolved on a preheated denaturing polyacrylamide gel. To control for the presence of any nonspecific cleavage, an equal amount of labeled RNAs, which are not treated with iodine, are also loaded on the same gel. Large RNAs are resolved on gels of different acrylamide concentrations, and are run for varying lengths of time to resolve the maximum portion of the molecule. For the *Tetrahymena* intron, which is approx 400 nucleotides long, 5% and 6% denaturing polyacrylamide gels are run. The 5% gels are typically run for 4 h 15 min and 3.5 h at 75 W, and the 6% gels are run for 2.5 h and 1 h at identical power. This procedure can resolve more than 98% of the nucleotides. The gels are dried and exposed to phosphorimager screens (Kodak, Rochester, NY). The intensities of the bands are quantitated by Bio-Rad Quantity One software. If a Molecular Dynamics machine is used for scanning the gel, the ImageQuant software is used for quantitation.

### 3.3. Selection of Active RNA

A primary requirement for successful NAIM assay is the ability to distinguish between active from inactive RNA population within a nucleotide analog-substituted pool *(5)*. Generally, there are two selection principles that have been used. The first approach requires physically separating the active variants from the inactive ones. This can be achieved by using a native gel-mobility, filter-binding, or denaturing PAGE, and have been successfully employed to study folding of ribozyme domain, RNA-protein interactions, and ribozyme cleavage activity *(11,24,25)*. The other approach is that in which a ribozyme-mediated ligation activity is employed, that selectively labels active variants from the pool of randomly doped RNAs *(7)*.

One of the most commonly used assays for NAIM is the 3'-exon ligation reaction executed by the L-21 G414 version of the *Tetrahymena* group I intron, which is the reverse of the second step of intron splicing (**Fig. 2A**; *26–28*). In this reaction, the terminal guanosine (G414) attacks the oligonucleotide substrate mimicking the 5'-3' ligated exon, and transfers the 3'-exon covalently to the 3'-end of the intron. By using a 3'-end-radiolabeled substrate, the active variants among the intron pool selectively label themselves. A typical ligation reaction with the L-21 G414 intron is performed as follows: The RNAs are prefolded by incubating in reaction buffer at 50°C for 10 min. After a quick spin, the RNAs are further heated for 2 min at 50°C, at this time the substrate dT(-1)S [CCCUC(dT)AAAAA], radiolabeled at 3'-end by yeast poly(A) polymerase (USB, Cleveland, OH) and [$\alpha$-$^{32}$P] cordycepin *(29)*, is dissolved in the same buffer and heated at 50°C for 2 min. After 2 min, dT(-1)S substrate, equal in volume to the ribozyme reaction, is added with proper mixing. The reaction

## A

**3'-exon ligation**

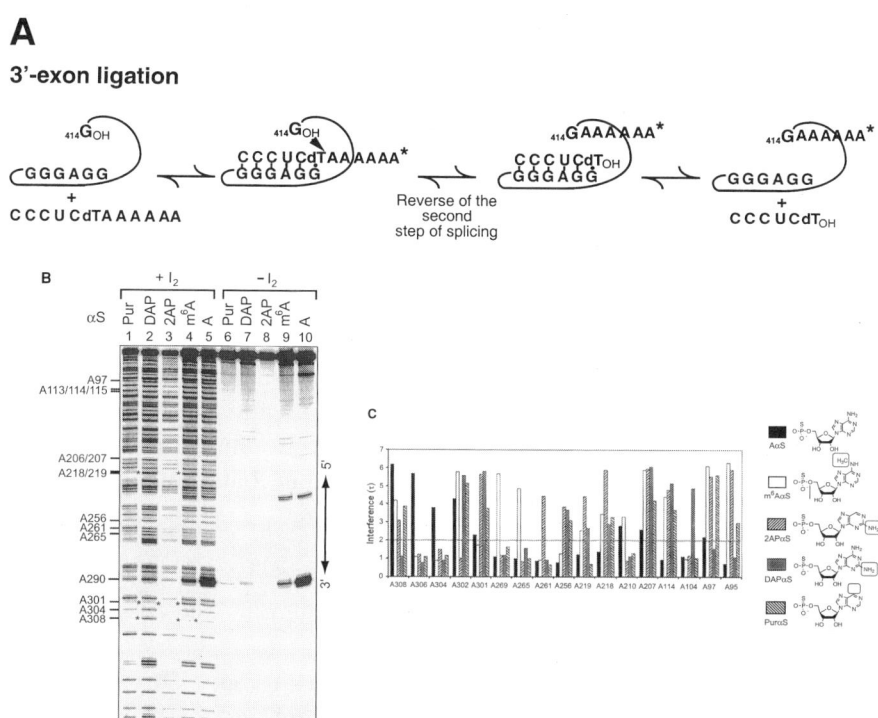

Fig. 2. Interference mapping on the *Tetrahymena* group I intron catalytic RNA. (**A**) Schematic of the 3'-exon ligation reaction for the L-21 G414 version of the group I intron with the 3'-radiolabeled substrate dT(−1)S. In the reaction, the terminal guanosine (G414) makes a nucleophilic attack on the substrate that mimics the 5'-3' ligated exon, and covalently transfers the 3'-exon portion on the 3'-end of the ribozyme. (**B**) Autoradiogram of the cleaved sequences of the ligated products that are doped with a subset of adenosine analogs. Some of the sites of interference are identified, and a few of those sites of interferences are marked by asterisks. Lanes 1–5 are iodine cleavage reactions of the ligated products. Lanes 6–10 are the corresponding lanes that were not treated with iodine, used to control for any nonspecific cleavage. The analogs corresponding to each lane are marked on top of the autoradiogram. A290 is a very consistent nonspecific cleavage site, and thus is always uninformative. Under our assay conditions, A302 and A306 show very strong phosphorothioate effect, and therefore are uninformative. (**C**) A quantitative depiction of the interference data derived from (B) via histogram. The 5'-radiobaled transcripts were also considered (as per Eq. 1) during calculation of the τ values.

is stopped after 10 min by adding 2 vol of loading buffer. The reaction is then split into two fractions and to one of the fractions $I_2$-ethanol solution is added (1/10th vol) to cleave the sulfur containing linkages. Both reactions are heated to 90°C for 2 min and loaded onto a 6% denaturing polyacrylamide gel (**Fig. 2B**), and after the completion of the run, the gel is dried and visualized by exposing the dried gel to a phosphorimager plate followed by scanning.

### 3.4. Data Quantitation

Quantitation of each band intensity from the iodine-cleavage sequence ladder is required to identify sites of interference as a result of analog substitution. The bands corresponding to cleavage products from the 5'-radiolabeled unselected RNA are compared with those from the selected RNA to control for incorporation differences at different sites, sites of phosphorothioate interference, and sites of nucleotide analog interference. Individual bands on the gels are selected, and the area under the peak corresponding to band intensity is calculated using the QuantityOne (Bio-Rad) software. Interference at each site is calculated by substituting the value of individual band intensities into the following equation:

$$\text{Interference } (\tau) = (N\alpha S_{\text{selected}})/(\delta\alpha S_{\text{selected}}) \div (N\alpha S_{\text{unselected}})/(\delta\alpha S_{\text{unselected}}) \tag{1}$$

where $N\alpha S$ is the parent phosphorothioate nucleotide and $\delta\alpha S$ is the modified nucleotide analog.

The equation normalizes the interference values for any site-specific incorporation differences or due to the phosphorothioate tag. For the sites of non-interference, the $\tau$ value is expected to be approx 1. To normalize the data, the average interference value for all positions that are within at least 2–3 standard deviations of the mean is calculated and the interference value at each position is divided by this value (**Fig. 2C**). The resulting set of values mostly hover around $1.0 \pm 0.2$. The $\tau$ value of above 2 is considered as interference, $\tau$ values below 0.5 are enhancement and values of approx 1 are non-interference. The cut-off point of 2.0 can be increased in case higher stringency of data is needed. But these criteria generally work quite well.

### 3.5. Nucleotide Analogs: Incorporation and Properties

Our laboratory and several others have successfully synthesized and tested phosphorothioate-tagged nucleotide analogs for various NAIM experiments. Here, we discuss incorporation properties within RNA of the published and a few unpublished analogs, as well as the information they provide on the chemical nature of RNA-RNA and RNA-ligand interactions. The ligands may include metal ions, small molecules, and proteins.

#### 3.5.1. Adenosine Analogs

Our discussion begins with purine nucleotides, starting with adenosine analogs, followed by pyrimidines, beginning with cytidine. Thus far, eleven

**A**

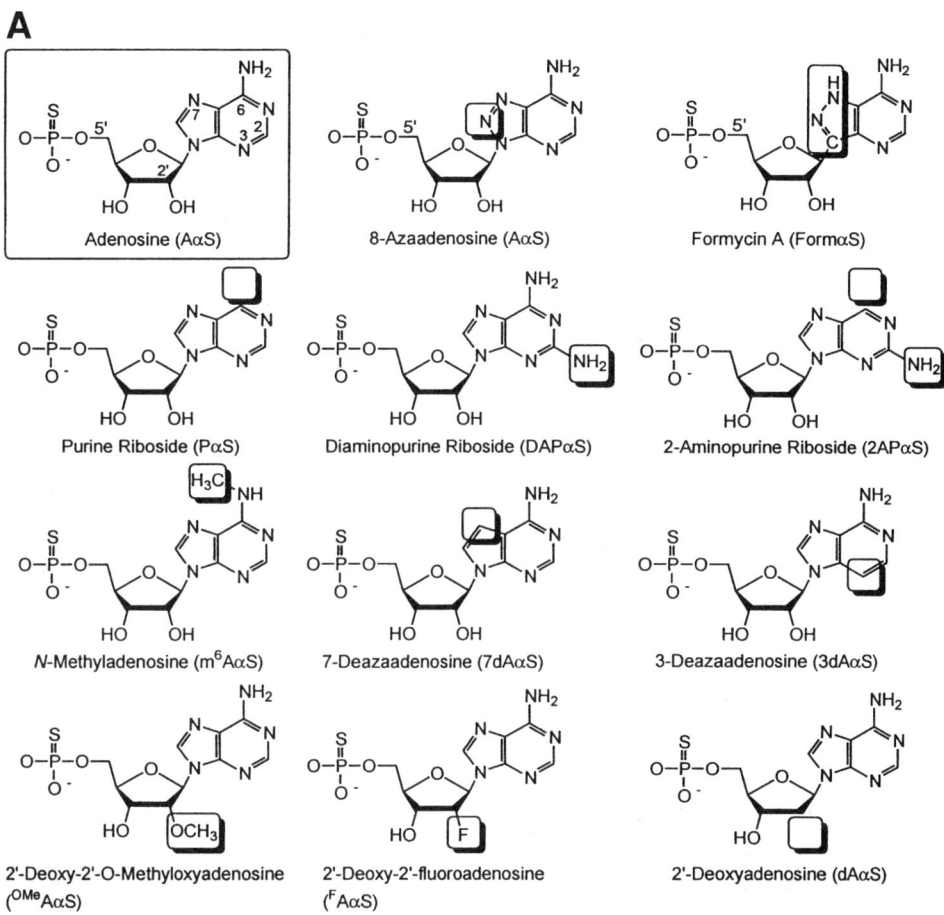

Fig. 3. Chemical structure of the modified nucleotide analogs. The site of chemical modification in an individual analog is marked by a box. (**A**) Adenosine analogs. (**B**) Guanosine analogs. (**C**) Cytidine analogs. (**D**) Uridine analogs.

adenosine analogs have been used for interference mapping (**Fig. 3A**). Eight among them are modified at various positions within the purine base. This subset includes purine riboside (PurαS), *N*-methyladenosine (m⁶AαS), 8-aza-adenosine (n⁸AαS), Formycin A (FormAαS), 7-deazaadenosine or tubercidin (7dAαS), 3-deazaadenosine (c³AαS), 2-aminopurine riboside (2APαS) and diaminopurine riboside (DAPαS). The analogs modified at the ribose sugar include 2'-deoxyadenosine (dAαS), 2'-deoxy-2'-fluoroadenosine (ᶠAαS), and 2'-*O*-methyladenosine (ᴼᴹᵉAαS), and all of these are modified at the 2'-OH position. Most of these analogs can be incorporated by the wild-type T7 RNA

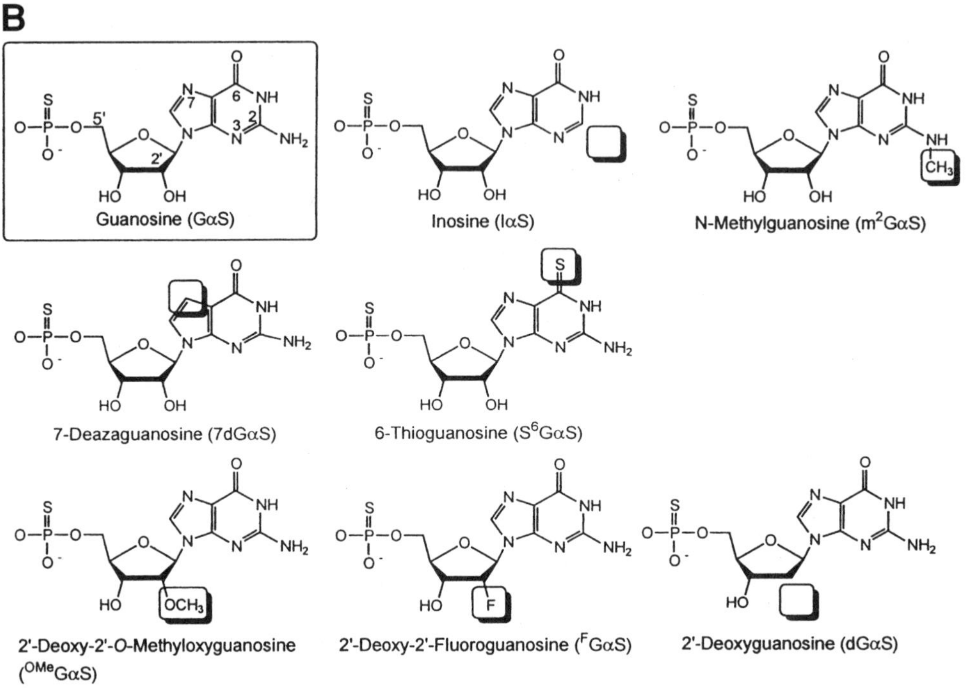

**B**

Fig. 3B.

polymerase. The 7dAαS and m⁶AαS require lower concentration than AαS for 5% level of incorporation, and DAPAαS incorporates more efficiently than AαS *(30)*. PurαS is a difficult analog to incorporate, and the intended 5% incorporation level is achieved by increasing the PurαS concentration and lowering the ATP concentration *(30)*. PurαS can sometimes incorporate unevenly, especially at sites with contiguous adenosines. 2APαS is incorporated by the wild-type polymerase, but requires higher concentrations than the parent nucleotide. Both n⁸AαS and FormAαS can be incorporated by the wild-type enzyme.

The minor-groove modified A-analogs c³AαS, dAαS, ᶠAαS, and ᴼᴹᵉAαS require the Y639F mutated form of T7 RNA polymerase for incorporation into RNA transcripts *(30,31)*. By far, the most difficult to incorporate of the adenosine analogs is c³AαS. The best reported incorporation level is about 1%. Both dAαS and ᶠAαS are incorporated efficiently, whereas ᴼᴹᵉAαS incorporation is poor, but generally even. It is difficult to obtain more than 2% incorporation level, although the incorporation level in some smaller transcripts may reach about 5%.

The set of A-analogs provide almost complete information on the chemical basis of adenosine function within RNAs *(30)*. The PurαS, m⁶AαS, and 2APαS

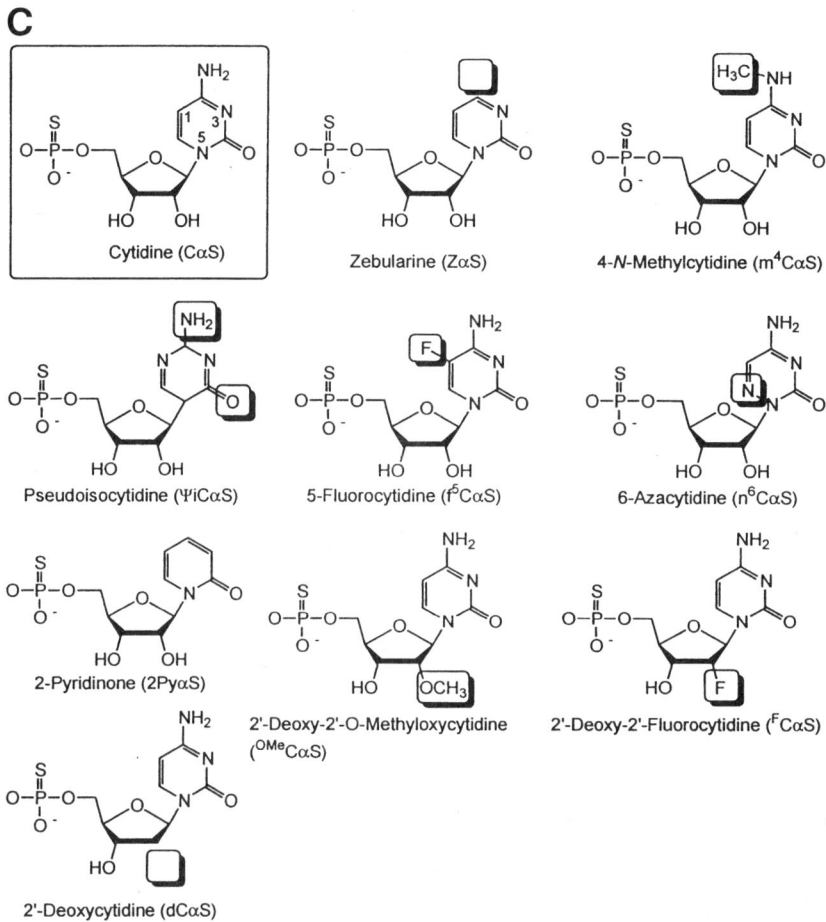

Fig. 3C.

provide information on the N-6 exocyclic amine of adenosine, either by deleting the amine group (as in the case of PuraS and 2APαS) or when a proton is replaced by a methyl group in m$^6$AαS *(6,8,30,32)*. Interference at any site resulting from PuraS or 2APαS incorporation indicates importance of the amine at that site. m$^6$AαS interference indicates that either both protons are necessary for H-bonding or the methyl group creates steric hindrance in a potentially cramped space. The effect may also be a result of modification of the electronic property of the purine ring—for example, pK$_a$s or pie-stacking.

Interference with 7dAαS, in which the N-7 nitrogen is replaced by a C—H group, indicates a critical major-groove contact with the ring nitrogen *(9,30)*. Interference with PuraS, m$^6$AαS, and 7dAαS strongly suggests Hoogsteen-type hydrogen bonding.

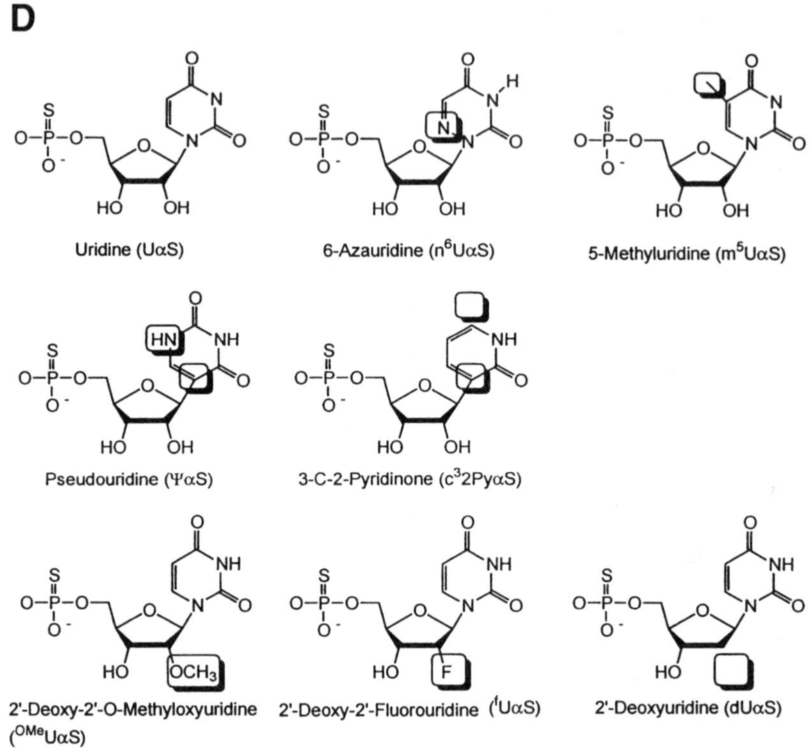

**D**

Uridine (UαS)                  6-Azauridine (n⁶UαS)              5-Methyluridine (m⁵UαS)

Pseudouridine (ΨαS)            3-C-2-Pyridinone (c³2PyαS)

2'-Deoxy-2'-O-Methyloxyuridine    2'-Deoxy-2'-Fluorouridine (ᶠUαS)    2'-Deoxyuridine (dUαS)
(ᴼᴹᵉUαS)

Fig. 3D.

The DAPαS and 2APαS have an extra amine at the C-2 position, and general interference with both of these analogs may indicate close minor-groove packing (6,9,30,33).

In c³AαS, the N-3 imino group of adenosine is replaced by a C-3, which allows an investigation of involvement of N-3 group in RNA function (31). This analog is particularly useful in mapping A-minor motifs within RNA.

n⁸AαS and FormAαS are analogs with perturbed N1 $pK_a$, and are used to examine adenosine base ionization within ribozymes (12,34).

### 3.5.2. Guanosine Analogs

Thus far, six guanosine analogs have been used for NAIM (**Fig. 3B**). Among these, four have modified functional groups within the base (inosine, IαS; $N^2$-methylguanosine, m²GαS; 7-deazaguanosine, 7dGαS; and 6-thioguanosine, S⁶GαS) and the remaining two at the 2'-OH of the ribose moiety (2'-deoxyguanosine, dGαS and 2'-O-methylguanosine, ᴼᴹᵉGαS).

Wild-type T7 RNA polymerase efficiently and uniformly incorporates both IαS and 7dGαS (7,35). S⁶GαS is incorporated reasonably well by the wild-type enzyme, but only in the presence of $Mn^{2+}$ in the transcription reaction *(24)*. The $S^6G\alpha S$ incorporated RNAs are highly unstable, which requires completion of all experiments within 48 h of transcription. $^{OMe}G\alpha S$ incorporation by the wild-type enzyme has been reported using a greatly increased ratio of analog to guanosine 5' triphosphate (GTP) *(3)*. But in several of the constructs we attempted including the L-21 G414 RNA, we found an inability to incorporate the analog. $m^2G\alpha S$ and $dG\alpha S$ are both minor-groove modified analogs. $dG\alpha S$ require the Y639F mutant enzyme for efficient incorporation *(21)*.

$I\alpha S$ and $m^2G\alpha S$ both help to reveal the role of N-2 exocyclic amine of guanosine *(7,36)*. Because of the lack of the amine, inosine may cause reduced duplex stability or missing tertiary contact, whereas replacement of a proton with the methyl group on the amine generally results only in loss of tertiary bonding. $7dG\alpha S$ lacks the N-7 ring nitrogen, and interference with this analog is an indication of major-groove contact at this position *(35)*. $S^6G\alpha S$, another major-groove modified analog has the O-6 replaced by a sulfur, and interference with this analog may probe RNA-RNA contact as well as sites of divalent or monovalent metal ion-binding site within RNA *(24)*.

The $dG\alpha S$ is a minor-groove ribose modified analog in which the 2'-OH is replaced by a hydrogen atom. Any critical H-bonding with 2'-OH of a guanosine can be examined with this analog.

### 3.5.3. Cytidine Analogs

Thus far, nine cytidine analogs have been synthesized and incorporated (**Fig. 3C**). These include six analogs that are modified within the pyrimidine base: 6-azacytidine ($n^6C\alpha S$), 5-fluorocytidine ($f^5C\alpha S$), pseudoisocytidine ($\Psi iC\alpha S$), zebularine ($Z\alpha S$), *N*-methylcytidine ($m^4C\alpha S$), 2-pyridinone ($2Py\alpha S$), and three analogs are modified at the 2'-OH of the ribose sugar: 2'-deoxycytidine ($dC\alpha S$), 2'-deoxy-2'-fluorocytidine ($^FC\alpha S$), 2'-*O*-methylcytidine ($^{OMe}C\alpha S$). $n^6C\alpha S$, $f^5C\alpha S$, and $\Psi iC\alpha S$ can be incorporated by the wild-type enzyme *(23,24)*. The other three base-modified analogs of cytidine require Y639F mutant polymerase and $Mn^{2+}$ for efficient incorporation *(11,15)* (Oyelere, A. et al., unpublished results). $dC\alpha S$ and $^FC\alpha S$ require Y639F mutant enzyme for incorporation (Oyelere, A. et al., unpublished results). Additionally, $^FC\alpha S$ requires $Mn^{2+}$ in the transcription buffer for incorporation into RNA transcripts. However, $^{OMe}C\alpha S$ could be incorporated by the wild-type enzyme *(3)*.

The $n^6C\alpha S$ has an extra-ring nitrogen in place of C-6. $f^5C\alpha S$ has a fluoro functionality at C-5, which introduces bulkiness in the major groove. $\Psi iC\alpha S$ is a C-linked nucleoside, in which the C-5 is also replaced by a nitrogen. In $Z\alpha S$, the N-4 exocyclic amine, which can be used to examine its importance in major-groove contacts, is missing. One interesting property of $n^6C\alpha S$, $f^5C\alpha S$, $\Psi iC\alpha S$, and $Z\alpha S$ is that they have different N-3 $pK_a$ relative to cytidine, and this characteristic has been utilized to determine $pK_a$ perturbation within RNA *(11,15)*. In $m^4C\alpha S$, a proton of the N-4 amine is replaced by a methyl group, whereas in $2Pyr\alpha S$ the N-4 amine (the N-3 ring nitrogen is replaced by a C) missing. They can be used to probe major-groove hydrogen bonding contacts (Oyelere, A. et al., unpublished results). The additional bulk of the methyl may cause interference if present in the context of insufficient space. The nature of information obtained from the 2'-OH derivatives of cytidine are similar to those obtained from similarly modified adenosine analogs.

### 3.5.4. Uridine Analogs

A total of seven analogs of uridine have been synthesized and incorporated into RNA (**Fig. 3D**). Four of them are base-modified: 6-azauridine ($n^6U\alpha S$), 3-C-2-pyridinone ($c^3-2Py\alpha S$), pseudouridine ($\Psi\alpha S$), 5-methyluridine ($m^5U\alpha S$) and the other three are modified in the ribose sugar: 2'-deoxyuridine ($dU\alpha S$), 2'-deoxy-2'-fluorouridine ($^FU\alpha S$), 2'-$O$-methyluridine ($^{Ome}U\alpha S$).

Except for $\Psi\alpha S$, the other base-modified analogs $n^6U\alpha S$ and $c^3-2Py\alpha S$ can only be incorporated into a transcript by the Y639F mutant enzyme (Oyelere, A. et al., unpublished results). Except for the $^{Ome}U\alpha S$ *(3)*, the other two ribose sugar-modified analogs also require the Y639F mutant enzyme *(37)*. $^FU\alpha S$ also needs $Mn^{2+}$ for effective incorporation.

$\Psi\alpha S$ is a C-linked nucleoside, and the C-5 is substituted by a nitrogen. This analog does not exhibit any interference within the L-21 G414 ribozyme ( Basu, S. and Strobel, S. A., unpublished results).

The 2'-OH modified analogs provide information similar in nature to identically modified adenosine analogs.

## 4. Notes

1. Solubility of certain nucleosides can adversely affect the yield of NTPαS synthesis. To dissolve the nucleoside, it is important to avoid using too much TEP. Excess TEP severely affects the reaction yield, so the volume should not exceed 1 mL. Most of the time, guanosine and its analogs are more difficult to dissolve than others. One can mildly heat the reaction vessel with an air-gun with continuous stirring to facilitate the dissolving process.

2. There are several incorporation-related issues that must taken into careful consideration. In cases in which transcription conditions require nucleotide triphosphate (NTP) concentration at some multiples of 1 m$M$, merely changing the analog concentration to that multiple may not achieve the desired level of incorporation. The conditions, especially the ratio of analog to NTP, may need to be reoptimized for each such system. Similarly, the concentration required for optimum incorporation when pure $S_p$ isomer is used must be reoptimized, but simply using one-half of the value used for the diastereomeric mixture may not work. This is true because the diastereomeric mixture may not contain equal amounts of the $S_p$ and $R_p$ isomers.

3. Sometimes, while attempting to obtain optimum incorporation, one may inadvertently create a condition for over-incorporation, exhibited by weaker bands compared to the parent analog. This may happen because an RNA that is over incorporated with an analog may be non-selectively detrimental to the activity and result in an overall weaker signal, falsely suggesting poor incorporation. In the case of a ribozyme ligation reaction, this can be detected by the topmost band in a lane, which represents the full-length uncleaved ligated product. An analog lane may have a substantially reduced amount of such product indicating poor ligation reaction, which may be a tell-tale sign of over incorporation. The analog concentration should be appropriately adjusted to verify that the problem is over-incorporation, and not under-incorporation.

4. We found that $S^6G\alpha S$ incorporated RNA transcripts are very unstable. All experiments must be concluded within 48 h of transcription. Storing the $S^6G\alpha S$ incorporated RNAs at $-80°C$ does not increase the half-life.

5. Sometimes, nonspecific cleavage may affect an experiment adversely, especially when the experiment requires cutting a band from a native gel to isolate the RNA. Under native conditions, the nonspecifically cleaved RNAs may still form a folded structure that may create problems during the sequencing step, as the non-full-length RNAs will create non-specific bands or bands of extra intensity. To solve this problem, the RNA is eluted into a SDS (1%) containing buffer to denature any extraneous nucleases, consequently limiting any nonspecific cleavage of the RNA. If SDS is used, the eluate must be phenol-chloroform-extracted to remove the SDS. Sometimes, an additional denaturing gel purification before the sequencing step may be necessary to have a pool of full-length RNA.

6. **Note:** many of the chemicals used for NAIM are hazardous, and proper precaution must be taken while handling them.

## Acknowledgments

We thank Dr. Scott Strobel for generous gift of some of the analogs and sharing with us unpublished results. We also thank Dr. A. Oyelere for his comments on the manuscript. The work is supported by a start-up fund to SB from the University of Pittsburgh.

## References

1. Conway, L. and Wickens, M. (1989) Modification interference analysis of reactions using RNA substrates. *Methods Enzymol.* **180**, 369–379.
2. Stern, S., Moazed, D., and Noller, H. F. (1988) Analysis of RNA structure using chemical and enzymatic probing monitored by primer extension. *Methods Enzymol.* **164**, 481–489.
3. Conrad, F., Hanne, A., Gaur, R. K., and Krupp, G. (1995) Enzymatic synthesis of 2'-modified nucleic acids: Identification of important phosphate and ribose moieties in RNase P substrates. *Nucleic Acids Res.* **23**, 1845–1853.
4. Gaur, R. K. and Krupp, G. (1993) Modification interference approach to detect ribose moieties important for the optimal activity of a ribozyme. *Nucleic Acids Res.* **21**, 21–26.

5. Ryder, S. P., Ortoleva-Donnelly, L., Kosek, A. B., and Strobel, S. A. (2000) Chemical probing of RNA by nucleotide analog interference mapping. *Methods Enzymol.* **317,** 92–109.
6. Strauss-Soukup, J. K. and Strobel, S. A. (2000) A chemical phylogeny of group I introns based upon interference mapping of a bacterial ribozyme. *J. Mol. Biol.* **302,** 339–358.
7. Strobel, S. A. and Shetty, K. (1997) Defining the chemical groups essential for *Tetrahymena* group I intron function by nucleotide analog interference mapping. *Proc. Natl. Acad. Sci. USA* **94,** 2903–2908
8. Boudvillain, M. and Pyle, A. M. (1998) Defining functional groups, core structural features and inter-domain tertiary contacts essential for group II intron self-splicing: a NAIM analysis. *EMBO J.* **17,** 7091–7104.
9. Siew, D., Zahler, N. H., Cassano, A. G., Strobel, S. A., and Harris, M. E. (1999) Identification of adenosine functional groups involved in substrate binding by the ribonuclease P ribozyme. *Biochemistry* **38,** 1873–1883
10. Ryder, S. P. and Strobel, S. A. (1999) Nucleotide analog interference mapping of the hairpin ribozyme: implications for secondary and tertiary structure formation. *J. Mol. Biol.* **291,** 295–311.
11. Oyelere, A. K., Kardon , J. R., and Strobel, S. A. (2002) pK(a) perturbation in genomic Hepatitis Delta Virus ribozyme catalysis evidenced by nucleotide analogue interference mapping. *Biochemistry* **41,** 3667–3675.
12. Jones, F. D. and Strobel, S. A. (2003) Ionization of a critical adenosine residue in the *neurospora* varkud satellite ribozyme active site. *Biochemistry* **42,** 4265–4276.
13. Gish, G. and Eckstein, F. (1988) DNA and RNA sequence determination based on phosphorothioate chemistry. *Science* **240,** 1520–1522.
14. Arabshahi, A. and Frey, P. A. (1994) A simplified procedure for synthesizing nucleoside 1-thiotriphosphates: dATPαS, dGTPαS, UTPαS, and dTTPαS. *Biochem. Biophys. Res. Commun.* **204,** 150–155.
15. Oyelere, A. K. and Strobel, S. A. (2000) Biochemical detection of cytidine protonation within RNA. *J. Am. Chem. Soc.* **122,** 10,259–10,267.
16. Eckstein, F. and Goody, R. S. (1976) Synthesis and properties of diastereoisomers of adenosine 5'-(*O*-1-thiotriphosphate) and adenosine 5'-(*O*-2-thiotriphosphate). *Biochemistry* **15,** 1685–1691.
17. Chen, J.-T. and Benkovic, S. J. (1983) Synthesis and separation of diastereomers of deoxynucleotide 5-*O*-(1-thio)triphosphates. *Nucleic Acids Res.* **11,** 3737–3751.
18. Chamberlain, M., Kingston, R., Gilman, M., Wiggs, J., and de Vera, A. (1983) Isolation of bacterial and bacteriophage RNA polymerases and their use in synthesis of RNA in vitro. *Methods Enzymol.* **101,** 540–568.
19. Griffiths, A. D., Potter, B. V. L., and. Eperon, I. C. (1987) Stereospecificity of nucleases towards phosphorothioate-substituted RNA: stereochemistry of transcription by T7 RNA polymerase. *Nucleic Acids Res.* **15,** 4145–4162.
20. Sousa, R. (2000) Use of T7 RNA polymerase and its mutants for incorporation of nucleoside analogs into RNA. *Methods Enzymol.* **317,** 65–74.
21. Sousa, R. and Padilla, R. (1995) A mutant T7 RNA polymerase as a DNA polymerase. *EMBO J.* **14,** 4609–4621.

22. Christian, E. L. and Yarus, M. (1992) Analysis of the role of phosphate oxygens in the group I intron from *Tetrahymena. J. Mol. Biol.* **228,** 743–758.
23. Milligan, J. F. and Uhlenbeck, O. C. (1989) Synthesis of small RNAs using T7 RNA polymerase. *Methods Enzymol.* **180,** 51–62.
24. Basu, S., Rambo, R. P., Strauss-Soukup, J., Cate, J. H., Ferre-D Amare, A. R., Strobel, S. A., and Doudna, J. A. (1998) A specific monovalent metal ion integral to the A-A platform of the RNA tetraloop receptor. *Nat. Struct. Biol.* **5,** 986–992.
25. Batey, R. T., Rambo, R. P., Lucast, L., Rha, B., and Doudna, J. A. (2000) Crystal structure of the ribonucleoprotein core of the signal recognition particle. *Science* **287,** 1232–1239.
26. Beaudry, A. A. and Joyce, G. F. (1992) Directed evolution of an RNA enzyme. *Science* **257,** 635–641.
27. Cech, T. R. (1990) Self-splicing of group I introns. *Ann. Rev. Biochem.* **59,** 543–568.
28. Mei, R. and Herschlag, D. (1996) Mechanistic investigations of a ribozyme derived from the *Tetrahymena* group I intron. Insights into catalysis and the second step of self-splicing. *Biochemistry* **35,** 5796–5809.
29. Lingner, J. and Keller, W. (1993) 3'-end labeling of RNA with recombinant yeast poly(A) polymerase. *Nucleic Acids Res.* **21,** 2917–2920.
30. Ortoleva-Donnelly, L., Szewczak, A. A., Gutell, R. R., and Strobel, S. A. (1998) The chemical basis of adenosine conservation throughout the *Tetrahymena* ribozyme. *RNA* **4,** 498–519.
31. Soukup, J. K., Minakawa, N., Matsuda, A., and Strobel, S. A. (2002) Identification of A-Minor Tertiary interactions within a bacterial group I intron active site by 3-Deazaadenosine interference mapping. *Biochemistry* **41,** 10,426–10,438.
32. Siew, D., Zahler, N. H., Cassano, A. G., Strobel, S. A., and Harris, M. E. (1999) Identification of adenosine functional groups involved in substrate binding by the ribonuclease P ribozyme. *Biochemistry* **38,** 1873–1883.
33. Strobel, S. A., Ortoleva-Donnelly, L., Ryder, S. P., Cate J. H., and Moncoeur, E. (1998) Complementary sets of noncanonical base pairs mediate RNA helix packing in the group I intron active site. *Nat. Struct. Biol.* **5,** 60–66.
34. Ryder, S. P., Oyelere, A .K., Padilla, J. L., Klostermeier, D., Millar, D. P., and Strobel, S. A. (2001) Investigation of adenosine base ionization in the hairpin ribozyme by nucleotide analog interference mapping. *RNA* **7,** 1454–1463.
35. Kazantstev, A. V. and Pace, N. R. (1998) Identification by modification-interference of purine N-7 and ribose 2'-OH groups critical for catalysis by bacterial ribonuclease P. *RNA* **4,** 937–947.
36. Ortoleva-Donnelly, L., Kronman, M., and Strobel, S. A. (1998) Identifying sites of RNA tertiary structure by Nucleotide Analog Interference Mapping with $N^2$-methylguanosine. *Biochemistry* **37,** 12,933–12,942.
37. Szewczak, A. A., Ortoleva-Donnelly, L., Ryder, S. P., Moncoeur, E., and Strobel, S. A. (1998) A minor groove RNA triple helix within the catalytic core of a group I intron. *Nat. Struct. Biol.* **5,** 1037–1042.

# 7

## Analysis of Global Conformational Transitions in Ribozymes

### David M. J. Lilley

#### Summary

The nucleolytic ribozymes are small, autonomously folding RNA species that require metal ions to adopt their active conformations. Three of these ribozymes contain one or more helical junctions, that act as key architectural elements in the structures. A combination of comparative gel electrophoresis and fluorescence resonance energy transfer (FRET) can be used to analyze the global conformation in solution, and to study the ion-induced folding processes. The FRET approach can also be extended into the time domain to analyze conformational equilibria, and increasingly to the single-molecule level.

**Key Words:** Comparative gel electrophoresis; fluorescence resonance energy transfer; single-molecules.

## 1. Introduction

RNA folding and catalytic activity are tightly linked. Just as a protein enzyme must fold into the conformation required for activity, ribozymes must also fold to create the local environment that supports catalysis. This may involve correctly oriented nucleobases, phosphate groups, or metal ions, or the environment that leads to the perturbation of the $pK_a$ of an important functional group. The folding could also create a precisely oriented arrangement at the reacting site, or perhaps introduce strain that can be used to accelerate the reaction by facilitating the trajectory into the transition state. RNA structure and folding processes are thus major keys to understanding the mechanisms of ribozyme action. RNA catalysis has been recently reviewed (*1,2*).

The nucleolytic ribozymes are relatively small RNA species that bring about site-specific cleavage of the backbone by a transesterification reaction in which a particular 2'-hydroxyl group attacks the 3'-phosphate in an $S_N2$ reaction, with

From: *Methods in Molecular Biology, vol. 252: Ribozymes and siRNA Protocols, Second Edition*
Edited by: M. Sioud © Humana Press Inc., Totowa, NJ

departure of the 5'-oxyanion to leave a cyclic 2'–3'phosphate *(3)*. The reaction is accelerated by a factor of $\geq 10^5$ in the ribozyme *(4)*. All that is required to bring about this reaction is the presence of divalent metal ions—in fact, even monovalent metal ions suffice for most ribozymes, although at very high concentrations *(5)*. The reverse (ligation) reaction is also efficiently catalyzed by some of the nucleolytic ribozymes.

The nucleolytic ribozymes are small, autonomously folding species. If the secondary structures of the hammerhead, hairpin, and varkud satellite (VS) ribozymes are compared, it is clear that a common feature is the presence of prominent helical junctions (**Fig. 1**). The hammerhead is built around a three-way junction, the VS has two three-way junctions that generate an H-shaped secondary structure, and the hairpin ribozyme contains a perfect four-way junction. Such junctions are very common in functional RNA molecules, and often have the effect of organizing the long-range architecture. In the case of the ribozymes, they play a central role in ensuring efficient folding into the active conformation.

In recent years, we have learned a great deal about the conformational properties of helical junctions (*see* **ref. 6**). Two principles are particularly important. First, stacking interactions are a major driving force in folding reactions, leading to pairwise coaxial stacking of helices. This often requires a thermodynamic choice between alternative stacking conformers to be made, and dynamic exchange between the conformers may occur in some systems *(7)*. Second, electrostatic interactions are extremely significant because of the polyelectrolyte character of nucleic acids, and thus, the binding of metal ions is frequently involved in mediating structural transitions.

We are now in the fortunate position of having crystal structures of the hammerhead *(8,9)*, hairpin *(10)*, and hepatitis delta virus (HDV) *(11)* ribozymes, leaving the VS as the only distinct member of this class of ribozyme that has not yielded to crystallography thus far. Crystal structures cast considerable light on possible origins of catalysis, yet such structures are usually not so much an end point as a new beginning. They are a tremendous stimulation for thinking about mechanisms and suggesting new lines of attack. In this laboratory, we have developed approaches to the study of the global structures of these species, and the folding processes by which the structures are formed. The mainstay of these has been comparative gel electrophoresis and fluorescence resonance energy transfer (FRET), coupled with detailed analysis of ribozyme action, mutational analysis, chemical modification, and molecular modeling.

## 2. Gel Electrophoresis

Gel electrophoresis provides a simple yet very powerful method to learn about the global shape of branched nucleic acids. The mobility of nucleic acids migrating in polyacrylamide gels under the influence of an electric field is very sensitive to

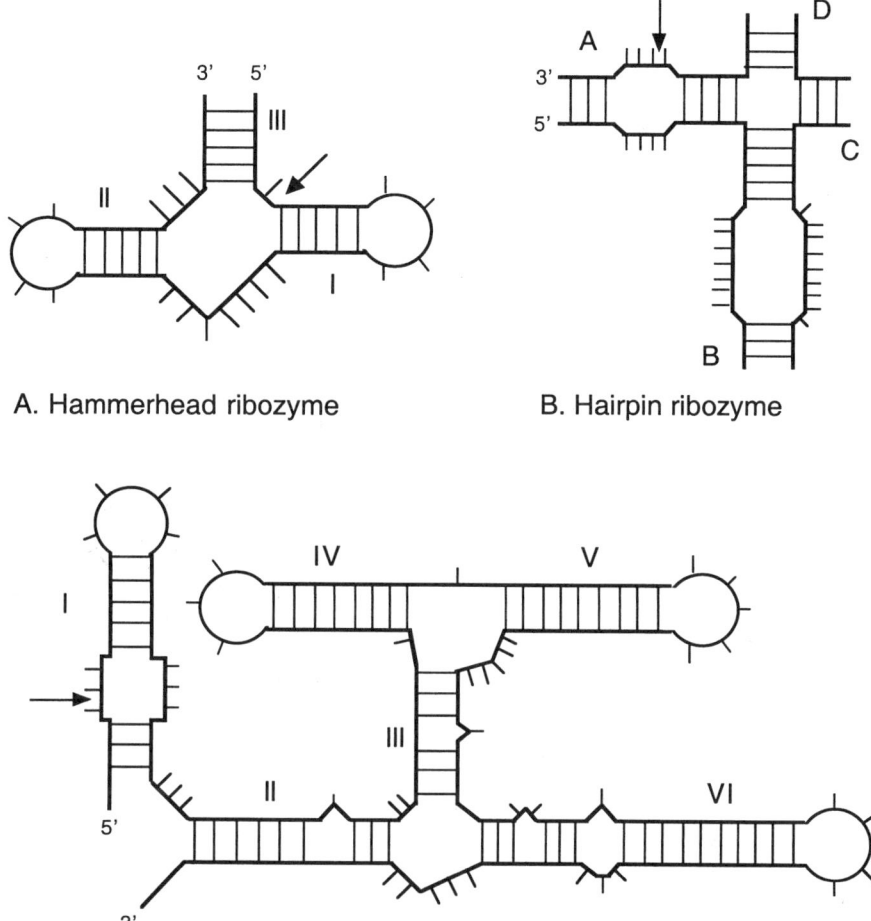

A. Hammerhead ribozyme

B. Hairpin ribozyme

C. VS ribozyme

Fig. 1. Schematic illustrations of the secondary structures of the hammerhead, hairpin, and VS ribozymes. In each case the helices are labeled conventionally. The positions of ribozyme cleavage are indicated by arrows.

the global shape of the molecule. The drawback of the method is the lack of a rigorous theoretical basis for interpreting mobilities. However, there are a number of theories that can provide at least qualitative agreement with experimental data *(12–14)*. Most of these are derived from the idea of the nucleic acid reptation *(15)*, in which the DNA moves through the pores of the gel matrix under the influence of the electric field. Lumpkin and Zimm *(13)* derived a relationship between the rate of migration ($\mu$) and the end-to-end distance of the molecule:

$$\mu = \frac{Q}{\zeta} \cdot \left\langle \frac{h_x^2}{L^2} \right\rangle \qquad \text{[Eq. 1]}$$

where Q is the charge on the molecule, $\zeta$ is the frictional coefficient for translation along the tube, L is the contour length of the molecule, and $h_x$ is the component of the end-to-end vector **h** in the direction of the electric field. The brackets indicate an average over an ensemble of configurations. Since branching, kinking, or bending of the nucleic acid will reduce the end-to-end distance, the theory predicts that such distortion will lead to a reduced mobility in the gel. This simple theory often provides a reasonable agreement with experiment, such as the retardation of DNA or RNA duplexes kinked by the presence of a central bulge *(16–19)*. Through many years of experience, this approach has been shown to work empirically. It has had a number of successes in which crystal structures have confirmed the global structure determined by electrophoresis (sometimes more than a decade later), including the Holliday junction in DNA *(20)* and the hammerhead *(21)* and hairpin *(22)* ribozymes. Perhaps more importantly, it has not yet led to any structures that have been proven incorrect.

## 3. Study of Junction Conformation by Comparative Gel Electrophoresis

This method is based on a comparison of the electrophoretic mobility of all the possible species derived from a given helical junction in which two arms are significantly longer (typically 40–60 bp) than the remaining arms (usually approx 15 bp). The number of combinations of *n* taken from m ($^mC_n$) is given by:

$$^mC_n = \frac{m!}{(m-n)!n!} \qquad \text{[Eq. 2]}$$

and thus there are three species that are derived from a three-way junction (each with two long and one short arm) and six that may be obtained from a four-way junction (each with two long and two short arms). The various mobilities of the species are analyzed on the assumption that the mobility will reflect the end-to-end distance of the two long arms (via **Eq. 1**), which is directly related to the angle subtended between these arms. Although we try to preserve the natural sequence of the helical arms as much as possible, particularly in the vicinity of the branchpoint, we remove any helical imperfections such as mismatches, and particularly base bulges. These might introduce kinks into the arms that would add vectorially to the angle between the long arms that we are trying to determine.

For RNA molecules, we have taken two approaches to the construction of the long-short species. The component strands can in principle be made by chemical synthesis, but in practice this becomes impractical considering that some of the longer component strands can be more than 100 nt in length. At present, RNA synthesis is generally not efficient beyond approx 60 nt, and is still very expensive when compared to DNA. In order to overcome these limitations, we have generally used the approach of synthesizing strands, in which only the central part (corresponding to the section around the junction itself) is RNA and the extended arm sections are made of DNA *(23)*. At first we were concerned that the resulting DNA-RNA junction might introduce a kink that could influence the results. However, by varying the length of the RNA helix and by introducing a hybrid DNA-RNA helix, and even by synthesizing a set of species composed entirely of RNA, we concluded that this did not affect our conformational conclusions in the case of the hammerhead ribozyme *(21)*. An alternative approach that avoids these concerns is to synthesize all the component strands by transcription, using T7 RNA polymerase together with a template generated by PCR from DNA oligonucleotides. In general, this is now our preferred approach.

We carry out the electrophoresis of the long-short RNA species typically in an 8% polyacrylamide (29:1 acrylamide:*bis*acrylamide ratio) in a buffer containing 90 m$M$ Tris borate (pH 8.3), 200 µ$M$ $MgCl_2$. With metal ion-containing buffers, it is important to recirculate at >1 L/h. Normally, the electrophoresis is carried out at room temperature, although temperature-dependent effects can be studied by the use of glass plates with serpentine water flow from a circulating bath. In our experiments, the RNA is radioactively labeled (either [5'-$^{32}$P]-labeling using T4 polynucleotide kinase [PNK], or by transcription in the presence of [α-$^{32}$P]-labeled cytidine 5' triphosphate [CTP]), and the bands are visualized by exposure of dried gels to storage phosphor screens and scanning in a phosphorimager.

Examples of the application of comparative gel electrophoresis applied to the two three-way junctions of the VS ribozyme are shown in **Fig. 2**. It can readily be seen from the patterns of mobility that both junctions change their global conformation on addition of magnesium ions, and the relative angles subtended between the long arms can be determined in each case.

## 4. Estimation of Dihedral Angles by Gel Electrophoresis

Joining together two RNA elements that kink the axis—such as base bulges or helical junctions—through a common duplex can result in a dihedral angle between outer helical arms. A simple example of this would be a duplex containing two base bulges, each of which kink the axis, in which the overall trajectory will be determined by the kink angles and the dihedral angle. The latter will be determined by the intrinsic structure of the elements generating the kinking, and the length of the duplex that connects them. If this length is systematically varied, this will generate a corresponding variation in the dihedral angle that can be revealed by the different electrophoretic mobility of the

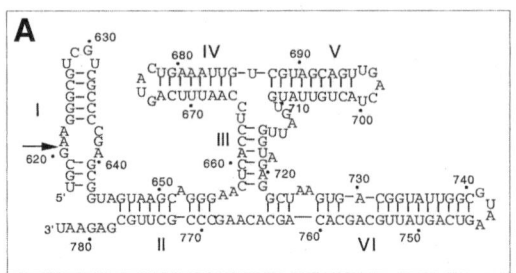

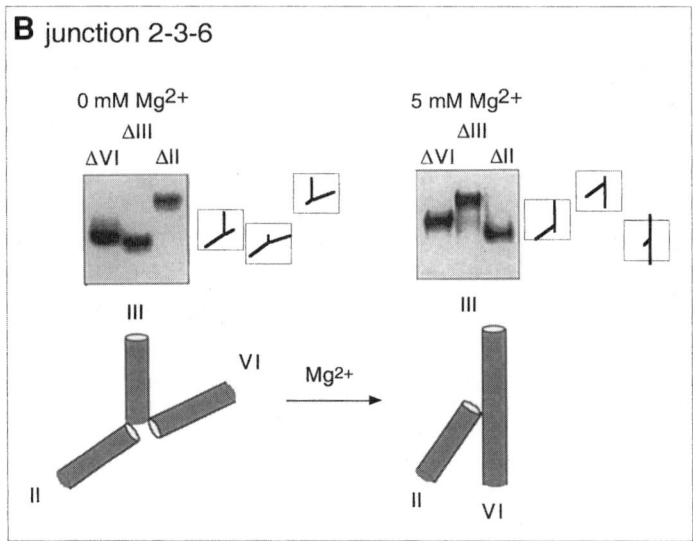

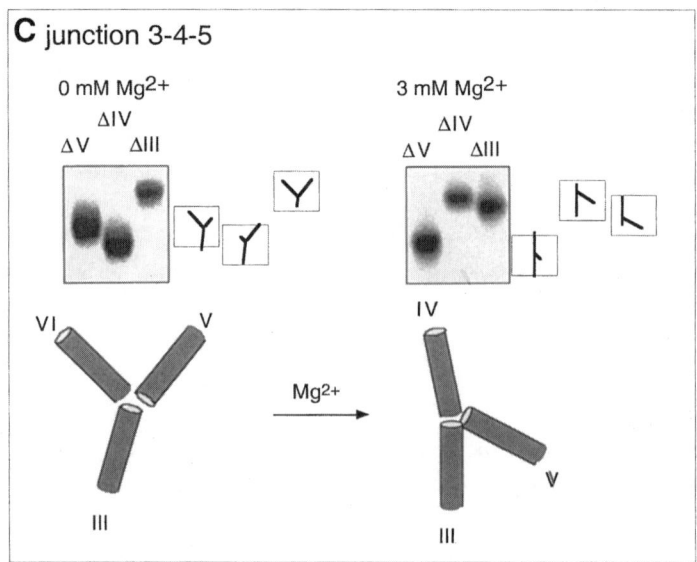

species. An early example of this was demonstrated for RNA duplexes with two base bulges *(16,17)*, in which variation in the length of the intervening duplex gave rise to the expected sinusoidal modulation of electrophoretic mobility, from which the helical periodicity of double-stranded RNA could be calculated. Such data can be modeled with simple geometric models or more sophisticated models using graphical modeling packages such as Insight, using the Lumpkin-Zimm equation to calculate the expected mobility.

A recent example of this approach arose in the analysis of the structure of the VS ribozyme. The structure of the two three-way helical junctions was determined using comparative gel electrophoresis and FRET (as discussed in the previous and following sections). The junctions could be put together through their common helix III, giving a coaxial arrangement of helices IV–III–VI, from which helices V and II radiated. However, there was no way of knowing the dihedral angle between helices II and V *a priori*, and we therefore used the method of variable phasing. We generated a series of ribozyme variants in which helix III was perfectly basepaired, and varied in length from 6–20 bp (**Fig. 3**). Helices VI and IV were shortened to 6-bp stem-loop structures, and helices II and V were extended by random sequences, removing the loops and bulges to create perfectly basepaired helices. The length of these helices was 40 bp in the species in which helix III was 6 bp, and were then both reduced by 1 bp for every 2 bp added to helix III. Thus, the contour length of the complete helix II–III–V path remained constant. We observe a sinusoidal modulation of the electrophoretic mobility (**Fig. 3**), as expected. The maximum mobility occurred in the species with a helix III of 10 bp; helices II and V should be approximately coplanar, and on opposite sides of stem III in this case. Using a simple geometric model, we could obtain a good fit to the experimental data, as shown in **Fig. 3**, from which we calculated a dihedral angle of approx 80° for the natural ribozyme. The analysis was extended using molecular graphics to generate an ensemble representation of the molecules used experimentally. A graphical search of torsional space around the axis of helix

---

Fig. 2. *(opposite page)* Comparative gel electrophoresis analysis of the two three-way junctions of the VS ribozyme. (**A**) The sequence of VS ribozyme in its natural, *cis*-acting form *(58)*. The position of ribozyme cleavage is indicated by the arrow. (**B** and **C**) Comparative gel electrophoretic analysis of the 2-3-6 (**B**) and 3-4-5 (**C**) junctions in the presence and absence of $Mg^{2+}$ ions *(41,47)*. Junctions having two long arms of 40 bp and one shorter arm of 10 (2-3-6) or 11 (3-4-5) bp were electrophoresed in a polyacrylamide gel in the presence of the indicated $Mg^{2+}$ concentrations. The arms are named according to the short arm—e.g., the species $\Delta VI$ has long II and III arms. The interpretation of the relative electrophoretic mobilities is indicated to the right of the gels, and by the schemes outlined at the bottom of each panel..

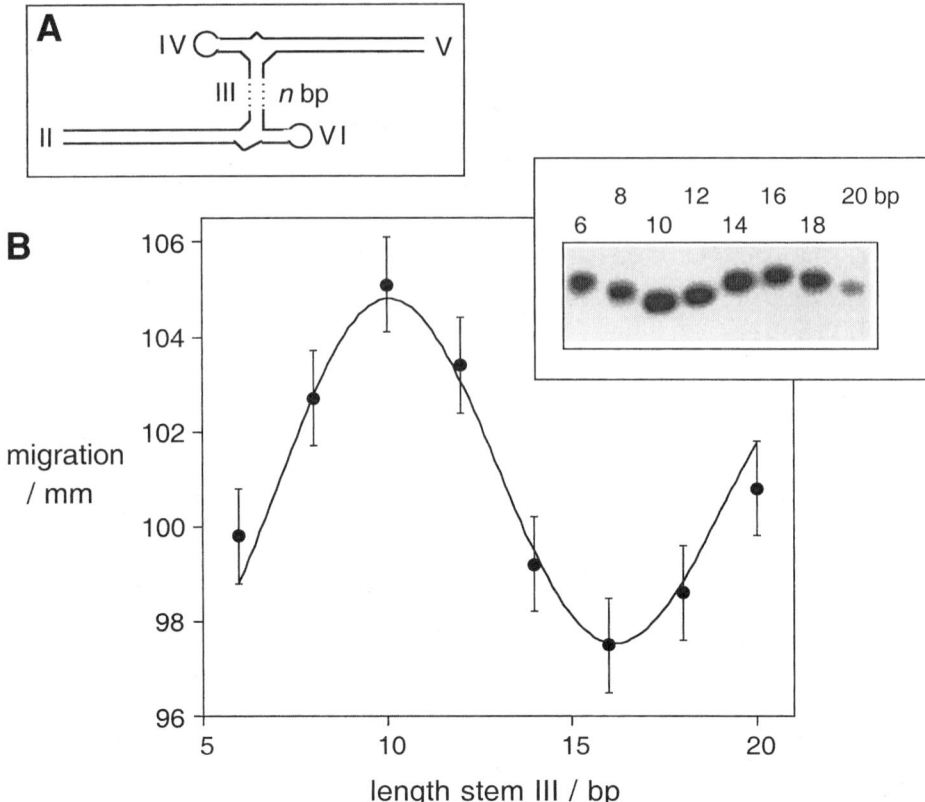

Fig. 3. Electrophoretic analysis of the dihedral angle subtended between helices II and V of the VS ribozyme. (**A**) The analysis employed a series of variant ribozymes in which helices II and V were elongated, and helices IV and VI shortened, and a perfectly basepaired helix III that varied in length from 6–20 bp. The dihedral angle between the two three-way junctions would vary with the length of the central helix III, and it would be expected that electrophoretic mobility would depend on this angle *(47)*. (**B**) The autoradiograph, and a plot of electrophoretic mobility as a function of helix III length. The electrophoretic analysis shows the expected sinusoidal modulation of mobility, and the measured mobilities (points) have been fitted (line) to a simple geometric model from which we can calculate the end-to-end distance for each length of stem III. A dihedral angle of approx 80° was calculated from this analysis. (**C**) Molecular graphics image of the global structure of the VS ribozyme *(47)*.

III enabled us to compare the end-to-end distances of the model to those derived from the electrophoretic mobility and **Eq. 1**. The best agreement of the measured end-to-end distances for all lengths of helix III gave a dihedral angle of 70°, in reasonable agreement with the value from the simple geometric model.

**C**

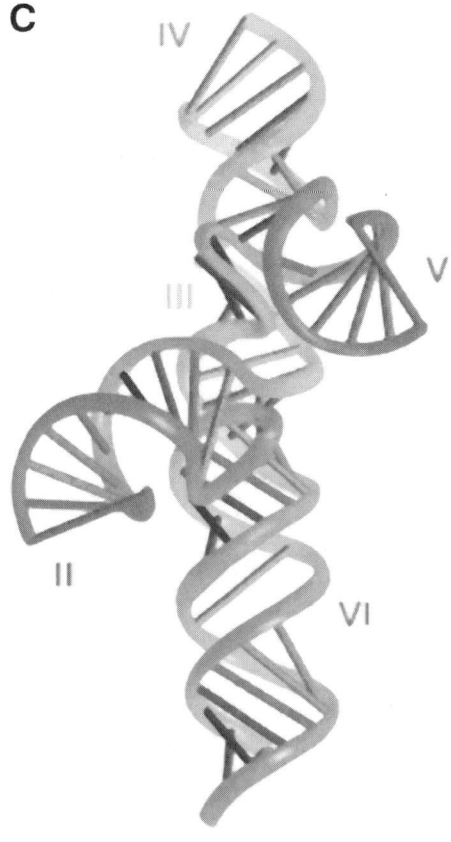

Fig. 3C.

## 5. Fluorescence Resonance Energy Transfer

FRET provides a powerful alternative approach to the study of global conformation in branched nucleic acids in free solution, based on the distance-dependent transfer of excitation between fluorophores attached at different positions *(24)*. With the usual fluorophores employed in these studies, we are most sensitive to distances in the range of 30–80 Å, which is perfect for studying the global conformation of ribozymes.

FRET can be studied in a variety of different ways. The simplest is in the steady-state, involving excitation by a constant light intensity. From this, the efficiency of energy transfer can be measured, but if multiple species are present, this will be an average property. Information on multiple species and dynamic processes can be obtained using time-resolved fluorescence.

In the most common approach, two different fluorophores (such as fluorescein and tetramethyl rhodamine) are attached to the termini of different helical arms. Excitation energy can be transferred from the fluorophore with the higher energy of excitation (donor) to that of lower energy (acceptor) in a radiationless process. The efficiency of energy transfer ($E_{FRET}$) between the two dyes is strongly distance-dependent, and by comparing these efficiencies for all the possible end-to-end vectors the global structure can be determined.

## 6. The Theoretical Basis of FRET

The theory of FRET as applied to nucleic acids has been described in detail by Clegg *(25)*. Upon the absorption of a photon, a fluorophore is excited into a higher electronic singlet state. Since the transition occurs in a time that is much more rapid than nuclear motion, the molecule is transiently in a higher vibrational level of this state, which then rapidly decays into the lowest vibrational level. The excited state can then become deactivated by the emission of a photon of lower energy; this is the Stokes shift of fluorescent emission. The excited state can also be depopulated by a variety of other processes, including collisional quenching and intersystems crossing to a triplet state with subsequent phosphorescent emission possible, and all of these lead to a reduction in the fluorescent quantum yield. However, in the presence of a second (acceptor) fluorophore with an absorption maximum to the red of the first (donor) fluorophore, resonant coupling between the emission dipole of the donor fluorophore and the excitation dipole of the acceptor fluorophore leads to a deactivation of the donor and an activation of the acceptor. The resonance energy transfer can be observed in a number of ways. The deactivation of the donor leads to a reduced fluorescent quantum yield ($\Phi$), and a shortened lifetime ($\tau$) of the excited state. The transfer of excitation to the acceptor results in increased fluorescent emission (sensitized emission), which provides the most sensitive way to measure FRET in steady-state experiments.

Förster *(24)* showed that the rate of energy transfer (rate constant $k_{FRET}$) is given by:

$$k_{FRET} = [9 \cdot \ln 10 \Phi^{D} \kappa^{2} J(\lambda)] / [128 \pi^{5} N n^{4} \tau_{D} R^{6}] \qquad \text{[Eq. 3]}$$

This contains the sixth power of the distance R between the fluorophores in the denominator, and this is the origin of the use of the method for obtaining distance information. $\tau_{D}$ is the excited state lifetime of the donor in the absence of an acceptor, $\Phi^{D}$ is the fluorescent quantum yield of the donor, $n$ is the refractive index of the medium, N is Avogadro's number, and $\kappa$ is related to the relative orientation of the two transition dipole moments. $J(\lambda)$ is the normalized spectral overlap integral, given by:

$$J(\lambda) = \frac{\int_0^\infty \phi^D(\lambda)\varepsilon^A(\lambda)\lambda^4 d\lambda}{\int_0^\infty \phi^D(\lambda)d\lambda} \qquad \text{[Eq. 4]}$$

where $\phi^D$ is the spectral shape of the fluorescent emission of the donor and $\varepsilon^A$ is the molar absorbance of the acceptor at each wavelength ($\lambda$). Thus, the rate of energy transfer depends on the extent of overlap between the emission spectrum of the donor and the excitation spectrum of the acceptor.

**Equation 3** can be simplified to:

$$k_{FRET} = \tau_D^{-1}(R_0/R)^6 \qquad \text{[Eq. 5]}$$

where $R_0$ is a characteristic distance for the donor-acceptor pair at which energy transfer is half-maximal efficient, given by:

$$(R_0)^6 = 8.8 \cdot 10^{28} \Phi^D \kappa^2 n^{-4} J(\lambda) \qquad \text{[Eq. 6]}$$

The main uncertainty in the interpretation of FRET data lies in the dependence on the relative orientation of the two transition moments. $\kappa^2$ can take values from 4 (colinear vectors) to 0 (orthogonal vectors). However, for the situation in which the dyes undergo flexible reorientation during the lifetime of the exited state $\kappa^2$ averages to two-thirds. Provided that at least one of the fluorophores is mobile, this is a good approximation, and this can be estimated by measurements of fluorescence anisotropy.

In general, what is measured in a FRET experiment is the efficiency of energy transfer ($E_{FRET}$). This corresponds to the quantum yield for energy transfer, and is the proportion of photons absorbed by the donor that lead to excitation of the acceptor by energy transfer.

$$E_{FRET} = \frac{k_{FRET}}{\sum k_{deact} + k_{FRET}} = \frac{k_{FRET}}{\tau_D^{-1} + k_{FRET}} \qquad \text{[Eq. 7]}$$

where $k_{deact}$ are the rate constants for all the processes leading to de-excitation of donors other than FRET. Substitution of **Eq. 5** leads to:

$$E_{FRET} = 1/[1 + (R/R_0)^6] \qquad \text{[Eq. 8]}$$

Thus, provided we can measure $E_{FRET}$, then in principle we can obtain molecular distance information in our system.

Although the focus is steady-state FRET methods in this chapter, it should be noted that $E_{FRET}$ can be obtained from the fluorescent lifetime of the excited state of the donor, since this is shortened by energy transfer to the acceptor (lifetime = $\tau_{DA}$) compared to that of the donor in the absence of the acceptor ($\tau_D$), and thus:

$$E_{FRET} = 1 - (\tau_{DA}/t_D) \qquad \text{[Eq. 9]}$$

## 7. The Normalized Acceptor Ratio Method

In our steady-state work we have found it most useful to measure efficiencies of energy transfer from the enhancement of acceptor fluorescence, using the acceptor normalization method of Clegg (Clegg, 1992 #814; Clegg, 1992 #1865). The emission at a given wavelength $(\nu_1)$ of a double-labeled sample excited primarily at the donor wavelength $(\nu')$ contains emission from the donor, emission from the directly excited acceptor and emission from acceptor excited by energy transfer from the donor:

$$F(\nu_1\nu') \propto \{\varepsilon^D(\nu')\Phi^A(\nu_1)E_{FRET}d\cdot a + \varepsilon^A(\nu')\Phi^A(\nu_1)\cdot a + \varepsilon^D(\nu')\Phi^D(\nu_1)\cdot d[(1 - E_{FRET})a + (1 - a)]\}$$

$$= F^A(\nu_1\nu') + F^D(\nu_1\nu') \qquad \text{[Eq. 10]}$$

where d and a are the molar fraction of molecules labeled with donor and acceptor respectively. Superscripts D and A refer to donor and acceptor, respectively. $\varepsilon^D(\nu')$ and $\varepsilon^A(\nu')$ are the molar absorption coefficients of donor and acceptor, respectively, and $\Phi^D(\nu_1)$ and $\Phi^A(\nu_1)$ are the fluorescent quantum yields of donor and acceptor, respectively. Thus, the spectrum contains the components resulting from donor emission $(F^D(\nu_1,\nu')$ ie the first term containing $\Phi^D(\nu_1))$ and those resulting from acceptor emission $(F^A(\nu_1,\nu')$ ie the latter two terms containing $\Phi^A(\nu_1))$.

The first stage of the analysis involves subtraction of the spectrum of RNA labeled only with donor, leaving just the acceptor components—i.e., $F^A(\nu_1,\nu')$. The pure acceptor spectrum thus derived is normalized to one from the same sample excited at a wavelength $(\nu'')$ at which only the acceptor is excited, with emission at $\nu_2$. We then obtain the normalized acceptor ratio:

$$(ratio)_A = F^A(\nu_1\nu')/F^A(\nu_2\nu')$$

$$= \{E_{FRET}\cdot d\cdot[\varepsilon^D(\nu')/\varepsilon^A(\nu'')] + [\varepsilon^A(\nu')/\varepsilon^A(\nu'')]\}[\Phi^A(\nu_1)/\Phi^A(\nu_2)] \qquad \text{[Eq. 11]}$$

$E_{FRET}$ is directly proportional to $(ratio)_A$ and can be easily calculated since $\varepsilon^D(\nu')/\varepsilon^A(\nu'')$ and $\varepsilon^A(\nu')/\varepsilon^A(\nu'')$ are measured from absorption spectra, and $\Phi^A(\nu_1)/\Phi^A(\nu_2)$ is unity when $\nu_1 = \nu_2$.

## 8. Instrumentation for Steady-State Fluorescence

Steady-state fluorescence measurements require a good fluorimeter. In this laboratory we use an SLM-Aminco 8100 (unfortunately, this is no longer manufactured), equipped with Glan-Thompson polarizers. Measurements of FRET efficiency are performed under photon counting conditions, with the polarizers crossed at the magic angle (54.7°) in order to remove polarization artifacts. Fluctuation of lamp intensity is corrected using a concentrated rhodamine B solution as a quantum counter.

## 9. Construction of Doubly Labeled Species and the Choice of Fluorophores

Fluorophores can be attached to nucleic acids internally (to bases or phosphates), or more commonly at 5' or 3' termini. They can be conjugated postsynthetically—typically as succinyl esters that are reacted with primary amine groups attached to the end of the nucleic acid via an alkyl chain—or they can be coupled as phosphoramidites as the final step of the synthesis. The former allows a much greater flexibility in terms of the fluorophores available, and with the advent of RNA synthesis based on *bis*(2-acetoxyethoxy)methyl (ACE)-protected phosphoramidites *(26)* this becomes more attractive.

For steady-state and time-resolved FRET experiments with nucleic acids in bulk solution, fluorescein is the most commonly used fluorescent donor (**Fig. 4A**). Its $\lambda_{max}$ is 495 nm, and it can be readily excited by the 488-nm line from an argon ion laser for time-resolved measurements. Fluorescein can be coupled during synthesis as a phosphoramidite, although some care in the choice structural isomer as isomeric purity is spectroscopically desirable. Fluorescein is very attractive for FRET experiments as the fluorophore is very mobile when it is attached to RNA at the 5' terminus via a six-carbon linker (with an anisotropy in the region of 0.1). This removes much of the uncertainty associated with the orientation of the transition moments, and using this donor, it is probably a safe assumption that $\kappa^2$ = two-thirds. In conjunction with experiments on the position of Cy3 (see below) we have deduced that the average position of fluorescein is maximally laterally extended from the helix *(27)*.

Using fluorescein as a donor, the two commonly chosen acceptors are rhodamine and cyanine 3 (**Fig. 4B,C**). Tetramethylrhodamine was used in many earlier studies *(28–38)*. It has a $\lambda_{max}$ of 560 nm and a good spectral overlap with fluorescein, yeilding an $R_0$ of approx 45 Å. A detailed photophysical study has shown that rhodamine experiences a number of distinct environments when attached at the 5' phosphate of a double-stranded DNA helix via a $C_6$ linker (Vamosi, 1996 #1773). In recent years, we *(22,39)* and others *(40)* have used indocarbocyanine-3 (Cy3) as the preferred acceptor for FRET measurements in bulk solution. It is spectrally quite similar to tetramethylrhodamine, although the overlap with the emission spectrum of fluorescein is better, and the $R_0$ is longer, at approx 56 Å *(27,49)*. The quantum yield is also higher than that of tetramethyl rhodamine, making the application of the *(ratio)$_A$* method easier. It is excited at 547 nm and has an emission maximum at 560 nm, and Cy3 emission can readily be deconvoluted from that of fluorescein. Cy3 can be attached to the 5'-terminus of RNA with good efficiency as a phosphoramidite, and there are no particular purification problems. When attached via a $C_3$ linker the attached dye is very constrained, with anisotropy values of approx 0.33, and using NMR, we have shown that the indole rings are stacked onto the end of the helix in the manner of

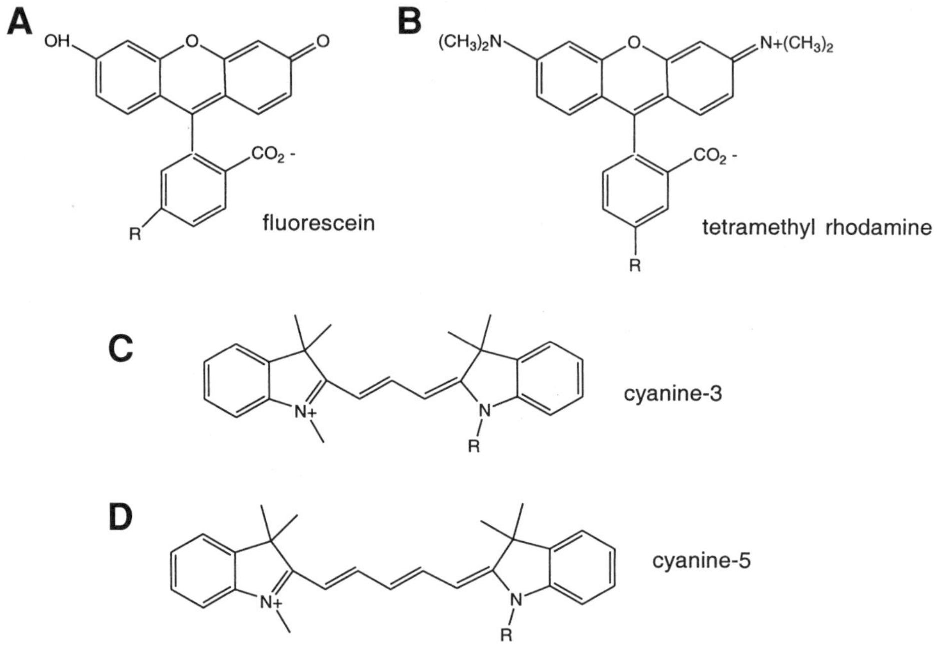

**A**

fluorescein

**B**

tetramethyl rhodamine

**C**

cyanine-3

**D**

cyanine-5

Fig. 4. Fluorophores commonly used in FRET experiments on nucleic acids. (**A**) Fluorescein. (**B**) Tetramethylrhodamine. (**C**) Cyanine3. (**D**) Cyanine5.

an additional basepair *(27)*. Knowledge of the position of this fluorophore is a major advantage in the structural interpretation of FRET data.

In single-molecule FRET studies of nucleic acids, the most common choice of fluorophores has been Cy3–Cy5 (**Fig. 4C,D**), in which Cy3 is now the donor and is excited by means of a frequency doubled YAG laser at 532 nm. This FRET pair has a number of advantages for these studies. The quantum yields of the dyes are very similar, and there is good spectral separation of donor and acceptor, making the separation of light emitted by the two fluorophores by means of dichroic mirrors relatively easy. Although the position of Cy5 on the end of the helix is not precisely known, the similarity in structure to Cy3 and the high anisotropy make it probable that it is terminally stacked in the same way as Cy3, and thus there is a potential uncertainty in the orientation factor $\kappa^2$.

In this laboratory, all RNA species used in FRET studies are generated by chemical synthesis of component strands followed by hybridization in appropriate combinations and purification. RNA species may be synthesized using ribonucleotide phosphoramidites with 2'-*tert* butyldimethylsilyl (*t*BDMS) protection. However, we have recently implemented ACE chemistry *(26)* (also available as custom synthesis by Dharmacon in Boulder), which has a number of advantages. Fluorophores are coupled to the 5' terminus as phosphoramidites, and the average efficiency of fluorophore conjugation is typically 97%. Oligoribonucleotides are deprotected in 25% ethanol/ammonia solution at 55°C for 6 h (dye-labeled) or 12 h (unlabeled). *t*BDMS groups are removed by treatment in 0.5 mL 1 *M* tetrabutylammonium fluoride in tetrahydrofuran for 16 h at 20°C in the dark with agitation. The deprotected RNA is desalted by gel filtration followed by ethanol precipitation. All RNA species are purified by gel electrophoresis in polyacrylamide gels (usually 20%) containing 7 *M* urea; fluorescently labeled species are significantly retarded in the gel system. Bands are excised, and the oligonucleotides electroeluted into 8 *M* ammonium acetate, and recovered by ethanol precipitation. Fluorescently labeled oligonucleotides are further purified by reversed-phase high-performance liquid chromatography (HPLC) (C18 column, eluted with a linear gradient of 100 m*M* ammonium acetate/acetonitrile, with a flow rate of 1 mL/min). The required stoichiometric combinations of fluorophore-labeled and unlabeled strands are then heated together in 90 m*M* Tris-borate (pH 8.3), 25 m*M* NaCl for 10 min at 80°C, followed by slow cooling over a period of several hours. The doubly labeled species are purified by electrophoresis in a polyacrylamide gel under nondenaturing conditions at 4°C at 150 V in 90 m*M* Tris-borate, pH 8.3, 25 m*M* NaCl with recirculation at >1 L/h. The fluorescent species are recovered by band excision and electroelution. Special precautions are necessary in the synthesis of RNA labeled with Cy5, as the fluorophore is significantly more sensitive than the related Cy3. It will not withstand the deprotection required for *t*BDMS synthesis, and thus ACE chemistry is necessary. Moreover, the labeled RNA should be protected from light at all stages.

## 10. Relative Distance Information From FRET

In principle, the distance information available from FRET may allow us to build a model of the global structure of interest. However, a great deal of useful information is available from steady-state FRET experiments without requiring the calculation of absolute distances.

FRET experiments can be performed in a comparative manner, somewhat analogous to the comparative gel electrophoresis discussed here. This is performed for a three- or four-way helical junction, with arms of equal length (if possible), and donor and acceptor fluorophores attached at the 5' termini. Just as with the electrophoresis, there are three and six different end-to-end distances that we can compare in a three- and four-way junction respectively (**Eq. 2**). Therefore, we prepare species from one strand 5'-labeled with donor (fluorescein), one strand 5'-labeled with acceptor (presently Cy3) and the remaining strands unlabeled. An example of such a fluorophore-labeled vector for the 2-3-6 junction of the VS ribozyme is shown in **Fig. 5A**. For best results,

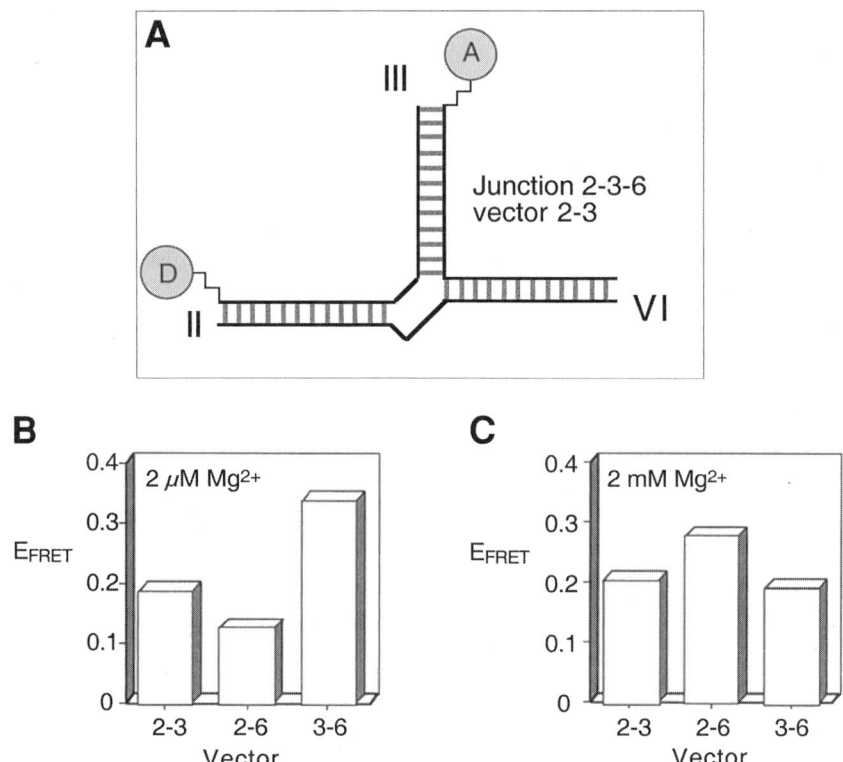

Fig. 5. FRET efficiencies compared for different end-to-end vectors of the 2-3-6 junction of the VS ribozyme as a function of $Mg^{2+}$ ion concentration. **(A)** Schematic illustration of the 2-3 vector of the junction, in which D represents the donor (fluorescein) and A the acceptor (Cy3). The vectors are each named according to the arms carrying the donor and acceptor, in that order. **(B and C)** Histograms of FRET efficiencies for three end-to-end vectors in the presence of 2 µ$M$ **(B)** and 2 m$M$ **(C)** $Mg^{2+}$ *(41)*. The relative FRET efficiencies can be interpreted in terms of the model illustrated in **Fig. 2B**.

the arm lengths will typically be in the range of 10–15 bp. The FRET efficiencies are compared for the three end-to-end vectors of the VS 2-3-6 junction at low (**Fig. 5B**) and high (**Fig. 5C**) magnesium-ion concentration *(41)*. Just as with the comparative gel electrophoresis experiments, the ion-induced structural transition is obvious from a comparison of the two histograms. At low magnesium ion concentrations the 3–6 vector is clearly the shortest end-to-end distance, and in the presence of 2 m$M$ magnesium ions, the 2–6 vector has become the shortest. The relative FRET efficiencies are fully consistent with the global structures of the junction determined from comparative gel electrophoresis, as shown earlier in **Fig. 2**.

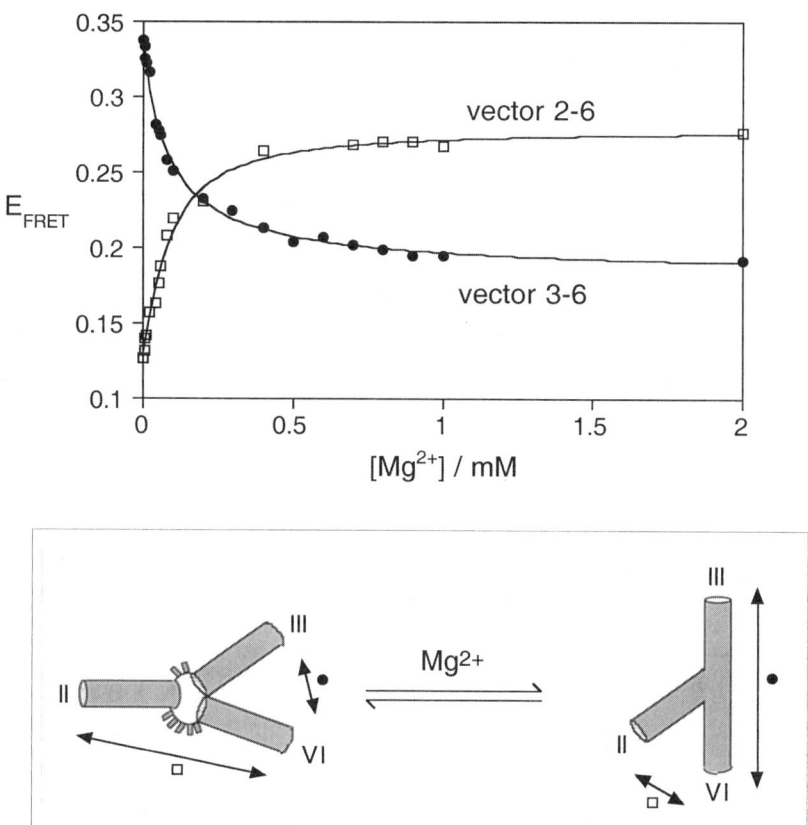

Fig. 6. Ion-induced folding of the 2-3-6 junction of the VS ribozyme analysed using two end-to-end vectors. Plot of FRET efficiency for the 2-6 (open squares) and 3-6 (closed circles) vectors as a function of $Mg^{2+}$ concentration *(41)*. The data are fitted (lines) to a simple two-state folding model. A scheme to illustrate the global conformational change is shown in the lower panel.

Using the acceptor normalization method, FRET efficiency can be easily measured over a range of magnesium ion concentrations, for different end-to-end vectors. An example is shown in **Fig. 6**, for the 2–6 and 3–6 vectors of the VS junction 2-3-6 *(41)*. The trends are quite clear, with the shortening of the 2–6 vector and the lengthening of the 3–6 vector as the junction undergoes ion-induced folding. Such titration data can be fitted to a two-state model, in which the unfolded ribozyme ($Rz_U$) is induced to fold (into ($Rz_F$)) by the binding of $x$ divalent metal ions:

$$Rz_U + x\, Mg^{2+} \rightleftharpoons Rz_F \cdot Mg^{2+}_x$$

The resulting FRET change will be:

$$\Delta E_{FRET} = [E_{FRET}(Rz_F) - E_{FRET}(Rz_U)] \cdot [K_A(Mg^{2+})^n/1 + K_A(Mg^{2+})^n] \qquad [Eq. 12]$$

where $K_A$ is the apparent binding constant for the metal ion and $n$ is a Hill coefficient that measures the cooperativity of binding. The magnesium concentration at which the transition is half complete ($[Mg]_{1/2}$) is given by $(1/K_A)^{1/n}$, and is more reliable than $K_A$ itself because high covariation. The data for the VS junction 2-3-6 have been fitted to this model (**Fig. 6**), giving values of $n = 1$ and $[Mg]_{1/2} = 100$ μ*M (41)*. This is a rather simple situation, in which the folding is induced by the non-cooperative binding of magnesium ions, which is likely to correspond to the binding of a single magnesium ion.

Not all ribozymes or junctions exhibit such simple folding transitions, and a more complex example is given by the hairpin ribozyme *(42)*. The natural form of this ribozyme comprises two loop-carrying arms that exist as adjacent arms of a perfect four-way junction (**Fig. 1B**). We used FRET to demonstrate that the ribozyme folds so as to bring the loops into close juxtaposition *(22)*, and this was later confirmed when the ribozyme was crystallized in the junction form *(10)*. The close contact between the loops creates the local environment in which catalysis can proceed. The folding is induced by the binding of magnesium ions, and in their absence there is no loop-loop interaction *(43)*. However, detailed analysis of the folding using steady-state FRET measurements showed that there are two parts to the transition *(42)*. The major part of the change in end-to-end FRET is brought about by highly cooperative ($n = 2.7$) metal ion binding. However, since this transition can be induced by monovalent ions, it is probable that only diffuse binding is involved. The remaining part of the change requires divalent metal ions, but is non-cooperative. It is likely that this corresponds to site-specific binding of magnesium ions. The isolated junction (in which the loops have been removed from the ribozyme) folds simply, by the non-cooperative binding ($n = 1$) of magnesium ions, and the complexity clearly arises from the multiple interactions between the loops.

FRET is also a very useful source of information of the progress of folding as a function of time. If the junction is removed from the hairpin ribozyme, folding is rather inefficient *(44)*, and thus is slow enough to be measured conventionally *(45)*. In the junction form, folding occurs in the millisecond range, and thus, stopped-flow techniques are necessary. In this case, we find that it is more sensitive to study the decrease in donor intensity as folding brings the acceptor closer. An example is shown in **Fig. 7**, in which folding is studied by the change in the fluorescein intensity following mixing with magnesium ions. These data can be fitted to one or more exponential functions to determine the rate constant(s) for the folding process.

**A**

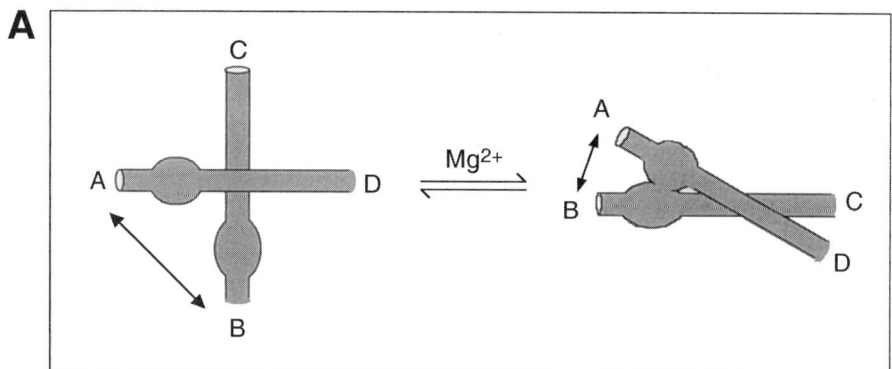

**B**

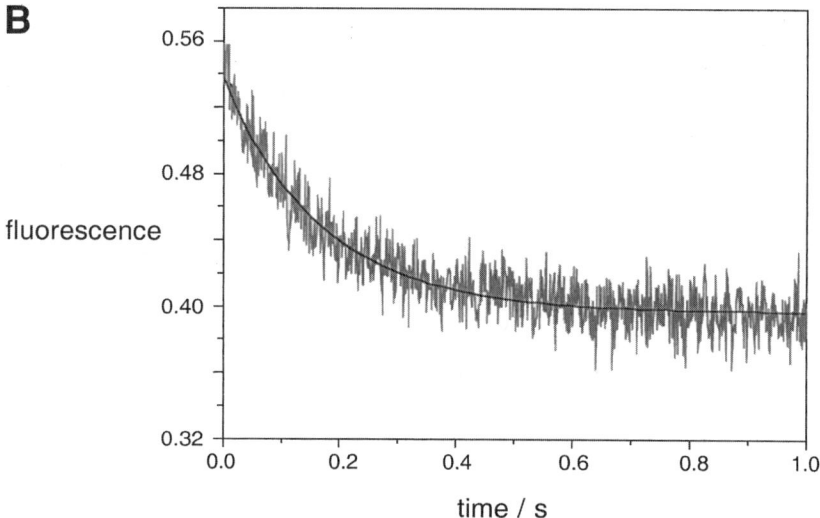

Fig. 7. Analysis of the kinetic process of the folding of the hairpin ribozyme by FRET, using stopped-flow mixing. **(A)** The hairpin ribozyme comprises two loop-carrying helices as adjacent arms of a four-way junction (*see* **Fig. 1B**). The arms are coaxially stacked in two pairs, A on D and B on C. At low $Mg^{2+}$ concentrations the loops are held away from contact, but upon addition of $Mg^{2+}$ ions the ribozyme undergoes a change in conformation that brings the loops into intimate contact *(22)*, as found in the crystal structure *(10)*. By placing donor and acceptor fluorophores on the B and A arms, we can follow the ion-induced folding process by the increase in FRET efficiency as the distance becomes shorter. **(B)** A time trace from the stopped-flow, showing the reduced donor (fluorescein) fluorescence after $Mg^{2+}$ addition to 500 $\mu M$, as a result of increased energy transfer as the ribozyme folds. The experimental data are plotted in gray, and the black line is an exponential fit. These data were kindly provided by Dr. T. Wilson.

## 11. Modeling RNA Structures Using FRET Information

The use of FRET-based information to build molecular models of RNA structures requires the derivation of absolute distance information. FRET is potentially extremely valuable in this process, because of the distance range it covers—typically 30–80 Å. This is not available from any other solution method, and may be complementary to the angular information available from residual dipolar couplings in nuclear magnetic resonance (NMR) experiments. A fairly limited number of distance constraints are usually available from FRET, compared to the hundreds of derived distances that a typical NMR experiment, but their length makes them highly valuable. In NMR experiments, distances are obtained from measurements of nuclear Overhauser enhancements that are limited to approx 5 Å, whereas the scale of FRET-based distances are on the global scale of the macromolecule. Since nucleic acids are generally rather extended structures, restricting the distance information to short range is a severe limitation that can leave structures under-determined, and even a relatively few long-range distance constraints from FRET can make a major difference. And in many cases, we can apply considerable additional information concerning the structure of interest, such as the geometry of A-form helices in double-helical RNA segments.

A number of limitations can potentially hinder the use of FRET data for structure building. One frequently discussed difficulty in the calculation of absolute distances from FRET arises from the orientation dependence of transfer efficiency. The term $\kappa$ in **Eq. 3** measures the relative orientation of donor and acceptor transition dipole moments, and can take a value between 0 and 4 for different orientations. In the case of flexible reorientation during the lifetime of the excited state, $\kappa^2$ averages to two-thirds, and as the interfluorophore distance increases this becomes a better approximation. Fluorescein attached via a six-carbon linker to the 5'-terminus of DNA is generally very mobile, with a low fluorescence anisotropy of approx 0.1, indicating that the fluorophore reorients rapidly during the lifetime of the excited state. Using fluorescein as the donor, $\kappa^2 = 2/3$ is generally a good approximation. Fortunately, the calculated distances are not as sensitive to the exact value of this parameter as it might appear because of the sixth root dependence (**Eq. 8**).

Assuming that we can dispose of the $\kappa^2$ (and other less worrying) problems, then we can measure a distance between the donor and acceptor fluorophores. However, in order to interpret this information in a manner that is useful for modeling the structure, we must know where the dyes are located relative to the nucleic acid. The cyanine dyes are clearly constrained by terminal attachment to double-stranded nucleic acids, with anisotropy values of 0.3 or higher. FRET studies of a series of DNA duplexes of different length, terminally

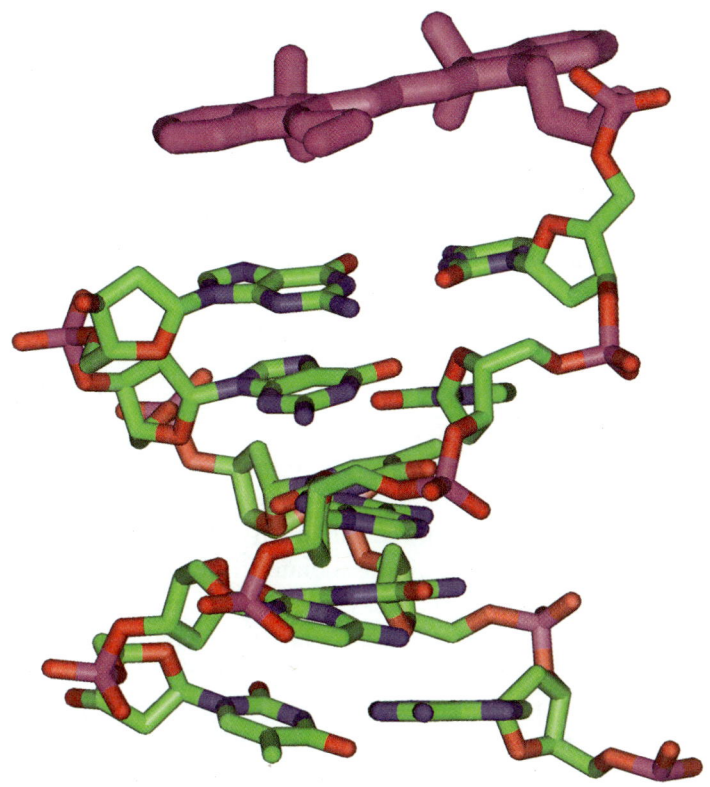

Fig. 8. The structure of Cy3 attached to the 5' end of double-stranded DNA, determined by NMR. The Cy3 is connected to the terminal phosphate via a three-carbon linker, and is thus equivalent to a ribose linkage. The NMR structure shows that the fluorophore (colored magenta) is stacked onto the end of the helix in the manner of a pseudo basepair *(27)*.

labeled with fluorescein and Cy3, indicated that the latter was located close to the helical axis, and this was demonstrated rigorously in an NMR study *(27)*, showing that the Cy3 was stacked onto the end of the helix in the manner of an additional basepair (**Fig. 8**). This is important, because it provides a known point from which distances are measured. By comparison with Cy3, terminally attached fluorescein is relatively flexible within a cone, and a single position cannot be ascribed to such a dynamic situation. However, using the known location of Cy3 together with interfluorophore distances for varying lengths of DNA duplexes from FRET, an effective position was determined. This showed that fluorescein is extended laterally from the DNA, as a result of a combination of charge repulsion and the flexible linker. So, despite the dynamic nature of the tethered fluorescein, we know the position from which it appears to act, and the combination of fluorescein-Cy3 is thus very useful for studying RNA structure. Moreover, this donor-acceptor pair is characterized by a relatively long $R_0$ of 56 Å *(27,40)*, providing distance information in the range of 35–80 Å with some reliability.

We applied this approach to analyse the structure of the 2-3-6 junction of the VS ribozyme folded in the presence of magnesium ions *(41)*, using end-to-end vectors labeled with the fluorescein-Cy3 combination. Independent distances were obtained by varying the lengths of the helical arms of the junction (11, 12, and 15 bp), and by reversing the fluorophore positions (e.g., vectors 2-6 and 6-2). After discarding vectors that yielded FRET efficiencies of less than 0.11 (considered unreliable), we were left with a total of 14 long-range distance measurements, in the range of 65.5–73.4 Å. These distances were then used to model the global structure of the junction. Activity data indicated that the arms were basepaired and helical for 3 bp on each arm beyond the formally unpaired core region of the junction, and canonical A-form helices were assumed; the structure of the helical arms and associated fluorophores was conserved during the subsequent modeling process. The central region was incorporated into the model by means of distance restraints between the junction-proximal ends of the duplex regions that allowed unpaired nucleotides to be in conformations ranging from extended to A-form. Simulated annealing and rigid body molecular mechanics were then used to determine the three-dimensional structure of the junction. Structures with the lowest violations of experimental restraints, derived from distance-restrained molecular dynamics, formed an extremely closely-defined group, as shown in **Fig. 9**. The structure obtained is almost planar, with arms III and VI being close to coaxial, arm II subtending an acute angle (65°) to arm VI, and an obtuse angle (136°) to arm III. The coaxial character of helices III and VI was checked by measuring the FRET efficiency of a 22-bp duplex corresponding to coaxially stacked 11-bp helices, obtaining a value of 70.1 Å; this is close to the value of 68.7 Å for the 6-3 vector of the junction. The global shape is fully consistent with the observations from comparative gel electrophoresis studies (**Fig. 2**).

A global structure of this type tells us little about the local structure in the core of the junction, and any progress in that regard requires additional information. In the case of the VS ribozyme 2-3-6 junction this came from an unexpected direction, using the crystal structure of the 50S ribosomal subunit *(46)* as a valuable structural resource. Inspection of the base sequence of the 23S rRNA revealed a three-way helical junction that was remarkably similar to that of the 2-3-6 junction, and replacement of the ribozyme junction with that from the ribosome yeilded a hybrid ribozyme that retained significant catalytic activity *(41)*. Therefore, it was probable that the two structures were very similar.

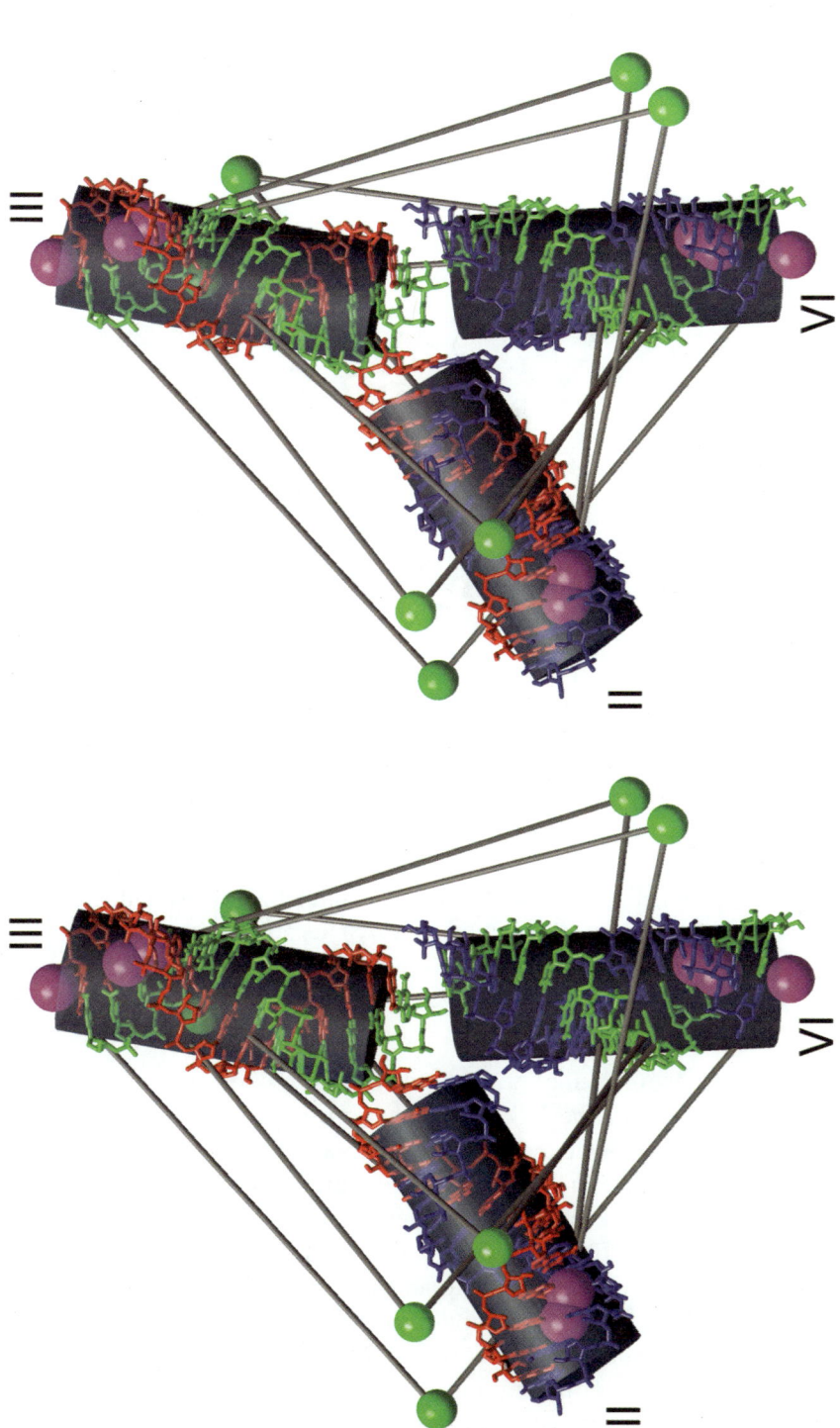

Fig. 9. The global structure of the 2-3-6 junction of the VS ribozyme determined by FRET. Parallel-eye stereographic represen-tation of the structure in 2 m*M* Mg$^{2+}$ ions (*41*), with the three strands differentiated by color. The positions of the fluorophores are indicated by single solid spheres (fluorescein, green, and Cy3, pink). Gray bars indicate the 14 FRET-derived distance constraints used in the calculations.

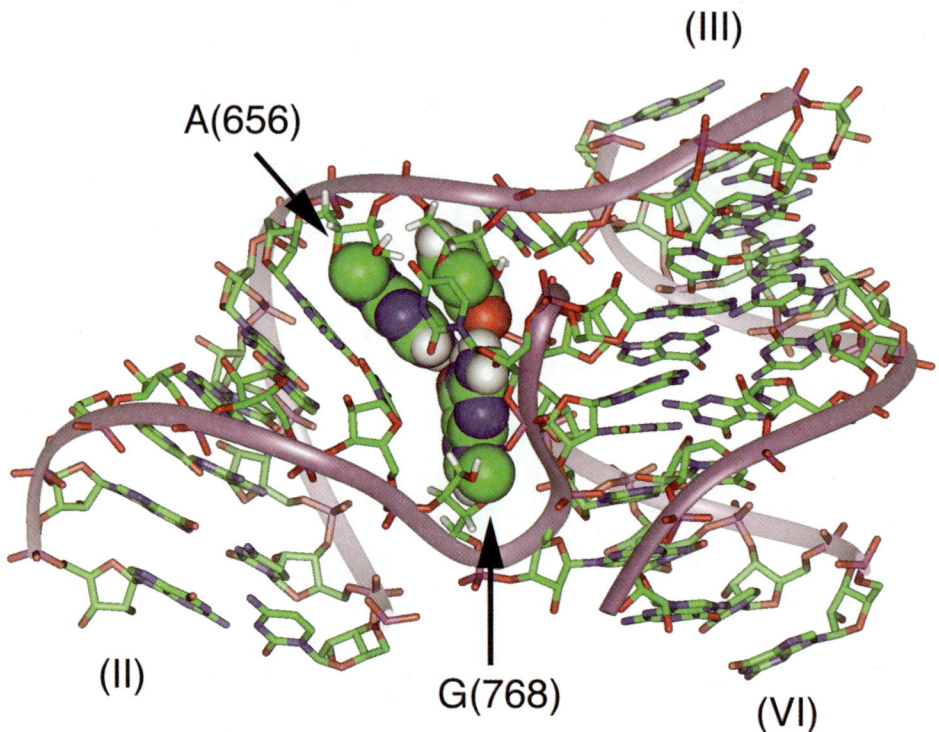

Fig. 10. The structure of a three-way RNA junction in 23S rRNA that is very similar in sequence to the VS 2-3-6 junction. The ribosomal junction can functionally substitute for the 2-3-6 junction in the ribozyme *(41)*. The structure of the ribosomal junction has been determined in the 50S ribosomal subunit by X-ray crystallography *(46)*. The equivalent helices in the VS junction are given in parentheses. The bases of the nucleotides corresponding to G768 and A656 are indicated.

The structure of the ribosomal junction is shown in **Fig. 10**. The global shape is the same as that deduced for the VS 2-3-6 junction, with the equivalents of helices III and VI coaxially stacked, and an acute angle subtended between helices II and VI. In this structure the long, single-stranded region connecting these helices is looped around, and the equivalent of G768 is packed against it, forming a wedge-like interaction with A656. This adenine is stacked onto the terminal basepair of the helix II equivalent, and thus this G-wedge interaction determines the trajectory of this arm of the ribozyme. Not surprisingly, G768 and A656 are the most important nucleotides of the VS ribozyme junction for catalytic activity, with mutations leading to up to 1000-fold reductions of activity. In later studies, we showed that the cleft formed between helices II and VI forms the principal binding site for the substrate, enabling the scissile phosphate to make contact with the A730 loop *(47)*. This loop is the probable active site of the ribozyme *(48)*, where a critical adenine A756 is believed to play a key role in the catalysis *(49)*.

## 12. Time-Resolved FRET

It is probable that many functional structures are in dynamic equilibria between alternative conformations in solution, and steady-state FRET measurements will only see an average of these. Thus, important information for the function of the RNA species may be missed. Time-resolved fluorescence provides a possible means to solve this problem. The details of this approach are beyond the scope of this chapter, and the reader who needs greater depth is directed to reviews of this subject *(50)*.

In brief, in time-resolved fluorescence we examine the lifetime ($\tau$) of the excited state of the fluorophore of interest. This is typically in the nanosecond range; for example the lifetime of fluorescein is of the order of 4 ns. The lifetime may be studied in either the time or frequency domain. In the time domain, the sample is excited by a brief pulse of light, generally produced using a mode-locked laser, and the decay of the resulting fluorescence is analyzed. Alternatively, in the frequency domain, the excitation is subjected to high-frequency amplitude modulation, and the phase shift and demodulation of the emission are analysed as a function of modulation frequency. In this laboratory, we make measurements in the frequency domain, using a K2 phase fluorimeter from ISS, where light from an argon ion laser is modulated using a Pockels cell driven by high frequency (up to 250 MHz) synthesizers. In either mode, the results can be analyzed in terms of multiple lifetimes, or distributions of lifetimes, thus providing information on different species present in solution. In a sample labeled with donor and acceptor, the lifetime of the donor will be shortened as a result of energy transfer to the acceptor (**Eq. 9**), and the data can be analyzed in terms of distributions of donor-acceptor lengths.

Time-resolved FRET measurements have been applied to the hairpin ribozyme *(51,52)*. We recently applied time-resolved FRET measurements to the analysis of a four-way junction derived from the IRES of hepatitis C virus *(61)*. This is somewhat different from the perfect junction of the hairpin ribozyme, for example, because of the presence of additional unpaired bases at the point of strand exchange. The time-resolved FRET data reveal that for the junction in the presence of $Mg^{2+}$ ions, although two vectors (corresponding to the two pairs of coaxially stacked helices) are well-fitted by a single Gaussian distribution of donor-acceptor lengths similar to a duplex of the same length, the remaining four vectors are fitted significantly better using two Gaussian

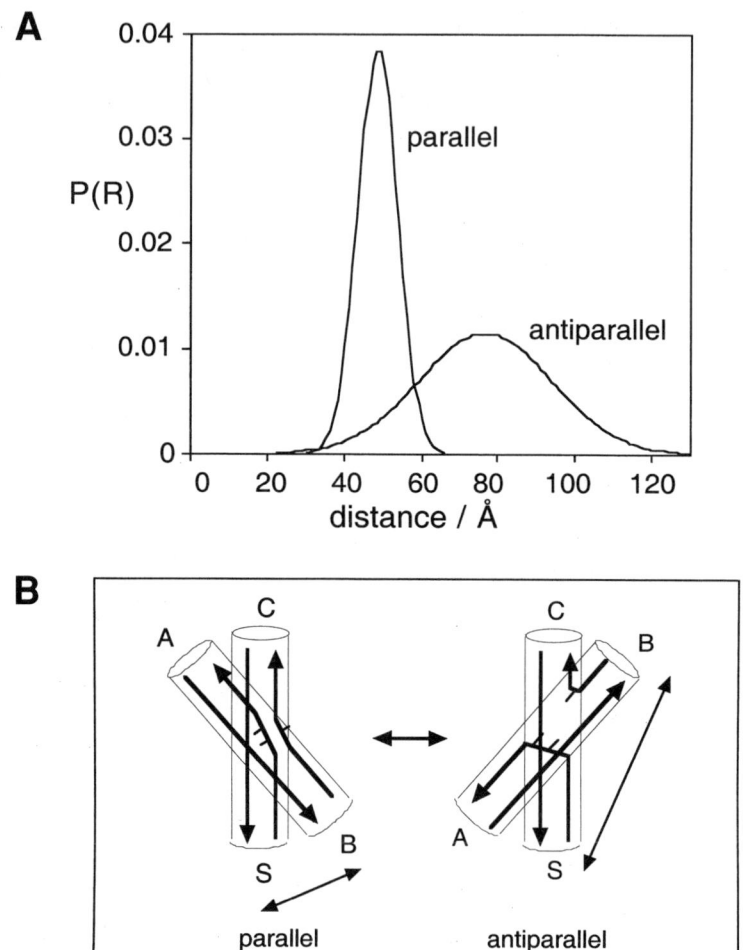

Fig. 11. Analysis of donor-acceptor distance distributions by time-resolved FRET measurements for the junction IIIabc of the HCV IRES *(61)*. This is a four-way junction in which there are three formally-unpaired bases at the point of strand exchange, forming a $2HS_2 2HS_1$ junction *(59,60)*. One end-to-end vector (BS) has been analyzed in terms of two Gaussian distributions of lengths **(A)**, corresponding to antiparallel and parallel conformations of the junction, illustrated in **(B)**. A single Gaussian distribution gave a significantly poorer fit to the experimental data.

distributions in each case **(Fig. 11)**. These correspond to parallel and antiparallel forms of the junction in rapid exchange, with similar populations. Clearly, such dynamic properties are likely to be important in the biological function of the RNA.

## 13. Single-Molecule FRET

Both steady-state and time-resolved fluorescence studies require $\geq 10^{12}$ molecules in the cuvet. One of the most exciting developments of the application of FRET to nucleic acids in the last few years is the ability to study molecules singly *(53,57)*. This has many advantages as compared to ensemble measurements. It is possible to analyze transitions that cannot be synchronized, and that therefore preclude stopped-flow analysis. A recent example is the study of transitions between stacking conformers of four-way DNA junctions *(7)*. Single-molecule studies provide a fresh perspective on the kinetic properties of molecules, in which the history and fate of states are known, and thus, transitions can be ordered in time. They have also revealed surprising heterogeneities between molecules in some systems *(57)*.

Although molecules can be studied in free solution *(54)*, a majority of the studies to date have been on molecules attached to glass surfaces *(55,56)*. This is readily achieved by means of biotin covalently attached to a 5'-terminus. If we wanted to study a three-way RNA junction, for example, we would assemble the junction from three strands separately 5'-labeled with Cy3 (donor), Cy5 (acceptor), and biotin (attachment). This can then be studied in a variety of ways. In one, the donor is excited by the evanescent wave arising from totally internally reflected light from a solid-state laser, and the surface is imaged through the objective lens of an inverted microscope (note that the molecules themselves cannot form an image as they are smaller than the wavelength of the light; they simply act as point sources of light, like a star). The light is split into donor and acceptor wavelengths by a series of dichroic mirrors, and the separate images are collected in an intensified CCD camera. The advantage is that data can be collected from many molecules simultaneously, but is limited in terms of time resolution to approx 100 ms. Time resolution can be substantially improved by collecting the light from individual molecules using a confocal microscope.

An example of single-molecule FRET data for the hairpin ribozyme in its natural, four-way junction form is shown in **Fig. 12** *(62)*. The construct is immobilized through the C arm, and carries Cy3 and Cy5 fluorophores on the A and B arms. In the open structure, which is predominant at low magnesium-ion concentrations, the fluorophores are relatively distant, and therefore the donor intensity is high. When the molecule folds into the conformation with intimate loop-loop interaction, efficient energy transfer occurs, which quenches the donor signal and provides greater emission intensity from the acceptor. Transitions between the two conformations are very clear, with anticorrelation of donor and acceptor signal as expected. A given molecule can be studied over multiple transitions, until one of the fluorophores (usually Cy5) undergoes photobleaching. Measurement of the dwell times in the two states yields the

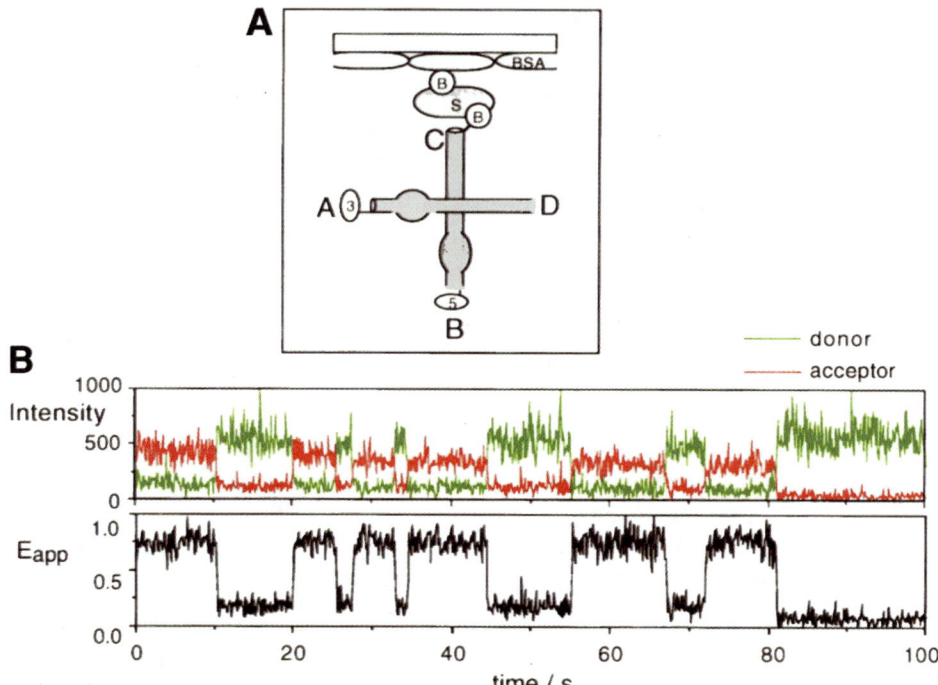

Fig. 12. Time traces for a single hairpin ribozyme molecule *(62)*. The ribozyme is undergoing transitions between folded and unfolded conformations (as shown in **Fig. 7A**). **(A)** The hairpin ribozyme is labeled with Cy3 (3, donor) on the A arm and Cy5 (5, acceptor) on the B arm. The molecule is attached via a biotin (B) attached to the terminus of the C arm, though a linkage with streptavidin (S) to biotinylated BSA on a quartz surface. Emitted light is divided into donor (Cy3) and acceptor (Cy5) and intensities are measured separately. **(B)** The upper traces show the separate donor (green) and acceptor (red) fluorescence as a function of time. In the folded state there is high energy transfer, yielding a higher fluorescence intensity for the acceptor. In the unfolded state, the donor shows higher intensity. Note the anticorrelation of donor and acceptor fluorescence in the folding/unfolding transitions. The lower trace shows a plot of apparent FRET efficiency over the same time range, calculated as $E_{app} = I_A/(I_D + I_A)$, where $I_D$ and $I_A$ are the measured fluorescence intensity of donor and acceptor, respectively.

rate of forward and backward transitions, which are found to be in good agreement with ensemble measurements. Temperature control is possible, and activation parameters have been measured in some systems. More rapid transitions can be analyzed using correlation analysis, and more complex systems can yield more than two levels of FRET efficiency.

## 14. Conclusion

The methods described in this chapter for the analysis of global structure and transitions in RNA are very powerful, providing long-range structural information in solution. In many ways, they are complementary to established methods for structural determination such as X-ray crystallography and NMR, providing an added perspective on folding processes and the dynamics of what are frequently structurally polymorphic species.

## Acknowledgments

I gratefully thank all my coworkers and collaborators over a number of years in the application of these methods to nucleic acid structure, including Daniel Lafontaine, David Norman, Tim Wilson, Sonya Melcher, Robert Clegg, Elliot Tan, and Taekjip Ha.

## References

1. Fedor, M. J. (2001) Structure and function of the hairpin ribozyme. *J. Mol. Biol.* **297,** 269–291.
2. Hammann, C. and Lilley, D. M. J. (2002) Folding and activity of the hammerhead ribozyme. *Chem. Bio. Chem.* **3,** 690–700.
3. Lilley, D. M. J. (1999) Structure, folding and catalysis in the small nucleolytic ribozymes. *Curr. Opin. Struct. Biol.* **9,** 330–338.
4. Soukup, G. A. and Breaker, R. R. (1999) Relationship between internucleotide linkage geometry and the stability of RNA. *RNA* **5,** 1308–1325.
5. Murray, J. B., Seyhan, A. A., Walter, N. G., Burke, J. M., and Scott, W. G. (1998) The hammerhead, hairpin and VS ribozymes are catalytically proficient in monovalent cations alone. *Chem. Biol.* **5,** 587–595.
6. Lilley, D. M. J. (2000) Structures of helical junctions in nucleic acids. *Quart. Rev. Biophys.* **33,** 109–159.
7. McKinney, S. A., Déclais, A.-C., Lilley, D. M. J., and Ha, T. (2003) Structural dynamics of individual Holliday junctions. *Nature Struct. Biol.* **10,** 93–97.
8. Pley, H. W., Flaherty, K. M., and McKay, D. B. (1994) Three-dimensional structure of a hammerhead ribozyme. *Nature* **372,** 68–74.
9. Scott, W. G., Finch, J. T., and Klug, A. (1995) The crystal structure of an all-RNA hammerhead ribozyme: a proposed mechanism for RNA catalytic cleavage. *Cell* **81,** 991–1002.
10. Rupert, P. B. and Ferré-D'Amaré, A. R. (2001) Crystal structure of a hairpin ribozyme-inhibitor complex with implications for catalysis. *Nature* **410,** 780–786.
11. Ferré-d'Amaré, A. R., Zhou, K., and Doudna, J. A. (1998) Crystal structure of a hepatitis delta virus ribozyme. *Nature* **395,** 567–574.
12. Lerman, L. S. and Frisch, H. L. (1982) Why does the electrophoretic mobility of DNA in gels vary with the length of the molecule. *Biopolymers* **21,** 995–997.
13. Lumpkin, O. J. and Zimm, B. H. (1982) Mobility of DNA in gel electrophoresis. *Biopolymers* **21,** 2315,2316.

14. Levene, S. D. and Zimm, B. H. (1989) Understanding the anomalous electrophoresis of bent DNA molecules: a reptation model. *Science* **245**, 396–399.
15. de Gennes, P. G. (1971) Reptation of a polymer chain in the presence of fixed obstacles. *J. Chem. Phys.* **55**, 572–578.
16. Bhattacharyya, A., Murchie, A. I. H., and Lilley, D. M. J. (1990) RNA bulges and the helical periodicity of double-stranded RNA. *Nature* **343**, 484–487.
17. Tang, R. S. and Draper, D. E. (1990) Bulge loops used to measure the helical twist of RNA in solution. *Biochemistry* **29**, 5232–5237.
18. Luebke, K. J. and Tinoco, I. (1996) Sequence effects on RNA bulge-induced helix bending and a conserved five-nucleotide bulge from the group I introns. *Biochemistry* **35**, 11,677–11,684.
19. Grainger, R. J., Murchie, A. I. H., Norman, D. G., and Lilley, D. M. J. (1997) Severe axial bending of RNA induced by the U1A binding element present in the 3' untranslated region of the U1A mRNA. *J. Molec. Biol.* **273**, 84–92.
20. Duckett, D. R., et al. (1988) The structure of the Holliday junction and its resolution. *Cell* **55**, 79–89.
21. Bassi, G., Møllegaard, N. E., Murchie, A. I. H., von Kitzing, E., and Lilley, D. M. J. (1995) Ionic interactions and the global conformations of the hammerhead ribozyme. *Nature Struct. Biol.* **2**, 45–55.
22. Murchie, A. I. H., Thomson, J. B., Walter, F., and Lilley, D. M. J. (1998) Folding of the hairpin ribozyme in its natural conformation achieves close physical proximity of the loops. *Molec. Cell* **1**, 873–881.
23. Duckett, D. R., Murchie, A. I. H., and Lilley, D. M. J. (1995) The global folding of four-way helical junctions in RNA, including that in U1 snRNA. *Cell* **83**, 1027–1036.
24. Förster, T. (1948) Zwischenmolekulare Energiewanderung und Fluoreszenz. *Ann. Phys.* **2**, 55–75.
25. Clegg, R. M. (1992) Fluorescence resonance energy transfer and nucleic acids. *Meth. Enzymol.* **211**, 353–388.
26. Scaringe, S. A. (2000) Advanced 5'-silyl-2'-orthoester approach to RNA oligonucleotide synthesis. *Methods Enzymol.* **317**, 3–18.
27. Norman, D. G., Grainger, R. J., Uhrin, D., and Lilley, D. M. J. (2000) The location of Cyanine-3 on double-stranded DNA; importance for fluorescence resonance energy transfer studies. *Biochemistry* **39**, 6317–6324.
28. Murchie, A. I. H., et al. (1989) Fluorescence energy transfer shows that the four-way DNA junction is a right-handed cross of antiparallel molecules. *Nature* **341**, 763–766.
29. Clegg, R. M., et al. (1992) Fluorescence resonance energy transfer analysis of the structure of the four-way DNA junction. *Biochemistry* **31**, 4846–4856.
30. Clegg, R. M., Murchie, A. I. H., Zechel, A., and Lilley, D. M. J. (1994) The solution structure of the four-way DNA junction at low salt concentration; a fluorescence resonance energy transfer analysis. *Biophys. J.* **66**, 99–109.
31. Clegg, R. M., Murchie, A. I. H., Zechel, A., and Lilley, D. M. J. (1993) Observing the helical geometry of double-stranded DNA in solution by fluorescence resonance energy transfer. *Proc. Natl. Acad. Sci. USA* **90**, 2994–2998.

32. Eis, P. S. and Millar, D. P. (1993) Conformational distributions of a 4-way DNA junction revealed by time-resolved fluorescence resonance energy transfer. *Biochemistry* **32,** 13,852–13,860.
33. Gohlke, C., Murchie, A. I. H., Lilley, D. M. J., and Clegg, R. M. (1994) The kinking of DNA and RNA helices by bulged nucleotides observed by fluorescence resonance energy transfer. *Proc. Natl. Acad. Sci. USA* **91,** 11,660–11,664.
34. Tuschl, T., Gohlke, C., Jovin, T. M., Westhof, E., and Eckstein, F. (1994) A three-dimensional model for the hammerhead ribozyme based on fluorescence measurements. *Science* **266,** 785–789.
35. Yang, M. S. and Millar, D. P. (1996) Conformational flexibility of three-way DNA junctions containing unpaired nucleotides. *Biochemistry* **35,** 7959–7967.
36. Stühmeier, F., Welch, J. B., Murchie, A. I. H., Lilley, D. M. J., and Clegg, R. M. (1997) The global structure of three-way DNA junctions with and without bulges: Fluorescence studies. *Biochemistry* **36,** 13,530–13,538.
37. Stühmeier, F., Lilley, D. M. J., and Clegg, R. M. (1997) The effect of bulges on the stability of three-way DNA junctions studied by fluorescence techniques. *Biochemistry* **36,** 13,539–13,551.
38. Miick, S. M., Fee, R. S., Millar, D. P., and Chazin, W. J. (1997) Crossover isomer bias is the primary sequence-dependent property of immobilized Holliday junctions. *Proc. Natl. Acad. Sci. USA* **94,** 9080–8084.
39. Bassi, G. S., Murchie, A. I. H., Walter, F., Clegg, R. M., and Lilley, D. M. J. (1997) Ion-induced folding of the hammerhead ribozyme: a fluorescence resonance energy transfer study. *EMBO J.* **16,** 7481–7489.
40. Jares-Erijman, E. A. and Jovin, T. M. (1996) Determination of DNA helical handedness by fluorescence resonance energy transfer. *J. Molec. Biol.* **257,** 597–617.
41. Lafontaine, D. A., Norman, D. G., and Lilley, D. M. J. (2001) Structure, folding and activity of the VS ribozyme: importance of the 2-3-6 helical junction. *EMBO J.* **20,** 1415–1424.
42. Wilson, T. J. and Lilley, D. M. J. (2002) Metal ion binding and the folding of the hairpin ribozyme. *RNA* **8,** 587–600.
43. Walter, F., Murchie, A. I. H., Thomson, J. B., and Lilley, D. M. J. (1998) Structure and activity of the hairpin ribozyme in its natural junction conformation; effect of metal ions. *Biochemistry* **37,** 14,195–14,203.
44. Zhao, Z.-Y., Wilson, T. J., Maxwell, K., and Lilley, D. M. J. (2000) The folding of the hairpin ribozyme: dependence on the loops and the junction. *RNA* **6,** 1833–1846.
45. Walter, N. G., Hampel, K. J., Brown, K. M., and Burke, J. M. (1998) Tertiary structure formation in the hairpin ribozyme monitored by fluorescence resonance energy transfer. *EMBO J.* **17,** 2378–2391.
46. Ban, N., Nissen, P., Hansen, J., Moore, P. B., and Steitz, T. A. (2000) The complete atomic structure of the large ribosomal subunit at 2.4 Å resolution. *Science* **289,** 905–920.
47. Lafontaine, D. A., Norman, D. G., and Lilley, D. M. J. (2002) The global structure of the VS ribozyme. *EMBO J.* **21,** 2461–2471.

48. Lafontaine, D. A., Wilson, T. J., Norman, D. G., and Lilley, D. M. J. (2001) The A730 loop is an important component of the active site of the VS ribozyme. *J. Molec. Biol.* **312,** 663-674.
49. Lafontaine, D. A., Wilson, T. J., Zhao, Z.-Y., and Lilley, D. M. J. (2002) Functional group requirements in the probable active site of the VS ribozyme. *J. Molec. Biol.* **323,** 23–34.
50. Lakowicz, J. R. (1999) *Principles of Fluorescence Spectroscopy.* Plenum Press, New York, NY, pp. 1–698
51. Walter, N. G., Burke, J. M., and Millar, D. P. (1999) Stability of hairpin ribozyme tertiary structure is governed by the interdomain junction. *Nature Struct. Biol.* **6,** 544–549.
52. Klostermeier, D. and Millar, D. P. (2001) Tertiary structure stability of the hairpin ribozyme in its natural and minimal forms: different energetic contributions from a ribose zipper motif. *Biochemistry* **40,** 11,211–11,218.
53. Ha, T., et al. (1996) Probing the interaction between two single molecules: fluorescence resonance energy transfer between a single donor and a single acceptor. *Proc. Natl. Acad. Sci. USA* **93,** 6264–6268.
54. Deniz, A. A., et al. (1999) Single-pair fluorescence resonance energy transfer on freely diffusing molecules: observation of Forster distance dependence and subpopulations. *Proc. Natl. Acad. Sci. USA* **96,** 3670–3675.
55. Ha, T., et al. (1999) Ligand-induced conformational changes observed in single RNA molecules. *Proc. Natl. Acad. Sci. USA* **96,** 9077–9082.
56. Zhuang, X., et al. (2000) A single molecule study of RNA catalysis and folding. *Science* **288,** 2048–2051.
57. Zhuang, X. W., et al. (2002) Correlating structural dynamics and function in single ribozyme molecules. *Science* **296,** 1473–1476.
58. Saville, B. J. and Collins, R. A. (1990) A site-specific self-cleavage reaction performed by a novel RNA in *Neurospora* mitochondria. *Cell* **61,** 685–696.
59. Brown, E. A., Zhang, H., Ping, L. H., and Lemon, S. M. (1992) Secondary structure of the 5' nontranslated regions of hepatitis C virus and pestivirus genomic RNAs. *Nucleic Acids Res.* **20,** 5041–5045.
60. Honda, M., et al. (1996) Structural requirements for initiation of translation by internal ribosome entry within genome-length hepatitis C virus RNA. *Virology* **222,** 31–42.
61. Melcher, S. E., Wilson, T. J., Lilley, D. M. (2003) The dynamic nature of the four-way junction of the hepatitis C virus JRES. *RNA* **9,** 809–820.
62. Tan, E., Wilson, T. J., Nahas, M. K., Clegg, R. M., Lilley, D. M., Ha, T. (2003) A four-way junction accelerates hairpin ribozyme folding via a discrete intermediate. *Proc. Natl. Acad. Sci. USA* **100,** 9308.-9313.

# 8

## In Vivo Detection of Ribozyme Cleavage Products and RNA Structure by Use of Terminal Transferase-Dependent PCR

Hsiu-Hua Chen, Daniela Castanotto, Jeanne LeBon, John J. Rossi, and Arthur D. Riggs

### Summary

Terminal transferase-dependent PCR (TDPCR) can be used after reverse transcription to analyze RNA. This method (RT-TDPCR) is able to provide in vivo information at nucleotide-level resolution, and has been used for study of ribozymes, RNA size, RNA structure, and RNA-protein interactions. A detailed protocol of RT-TDPCR is presented here with examples of its use in detecting ribozyme cleavage intermediates in yeast and a RNA transcription start site in mammalian cells.

**Key Words:** RT-TDPCR; ribozyme; RNA structure; mammalian RNA; transcription start.

## 1. Introduction

Reverse transcription (RT) followed by terminal transferase-dependent polymerase chain reaction (RT-TDPCR) was developed in our laboratory for analyzing yeast and mammalian RNA molecules (*1–5*). With RT-TDPCR, much information can be obtained about in vivo RNA size, structure and protein binding, and the procedure has been applied to the detection of ribozyme cleavage (*1–3*). RT-TDPCR is derived from TDPCR (*6*), which had previously been used for the study of DNA footprints and chromatin structure (*7–10*). When preceded by RT, TDPCR provides an extremely sensitive, versatile, and quantitative assay with nucleotide-level resolution, and can be used for in vivo detection of mRNA lesions, stem-loop structures, splicing sites, protein footprints, and transcriptional start sites (*1–3*). The procedure has adequate sensitivity and specificity for studies of eukaryotic cells; yeast and mammalian

From: *Methods in Molecular Biology, vol. 252: Ribozymes and siRNA Protocols, Second Edition*
Edited by: M. Sioud © Humana Press Inc., Totowa, NJ

ribozyme cleavage intermediates and final products have been studied both in vitro and in vivo by RT-TDPCR *(1,3–5)*. For example, RT-TDPCR was used in conjunction with oligodeoxynucleotides to identify accessible sites in an RNA target—NCOA3 mRNA—which often is upregulated in breast cancer. Ribozymes targeted to accessible sites identified in this way were effective in downregulating the target message in vivo *(5)*.

As illustrated in **Fig. 1**, RT-TDPCR includes the following steps: i) First-strand cDNA synthesis using a gene-specific, biotinylated first primer and a reverse transcriptase; ii) capture and enrichment of the biotinylated-strand cDNA using magnetic streptavidin beads; iii) riboG-tailing of cDNA on the beads using terminal deoxynucleotidyl transferase (TdT); iv) ligation of the riboG-tailed cDNA on the beads to a double-stranded DNA linker; v) PCR amplification using a nested second primer and a universal linker-primer; vi) labeling of PCR products using a $^{32}$P-$\gamma$-adenosine 5' triphosphate (ATP)- or fluorescent-dye-labeled nested-third primer; and vii) sequencing gel electrophoresis to resolve the labeled products. Gel bands are seen where primer extension by RT reaches the end of an RNA molecule or is terminated (or strongly paused) by lesions in the RNA, or by stable secondary structure. Lesions or breaks in the RNA can be introduced by various agents; for example, by treatment of permeabilized cells with a ribonuclease, and RNA footprint studies using RNase T1 have identified specific in vivo protein-binding sites *(1,3)*. **Figure 2** shows two typical results obtained using RT-TDPCR for analysis of total cellular RNA. Panel A depicts ribozyme cleavage products from the expression of a ribozyme in yeast. For this study, AMV RT was used along with a conventional sequencing gel and radioactive detection *(1,3)*. Panel B shows the location of a transcription start site in mouse X*ist* RNA *(11,12)* determined by RT-TDPCR using *C. therm.*™ polymerase, a LI-COR IRD-700 fluorescent-dye labeled primer and a LI-COR sequencer (a non-radioactive detection system) (Chen, H. -H. and Riggs, A. D., unpublished data).

Results shown in **Fig. 2A** illustrate several aspects of RT-TDPCR. Total RNA was prepared from a *Saccharomyces cerevisiae* strain JM43 that had been transformed either with a *cis*-active (self-cleaving) ribozyme construct, pWC1Rz (lanes 4 and 5), or an inactive mutant ribozyme construct, pWC1MtRz (lanes 2 and 3). Both pWC1Rz and pWC1MtRz plasmids contain actin promoters to drive transcription, and in each construct the first exon of the actin gene is followed by a 192-nt fragment, without any splice sites, containing an open reading frame fused either with a cis-ribozyme-encoding sequence (Rz) or an inactive ribozyme mutant sequence (MtRz). The MtRz is identical to the active Rz, except that it contains a G5 to A5 transition mutation in the catalytic core of ribozyme sequence, which abolishes its self-cleavage activity. The Rz or MtRz sequence is followed by the actin exon II and a β-Gal

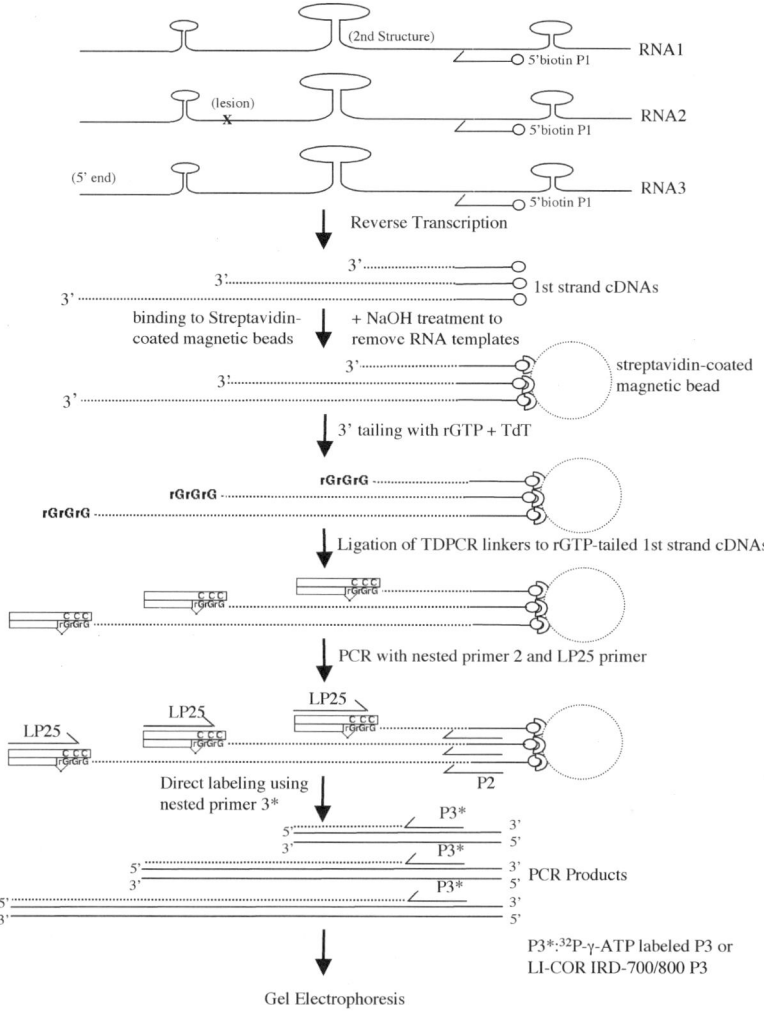

Fig. 1. Schematic outline of the RT-TDPCR procedure. i) RNA is reverse tran-scribed using a biotin-labeled, gene-specific primer (P1); ii) the newly synthesized cDNA strand is captured by streptavidin-coated magnetic beads, and is iii) ribo-tailed using TdT and riboGTP. iv) An oligonucleotide linker with a blocked 3'-terminus is ligated to the tailed, 3' end of the cDNA strand. The cDNA molecules, now having a defined sequence on both the 5' and 3' ends, are v) PCR amplified using a nested gene-specific primer (P2) and a linker-specific primer (LP25). vi) The amplified DNA frag-ments are directly primed by using a third gene-specific primer (P3) with either radioactive ($^{32}$P) or fluorescent-dye (LI-COR IRD-700/800) labeled at its 5' end. vii) The labeled products are separated by use of a DNA sequencing gel and visualized by autoradiography or phosphorimaging ($^{32}$P) or by direct read out from a LI-COR sequencer.

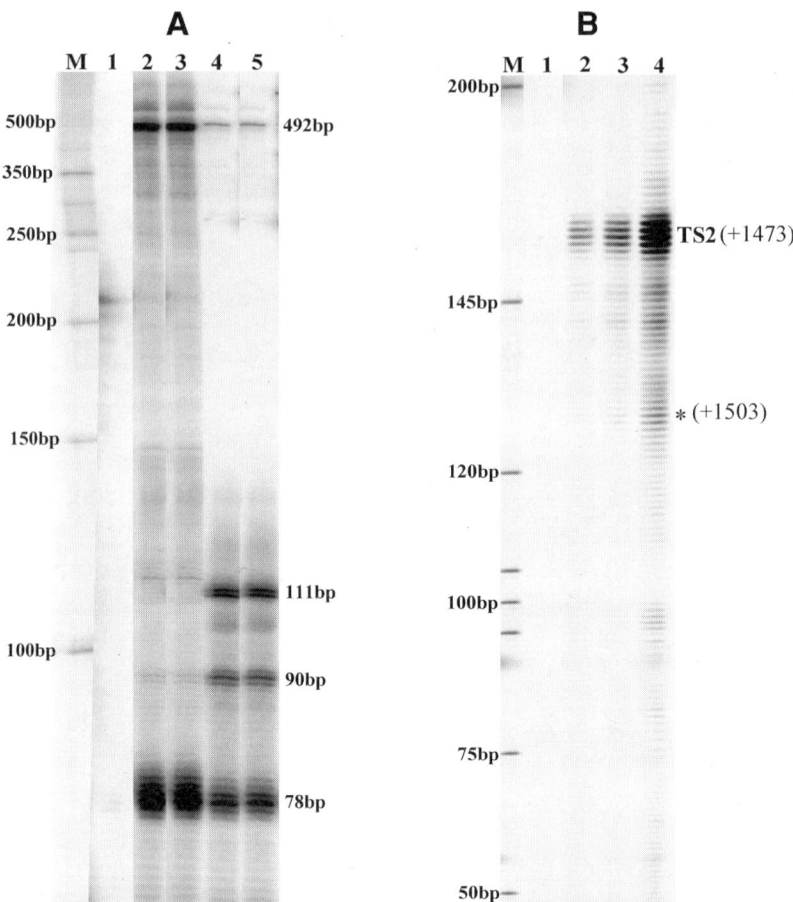

Fig. 2. **(A)** RT-TDPCR analysis of ribozyme self-cleavage products in yeast. Total RNA isolated from untransformed *S. cerevisiae* strain JM43 (lane 1), and transformants of pWC1MtRz (lanes 2 and 3) and pWC1Rz (lanes 4 and 5), were analyzed by RT-TDPCR. Details of the vector construct are given in the text. Lane M shows a 50-bp DNA ladder. Two nested primers were used: primer 1 (β-Gal 23) 5'-biotin-TTAAGTTGGGTAACGCCAGGGTT; primer 2 (β-Gal Nest), 5'-ACGCCAAGGTTT TCCCAGTCACGAC. Primer 2 was $^{32}$P-labeled by kination and used for visualization of the PCR product fragments by the primer extension procedure (**Subheading 3.3.7.**). A phosphorimager was used for detection of bands; exposure was for only a few minutes. **(B)** RT-TDPCR analysis of mammalian RNA with nonradioactive detection using infrared dyes (**Subheading 3.3.8.**) . Total RNA from mouse embryonic stem cells was used to detect a second transcription start site in mouse *Xist* RNA. A LI-COR sequencing system was used for separation and detection. RNA used for analysis is 0, 50, 500, and 1000 ng for lanes 1, 2, 3, and 4, respectively. Lane M is LI-COR IRD-700 labeled DNA markers (50–350 bp).

reporter gene *(1,13)*. The RNA samples were subjected to RT-TDPCR using β-Gal-specific primers *(1)*. As shown in lanes 2 and 3, RT-TDPCR analysis of RNA from pWC1MtRz, which contains the mutant, inactive ribozyme, shows a strong 492-nt band, which corresponds to the full-size transcript. pWC1Rz (lanes 4 and 5) contains an active *cis*-cleaving ribozyme, and shows a strong 111-nt band corresponding to the expected size of the 3' ribozyme cleavage product. RT-TDPCR can detect the remaining uncleaved transcript, although the intensity of the band representing full-length transcripts is only 4% of that observed for the mutant ribozyme construct. The identity of the 111-nt cleavage product was confirmed by isolating the band from the gel and determining its sequence *(1)*. This is generally a useful feature of RT–TDPCR; any band of interest can be isolated from the gel and its sequence determined by sequencing either directly or after cloning. In addition to a strong 492-nt band representing extension to the 5' end of the mRNA, all lanes with both the active and inactive ribozyme (lanes 2–5) have a cluster of bands centered at 78 nt. This band cluster is at the site of insertion of the ribozyme, and probably results from structure-induced pausing of RT during primer extension. This is to be expected, since previous work has shown ribozyme structure to cause pausing of AMV RT *(14)*. A cluster of bands is seen, rather than ~~not~~ a single band, because of variable addition of Gs by TdT. The bands centered at 90 nt, which are present in pWC1Rz but not in pWC1MtRz, are of interest because they demonstrate that RT-TDPCR is detecting subtle RNA structural changes that occurr after ribozyme cleavage. A pause site at 90 nt is consistent with structural predictions based on nucleotide-base interactions detected by nuclear magnetic resonance (NMR) in the cleaved ribozyme *(15,16)*. Lane 1 shows that total cellular RNA isolated from untransformed strain JM43 is free of any of the bands described here; there is a low background.

**Figure 2B** shows an experiment done to determine a transcription start site in mammalian embryonic stem (ES) cells. Total RNA was isolated from male mouse ES cells, and various amounts of RNA (0, 50, 500, and 1000 ng for lane 1, 2, 3, and 4, respectively) were analyzed by RT-TDPCR using *C. therm.*™ polymerase (Roche) and three *Xist*-specific primers. The band marked TS2 (+1473), which is 155 nt in size, shows an apparent transcription start site, and its band intensity of which increases with the amount of RNA used at RT step (lanes 1–4). The asterisk (*) marks a band that is 126 nt in size, which corresponds to +1503 nt in the *Xist* cDNA sequence *(11)* and was reported by Johnston, et. al *(12)* as the tentative second transcription site. The sensitivity, data quality, and resolution of the RT-TDPCR method are apparent, even with nonradioactive detection.

The LI-COR system is now routinely used in our laboratory becuase of its advantages, which include nonradioactive, non-biohazardous detection and evenly spaced band resolution. However, the LI-COR instrument is expensive and may not be available, so the conventional, radioactive detection system is also described here.

## 2. Materials
### 2.1. Reagents and Chemicals

1. $^{32}$P-γ-ATP.
2. Dynabeads M-280 streptavidin (Dynal).
3. SequaGel Solutions (National Diagnostics).
4. 100 m*M* guanosine 5' triphosphate (GTP), 100 m*M* ATP, and 100 m*M* dNTP.
5. Dithiothreitol (DTT).
6. *N,N,N',N'*-tetramethyl-ethylenediamine (TEMED) (Sigma).
7. Ammonium persulfate.
8. Xylene cyanol.
9. Bromophenol blue.
10. LI-COR IR$^2$ stop solution.
11. LI-COR mol wt STR markers (50–350 bp, either IRD-700- or IRD-800-labeled) (LI-COR, Inc).
12. 25-bp and 50-bp DNA Ladders (Gibco BRL).
13. Betaine (Sigma).
14. RNA STAT-60 (Tel-Test, Inc).
15. 2X BW buffer: 10 m*M* Tris-HCl, pH 7.5, 1 m*M* EDTA, 2 *M* NaCl.

### 2.2. Enzymes

1. AMV reverse transcriptase (20 U/μL).
2. Moloney murine leukemia virus (MMLV) reverse transcriptase (200 U/μL).
3. C. therm. polymerase (4 U/μL).
4. RNasin (20 U/μL).
5. TdT (15 U/μL).
6. T4 DNA ligase (3 U/μL).
7. AmpliTaq polymerase (5 U/μL).
8. T4 polynucleotide kinase (PNK) (10 U/μl).
9. *E. coli* Exonuclease I (10 U/μL).
10. T4 DNA polymerase (3 U/μL).
11. Expand Long PCR system (3.5 U/μL, Roche).

### 2.3. Equipment

1. A thermocycler such as a PTC-100 programmable Thermal.
2. PhosphorImager 425S (Molecular Dynamics).
3. Electrophoresis Model SA system (Gibco-BRL).
4. 6000 V power supply.
5. LI-COR IR$^2$ LONG READIR DNA Sequencer 4200 system (LI-COR, Inc.).

## 2.4. Primers and TDPCR Linker

1. Gene-specific primers: Either the Oligo 4™ or the Oligo 5.1™ program (National Biosciences, Inc.) is used to design the primers.

   The first gene-specific primer (Bio-P1) is 5'-biotinylated and designed to have a Tm of approx 50°C for use with AMV or Moloney MMLV transcriptase or approx 70°C for use with *C. therm.* polymerase.

   Since the second primer (P2) is used together with LP25 (the linker-primer) at the PCR step, it is designed to have a Tm of approx 63°C, and the Tm of the third primer (P3) is around 65°C.

2. Universal linker primer (LP25) (25 nt): 5'-GCGGTGACCCGGGAGATC TGAATTC-3'.

3. TDPCR linker primers: Upper primer (LP27): 5'-GCGGTGACCCGGGAG ATCTGATTCCC-3', (27mer). Lower primer (LP24*$C_5$): 5'-AATTCAGAT CTCCCGGGTCACCGC-pentylNH$_2$-3', (24mer). Note that this primer is made with an aminopentyl blocking group at its 3' terminus *(6)*.

## 3. Methods

### 3.1. Preparation of the TDPCR Linkers

1. Kinasing the lower primer: Prepare the kinase mix in a 1.7-mL microfuge tube. (Usually 10 tubes are done at the same time.) Incubate at 37°C for 1.5–2 h followed by 20 min at 65°C *(6)*. Chill on ice (*see* **Table 1**).

2. Annealing with the upper primer: Add 44 µL of 200 µ*M* upper primer directly to the kinased mix (the final concentration of the linkers is 20 µ*M*), mix well, quick-spin, add a lid lock to each tube, and then denature at 95°C for 5 min using a heat block. Turn off the power but leave tubes in the block, and allow gradual cooling to room temperature. Leave at 4°C overnight and store at –20°C.

3. Always thaw the linkers on ice before preparing the ligation mix.

### 3.2. Labeling DNA Mol-Wt Standards

1. Label a 25-bp or 50-bp DNA ladder with $^{32}$P-$\alpha$-deoxycytidine 5' triphosphate (dCTP) using T4 DNA polymerase according to instructions from BRL.

2. Purify the labeled DNA ladder on a G-25 spin column. Add a 2X-vol formamide loading dye and denature for 2 min at 95°C before loading along side RT-TDPCR samples on a sequencing gel.

3. When the $^{32}$P markers are fresh, 0.5–1.0 µL is sufficient for an overnight exposure. Markers can be used for up to 3 mo, compensating for radioactive decay by increasing the amount loaded.

### 3.3. RT-TDPCR Procedure

#### 3.3.1. RT Step

1. First strand cDNA Synthesis using *C. therm.*: Usually 0–1 µg of RNA in a volume of 5 µL is used per reaction. Prepare *C. therm.* RT mix, 15 µL per sample. Prepare enough for total number of samples + 2 extra (*see* **Table 2**).

**Table 1**
**Kinase Mix**

| Component | 1X μL |
|---|---|
| H₂O | 292 |
| 10X kinase buffer (NEB) | 40 |
| 100 mM ATP (BMB) | 4 |
| 200 μM lower primer | 44 |
| T4 DNA kinase (NEB) | 20 |
| Total | 400 μL |

**Table 2**
**RT Mix**

| Component | 1X μL |
|---|---|
| DEPC H₂O | 5.76 |
| 5X RT buffer | 4.0 |
| 100 mM DTT | 1.0 |
| 25 mM dNTP | 0.64 |
| RNasin | 0.5 |
| DMSO | 0.6 |
| 20 μM Bio-P1 | 1.0 |
| *C. therm.* polymerase* | 1.5 |
| Total | 15 μL |

2. Add 15 μL RT mix to each 5 μL RNA sample. Mix well and do a quick spin (*see* **Note 1**).
3. Add 10–20 μL mineral oil on top (omit if the thermocycler has a hot-bonnet). Primer extension is carried out at 70°C or at the Tm of Bio-P1 for 30 min.

### 3.3.2. Magnetic Bead Capture and Enrichment

1. Prepare Dynabeads using a magnetic particle concentrator (MPC): Swirl the Dynabead bottle to completely resuspend the beads and then take out enough for all of the samples. Usually, 20 μL/sample is adequate for 20 pmol of the biotinylated primer 1.
2. Transfer 200 μL (e.g., 20 μL per sample × 10 samples) of well-resuspended bead solution to a 1.5-mL tube; use the MPC to separate the beads from the supernatant and remove the supernatant.
3. Wash beads twice with 400 μL 2X BW (10 mM Tris-HCl, pH 7.5, 1 mM EDTA, 2–2.5 M NaCl) by pipetting up and down, and then use the MPC to remove supernatant.
4. Resuspend beads in 200 μL 2X BW. After RT, add 20 μL of the bead solution to each sample. (**Note:** DO NOT vortex while the Dynal beads are present; a quick spin at low speed is OK—e.g., few seconds at 1000 rpm).

**Table 3**
**TdT Mix**

| Component | 1X μL |
|---|---|
| Dd $H_2O$ | 4.93 |
| 5X TdT buffer | 4.0 |
| 100 m$M$ GTP (Roche) | 0.4 |
| TdT (15 U/μL, BRL) | 0.67 |
| Total | 10 μL |

5. Immobilize DNA on the beads by rotating the mixture at room temperature for 15–60 min (see the instructions by the manufacturer). Remove supernatant using the MPC.
6. Wash once with 50 μL of 2X BW and remove supernatant using the MPC.
7. Remove the RNA from the RNA-DNA hybrid by addition of 50 μL 0.15 $M$ NaOH to the beads, mix, and incubate at 37°C for 5–10 min. Remove supernatant, using the MPC. The cDNA will remain on the beads.
8. Wash beads once with another 50 μL 0.15 $M$ NaOH. Remove supernatant using the MPC.
9. Wash beads twice with 100 μL of TE, pH 7.5. Remove supernatant using the MPC.
10. Resuspend beads in 10 μL 0.1X TE, pH 7.5.

### 3.3.3. Terminal Transferase (TdT) Step

1. Prepare TdT mix (10 μL each): Prepare enough for total number of samples + 2 extra (*see* **Table 3**).
2. Add 10 μL TdT mix to each sample, mix well by pipetting, and incubate at 37°C for 15 min.
3. Wash twice with 100 μL TE, pH 7.5. Remove supernatant using the MPC.
4. Resuspend in 15 μL of 0.1X TE, pH 7.5.

### 3.3.4. Ligation

1. Prepare the ligation mix (15 μL each): Prepare enough for total number of samples + 2 extra (*see* **Table 4**).
2. Add 15 μL of ligation mix to each sample, and mix well by pipet.
3. Add a lid lock to each tube and incubate at 17°C overnight.

### 3.3.5. PCR Amplification

1. Do a quick spin to bring down the ligation mix, and remove supernatant using the MPC.
2. Wash the beads twice with 100 μL TE, pH 8.0, remove supernatant using the MPC, and resuspend each bead sample in 10–30 μL 0.1X TE, pH 8.0. If 20 or 30 μL of 0.1X TE, pH 8.0 was used to resuspend the beads, keep the leftover at 4°C. **Note:** Do not freeze when the magnetic beads are present!

**Table 4**
**Ligation Mix**

| Component | 1X µL |
|---|---|
| dd H$_2$O | 7.95 |
| 1 *M* Tris-HCl | 1.5 |
| 1 *M* MgCl$_2$ | 0.3 |
| 1 *M* DTT | 0.3 |
| 100 m*M* ATP | 0.3 |
| 10 mg/mL BSA | 0.15 |
| 20 µ*M* TDPCR linker | 3.0 |
| T4 DNA ligase | 1.5 |
| Total | 15 µL |

**Table 5**
**PCR Mix**

| Component | 1X µL |
|---|---|
| dd H$_2$O | $x$ |
| 5X Taq buffer * | 10 |
| 25 m*M* MgCl$_2$* | 3 or 4 |
| 25 m*M* dNTPs | 0.5 |
| 20 µ*M* primer 2 (P2) | 0.5 |
| 20 µ*M* LP25 | 0.5 |
| AmpliTaq* | 1.0 |
| Total | 40 µL |

*See **Notes 2–5**.

3. Prepare the PCR mix, 40 µL per sample: Prepare enough for total number of samples + 2 extra (*see* **Table 5**).
4. Add 40 µL of PCR mix to each 10 µL resuspended sample. Mix well by pipetting (total volume should be 50 µL).
5. Add 30 µL of mineral oil on top. The oil can be omitted if the thermocycler has a hot bonnet; in this case remember to prestart the program and set at pause to allow the hot bonnet to equilibrate to temperature.
6. After an initial 3 min at 95°C, perform 20 cycles of 45 s at 95°C, 2 min at 63°C or the Tm of the second primer, and 3 min at 72°C. Do 30–40 cycles if band isolation is desired.
7. Keep the PCR products at 4°C (do not freeze!) for more labeling reactions. Although most of the thermostable DNA polymerases we have used remained active for months if PCR mixtures were stored properly at 4°C, we add 0.1 µL additional polymerase into the final labeling reaction.

**Table 6**
**Primer Labeling Mix**

|  | 1X μL | 5X μL |
|---|---|---|
| $H_2O$ | 5.9 | 29.5 |
| 10X kinase buffer | 1.0 | 5.0 |
| 0 μ*M* P3 | 1.1 | 5.5 |
| T4 DNA kinase | 1.0 | 5.0 |
| Total | 9.0 μL | 45.0 μL |

### 3.3.6. Primer Labeling Using $^{32}P$

1. Labeling of Primer 3. Prepare the reaction mixture (*see* **Table 6**).
2. Mix well, on ice, quick-spin on ice, take to the designated radioactivity area, add 1 μL (1X) or 5 μL (5X) $^{32}$P-γ-ATP, mix well, and spin.
3. Incubate at 37°C for 1–1.5 h, then at 65°C 15 min. Purify using a G-25 spin column (suitable for 50–100 μL per column). Take 0.5–1.0 μL to count. Use immediately or store at –20°C.

### 3.3.7. Primer Extension

1. Take 10 μL from each PCR reaction and add 1 μL or more (depending on the counts/μL) of $^{32}$P-γ-ATP-labeled primer 3, spin, and add 10 μL mineral oil.
2. Perform primer extension using a thermal cycler: pause at 95°C and put the tubes in, then denature at 95°C for 2 min, followed by 3–9 cycles at 95°C for 45 s, annealing for 2 min at Tm of primer 3, 72°C for 3 min, and another 10 min at 72°C.
3. If *Exo*I treatment was used (*see* **Note 5**), add the labeled primer 3 directly to the Exo-treated sample and perform the primer extension reaction.
4. Preparation for gel loading: Add an equal volume of formamide loading solution (95% formamide, 20 m*M* ethylenediaminetetraacetic acid [EDTA], pH 8.0, 0.05% Xylene cyanol, 0.05% bromphenol blue) to each sample, denature at 95°C for 2 min, and then cool on ice before loading (usually 10 μL/lane) onto a 6% or 8% acrylamide/8 *M* urea-sequencing gel. Labeled samples can be stored at –20°C.

### 3.3.8. Labeling With LI-COR Dyes IRD-700 or 800

1. Order primer 3 from LI-COR tagged at the 5'-end with either the fluorescent infrared dye IRD-700 or IRD-800 (*see* **Notes 3** and **4**).
2. Resuspend the primer (which arrives as a dry pellet from LI-COR) according to LI-COR's instruction to obtain 1 μ*M* final concentration.
3. After PCR, transfer 9 μL from each reaction to a new tube, add 1 μL of 1 μ*M* LI-COR primer 3 and perform primer extension as above for $^{32}$P labeling. If Exo I treatment was done (*see* **Note 5**), add the LI-COR primer 3 directly to the Exo-treated sample and perform primer extension.

4. Add 3 µL of LI-COR gel loading dye (IR$^2$ Stop solution, LI-COR) to each sample, denature at 95°C for 2 min, cool on ice and load (usually 2 µL/lane) onto a 5% or 6 % LI-COR sequencing gel. Store labeled samples in the dark at –20°C.

### 3.3.9. Gel Electrophoresis

1. $^{32}$P-γ-ATP-labeled samples: Any DNA sequencing electrophoresis apparatus and power supply (6000 V) is suitable. Usually the gel is run at 75 W for 4–5 h until the xylene cyanol reaches the bottom of the gel. The gel is transferred to a 3-mm Whatman paper, covered with a piece of Saran wrap, and dried under vacuum at 80°C. The gel is exposed overnight using a PhosphoImager cassette and scanned the next morning (**Fig. 2A**).
2. IRD-700 or -800 labeled samples: To run LI-COR IRD-primer-labeled samples, we use a IR$^2$ Long Reader DNA Sequencer from LI-COR. The data is automatically saved as a TIFF file during the run, and the TIFF file can be opened directly by a PhotoShop program (**Fig. 2B**) or analyzed quantitatively by programs that accept TIFF files.
3. Both PhosphorImager and LI-COR gel files can be read and quantitated using Gene ImagIR (an upgraded version of RFLP) from Scanalytics, Inc.

## 4. Notes

1. Choice of reverse transcriptase. All RTases should be tested as described in **ref.** *1* to obtain reliable results. AMV and Moloney MMLV RTases, which have a temperature optimum of at 42–45°C, often work well, but can pause at any secondary structures in RNA molecules. This can provide useful information. Reverse transcriptases, such as *C. therm.*™, that have higher temperature optimums can transcribe through most secondary structures. Routinely, C. therm. has been used in our laboratory for RT-TDPCR for its better extension ability at higher temperature to overcome secondary structures present in RNA molecules. For determining secondary structures of RNA, AMV RT or Moloney MMLV RT in addition to *C. therm.* should be used *(1)*.
2. 5X Taq buffer: 200 m*M* NaCl, 50 m*M* Tris-HCl, pH 8.9, 0.05% (w/v) gelatin. 10X Taq buffer from Perkin-Elmer is also suitable, but our homemade buffer with pH at 8.9 works better in our experience. The final [Mg$^{2+}$] should be between 1.5–2.0 m*M* to optimize PCR depending on primer and template.
3. Use of Betaine to a final concentration of 1.5 *M* during the PCR amplification step (**Subheading 3.3.5.**) has proved to be advantageous for high G + C regions with apparent secondary structure problems *(1)*.
4. Expand Long PCR System (Roche) has been used recently in our laboratory for PCR step and works well to allow for longer readouts when using the LI-COR long-ranger gel electrophoresis system. We follow the instructions from BMB to prepare PCR mixtures.
5. (Optional) *E. coli* exonuclease I treatment after PCR amplification: This step eliminates all primers left after the PCR reaction to allow primer 3 to be the only primer present in the direct labeling step. Exo I treatment: i) Use the MPC to

attract the beads and transfer 9 µL each PCR solution to a new tube and add 1 µL of Exo I (diluted to 1 U/µL with water or 1X PCR buffer from the Amersham stock, which is 10 U/µL). ii) Incubate at 37°C for 30 min to eliminate unincorporated primers (P1, P2, and LP25) left from the previous steps. To inactivate Exo I, incubate at 72°C for 15 min.

6. RT-TDPCR is a procedure that involves many steps, and numerous pipettings. Each step is relatively robust, but take special care to make sure that all components of the reaction mixtures are at the correct concentrations and active, especially the enzymes. Always add all the components together except the enzyme, mix well, then add the enzyme and mix well again.

7. To ensure the reproducibility of results, performing experiments in duplicate is highly recommended (*see* **Fig. 2A**). In general, thin-walled tubes give better results.

8. The LI-COR primers IRD-800 and IRD-700 are light-sensitive; handling under dimmed or yellow light is recommended.

9. Both IRD-700 and IRD-800 provide similar sensitivity. A useful fact is that both can be detected simultaneously, allowing two differently labeled samples to be run together in the same lane.

10. LI-COR gels often show evenly spaced bands to 1000 bp, although good-quality data usually does not go beyond 500 bp.

11. RT-TDPCR theoretically and experimentally requires little RNA. We routinely find that 50 ng of total mammalian RNA provides good signals in our experience.

12. If band isolation is desired, the PCR step can be done with 30–40 cycles. When quantitation is desired, one must stay in the linear, quantitative range of PCR, as is usually the case for the first 20 cycles. In general, if more than 20 cycles are needed, it is likely that at least one of the primers is poor or at least one step is inefficient. If this is the case, the patterns seen may not be reproducible.

13. TDPCR generally shows a weak band at every position. This is normal and is caused by polymerase pausing, as well as variability in tailing by TdT. When Taq polymerase is used for PCR, every LMPCR or TDPCR band has a shadow band. This is the result of variability in the addition of an extra base by Taq, and indicates that the enzyme was no longer adequately active in the last few cycles of PCR. Also note that the size purities of the linker primer and labeling primers are important, because the final labeled DNA fragments must be separated with single-nucleotide resolution.

## Acknowledgments

We thank Louise Shively for critical reading of this manuscript. This work was supported by NIH grant GM50575 to A.D.R. and AI29329 to J.J.R.

## References

1. Chen, H.-H., Castanotto, D., LeBon, J. M., Rossi, J. J., and Riggs, A. D. (2000) In vivo, high-resolution analysis of yeast and mammalian RNA-protein interactions, RNA structure, RNA splicing and ribozyme cleavage by use of terminal trnsferase-dependent PCR. *Nucleic Acids Res.* **28,** 1656–1664.

2. Buettner, V. L., LeBon, J. M., Gao, C., Riggs, A. D., and Singer-Sam, J. (2000) Use of terminal transferase dependent antisense RNA amplification to determine the transcription start site of the Snrpn gene in individual neuron. *Nucleic Acids Res.* **28,** E25.

3. Chen, H. H., Castanotto, D., Rossi, J. J., and Riggs, A. D. (1999) RNA analysis by terminal transferase-dependent PCR. In: *Intracellular Ribozyme Applications: Principles and Protocols.* (Rossi, J. J. and Couture, L., eds.), Horizon Scientific Press, Norfolk, UK, pp. 217–229.

4. Scherr, M., LeBon, J., Riggs, A. D., and Rossi, J. J. (1999) The use of cell extracts and antisense deoxyribo-oligo nucleotides for identifying ribozyme cleavage sites on messager RNAs. In: *Interacellular Ribozyme Applications, Principles, and Protocols.* (Rossi, J. J. and Couture, L., eds.), Horizon Scientific Press, Norfolk, UK, pp. 47–56.

5. Scherr, M., LeBon, J., Castanotto, D., Cunliffe, H. E., Meltzer, P. S., Ganser, A., et al. (2001) Detection of antisense and ribozyme accessible sites on native mRNAs: application to NCOA3 mRNA. *Mol. Therapy* **4,** 454–460.

6. Komura, J. and Riggs, A. D. (1998). A sensitive and versatile method of genomic sequencing: ligation-mediated PCR with ribonucleotide tailing by terminal deoxynucleotidyl transferase. *Nucleic Acids Res.* **26,** 1807–1811.

7. Komura, J., Ikehata, H., Hosoi, Y., Riggs, A. K., and Ono, T. (2001) Psoralen cross-links at the nucleotide level in mammalian cells: suppression of cross-linking by transcription factor- or nucleosome-binding. *Biochemistry* **40,** 4096–4105.

8. Kontaraki, J., Chen, H.-H., Riggs, A. D., and Bnifer, C. (2000) Chromatin fine structure profiles for a developmentally regulated gene: reorganization of the lysozyme locus before trans-activator binding and gene expression. *Genes Develop.* **15,** 2106–2122.

9. Chen, H.-H., Kontaraki, J., Bonifer, C., and Riggs, A. D. (2001) Terminal transferase-dependent PCR (TDPCR) for in vivo UV photofootprinting of vertebrate cells. *Sci. STKE* **77,** PL1.

10. Besaratinia, A., Bates, S. E., and Pfeifer, G. P. (2002) Mutational signature of the proximate bladder carcinogen N-hydroxy-4-acetylaminobiphenyl: inconsistency with the p53 mutational spectrum in bladder cancer. *Cancer Res.* **62,** 4331–4338.

11. Brockdorff, N., Ashworth, A., Kay, G. F., McCabe, V. M., Norris, D. P., Cooper, P. J., et al. (1992) The product of the mouse *Xist* gene is a 15 kb inactive X-specific transcript containing no conserved ORF and located in the nucleus. *Cell* **71,** 515–526.

12. Johnston, C. M., Nesterova, T. B., Formston, E. J., Newall, A. E. T., Duthie, S. M., Sheardown, S. A., and Brockdorff, N. (1998) Developmentally regulated *Xist* promoter switch mediates initiation of X inactivation. *Cell* **94,** 809–817.

13. Castanotto, D., Chow, W. A., and Rossi, J. J. (1998) Unusual interactions between cleavage products of a cis-cleaving hammerhead ribozyme. Antisense nucleic *Acid Drug Dev.* **8,** 1–13.

14. Lin, J. and Rossi, J. J. (1998) Identification and characterization of yeast mutants that overcome an experimentally introduced block to splicing at the 3' splice site. *RNA* **2,** 835–848.

15. Simorre, J. P., Legault, P., Hangar, A. B., Michiels, P., and Pardi, A. (1997) A conformational change in the catalytic core of the hammerhead ribozyme upon cleavage of an RNA substrate. *Biochemistry* **36,** 518–525.
16. Murray, J. B., Terwey, D. P., Maloney, L., Karpeisky, A., Usman, N., Beigelman, L., and Scott, W. G. (1998) The structural basis of hammerhead ribozyme self-cleaage. *Cell* **92,** 665–673.

# 9

## Identification of Efficient Cleavage Sites in Long-Target RNAs

### Wei-Hua Pan and Gary A. Clawson

#### Summary

In this chapter, we describe a procedure for identification of efficient hammerhead ribozyme (hRz) cleavage sites in target RNAs. An active hRz library, containing randomized recognition sequences flanked by fixed 5' and 3' regions, is designed to generate enormous diversity. The library is incubated with target RNA at an elevated temperature in the absence of magnesium, and bound library pools are isolated, reamplified, and rebound to target RNA. After two rounds, the active preselected library pool is incubated at 37°C with target RNA in the presence of magnesium, and cleavage products are directly identified on sequencing gels. The protocol identifies highly active hRz, which typically have $K_m$s of 20–80 n$M$, and $k_{cat}/K_m$ values of $10^6$.

**Key Words:** Ribozyme; library selection; selex.

## 1. Introduction

RNA presents unique problems for molecular targeting. Although RNA is considered to be a "linear" molecule, long RNAs actually form complex secondary and tertiary structures. The assumed structures mask various regions of a particular RNA, often appearing to leave relatively few "accessible" regions. Although computer modeling programs such as mFold *(1)* have been devised, many important parameters cannot yet be factored in, and thus the program's predictive power is still relatively limited.

A number of investigators have devised protocols for identifying accessible sites experimentally *(2–11)*. These have employed "libraries" of random oligonucleotides or various ribozymes, especially hammerhead Rz (hRz). In general, however, these protocols are complex and technically difficult, and consequently they have not been widely employed. We previously described a

From: *Methods in Molecular Biology, vol. 252: Ribozymes and siRNA Protocols, Second Edition*
Edited by: M. Sioud © Humana Press Inc., Totowa, NJ

systematic evolution of ligands by exponential enrichment ("SELEX") approach with a library of oligonucleotides containing random flanking sequences surrounding a potential hRz cleavage site *(12)*. hRz targeted to accessible sites identified using this protocol were generally 2–3 orders of magnitude more active than hRz targeted to sites that were selected using computer programs. Although this protocol is effective in at identifying good sites for targeting, it is also labor-intensive and time-consuming. Here, we describe a second protocol, utilizing a library of catalytically active hRz with randomized regions flanking the catalytic core. After initial binding under inactive conditions, the preselected hRz are incubated with target under active conditions, and sites of cleavage are directly identified on sequencing gels. As may be anticipated, hRz targeted to sites identified using the hRz library are generally more active than those targeted to sites identified using the oligonucleotide library (*see* **Note 1**).

### 1.1. Selection Protocol Using a Library of Active hRz

A double-stranded DNA library is used to generate an hRz-library with multiple copies of approx $10^{10}$ different RNA sequences. Each transcript is 79-nt in length, with a central ribozyme core flanked on each side by random sequences of 9Ns and by defined 5'/3'-end sequences (**Fig. 1**). The fixed 5' and 3' sequences allow a polymerase chain reaction (PCR)-based iterative protocol for regeneration of bound species. They also decrease (by 20X) the catalytic activity of the active ribozyme species. As a result, hRz targeted to identified cleavage sites are highly active when they are subsequently tested without the flanking regions.

DNA templates of targeted RNA are generated by reverse transcriptase (RT)-PCR with a T7 promoter in the 5'-primers. To circumvent the problem of microheterogeneity of transcripts at the 3'-ends, a 3'-primer encoding a self-cleaving ribozyme is utilized, so that transcripts with precise 3'-GUC ends are produced during in vitro runoff transcription. This allows labeling of target RNAs at either the 5' or 3' end.

Incubating the active hRz library with target RNA does not yield identifiable cleavage products, because of the huge diversity engineered into the library. To circumvent this, the ribozyme library is first subjected to selection at 85°C in magnesium-free buffer (*see* **Note 2**), to allow isolation of RNA molecules that anneal to the denatured target RNA (**Fig. 2A, [a]** and **[b]**). The bound hRz-library RNA pool is subsequently amplified (**Fig. 2A, [c]**) by RT-PCR, transcribed, and is then subjected to a second round of selection at a lower target RNA ratio; this serves to increase the selection stringency and to decrease background.

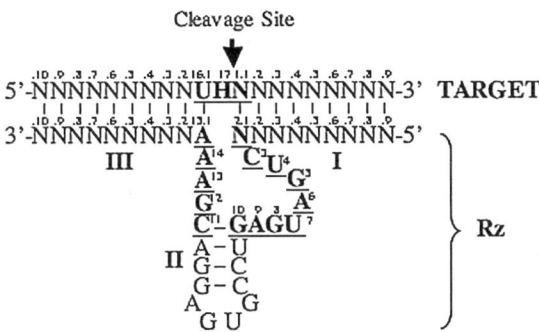

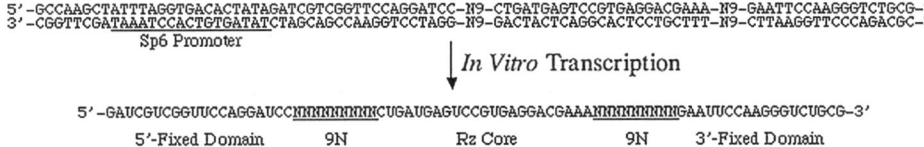

Fig. 1. The Random hRz Selection Library. (**A**) Drawing of a *trans*-acting hRz bound to its target RNA, showing a central domain (the catalytic core nucleotides are underlined) for an hRz, flanked by two random 9-nt 5'/3'-flanking regions (I and III). Arrow shows the site of cleavage in the target RNA, which is just 3' to the NUH triplet in the target RNA. (**B**) Generation of the library of random hRz-RNA transcripts. Primers (as described) are annealed together, and used for PCR amplification to yield a double-stranded DNA library. The Sp6 RNA polymerase promoter (underlined) is then utilized for transcription of the 79-nt random hRz library. Rz, ribozyme. Modified from **ref. *14***.

Next, the reamplified second-round selected hRz library RNA pool is used to cleave 5'- or 3'-end-labeled target-RNA (**Fig. 2A**, [**d**] and [**e**]). The cleaved products are then analyzed on sequencing gels, in comparison with G-, A-, and/or base-hydrolysis ladders (**Fig. 2A**, [**f**]), and the cleavage sites are precisely identified (**Fig. 2B**).

## 1.2. Representative Parameters for Selected hRz

Essentially all hRz targeted to sites identified using this protocol are highly active in vitro. For example, all six selected sites in Akt-3 (*13*) target RNA show excellent cleavage activity against full-length Akt-3 mRNA in vitro (**Fig. 3**).

For a number of diverse targets, $K_m$s for library selected hRz are generally in the 20–80-n$M$ range, and $k_{cat}/K_m$ values are on the order of $10^6$. In many

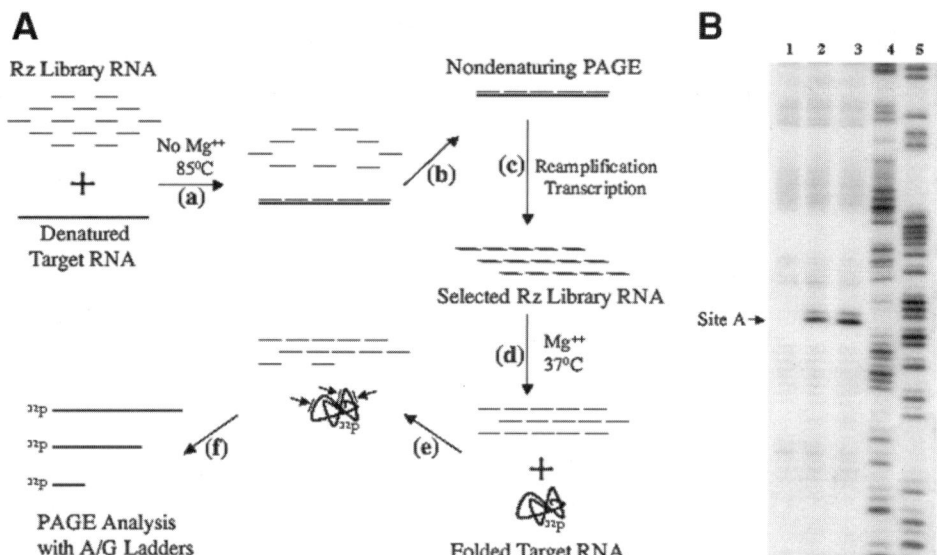

Fig. 2. (A) Schematic overview of the library selection procedure. The hRz-library RNA and target RNA are annealed (at elevated temperature) to form RNA-RNA complexes (A.a). The complexes are then isolated (A.b) and re-generated (A.c) by RT-PCR and in vitro transcription. After a second round with a reduced concentration of target RNA, the selected hRz-library RNA and the 5'- (or 3') end 32P-labeled target RNA are incubated at 37°C in the presence of magnesium (A.d), to allow target cleavage (A.e). The cleaved products are then separated on 6% sequencing gels under standard conditions (A.f). Modified from ref. 14. (B) Section of a gel following this library selection protocol with Akt-3 target RNA. Lanes 1, 2, and 3 show cleavage products after 0, two, or three rounds of preselection. Lanes 4 and 5 show A and G hydrolysis ladders generated from target RNA by RNase U2 and T1 digestions (respectively). This cleavage site was designated as A. Six efficient cleavage sites were identified on the various gels. Rz, ribozyme.

studies with HPV 16 E6/E7 target RNA (14), 8 (of 11) hRz selected with this procedure showed greater activity than the best hRz identified using our SELEX protocol (12), yet two others showed equivalent activity under standard conditions. All of the HPV 16 E6/E7-targeted hRz selected using this procedure were active in cell-culture studies (14). Similarly, all of the HPV 11 E6/E7-targeted hRz constructs tested showed excellent activity in cell-culture studies, with those containing selected hRz in our SNIP cassette producing 80–90% reductions in E6/E7 levels at 5 d.

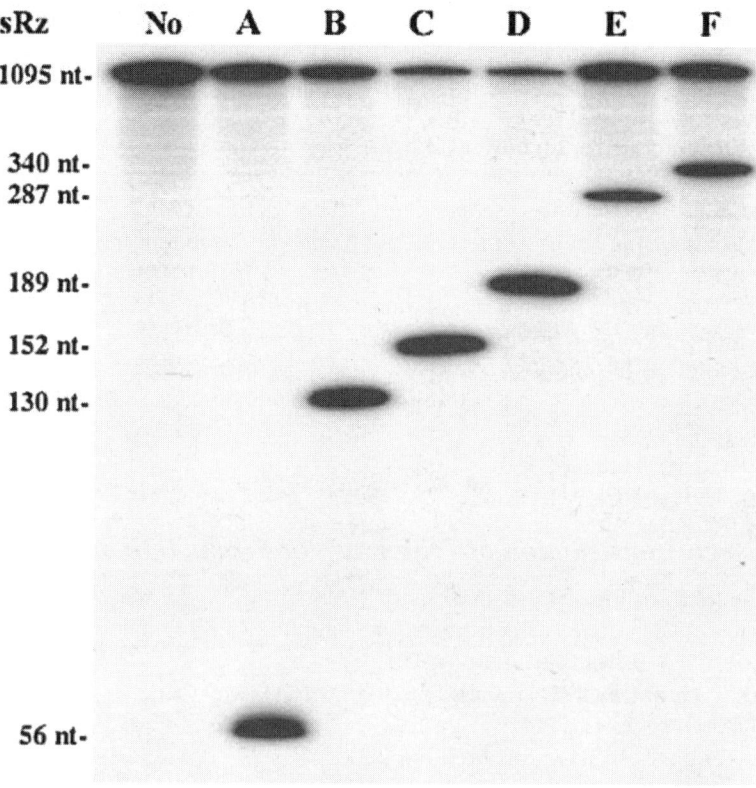

Fig. 3. In vitro cleavage analysis of sRz targeted to Akt-3 RNA. Library selection was performed using Akt-3 RNA as target, and sRz was targeted to the identified cleavage sites were constructed and tested. Reactions contained a 1:1 ratio of sRz to target RNA, and were for 2 h at 37 C. sRz designations (A–F) refer to the identified cleavage sites.

## 2. Materials

### 2.1. Design/Construction of a DNA Library for Expression of the hRz Library

1. hRz library primer: 5'-CGCAGACCCTTGGAATTC-NNNNNNNNN-TTTCGTCCTCACGGACTCATCAG-NNNNNNNNN-GGATCCTGGAACCGACGAT-3.'
2. Sp6 promoter primer: 5'-GCCAAGCTATTTAGGTGACACTATAGATCGTCGGTTCCAGGATCC-3'.
3. 5'-end primer: 5'-GCCAAGCTATTTAGGTGA-3'.
4. 3'-end primer: 5'-CGCAGACCCTTGGAATTC-3'.

5.  4 m*M* deoxynucleotide 5' triphosphate (dNTP): 4 m*M* deoxyadenosine triphosphate (dATP), 4 m*M* deoxycytidine 5' triphosphate (dCTP), 4 m*M* deoxyguanosine 5' triphosphate (dGTP), 4 m*M* deoxythymidine 5' triphosphate (dTTP) in 20 m*M* Tris-HCl, pH 7.4.
6.  10X Pfx amplification buffer (Gibco-BRL).
7.  50 m*M* MgSO$_4$.
8.  Platinum Pfx DNA polymerase (2.5 U/μL, Gibco-BRL).
9.  Non-denaturing loading buffer: 0.05% phenol blue and 0.05% xylene cynol FF in 20% glycerol.
10. 10X TBE buffer: 900 m*M* Tris-borate, 20 m*M* EDTA, pH 8.5.
11. 8% polyacrylamide urea-free TBE-gel.
12. 1% ethidium bromide.
13. 25:24:1 (v:v:v) Phenol:chloroform:isoamyl alcohol (IAA), pH 7.9 (Ambion).
14. 5 *M* NaCl.
15. 100 and 70% ethanol.
16. 20 m*M* Tris-HCl, pH 7.4.

## *2.2. Design/Construction of Templates for Production of Target RNAs*

1.  Primers from **Fig. 4** (5'-primers and 3'-primers).
2.  Pretemplate target-RNA constructs(for example, subcloned in an expression vector).
3.  Reagents 5–9 from **Subheading 2.1.**
4.  50X TAE buffer 2 *M* Tris-acetate, 50 m*M* EDTA, pH 8.0.
5.  1% Agarose TAE-gel.
6.  Reagents 12–16 from **Subheading 2.1.**

## *2.3. Transcription and Isolation of Library RNA and Target RNA*

1.  Templates from **Subheading 2.1.–2.2.**
2.  10 m*M* NTP, 10 m*M* ATP, 10 m*M* cytidine 5' triphosphate (CTP) 10 m*M* GTP, 10 m*M* uridine 5' triphosphate (UTP) in 20 m*M* Tris-HCl, pH 7.4.
3.  Optimized 5X transcription buffer (Promega).
4.  100 m*M* dithiothreitol (DTT).
5.  α-$^{32}$P-CTP (3000 Ci/mmol, 10.0 mCi/mL = 3.33 pmol/mL).
6.  Sp6 RNA polymerase (15 U/μL, Promega).
7.  T7 RNA polymerase (20 U/μL, Promega).
8.  RNase inhibitor (40 U/μL, Promega).
9.  Dimethyl sulfoxide (DMSO).
10. Denaturing loading buffer: 0.05% phenol blue and 0.05% xylene cynol FF in 80% formamide (Super pure-grade, Fisher Scientific Co.) and 20% 0.5 *M* EDTA, pH 8.0.
11. 6% Polyacrylamide 7 *M* urea TBE-gel.
12. Elution buffer: 250 m*M* NaCl, 20 m*M* Tris-HCl, pH 7.4.
13. 100 and 70% ethanol.
14. 20 m*M* Tris-HCl, pH 7.4.

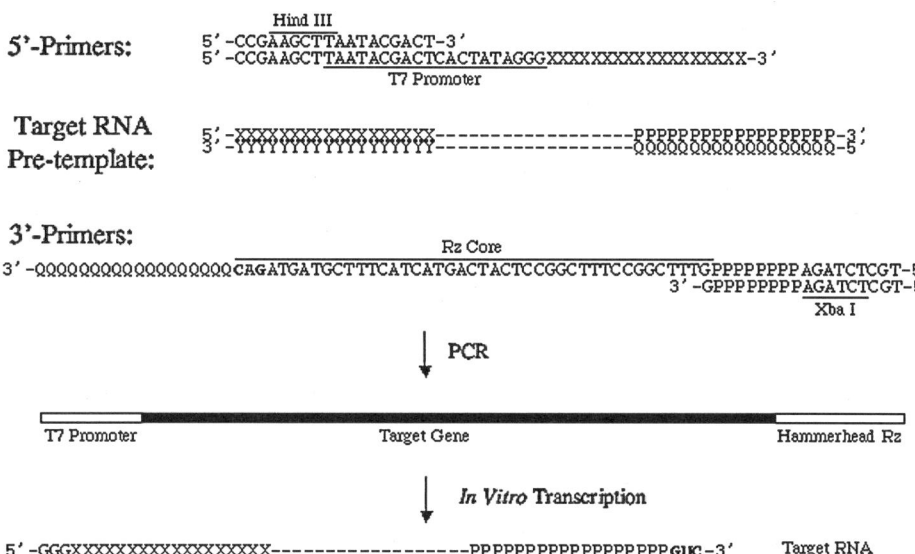

Fig. 4. Schematic representation of a template for expression of target RNA. The 5'-primers were designed to have a 3'-region ($X_{18}$) matching the 5' region from the chosen target, as well as a T7 promoter and an upstream *Hind*III site (as designated). The target RNA pre-template is shown with the X-region, as well as the most 3'-region ($P_{18}$). The 3'-primers contain: i) a Q-region, which is reverse-complementary to the 3'P-region in the pre-template; ii) a central *cis*-acting hRz core. This includes the reverse complement of the targeted GUC (shown in bold at the 3' end of the ribozyme core, which is underlined), and a $P_8$ region that matches a portion of the P region in the pre-template; and iii) an additional primer at the 5'-end, which adds a *Xba*I restriction endonuclease site. Following PCR, and target RNA expression construct is produced, which contains a 5' T7 promoter for transcription, and a 3' *cis*-acting hRz, which produces a uniform 3' end to target RNA transcripts. Rz, ribozyme. Modified from **ref. 14**.

## 2.4. hRz-Library Pre-Selection

### 2.4.1. Selection of Bound-hRz Library RNA Pool

1. hRz library RNA.
2. Target RNA.
3. 20 m$M$ Tris-HCl, pH 7.4.
4. Non-denaturing loading buffer: 0.05% phenol blue and 0.05% xylene cyanol FF in 20% glycerol.
5. 3 $M$ Na Acetate, pH 5.2.
6. 100 and 70% ethanol.

## 2.4.2. Re-Amplification of Bound-hRz Library RNA Pool

1. hRz library RNA/target RNA complex.
2. 3'-end primer: 5'-CGCAGACCCTTGGAATTC-3'.
3. 5 m$M$ dNTP (Qiagen)
4. 10X Reverse transcription buffer (Qiagen).
5. RNase inhibitor (40 U/µL, Promega).
6. Omniscript reverse transcriptase (4 U/µL, Qiagen).
7. PCR reagents (*see* **Subheading 2.1.**), but the hRz library primer is replaced by reverse transcribed DNA.
8. Transcriptional reagents (*see* **Subheading 2.3.**).

## 2.5. Production of $^{32}$P-Labeled Target RNA

### 2.5.1. 5'-End Labeling of Target RNA

1. Target RNA.
2. 10X NEBuffer 3 (New England Biolabs).
3. Alkaline phosphatase, calf intestinal (CIP, 10 U/µL, New England Biolabs Inc.).
4. 10X NEBuffer for T4 polynucleotide kinase (PNK) (New England Biolabs).
5. T4 polynucleotide kinase (10 U/µL, New England Biolabs).
6. γ-$^{32}$P-ATP (3000 Ci/mmol, 10.0 mCi/mL = 3.33 pmol/µL).
7. 25:24:1 Phenol:chloroform:IAA, pH 7.9 (Ambion).
8. 5 $M$ NaCl.
9. Reagents 10–14 from **Subheading 2.3**.

### 2.5.2. 3'-End Labeling of Target RNA

1. Target RNA.
2. 10X NEBuffer for T4 PNK (New England Biolabs).
3. 1 $M$ Tris-HCl, pH 8.1.
4. 1 $M$ MgCl$_2$.
5. T4 PNK (10 U/µL, New England Biolabs).
6. 5X Poly (A) polymerase reaction buffer (Amersham Life Science)
7. Poly (A) polymerase (500 U/µL, Amersham Life Science).
8. α-$^{32}$P-CoTP (3000 Ci/mmol, 10.0 mCi/mL = 3.33 pmol/µL).
9. 25:24:1 Phenol:chloroform:IAA, pH 7.9 (Ambion).
10. 5 $M$ NaCl.
11. Reagents 10–14 from **Subheading 2.3**.

## 2.6. Active hRz Library Selection

### 2.6.1. Generation of A-, G-, and Limited Alkaline Hydrolysis RNA Ladders

1. $^{32}$P-labeled target RNAs.
2. U2/T1 cocktail: 10 $M$ urea, 30 m$M$ NaCitrate, pH 3.5, 1.5 m$M$ EDTA, pH 8.0, 0.3 µg/µL tRNA, 0.02% phenol blue, 0.02% xylene cynol FF.
3. Ribonuclease U2 (A-specific, USB).

4. Ribonuclease T1 (G-specific, USB).
5. 2X alkaline hydrolysis buffer: 50 m$M$ NaHCO$_3$/Na$_2$OH, pH 9.0, 3.0 m$M$ EDTA, pH 8.0, 0.3 μg/μL tRNA.
6. Reaction stop buffer: 10 $M$ urea, 25 m$M$ EDTA, 0.02% phenol blue, 0.02% xylene cyanol FF.
7. 6% Polyacrylamide 7 $M$ urea TBE-sequencing gel.

## 2.6.2. End-Labeled 32P-Target RNA Cleavage by Selected hRz Library RNA Pool

1. $^{32}$P-labeled target-RNAs.
2. Random hRz library RNA.
3. Second round selected hRz library RNA.
4. 1 $M$ MgCl$_2$.
5. RNase inhibitor (40 U/μL, Promega).
6. Reagents 6–7 from **Subheading 2.6.1.**

## 2.7. Procedure for Making Templates for Transcription of Selected hRz (sRz)

1. sRz primers: 5'-GACCCTTGGAATTC-9N-TTTCGTCCTCACGGACTCATC AG-6N-GGATCCTGGAACCTATAG-3', where the Ns are specific for the target RNA region selected.
2. Sp6 primer: 5'-GCCAAGCTATTTAGGTGACACTATAGGTTCCAGGATCC-3'.
3. 5'-end primer: 5'-GCCAAGCTATTTAGG-3'.
4. 3'-end primer: 5'-GACCCTTGGAATTC-3'.
5. Reagents 5–16 from **Subheading 2.1.**

## 2.8. Cleavage Tests of Selected Ribozymes

1. 5'-end $^{32}$P-labeled target RNA.
2. Selected ribozyme-RNAs.
3. Reagents 3–5 from **Subheading 2.6.2.**

## 2.9. Quantitation of Target RNA Transcripts for Assessment of sRz Activity in Cell Culture

### 2.9.1. Total RNA Isolation

1. RNAqueous-4PCR kit (Ambion).
2. RNase-free DNase.

### 2.9.2. Reverse Transcription of Target RNA and TBP RNA

1. Total RNA, isolated from cells.
2. 3'-primer for target-RNA.
3. 3'-primer for TBP-RNA: 5'-CTGGAAAACCCAACTTCTGTACAA-3'.
4. 5 m$M$ dNTP (Qiagen).
5. 10X reverse transcription buffer (Qiagen).

6. RNase inhibitor (40 U/μL, Promega).
7. Sensiscript reverse transcriptase (4 U/μL, Qiagen).

## 2.9.3. Quantitation of Target RNA

1. Reverse-transcribed cDNA.
2. 5'-primer for the target RNA.
3. 5'-primer for TBP RNA: 5'-ACCACGGCACTGATTTTCAGT-3'.
4. TaqMan probe for the target RNA: $(Cy5)-N_{20}-(BHQ3)$.
5. TaqMan probe for TBP RNA: (VIC)-TGTGCACAGGAGCCAAGAGTG AAGA-(BHQ1).
6. 3'-primer for the target RNA.
7. 3'-primer for TBP RNA: 5'-CTGGAAAACCCAACTTCTGTACAA-3'
8. ROX (carboxy-X-rhodamine, succinimidyl ester).
9. 10X PCR buffer.
10. 25 m$M$ MgCl$_2$.
11. 4 m$M$ dNTP: 4 m$M$ dATP, 4 m$M$ dCTP, 4 m$M$ dGTP, 4 m$M$ dTTP in 20 m$M$ Tris-HCl, pH 7.4.
12. HotStarTaq DNA polymerase (Qiagen).

# 3. Methods

## 3.1. Design/Construction of a DNA Library for Expression of the hRz Library

The procedure starts with construction of a single-stranded DNA library containing $6.87 \times 10^{10}$ sequences (1.5 mg of DNA) by automated solid-state synthesis. The sequence diversity is created by randomizing two domains totaling 18-nt (9 Ns and 9 Ns) flanking a functional ribozyme core (23-nt), and using fixed sequences for both 5'/3'-ends. These fixed sequences are designed to facilitate cloning into a ribozyme expression cassette and to allow for PCR amplification and in vitro transcription. The library sequence used is 5'-CGCAGACCCTTGGAATTC-NNNNNNNNN-TTTCGTCCTCACGGACT CATCAG-NNNNNNNNN-GGATCCTGGAACCGACGAT-3'. The Sp6 primer (containing an Sp6 RNA polymerase promoter), 5'-end primer, and 3'-end primer are designed to utilize PCR amplification of the randomized sequence in order to construct the double-stranded DNA library (**Fig. 1**; the library should be sequenced to confirm its composition). The library is then transcribed using Sp6 RNA polymerase to generate a random pool of multiple copies of approx 70 billion different ribozyme sequences.

1. In 100 μL of reaction:
   | | |
   |---|---|
   | H$_2$O | 75 μL |
   | 10X Pfx amplification buffer | 10 μL |
   | 50 m$M$ MgSO$_4$ | 2 μL |
   | 4 m$M$ dNTPs | 5 μL |

| hRz library primer (10 pmol/µL) | 1 µL |
| Sp6 promoter primer (10 pmol/µL) | 1 µL |
| 5'-end primer (100 pmol/µL) | 2 µL |
| 3'-end primer (100 pmol/µL) | 2 µL |
| Platinum Pfx DNA polymerase (2.5 U/µL) | 2 µL |

2. Aliquot to five PCR tubes, perform PCR as follows: 94°C, 3 min; (94°C, 30 s, 52°C, 40 s, 72°C, 60 s) for 6, 9, 13, 15, 18 cycles; 72°C, 5 min.

3. Mix 5 µL PCR products with 2 µL non-denaturing loading buffer, and then load on 8% polyacrylamide urea-free TBE-gel.

4. Run the gel at 10 V/cm for 10 min, and then increase to 20 V/cm for 1 h.

5. Stain the gel for 3 min in ethidium bromide working solution (100 µL of 1% ethidium bromide diluted in 200 mL $H_2O$), rinse once with $H_2O$ and analyze the results under ultraviolet (UV) light.

6. According to the previously described data, perform an additional 100 µL reaction under optimal conditions.

7. Extract the optimal PCR products once by 25:24:1 phenol:chloroform:IAA, pH 7.9, then precipitate by adding NaCl to 250 m$M$, and 2.5X vol of 100% ethanol, with incubation at –80°C for 15 min.

8. Centrifuge the PCR products at 20,000$g$ for 20 min at 4°C, discard the supernate.

9. Wash the pellet once with 70% ethanol, centrifuge for 10 min and discard the supernate.

10. Air-dry the pellet and resuspend in 50 µL of 20 m$M$ Tris-HCl, pH 7.4.

### 3.2. Design/Construction of Templates for Production of Target RNAs

We have generally employed RT-PCR for cloning of full-length target RNA (or fragments thereof), although the DNA constructs can be obtained by other means. We refer to these constructs as pre-templates.

Double-stranded DNA templates for production of target RNA transcripts are constructed by adding the T7 RNA polymerase promoter and a 3'-ribozyme tail (**Fig. 4**) to the pre-template constructs. The T7-promoter primer adds a T7 RNA polymerase promoter to the 5'-end of the pre-template, and the ribozyme-primer adds an additional tail at the 3'-end. The X-part of the T7 primer (18 nt in length) represents the sense sequence of the 5'-end of the pre-template (*see* **Fig. 4**). The Q-part (the 3'-end of the *cis*-acting ribozyme primer, 18 nt), is reverse-complementary to the 3'-end of the sense target RNA pre-template. The P-part (5'-end of the *cis*-acting ribozyme primer, 8 nt) forms the 3'-end (Helix III) of the *cis*-acting ribozyme (**Fig. 4**). The *Hin*dIII and *Xba*I restriction endonuclease sites are added for cloning purposes, and are placed 5' to the T7 promoter and 3' to the *cis*-acting ribozyme flanking sequence, respectively.

Primers are synthesized as indicated (**Fig. 4**). The protocol for production of target RNA templates follows that described previously (*see* **Subheading 3.1.**)

### *3.3. Transcription and Isolation of Library RNA and Target RNA*

In vitro transcription reactions (for both the ribozyme-library RNA pool and target RNA) are transcribed in vitro using the Riboprobe System (Promega, Madison, WI) with $^{32}$P-CTP (when labeled transcripts are desired). Reactions employ Sp6 (for transcription of ribozyme-library RNA) or T7 (for transcription of target RNA) RNA polymerases; after transcription, RNAs are gel-purified.

In 100 µL of reaction:

| | |
|---|---|
| H$_2$O | 44 µL |
| DMSO | 5 µL |
| Optimized 5X transcription buffer | 20 µL |
| 100 m*M* DTT | 10 µL |
| 10m*M* NTP | 10 µL |
| Templates from **Subheading 3.1.** or **3.2.** | 5 µL |
| Sp6 or T7 RNA polymerase | 5 µL |
| RNase inhibitor | 1 µL |

If $^{32}$P-labeled RNA is desired, 1–10 µL of $\alpha$-$^{32}$P-CTP may be added to the reaction described here.

1. Incubate the samples for 2 h at 37°C, then add 5 µL of RNase-free DNase and follow with an additional 30 min incubation at 37°C.
2. Add 100 µL of denaturing loading buffer, heat 5 min at 85°C and then chill on ice.
3. Load on a 6% polyacrylamide 7 *M* urea TBE-gel, and run at 1 W/cm until the samples have migrated into the gel, and then increase the power to 2 W/cm for 1 h (for hRz library RNA) or 3–4 h (for target RNA) (*see* **Note 3**). Put the nonradioactive gel on kitchen-wrap, and then transfer onto a thin-layer chromatography (TLC) plate (radioactive gel strips may be identified by using autoradiography film).
4. Identify the gel strip containing the RNA under 254 nm UV light, and excise with a flamed razor blade.
5. Transfer the gel-strip into a 1.5-mL tube and homogenize with a sealed pipet tip (they can be sealed by burning).
6. Add 700 µL of elution buffer, and shake 2 h at 4°C.
7. Heat 5 min at 85°C, and then spin 5 min at 20,000*g*.
8. Recover 400 µL of the supernate, and precipitate the RNA by adding 1 mL 100% ethanol at –80°C for 15 min.
9. Perform **steps 8–9** from **Subheading 3.1.**
10. Resuspend the RNA in 20 µL of 20 m*M* Tris-HCl, pH 7.4.
11. Quantitate by UV absorbance.

### *3.4. hRz-Library Pre-selection*

Two rounds of pre-selection are performed (*see* **Note 4**). Each round is performed as follows: 100 m*M* ribozyme-library RNA pool and 1 µ*M* target RNA are mixed in 20 m*M* Tris-HCl, pH 7.4, heated to 85°C for 3 min, and then cooled to 37°C over a 30-min period, allowing RNA-RNA complexes to form. Bound complexes are separated from the unbound ribozyme-library RNA pool in a non-denaturing, 8% (urea-free) polyacrylamide-TBE gel at room temperature, and RNA-RNA complexes (containing the target with bound species from the ribozyme library) are isolated and purified, and these selected hRz-library RNAs are reverse-transcribed to produce their cDNAs using 3'-end primer (as described previously for construction of hRz-library template). The product is then amplified using PCR, with the Sp6 RNA polymerase promotor primer and 5'-/3'-end primers, and subsequently transcribed using Sp6 RNA polymerase. This produces a new hRz library RNA pool that is enriched for library constituents that can bind to denatured target RNA (**Fig. 2**).

This hRz-library RNA pool is then subjected to this same procedure for one round. However, for this second-round selection, stringency is increased by reducing (by half) the target RNA concentration. After two rounds, the selected pool of hRz-library RNA is used to cleave the corresponding target RNA.

### 3.4.1. Selection of Bound-hRz Library RNA Pool

1. In 100 μL of reaction, 100 μ*M* hRz library RNA and 1 μ*M* target RNA are mixed in 20 m*M* Tris-HCl, pH 7.4.
2. Heat to 85°C for 3 min and then cool to 37°C over a 300min period.
3. Chill on ice and add 25 μL non-denaturing loading buffer.
4. Load on 8% polyacrylamide urea-free TBE-gel (*see* **Note 5**) run the gel at 10 V/cm for 10 min, and then increase to 20 V/cm for 1 h (*see* **Note 6**).
5. Isolate the RNA-RNA complexes. (*See* **Subheading 3.3., steps 3–5**. If the target-RNA is greater than 1000 nt, the complexes may be isolated by excising a gel strip of 3-mm width, adjacent to the bottom of the well.)
6. Add 650 μL of 20 m*M* Tris-HCl, pH 7.4, and shake for 2 h at 4°C.
7. Heat 5 min at 85°C, and then spin 5 min at 20,000*g*.
8. Recover 360 μL of the supernatant and add 40 μL 3 *M* Na acetate, pH 5.2, and then precipitate the complex by 1 mL 100% ethanol at –80°C for 15 min.
9. Perform **steps 8** and **9** from **Subheading 3.1.**
10. Resuspend the complex in 10 μL of 20 m*M* Tris-HCl, pH 7.4.

### 3.4.2. Reamplification of Bound-hRz Library RNA Pool

1. In 20 μL of reaction:

| | |
|---|---|
| H₂O | 7 μL |
| 10X reverse transcription buffer | 2 μL |
| 3'-end primer (10 pmol/μL) | 2 μL |
| 5 m*M* dNTP | 2 μL |
| hRz library RNA/target RNA complex | 5 μL |
| RNase inhibitor (10 U/μL) | 1 μL |
| Omniscript reverse transcriptase (4 U/μL) | 1 μL |

2. Incubate 1 h at 37°C.
3. PCR-amplify the cDNA (*see* **Subheading 3.1.** and **Note 7**). Perform transcription as described in **Subheading 3.3.**).
4. This first-round selected hRz library RNA pool is ready for next-round selection.
5. Repeat **Subheading 3.4.1.**, with a 50% reduction in target RNA concentration, to obtain the second round sRz library RNA pool.

### 3.5. Production of $^{32}$P-Labeled Target RNA

To produce 5'-end $^{32}$P-labeled target RNA, CIP is used to remove the triphosphate groups from the 5'-end of unlabeled transcripts. Dephosphorylated transcripts are labeled using T4 PNK with $\gamma$-$^{32}$P-ATP.

To produce 3'-end $^{32}$P-labeled target RNA, T4 PNK is used to cleave the 2', 3' cyclic phosphate bond and remove the phosphate group *(15)*. Dephosphorylated transcripts are labeled using Poly (A) polymerase with $\alpha$-$^{32}$P-CoTP.

#### 3.5.1. 5'-End Labeling of Target RNA

1. In 40 μL of reaction:
   | | |
   |---|---|
   | $H_2O$ | 30 μL |
   | 10X NEBuffer 3 | 4 μL |
   | Non-radioactive target RNA (10 pmol/mL) | 4 μL |
   | Alkaline phosphatase (10 U/mL) | 2 μL |
2. Incubate 1 h at 37°C, and increase the volume to 100 μL with 20 m*M* Tris-HCl, pH 7.4.
3. Extract once with 25:24:1 phenol:chloroform:IAA, pH 7.9 and precipitate (*see* **Subheading 3.1., steps 7–9**).
4. Air-dry the pellet and resuspend the 5'-dephosphorylated target-RNA in 10 μL 20 m*M* Tris-HCl, pH 7.4.
5. Perform 40 μL 5'-end-labeling reaction:
   | | |
   |---|---|
   | $H_2O$ | 22 μL |
   | 10X NEBuffer for T4 polynucleotide kinase | 4 μL |
   | 5'-dephosphorylated target-RNA | 5 μL |
   | $\gamma$-$^{32}$P-ATP | 7 μL |
   | T4 PNK (10 U/mL) | 2 μL |
6. Incubate 1 h at 37°C, and add 40 μL non-denaturing loading buffer.
7. Heat to 85°C for 5 min, and then chill on ice.
8. Perform **steps 3–11** from **Subheading 3.3.**

#### 3.5.2. 3'-End-Labeling of Target RNA

1. In 40 μL of reaction:
   | | |
   |---|---|
   | $H_2O$ | 31.2 μL |
   | 1 *M* Tris-HCl, pH 8.1 | 2 μL |
   | 1 *M* $MgCl_2$ | 0.8 μL |
   | Non-radioactive target-RNA (10 pmol/mL) | 4 μL |
   | T4 PNK (10 U/mlL) | 2 μL |

2. Incubate 1 h at 37°C, and increase the volume to 100 μL by 20 m*M* Tris-HCl, pH 7.4.
3. Extract once by 25:24:1 phenol:chloroform:IAA, pH 7.9 and precipitate (*see* **Subheading 3.1., steps 7–9**).
4. Air-dry the pellet and resuspend the 3'-dephosphorylated target RNA in 10 μL 20 m*M* Tris-HCl, pH 7.4.
5. Perform 40 μL 5'-end-labeling reaction:
   | | |
   |---|---|
   | H$_2$O | 19 μL |
   | 5X poly (A) polymerase reaction buffer | 8 μL |
   | 3'-dephosphated target RNA | 5 μL |
   | α-$^{32}$P-CoTP (3000 Ci/mmol) | 7 μL |
   | Poly (A) polymerase (500 U/mL) | 1 μL |
6. Incubate 1 h at 37°C, and add 40 μL non-denaturing loading buffer.
7. Heat to 85°C for 5 min, and then chill on ice.
8. Perform **steps 3–11** from **Subheading 3.3.**

### 3.6. Active hRz Library Selection

A trace amount of end-labeled target RNA (approx 50,000 cpm) is incubated with 40 μ*M* of the selected ribozyme-library RNA pool in 5 μL of 20 m*M* Tris-HCl, pH 7.4, and 25 m*M* MgCl$_2$. Samples containing the various cleaved transcripts are then analyzed by polyacrylamide gel electrophoresis (PAGE) using a 6% urea gel, in comparison with A-, G-, and limited alkaline hydrolysis ladders *(16)*.

#### 3.6.1. Generation of A-, G-, and Limited Alkaline Hydrolysis RNA Ladders

1. A-specific hydrolysis ladder:
   | | |
   |---|---|
   | U2/T1 cocktail | 7 μL |
   | End-$^{32}$P-labeled target-RNA (50,000 cpm/μL) | 2 μL |
   | Ribonuclease U2 (0.1 U/μL) | 3 μL |
2. G-specific hydrolysis ladder:
   | | |
   |---|---|
   | U2/T1 cocktail | 7 μL |
   | End-$^{32}$P-labeled target RNA (50,000 cpm/μL) | 2 μL |
   | Ribonuclease T1 (0.1 U/μL) | 3 μL |
3. Incubate 15 min at 50°C, and then chill on ice.
4. Limited alkaline hydrolysis RNA ladder:
   | | |
   |---|---|
   | 20 m*M* Tris-HCl, pH 7.4 | 0.5 μL |
   | End-$^{32}$P-labeled target-RNA (50,000 cpm/μL) | 2 μL |
   | 2X alkaline hydrolysis buffer | 2.5 μL |
5. Boil for 5–10 min and chill on ice, and then add 7 μL reaction stop buffer.

#### 3.6.2. End-Labeled $^{32}$P-Target RNA Cleavage by Selected hRz Library RNA Pool

1. Non-selected hRz library RNA (control):
   | | |
   |---|---|
   | 250 m*M* MgCl$_2$ (in 20 m*M* Tris-HCl, pH 7.4) | 0.5 μL |
   | 10 U/μL RNase inhibitor (same as above) | 0.5 μL |

| End-$^{32}$P-labeled target RNA (50,000 cpm/µL) | 2 µL |
|---|---|
| hRz (non-selected) library RNA (100 µ*M*) | 2 µL |

2. Second-round selected hRz library RNA:

| 250 m*M* MgCl$_2$ (in 20 m*M* Tris-HCl, pH 7.4) | 0.5 µL |
|---|---|
| 10 U/µL RNase inhibitor (same as above) | 0.5 µL |
| End-$^{32}$P-labeled target RNA (50,000 cpm/µL) | 2 µL |
| Second-round selected hRz library RNA (100 µ*M*) | 2 µL |

3. Incubate 2 h at 37°C and add 7 µL reaction stop buffer.
4. Heat to 85°C for 5 min, and then chill on ice.
5. Load the samples, 3 µL for each (from **Subheadings 3.6.1., steps 1**, **2**, and **4** and **3.6.2., steps 1** and **2**) on a 6% polyacrylamide 7 *M* urea TBE-sequencing-gel. Run the gel at 30 W until the samples have migrated into the gel, and then increase the power to 60 *W* for 1.5, 3, and 6 h.
6. Dry the gels and analyze the results by autoradiography.

## 3.7. Procedure for Making Templates for Transcription of Selected hRz (sRz)

Once selected target cleavage sites have been identified, overlapping oligo-nucleotides are synthesized, with fixed nucleotides in place of the random the recognition sequences present in the original library; these will basepair in a reverse-complementary manner to the target RNA at the selected regions. The 3' recognition sequence is shortened to six nucleotides, because in vitro speci-ficity testing has shown little or no effect for nucleotide substitutions after six positions, and because transcription yields were routinely low when nine nucle-otides were utilized. The Sp6, 5'end, and 3'end primers are those specified in **Subheading 2.7.**

1. In 100 µL of reaction:

| H$_2$O | 75 µL |
|---|---|
| 10X Pfx amplification buffer | 10 µL |
| 50 m*M* MgSO$_4$ | 2 µL |
| 4 m*M* dNTPs | 5 µL |
| sRz primer (10 pmol/µL) | 1 µL |
| Sp6 primer (10 pmol/µL) | 1 µL |
| 5'-end primer (100 pmol/µL) | 2 µL |
| 3'-end primer (100 pmol/µL) | 2 µL |
| Platinum Pfx DNA polymerase (2.5 U/µL) | 2 µL |

2. Perform **steps 2–10** from **Subheading 3.1.**

## 3.8. Cleavage Tests of sRz

sRz targeted to individual library-selected sites are transcribed from double-stranded DNA oligonucleotides, using Sp6 RNA polymerase as described pre-viously for hRz-library RNAs. For standard screening of sRz activity, incubations contain trace amounts of 5'-$^{32}$P-labeled target RNA and 40 n*M* ribozyme RNA, and are for 30 min at 37°C in 20 m$M$ Tris-HCl, pH 7.4, 25 m$M$ MgCl$_2$. After incubations, samples are separated by PAGE in a 6% urea gel; the gels are dried and radioactivity is analyzed using a PhosphorImager.

For kinetic analyses, a trace amount of $^{32}$P-5'-labeled target RNA is mixed with unlabeled target RNA (to yield final concentrations of 10, 100, 333, and 1000 n$M$ target RNA) and ribozyme-RNA (40 n$M$ final concentration), and incubations are performed using the same conditions as described previously *(12)*, except that incubation times are varied (for 40 s, 1, 2, 5, 10, 30 min, and 1 h). Samples are separated by PAGE in a 6% urea gel, and then dried and analyzed using a PhosphorImager.

1. Transcribe sRz from the sRz template (**Subheading 3.7.**), performed as described in **Subheading 3.3.**
2. In 5 µL of reaction:

   | | |
   |---|---|
   | 250 m$M$ MgCl$_2$ (in 20 m$M$ Tris-HCl, pH 7.4) | 0.5 µL |
   | 10 U/µL RNase inhibitor (same as above) | 0.5 µL |
   | End-$^{32}$P-labeled target-RNA (10,000 cpm/µL) | 2 µL |
   | sRz RNA (0.1 µ$M$) | 2 µL |

3. Incubate 30 min at 37°C and add 7 µL denaturing loading buffer.
4. Heat to 85°C for 5 min, and then chill on ice.
5. Load 3 µL on 6% polyacrylamide 7 $M$ urea TBE-sequencing-gel, run at 30 W until the samples have migrated into the gel, and then increase the power to 60 W for 1.5 h.
6. Dry the gel and analyze the results by autoradiography and PhosphorImager.

## 3.9. Quantitation of Target RNA Transcripts for Assessment of sRz Activity in Cell Culture

Transfection/co-transfection protocols can be tailored as appropriate for specific targets. We then employ quantitative real-time PCR (QPCR) to quantitate target RNA levels. We use a Stratagene (La Jolla, CA) Mx4000 machine and TaqMan 5'-nuclease methodology, with 6-carboxy-fluorescein (FAM), Cy5, and VIC dyes, in conjunction with Black Hole Quencher 1 (BHQ1) or 3 (BHQ3), as appropriate.

For target RNAs, the initial 3'-primer for RT is a 30mer, reverse-complementary to a chosen region of the target RNA. For QPCR amplification, the 5' and 3' primers are chosen as 20–25mers (we routinely search databases for selected primers). The TaqMan probe is generally a 20mer, (Cy5)-N20-(BHQ3), matching a selected region lying between the 5' and 3' primers on the target RNA (*see* **Note 9**).

As an internal control, QPCR amplification of Tata-box-binding protein (TBP, an integral component of TFIID) is performed in the same RT reactions, using VIC-labeled primer. For TBP, the initial 3'-primer we use is CTG GAA

AAC CCA ACT TCT GTA CAA, which is reverse-complementary to nt 718–742. For QPCR, the same 3'-primer was used. The 5'-primer is ACC ACG GCA CTG ATT TTC AGT, which matches nt 625–646. The TaqMan probe is (VIC)-TGT GCA CAG GAG CCA AGA GTG AAGA – (BHQ1), which matches nt 659-683. The $C_T$ value for TBP RNA is approx $25.90 \pm 0.10$ (for comparative purposes, that for 18S rRNA is $16.16 \pm 0.04$). Finally, ROX (carboxy-X-rhodamine, succinimidyl ester) was used as dye for a volume control.

QPCR amplifications are performed with the Sensicript Reverse Transcriptase and HotStarTaq reagents/protocols as described previously, except that 7% DMSO is included in the RT mix. Final concentrations of TBP and target RNA primers are 100 n$M$, although standard curves (25–100 n$M$) are routinely run (*see* **Note 8**). Denaturation cycles are at 94°C, annealing cycles are at 52°C, and extension cycles are at 72°C.

QPCR results are analyzed using the REST© program *(18)*. Analyses are reported for pair-wise fixed-reallocation randomization tests. Results are generally similar using randomization results not normalized by TBP expression.

## 4. Notes

1. One surprising finding has been that DNAzymes (Dz) targeted to sites identified using the hRz library protocol have almost uniformly shown no cleavage activity. For example, we identified a number of regions in HPV 16 E6/E7 mRNA using the hRz library protocol. Twelve of 13 Dz targeted to sites within these regions showed no cleavage of target RNA (the other showed only weak activity). In contrast, Dz targeted to regions identified using the oligonucleotide selection protocol have routinely shown good cleavage activity. Thus, both approaches can be useful, depending upon the application. We are currently developing/refining a library selection protocol for a random Dz library.

2. When annealing is performed at 37°C, cleavage products are not seen. The annealing temperature of 85°C allows selection of a subset of the hRz library that is capable of binding to denatured target. Upon reamplification, this subset is therefore enriched for ribozyme targeted to accessible sites relative to the starting library (which contained large numbers of species not represented in the linear target sequence), allowing cleavage products to be identified.

3. The surface temperature of the glass plate should exceed 50°C.

4. One round of preselection generally results in relatively infrequent and faint bands. Three rounds of preselection produces results similar to results observed for two rounds (*see* **Fig. 3**).

5. The bottom footprint of the sample-well should be 50 mm $^2$, for example, $1 \times 50$ mm. Otherwise, nonspecific complexes form.

6. The surface temperature of the glass plate should not exceed 35°C.

7. The hRz library primer is replaced by 5 µL of the reverse-transcribed DNA, and a non-RT control should be performed for the PCR amplification to ensure that the isolated RNA-RNA complexes are without template-DNA contamination.

8. Primer concentrations can be adjusted to yield cycle threshold ($C_T$) values in the same range as TBP.
9. Choosing primers to known library-selected sites can significantly increase amplification efficiency.

## Acknowledgments

This work was supported by a Research and Development Agreement with BioSan, Inc.

## References

1. Zuker, M. and Jacobson, A. B. (1998) Using reliability information to annotate RNA secondary structures. *Rna* **4**, 669–679.
2. Jarvis, T., Wincott, F., Alby, L., McSwiggen, J., Beigelman, L., Gustofson, J., et al. (1996) Optimizing the cell efficacy of synthetic ribozymes. *J. Biol. Chem.* **271**, 29,107–29,112.
3. Mir, A., Lockett, T., and Hendry, P. (2001) Identifying ribozyme-accessible sites using NUH-triplet targeting gapmers. *Nucleic Acids Res.* **29**, 1906–1914.
4. Pierce, M. and Ruffner, D. (1998) Construction of a directed hammerhead ribozyme library: towards the identification of optimal target sites for antisense-mediated gene inhibition. *Nucleic Acids Res.* **26**, 5093–5101.
5. Bramlage, B., Luzi, E., and Eckstein, f. (2000) HIV-1 LTR as a target for synthetic ribozyme-mediated inhibitionn of gene expression: site selection and inhibition in cell culture. *Nucleic Acids Res.* **21**, 4059–4067.
6. Putlitz, J., Yu, Q., Burke, J., and Wands, J. (1999) Combinatorial screening and intracellular antiviral activity of hairpin ribozymes directed against hepatitis B virus. *J. Virol.* **73**, 5381–5387.
7. Yu, Q., Pecchia, D., Kingsley, S., Heckman, J., and Burke, J. (1998) Cleavage of highly structured viral RNA molecules by combinatorial libraries of hairpin ribozymes. *J. Biol. Chem.* **273**, 23,524–23,533.
8. Birikh, K. R., Berlin, Y. A., Soreq, H., and Eckstein, F. (1997) Probing accessible sites for ribozymes on human acetylcholinesterase RNA. *RNA* **3**, 429–437.
9. Cairns, M., Hopkins, T., Witherington, C., Wang, L., and Sun, L. (1999) Target site selection for an RNA-cleaving catalytic DNA. *Nature Biotechnol.* **17**, 480–486.
10. Lima, W., Brown-Driver, V., Fox, M., Hanecak, R., and Bruice, T. (1997) Combinatorial screening and rational optimization for hybridization to folded hepatitis C virus RNA of oligonucleotides with biological antisense activity. *J. Biol. Chem.* **272**, 626–638.
11. Lieber, A. and Strauss, M. (1995) Selection of efficient cleavage sites in target RNAs by using a ribozyme expression library. *Mol. Cell. Biol.* **15**, 540–551.
12. Pan, W.-H., Devlin, H., Kelley, C., Isom, H., and Clawson, G. (2001) A selection system for identifying accessible sites in target RNAs. *RNA* **7**, 610–621.
13. Masure, S., Haefner, B., Wesselink, J., Hoefnagel, E., Mortier, E., Verhasselt, P., et al. (1999) Molecular Cloning, expression and characterization of the human serine/threonine kinase Akt-3. *Eur. J. Biochem.* **265**, 353–360.

14. Pan, W. H., Xin, P., Bui, V., and Clawson, G. (2003) Rapid identification of efficient target cleavage sites using a hammerhead ribozyme library in an iterative manner. *Mol. Therapy* **7,** 1–11.

15. Loria, A. and Pan, T. (2000) The 3' substrate determinants for the catalytic efficiency of the Bacillus subtilis RNase P holoenzyme suggest autolytic processing of the RNase P RNA in vivo. *RNA* **6,** 1413–1422.

16. Donis-Keller, J. (1980) Phy M: An RNase activity specific for U and A residues useful in RNA sequence analysis. *Nucleic Acids Res.* **8,** 3133–3142.

17. Herasse, M., Ono, Y., Fougerousse, F., Kimura, E., Stockholm, D., Beley, C., et al. (1999) Expression and functional characteristics of calpain 3 isoforms generated through tissue-specific transcriptional and posttranscriptional events. *Mol. Cell Biol.* **19,** 4047–4055.

18. Pfaffl, M., Horgan, G., and Dempfle, L. (2002) Relative expression software tool (REST) for group-wise comparison and statistical analysis of relative expression results in real-time PCR. *Nucleic Acids Res.* **30,** e36.

# 10

## Selection In Vitro of Allosteric Ribozymes

### Adam Roth and Ronald R. Breaker

### Summary

The capacities of RNA for both catalysis and molecular recognition have been appreciated for some time. Recently, a number of studies have shown that these distinct functional classes may be combined to generate multipartite ribozymes in which the catalytic and ligand-binding elements act coordinately. In this chapter, we describe a procedure known as allosteric selection, in which an RNA catalyst in conjunction with a random sequence domain is subjected to iterative in vitro selection in order selectively to recover ligand-dependent ribozymes.

**Key Words:** Allosteric ribozyme; in vitro selection; aptamer; RNA engineering; hammerhead ribozyme.

## 1. Introduction

Among the multiplicity of tasks attended by RNA, in its natural roles as well as in experimentally derived ones, is the assumption of tertiary structures capable of catalyzing an impressive array of chemical transformations and of forming specific binding sites for a wide range of ligands (1–5). These two functional classes—RNA catalysts (ribozymes) and ligand-binding RNAs (aptamers)—have been studied largely in isolation from one another. However, in recent years, molecular engineering efforts have demonstrated that aptamer elements may be appended to ribozyme domains in such a way that the entities act coordinately (6,7). Specifically, the catalytic activity of a chimeric molecule containing spatially distinct ribozyme and ligand-binding domains may be modulated in response to the cognate ligand of the attached aptamer. This interdomain communication harks back to an ancient era of complex, RNA-centered metabolism, but also heralds a future in which modular ribozymes may be harnessed as biosensors, diagnostic tools, and agents for the control of gene expression in vivo.

From: *Methods in Molecular Biology, vol. 252: Ribozymes and siRNA Protocols, Second Edition*
Edited by: M. Sioud © Humana Press Inc., Totowa, NJ

Several strategies may be applied toward the construction of ligand-sensitive, or allosteric ribozymes. Although these strategies share the same directive—to assemble a multidomain RNA in such a way that a ligand-induced conformational change within one element culminates in an altered conformation in a separate catalytic component—they differ primarily in the extent to which they employ combinatorial methods. On one end of the spectrum is modular rational design, which eschews combinatorial techniques in favor of allosteric ribozyme assembly wholly from pre-existing functional modules. Although the hammerhead self-cleaving motif has thus far been the most commonly used catalytic platform for allosteric ribozyme construction *(8)*, other naturally occurring ribozymes have also been brought under allosteric control using rational design principles *(9,10)*. Furthermore, catalysts selected in vitro, including DNA enzymes, have been shown to be equally susceptible to engineered ligand sensitivity *(9,11)*. Similarly, effector-binding elements targeting an impressive variety of ligands may be drawn from a substantial repository of aptamers derived from both in vitro selections and natural sources *(4,12)*. Finally, highly focused in vitro selections have yielded a third class of RNA elements—communication modules—which serve to relay structural changes between discrete RNA domains *(9,13–15)*. With these three basic classes of functional RNAs at their disposal, molecular engineers have shown the feasibility of functionally linking disparate aptamer, catalyst, and relay domains in an interchangeable fashion while relying only on principles of rational design.

Also in the area of rational design are catalysts designed to be responsive to specific oligonucleotide effectors. Because of the highly predictable manner in which nucleic acid strands associate to form secondary structure, modulation of catalysis may be achieved simply by supplying oligonucleotide effectors that serve to stabilize or disrupt critical domains. Based on this principle, a number of generalizable strategies have been implemented for the control of nucleic acid enzymes with specific oligonucleotide effectors *(16–21)*.

On the opposite end of the spectrum from rational design is the selection in vitro of allosteric ribozymes from completely random sequence pools. In contrast to rational design, in which the functional domains are predefined and need only to be productively linked, selection from a random sequence population offers, in principle, the ability to bring any readily selectable catalytic activity under the allosteric control of almost any ligand. Although this approach has been successfully used *(18)*, it is anticipated to be a relatively inefficient route to the generation of ligand-responsive ribozymes, due to the substantial complexity of the demands imposed on such a selection. RNA molecules that contain more than one discrete functional domain will be exceedingly rare in sequence space, especially if those domains are in dialog with one another.

It is perhaps not surprising that the combination of rational design with in vitro selection has emerged as one of the most powerful tools for the creation of allosteric ribozymes *(22–26)*. The union of a defined catalytic platform with a judiciously inserted random sequence domain releases the random portion from the obligation to evolve multiple (catalyst and aptamer) domains simultaneously. Rather, this construction allows the sampling of sequence space to be more singularly applied to the task of ligand binding when subjected to an appropriate iterative selection protocol. (Although the converse of this arrangement—the joining of random sequence to a defined aptamer element for the purpose of bringing a novel catalyst under the allosteric control of a particular ligand—may also be employed, this variation has typically been used to bias catalytic activities toward a desired substrate *[27]*.) One of the primary advantages of allosteric selection, as this technique is known, is the ability to isolate ribozymes that are regulated by virtually any effector. Moreover, such ribozymes may be generated in parallel, allowing a panoply of allosteric catalysts, each responsive to a distinct effector, to emerge from the same in vitro selection. A procedure for the in vitro selection of allosteric hammerhead ribozymes is described in this chapter.

As mentioned previously, a variety of naturally occurring and artificially derived motifs have been enlisted as catalytic platforms in the pursuit of allosteric ribozymes, although none as predominantly as the hammerhead self-cleaving motif. Part of the reason for the abundance of allosteric hammerhead ribozymes is historical—the first example of an engineered catalyst responding to a small organic effector used this motif *(28)*. Other factors that contribute to the frequent usage of this catalytic platform are its small size and thorough structural characterization—features that facilitate experimental manipulation and relieve investigators of the task of identifying structural elements amenable to allosteric modulation. Although the allosteric selection protocol detailed in this chapter specifies the hammerhead catalytic domain, it is important to note that the concepts apply broadly, and that nearly any ribozyme may be substituted for the hammerhead provided that proper consideration is given to the attachment site of the random sequence domain.

In the case of the hammerhead, the random sequence domain is routinely affixed to stem II (**Fig. 1**), a structural element whose stability is critical for catalytic activity *(29)*. By exchanging this crucial secondary structure element with random sequence and subjecting the resulting pool to an appropriate in vitro selection protocol, the random portion is interrogated for solutions to a particular query posed by the investigator (**Fig. 2**). Accordingly, if an allosteric ribozyme is sought whose catalytic activity is positively regulated by a particular ligand, the first step is to cull from the population those molecules that are constitutively active in the absence of effector because of

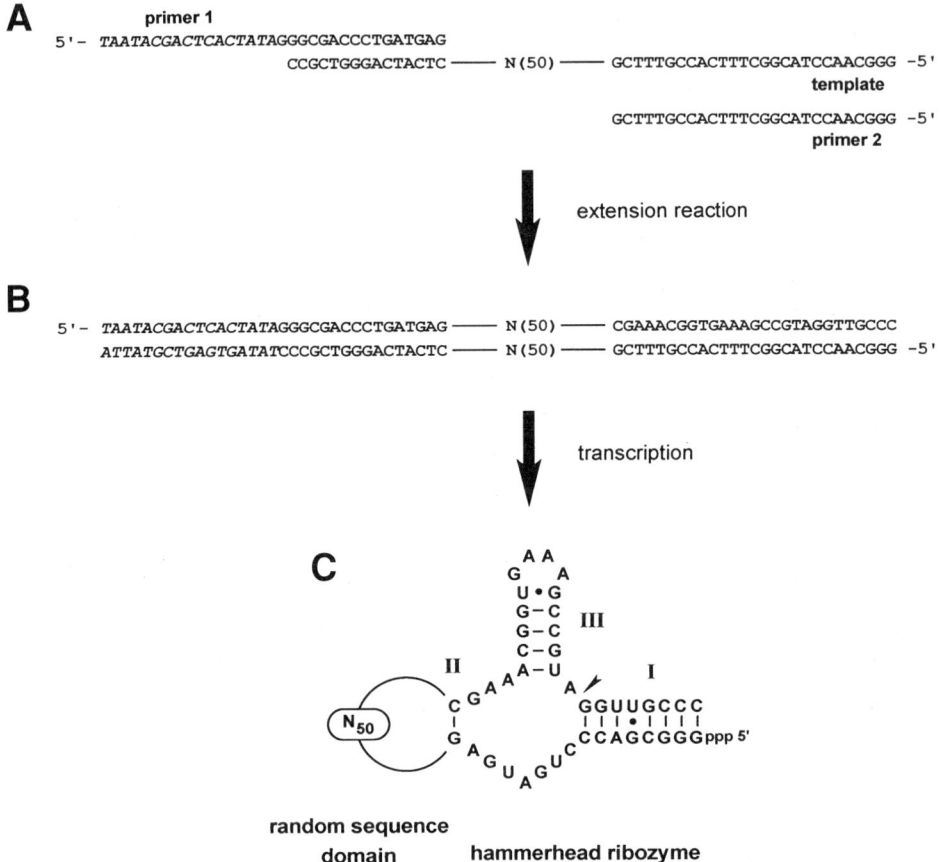

Fig. 1. Construction of the starting pool. The oligonucleotides required for this allosteric selection protocol are shown in (**A**). Primer 1 and the template oligonucleotide are included in an extension reaction, generating a double-stranded DNA (**B**), which in turn, is employed as a template for transcription of the initial RNA pool. Sequences corresponding to the T7 promoter are italicized. The secondary structure of the resulting hammerhead ribozyme-based population is shown in (**C**). Roman numerals denote stem structures of the ribozyme domain. With the exception of a single C-G basepair, stem II is replaced by random sequence ($N_{50}$). The arrowhead indicates the expected cleavage site.

stable stem II-like elements in the random domain. This is referred to as negative selection. Following incubation of the pool in the absence of the designated ligand, active ribozymes that have catalyzed their own self-cleavage are separated electrophoretically, by virtue of their smaller size, from inac-

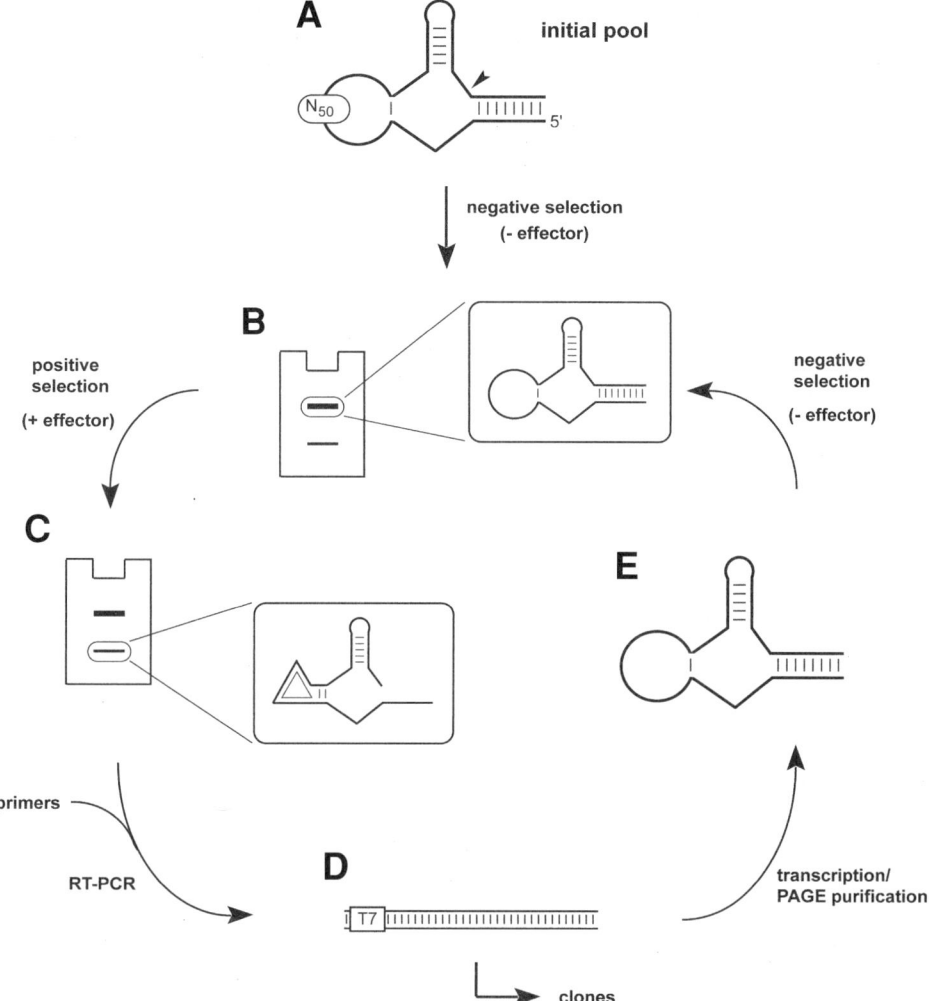

Fig. 2. Outline of the basic allosteric selection scheme. The gel-purified, hammer-head ribozyme-based starting population (**A**) is subjected to negative selection in the absence of effectors. Molecules remaining uncleaved during this step are gel-purified (**B**) and subsequently included in a positive selection in the presence of an effector (represented by a triangle in [**C**]). The 5' cleavage products generated during this step (depicted in ligand-bound form) are isolated from a gel (**C**), converted to cDNAs and amplified by PCR. Transcription of the resulting DNA pool (**D**) provides RNA (**E**) for the next round of negative and positive selections.

tive, uncleaved molecules. The purified subpopulation of catalysts that are largely inactive in the absence of ligand is recovered and subsequently incubated in the presence of ligand, in what is known as the positive selection step. Gel electrophoresis is employed again to isolate those molecules that are cleaved during this step. Following conversion of the selected RNAs into cDNAs, the sequences are amplified using polymerase chain reaction (PCR) to generate templates for the next round of selection. With multiple iterations of these steps, populations may become progressively more enriched for discrete RNA domains that, upon ligand binding, stabilize or functionally replace a crucial ribozyme structural element in order to confer ligand-mediated modulation of catalytic activity.

## 2. Materials

### 2.1. Construction of the Starting Pool

1. Appropriate chemically synthesized DNA oligonucleotides.
2. 5X reverse transcriptase buffer: 250 m$M$ Tris-HCl, pH 8.3, 375 m$M$ KCl, 15 m$M$ MgCl$_2$.
3. 100 m$M$ dithiothreitol (DTT).
4. 10X deoxynucleotide 5' triphosphate (dNTP) mix: 10 m$M$ each of deoxyguanosine 5' triphosphate (dGTP), deoxyadenosine 5' triphosphate (dATP), deoxythymidine 5' triphosphate (dTTP), and deoxycytidine 5' triphosphate (dCTP).
5. Reverse transcriptase (e.g., SuperScript II, Invitrogen).
6. 10X transcription buffer: 500 m$M$ Tris-HCl, pH 7.5, 150 m$M$ MgCl$_2$, 50 m$M$ DTT, 20 m$M$ spermidine.
7. 5X NTP mix: 10 m$M$ each of guanosine 5' triphosphate (GTP), ATP, UTP, and CTP.
8. [α-$^{32}$P]UTP (3000 Ci/mmol, Amersham Biosciences).
9. T7 RNA polymerase.
10. Crush/soak buffer: 200 m$M$ NaCl, 10 m$M$ Tris-HCl, pH 7.5, 1 m$M$ ethylenediaminetetraacetic acid (EDTA).
11. 100% ethanol.
12. 2X denaturing gel-loading buffer: 90 m$M$ Tris, 90 m$M$ borate, 18 $M$ urea, 20% sucrose (w/v), 1 m$M$ EDTA, 0.1% sodium dodecyl sulfate (SDS), 0.05% xylene cyanol FF, 0.05% bromophenol blue.
13. Reagents and apparatus for denaturing polyacrylamide gel electrophoresis (PAGE).

### 2.2. Negative Selection

1. 2X reaction buffer: 100 m$M$ Tris-HCl, pH 7.5, and 40 m$M$ MgCl$_2$.
2. 100% ethanol.
3. Crush/soak buffer: *see* **Subheading 2.1., item 10**.
4. 2X denaturing gel-loading buffer: *see* **Subheading 2.1., item 12**.
5. Reagents and apparatus for denaturing PAGE.

### 2.3. Positive Selection

1. 2X reaction buffer: *see* **Subheading 2.2., item 1**.
2. 10X effector solution.
3. 100% ethanol.

4. Crush/soak buffer: *see* **Subheading 2.1., item 10**.
5. 2X denaturing gel-loading buffer: *see* **Subheading 2.1., item 12**.
6. Reagents and apparatus for denaturing PAGE.
7. X-ray film.

### 2.4. Reverse Transcription and PCR Amplification

1. 5X reverse transcriptase buffer: *see* **Subheading 2.1., item 2**.
2. 100 m$M$ DTT.
3. dNTP mix: 2 m$M$ each of dGTP, dATP, dTTP, and dCTP.
4. Reverse transcriptase.
5. 10X PCR buffer: 100 m$M$ Tris-HCl, pH 8.3, 500 m$M$ KCl, 15 m$M$ MgCl$_2$, 0.1% gelatin.
6. Appropriate 5' and 3' DNA oligonucleotide primers.
7. *Taq* DNA polymerase.
8. 100% ethanol.
9. Reagents and apparatus for agarose gel electrophoresis.

### 2.5. Evaluation and Control of Allosteric Selection Progress

1. Reagents for the cloning and sequencing of PCR products.
2. 3 $M$ NaOAc, pH 5.2.
3. 1 $M$ NaOH.
4. 0.5 $M$ EDTA.
5. 10 mg/mL glycogen.
6. 100% ethanol.

(**Items 2–6** are required only if a more stringent negative selection is necessary.)

## 3. Methods

The following protocol may be applied in the search for entirely new aptamer sequences that modulate hammerhead ribozyme activity (*see* **Note 1**). The use of a random region approx 50 nucleotides in length is recommended when selecting for these novel aptamer domains. Although this allows the sampling of only a minute fraction of possible sequences, the number of unique solutions existing in sequence space is predicted to be so vast that some are expected to occur with reasonable frequency even in a comparatively small starting population. Indeed, ribozyme populations containing insertions of 25–50 random positions have proved to be sufficient for the selection of catalysts responsive to a wide assortment of effectors including metals, small organic molecules, oligonucleotides, and proteins (*see* **Note 2**).

### 3.1. Construction of the Starting Pool

The structural element that is most commonly engineered in the pursuit of allosteric hammerhead ribozymes is stem II. Typically, the entire stem II structure, save for a critical C-G basepair, is exchanged for random sequence (**Fig. 1C**). The double-stranded DNA template encoding such a population is conveniently manufactured by using a DNA polymerase to extend synthetic oligonucleotides complementary at their 3' ends (**Fig. 1A,B**; *see* **Note 3**).

1. Employing standard solid phase methods, chemically synthesize the appropriate pair of DNA oligomers with overlapping complementary 3' ends (an overlap of 15–20 nucleotides is sufficient) (**Fig. 1A**). The oligonucleotides may be purified by denaturing PAGE or high-performance liquid chromatography (HPLC) to ensure length homogeneity.
2. In order to prepare a double-stranded DNA template for transcription, incubate 300 pmol of primer and 200 pmol of the oligonucleotide bearing random sequence (primer 1 and template, respectively; *see* **Fig. 1** and **Note 4**) in a 100 µL extension reaction containing 50 m$M$ Tris-HCl, pH 8.3, 75 m$M$ KCl, 3 m$M$ MgCl$_2$, 10 m$M$ DTT, 1 m$M$ each of dGTP, dATP, dTTP, and dCTP, and 8 U/µL reverse transcriptase (e.g., SuperScript II, Invitrogen). The mixture may be heated to 90°C for 1 min and then cooled to room temperature before adding reverse transcriptase to promote productive annealing.
3. Incubate for 1 h at 42°C.
4. Precipitate the double-stranded DNA with 2.5 vol ethanol and pellet by centrifugation at a minimum of 10,000$g$ for 15 min.
5. Resuspend in a 200 µL transcription reaction containing 50 m$M$ Tris-HCl, pH 7.5, 15 m$M$ MgCl$_2$, 5 m$M$ DTT, 2 m$M$ spermidine, 2 m$M$ each of GTP, ATP, UTP, and CTP, 10 µCi [$\alpha$-$^{32}$P]UTP, and 1000 U T7 RNA polymerase.
6. Allow transcription to proceed for 1–2 h at 37°C (*see* **Note 5**).
7. To concentrate the sample, precipitate with ethanol and centrifuge as in **step 4**.
8. Heat the resuspended sample to 90°C for 1 min in the presence of denaturing gel-loading buffer to promote melting of DNA template strands. Snap-cool on ice.
9. Resolve full-length precursor RNAs by denaturing (8 $M$ urea) 6% PAGE (*see* **Note 6**).
10. Excise from the gel the band containing precursor RNAs. The quantity of RNA should be sufficient for detection by shadowing with ultraviolet (UV) light.
11. Elute RNA from the crushed or diced gel slice in 200 m$M$ NaCl, 10 m$M$ Tris-HCl, pH 7.5, 1 m$M$ EDTA for ~2 h at 37°C or overnight at 4°C.
12. Transfer the solution containing eluted material to a fresh tube, taking care to exclude polyacrylamide debris to the extent possible.
13. Precipitate the size-purified RNA population with ethanol and pellet by centrifugation.
14. After resuspending in sterile deionized water, determine the RNA concentration with a spectrophotometer or by measuring the extent of incorporation of radiolabeled UTP (*see* **Note 7**).

### 3.2. Negative Selection

1. Incubate approx 1000 pmol of the gel-purified starting pool at 23°C for 2–5 h in a 200-µL reaction mixture containing 50 m$M$ Tris-HCl, pH 7.5, and 20 m$M$ MgCl$_2$ in the absence of the designated effector molecules (*see* **Notes 8** and **9**).

2. Precipitate with ethanol and centrifuge as in **Subheading 3.1.**
3. Resuspend the sample in a convenient volume of water and denaturing gel-loading buffer.
4. Separate the cleaved products from the uncleaved precursor RNAs by electrophoresis through a denaturing 6% polyacrylamide gel.
5. Using a phosphorimager, record the fraction of RNA that was cleaved in order that the efficacy of the negative selection step may be monitored by comparing this value with ones determined for subsequent populations.
6. Excise the uncleaved RNAs from the gel and elute as in **Subheading 3.1.** (*see* **Note 10**).
7. Concentrate the sample by precipitation with ethanol and centrifugation.

### 3.3. Positive Selection

1. Resuspend the uncleaved RNAs in water. Include one-half of this sample in a 100-µL (final volume) reaction mixture with a buffer composition identical to that of the negative selection buffer used in **Subheading 3.2., step 1**. This reaction, which contains no added effector molecules, serves as a negative control. The other half of the sample is included in a reaction mixture identical to the first, except that it also contains a defined concentration of the designated effector molecule(s) (*see* **Notes 11** and **12**). This is known as the positive selection reaction (*see* **Note 13**). Splitting the pool in this manner allows the experimenter to monitor accurately any ligand-dependent activity that might emerge as the selection progresses.
2. Incubate both reactions for 30 min at 23°C (*see* **Note 14**).
3. Concentrate each sample by precipitation with ethanol and centrifugation.
4. Dissolve the pellets in water and add denaturing gel-loading buffer.
5. Separate the cleaved products from the precursor RNAs by electrophoresis through a denaturing 6% polyacrylamide gel. Leave a space of at least one lane between the baseline control (minus effector) and positive selection reaction (plus effector) samples to avoid cross-contamination.
6. Calculate the fraction of cleaved RNA in each lane using a phosphorimager (*see* **Note 15**).
7. Excise the 5' cleavage products from the positive selection lane, using an autoradiogram to locate the corresponding zone (*see* **Note 16**).
8. Elute the selected RNAs from the gel slice as in **Subheading 3.1.**
9. Precipitate with ethanol and pellet by centrifugation (*see* **Note 17**).

### 3.4. Reverse Transcription and PCR Amplification

1. Resuspend the pellet in 20 µL water.
2. Add 50 pmol 3' primer (primer 2; unless this was already added as a carrier in **Subheading 3.3., step 9**).
3. Heat to 80°C for 1 min and then snap-cool on ice.
4. Using the pre-annealed mixture of selected RNA and 3' primer (primer 2), assemble a 50-µL reverse-transcription reaction containing 50 m$M$ Tris-HCl, pH 8.3, 75 m$M$ KCl, 3 m$M$ MgCl$_2$, 10 m$M$ DTT, 0.5 m$M$ each dNTP, and 8 U/µL reverse transcriptase (SuperScript II).

5. Incubate for 1 h at 42°C.
6. Include a 5–10-μL aliquot of the reverse transcription as template in a 100-μL PCR containing 10 m$M$ Tris-HCl, pH 8.3, 50 m$M$ KCl, 1.5 m$M$ MgCl$_2$, 0.01% gelatin, 200 μ$M$ each dNTP, 0.5 μ$M$ each of the 5' and 3' primers (primers 1 and 2, respectively), and 0.05 U/μL *Taq* DNA polymerase (*see* **Note 18**).
7. Amplify using standard cycling parameters: 94°C for 30 s, 50°C for 30 s, and 72°C for 1 min (*see* **Note 19**).
8. Confirm the full amplification of products of the expected length by subjecting a 5-μL aliquot to 3% agarose gel electrophoresis.
9. Precipitate the PCR products with ethanol and pellet by centrifugation.
10. Employ these DNA molecules as templates to transcribe the RNA pool for the next round of selection, as in **Subheading 3.1., step 5** (*see* **Note 20**). Reserve a portion of this DNA population for archiving purposes.

### 3.5. Evaluation and Control of Allosteric Selection Progress

Continue with iterations of the previous procedure until the enrichment of ligand-responsive sequences is indicated by RNA cleavage levels in the positive selection reaction that are significantly elevated relative to the baseline (no effector) control (*see* **Note 21**). Once this signal has been detected, the duration of positive selection reactions in subsequent rounds may be safely reduced to apply stronger selective pressure for ligand-dependent activity (*see* **Notes 22** and **23**). In response to this pressure, the ligand-dependent cleavage rate of the population as a whole is expected to increase over the course of several generations until it reaches a plateau, thereby signaling a relatively static pool composition likely to be predominated by a small number of unique ligand-sensitive sequences. Cloning and sequencing of representative molecules may be performed at this stage to identify prevailing ribozymes (*see* **Notes 24** and **25**). The ligand sensitivity of individual sequences may then be assayed, allowing specific features of these molecules, such as fold activation in the presence of effector and apparent $K_D$ of the RNA-ligand interaction, to be evaluated and compared (*see* **Note 26**).

Ribozymes derived from such allosteric selections typically respond to the intended targets with impressive specificities and with $K_D$ values commensurate with the employed concentrations. However, when allosteric ribozymes with improved performance characteristics are desired, sequences may be subjected to reselection pending the influx of mutation that is required for continued evolution to occur. A pool of variants based on a single parental sequence may be generated by synthesizing a heterogeneous template population using defined mixtures of phosphoramidite monomers at appropriate positions. The introduction of mutations via doped chemical synthesis affords the investigator a high degree of control over the level of mutagenesis and allows the confinement of sequence variability to the allosteric domain. Resulting mutagenized ribozyme pools may be subjected to reselection, with the stringency of the selection parameters adjusted accordingly to search for optimized variants (*see* **Note 27**). At the outset of a reselection, for example, the concentration of target ligand in the positive selection reaction is commonly reduced by an order of magnitude in an effort to isolate variants with improved affinities (*see* **Note 28**).

The careful application of negative and positive selection pressure to RNA pools, as described in the preceding sections, is a powerful approach that may lead directly to ligand-dependent ribozymes. However, the investigator should be cautioned that there are a number of strategies RNA can adopt that will allow ligand-independent sequences to survive and even flourish despite the constraints imposed by the selection. Once such sequences establish a firm presence in the population, they provide significant obstacles to the enrichment of ligand-responsive molecules. Furthermore, their selective removal from the population to the point where ligand-sensitive sequences can compete effectively often poses a considerable challenge.

One example of a feature conducive to the perseverance of ligand-independent ribozymes is a slow rate of self-cleavage. Although molecules that exhibit this characteristic certainly suffer a reduction in number during the negative selection step, there is also a significant quantity that survives intact, and these impostors are subsequently deposited into the positive selection step. Ribozymes that employ this survival mechanism are likely to occur relatively more frequently in the initial pool, and thus enjoy a numerical advantage that allows them to effectively overwhelm their ligand-responsive counterparts. The most effective means to combat such sequences is to exploit their liability—slow reaction speed—by instituting more aggressive negative and positive selection steps. Accordingly, the duration of the negative selection reaction is extended to deplete the population more thoroughly of slow catalysts (*see* **Note 29**) and that of the positive selection is reduced in order to favor fast, ligand-dependent enzymes.

There are a number of indicators that a population has become overrun with effector-independent ribozymes such as the slowly processing type described here. After several rounds of allosteric selection (typically between 3 and 10 cycles), the levels of RNA cleavage occurring during the negative selection steps (**Subheading 3.2., step 5**), rather than describing an expected downward trend, might instead be observed to remain constant or even increase over successive rounds. This rising trend in the fraction cleaved during the negative selection reaction is usually accompanied by an effector-independent response during the positive selection reaction (marked by equivalent amounts of cleavage in the positive selection reaction and the negative [no effector added] control [**Subheading 3.3., step 6**]). If patterns similar to these are observed during the course of a selection, they should alert the investigator to the likelihood that ligand-insensitive catalysts are being preferentially amplified. However, the dominant survival strategy employed by such a population may not be immediately apparent, since ligand-independent ribozymes that rely on any of several alternative mechanisms would yield similar profiles.

In addition to the slow reaction rate example cited previously, one particularly noteworthy strategy permitting ligand-independent ribozymes to survive is distribution between active and inactive conformations. Although the active ribozyme conformation is susceptible to negative selection, the inactive form enables a proportion of sequences effectively to bypass the negative selection step and be routed to the positive selection, where those molecules assuming the active structure upon refolding will be amplified. Ribozymes that employ this tactic are impervious to conventional negative selections, since extended incubation times are ineffective in combating species that remain in their inactive conformations indefinitely. Instead, more aggressive countermeasures must be undertaken, which are based on iterations of denaturation and refolding. The object of this repetitive cycling is to improve the penetration of the negative selection step by providing multiple opportunities for the misfolded form to redistribute to the active structure. Three stringent negative selection protocols based on this strategy are described here:

1. Multiple, consecutive negative selection steps may be employed. Specifically, the procedure outlined in **Subheading 3.2.** may be performed several times in series, with the uncleaved RNA obtained in **step 7** resubmitted to **step 1** of the same section (*see* **Note 30**). During each negative selection step, the fraction of ligand-independent catalysts adopting the active conformation is eliminated, leading to a substantial cumulative reduction.

2. The negative selection reaction may be subjected to periodic temperature spikes in order to promote denaturation and subsequent refolding. The incubation detailed in **Subheading 3.2., step 1** is performed as described, except that it is punctuated at regular intervals (0.5–5 h, depending on the total duration of the negative selection reaction) by heating to 65°C for 1 min (*see* **Note 31**).

3. Periods of negative selection are interspersed with episodes of chemically induced denaturation (mild alkali treatment) (*see* **Note 32**). Essentially, the negative selection reaction described in **Subheading 3.2., step 1** is substituted with the procedure listed here. A reduced reaction scale is employed here since, presumably, at least several rounds of selection have been completed at this stage (*see* **Note 20**).

   a. Incubate 50–200 pmol of gel-purified RNA for 30 min at 23°C in a 40-μL volume containing 50 m$M$ Tris-HCl, pH 7.5, and 20 m$M$ MgCl$_2$ in the absence of effector.

   b. Add to this sample 10 μL 3 $M$ NaOAc, pH 5.2, 1.5 μL 0.5 $M$ EDTA, 0.5 μL glycogen (10 mg/mL), and 48 μL water. Mix well.

   c. Precipitate RNA with 250 μL ethanol and pellet by centrifugation.

   d. Resuspend the pellet in 90 μL water.

   e. Denature the RNA by adding 10 µL 1 *M* NaOH. Mix well.

   f. Neutralize immediately by adding 11 µL 3 *M* NaOAc, pH 5.2.

   g. Precipitate RNA with 275 µL ethanol and pellet by centrifugation.

   h. Resuspend pellet in 20 µL water.

   i. Repeat **steps a–h** two more times.

   j. Repeat **step a** for a fourth and final time. Proceed to **Subheading 3.2., step 2**.

## 4. Notes

1. This same protocol may also be employed to select communication modules that will report to the hammerhead ribozyme the occupation state of a pre-existing aptamer *(13)*. The construction of this type of starting pool involves linking pre-defined catalytic and aptamer domains with a short bridge of random sequence (~10 nucleotides). In a typical design, each domain is divested of a functionally critical stem structure and conjoined to another at these modified sites by intervening random sequence with the expectation that selected sequences may convey a ligand-binding event to the catalytic domain. This variation of allosteric selection requires starting pools of relatively low complexity because of the limited number of randomized positions typically introduced. As a result, the investigator is afforded comprehensive coverage of sequence space in the search for functional modules. This method may therefore be particularly suitable if aptamers already exist for the ligand in question. However, it should be noted that ribozymes resulting from communication module selections have generally exhibited lower levels of ligand-induced ribozyme activation than ones produced by allosteric selections employing completely random ligand-binding domains.

2. If a catalytic domain other than the hammerhead self-cleaving motif is to be brought under the control of especially large effector molecules, such as proteins, it may be advisable to extend the substitution of random sequence to portions of the conserved catalytic core. Robertson and Ellington *(23)* have observed that, without this provision, effector-mediated activation is quite modest, probably resulting from steric inhibition caused by interaction of the effector with multiple sites on the ribozyme. However, such an approach is not recommended for the hammerhead catalytic core or other small, highly conserved motifs that are expected to be quite intolerant of mutations.

3. In designing any population for allosteric selection (regardless of the identity of the catalytic domain), ensure that the lengths of the fixed sequences flanking the random domain are adequate (~15 nucleotides) to permit amplification of the expected cleavage product with the appropriate primers. Also, confirm that the anticipated difference in size between the precursor and product RNAs will allow the efficient separation of these molecules by denaturing polyacrylamide gel electrophoresis.

4. The decision to initiate allosteric selection with 200 pmol of different sequences balances the desire for a highly diverse sequence population with the ability to perform laboratory manipulations easily. Scaling up the complexity of the starting pool by a factor of ten may be reasonably accomplished, but the use of still

larger populations will be unwieldy and quite expensive. Conversely, the use of libraries containing a relatively small number of randomized positions obviates the need for large pools. For example, in a typical selection for a communication module in which ten random positions are incorporated, there will be millions of copies of every possible sequence variant in a starting pool of only 20 pmol.

5. During the course of the transcription a white precipitate may be observed, which is caused by the formation of an insoluble complex between $Mg^{2+}$ ions and inorganic phosphate. The appearance of such a precipitate usually indicates a robust yield, and further incubation at 37°C beyond this point will not result in the significant production of additional RNA because of the limited availability of free $Mg^{2+}$ ions. If extraordinarily high yields are required, inorganic pyrophosphatase may be added to the reaction at 0.1 U/μL.

6. In addition to separating precursor molecules from prematurely terminated transcripts and DNA template strands, this step serves as an extra layer of negative selection, as it disfavors the isolation of ribozymes that undergo ligand-independent self-cleavage during the transcription.

7. Since 200 pmol of template DNA were included in the transcription, a yield of at least that amount of purified RNA is required to ensure that there is no significant loss in pool complexity in the conversion from DNA to RNA. However, run-off transcriptions routinely produce 10–20 copies of each sequence, effectively amplifying the pool. This redundancy in sequence representation makes it less likely that ribozymes with the desired characteristics will escape detection.

8. The duration of the negative selection step should be long enough to strongly disfavor those ribozymes that assume relatively stable structures conducive to catalysis in the absence of effectors. However, the temptation should be avoided specifically during the early rounds of selection, to extend this incubation for even longer periods in an attempt to further reduce the background. Because allosteric ribozymes inherently possess the capacity to assume at least two distinct structural states, ribozymes with active conformations that are stabilized through ligand binding may sample the active state, albeit infrequently, even in the absence of ligand and run the risk of being depleted from the population as a result of an overly aggressive negative selection. In other words, a certain load of ligand-independent ribozymes must be carried in the early stages of the selection in order to ensure that ligand-dependent catalysts are not inadvertently discarded. Only during later rounds, at which time ligand-responsive sequences presumably have been greatly amplified, should extended negative selections be considered.

9. The buffer conditions used in the negative (and positive) selection step are selected mainly with regard to performance of the hammerhead ribozyme motif. The composition of the selection buffer may be altered (to accommodate the structural requirements of a particular ligand, for example) as long as it remains compatible with near optimal performance of the catalytic platform.

10. It is often convenient to isolate the cleaved product band as well, for use as a size marker in the ensuing positive selection steps. Of course, caution must be exercised to avoid the cross-contamination of these samples.

11. It is imperative that the negative selection and the positive selection are identical in all respects, apart from the presence of the effector compound(s) in the positive selection. Even the order of addition of reaction components must be strictly maintained; lapses in attention to this detail may result in the isolation of ribozymes responsive to unintended stimuli *(22)*. It is recommended that the RNA is added last to the negative selection reaction, and as the penultimate component (immediately before the effector[s]) to the positive selection reaction.

12. The decision regarding which concentration of a particular ligand to employ in the positive selection reaction is influenced generally by the ligand's anticipated potential for interaction with RNA. Compounds such as aminoglycoside antibiotics, for example, which associate readily with RNA because of the abundance of functional groups that may participate in hydrogen bonding, might be added to selection reactions at micromolar to nanomolar concentrations. In contrast, for ligands such as aliphatic amino acids, which are not expected to bind RNA with the same facility, concentrations in the low millimolar range might be advisable. Note that while it is certainly possible to recover allosteric ribozymes with $K_D$ values significantly below the concentration of ligand used during the selection, the probability of such an outcome is low. In a particular random population, sequences that avidly bind a target will occur far more rarely than those exhibiting more modest affinities.

13. Although the procedure outlined here describes the generation of ligand-activated allosteric ribozymes, catalysts that are inactivated in response to a particular ligand may be similarly isolated by making simple adjustments to the protocol. Specifically, a selection for allosteric ribozymes inhibited by a particular ligand would require including the ligand not just in the negative selection reaction but also during transcription of the pool. The ligand would then be omitted from the positive selection reaction in order to permit the self-processing of those ribozymes relieved of inhibition in the absence of ligand.

14. As a corollary to **Note 8**, the duration of the positive selection reaction should not be shortened appreciably beyond the recommended time during the earliest rounds of the selection. Although an abbreviated incubation would indeed serve to lower the background by reducing the number of ligand-independent ribozymes that are permitted to self-cleave, the time constraint would apply equally to ligand-responsive ribozymes, for which a failure to self-cleave at this critical stage could result in elimination from the pool. It is best to adopt a conservative approach in the initial rounds of the selection, tolerating relatively high levels of effector-independent catalysis in order to ensure the recovery of sparsely represented molecules with desired allosteric properties. Increasingly stringent positive selection conditions are more appropriate in later rounds, once ligand-dependent sequences have been amplified to the point at which their inefficient recovery is acceptable, and, in fact, even beneficial, to the progress of the selection.

15. In the initial round of the allosteric selection (and typically for the next several rounds), the fractions of cleaved ribozymes in the negative control and positive selection lanes should be equal. (Despite the preceding negative selection step,

there may still be a substantial background of ligand-independent ribozymes.) If the extent of cleavage in the positive selection lane is markedly reduced compared to the negative control, it could be an indication that the selected ligand or ligand cocktail itself inhibits the function of the catalytic domain, which would interfere with the isolation of an allosteric ribozyme population. Conversely, if significantly more cleavage is observed in the positive selection lane during the initial round, when no amplification of ligand-dependent activity has yet occurred, it suggests a nonspecific enhancement of enzyme function by the ligand, which may also prove to be problematic.

16. Especially during the early rounds of the selection, persisting ligand-independent ribozymes may be abundant enough to permit the detection of a cleavage product band on an autoradiogram. (If an overnight exposure is required, the film cassette should be stored at 4°C to slow the diffusion rate of bands in the gel.) If detection of the product band ever proves to be difficult, however, the investigator may rely on a size marker to infer its location (*see* **Note 10**). To avoid contaminating the positive selection sample with size-marker RNAs, ensure that these samples are not loaded in adjacent lanes.

17. The efficient recovery of cleaved RNA molecules from the positive selection is critical during early selection rounds, when the occurrence of ligand-responsive sequences is rare. When the amount of cleaved RNA is very low or undetectable, there is a more pronounced risk of losing molecules during the ethanol-precipitation step. To improve the yield of ethanol precipitation in such cases, the addition of a carrier is recommended. Suggested carrier molecules are either glycogen (20 μg) or the 3' primer (primer 2; 50 pmol), which is quite convenient, as it is subsequently required for the reverse transcription step.

18. During the first one or two rounds of the selection, it is recommended to amplify most or all of the cDNA by scaling up the number of 100-μL PCRs. This will favor the propagation of even the rarest selected sequences in subsequent generations.

19. The number of cycles required for complete amplification (full incorporation of primers) varies inversely with the amount of template DNA. For each individual round of the in vitro selection, the appropriate number of PCR cycles may be estimated from the quantity of input cDNA, which in turn, may be approximately derived from the fraction of cleaved RNA in the positive selection reaction. Alternatively, if the fraction of cleaved RNA was below the level of detection, the number of PCR iterations necessary for complete amplification may be determined empirically by analyzing aliquots withdrawn at defined intervals from a scout PCR. (Do not subject the amplification to more cycles than are required, as continued cycling past the point of primer availability may result in undesired higher mol-wt products derived from mispriming events.)

20. Depending on the complexity of the starting pool, it may be convenient to use smaller RNA populations (with reaction volumes scaled down accordingly) once the initial round of the selection has been completed.

21. If the selection has included multiple potential allosteric effectors in parallel rather than only one, it is likely that the first allosteric ribozymes to emerge will

be controlled by a small subset of these effectors. Since the presence of a dominant class of sequences responding to this effector subset provides a barrier to the recovery of ribozymes sensitive to the remaining ligands, this dominant class must be removed selectively before an efficient and productive search for other classes can proceed. Therefore, after the first robust ligand-dependent cleavage signal appears, the investigator should perform tests to identify the effector(s) to which these earliest emerging sequences are responding. In subsequent rounds, this effector is omitted from the positive selection reaction and included in the negative selection reaction as a means to cull its dependent sequences from the pool so that other ligand-sensitive classes may gain a foothold. Once a new dominant population is established that is activated by one or more of the remaining effectors, the process is repeated to clear the way for yet another class of ligand-sensitive ribozymes to expand into a newly vacated niche. This successive dropping out of ligands is continued until the population ceases to respond, which would suggest that all major subclasses sensitive to effectors in the initial cocktail have been retrieved. (Note that to optimize the response of a population to a particular ligand, selection under more stringent incubation conditions may be resumed using the relevant archived pool.)

22. A conservative regimen is advised, in which the incubation time is reduced gradually enough to maintain a detectable cleavage signal in each round. Reaction times that are too drastically reduced may result in an evolutionary bottleneck that is detrimental to the selection.

23. Concomitantly with the introduction of stronger positive selection pressure, a more stringent negative selection may be applied by increasing the duration of the negative selection reaction.

24. For the cloning of PCR products, it is quite convenient to employ a TOPO TA Cloning Kit (Invitrogen), as this method does not require restriction sites at amplification product termini.

25. Populations in the plateau phase of the selection are usually relatively homogeneous in composition, because of aggressively shortened incubation times that permit the survival of only the fittest allosteric ribozymes. In the event that a more diverse sampling of ligand-sensitive molecules is required, it is recommended to isolate sequences from ligand-responsive populations that have not yet been subjected to stringent reaction-time constraints.

26. Templates for the transcription of individual clones may be prepared by PCR amplification of plasmid DNA or with appropriate chemically synthesized oligonucleotides.

27. Aside from serving as reservoirs for high performance allosteric ribozymes, mutagenized pools may also be employed for an alternative purpose. If the positive selection reaction, rather than being increased in stringency, is instead maintained under permissive conditions (defined here as supporting robust activity of the parent molecule), then a coarse structure/function assignment of the ligand-binding sequence may emerge from analysis of the resulting constellation of ligand-sensitive mutants. The completion of only a few rounds of allosteric

selection under mild conditions should be sufficient to recover a strong ligand-dependent signal, at which point the sequences of representative clones may be examined. Since selection in this relatively noncompetitive environment allows ligand-sensitive sequences of varying efficiencies to persist, a large panel of non-disruptive mutations may be sampled. In effect, these mutants provide a type of artificial phylogeny in which the distribution of mutations can yield important insights in the structure of the pertinent domain. As a general rule, highly conserved regions are essential to function, and those that tolerate high levels of mutation are less critical. Also, the presence of covarying residues (in which a pattern of mutations is observed that would preserve basepairing between two elements) may sometimes be interpreted in support of secondary structure elements.

28. Note that the initial reselection pool is a diverse population and, for the reasons discussed in **Note 14**, should not be immediately challenged with the increased stringency of multiple parameters simultaneously. Thus, if the target concentration in the positive selection reaction has been reduced to isolate sequence variants exhibiting improved $K_D$ values, the reaction time constraint should be kept relatively relaxed until a ligand-dependent signal is detected.

29. Negative selection reactions of up to 48 h have been used successfully to combat ligand-independent ribozymes. (Since, by this point in the allosteric selection, it is likely that multiple iterations have been performed, the caveats issued in **Note 8** no longer apply. If ligand-dependent ribozymes exist in the population, they have probably been amplified sufficiently to eliminate their risk of extinction during an extended negative-selection reaction.)

30. Since this method requires extensive manipulation of the RNA population, there may be a significant overall reduction in yield. For this reason, ensure that adequate amounts of the relevant pool are used to initiate this negative-selection protocol.

31. The number of thermal spikes should be kept reasonably low (no more than four) because of the increased background rate of RNA transesterification at elevated temperatures.

32. Alkaline denaturation has proven to be the most effective of these three techniques in selecting against ligand-independent sequences that distribute between active and inactive structures.

## References

1. Doudna, J. A. and Cech, T. R. (2002) The chemical repertoire of natural ribozymes. *Nature* **418,** 222–228.
2. Wilson, D. S. and Szostak, J. W. (1999) In vitro selection of functional nucleic acids. *Annu. Rev. Biochem.* **68,** 611–647.
3. Breaker, R. R. (1997) In vitro selection of catalytic polynucleotides. *Chem. Rev.* **97,** 371–390.
4. Gold, L., et al. (2002) One, two, infinity: genomes filled with aptamers. *Chem. Biol.* **9(12),** 1259–1264.
5. Hermann, T. and Patel, D. J. (2000) Adaptive recognition by nucleic acid aptamers. *Science* **287,** 820–825.

6. Breaker, R. R. (2002) Engineered allosteric ribozymes as biosensor components. *Curr. Opin. Biotechnol.* **13(1),** 31–39.
7. Kuwabara, T., Warashina, M., and Taira, K. (2000) Allosterically controllable ribozymes with biosensor functions. *Curr. Opin. Chem. Biol.* **4(6),** 669–677.
8. Soukup, G. A. and Breaker, R. R. (2000) Allosteric nucleic acid catalysts. *Curr. Opin. Struct. Biol.* **10,** 318–325.
9. Kertsburg, A. and Soukup, G. A. (2002) A versatile communication module for controlling RNA folding and catalysis. *Nucleic Acids Res.* **30(21),** 4599–4606.
10. Thompson, K. M., et al. (2002) Group I aptazymes as genetic regulatory switches. *BMC Biotechnol.* **2(1),** 21.
11. Levy, M. and Ellington, A. D. (2002) ATP-dependent allosteric DNA enzymes. *Chem. Biol.* **9,** 417–426.
12. Osborne, S. E. and Ellington, A. D. (1997) Nucleic acid selection and the challenge of combinatorial chemistry. *Chem. Rev.* **97,** 349–370.
13. Soukup, G. A. and Breaker, R. R. (1999) Engineering precision RNA molecular switches. *Proc. Natl. Acad. Sci. USA* **96(7),** 3584–3589.
14. Soukup, G. A., Emilsson, G. A., and Breaker, R. R. (2000) Altering molecular recognition of RNA aptamers by allosteric selection. *J. Mol. Biol.* **298(4),** 623–632.
15. Robertson, M. P. and Ellington, A. D. (2000) Design and optimization of effector-activated ribozyme ligases. *Nucleic Acids Res.* **28(8),** 1751–1759.
16. Porta, H. and Lizardi, P. M. (1995) An allosteric hammerhead ribozyme. *Biotechnology* **13(2),** 161–164.
17. Burke, D. H., Ozerova, N. D., and Nilsen-Hamilton, M. (2002) Allosteric hammerhead ribozyme TRAPs. *Biochemistry* **41(21),** 6588–6594.
18. Robertson, M. P. and Ellington, A. D. (1999) In vitro selection of an allosteric ribozyme that transduces analytes to amplicons. *Nat. Biotechnol.* **17(1),** 62–66.
19. Warashina, M., Kuwabara, T., and Taira, K. (2000) Working at the cutting edge: the creation of allosteric ribozymes. *Structure Fold Des.* **8(11),** R207–212.
20. Wang, D.Y., et al. (2002) A general approach for the use of oligonucleotide effectors to regulate the catalysis of RNA-cleaving ribozymes and DNAzymes. *Nucleic Acids Res.* **30(8),** 1735–1742.
21. Komatsu, Y., et al. (2000) Construction of new ribozymes requiring short regulator oligonucleotides as a cofactor. *J. Mol. Biol.* **299(5),** 1231–1243.
22. Koizumi, M., et al. (1999) Allosteric selection of ribozymes that respond to the second messengers cGMP and cAMP. *Nat. Struct. Biol.* **6(11),** 1062–1071.
23. Robertson, M. P. and Ellington, A. D. (2001) In vitro selection of nucleoprotein enzymes. *Nat. Biotechnol.* **19,** 650–655.
24. Piganeau, N., et al. (2001) An allosteric ribozyme regulated by doxycyline. *Angew Chem. Int. Ed. Engl.* **40(19),** 3503.
25. Piganeau, N., et al. (2001) Corrigendum: an allosteric ribozyme regulated by doxycycline. *Angew Chem. Int. Ed. Engl.* **40,** 3503.
26. Piganeau, N., Thuillier, V., and Famulok, M. (2001) In vitro selection of allosteric ribozymes: theory and experimental validation. *J. Mol. Biol.* **312(5),** 1177–1190.

27. Baskerville, S. and Bartel, D. P. (2002) A ribozyme that ligates RNA to protein. *Proc. Natl. Acad. Sci. USA* **99(14),** 9154–9159.
28. Tang, J. and Breaker, R. R. (1997) Rational design of allosteric ribozymes. *Chem. Biol.* **4(6),** 453–459.
29. Tuschl, T. and Eckstein, F. (1993) Hammerhead ribozymes: importance of stem-loop II for activity. *Proc. Natl. Acad. Sci. USA* **90,** 6991–6994.

# 11

## Regulation of Ribozyme Cleavage Activity by Oligonucleotides

### Yasuo Komatsu and Eiko Ohtsuka

### Summary

Here, we describe allosteric ribozymes, which are activated by the addition of a short regulator oligonucleotide. The allosteric hammerhead ribozyme, which contains a single-stranded loop instead of stem II, exhibited minimal cleavage of the target RNA; however, it becomes active by the addition of oligonucleotides that bind to the ribozyme. We have also carried out in vitro selection to obtain an allosteric hairpin ribozyme, which is activated with a short oligonucleotide. The ribozyme contains the characteristic hairpin loop with the structure that changes upon regulator oligonucleotide binding. Since both regulations are sequence-specific, ribozymes containing different sequences can be positively and independently controlled by a specific oligonucleotide. Furthermore, these allosteric hairpin loops have the ability to control the activities of other functional RNAs.

**Key Words:** RNA; ribozyme; hammerhead ribozyme; hairpin ribozyme; catalytic RNA; allosteric control; oligonucleotide; sensor.

## 1. Introduction

Both hammerhead and hairpin ribozymes, which are well-known as small, catalytic RNAs, can catalyze either cleavage or ligation of RNA with sequence specificity. Although they have been applied to gene therapy *(1,2)* and gene discovery *(3)*, by cutting the mRNAs of target genes, these ribozymes—which compose the catalytic core of viral satellite RNA—usually do not have the ability to regulate their catalytic activities. Therefore, allosteric control has been applied to the ribozymes *(4,5)*.

Breaker and colleagues (*see* Chapter 10) carried out an in vitro selection of stem II cells of the hammerhead ribozyme, and constructed a series of allosteric ribozymes responding to small molecules, such as adenosine 5' triphosphate (ATP) *(6)*, cyclic nucleotides *(7)*, or flavin mononucleotide (FMN) *(8)*.

From: *Methods in Molecular Biology, vol. 252: Ribozymes and siRNA Protocols, Second Edition*
Edited by: M. Sioud © Humana Press Inc., Totowa, NJ

Herschlag and colleagues constructed a hammerhead ribozyme that contained an abasic residue at a specific position within the catalytic core *(9,10)*. Although the cleavage was reduced, the activity was rescued 300-fold by the addition of the missing base. Taira and colleagues described four pieces of ribozyme that could be activated with an oligoribonucleotide *(11)*, and used this system to investigate gene function in vivo *(2)*. Porta and Lizard also reported an allosteric hammerhead ribozyme, which contained both an inherent ribozyme sequence and a long, extra tail *(12)*. By the addition of an oligonucleotide complementary to the extra sequence, the ribozyme folded to form a proper secondary structure. The oligonucleotide facilitated the ribozyme binding to the substrate, but did not take part in the formation of the catalytic core.

General methods to control ribozyme activity have not been established. Therefore, we sought to develop a new method to control the ribozyme activity. A hairpin loop and an oligonucleotide bound to the loop form one-half of the pseudo-knot structure (**Fig. 1**). We designed an allosteric hammerhead ribozyme, which was activated by the introduction of this motif by using a short, complementary oligonucleotide as a cofactor *(13)*. Stem II of the wild type hammerhead ribozyme (**Fig. 2A**) was substituted with a non-self-complementary loop sequence (loop II) to abolish the cleavage activity (**Fig. 2B**). The new ribozyme had almost no cleavage activity with the target RNA. However, it exhibited the cleavage activity in the presence of a regulator oligoribonucleotide (RLO), which was complementary to loop II of the ribozyme.

Loop II consisted of 17 or 19 non-self-complementary bases, as shown in **Fig. 2B**. The loop II length restricted that of the RLO. Although the ribozyme (H43C) with a 17-base loop II showed weak cleavage activity in the presence of a 9- or 11-base RLO, a 7- or 8-base RLO activated the ribozyme efficiently. H45C, with a 19-base loop II became more active by the addition of a 9-base RLO. We showed that the binding of the long RLO promoted the release of the substrate RNA from the ribozyme/substrate complex *(13)*. A short loop II is insufficient to form a stable active conformation after a long RLO binds. However, both ribozymes were most efficiently activated in the presence of an 8-base 2'-*O*-methyloligoribonucleotide (m7G) by as much as about 750-fold, as compared with their activities in the absence of the oligonucleotide (m7G: 2'-OMe(GAGUGAG)rG, **Fig. 2B**). We also showed that an RLO with a 3'-dangling purine base (circled in **Fig. 2C**) could serve as an efficient cofactor of the ribozyme (*see* **Note 1**).

The hairpin ribozyme has two stem-loop domains, and both domains involve two helices and one internal loop (helices 1, 2, and loop A in domain I; helices 3, 4, and loop B in domain II) *(14–16)*. Essential bases for the cleavage activity are located in the two internal loops. It has been proved that the bent structure is an active conformation *(17–19)*.

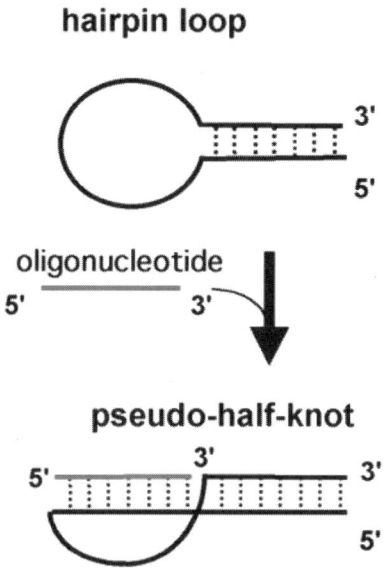

Fig. 1. Scheme of formation of pseudo-half-knot structure. The gray bar in the pseudo-half-knot indicates a complementary oligonucleotide binding with a hairpin loop. Dotted lines indicate basepairs.

The hairpin ribozyme had not been modified into an allosteric enzyme that responds to small molecules, in contrast to the hammerhead ribozyme. Recently, we reported the construction of an allosteric hairpin ribozyme, which is activated by a short oligonucleotide *(20)*. We carried out an in vitro selection to obtain an allosteric hairpin ribozyme, which has cleavage activity in the presence of an exogenous short oligonucleotide as a regulator. Random sequences were inserted in the region that corresponds to the hairpin loop of the ribozyme (**Fig. 3B**). The hairpin ribozymes produced exhibited the cleavage activity specifically in the presence of the RLO, and they contained a characteristic hairpin loop, with a structure that changed upon binding with the RLO. The ribozymes with high cleavage activities gained similar characteristic hairpin loops at the random domain (**Fig. 4**). In the absence of the oligonucleotide, the loop domain within the allosteric ribozyme probably forms a slipped hairpin loop, and the complementary sequence, with the RLO located at the single-stranded loop, would allow easy access for the oligonucleotide. The binding of the RLO triggers a structural change of the hairpin loop to form an active conformation, as shown in **Fig. 4**. Furthermore, we constructed an allosteric hammerhead ribozyme by introducing the characteristic hairpin loop (**Fig. 3C**). The modified hammerhead ribozyme was also changed to an allosteric ribozyme, which was activated by the addition of the RLO. The characteristic

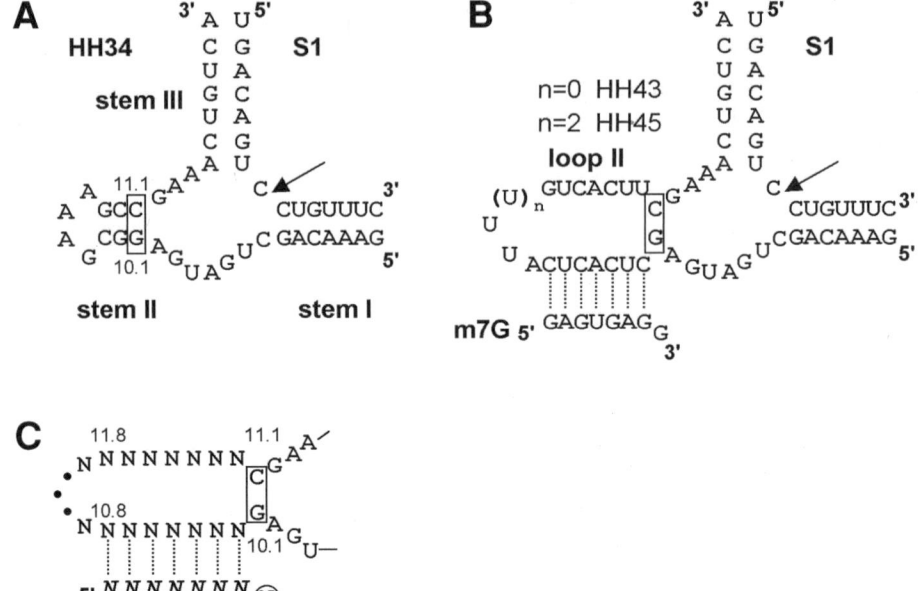

Fig. 2. (**A**) Secondary structure of the complex of a hammerhead ribozyme (HH34) and a substrate (S1). HH34 consists of 34 bases and S1. (**B**) Secondary structure of the complex of allosteric hammerhead ribozymes (H43C and H45C) and S1. H43C and H45C consist of 43 and 45 bases, respectively and both involve a cytidine base at position 11.1 (C11.1). Stem II is substituted for loop II (17 or 19 bases). The arrow indicates the cleavage site. (**C**) Secondary structure of the loop II domain and RLO. Bases $N_{10.2-10.8}$ are complementary with bases $N_{1-7}$ (italic characters) of the RLO. The dangling base at the 3'-end of the RLO is circled.

hairpin loop, which has proven to be regulated by an exogenous oligonucleotide, may be used to control RNA functions in various fields.

## 2. Materials

### 2.1. Oligoribonucleotides

Oligoribonucleotides can be prepared, as previously reported *(21,22)*. After chemical synthesis of the oligoribonucleotides, they are purified using reverse-phase and anion-exchange column chromatography (*see* **Note 2**).

#### 2.1.1. Reverse-Phase High-Performance Liquid Chromatography (HPLC)

1. Buffer A: 5% acetonitrile, 0.1 *M* triethylammonium acetate (TEAA; pH 7.0).
2. Buffer B: 25% acetonitrile, 0.1 *M* TEAA, pH 7.0.
3. Column: μ-Bondasphere (Waters; φ3.9 × 150 mm).

**A**

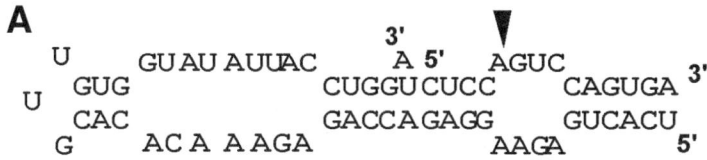

**B**

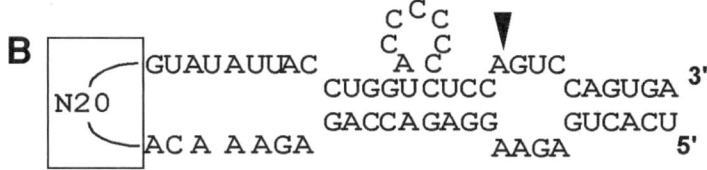

**C**

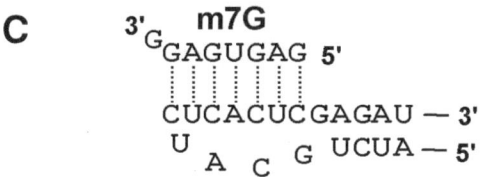

Fig. 3. (**A**) Secondary structures of the wild-type hairpin ribozyme (*trans*-cleavage system) and the ribozyme with a 20-nucleotide random-sequence domain (N$_{20}$) (**B**, *cis*-cleavage system). Arrows indicate cleavage sites. (**C**) Predicted secondary structure of the hairpin loop region of a selected allosteric ribozyme in the presence of a cofactor oligonucleotide (m7G). Dotted lines indicate basepairs between ribozymes and m7G.

### 2.1.2. Anion-Exchange HPLC

1. Buffer A: 20% acetonitrile.
2. Buffer B: 20% acetonitrile, 2 $M$ ammonium formate.
3. Column: TSKgel DEAE 2SW (Toso; $\phi$4.6 × 350 mm).

## 2.2. 5'-End-Labeling of Oligoribonucleotide

1. Kination buffer 1X: 50 m$M$ Tris-HCl, pH 7.6, 10 m$M$ MgCl$_2$, and 10 m$M$ 2-mercaptoethanol.
2. [$\gamma$-$^{32}$p]ATP.
3. T4 polynucleotide kinase (PNK) (*E. coli* A19).
4. NENSORB 20 (DuPont).
5. Sep-Pak cartridge (C8 or C18).
6. Oligoribonucleotide (100 pmol).

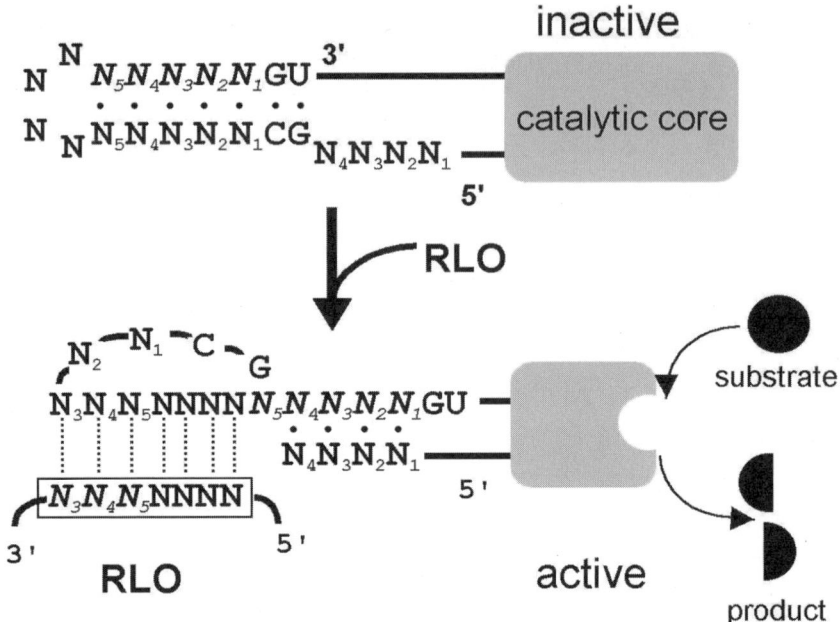

Fig. 4. Structural change of the allosteric hairpin motif. Bases $N_{1-5}$ in the motif are complementary with bases $N_{1-5}$ (italic characters). The gray box and black circle indicate the catalytic core of the hairpin ribozyme and the substrate, respectively.

## 2.3. Cleavage Reaction of Hammerhead Ribozyme

### 2.3.1. Cleavage Reaction

1. Annealing buffer: 100 m$M$ NaCl, 50 m$M$ Tris-HCl, pH 7.5.
2. Magnesium ion buffer: 50 m$M$ MgCl$_2$, 50 m$M$ Tris-HCl, pH 7.5.
3. Stop solution: 10 $M$ urea, 50 m$M$ Na$_2$ ethylenediaminetetraacetic acid (EDTA), 0.05% bromophenol blue, 0.05% xylene cyanol.
4. 20% polyacrylamide gel (20 × 30 × 0.05 cm): Total volume 50 mL, 40% acrylamide (acrylamide:*bis*-acrylamide, 19:1), 25 mL; 10X TBE, 5 mL; urea, 24.03 g; ammonium persulfate, 20 mg; $N,N,N',N'$-tetramethylethylenediamine (TEMED), 15 µL.

### 2.3.2. Gel Mobility-Shift Assay

1. Reaction buffer: 50 m$M$ Tris-HCl, pH 7.5, 50 m$M$ NaCl, 25 m$M$ MgCl$_2$, 3% glycerol, 0.05% bromophenol blue, 0.05% xylene cyanol.
2. Electrophoresis buffer: 50 m$M$ Tris-acetate, pH 7.5, 25 m$M$ magnesium acetate, 50 m$M$ NaCl.
3. 15% polyacrylamide gel (20 × 30 × 0.05 cm): 40% acrylamide (acrylamide:*bis*-acrylamide, 19:1), 18.7 mL; 10X TBE, 5 mL; ammonium persulfate, 20 mg/mL; TEMED, 15 mL.

## 2.4. Cleavage Reaction of Hairpin Ribozyme

### 2.4.1. Transcription

1. AmpliScribe™ transcription kit (EPICENTRE TECH.).
2. [α-$^{32}$P]UTP.
3. Template DNA.
4. NAP-5 (Amersham Pharmacia).

### 2.4.2. Cleavage Reaction

1. Cleavage buffer 1X: 40 m$M$ Tris-HCl, pH 7.5, 12 m$M$ MgCl$_2$, 2 m$M$ spermidine. 5X: 200 m$M$ Tris-HCl, pH 7.5, 60 m$M$ MgCl$_2$, 10 m$M$ spermidine.
2. Stop solution: 50 m$M$ Na$_2$EDTA containing 10 $M$ urea, 0.05% bromophenol blue, 0.05% xylene cyanol.
3. 10% polyacrylamide gel containing 8 $M$ urea (40 × 30 × 0.05 cm): 80 mL, 40% acrylamide (acrylamide:*bis*-acrylamide, 19:1), 20 mL; 10X TBE, 8 mL; urea, 38.44 g; ammonium persulfate, 20 mg/mL; TEMED, 20 µL.
4. Mineral oil (Pharmacia).

## 3. Methods

### 3.1. 5'-End-Labeling of Oligoribonucleotide

1. Incubate purified oligoribonucleotide (100 pmol) in 50 m$M$ Tris-HCl, pH 7.6, 10 m$M$ MgCl$_2$, 10 m$M$ 2-mercaptoethanol with [γ-$^{32}$p]ATP, and T4 PNK (*E. coli* A19) for 1 h at 37°C.
2. Desalt the reaction solution using a NENSORB 20 column or a Sep-Pak cartridge. The labeled oligonucleotide can be eluted with 50% aqueous ethanol (800 µL).
3. Evaporate the solvent under reduced pressure.
4. Add appropriate amount of distilled water to the labeled oligonucleotide (*see* **Note 3**).

### 3.2. Cleavage Reaction Procedure With the Hammerhead Ribozyme

#### 3.2.1. Excess Enzyme Conditions

When a cleavage reaction under enzyme excess conditions is carried out, the concentration of the RLO should be sufficiently higher than that of the ribozyme. The reaction using 1.5 or 4 Molar equivalents of RLO to the ribozyme is described below (*see* **Note 4**).

1. Dissolve the 5'-end-labeled substrate RNA, the ribozyme, and the RLO in annealing buffer (20 µL). The concentrations of the substrate and the ribozyme are 20 n$M$ and 3 µ$M$, respectively. The RLO concentration is 4.6 or 12 µ$M$.
2. Heat the combined solution at 65°C for 2 min, and then transfer it to an ice bath.
3. Distribute aliquots (2 µL) from the complex solution into nine 0.5-mL tubes, and then add an equal volume of the magnesium ion buffer to the aliquots.

4. Mix the solution by vortexing (the final concentrations of the substrates, ribozyme, and RLO are 10 nM, 1.5 µM, and 2.3 or 6 µM, respectively) and incubate it at 37°C.
5. Add mineral oil (7 µL), which is preincubated at 37°C, to each reaction tube just after the reaction starts (*see* **Note 5**).
6. Add the stop solution (5 µL) to the reaction solutions at time intervals, and vigorously mix the solutions.
7. Cleavage products can be fractionated on 20% polyacrylamide gels containing 8 *M* urea, and the percentages of the cleaved products are estimated by measuring the radioactivities with a Bioimaging analyzer (FUJIX BAS2000). The $k_{obs}$ values can be determined as previously reported *(23)*.

### 3.2.2. Measurement of Dissociation and Cleavage-Rate Constants

The dissociation constant between the RLO and the ribozyme/substrate complex can be obtained from the cleavage reactions with various RLO concentrations. After the observed rate constants ($k_{obs}$) are measured for each reaction, the dissociation constant (*K*d) of the RLO can be obtained from curve-fitting of plots of *k*obs vs each concentration of RLO. Both the highest chemical step ($k_{RLO(+)}$) and the rate constant ($k_{RLO(-)}$) without the RLO are also calculated from the curve-fitting. Cleavage reactions are carried out as described for the cleavage reaction under enzyme excess conditions. *K*d, $k_{RLO(+)}$ and $k_{RLO(-)}$ are determined from the equation $k_{obs} = (k_{RLO(-)} * Kd + k_{RLO(+)} [RLO])/(Kd + [RLO])$, as reported previously *(13,24)*. The $k_{RLO(-)}$ is also determined by the cleavage reaction of the ribozyme/substrate complex in the absence of the RLO.

### 3.2.3. Multiple-Turnover Reactions Using the Ribozyme/RLO Complex

1. Dissolve the 5'-end-labeled substrate RNA, the ribozyme, and the RLO in the annealing buffer (8 µL). Heat the mixture of the three pieces of the complex at 65°C for 2 min, and then immediately transfer it into an ice bath.
2. Add an equal volume of magnesium ion buffer (8 µL) to the complex solution (8 µL) to initiate the cleavage reaction. After flash-vortexing and spinning the tube down, incubate the reaction at 37°C.
3. Take an aliquot (1–2 µL) from the solution at the time intervals and add it to the stop solution (5 µL). After vortexing and spinning the tube down, leave it on ice.
4. The cleavage products can be fractionated on 20% polyacrylamide gels containing 8 *M* urea. The initial velocities are estimated from the percentages of the cleavage products. We obtained kinetic parameters using the Hanes-Woolf plot.

### 3.2.4. Gel Mobility-Shift Assay

#### 3.2.4.1. ABSENCE OF RLO

1. Dissolve the 5'-end-labeled substrate RNA (4 n*M*, 5 µL: *see* **Note 6**) and the ribozyme (10 µL) in the reaction buffer, respectively. Prepare various concentrations of the ribozyme solutions (*see* **Note 7**).

2. Heat the ribozyme and the substrate solutions separately at 65°C for 2 min, and then transfer them into an ice bath. Add an equal volume of the ribozyme solution (5 μL) to the substrate solution. Mix the solution by vortexing, and incubate the combined solution (10 μL), containing the substrate (2 n*M*) and the ribozyme at 37°C for 1 h.
3. Perform electrophoresis on a nondenaturing 15% polyacrylamide gel (acrylamide:*bis*-acrylamide, 19:1), maintaining the temperature of the gel at room temperature (*see* **Note 8**).

3.2.4.2. PRESENCE OF RLO

The gel mobility-shift analyses in the presence of the RLOs are the same as those described previously, except for the addition of the RLO, which is premixed with the ribozyme (*see* **Note 9**).

### 3.3. Cleavage Reaction Procedure With the Hairpin Ribozyme

### 3.3.1. Transcription

1. Mix the following components:
   Transcription buffer (AmpliScribe™ transcription kit)    1 μL
   25 m*M* ATP                                               1 μL
   25 m*M* cytidine 5' triphosphate (CTP)                    1 μL
   25 m*M* guanosine 5' triphosphate (GTP)                   1 μL
   25 m*M* uridine 5' triphosphate (UTP)                     1 μL
   [α-$^{32}$P]UTP                                           0.5 μL
   100 m*M* DTT                                              1 μL
   DNA template (20 ng)                                      x μL
   T7 RNA polymerase                                         1 μL
2. Complete to 10 μL with water and incubate the mixture at 37°C for 2 h.
3. Add DNase I (0.5 μL; 1 MBU; AmpliScribe™) to the solution, and incubate it at 37°C for 15 min.
4. Add distilled water to the solution to a total volume of 500 μL, and load the reaction mixture on a NAP-5 column for desalting. Transcripts are eluted with distilled water (1000 μL).
5. After phenol:chloroform and chloroform extractions, ethanol precipitation is carried out, and the transcribed RNA is dissolved in distilled water (20 μL; *see* **Note 10**).

### 3.3.2. Cleavage Reaction (see **Note 11**)

1. Dissolve the transcribed RNA (16 pmol) and the RLO (200 pmol) separately in 8 μL and 10 μL of reaction buffer, respectively. Heat these solutions at 90°C for 2 min and then immediately transfer them to an ice bath to denature each molecule.
2. After mixing the RLO by vortexing, add the RLO solution (8 μL) to the ribozyme transcripts.
3. Mix the combined solution by vortexing, and immediately incubate the reaction mixture at 37°C.

4. Take an aliquot (1–2 µL) from the reaction solution at the time intervals, and then add it to the stop solution.

5. Cleavage products can be fractionated by electrophoresis on a 10% polyacrylamide gel containing 8 $M$ urea. From the radioisotope activity measured by an imaging analyzer, the observed rate constants can be calculated from curve fitting, using the equation $S/(L + S) = a(1 - \exp(-k_{obs}t))$, in which $S$ is the concentration of cleaved products, $L$ is the concentration of the full-length transcripts, $t$ is time, and $a$ is the percentage of active molecules vs total RNA molecules.

6. The dissociation constant ($K_d$) of the ribozyme/RLO complex and both $k_{RLO(+)}$ and $k_{RLO(-)}$, which are the cleavage-rate constants with a saturating amount of RLO and without RLO, respectively can be calculated from the observed rate constant ($k_{obs}$) values for various RLO concentrations (We obtained the $K_d$ value of the ribozyme/RLO complex using 1 µ$M$ of the ribozyme, and 0–100 µ$M$ RLO). The $K_d$ and the highest chemical step ($k_{RLO(+)}$) are determined from the equation $k_{obs} = (k_{RLO(-)}*K_d + (k_{RLO(+)})*[RLO])/(K_d + [RLO])$, as reported previously *(13,24)*. Alternatively, the $k_{RLO(-)}$ value is determined from the cleavage reaction of the ribozyme in the absence of RLO.

### 3.3.3. Cis-Cleavage Reaction During Transcription from the DNA Template

1. Dissolve both the DNA template (50 nmol) and the RLO (600 pmol) in the cleavage buffer that contains the following components:

   | | |
   |---|---|
   | 5X cleavage buffer | 4 µL |
   | 2.5 m$M$ nucleotide triphosphates (NTPs) (ATP, GTP, CTP, UTP) | 4 µL |
   | 100 m$M$ DTT | 1 µL |
   | [α-$^{32}$P]-UTP | 1 µL |
   | T7 RNA polymerase | 2 µL |

2. Complete with water to 20 µL, and then incubate the reaction solution at 37°C.

3. Take an aliquot (1–2 µL) at time intervals from the reaction solution (*see* **Note 12**), and transfer the aliquots to the stop solution (5 µL). After the reaction is stopped, place the tubes on ice.

4. Analyze the cleavage products by 10% polyacrylamide gel containing 8 $M$ urea.

5. The cleavage rate constants are determined by fitting the data by a least-squares method to the following equation, as described previously *(25,26)*, $L/(L + S) = (1 - b)[[1 - \exp(-k_{cis}t)]/k_{cis}t] + b$, where $L$ is the concentration of the full-length transcripts, $S$ is the concentration of cleaved transcript, $t$ is time, $b$ is the percentage of the inactive molecule, and $k_{cis}$ is the rate constant for *cis*-cleavage.

## 4. Notes

1. To make the ribozyme inactive in the absence of the RLO, not less than bases $N_{10.2-10.7}$ should not be complementary with bases $N_{11.2-11.7}$, respectively.

2. To highly purify chemically synthesized RNA, reverse-phase as well as anion-exchange column chromatographies are required. After the purification with an anion-exchange column, we desalt the RNA by gel filtration using Sephadex resin (G10,

G25, or G50: Pharmacia). The gel filtration is carried out using distilled water. The concentrations of the purified RNA are determined by ultraviolet measurements.

3. If the labeled oligonucleotide contains degraded products, we carry out an additional purification using a denaturing polyacrylamide gel (either 15% or 20% polyacrylamide, prepared as described in the analysis of the cleavage reaction).

4. In our experiments, the final concentrations of the ribozyme and the RLO were 40 n$M$ and 30 µ$M$, respectively, and the concentrations of the substrate RNA were varied from 0.2 µ$M$–2 µ$M$.

5. When the reaction volume is lower than 10 µL or the reaction time exceeds 60 min, we add mineral oil to the reaction solution to prevent evaporation of the solvent. On the other hand, in a reaction in which the volume is more than 10 µL and the incubation is within 60 min, aliquots are taken from the solution at time intervals and then transferred into the stop solution, without adding mineral oil.

6. We usually use a 5'-end-labeled substrate analog, which has a 2'-$O$-methyl nucleotide at the cleavage site to prevent the cleavage reaction during electrophoresis.

7. We prepared 2–280 n$M$ of the ribozymes in our experiments.

8. To maintain the pH of the electrophoresis buffer, it is advisable to circulate the buffer slowly, using a pump.

9. We prepared the final reaction solution to contain 140 n$M$ of the ribozyme, 2 n$M$ of the substrate, and 560 n$M$ of the RLO.

10. Further purification of the transcripts can be achieved by gel electrophoresis using denaturing polyacrylamide gel. After the transcripts are mixed with the stop solution, the transcription products are fractionated by electrophoresis on a 10% polyacrylamide gel containing 8 $M$ urea. The ribozyme bands are excised from the gel, and the transcripts are eluted from the gel slice in distilled water (800–1000 µL) at 4°C overnight. The eluted products are purified on a NENSORB™20 column (DU PONT) or Sep-Pak cartridge (Waters; C8 or C18). After the eluants are evaporated under reduced pressure, the transcripts are dissolved in distilled water.

11. A *trans*-cleavage reaction of the hairpin ribozyme can be carried out using the same method described in the hammerhead ribozyme reaction.

12. After tapping and gently spinning down the tube, an aliquot is removed. Put the tube immediately back in the incubator. These operations should be done as quickly as possible so that the data do not vary.

## Acknowledgments

The authors thank Dr. K. Matsubara for his support and Dr. M. Koizumi for helpful discussions. We are grateful to Shigeko Yamashita, Kaoru Nobuoka, and Naoko Karino-Abe for data analysis and technical support.

## References

1. Michienzi, A., Cagnon, L., Bahner, I., and Rossi, J. J. (2000) Ribozyme-mediated inhibition of HIV 1 suggests nucleolar trafficking of HIV-1 RNA. *Proc. Natl. Acad. Sci. USA* **97,** 8955–8966.

2. Kawasaki, H., Schiltz, L., Chiu, R., Itakura, K., Taira, K., Nakatani, Y., and Yokoyama, K. K. (2000) ATF-2 has intrinsic histone acetyltransferase activity which is modulated by phosphorylation. *Nature* **405,** 195–200.
3. Beger, C., Pierce, L. N., Kruger, M., Marcusson, E. G., Robbins, J. M., Welcsh, P., et al. (2001) Identification of Id4 as a regulator of BRCA1 expression by using a ribozyme-library-based inverse genomics approach. *Proc. Natl. Acad. Sci. USA* **98,** 130–135.
4. Soukup, G. A. and Breaker, R. R. (2000) Allosteric nucleic acid catalysts. *Curr. Opin. Struct. Biol.* **10,** 318–325.
5. Robertson, M. P. and Ellington, A. D. (1999) In vitro selection of an allosteric ribozyme that transduces analytes to amplicons. *Nat. Biotechnol.* **17,** 62–66.
6. Tang, J. and Breaker, R. R. (1997) Rational design of allosteric ribozymes. *Chem. Biol.* **4,** 453–459.
7. Koizumi, M., Soukup, G. A., Kerr, J. N., and Breaker, R. R. (1999) Allosteric selection of ribozymes that respond to the second messengers cGMP and cAMP. *Nat. Struct. Biol.* **6,** 1062–1071.
8. Soukup, G. A. and Breaker, R. R. (1999) Engineering precision RNA molecular switches. *Proc. Natl. Acad. Sci. USA* **96,** 3584–3589.
9. Peracchi, A., Beigelman, L., Usman, N., and Herschlag, D. (1996) Rescue of abasic hammerhead ribozymes by exogenous addition of specific bases. *Proc. Natl. Acad. Sci. USA* **93,** 11,522–11,527.
10. Peracchi, A., Matulic-Adamic, J., Wang, S., Beigelman, L., and Herschlag, D. (1998) Structure-function relationships in the hammerhead ribozyme probed by base rescue. *RNA* **4,** 1332–1346.
11. Kuwabara, T., Warashina, M., Tanabe, T., Tani, K., Asano, S., and Taira, K. (1998) A novel allosterically trans-activated ribozyme, the maxizyme, with exceptional specificity in vitro and in vivo. *Mol. Cell* **2,** 617–627.
12. Porta, H. and Lizardi, P. M. (1995) An allosteric hammerhead ribozyme. *Bio/ Technology (NY)* **13,** 161–164.
13. Komatsu, Y., Yamashita, S., Kazama, N., Nobuoka, K., and Ohtsuka, E. (2000) Construction of new ribozymes requiring short regulator oligonucleotides as a cofactor. *J. Mol. Biol.* **299,** 1231–1243.
14. Earnshaw, D. J. and Gait, M. J. (1997) Progress toward the structure and therapeutic use of the hairpin ribozyme. *Antisense Nucleic Acid Drug Dev.* **7,** 403–411.
15. Walter, N. G. and Burke, J. M. (1998) The hairpin ribozyme: structure, assembly and catalysis. *Curr. Opin. Chem. Biol.* **2,** 24–30.
16. Fedor, M. J. (2000) Structure and function of the hairpin ribozyme. *J. Mol. Biol.* **297,** 269–291.
17. Feldstein, P. A. and Bruening, G. (1993) Catalytically active geometry in the reversible circularization of 'mini-monomer' RNAs derived from the complementary strand of tobacco ringspot virus satellite RNA. *Nucleic Acids Res.* **21,** 1991–1998.
18. Komatsu, Y., Koizumi, M., Nakamura, H., and Ohtsuka, E. (1994) Loop-Size Variation to Probe a Bent Structure of a Hairpin Ribozyme. *J. Am. Chem. Soc.* **116,** 3692–3696.

19. Rupert, P. B. and Ferre-D'Amare, A. R. (2001) Crystal structure of a hairpin ribozyme-inhibitor complex with implications for catalysis. *Nature* **410,** 780–786.

20. Komatsu, Y., Nobuoka, K., Karino-Abe, N., Matsuda, A., and Ohtsuka, E. (2002) In vitro selection of hairpin ribozymes activated with short oligonucleotides. *Biochemistry* **41,** 9090–9098.

21. Komatsu, Y., Koizumi, M., Sekiguchi, A., and Ohtsuka, E. (1993) Cross-ligation and exchange reactions catalyzed by hairpin ribozymes. *Nucleic Acids Res.* **21,** 185–190.

22. Komatsu, Y., Kanzaki, I., Koizumi, M., and Ohtsuka, E. (1995) Modification of primary structures of hairpin ribozymes for probing active conformations. *J. Mol. Biol.* **252,** 296–304.

23. Ruffner, D. E., Stormo, G. D., and Uhlenbeck, O. C. (1990) Sequence requirements of the hammerhead RNA self-cleavage reaction. *Biochemistry* **29,** 10,695–10,702.

24. Araki, M., Okuno, Y., Hara, Y., and Sugiura, Y. (1998) Allosteric regulation of a ribozyme activity through ligand-induced conformational change. *Nucleic Acids Res.* **26,** 3379–3384.

25. Siwkowski, A., Shippy, R., and Hampel, A. (1997) Analysis of hairpin ribozyme base mutations in loops 2 and 4 and their effects on cis-cleavage in vitro. *Biochemistry* **36,** 3930–3940.

26. Shippy, R., Siwkowski, A., and Hampel, A. (1998) Mutational analysis of loops 1 and 5 of the hairpin ribozyme. *Biochemistry* **37,** 564–570.

# 12

## Tetracycline-Regulated Expression of Hammerhead Ribozymes In Vivo

### Emma T. Bowden and Anna T. Riegel

#### Summary

A major obstacle to achieving constitutive ribozyme expression in cells is that expression or elimination of the target gene may provide either a growth advantage or disadvantage to the cells that express ribozyme. Many approaches have been used to overcome this problem, mostly based on the effort to create conditional or inducible expression systems. In this chapter, we describe the most common choice for overcoming this problem, tetracycline-regulated ribozyme expression. This system consists of two central components: transcriptional transactivators that interact specifically with bacterial *cis*-regulatory elements and antibiotics that can modulate the binding of the transactivators at low and nontoxic doses. Here, we summarize protocols to generate cell lines expressing tetracycline-regulated ribozyme constructs.

**Key Words:** Tetracycline; regulatable; ribozymes.

## 1. Introduction

In the early 1980s, it was discovered that RNA had catalytic activity, and this changed the traditional view of the role of these molecules as exclusively protein-encoding molecules. Catalytic RNA (ribozyme) activity was demonstrated for self-cleaving introns and the RNA component of the tRNA-processing enzyme RNase *(1,2)*. After this finding, studies of the replication of plant viruses and virusoids revealed examples of other classes of ribozymes *(3–5)*. At this point, there are five major RNA based motifs that have been derived from naturally occurring ribozymes. These include hairpin, hammerhead, group I intron, ribonuclease P, and hepatitis delta virus (HDV) ribozymes. From these, the two perhaps most studied and utilized are the hammerhead and hairpin ribozymes, mostly because of their inherent simplicity and relatively small size.

From: *Methods in Molecular Biology, vol. 252: Ribozymes and siRNA Protocols, Second Edition*
Edited by: M. Sioud © Humana Press Inc., Totowa, NJ

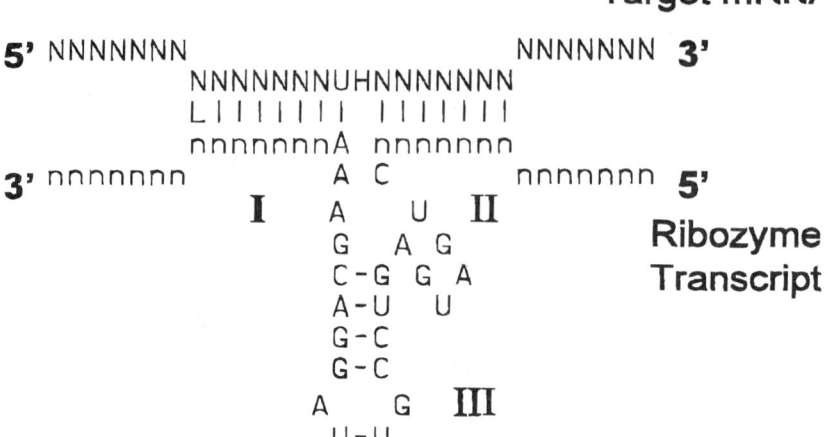

Fig. 1. General structure of a hammerhead ribozyme RNA hybridized to the target RNA. Stems I and III are antisense sequences specific to the target RNA. Stem II is the catalytically active region of the ribozyme.

The most frequently used hammerhead ribozyme is based on that described by Haseloff and Gerlach *(3)*.This has been modified using mutational analysis to identify the minimal essential elements for catalytic activity (**Fig. 1**). These include a highly conserved, 22-nucleotide catalytic domain (stem II), a basepairing sequence flanking the susceptible 3',5'-phosphodiester bond (stems I and III), and a recognition sequence on the target RNA such as GUC. Hammerhead ribozymes can cleave any 5'-NUH-3' triplets of an RNA molecule, in which U is conserved, N is any nucleotide, and H can be a C, U, or A, but not G *(6)*. Comparative studies have extended and refined this rule, and revealed that the reaction rate decreases in the following order: AUC, GUC>GUA, AUA, CUC>AUU, UUC, UUA>GUU, CUA>UUU, CUU *(7,8)*. The minimized, enzymatically active core and the two targeting arms of 4–9 nucleotides result in a ribozyme of 30–40 nucleotides total. The flanking sequences of stems I and III hybridize as antisense sequences with the targeted-sense RNA, according to Watson and Crick basepairing *(6)*. There has been the suggestion that the antisense sequences flanking the ribozyme core could block translation of the targeted RNA *(9)*. However, in our experience, point mutations in our core sequence internal to stem II eliminate enzymatic activity, but have no effect on "antisense" binding.

As they exist in nature, hammerhead ribozymes act in *cis (3,4)*. Thus, they are self-cleaving molecules. It has been demonstrated that they could be made to function in trans—e.g., on exogenous substrates—and thus could be targeted to virtually any RNA *(10)*. This makes them extremely attractive tools for specific gene inactivation, and has given rise to the idea that the mRNA coding for any protein associated with disease can be selectively cleaved by ribozymes. This belief has fueled the study of the catalytic RNAs with the idea that they can be developed as potential therapeutic agents.

The first major problem when considering the use of ribozymes against in vivo targets is that the biophysical parameters governing RNA folding in vivo have not yet been clearly determined. Therefore, there is no clear way to evaluate the accessibility of a potential RNA substrate cleavage site. As a first approximation, the gross topography of the substrate RNA can be simulated using an RNA secondary structure-folding program. This allows at least an approximate idea of whether a target cut site is buried within a structural element of the RNA. However, there are limitations in the size of the RNA that can be evaluated using these models, and it is clear that these predictions are a gross approximation of RNA secondary structure and that they also do not consider either tertiary structure or RNA-protein interactions. Some methods have been described that will allow the selection of target-specific efficient ribozymes from a pool in an in vivo system *(11)*. Briefly, a library of ribozymes containing a constant ribozyme core sequence, but random hybridizing arm sequences, is mixed with the target RNA in a cellular extract. The various ribozymes will produce cleavage products of the target, but the most efficient will produce the most abundant cleavage products. Several of these abundant cleavage products are cloned and sequenced; this provides knowledge about the cleavage sites recognized by the most efficient ribozymes in the library and facilities the design of ribozymes for in vivo uses. However, we frequently perform random selection of ribozyme targets, and in our experience the majority of predicted sites yield satisfactory results. Therefore, from the initial selection of 3–5 ribozymes, we are usually able to find at least one with excellent in vitro and in vivo efficacy *(12–16)*.

The next major problem is that of translating expected in vitro activities of ribozymes to in vivo targets. This encompasses questions about the intra- and extracellular stability of the ribozymes, delivery to target cells, target accessibility, and optimal catalytic activity and specificity of the ribozyme. Although the RNAs have catalytic activity, they must present in sufficient amounts to affect both a significant proportion of the cell population, and within those cells, a significant proportion of the target RNA. Delivery can be achieved by exogenous application of synthetic ribozymes or through a gene-therapy approach relying on the intracellular expression of ribozyme transcripts from plasmid or viral vectors. Each of these approaches has different considerations in regard to delivery, stability, activity, and toxicity. The issues of stability, toxicity, cellular pharmacokinetics, and delivery agents for exogenously added synthetic ribozymes are not further discussed here. However, these questions have been examined extensively elsewhere, particularly in the field related to the delivery of antisense oligonucleotides *(17)*.

We have chosen standard eukaryotic expression plasmids to provide an endogenous ribozyme delivery system. This approach overcomes some of the problems inherent to the exogenous delivery system. For constitutive ribozyme expression, we routinely use the commercially available pRc/CMV plasmid from Invitrogen (San Diego, CA). The CMV promoter in this construct provides high transcription levels in all the cell lines we have examined to date. Furthermore, the BMG polyadenylation (poly [A]) signal facilitates transcriptional termination, the addition of the poly A tail adds stability to the ribozyme, and the T7 and SP6 promoter binding sites that flank the multiple cloning site enable the generation of in vitro runoff transcripts of the ribozyme inserts if needed for in vitro ribozyme assays. Finally, a eukaryotic transcription unit for G418 resistance allows selection of stably transfected cells. However, it is important to consider that a major obstacle to achieving constitutive ribozyme expression in cells is that expression or elimination of the target gene may provide either a growth advantage or disadvantage to the cells that express ribozyme. Several approaches have been used to overcome this problem, mostly based in the effort to create conditional or inducible expression systems. In this chapter, we describe the most common choice for overcoming this problem—tetracycline-regulated ribozyme expression *(18)*.

An inducible system requires three components: a regulatory unit, a responsive element, and an induction agent. Ideally, the system should allow rapid initiation and/or termination of expression with the addition of the appropriate external stimulus. There should be no basal expression of the target gene when expression is not induced, and there should be a high level of expression of the target gene upon induction. Under these conditions, it should be possible to generate variable levels of expression of the target gene while avoiding the long-term toxic effects of ribozyme expression. At this point, the tetracycline-regulated expression system represents the most widely used example of regulated expression constructs. The basis of the original tetracycline-regulated expression system used the tetracycline transactivator protein (tTa), a chimeric protein containing the VP16 activation domain of herpes simplex virus fused to the tet repressor protein of *E. coli* as the regulatory unit. In the absence of tetracycline, tTa exhibits a high binding affinity for the tetracycline resistance operator (tetO), the responsive element. In the presence of tetracycline (the inducer) a conformational change in the repressor domain prevents tTa from binding to tetO. By placing multiple copies of tetO in front of a minimal promoter linked to a target gene, tTa can promote target-gene expression in the absence of tetracycline. Upon addition of tetracycline, the ability of tTa to bind to tetO is inhibited, and expression of the target gene is significantly reduced

(Tet-off). In general, the tTa system will exhibit a high level of induction when at the "on" position and will exhibit tight control when turned "off" by the addition of low levels of tetracycline.

In a variation of the Tet-off system, a reverse transactivator (rtTa) that will bind to tetO only in the presence of tetracycline was developed. Multiple copies of tetO upstream of a minimal promoter and target gene result in target-gene expression by the addition of tetracycline (Tet-on). Removal of tetracycline results in a significant reduction in target gene expression. Therefore, this system works exactly opposite to the tTa. In the absence of the inducing agent tetracycline, tTa does not bind or only binds very weakly to the tetO sequences therefore no target gene is transcribed. The induction of transgene expression by tetracycline is extremely rapid, and can occur within a period of hours.

The choice of the Tet-on vs Tet-off system will depend on the ultimate applications, and perhaps the target, of the ribozyme in the individual system being examined. Both systems have several technical problems that can influence the choice of a particular application. For example, the Tet-off system relies on the constitutive expression of the VP16-tTa regulatory unit, and there has been some concern about cellular toxicity of this protein when highly expressed. Furthermore, if expression of the target gene is to be repressed for any significant period of time (perhaps because of the effects of downregulation of the ribozyme target), continuous culture in the presence of tetracycline can become burdensome, and may produce unwanted phenotypic changes. The major disadvantage of the Tet-on system is that in the absence of tetracycline, the rtTa can bind—albeit weakly—to the tetO, inducing a basal level of expression of the ribozyme (leakiness). However, both systems have proved to be successful, as a review of the literature will demonstrate, and as with many experimental systems, careful preparation and controls can identify potential problems.

We usually use the Tet-off system for ribozyme studies, and this requires the following plasmids, all of which are available from Dr. Hermann Bujard at the University of Heidelberg. Detailed information on this system is available at http://www.zmbh.uni-heidelberg.de/Bujard and from his numerous publications that document the development of this system. The tetracycline transactivator (tTa) plasmid (pUHD15-1) contains the human cytomegalovirus (CMV) promoter that drives expression of tTa. The tetracycline-controlled (TetO) luciferase plasmid (pUHC13-3) has seven copies of the 42-nucleotide-long inverted repeats of TetO localized upstream from the minimal CMV promoter and the luciferase gene of Photinus pyralis fused to the SDV40 small-t intron and the SV40 polyadenylation signal. Finally, the tetracycline-controlled (TetO) modified cloning vector (pUHD10-3) also has seven copies of the 42-nucleotide-long inverted repeats of TetO localized upstream from the minimal

CMV promoter, as well as a multiple cloning site containing unique *Sac*II, *Eco*RI, and *Xba*I sites followed by an SV40 polyadenylation signal.

We usually create the Tet-regulated system in a cell line of interest in two steps. First, we isolate cloned and stable cell lines that express either the tTa or the rtTa. This provides cell lines with an established and defined genetic background, and this allows a direct comparison when isolating stable subclones from the second round of transfection in which the ribozyme regulated by a responsive element is introduced. We advise against the introduction of the regulator and the response element together since the two plasmids frequently integrate at the same site, bringing the enhancer region of the regulatory plasmid into close enough proximity with the minimal promoter of the response element to drive an elevated basal activity of the ribozyme. Once tTa-expressing clones are isolated, we identify expression levels and tetracycline regulatability using an indirect method in which a tTa-sensitive reporter is transiently introduced into the clonal cell lines. Finally, we stably transfect the tetracycline-responsive (tetO) plasmid into clonal-cells in which high levels of tetracycline regulatability have been established.

Once these stable cell lines have been isolated and characterized in terms of their tetracycline regulatability, they can be examined closely for the effects of temporal regulation of the target gene by ribozyme expression. In this regard, our laboratory is particularly interested in models of tumor-cell growth. Tetracycline regulation of tumor cell-related targets allows us to examine phenotypic changes in the behavior of tumor cells in both in vitro and in vivo tumor models. In particular, we find this approach extremely useful for in vivo models of tumor growth in which expression of the ribozyme can be modulated at different experimental phases of tumor development—for example, tumor "take" vs a block or reversal of tumor growth. In fact, it should be noted that we have observed that ribozyme expression in vitro for several factors produced no major changes in proliferation. However, when these cells were implanted as a xenograft in an in vivo model, the target being modulated by the ribozyme was found to have a profound role in tumor growth. This effect could be titrated, and even very small decreases of the levels of the target in this model could result in profound phenotypic responses in terms of tumor development *(9,19,20)*. Therefore, in this chapter, we also include a some details of the considerations for this type of experiment.

## 2. Materials

All materials and chemicals used, even during standard molecular biology procedures, should be considered potentially hazardous. Particular care should be taken when handling and disposing of antibiotic solutions, from an environmental standpoint.

## 2.1. Preparation for Stable Transfections

1. Complete growth media (*see* **Note 1**).
2. Selection media for selection plasmids 1 and 2 (*see* **Note 2**).

## 2.2. Isolation and Characterization of Stable tTa Expressing Cells

1. Serum-free growth media: this is the basal growth media for the cell line of choice minus all additions, including serum and any supplements.
2. Fugene 6 (Roche, Indianapolis, IN).
3. tTa expression plasmid (*see* **Note 3**).
4. Selection plasmid 1.
5. Selection media for selection plasmid 1.

## 2.3. Screen Clonal Cell Lines for Expression of tTa and Tet Regulation

1. Tet-regulated luciferase-reported plasmid.
2. Tetracycline 1 μg/μL (*see* **Note 4**).
3. Phosphate-buffered saline (PBS): 137 m$M$ NaCl, 2.7 m$M$ KCl, 4.3 m$M$ Na$_2$HPO$_4$, 1.4 m$M$ KH$_2$PO$_4$, pH 7.3.
4. Lysis buffer was provided in kit as 5X solution (*see* **Note 5**).

## 2.4. Clone Ribozymes Into the Tetracycline-Controlled (TetO) Modified Cloning Vector

1. 10 μg highly purified oligonucleotides.
2. Annealing buffer: 100 m$M$ KCl, 10 m$M$ KPO$_4$.
3. Hybridization buffer: 6 X standard saline citrate (SSC) (0.9 $M$ NaCl, 0.09 $M$ sodium citrate, pH 7.0), 0.5% sodium dodecyl sulfate (SDS), 5X Denhardt's solution, 100 μg/μL denatured salmon-sperm DNA.

## 2.5. In Vitro Testing for Ribozyme Activity

1. 1 $M$ Tris-HCl, pH 8.5.
2. 400 m$M$ MgCl$_2$.

## 2.6. Transfection, Isolation, and Characterization of Double Transfectants

1. Ribozyme construct in Tet-responsive plasmid.
2. Selection plasmid 2.
3. Selection media for selection plasmid 2.

## 3. Methods

We generally use certain lab manuals for standard molecular biology techniques. We recommend *Current Protocols in Molecular Biology* by Frederick M. Ausubel (Greene & Wiley Interscience) and *Molecular Cloning: A Lab Manual* by T. Maniatis, Joseph Sambrook, and E. F. Fritsch (Cold Spring

Harbor Laboratory). Unless otherwise noted, all of our standard protocols are based on these publications.

## 3.1. Isolation and Characterization of Stable tTa-Expressing Cells

### 3.1.1. Preparation for Stable Transfections

There are a number of steps that are prerequisite to the transfection procedure.

1. Ensure that the cell line to be transfected can be grown as single colonies. For adherent cells, the best way to examine this is to plate ~100 cells in a 10-cm tissue culture-treated dish, and feed with the optimal growth media for that cell line. Media should be changed every 4 d for approx 2 wk before the number of viable colonies is counted (*see* **Note 6**).
2. Determine the selection conditions for the parental cell line. There will be two independent selection steps, and therefore, this should be performed for both of the selection agents to be used. For example, if G418 selection is to be used, the minimum levels of this drug that will kill all the cells in 4–8 d should be determined (*see* **Note 7**).
3. Determine the most efficient method of transfecting the parental cell type (*see* **Subheading 3.1.2.**).

### 3.1.2. Transfection and Isolation of Clonal-Cell Lines Expressing tTa

There are a variety of methods available for successful transfection of mammalian cell lines. These methods have been described in detail elsewhere, including several of the standard laboratory manuals. We frequently use Fugene 6 (Roche), a non-liposomal transfection agent. In our experience, this provides excellent transfection efficiencies in a variety of cell lines, with very low toxicity. As with most of these proprietary preparations, the manufacturer provides extremely detailed protocols for a variety of cell types and conditions. Here, we will briefly describe our usual protocol, which we use for adherent cells.

1. Plate the cells 1 d before the transfection experiment. The appropriate plating density will depend upon the growth rate of the cells being used. Cells should be 50–80% confluent on the day of transfection. We use $1–3 \times 10^5$ cells per 35-mm culture dish.
2. Dilute 3 µL Fugene into 100 µL of serum free medium per 35 mm culture dish. Add 1–2 µg DNA to the pre-diluted Fugene. Mix and incubate for 15 min at room temperature.
3. Add the DNA Fugene mixture to the cells dropwise. Swirl the plate to ensure an even distribution of the complex.
4. Return cells to the appropriate incubator growth conditions for 24 h (*see* **Note 8**).
5. Trypsinize cells and transfer to a 10-cm dish in the presence of selection media.
6. Change selection media every 3–4 d to maintain selection. Depending on the growth rate of the cells, clones should begin to appear after 2 wk.

7. At this point, clonal-cell lines can be isolated by using cloning rings or by a limiting dilution procedure. Briefly, the fastest way to isolate single clones is using cloning rings. In order to do this, as the clones grow, they must be very well-separated from each other. This method cannot be used for every cell line, and it introduces a significant risk of cross-contamination. A more conservative approach has been addressed by limiting dilution cloning so that after drug-resistant cells have been isolated, they are trypsinized and diluted to a final concentration of one cell per 100 μL of selection media and plated into 96-well plates. After 2–6 wk, individual clones can be grown up for testing.

8. Clonal-cell lines should be expanded under selection until there are sufficient cells to screen tTa expression and functionality.

### 3.1.3. Screen Clonal-Cell Lines for Expression of tTa and Tetracycline Regulation (see **Note 9**).

1. Transiently transfect tTa clonal-cell lines with the tetracycline-regulatable luciferase reporter construct, using the protocol described in **Subheading 3.1.1.**
2. 24 h after transfection, aspirate growth medium from cells to be assayed.
3. Rinse cells with PBS. Care must be taken to avoid dislodging attached cells. Aspirate as much of the PBS as possible.
4. Add sufficient lysis buffer, pre-equilibrated to room temperature, to just cover the cells (approx 300 μL for a 35-mm dish).
5. Rock the 35-mm dishes to ensure that the cells are completely covered with the lysis buffer. Scrape the dishes to ensure that the cells are released from the plate. Transfer the cell suspension to a microcentrifuge tube and place on ice.
6. Vortex microcentrifuge tube 10–15 s. Centrifuge at $12,000g$ for 15 s at room temperature. Transfer the supernatant to a new tube.
7. Cell lysates can be stored at $-70°C$ or used immediately to assay luciferase activity.
8. If using a manual luminometer, dispense 100 μL of luciferase assay reagent at room temperature into luminometer tubes. Program the luminometer to give a 2-s delay before a 10-s measurement reading. Add 20 μL of cell lysate to the luminometer tube, and vortex briefly before reading.
9. Choose several clones that exhibit high levels of regulatability of luciferase activity (*see* **Note 10**).

### 3.2. Clone Ribozymes Into the Tetracycline-Controlled (TetO) Modified Cloning Vector

Synthesizing the respective sense and antisense DNA-oligonucleotides generates ribozyme expression cassettes. These cassettes code for binding regions in the target RNA molecule, the catalytic core, and additional sequences to generate appropriate restriction enzyme overhangs and allow easy ligation into the expression plasmid. We usually proceed with this while clonal cells from the first round of transfection are growing up to sufficient numbers to allow screening for tetracycline regulatability. Although these protocols are used to describe the generation of a tetracycline-inducible system, we usually try to use restriction enzyme overhangs that will allow us to clone into the tetracycline-controlled cloning vector (pUHD10-3) as well as a commercial constitutive expression vector such as pCDNA3 (Invitrogen). EcoRI and XbaI sites in the multiple cloning sites of these two vectors allow appropriate directional cloning. The commercial constitutive expression vector is useful because it contains the SP7 and T7 (or equivalent) promoter binding sites flanking the multiple cloning site. This enables us to generate in vitro runoff transcripts of our ribozyme inserts. We have also noted that the oligonucleotides used to generate the ribozyme expression cassette must be generated from high-quality high-performance liquid chromatography (HPLC) or gel-purified oligonucleotides to ensure that the majority are full-length.

1. Mix 10 µg of each oligonucleotide in 100 µL of annealing buffer and heat to 85°C. Shut off the heat block and allow the reaction mixture to cool down to room temperature.

2. Ligate an aliquot of the annealing mix into the pCDNA3 and pUHD10-3 vectors that have been cut with *Eco*RI and *Xba*I and purified, with both procedures using standard molecular biology protocols.

3. Transform the ligation reactions into transformation-competent *E. coli* and after growth on plates under selective conditions, pick and transfer approx 50 single isolated colonies to fresh master plates in duplicate and again under the appropriate selection conditions.

4. Transfer the bacterial colonies from the master plates onto nitrocellulose filters. Denature filters in 0.5 *M* NaOH and 1.5 *M* NaCl solution for 2 min. Neutralize filters for 5 min in 0.5 *M* Tris-HCl, pH 7.5, 1.5 *M* NaCl, and rinse in 5X SSC for 5 min. Dry filters between two sheets of Whatmann paper and fix DNA onto the filters by crosslinking.

5. Prehybridize filters for 2 h in hybridization buffer and then add the hybridization probe (100 µg of single-stranded oligonucleotide used to generate the ribozyme cassette labeled with [$^{32}$P-ATP]-T4-kinase with a minimal activity of $1 \times 10^6$ cpm/mL hybridization buffer). Hybridize at 42°C overnight. The hybridization strength and washing conditions will depend on the length of the oligonucleotide used as a probe. With a probe length of 40–45 nucleotides, we add 25% formamide to the hybridization buffer, and the final wash of the filter is in 0.5X SSC/0.1% SDS at 42°C for 30 min.

6. Positive clones are identified by autoradiography. Pick positive clones from the master plate and set up small cultures for plasmid DNA preparation.

7. Prepare a small amount of plasmid DNA. We frequently use the Quiprep Spin Plasmid Miniprep Kit from Qiagen (Chatsworth, CA). We use these preparations to verify sequence and orientation of the ribozyme cassette inserts by DNA sequencing.

8. Prepare a large-scale batch of plasmid DNA that has been proven by sequencing to contain full-length insert in the correct orientation.

## 3.3. In Vitro Testing for Ribozyme Activity

We have discussed that the efficiency of the ribozymes can be examined prior to in vivo experiments. This is one of the reasons to clone the ribozyme cassette into a constitutive eukaryotic expression system such as the one we have described. However, this procedure requires that the target gene must also be cloned into a similar expression vector system.

1. Substrate and ribozyme transcripts are generated using a standard *in vitro* transcription kit and protocol (Riboprobe® System-T7, Promega, Madison, WI). Substrate RNA is transcribed, "hot" in the presence of $[^{32}P\text{-CTP}]$ at 50 μCi/reaction. The ribozyme-RNA for the cleavage reaction is transcribed "cold" (non-radioactive) (**Note:** we also recommend that the ribozyme RNA should be transcribed in a separate "hot" reaction to generate a positive control for the generation of full-length, non-degraded ribozyme RNA. This reaction will also provide information about the molar ratios of ribozyme RNA to substrate RNA).
2. Remove free nucleotide from the transcripts using Chromaspin-100 columns from Clontech (Palo Alto, CA) to isolate transcript sizes from 80–1000 nucleotides. Measure activity of the transcripts and add 1 $M$ Tris-HCL, pH 8.5 to yeild a final concentration of 50 m$M$. Heat transcripts for 2 min at 85°C and then cool to room temperature (*see* **Note 10**).
3. Add 1 U RNAse inhibitor and $MgCl_2$ (20mM final concentration) to each of the transcripts. Mix the "hot" substrate and cold ribozyme transcripts at two molar ratios (we usually try 1:10 and 1:100). Incubate at 50°C (final reaction volume should be approx 20 μL. Remove 5-μL aliquots at 5, 15, 60, and 120 min. Add 2 μL of loading buffer (for example, the stop solution from a USB-sequencing kit) and store each sample on ice until the end of the reaction. **Note:** there should be approx 50,000 cpm in each aliquot. Heat the samples to 85°C for 2 min and run out the samples on a 6% urea/polyacrylamide gel. Specific cleavage products can be detected by autoradiography.

## 3.4. Transfection, Isolation, and Characterization of Double Transfectants

This procedure follows the protocol described in **Subheading 3.1.1.**, but uses the ribozyme construct in the Tet-responsive plasmid. A second plasmid with a selection marker distinct from that used in the isolation of tTa clones is used. For example, if in the first round of isolating tTa-expressing clones the selection plasmid introduced resistance to G418, in the second round when the Tet-responsive plasmid is introduced, a different antibiotic selection marker must be introduced on the selection plasmid (e.g., zeocin).

### 3.4.1. Transfection and Isolation of Clonal-Cell Lines Expressing Tet-Responsive Ribozyme Expression Construct

1. Transfection can be performed as described in **Subheading 3.1.1.**

2. Clonal-cell lines should be isolated either using cloning rings or by a limiting dilution procedure.

3. Clonal-cell lines should be expanded under selection until there are enough cells to screen tTa expression and functionality, as described in **Subheading 3.1.2.** (**Note 12**). Clonal-cell lines that have maintained a good regulatability of tTa expression can be screened for ribozyme expression (*see* **Subheading 3.4.2.**).

### 3.4.2. Screen Clonal-Cell Lines for Tetracycline Regulation of Ribozyme Expression

1. Plate equal numbers of cells from one clone into 35-mm dishes in the presence and absence of Tet. Set up sufficient dishes for a time-course. Initially, we suggest 6, 24, 48, and 72 h.

2. Ribozyme activity can be measured both directly and indirectly by examining the RNA and protein levels of the ribozyme target gene.

### 3.5. Tetracycline Regulation of Ribozyme Expression in Mouse-Tumor Xenograft Models

As previously mentioned, our laboratory is particularly interested in models of tumor cell growth. Tetracycline regulation of ribozymes against potential tumor-related gene targets is extremely useful for in vivo models of tumor growth in which expression of the ribozyme can be modulated at various experimental phases of tumor development. Although we do not provide an in-depth discussion of all the technical issues surrounding these experiments here, we present our standard lab protocol for this procedure as it is performed with the MCF-7 breast cancer-cell line.

1. Twenty million cells are needed per injection site (*see* **Note 13**).

2. Suspend cells in 0.2 mL of sterile PBS and inject subcutaneously into the flanks of athymic female nude mice.

3. One day before the injection, the mice receive a 0.72-mg 17 β-estradiol pellet (**Note 14**).

4. Mice are fed either a doxycycline-containing diet (Bios-Serv, Frenchtown, NJ) or normal food for the duration of the study, and tumor measurements are taken every 2–3 d (*see* **Note 15**).

## 4. Notes

1. We have assumed a general working knowledge of tissue-culture skills. This includes the basic tenets of maintaining cell lines in culture, and that individual cell lines all have their own optimal growth conditions. For a cell line with which the researcher has little experience, the source or provider of the cell line will usually provide this information. Pertinent information includes growth media, serum requirements, supplements, cell-density restrictions, and approximate growth rates.

2. This refers to the antibiotic that the selection plasmid confers resistance to. The dose of antibiotic required must be determined empirically, as discussed in **Subheading 3.1.1.**

3. All plasmids used as reagents in these procedures must be high-quality preparations, consisting predominantly of supercoiled DNA, and isolated using methods in which extremely low levels of contaminants are present. In particular, some cell lines are extremely sensitive to the presence of bacterial endotoxins. We routinely use Qiagen Maxiprep kits that provide high-quality plasmid DNA and can be used to generate endotoxin-free plasmid preparations if necessary for the cell lines being used for transfections.

4. Instead of tetracycline hydrochloride, some groups—including ours—use the tetracycline analog, doxycycline hydrochloride which functions with tTa as well as rtTa. In fact, it is reported to be the most potent effector for rtTa. As with tetracycline, the doxycycline working solution should be stored at 4°C in the dark and remade every 2 wk. Stock solutions can be stored in aliquots at –20°C for up to 6 mo.

5. We usually use a luciferase assay kit from Promega that provides proprietary lysis buffer as a component. They have a choice of buffers for this purpose, and the choice of buffers depends on the cell type being examined. We usually use the passive lysis buffer (PLB), the use of which allows standardization techniques including protein determination. This allows luciferase activity to be expressed per mg of protein.

6. Selection of stable colonies requires the survival of single cells to grow in isolation during the selection procedure. Conditions that permit this requirement are necessary for the establishment of clonal-cell lines. The easiest way to determine how many colonies are present on the tissue-culture dish is to stain the dish with a 2% methylene blue solution after aspiration of growth media. After removing excess stain following a 2-min incubation, a brief rinse in water will reveal the localization of established colonies.

7. It is critical that cells are subconfluent while under selection. The selection process is usually effective only when the cells are undergoing logarithmic growth. Furthermore, it should be noted that for many of the commonly used antibiotics that the cells can undergo several rounds of division in the presence of effective lethal doses of drug.

8. Unlike many transfection reagents, Fugene does not need to be washed from the cells after transfection. Prolonged exposure does not induce cytotoxicity in most cell lines.

9. We routinely use the luciferase assay system from Promega. This kit provides everything necessary for these assays, and explains in great detail how to set up assays on different scales and for use with different luminometers such as manual, single-tube, and plate-reading types. We also describe an indirect method to examine the levels of tTa in clones. This is based on transient transfection of the clonal-cell lines with a tetracycline-sensitive luciferase reporter plasmid. We have found it difficult to detect the levels of expression of tTa protein in stable cell lines. This indirect method, as described allows both functionality and regulatability to be determined.

10. Multiple studies have shown that after administration in feed and water, antibiotics can be detected in the serum of cattle and calves at concentrations sufficiently high to affect the tetracycline-regulated elements described in this chapter. Therefore, it is crucial to screen batches of serum prior to use when attempting these experiments. We test serum lots using the Hela-cell-derived X1/5 cell line. This line stably expresses the tTa sequence and a tetracycline-responsive element to drive luciferase expression. It is extremely sensitive to tetracycline, and can be used to compare basal activity stimulated by various batches of serum. This cell line is available from the European Collection of Cell Cultures (ECACC). It should be noted that in the presence of 1 µg/µL doxycycline, these cells exhibit levels of luciferase activity below the level of detection, and in the complete absence of antibiotic levels of activity will increase up to five orders of magnitude. A batch and lot of serum demonstrated to show negligible effects should be kept as a control for testing against new lot. When beginning these studies, it is advisable to order multiple test samples of serum of different lot from several companies and to compare these directly using this assay. The "best" serum batch isolated from this test then becomes the control for testing other batches.

11. The in vitro transcription of a full-length non-degraded substrate RNA is critical. Always generate a substrate RNA smaller than 1000 nucleotides in length. To control for nonspecific degradation during the cleavage reaction, incubate the substrate RNA without ribozyme in parallel with the other reactions. Sometimes it may be necessary to optimize the incubation time and temperature, the magnesium concentration, and the molar ratios of substrate to ribozyme.

12. We have found that maintenance of cell lines in tetracycline-containing media can cause these cells to lose the ability to regulate the Tet operon. The mechanism for this process is unknown. Therefore, Tet regulatability of luciferase, as described in **Subheading 3.1.2.**, should be tested frequently to ensure that the regulatability of tTa has been maintained. An essential precaution in order to maintain this phenotype is that immediately after we have characterized our cell lines as exhibiting high levels of regulatability of ribozyme expression, we expand them to make many freeze-downs of cells that can be returned to if this problem arise.

13. The number of cells per site will depend on the model cell line chosen. This is the number of cells we usually use for the breast cancer-cell line MCF-7, in which many of our studies are performed. Plan to have tTa single-transfected cells as well as tTa-ribozyme double-transfected cells for these experiments. The single-transfected cell line will provide a control for any possible effects of tetracycline on tumor growth. The tTa-ribozyme cells will be used in the presence and absence of tetracycline, yeilding a total of three experimental groups.

14. This is a necessary supplement for the growth of this particular breast cancer-cell line.

15. In earlier studies, mice were given tetracycline in their water supply. We have found that levels of the antibiotic are more stable when added to the food supply. We routinely use pellets from Bio-Serv Inc. (Frenchtown, NJ). These are particularly useful because they contain a dye that can be monitored. We have found

that removal of the doxycycline-supplemented feed for 24 h will reverse the tetracycline-induced changes in gene expression. Changes in gene expression can be directly monitored from extracts of tumor tissue by Western blot or Northern analysis.

## Acknowledgments

We would like to thank Dr. Anton Wellstein for critical reading and discussion of this manuscript.

## References

1. Kruger, K., Grabowski, P. J., Zaug, A. J., Sands, J., Gottschling, D. E., and Cech, T. R. (1982) Self-splicing RNA: autoexcision and autocyclization of the ribosomal RNA intervening sequence of Tetrahymena. *Cell* **31,** 147–157.
2. Peebles, C. L., Perlman, P. S., Mecklenburg, K. L., Petrillo, M. L., Tabor, J. H., Jarrell, K. A., and Cheng, H. L. (1986) A self-splicing RNA excises an intron lariat. *Cell* **44,** 213–223.
3. Gerlach, W. L. and Haseloff, J. (988) Simple RNA enzymes with new and highly specific endoribonuclease activities. *Nature* **334,** 585–591.
4. Hampel, A. and Tritz, R. (1989) RNA catalytic properties of the minimum (-)sTRSV sequence. *Biochemistry* **28,** 4929–4933.
5. Branch, A. D. and Robertson, H. D. (1991) Efficient trans cleavage and a common structural motif for the ribozymes of the human hepatitis delta agent. *Proc. Natl. Acad. Sci. USA* **88,** 10,163–10,167.
6. Symons, R. H. (1992) Small catalytic RNAs. *Annu. Rev. Biochem.* **61,** 641–671.
7. Perriman, R., Delves, A., and Gerlach, W. L. (1992) Extended target-site specificity for a hammerhead ribozyme. *Gene* **113,** 157–163.
8. Zoumadakis, M. and Tabler, M. (1995) Comparative analysis of cleavage rates after systematic permutation of the NUX consensus target motif for hammerhead ribozymes. *Nucleic Acids Res.* **23,** 1192–1196.
9. Wellstein, A., Schulte, A. M., Malerczyk, C., Tuveson, A. T., Aigner, A., Czubayko, F., and Riegel, A. T. (1998) Ribozyme targeting of angiogenic molecules. *Antiangiogenic Agents in Cancer Therapy*, Humana Press, Totowa, NJ, pp. 423–441.
10. Uhlenbeck, O. C. (1987) A small catalytic oligoribonucleotide. *Nature* **328,** 596–600.
11. Lieber, A. and Strauss, M. (1995) Selection of efficient cleavage sites in target RNAs by using a ribozyme expression library. *Mol. Cell Biol.* **15,** 540–551.
12. Czubayko, F., Riegel, A. T., and Wellstein, A. (1994) Ribozyme-targeting elucidates a direct role of pleiotrophin in tumor growth. *J. Biol. Chem.* **269,** 21,358–21,363.
13. Juhl, H., Downing, S. G., Wellstein, A., and Czubayko, F. (1997) HER-2/neu is rate-limiting for ovarian cancer growth. Conditional depletion of HER-2/neu by ribozyme targeting. *J. Biol. Chem.* **272,** 29,482–29,486.
14. Czubayko, F., Liaudet-Coopman, E. D., Aigner, A., Tuveson, A. T., Berchem, G. J., and Wellstein, A. (1997) A secreted FGF-binding protein can serve as the angiogenic switch in human cancer. *Nat. Med.* **3,** 1137–1140.

15. Czubayko, F., Downing, S. G., Hsieh, S. S., Goldstein, D. J., Lu, P. Y., Trapnell, B. C., and Wellstein, A. (1997) Adenovirus-mediated transduction of ribozymes abrogates HER-2/neu and pleiotrophin expression and inhibits tumor cell proliferation. *Gene Ther.* **4**, 943–949.

16. Czubayko, F., Schulte, A. M., Berchem, G. J., and Wellstein, A. (1996) Melanoma angiogenesis and metastasis modulated by ribozyme targeting of the secreted growth factor pleiotrophin. *Proc. Natl. Acad. Sci. USA* **93**, 14,753–14,758.

17. Kopper, L. and Kovalszky, I. (1994) Antisense tumor therapy (a dream under construction). *In Vivo* **8**, 781–786.

18. Gossen, M. and Bujard, H. (1992) Tight control of gene expression in mammalian cells by tetracycline- responsive promoters. *Proc. Natl. Acad. Sci. USA* **89**, 5547–5551.

19. Czubayko, F., Liaudet-Coopman, E. D., Aigner, A., Tuveson, A. T., Berchem, G. J., and Wellstein, A. (1997) A secreted FGF-binding protein can serve as the angiogenic switch in human cancer [see comments]. *Nat. Med.* **3**, 1137–1140.

20. List, H. J., Lauritsen, K. J., Reiter, R., Powers, C., Wellstein, A., and Riegel, A. T. (2001) Ribozyme targeting demonstrates that the nuclear receptor coactivator AIB1 is a rate-limiting factor for estrogen-dependent growth of human MCF-7 breast cancer cells. *J. Biol. Chem.* **276**, 23,763–23,768.

# 13

## Ribozyme Expression Systems

### Masayuki Sano and Kazunari Taira

#### Summary

We have succeeded in constructing an effective system for the expression of ribozymes under the control of a human tRNA$^{Val}$ promoter, which ensures a high level of production of ribozymes in vivo. The engineered tRNA$^{Val}$-driven ribozymes, based on computer-predicted secondary structure, were relatively stable and were transported to the cytoplasm, where they could be colocalized with their target RNA. The activity of the exported ribozymes was significantly higher than that of ribozymes that remained in the nucleus. This chapter examines the methods for construction of the appropriate vector that can produce tRNA-driven ribozymes with high-level intracellular activity and for analysis of the constructs in cells.

**Key Words:** Ribozyme; tRNA$^{Val}$ promoter; cloverleaf structure; cytoplasmic localization.

## 1. Introduction

Hammerhead ribozymes are relatively small and versatile catalytic RNA molecules that can cleave their target RNAs with high specificity. The potential of sequence-specific inhibition of gene expression by these ribozymes has been demonstrated (1–3). Despite many successful studies based on sophisticated construction for the expression of ribozymes, the efficacy of ribozymes in vivo is still a matter of trial and error. Potential reasons for the limited ability of ribozymes to function in living cells might be: i) the substrate mRNA is likely to be in a highly folded structure and may also be protected by proteins bound to the substrate; ii) a large excess of ribozyme molecules is usually required to achieve a significant decrease in the level of a target RNA; iii) several cellular RNases cause the degradation of ribozymes; and iv) since the spliced mRNAs are transported rapidly from the nucleus to the cytoplasm, active ribozymes must also be transported to the cytoplasm to

From: *Methods in Molecular Biology, vol. 252: Ribozymes and siRNA Protocols, Second Edition*
Edited by: M. Sioud © Humana Press Inc., Totowa, NJ

ensure colocalization with their targets. Thus, appropriate ribozyme-expression systems require a high level of production of a ribozyme with high stability, as well as cytoplasmic localization of the ribozymes *(4–6)*.

The RNA polymerase III (pol III) system is very attractive for the expression of ribozymes because a high level of transcription can be achieved. The level of expression of ribozyme transcripts produced from a pol III promoter is at least 1–3 orders of magnitude higher than that produced from a pol II promoter. Thus, a tRNA-based system for the expression of ribozymes was established that resulted in both high levels of expression and of intracellular stability *(1–3,6–8)*. In our tRNA-based system, the sequence encoding the ribozyme is ligated downstream of the sequence for a partially modified human tRNA$^{Val}$, with its last seven bases removed, to block 3'-end processing and replaced by a short linker of less than 20 nucleotides. Since the promoter is located within the tRNA sequence being transcribed, part of the tRNA is incorporated into the ribozyme when transcription occurs. Furthermore, a major advantage of our tRNA-based system is the ability to colocalize the expressed ribozyme in the cytoplasm with its target mRNA. The tRNA$^{Val}$-ribozymes— which exhibit a secondary structure that resembles the cloverleaf structure of a natural tRNA, as predicted by a computer folding program—are transported efficiently from the nucleus to the cytoplasm via the action of exportin-t (Xpo-t) *(9)*, a tRNA-binding protein that functions with Ran GTPase *(10,11)*, which regulates transport by catalyzing the hydrolysis of guanosine 5' triphosphate (GTP). We previously demonstrated that the intracellular activities of ribozymes that were transported to the cytoplasm were significantly higher than those of ribozymes that remained in the nucleus *(12)*.

This chapter describes how to design the appropriate ribozyme that exhibits higher activity in cells and how to construct our ribozyme-expression vector. Methods for analysis in cells of constructs are also described.

## 2. Materials

### *2.1. Construction of the Ribozyme-Expression Vector*

1. tRNA$^{Val}$-expression vector: pUC-KE (a plasmid that contains the promoter of a human gene for tRNA$^{Val}$, *Kpn*I and *Eco*RV sites, between the *Eco*RI and *Sal*I sites of pUC 19).
2. Custom-synthesized DNAs:
   a. Template DNA.
   b. Primer A: 5'-CCG GTT CGA AAC CGG GCA C-3'.
   c. Primer B: 5'-AAA AAA AGA TAT CCG GGT ACC T-3'.
   d. M13 primer P7: 5'-CGC CAG GGT TTT CCC AGT CAC GAC-3'.
   e. M13 primer P8: 5'-AGC GGA TAA CAA TTT CAC ACA GGA AAC-3'.
3. 10X *Taq* polymerase buffer: 100 m$M$ Tris-HCl, pH 8.3, 15 m$M$ MgCl$_2$, 500 m$M$ KCl.

4. Deoxynucleotide 5' triphosphate (dNTP) mixture: deoxyadenosine 5' triphosphate (dATP), deoxycytidine 5' triphosphate (dCTP), deoxyguanosine 5' triphosphate (dGTP), and deoxythymidine 5' triphosphate (dTTP), each at 2.5 m$M$.

5. 10X L (low-salt) buffer: 100 m$M$ Tris-HCl,pH 7.5, 100 m$M$ MgCl$_2$, 10 m$M$ dithiothreitol (DTT).

6. *Taq* DNA polymerase (5 U/μL).

7. Restriction enzymes: *Csp* 45I (8 U/μL), *Kpn*I (10 U/μL).

8. 50X TAE buffer: 2 $M$ Tris-acetate, 0.05 $M$ ethylenediaminetetraacetic acid (EDTA), pH 8.0.

9. Ethanol: both 100 and 70% in distilled water.

10. Phenol and chloroform.

11. Agarose.

12. Luria-Bertani (LB) medium: 10 g bacto tryptone, 5 g bacto yeast extract, 10 g NaCl in 1 L distilled water.

13. *E. coli* (JM109, DH-5) or other competent cells.

14. Kit for purification of DNA fragments from agarose gels (e.g., QIAquick Gel Extraction Kit; Qiagen, Hilden, Germany).

15. Kit for ligation of DNA fragments to the vector (e.g., DNA Ligation kit Ver.2; Takara Shuzo Co., Kyoto, Japan).

## *2.2. Confirmation of the Expression of Ribozymes and Their Activities in Cells*

1. Target gene-expression plasmid for luciferase assay.

2. β-galactosidase control plasmid for β-galactosidase assay (e.g., PSV-β-galactosidase control vector, Promega, Madison, WI).

3. Cell lines (e.g., HeLa S3 cell).

4. Cell-growth medium (e.g., Dulbecco's modified Eagle's medium [DMEM], Gibco-BRL, Gaithersburg, MD).

5. Serum-free medium (e.g., OPTI-MEM, Gibco-BRL).

6. Fetal bovine serum (FBS).

7. Antibiotic mixture (e.g., Antibiotic-Antimycotic, Gibco-BRL).

8. Transfection reagents (e.g., TransIT-LT1, Pan Vera, Madison, WI; PolyFect, Qiagen).

9. PBS buffer: 8 g NaCl, 0.2 g KCl, 1.15 g Na$_2$HPO$_4$ (anhydrous), 0.2 g KH$_2$PO$_4$ (anhydrous) in 1 L distilled water.

10. Trypsin-EDTA.

11. Digitonin lysis buffer: 50 m$M$ HEPES/KOH, pH 7.5, 50 m$M$ potassium acetate, 8 m$M$ MgCl$_2$, 2 m$M$ EGTA, and 50 μg/mL digitonin.

12. NP-40 lysis buffer: 20 m$M$ Tris-HCl, pH 7.5, 50 m$M$ KCl, 10 m$M$ NaCl, 1 m$M$ EDTA, and 0.5% NP-40.

13. RNA extraction reagent (e.g., ISOGEN reagent, Wako, Osaka, Japan).

14. Diethylpyrocarbonate (DEPC)-treated distilled water.

15. 20X MOPS buffer: 0.4 $M$ MOPS (pH 7.0), 100 m$M$ sodium acetate, 20 m$M$ EDTA.

16. 12.3 $M$ formaldehyde.

17. Deionized formamide.
18. 200 µg/mL ethidium bromide.
19. 10X formamide gel-loading buffer: 50% glycerol, 10 m$M$ EDTA, 0.25% (w/v) bromophenol blue, 0.25% (w/v) xylene cyanol FF.
20. Hybridization buffer (e.g., ultrasensitive hybridization buffer, Ambion, Austin, TX).
21. 10X T4 polynucleotide kinase (PNK) buffer: 500 m$M$ Tris-HCl, pH 8.0, 100 m$M$ MgCl$_2$, 50 m$M$ DTT.
22. T4 PNK (10 U/µL).
23. Nylon membrane (e.g., Hybond-N nylon membrane, Amersham Co., Buckinghamshire, UK).
24. $\gamma^{32}$P-ATP (10 mCi/mL).
25. Custom-synthesized DNA probe for Northern hybridization.
26. 2X SSPE buffer with 0.1% sodium dodecyl sulfate (SDS): 1.5 $M$ NaCl, 17.3m$M$ NaH$_2$PO$_4$, 2.5 m$M$ EDTA, 0.1% SDS.
27. Cell lysis buffer (e.g., Reporter lysis buffer, Promega, Madison, WI).
28. Kit for luciferase assay (e.g., PicaGene Kit, Tokyo-inki, Tokyo, Japan).

## 3. Methods

### 3.1. Prediction of the Secondary Structure of the Substrate and Selection of the Target Sites

The selection of target sites on the substrate RNA is critically important, and must be carried out with considerable forethought. Although some of experimental approaches seem to give more reliable results *(13)*, RNA-folding programs such as Mfold are conveniently available for the search of ribozyme-accessible sites.

1. Predict secondary structures of the substrate RNA by Mfold program (which can be freely available at Website; http://www.bioinfo.rpi.edu/applications/mfold/old/rna/) (*see* **Note 1**).
2. Search any sequences for an NUH triplex (where N can be A, U, G, or C, and H can be A, U or C) that can be cleaved by hammerhead ribozymes. Among the NUH triplets, GUC is the most efficiently cleaved, and CUC and AUC are cleaved somewhat less efficiently in vitro. Therefore, when target sites are to be chosen, GUC, CUC, or AUC may be preferable.
3. If possible, search the target sequences within 500 bases downstream of a start codon.
4. Identify extended single-stranded or unstructured regions, which may be preferable target sites for ribozymes (**Fig. 1**). In particular, loop regions have proven to be highly efficacious in some cases. In contrast, since stable stem regions are likely to be inaccessible for ribozymes, there are many examples of failures *(8)*.
5. Select about 10 sites for the target of ribozymes according to the previous criteria.

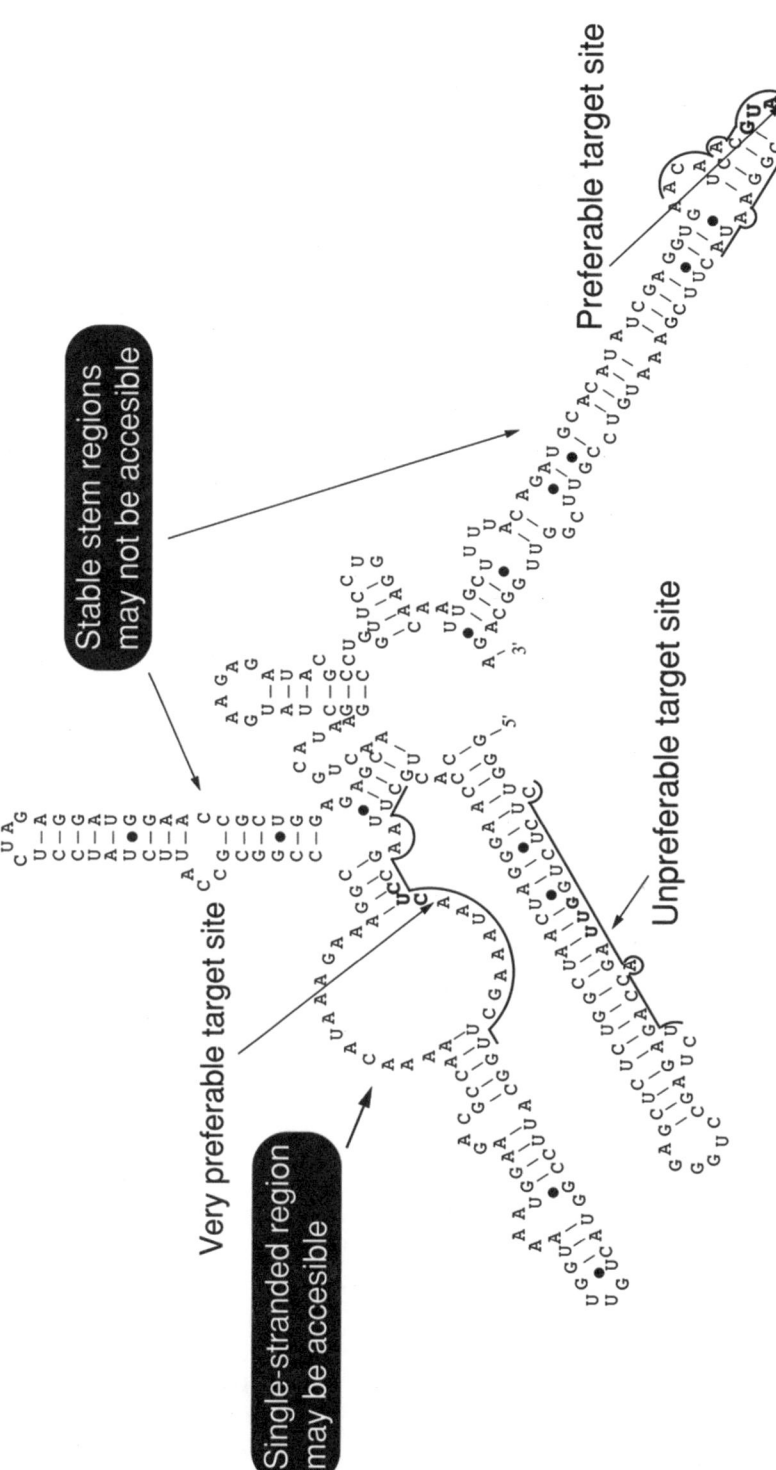

Fig. 1. Selection of suitable target sites to be cleaved by ribozymes. The secondary structure of a mRNA, as predicted by Mfold, is shown. An extended single-stranded or an unstructured region, especially a loop region, is relatively accessible for a ribozyme. In contrast, targets within stable stem structures are inaccessible.

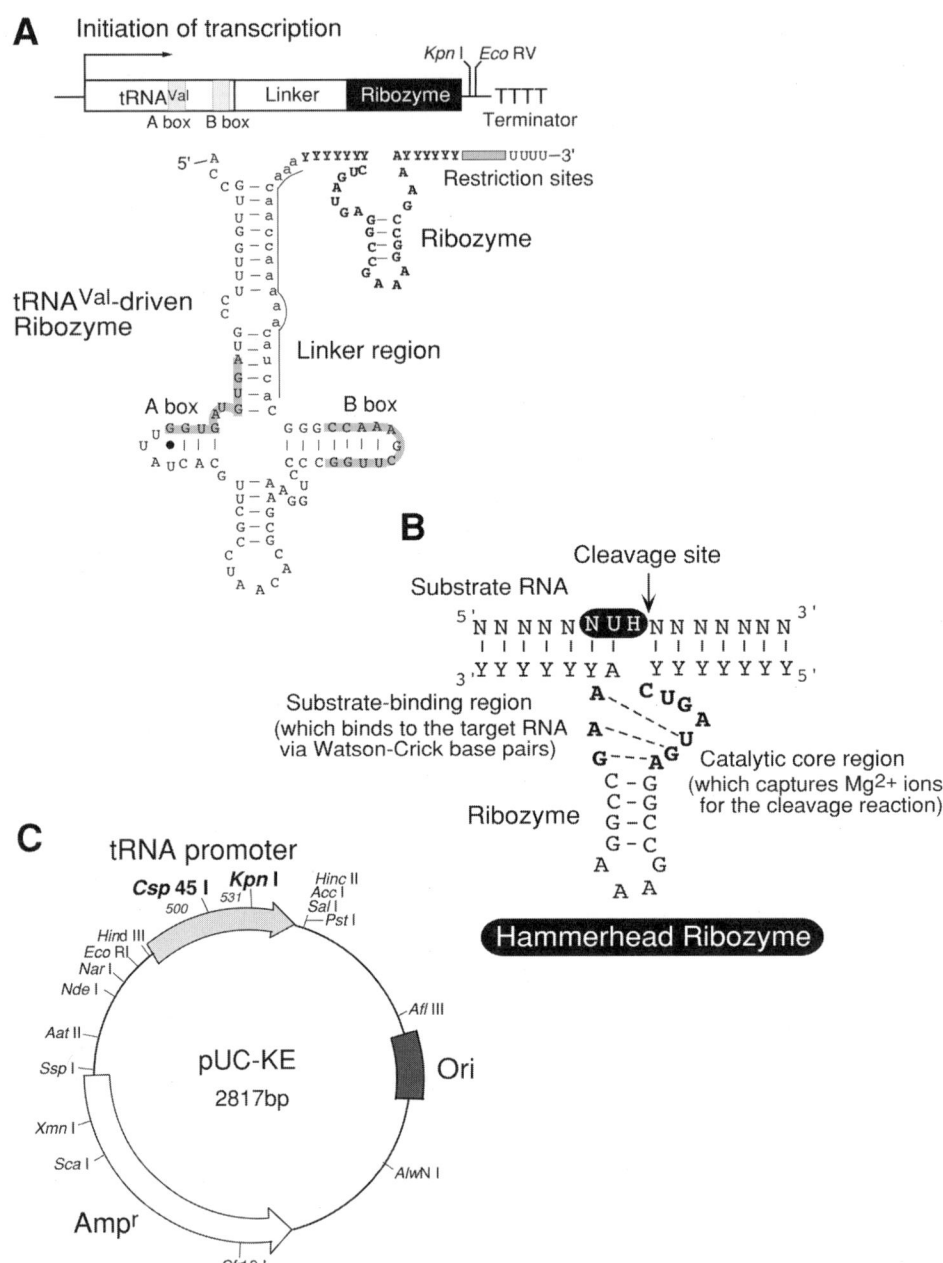

Fig. 2. Design of a hammerhead ribozyme. (**A**) The ribozyme-expression system. The tRNA^Val portion is attached to the ribozyme (indicated by the bold letters) because the promoters of the gene, namely A and B boxes, are internal. (**B**) Secondary

## 3.2. Design of the Ribozyme

In the tRNA$^{Val}$-based system that we use, since the sequence encoding the ribozyme is ligated downstream of the sequence for the human tRNA$^{Val}$ through a linker, it is inevitable that part of the tRNA$^{Val}$ becomes incorporated into the ribozyme (**Fig. 2A**). Therefore, we must ensure that the entire secondary structure of ribozyme transcripts will form the appropriate conformation when we design the functional ribozymes.

1. Design seven nucleotides, complementary to the sequence of the substrate RNA, both upstream and downstream of the chosen target triplets as the substrate-binding sites (**Fig. 2B**).
2. Predict the entire secondary structure of the tRNA$^{Val}$-ribozyme transcript (approx 140 nucleotides) by the Mfold program (**Fig. 2A**). The sequence of the transcript is described as follows: 5'-ACC GUU GGU UUC CGU AGU GUA GUG GUU AUC ACG UUC GCC UAA CAC GCG AAA GGU CCC CGG UUC GAA ACC GGG CAC UAC AAA AAC CAA CAA **AYY YYY YYC UGA UGA GGC CGA AAG GCC GAA AYY YYY Y**AG GUA CCC CGG AUA UCU UUU UUU-3'
   Bold letters correspond to the sequence of the ribozyme, and polyY corresponds to sequences complementary to the target site—namely, the binding site of the ribozyme that is adjacent to the catalytic domain. PolyY regions should be replaced by the chosen sequence of interest.
3. Confirm that the predicted secondary structure maintains a cloverleaf structure and that the predicted substrate-binding sites of the ribozyme are free and not embedded in any stem-structures (see **Note 2**).
4. Choose about five ribozymes according to the previous criteria.

## 3.3. Construction of the Ribozyme-Expression Vector

### 3.3.1. Preparation of the Insert DNA

1. To insert the sequence of the designed ribozyme into the tRNA$^{Val}$-expression vector, pUC-KE (**Fig. 2C**), prepare a synthetic oligonucleotide as described here:
   5'-CCG G<u>TT CGA A</u>AC CGG GCA CTA CAA AAA CCA ACA
       *Csp*45I
   **AAY YYY YYY CTG ATG AGG CCG AAA GGC CGA AAY YYY YY**A <u>GGT ACC </u>CCG GAT ATC TTT TTT T-3'
           *Kpn*I
   Bold letters correspond to the sequence of the ribozyme, and polyY corresponds

---

Fig. 2. *(continued)* structure of a *trans*-acting hammerhead ribozyme. The polyY corresponds to the binding site of the ribozyme. Bold letters correspond to the catalytic core region. (**C**) A plasmid for cloning of a ribozyme. The promoter of a human gene for tRNA$^{Val}$ is inserted into pUC19 plasmid. The restriction enzyme sites, *Csp*45I and *Kpn*I, are used for cloning of a ribozyme.

to the binding site of the ribozyme. The underlined letters are the sequences that correspond to restriction sites.

2. For PCR, combine the following reagents:
   a. 1 μL (0.05 μmol of template DNA) solution of template DNA.
   b. 1 μL 100 μM primer A.
   c. 1 μL 100 μM primer B.
   d. 10 μL 10X *Taq* polymerase buffer.
   e. 10 μL dNTP mix.
   f. 0.5 μL *Taq* polymerase (5 U/μL).
   g. 76.5 μL distilled water.
3. Place the reaction mixture in a thermal cycler. Program the cycler to execute 25 cycles with the following program: 94°C, 30 s; 55°C, 30 s; 72°C, 1 min. After amplification, check an aliquot for the polymerase chain reaction (PCR) products by electrophoresis on 2.0% agarose minigel.
4. Purify the fragment by phenol-chloroform extraction and ethanol-precipitation. Resuspend the dried pellet in a proper volume of distilled water.
5. To digest PCR fragments, adjust the volume to 100 μL while adding 1 μg of the fragments, 10 μL of 10X L buffer, 2 U of *Csp*45I, and 2 U of *Kpn*I, and incubate for 2 h at 37°C.
6. Extract the digested fragments with phenol and chloroform, and concentrate them in a proper volume of TE buffer after ethanol-precipitation.

### 3.3.2. Preparation of a Vector

1. Digest 2 μg of the vector plasmid "pUC-KE" with *Csp*45I and *Kpn*I. In a microcentrifuge tube, digest the plasmid with each 5 U of *Csp*45I and *Kpn*I in 1X L buffer for 2 h at 37°C in a volume of 100 μL.
2. After digestion, load the reaction mixture onto a 1.0% agarose gel and electrophorese the sample in 1X TAE buffer.
3. Excise gel pieces that contain the fragment of the vector and purify them using a column (e.g., QIAquick Gel Extraction kit; Qiagen).
4. Mix 0.03 pmol of the digested vector and about 0.1–0.3 pmol of the digested insert DNA in a final volume of 5 μL. Add 5 μL of solution I of DNA Ligation kit Ver. 2 to the reaction mixture and incubate for at least 30 min at 16°C (*see* **Note 3**).
5. Transform *E. coli* host cells with the ligation mixture and plate the cells on LB agar plates that contain ampicillin at 100 μg/mL.
6. After colonies have become apparent (12–16 h after plating), confirm the positive clones that have ribozyme-containing vector using PCR. Pick up a colony and suspend it in distilled water. Adjust the final volume of 10 μL while adding 1 μL of 10X *Taq* buffer, 1 μL of dNTP mix, each 1 μL of 10 μM M13 primers P7 and P8, and 1 U of *Taq* polymerase. Perform PCR with the following program: 94°C for 1 min, then 25 cycles of 94°C for 30 s, 55°C for 30 s, and 72°C for 1 min.
7. Check the positive clones by electrophoresis on 2.0% agarose minigel. As a control, a negative clone must be also checked.
8. Perform standard plasmid minipreps from approx 5 colonies.

9. Determine the nucleotide sequence of the positive clones to confirm the nature of the construct.

## 3.4. Confirmation of the Expression of Ribozymes and Their Activities in Cells

To inhibit the expression of the gene of interest by ribozymes, the constructed expression vector must be transfected into cells. Two different approaches are commonly used to introduce the vector into cells—namely, transient transfection and stable transfection. In general, depending on the purpose, one approach should be chosen. Since our transiently transfected tRNA$^{Val}$-driven ribozymes are often very active in cells, we will describe the method for transient transfection that we use (*see* **Note 4**), and for fractionation of the cells. After fractionation, Northern blotting analysis reveals the level of expression of the transcripts as well as their intracellular localization.

### 3.4.1. Transient Transfection With the Vector to HeLa Cells and Isolation of the Nuclear and Cytoplasmic RNAs

1. The day before transfection, seed cells in 10-cm dish at density of $1 \times 10^7$ cells per 10 mL of cell-growth medium. Cells were grown to ~80% confluence on the day of transfection.
2. In a 1.5-mL microcentifuge tube, add 40 μL of TransIT regent into 1 mL of serum-free medium (e.g., OPTI-MEM) and mix thoroughly by vortexing. Incubate at room temperature for 5–20 min.
3. Add 20 μg of the plasmid to the diluted TransIT reagent and mix by gentle pipetting. Incubate at room temperature for 5–20 min.
4. While complex formation takes place, wash cells with phosphate-buffered saline (PBS) and replace with 9.0 mL of fresh cell-growth medium.
5. Apply the complexes to cells and incubate for 24–36 h.
6. Harvest trypsinized cells and wash twice with PBS.
7. To prepare the cytoplasmic fraction, resuspend collected cells in 200 μL of digitonin lysis buffer on ice for 10 min.
8. Centrifuge the lysate at 1000g for 5 min and collect the supernatant as the cytoplasmic fraction.
9. Resuspend the pellets in 100 μL of NP-40 lysis buffer and incubate on ice for 10 min.
10. Centrifuge the lysate at 1000g for 5 min and remove the supernatant.
11. Resuspend the pellets in 100 μL of NP-40 lysis buffer and then collect the resultant lysate as the nuclear fraction.
12. Add 1 mL of ISOGEN reagent and 200 μL of chloroform to each lysate, and mix vigorously for 15 s. Centrifuge the mixture at 12,000g for 15 min at 4°C (*see* **Note 5**).
13. Carefully collect the aqueous phase and add 500 μL of isopropanol. Centrifuge the mixture at 12,000g for 15 min at 4°C. Rinse the pellet with 1 mL of 70% ethanol.

14. Remove the supernatant and resuspend the dried pellet in 30 μL of diethylpyrocarbonate (DEPC)-treated water. Measure the concentration of RNA with a spectrophotometer.

## 3.4.2. Confirmation of Steady-State Levels of Expression and Intracellular Localization of tRNA$^{Val}$-Driven Ribozymes by Northern Blotting Analysis

1. Set up the denaturing reaction mixture:
   a. 5.5 μL of RNA (up to 20 μg) solution
   b. 1.0 μL of 20X MOPS electrophoresis buffer
   c. 3.5 μL of 12.3 *M* formaldehyde
   d. 10 μL of formamide
   e. 1.0 μL of ethidium bromide (200 μg/mL)
2. Incubate the RNA solutions for 15 min at 65°C. Chill the samples for 5 min on ice water.
3. Add 2 μL of 10X formamide gel-loading buffer to each sample.
4. Load the samples onto a 2.0% agarose gel containing 2.2 *M* of formaldehyde and electrophorese the sample in 1X MOPS electrophoresis buffer.
5. To check the quality of preparations of RNA, visualize the RNAs by placing the gel on an ultraviolet (UV) transilluminator.
6. Transfer the RNAs to a nylon membrane (e.g., Hybond-N nylon membrane, Amersham Co.) for about 12–16 h. After transferring, irradiate the membrane by UV light to crosslink.
7. Incubate the membrane for 1 h at 42°C in a proper volume of hybridization buffer (e.g., ultrasensitive hybridization buffer, Ambion; *see* **Note 6**).
8. While prehybridizing the membrane, radiolabel the synthetic oligonucleotide complementary to the sequence of the ribozyme in a 100-μL reaction mixture that includes 10 μL of 10X T4 PNK buffer, 5 μL of γ$^{32}$P-ATP and 2 μL (10 U) of T4 PNK for 30 min at 37°C.
9. Denature the $^{32}$P-labeled probe by heating for 5 min at 95°C. Chill the probe rapidly on ice water.
10. Add the denatured radiolabeled probe directly to the prehybridization solution. Continue incubation for 12–16 h at 42°C.
11. Discard the hybridization buffer and wash the membrane twice for 10 min in 2X SSPE with 0.1% SDS at 42°C.
12. Dry the membrane on blotting paper and detect an image of the membrane by scanning in a phosphorimager (**Fig. 3B**).

## 3.4.3. Measurement of Luciferase Activity

The luciferase assay is more sensitive than other commonly used reporter assays. In studies to evaluate the intracellular activity of ribozyme transcripts, a plasmid that encodes an appropriate ribozyme unit(s), and a plasmid that encodes the target sequence fused with a gene for luciferase are cotransfected into cells. After transient expression of the ribozyme and its substrate fused

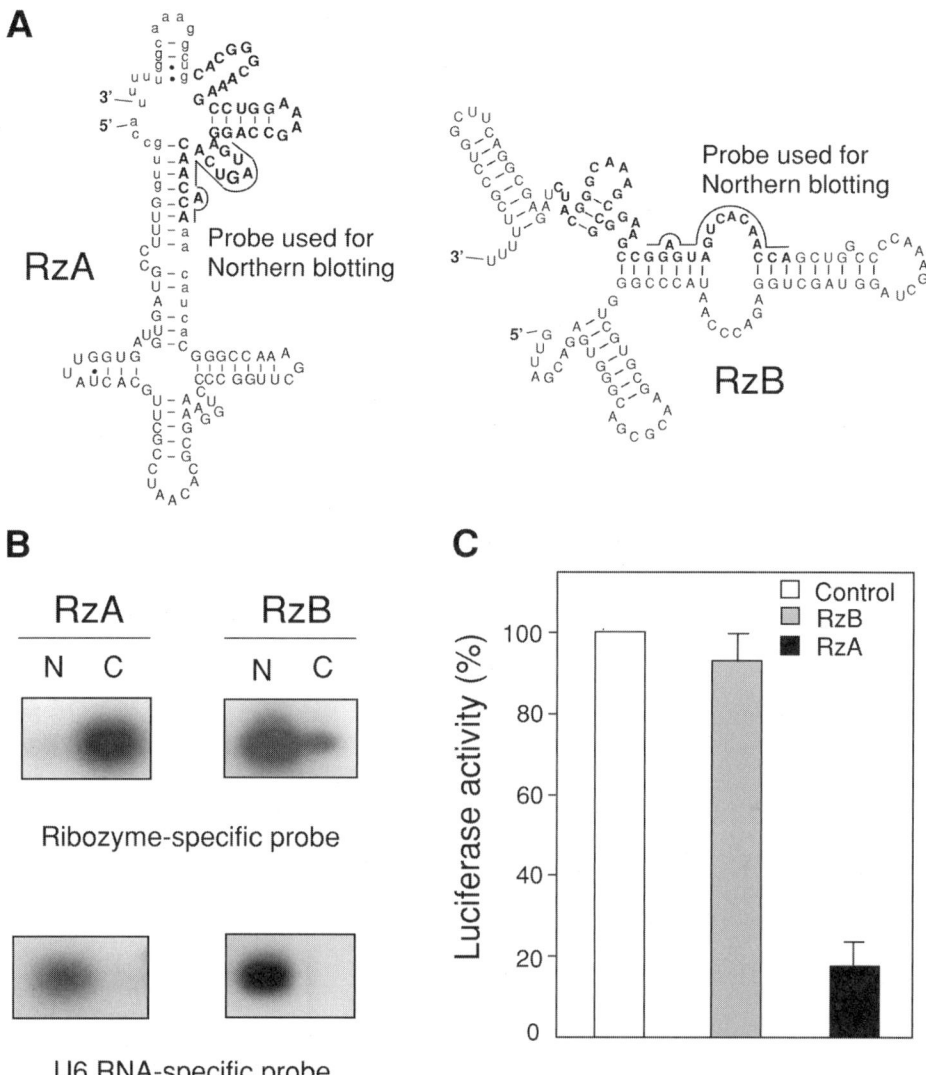

Fig. 3. Relationship between the activities of ribozymes and their intracellular localization. (A) Secondary structures of tRNA-driven ribozymes. RzA is produced by tRNA$^{Val}$-promoter. An appropriate design makes tRNA$^{Val}$-ribozyme to show a cloverleaf structure at the portion of tRNA$^{Val}$. Rz-B is produced by tRNA$_i^{Met}$-promoter. (B) Northern blotting analysis performed with the RNA from the nuclear (indicated as N) and cytoplasmic (indicated as C) fractions 24 h after transfection. As a control, to demonstrate successful separation, intracellular U6 snRNA, which is known to remain in the nucleus, was also analyzed. (C) Activities of the RzA and RzB in cells at 24 h after transfection, which were estimated by luciferase assay.

with the luciferase, the activity of the ribozyme is estimated by measuring luciferase activity in individual cell lysates. Here, we describe the method for transfection by PolyFect reagent (Qiagen).

1. The day before transfection, seed cells in 12-well plates at density of $4 \times 10^5$ cells per 1.0 mL of cell-growth medium. Cells are grown to approx 80% confluence on the day of transfection.
2. Dilute 1–3 μg of the ribozyme-expression vector, 0.1–0.5 μg of target gene-expressing plasmid, which encodes the chimeric luciferase gene, and 0.1 μg of the β-galactosidase control vector with serum-free medium to a total volume of 100 μL.
3. Add 10–25 μL of PolyFect reagent to the mixture. Mix by gentle pipeting and incubate at room temperature for 5–10 min.
4. While complex formation takes place, wash cells with PBS and replace by 600 μL of fresh cell-growth medium.
5. Add 650 μL of cell-growth medium to the reaction mixture. Mix by gentle pipeting and transfer the total volume to the cells.
6. Incubate cells for 24–48 h at 37°C.
7. After incubation, wash the cells twice with PBS.
8. Add 150 μL of cell lysis buffer and incubate for 20 min at room temperature. Transfer the cell lysate to a 1.5-mL microcentrifuge tube.
9. Centrifuge the cell lysate at maximum speed for 1 min at 4°C. Carefully transfer the supernatant to a fresh 1.5-mL microcentrifuge tube.
10. Add 20 μL of the supernatant to 100 μL of luciferin solution in the luminometer tube and immediately measure the light output for a period of 10–15 s at room temperature. An example of the result is shown in **Fig. 3C.**
    To normalize the efficiency of transfection by reference to β-galactosidase activity, the chemiluminescence signal resulting from the β-galactosidase should be determined with β-galactosidase assay kits such as a luminescent β-galactosidase Genetic Reporter System (Clontech, Palo Alto, CA).

## 4. Notes

1. Other programs are also freely available, such as RNA structure (http://rna.chem.rochester.edu/RNAstructure.html).
2. If the predicted secondary structure of the transcripts does not form a cloverleaf structures, it is likely that those transcripts will accumulate in the nucleus without being transported into the cytoplasm (**Fig. 3A,B**).
3. The ligation reaction is completed within 30 min using DNA Ligation kit Ver.2, according to the manufacturer's protocol.
4. We often use the TransIT-LT1 reagent (Pan Vera, Madison, WI) as the transfection reagent. However, depending on cell type, an appropriate transfection reagent must be chosen for high efficiency of the transfection according to the manufacturer's protocol.
5. In general, the RNA is extracted from cells by the guanidinium thiocyanate phenol-chloroform method.

6. The ultrasensitive hybridization buffer can be used either in overnight hybridization, or in 2 h hybridization. A more effective hybridization increases signal 20- to 50-fold over traditional hybridization buffers, according to the manufacturer's protocol.

## References

1. Kawasaki, H., Eckner, R., Yao, T.-P., Taira, K., Chiu, R., Livingston, D. M., and Yokoyama, K. K. (1998) Distinct roles of the co-activators p300 and CBP in retinoic-acid-induced F9-cell differentiation. *Nature* **393,** 284–289.
2. Kuwabara, T., Warashina, M., Tanabe, T., Tani, K., Asano, S., and Taira, K. (1998) A novel allosterically *trans*-activated ribozyme, the maxizyme, with exceptional specificity *in vitro* and *in vivo*. *Mol. Cell* **2,** 617–627.
3. Tanabe, T., Kuwabara, T., Warashina, M., Tani, K., Taira K., and Asano, S. (2000) Oncogene inactivation in a mouse model. *Nature* **406,** 473–474.
4. Bertrand, E., Castanotto, D., Zhou, C., Carbonnelle, C., Lee, G. P., Chatterjee, S., et al. (1997) The expression cassette determines the functional activity of ribozymes in mammalian cells by controlling their intracellular localization. *RNA* **3,** 75–88.
5. Good, P. D., Krikos, A. J., Li, S. X., Lee, N. S., Gilver, L., Ellington, A., et al. (1997) Expression of small, therapeutic RNAs in human cell nuclei. *Gene. Ther.* **4,** 45–54.
6. Koseki, S., Tanabe, T., Tani, K., Asano, S., Shioda, T., Nagai, Y., et al. (1999) Factors governing the activity *in vivo* of ribozymes transcribed by RNA polymerase III. *J. Virol.* **73,** 1868–1877.
7. Kato, Y., Kuwabara, T., Warashina, M., Toda, H., and Taira, K. (2001) Relationships between the activities *in vitro* and *in vivo* of various kinds of ribozyme and their intracellular localization in mammalian cells. *J. Biol. Chem.* **276,** 15,378–15,385.
8. Warashina, M., Kuwabara, T., Kato, Y., Sano, M., and Taira, K. (2001) RNA-protein hybrid ribozymes that efficiently cleave any mRNA independently of the structure of the target RNA. *Proc. Natl. Acad. Sci. USA* **98,** 5572–5577.
9. Kuwabara, T., Warashina, M., Sano, M., Tang, H., Wong-Staal, F., Munekata, E. and Taira, K. (2001) Recognition of engineered tRNAs with an extended 3' end by Exportin-t (Xpo-t) and transport of tRNA-attached ribozymes to the cytoplasm in somatic cells. *Biomacromolecules* **2,** 1229–1242.
10. Arts, G.-J., Fornerod, M., and Mattaj, I. W. (1998) Identification of a nuclear export receptor for tRNA. *Curr. Biol.* **8,** 305–314.
11. Kutay, U., Lipowsky, G., Izaurralde, E., Bischoff, F. R., Schwarzmaier, P., Hartmann, E., and Görlich, D. (1998) Identification of a tRNA-specific nuclear export receptor. *Mol. Cell* **1,** 359–369.
12. Kuwabara, T., Warashina, M., Koseki, S., Sano, M., Ohkawa, J., Nakayama, K., and Taira, K. (2001) Significantly higher activity of a cytoplasmic hammerhead ribozyme than a corresponding nuclear counterpart: engineered tRNAs with an extended 3' end can be exported efficiently and specifically to the cytoplasm in mammalian cells. *Nucleic Acids Res.* **29,** 2780–2788.
13. Eckstein, F. (1998) Searching for the ideal partner. *Nature Biotechnol.* **16,** 24.

# 14

## Design and Expression of Chimeric U1/Ribozyme Transgenes

Roger Abounader, Robert Montgomery, Harry Dietz, and John Laterra

### Summary

The U1snRNA/ribozyme/antisense construct (designated U1/ribozyme) is a chimeric transgene that has proven to be very useful for inhibiting the expression of targeted genes in vitro and in vivo. It consists of a combination of hammerhead ribozyme flanked by target-specific antisense sequences that are in turn flanked by the loops and promoter of U1 snRNA. To construct U1/ribozymes, antisense/ribozyme sequences are first designed corresponding to ribozyme-cleavage consensus "GUC" sequences that are present in the targeted mRNA. Antisense/ribozyme sequences are then inserted between the U1 snRNA loops, and the conceptual secondary structure of the encoded regulatory RNA is analyzed to ensure proper folding. Appropriate antisense/ribozymes are then synthesized as oligonucleotides, annealed, and ligated into the pU1 vector containing the U1 snRNA promoter and loops to yield the pU1/ribozyme expression vector. Constructs can then be transiently or stably expressed in vitro and in vivo to inhibit the expression of target genes. U1/ribozymes can also be expressed in viral vectors for more efficient transfection, or complexed to liposomes for systemic delivery.

**Key Words:** U1/ribozyme; U1 snRNA; hammerhead ribozyme; antisense; knock-down; inhibition of gene expression; stable transfection; transient transfection; liposomes; adenovirus.

## 1. Introduction

The U1snRNA/ribozyme/antisense construct (designated U1/ribozyme) is a recently developed chimeric transgene that has proven to be very useful for inhibiting the expression of targeted genes in vitro and in vivo *(1–4)*. The construct is designed to incorporate many features that enhance the inhibitory effects of naturally occuring antisense cRNAs *(4)*. It consists of the combination of a hammerhead ribozyme flanked by target-specific antisense sequences

From: *Methods in Molecular Biology, vol. 252: Ribozymes and siRNA Protocols, Second Edition*
Edited by: M. Sioud © Humana Press Inc., Totowa, NJ

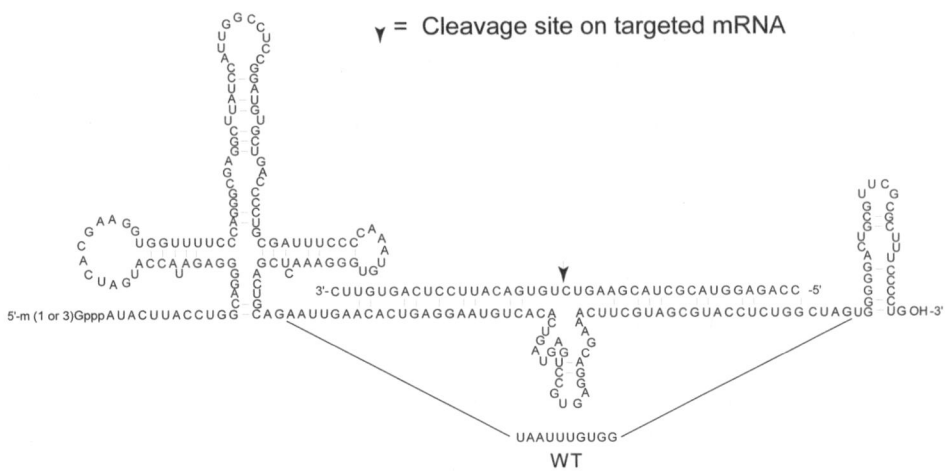

Fig. 1. Final U1/ribozyme structure and sequence. Example that was successfully used to inhibit scatter-factor gene expression in cells after both transient and stable expression in vitro and in tumor xenografts after in vivo transgene delivery.

that are in turn flanked by the loops and promoter of U1 snRNA (**Fig. 1**). The 22-nucleotide autocatalytic hammerhead ribozyme cleaves the targeted message at the consensus sequence 5'-GUC-3'. The two (~20 nucleotides each) antisense sequences that flank the ribozyme are complementary to regions of the targeted mRNA immediately 5' and 3' to the GUC cleavage site. The hammerhead ribozyme and antisense sequences are flanked by the U1 snRNA stem loops, and expression is driven by the U1 promoter (**Fig. 1**). The whole construct is subcloned into a mammalian expression vector that contains an antibiotic-resistance selectable marker for stable expression in cells or tissues. In the presence of flanking antisense sequences, the hammerhead ribozyme can cleave any RNA that contains the triplet GUC sequence. The antisense sequences provide the specificity for targeting the mRNA of interest. The U1 snRNA backbone, which constitutes the novel part of the construct, is believed to significantly facilitate and increase the efficiency of the inhibitory effects of the construct. snRNAs are essential components of the spliceosome complex, and are abundant and stable in the nucleus of mammalian cells (*5*). They are characterized by stable stem-loops, and the 3' loop generally has a high GC nucleotide content. These structures enhance the stability of the targeting molecules by conferring resistance to the activity of exonucleases (*6*). U1 snRNA has other favorable attributes, including a potent and constitutively active promoter, the ability of the unusual trimethylguanosine 5' cap to signal transport back into the nucleus where ribozyme action may be most effective, and the lack of polyadenylation, a factor that may influence transcript trafficking and localization *(7–9)*. Moreover, unlike other spliceosome components, U1 snRNA is widely dispersed in the nucleoplasm *(10)*. Expression of chimeric U1/ribozyme cRNAs can achieve potent and specific cleavage of targeted mammalian mRNA. Plasmids that express the construct can be used for either transient or stable transfection of cultured cells to inhibit the expression of genes of interest. U1/ribozymes have also been successfully expressed in adenoviruses and complexed to liposomes for localized or systemic inhibition of in vivo gene expression *(1)*. This chapter focuses on the design and construction of U1/ribozymes, as well as their stable expression in cells for the inhibition of gene expression.

## 2. Materials

### 2.1. Design of U1/Ribozymes

1. Access to GenBank (www.ncbi.nlm.nih.gov) for sequence retrieval and blast comparisons.
2. Nucleic acid analysis software for sequence manipulations, such as finding of antiparallel, complementary sequences (e.g., DNA strider, OMIGA, or other).
3. Software for analysis of RNA secondary structures (e.g., MulFold, available on the web for free download at [www.cgal.icnet.uk/macsoft.html]).
4. Software for visualization of RNA secondary structure (e.g., LoopDloop, available on the web for free download at [www.cgal.icnet.uk/macsoft.html]).

### 2.2. Construction of U1/Ribozyme Plasmid Expression Vector

1. Oligonucleotide synthesis facility.
2. DNA-sequencing facility.
3. pU1 vector (containing U1 snRNA promoter and loop sequences with appropriate restriction sites for subcloning of ribozyme/antisense oligos [**Fig. 2**]) (available upon request from corresponding author) (*see* **Note 1**).
4. Restriction enzymes *Eco*RI, *Spe*I, and *Bam*HI and corresponding buffers.
5. dH$_2$O.
6. Agarose powder (preferably CTG type).
7. 5X TBE buffer: 54 g Tris-base, 27.5 g boric acid, and 20 mL of 0.5 *M* ethylenediaminetetraacetic acid (EDTA), pH 8.0, in 1 L of dH$_2$O.
8. Ethidium bromide (0.5 μg/mL TBE) (light sensitive).
9. DNA-loading buffer.
10. Gel extraction kit (e.g., QIAEX II from Qiagen, Valencia, CA).
11. 100% ethanol.
12. Sodium acetate (3 *M*).
13. T4 DNA ligase.
14. 10 X ligase buffer: 200 m*M* Tris-HCl, pH 7.6, 50 m*M* MgCl$_2$, 50 m*M* dithiotheritol (DTT), 0.5 mg/mL bovine serum albumin (BSA), 1 m*M* adenosine 5' triphosphate (ATP).
15. Competent *E. coli* (e.g., DH5α).

# A
## U1snRNA Sequence

<u>GGATCC</u>*GCCAACCGAAAGTTGCTCCTTAACACAGGCTAAGGACCAGCTTCTTTGG*
*GAGAGAACAGACGCAGGGGCGGGAGGGAAAAAGGGAGAGGCAGACGTCACTTC*
*CCCTTGGCGGCTCTGGCAGCAGATTGGTCGGTTGAGTGGCAGAAAGGCAGACG*
*GGGACTGGGCAAGGCACTGTCGGTGACATCACGGACAGGGCGACTTCTATGTAG*
*ATGAGGCAGCGCAGAGGCTGCTCGTTCGCCACTTGCTGCTTCGCCACGAAGGAG*
*TTCCCGTGCCCTGGGAGCGGGTTCAGGACCGCGGATCGGAAGTGAGAATCCCA*
*GCTGTGTGTCAGGGCTGGAAAGGGCTCGGGAGTGCGCGGGGCAAGTGACCGTG*
*TGTGTAAAGAGTGAGGCGTATGAGGCTGTGTCGGGGCAGAGGCCCAAGATCT*GA
TACTTACCTGGCAGGGGAGATACCATGATCACGAAGGTGGTTTTCCCAGGGC
GAGGCTTATCCATTGCACTCCGGATGTGCTGACCCCTGCGATTTCCCCAAATG
TGGGAAACTCGACTGCAG<u>AATT</u>CGTA<u>CTAG</u>TGGGGGACTGCGTTCGCGCTTT
CCCCTGA*CTTTCTGGAGTTTCAAAAACAGACCGTACGCCAAGGGTCATGTCTTTTT*
*TCGTATTGGTTTGTGTCTTAGTTGTTAATCCTACAGT*<u>GGATCC</u>

# B
## Hammerhead Ribozyme sequence

cugaugaguccgugaggacgaa

Fig. 2. Sequences of U1 snRNA and hammerhead ribozyme. (**A**) 682-nucleotide sequence of U1 snRNA cassette that was inserted into the *Bam*HI site (double-underlined) of the modified pU1 plasmid. The non-italicized sequences represent the loops of U1 snRNA. The underlined sequences represent the added *Eco*RI and *Spe*I restriction sites for cloning oligonucleotides containing antisense and ribozyme sequences. The italicized sequences contain the U1 promoter of U1 snRNA. (**B**) RNA sequence of the hammerhead ribozyme.

16. SOC medium: 20 g tryptone, 5 g yeast extract, 0.5 g NaCl, and 20 mL of sterile 1 *M* glucose solution (added after autoclaving the former for 20 min) in 1 L of dH$_2$O.
17. Low-salt LB medium + Zeocin: autoclave 20 g tryptone, 5 g yeast extract, and 5 g NaCl in 1 L dH$_2$O, cool down to room temperature and add Zeocin (100 µg/mL).
18. Low-salt LB agar plates + Zeocin: autoclave 10 g bactotryptone + 5 g yeast extract + 5 g NaCl + 15 g bactoagar in 1 L dH$_2$O; let it cool to approx 60°C, add Zeocin (100 µg/mL), and pour in bacterial plates.
19. Plasmid extraction and purification Mini and Maxi kits (e.g., Qiagen).
20. pU1 sequencing 3' primer (TCCACTGTAGGATTAACAAC).

## *2.3. Generation of Stable Mammalian-Cell Transfectants*

1. Liposomes for transfection (e.g., Fugene from Roche, Indianapolis, IN).
2. Zeocin or other selection antibiotic.
3. pU1/ribozyme construct.
4. Colony-picking filters.

## 2.4. Screening of Transfected Cells for Gene-Expression Inhibition

1. RNA extraction kit (e.g., RNeasy from Qiagen).
2. Cell lysis buffer (e.g., RIPA buffer).
3. Northern analysis reagents (*see* standard manuals for protocols).
4. Western analysis reagents (*see* standard manuals for protocols).

## 3. Methods

### 3.1. Design of U1/Ribozymes (see Fig. 3)

1. Download cDNA sequence of your gene of interest from GenBank into a DNA analysis program.
2. Find all the (GTC) sequences in your cDNA, giving preference to those closest to the 5' end. Proximity to the 5' end gives a theoretical advantage, but is not an absolute necessity.
3. Extend approx 20 nucleotides from the "C" of GTC on either side.
4. Convert the sequence above into antiparallel sequence and then into RNA.
5. Replace the "G" of GAC with the hammerhead ribozyme sequence (**Fig. 2**) (this is the ribozyme/antisense sequence).
6. Add U1 snRNA sequences onto 5' and 3' ends of the ribozyme/antisense sequence.
7. Put the U1 snRNA/ribozyme/antisense sequence (~221 nucleotides) in a simple text file. Open it in Mulfold software and do the folding. Visualize the resulting analysis in LoopDloop to check RNA secondary structure. The criteria for a good secondary structure are not absolute, and include: i) undisrupted U1 snRNA loops, ii) antisense sequence with few bonds (four or less in a row), and iii) conserved hammerhead structure (*see* **Note 2**; **Fig. 4**).
8. Choose two or three promising sequences (if possible), and proceed to oligonucleotide synthesis of antisense/ribozyme and *Eco*RI and *Spe*I overhangs as described below.
9. Transform ribozyme/antisense sequence back to DNA and add AATT (*Eco*RI) to 5' end (sequence I).
10. Find out complementary sequence (without 5'-AATT) and add CTAG (*Spe*I) to its 5' end (sequence II) (*see* **Note 3**).

### 3.2. Construction of U1/Ribozyme Plasmid Expression Vector (see Fig. 2)

1. Synthesize oligonucleotides corresponding to both sequences I and II (~61 nucleotides each).
2. Anneal oligonucleotides I and II: add equimolar amounts of each oligonucleotide, bring to 95°C for 5 min and then cool slowly at room temperature to anneal.
3. Restriction digest pU1 plasmid (~10 µg) with *Eco*RI and *Spe*I (do not dephosphorylate).
4. To avoid background resulting from uncut plasmid, cast 1% agarose gel in TBE, run restriction-cut pU1 (also run 1 µg uncut plasmid as control), extract cut plasmid from the gel using gel-extraction kit, and measure concentration.
5. Add cut pU1 to annealed oligos (molar ratio 1:5, respectively), T4 DNA ligase, ligase buffer, and appropriate volume of dH$_2$O, and ligate overnight at 16°C.

Your target cDNA sequence with
20 nucleotides on either side of **C**

5'-CCAGAGGTACGCTACGAAGTCTGTGACATTCCTCAGTGTTC-3'

            ↓   Transform to anti parallel
                and then to RNA

5'-GAACACUGAGGAAUGUCACAGACUUCGUAGCGUACCUCUGG-3'

            ↓   Replace **G** by *ribozyme* sequence

5'-GAACACUGAGGAAUGUCACA*cugaugagucc
gugaggacgaa*ACUUCGUAGCGUACCUCUGG-3'

            ↓   Insert into U1snRNA loops at
                EcoRI/SpeI <u>restriction sites</u>

5'-GAUACUUACCUGGCAGGGGAGAUACCAUGAUCACGAAGGUGGUU
UUCCCAGGGCGAGGCUUAUCCAUUGGCCUCCGGAUGUGCUGACCC
CUGCGAUUUCCCCAAAUGUGGGAAACUCGACUGCAG<u>AAUU</u>
GAACACUGAGGAAUGUCACAcugaugaguccgugaggacgaaACUUCGUA
GCGUACCUCUGG<u>CUAG</u>UGGGGGGCUGCGUUCGCGCUUCCCCUG-3'

            ↓   Check RNA secondary structure

            ↓   Transform antisense/ribozyme to DNA
                and add <u>EcoR1 site</u> to 5'  (Sequence I)

5'-<u>AATT</u>GAACACTGAGGAATGTCACActgatgagtc
cgtgaggacgaaACTTCGTAGCGTACCTCTGG-3'

            ↓   Add <u>Spe I site</u> to 5' of complementary
                antisense/ribozyme sequence (Sequence II)

5'-<u>CTAG</u>CCAGAGGTACGCTACGAAGTttcgtcctc
acggactcatcagTGTGACATTCCTCAGTGTTC-3'

            ↓

Synthesize sequences I and II

6. Transform ligation product using DH5α competent cells.
7. Grow bacteria overnight at 37°C on low-salt LB-agar plates supplemented with Zeocin (100 μg/mL).
8. Pick colonies (~10 per construct) and grow in LB (low-salt) medium overnight.
9. Collect bacteria and extract plasmids using plamid extraction kit (Mini preps).
10. Check for successful ligation by restriction-digesting ~3 μg of each plasmid with *Bam*HI and electrophoresing on a 2% agarose gel along with *Bam*HI-cut pU1 control plasmid. pU1 plasmids with inserted ribozyme/antisense will yield slightly larger *Bam*HI cut bands than control pU1. Checking for ligation product by cutting with *Eco*RI/*Spe*I is not recommended, as the inserts are difficult to detect because of their small size.
11. Re-transform pU1 plasmids containing inserted ribozyme/antisense into DH5α, grow bacteria, pick colonies, and extract plasmids (Maxi preps) as described previously.
12. Send plasmids for sequencing to verify correct orientation and sequence. Correct sequences are your ready-to-use pU1/ribozymes.
13. U1/ribozymes can also be expressed in adenoviruses for more efficient transient delivery and complexed with liposomes for systemic in vivo delivery (*see* **Notes 6** and **7**).

### 3.3. Generation of Stable Mammalian-Cell Transfectants

1. Before starting stable transfections, check cells of interest for Zeocin resistance and determine the minimal Zeocin concentration that kills 100% of untransfected cells. Most eukaryotic cells are zeocin-sensitive and will die after exposure to at 50–500 μg/mL Zeocin for about 1 wk. (*See* **Note 4** if your cells are resistant to Zeocin.)
2. Grow cells to approx 30% confluency in medium free from selection antibiotic (more confluent cells more poorly).
3. Transfect cells with pU1/ribozyme plasmids using liposomes. We use Fugene reagent for transfections. Also transform with pU1 to generate control-transfected clones.
4. Grow cells to confluency (without zeocin).
5. Passage cells into zeocin-containing medium. Depending on the transfection efficiency of the cells, splitting ratios at this passage can vary widely between 1:10 (low transfection efficiency) and 1:500 (high transfection efficiency). The goal is to disperse the cells so that sufficiently large single colonies can grow on the plates without overlapping. Therefore, it is advisable to split the cells in various dilutions to ensure an appropriate number and distribution of colonies on the plates.

---

Fig. 3. *(opposite page)* Example transformation and analysis of targeted gene and U1/ribozyme construct leading to the antisense/ribozyme sequences that must be synthesized to generate specific U1/ribozymes. Gray shaded sequences: gene of interest; capital letters: U1 snRNA loops; small letters: hammerhead ribozyme; underlined sequences: restriction sites (refer to text for more details).

**A**

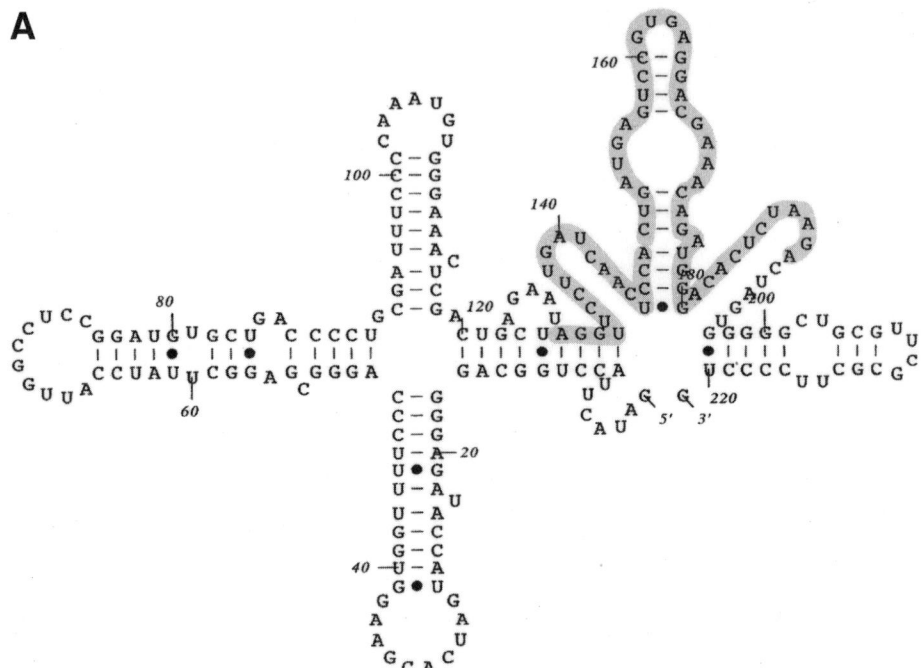

**B**

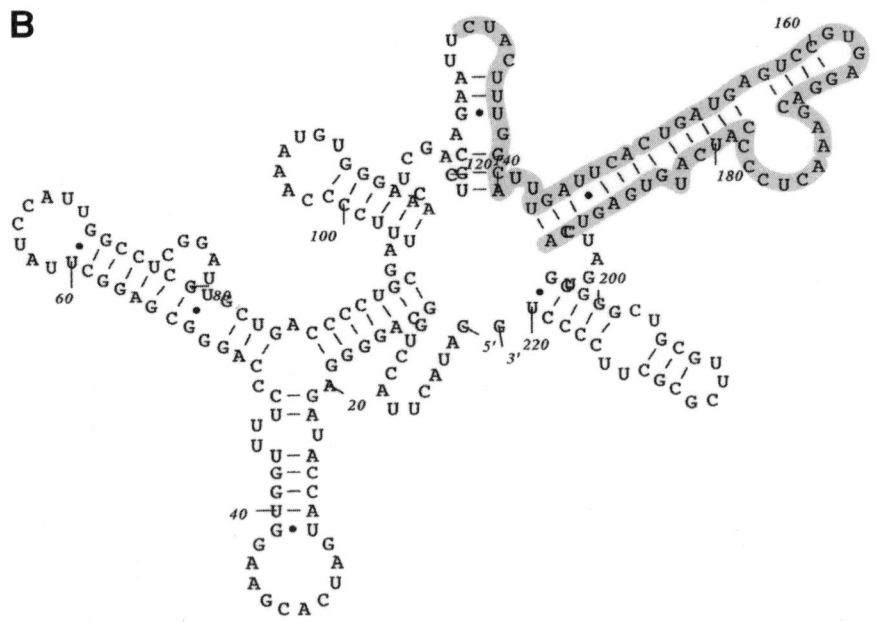

6. Regularly check the plates for colonies while replenishing medium and zeocin. Depending on the cell type, colonies can take up to several weeks to grow. Allow colony size to reach >1-mm diameter).

7. Pick colonies when they reach appropriate size. An easy way to pick colonies is with commercially available, small, sterile round filters that are dipped in trypsin and placed for a few minutes on the colonies after medium removal and after washing with phosphate-buffered saline (PBS). The filters can then be shaken in the medium of the destination plates to release the colony-derived cells. Alternatively, colonies can be scraped and aspirated with a sterile pipet tip after short exposure to a droplet of trypsin. Make sure that plates do not dry out during this procedure, as this will kill the cells.

8. Transfer each colony to a separate well of a 6-well plate and grow to confluency in zeocin-containing medium.

9. Trypsinize and transfer cells from the 6-well plates into two or three 10-cm dishes and grow to confluency. One of these plates will be used for screening gene expression by Northern analysis, and cells from the other two will be cryopreserved to be regrown if the clones display adequate expression knock-down of the gene of interest.

## 3.4. Screening of Transfected Cells for Gene-Expression Inhibition

1. We use Northern analysis and western blotting to screen for mRNA and protein levels in candidate knock-down clones. Alternative methods such as quantitative polymerase chain reaction (PCR) or enzyme-linked immunosorbent assay (ELISA) can also be used, if preferred.

2. To screen for mRNA of interest, isolate total RNA from cells and perform quantitative Northern analysis comparing mRNA levels with those of pU1 control-transfected cells. Knock-down levels can vary considerably, and depend on various factors, including levels of expression of gene in question, U1/ribozyme design, and type of cells. We have achieved up to 90–95% inhibition of expression for most genes in most cell lines.

3. Regrow clones that show adequate gene-expression inhibition based on mRNA levels and screen for protein levels, by Western blotting—again compared to pU1(control) clones run on the same gels.

4. *See* **Note 5** for clone growth and maintenance.

---

Fig. 4. *(opposite page)* Example analysis of the secondary structure of U1/ribozyme RNA using MulFold. **(A)** Promising secondary structure showing well-preserved U1 snRNA and ribozyme loops (gray shaded) and an antisense sequence (gray shaded) with a large number of bond-free nucleotides. **(B)** Less promising secondary sturcture showing partial loss of the ribozyme loop and an antisense sequence that is compromised by several bonds.

## 4. Notes

1. The pU1 vector was constructed on the backbone of the pZeoSV prokaryotic/ eukaryotic expression vector (Invitrogen; sequence available on the Invitrogen website: www.invitrogen.com). The SV40 promoter, polyadenylation site, and polylinker were excised from pZeoSV at the *Bam*HI sites, and a U1 snRNA expression cassette was cloned at this restriction site. Two rounds of site-directed mutagenesis were then performed to change four nucleotides flanking the Sm protein-binding site of U1 snRNA, creating unique *Eco*RI and *Spe*I restriction sites for insertion of the antisense/ribozyme sequences *(4)*.

2. The secondary structure requirements for the design and construction of the U1/ ribozymes do not represent strict criteria for the success of gene-expression inhibition. They are theoretical criteria that may improve the probability of successful results.

3. It is advisable to design and construct three different ribozymes that target the mRNA at different places for each gene in order to increase the chances of obtaining a construct that achieves a successful knock-down.

4. If cells are not sensitive to zeocin or if another selection antibiotic is preferred, the entire U1/ribozyme construct can be removed from pU1/Zeo by restriction digestion with BamHI and subcloned in any other mammalian-expression vector. The new vector should not have a promoter to drive the expression of the U1/ ribozyme, as this promoter could interfere with the U1 promoter that drives the expression of the transgene.

5. If inhibition of the targeted genes makes the cells develop a growth disadvantage relative to cells with normal gene expression, knock-down might be gradually lost in stable transfectants. This results from transfection revertants, which grow faster than knock-down cells and overtake the culture. This potential problem can be diminished by constantly keeping transfected cells in the selection antibiotic (zeocin) and by keeping early passage stocks and using early-passages of transfected clones.

6. pU1/ribozymes can be used for lipid-mediated transient transfections. The results can be compromised by transfection efficiency problems. Much better transient transfection results can be achieved with adenoviruses. We have successfully expressed U1/ribozymes in adenoviruses (driven by the U1 promoter) and have achieved high levels of gene-expression inhibition at mRNA levels. Effects on protein levels in transiently transfected cells will depend upon rates of cell division and half-life of the targeted protein. We have successfully used U1/ ribozyme-expressing adenoviruses for both in vitro and in vivo experiments *(1)*.

7. U1/ribozyme-expressing plasmids can also be complexed to liposome fractions for systemic in vivo delivery. We have successfully complexed pU1/ribozymes to liposomes composed of a DOTIM/cholesterol mixture. We delivered these intravenously to nude mice and achieved significant inhibtion of targeted gene expression in the animals *(1)*.

## Acknowledgments

Supported by Elsa U. Pardee Foundation and NIH Grant RO1 NS32148.

## References

1. Abounader, R., Lal, B., Luddy, C., Koe, G., Davidson, B., Rosen, E. M., et al. (2002) In vivo targeting of SF/HGF and c-met expression via U1snRNA/ ribozymes inhibits glioma growth and angiogenesis and promotes apoptosis. *FASEB J.* **16(1),** 108–110.
2. Abounader, R., Ranganathan, S., Lal, B., Fielding, K., Book, A., Dietz, H., et al. (1999) Reversion of human glioblastoma malignancy by U1 small nuclear RNA/ ribozyme targeting of scatter factor/hepatocyte growth factor and c- met expression. *J. Natl. Cancer Inst.* **91(18),** 1548–1556.
3. Jiang, W. G., Grimshaw, D., Lane, J., Martin, T. A., Abounader, R., Laterra, J., et al. (2001) A hammerhead ribozyme suppresses expression of HGF/SF receptor, cMET, and reduces migration and invasiveness of breast cancer. *Clin. Cancer Res.* **7,** 2555–2556.
4. Montgomery, R. A., and Dietz, H. C. (1997) Inhibition of fibrillin 1 expression using U1 snRNA as a vehicle for the presentation of antisense targeting sequence. *Hum. Mol. Genet.* **4,** 519–525.
5. Guthrie, C. and Patterson, B. (1988).] Spliceosomal snRNAs. *Annu. Rev. Genet.* **22,** 387–419.
6. Hjalt, T. A. and Wagner, E. G. (1995).] Bulged-out nucleotides protect an antisense RNA from RNase III cleavage. *Nucleic Acids Res.* **23(4),** 571–579.
7. Fischer, U., Darzynkiewicz, E., Tahara, S. M., Dathan, N. A., Luhrmann, R., and Mattaj, I. W. (1991).] Diversity in the signals required for nuclear accumulation of U snRNPs and variety in the pathways of nuclear transport. *J. Cell Biol.* **113(4),** 705–714.
8. Hamm, J., Darzynkiewicz, E., Tahara, S. M., and Mattaj, W. I. (1990).] The trimethylguanosine cap structure of U1 snRNA is a component of a bipartite nuclear signal. *Cell* **62,** 569–577.
9. Zhong, L., Batt, D. B., and Carmichael, G. G. (1994) Targeted nuclear antisense RNA mimics natural antisense-induced degradation of polyoma virus early RNA. *Proc. Natl. Acad. Sci. USA* **91,** 4258–4262.
10. Carmo-Fonseca, M., Tollervey, D., Pepperkok, R., Barabino, S. M., Merdes, A., Brunner, C., et al. (1991) Mammalian nuclei contain foci which are highly enriched in components of the pre-mRNA splicing machinery. *EMBO J.* **10(1),** 195–206.

# 15

## Design and Validation of Therapeutic Hammerhead Ribozymes for Autosomal Dominant Diseases

Jason J. Fritz, Marina Gorbatyuk, Alfred S. Lewin, and William W. Hauswirth

### Summary

   Hammerhead ribozymes are small, catalytic RNAs that can be designed to effectively inhibit gene expression in an allele-specific manner. It is the high level of sequence discrimination, coupled with the minimal cleavage-site requirements of hammerhead ribozymes, that makes these catalytic RNAs so amenable for use as therapeutic agents for autosomal dominant diseases. Here, we present a detailed set of protocols for the design and validation of hammerhead ribozymes for the treatment of autosomal dominant disease, with specific examples of hammerhead ribozymes targeted against human P23H rod opsin mRNA, a major cause of dominant retinitis pigmentosa.

   **Key Words:** Hammerhead ribozyme design; gene therapy; kinetics; recombinant adeno-associated virus; retinitis pigmentosa; autosomal dominant diseases.

## 1. Introduction

   Dominant diseases are of two types: those that lead to insufficient production of essential proteins (haploinsufficiency) and those that result in the accumulation of nonfunctional or cytotoxic proteins (dominant-negative mutations). Autosomal dominant retinitis pigmentosa (ADRP) is an example of the latter type, in which the production of mutant proteins (e.g., rhodopsin or peripherin/rds) results in the apoptotic death of photoreceptor cells in a progressive manner, eventually leading to loss of vision *(1,2)*. There are more than 100 point mutations in the rhodopsin gene alone that cause ADRP *(3)*, and thus, much of our research has focused on the development of mutation-specific ribozymes for use as therapeutic agents for gene therapy of rhodopsin-linked ADRP. Our hypothesis has been that we should be able to preserve

From: *Methods in Molecular Biology, vol. 252: Ribozymes and siRNA Protocols, Second Edition*
Edited by: M. Sioud © Humana Press Inc., Totowa, NJ

vision by using mutation-specific ribozymes to selectively reduce the cellular population of aberrant rhodopsin mRNAs, without significantly affecting the level of mRNA molecules that encode the wild-type protein. By cleaving the targeted mRNA molecules in a sequence-specific manner between their 5' 7-methylguanosine cap and 3' poly A tail, it is expected that the two RNA cleavage products will be quickly and efficiently degraded by cellular nucleases *(4,5)*. The in vitro characterization system outlined in this chapter is sufficient for the rapid screening of therapeutic hammerhead ribozymes prior to cloning and production of recombinant adeno-associated viral vectors for work in cell-culture and/or in vivo model systems.

## 1.1. The Hammerhead Ribozyme

The hammerhead ribozyme is the smallest naturally occurring ribozyme identified capable of performing sequence-specific cleavage of a phosphodiester bond *(6)*. The hammerhead ribozyme can be designed to cleave in *trans* any substrate RNA containing a $^{5'}NUX^{3'}$ triplet, where N is any nucleotide and X is any nucleotide except guanosine. By virtue of its minimal target-sequence requirement, the hammerhead ribozyme can be utilized for the specific downregulation of many mutant alleles, since variations of the NUX triplet occur frequently throughout all RNA messages. It should be noted, however, that not all NUX triplets are cleaved with the same efficiency. GUCUU and GUCUA sites have been shown to be the most efficiently cleaved variants of the NUX sites in vitro *(7)*, followed by GUC, CUC, UUC, and then the remaining NUX combinations *(8)*.

In addition to its minimal target-sequence requirements, the hammerhead ribozyme is also capable of discriminating in vivo between two target molecules that differ by just one nucleotide *(9)*. Hormes et al. performed a head-to-head comparison in which traditional antisense RNA molecules and hammerhead ribozymes were directed against identical human immunodeficiency virus-1 (HIV-1) target sequences and assayed for their abilities to inhibit replication *(10)*. Hammerhead ribozymes inhibited viral replication two- to 10-fold better than antisense RNA molecules that recognized the same sequences. In addition, the allele specificity afforded by hammerhead ribozymes may prove to be an advantage over small interfering RNAs (siRNAs) *(11)*, as well as antisense RNA approaches to selectively downregulate aberrant gene expression.

## 1.2. Target-Site Selection

For a mutation-specific hammerhead ribozyme to distinguish between wild-type and mutant transcripts, the point mutation must be within four bases of the scissile phosphodiester bond *(12)*. This consideration greatly limits the targeted region of the mutant allele to a fixed domain immediately flanking the point mutation. In addition, this target sequence of the mRNA containing the putative cleavage triplet must be accessible for base pairing with the flanking arms of the ribozyme in order to allow proper folding of the catalytic core so that sequence-specific cleavage can take place. Because of intramolecular basepairing, long mRNAs characteristically exhibit complex secondary and tertiary structures *(13)*. The reliable prediction of an intended target site's accessibility, within the full-length mRNA molecule, thus poses the greatest barrier to the success of a ribozyme-mediated therapy. Several in vitro strategies exist to predict such target-site accessibility, including the functional screening of random ribozyme and DNAzyme libraries *(14; see* Chapters 20 and 23), accessibility mapping of RNA using enzyme or chemical reagents *(15)*, or RNase H *(16)*, and computer-aided RNA-folding programs *(17)*. Because none of these strategies can account for the presence of RNA binding proteins that appear to alter mRNA structure in vivo *(18)*, one should be careful to avoid assigning too much confidence to these technologies. Novel methods for target site selection are described in Chapters 8, 9, and 19. However, testing of hammerhead ribozymes in tissue-culture cells is a recommended step, since this permits evaluation of activity in an intracellular environment. Nevertheless, testing in vivo in the relevant tissue is the best assay to verify the accessibility of a target site.

### 1.3. Therapeutic Hammerhead Ribozyme Design

Attempts to design allele-specific therapeutic hammerhead ribozymes have traditionally exploited situations in which a point mutation(s) generates a novel ribozyme cleavage site that is not present in the wild type allele *(19–25)*. Grassi and colleagues introduced two point mutations into the mouse *COL1A1* gene: i) a G-to-T transversion causing an amino acid substitution (Gly349Cys) responsible for osteogenesis imperfecta, and ii) a C-to-T transition that generated a novel GUC cleavage triplet *(20)*. Hammerhead ribozymes designed to target the novel GUC site in the mutant transgene selectively reduced the amount of mutant mRNA, yet did not reduce mRNA levels from transgenes lacking the GUC site, but containing the G-to-T transversion. More commonly, however, the targeted point mutation does not generate a novel cleavage site, but is within a few nucleotides of an NUX site common to both the mutant and wild-type alleles. In this case, it is recommended to keep the hammerhead ribozyme's hybridizing arms (helices I and III) relatively short to ensure specificity for the mutant mRNA substrate. The combined length of the sequence targeted by helices I and III can be up to 12 nucleotides and still discriminate a single base mismatch in either of the arms *(26)*. Consequently, after identifying a potential NUX sequence that lies within requisite proximity of a desired

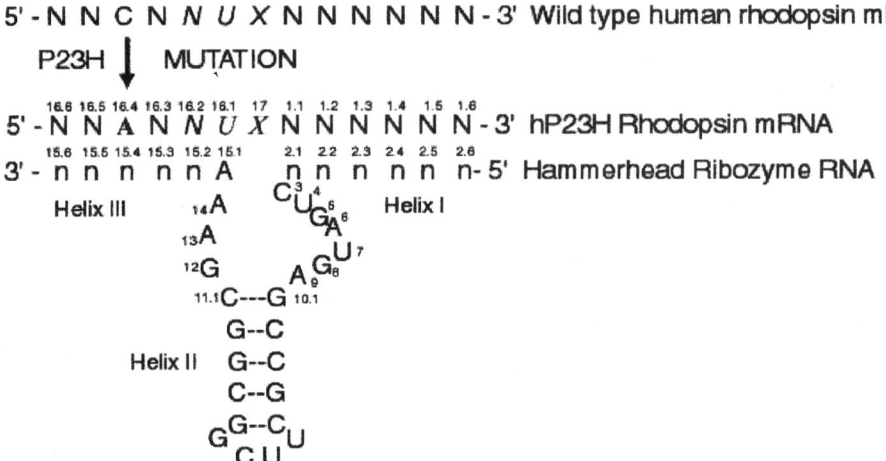

Fig. 1. Secondary structure of a hammerhead ribozyme designed to target the human P23H rhodopsin mRNA. The wild-type and mutant targets are in upper-case "N" and the ribozyme is in lower-case "n." Cleavage occurs just downstream of the X in the *NUX* target site. The targeting arms of the ribozyme anneal to the target to form helices I and III while helix II is formed internal to the ribozyme. The location of the C to A point mutation responsible for P23H has been indicated with an arrow. The numbering system is that of Hertel et al. *(29)*.

point mutation, hybridizing arms (helices I and III) are designed to basepair with a stretch of 12 nucleotides surrounding the "X" nucleotide in the NUX triplet site. Helices I and III are connected through the conserved nucleotides forming the catalytic core and a four basepaired stem (helix II), which is closed by an RNA tetra-loop (GNRA or UUCG) for increased stability. **Figure 1** shows a schematic of a hammerhead ribozyme designed to target mutant human P23H rhodopsin. The allele-specificity of a hammerhead ribozyme that targets a NUX site common to both the wild-type and mutant mRNA species can be improved by varying the lengths of the hybridizing arms and adopting an asymmetric design (Fritz, Lewin, Hauswirth, unpublished data). It has also been shown that hammerhead ribozymes with asymmetric hybridizing arms have higher catalytic turnover numbers than traditional hammerhead ribozymes *(27)*.

## 2. Materials

Always wear sterile laboratory gloves when working with RNA, and change them often during these protocols to prevent contamination with RNases. All water used in these protocols is deionized, sterile, and tested to be nuclease-free, and will be referred to hereafter as $dH_2O$.

## 2.1. Deprotection of Chemically Synthesized Ribozyme and Substrate RNAs

1. Ribozyme and substrate RNA oligonucleotides are chemically synthesized with an acid-labile orthoester-protecting group on the 2'-hydroxyl (2'-angiotensin-converting enzyme [ACE]). They are purchased from Dharmacon Research, Inc. (Boulder, CO).
2. 100 m$M$ acetic acid (adjusted to pH 3.8 with $N,N,N',N'$-tetramethyl-ethylenediamine [TEMED]).
3. dH$_2$O.

## 2.2. 5'-End-Labeling of Deprotected Substrate RNAs

1. 50 pmols of freshly deprotected substrate RNA (*see* **Note 1**).
2. 10X T4 polynucleotide kinase (PNK) buffer: 700 m$M$ Tris-HCl, pH 7.6 at 25°C, 100 m$M$ MgCl$_2$,5 m$M$ dithiothreitol (DTT). T4 PNK (10 U/μL).
3. RNAsin 40 U/μL (diluted 1:10 from stock in 0.1 M dithiothreitol [DTT]).
4. [γ$^{32}$P]-adenosine 5' triphosphate (ATP) (150 μCi/μL). **Caution:** radiation hazard.
5. dH$_2$O.
6. Sephadex G-25 fine resin (Pharmacia, Piscataway, NJ).

## 2.3. Hammerhead Ribozyme Cleavage Time-Course Analysis

1. 25 pmol/μL stock of deprotected non-radiolabeled substrate RNA, both mutant and wild-type sequences.
2. 0.5 pmol/μL stock of freshly 5' end-labeled substrate RNA, both mutant and wild-type sequences (*see* **Note 2**).
3. 5 pmol/μL stock of deprotected hammerhead ribozyme.
4. 400 m$M$ Tris-HCl (pH 7.5 at 37°C).
5. 200 m$M$ MgCl$_2$.
6. RNasin (diluted 1:10 from stock in 0.1 $M$ DTT).
7. RNA gel-loading dye: 90% formamide, 50 m$M$ ethylenediaminetetraacetic acid (EDTA), 0.05% bromophenol blue, and 0.05% xylene cyanol.
8. 8 $M$ urea, 10% acrylamide (w/v) gel solution for denaturing polyacrylamide gel electrophoresis (PAGE).
9. 1X TBE buffer: 89 m$M$ Tris borate, pH 8.3, 20 m$M$ EDTA.
10. Gel fix solution: 40% methanol, 10% acetic acid, and 3% glycerol.
11. dH$_2$O.

## 2.4. Cleavage Time-Course at Physiological Magnesium

1. All of the reagents from **Subheading 2.3.** will be needed.
2. The exception is that a stock of 20 m$M$ MgCl$_2$ will be used in this protocol in place of the 200 m$M$ MgCl$_2$ stock to yield a final (Mg$^{2+}$) in solution of 2 m$M$.

## 2.5. Competitive Cleavage Time-Course Analysis

You will need all of the reagents listed in **Subheading 2.3.**

## 2.6. Multiple-Turnover Kinetic Analysis

1. 30 pmol/µL substrate solution contains: 120 µL of RNase-free water, 15 µL of a 300-pmol/µL solution of unlabeled substrate, and 15 µL of the $^{32}$P-labeled substrate produced in **Subheading 3.2.**
2. 3 pmol/µL substrate solution contains: 135 µL of RNase-free water and 15 µL of the 30 pmol/µL substrate solution.
3. 0.3 pmol/µL substrate solution contains: 99 µL of RNase-free water and 1 µL of the 30 pmol/µL substrate solution.
4. 0.3 pmol/µL ribozyme solution.
5. 400 m*M* Tris-HCl, pH 7.5 at 37°C.
6. 200 m*M* MgCl$_2$.
7. RNasin (diluted 1:10 from stock in 0.1 *M* DTT).
8. RNA gel-loading dye: 90% formamide, 50 m*M* EDTA, 0.05% bromophenol blue, and 0.05% xylene cyanol.
9. 8 *M* urea, 10% acrylamide (w/v) gel solution for denaturing PAGE.
10. 1X TBE buffer: 89 m*M* Tris-borate, pH 8.3, 20 m*M* EDTA.
11. Gel-fix solution: 40% methanol, 10% acetic acid, and 3% glycerol.
12. dH$_2$O.

## 2.7. Calibration Curve

1. 3 pmol/µL substrate and 0.3 pmol/µL substrate solutions prepared in **Subheading 2.6.**
2. HybondN$^+$ membrane (Amersham Pharmacia Biotech, Piscataway, NJ).

## 2.8. Transient Cotransfection of HEK 293 Cells With Plasmids Expressing Target and Ribozyme

1. Choice of cell line. To eliminate expressed target background, use a cell line in which the gene of your interest is not expressed (in case of human P23H rhodopsin HEK 293 cell line). Transfection should be done in a laminar flow hood in a dedicated tissue-culture room (BL2 conditions).
2. Coding sequences for the target mRNA (in our case, rod-cell opsin) and the ribozyme are inserted into expression vectors employing promiscuous non-regulated promoters—e.g., the cytomegalovirus (CMV) immediate early promoter or the chicken β-actin-CMV hybrid promoter. It is very important to use a highly purified preparation of DNA for transfection—for example, CsCl-banded plasmid DNA.
3. Dulbecco's modified Eagle's medium (DMEM, Sigma-Aldrich Co.) supplemented with 10% fetal bovine serum (FBS, Sigma-Aldrich Co.), 100 U/mL of penicillin, 100 mg/mL of streptomycin (10X PenStrep, Mediatech, Cellgrow, VA) are needed to grow cells. 1X trypsin (Sigma-Aldrich Co.) phosphate-buffered saline (PBS) is used to remove adherent cells from dishes for subculturing splitting.
4. Opti-Mem I medium (Sigma-Aldrich Co.) (without serum) is used to dilute plasmid DNA and transfection LIPOPHECTAMIN 2000 Reagent (Life Technologies).

5. Tissue-culture dish $100 \times 20$ mm (Corning Inc.)
6. Incubator equilibrated to 37°C with 5% $CO_2$ to maintain growing cells.

## 2.9. Analysis of Ribozyme Activity by Quantitative Reverse Transcriptase-Polymerase Chain Reaction (RT-PCR)

1. RNA preparation kits are available from Qiagen or Sigma.
2. First-strand cDNA synthesis kit (Amersham Bioscience) to perform RT-PCR.
3. Commercially-synthesized sequence-specific primers for control as well as for rod opsin.
4. $\alpha$-[$^{32}$P]-ATP (167 µCi/µL) (NEN, Boston, MA). **Caution:** Radiation hazard.

## 3. Methods

### 3.1. Deprotection of Chemically Synthesized Ribozyme and Substrate RNAs

Chemically synthesized ribozyme and substrate RNAs are usually ordered on a 0.05 micromol synthesis scale from Dharmacon Research, Inc., *see* **Note 1**).

1. Quick-spin the dried RNA pellet to ensure that all of the vacuum-dried material is at the bottom of the tube. The acid-labile orthoester-protecting group on the 2'-hydroxyl of the synthesized RNA oligonucleotides is removed by incubation in 100 µL of 100 m$M$ acetic acid (adjusted to pH 3.8 with TEMED) for 30 min at 60°C. The volume of the deprotection step is decreased 75% from the amount suggested by the manufacturer.
2. Deprotected RNA oligonucleotides are dried under vacuum and resuspended in an appropriate volume of dH$_2$O to yield a 300-pmol/µL working stock and stored at –80°C. In most instances, we have found it unnecessary to electrophoretically purify synthetic ribozymes or targets purchased from Dharmacon Research, Inc. on gels.

### 3.2. 5'-End-Labeling of Deprotected Substrate RNAs

1. Combine the following reagents in a sterile 1.5-mL microcentrifuge tube for a 10-µL end-labeling reaction: 2 µL RNA oligo (25 pmol/µL), 1 µL 10X T4 PNK buffer, 1 µL RNAsin (diluted 1:10 in 0.1 $M$ DTT), 4 µL dH$_2$O, 1 µL [$\gamma^{32}$P]-ATP (150 µCi/µL) (ICN), 1 µL T4 PNK (10 U/µL).
2. Incubate the tube at 37°C for 30 min.
3. After 30 min, add 90 µL dH$_2$O and heat-inactivate the kinase at 65°C for 3 min.
4. Extract twice with 100 µL phenol-chloroform and once with 100 µL chloroform.
5. Purify the aqueous layer by filtration through a 1-mL Sephadex G-25 (Pharmacia) spin column. The resulting stock solution will be at a concentration of 0.5 pmol/µL, and should be stored at –20°C (*see* **Note 2**).

### 3.3. Hammerhead Ribozyme-Cleavage Time-Course Analysis

Initial cleavage assays use short synthetic RNA oligonucleotide targets ranging in length from 13–16 bases and synthetic hammerhead ribozymes. Use of commercial reagents allows a quick and cost-effective way to screen potential target sites and ribozymes before beginning the labor-intensive cloning and packaging protocols needed to make recombinant adeno-associated viral vectors for in vivo experiments. We initially assay the ability of the ribozyme(s) to independently cleave the mutant or wild-type substrates. Cleavage time-course analysis is performed on each substrate separately to identify the time-point(s) when 10–15% of the target has been cleaved by the ribozyme. During this window of time, the cleavage reaction is still nearly a linear function of time, and can be fitted to the Michalis-Menten equation to determine the catalytic properties of the ribozyme as described in **Subheading 3.6.** Initial cleavage time-course reactions are carried out at a molar excess of substrate (multiple-turnover conditions), so that the ratio of ribozyme to substrate is 1:20. The final volume for each cleavage time-course reaction is 100 μL, and are performed in triplicate.

1. In a 1.5-mL microcentrifuge tube combine the following reagents: 38 μL $H_2O$ 4 μL (25 pmol/μL) unlabeled mutant substrate, and 2 μL (0.5 pmol/μL) $^{32}P$-end-labeled mutant substrate.
2. Combine the following reagents in a second sterile 1.5-mL microcentrifuge tube: 25 μL $dH_2O$, 10 μL 400 m$M$ Tris-HCl, pH 7.5 at 37°C, and 1 μL ribozyme (5 pmol/μL).
3. Equilibrate the mutant substrate solution (prepared in **step 1**) to 37°C.
4. Heat the ribozyme solution (prepared in **step 2**) at 65°C for 2 min, then remove it from heat source and allow the tube to cool at room temperature for 10 min. Add 10 μL 200 m$M$ $MgCl_2$ and 10 μL RNasin (diluted 1:10 from stock in 0.1 $M$ DTT) to the ribozyme solution, mix gently, and equilibrate the tube at 37°C for 10 min.
5. Following the 10-min equilibration step, add the target solution to the ribozyme solution, mix gently, and immediately remove a 10-μL aliquot from the tube. Add this aliquot to a prelabeled sterile 1.5-mL microcentrifuge tube containing 10 μL of RNA gel-loading dye, and place the tube on ice for 1 min.
6. Continue to remove eight more 10-μL aliquots at various time-points thereafter (e.g., $t = 1, 2, 5, 10, 20, 30, 60,$ and 120 min). To account for pipetting/transfer errors, always prepare at least 25% more reaction volume than needed. Store the reaction aliquots at –20°C until ready to analyze by denaturing PAGE.
7. Cleavage products for short synthetic RNA oligonucleotide substrates are analyzed by electrophoresis on 8 $M$ urea, 10% acrylamide (w/v) denaturing gels run in 1X TBE buffer. Once gels have been poured and allowed to polymerize, they are pre-run at to warm the gel to approx 45°C.
8. Denature reaction aliquots at 85°C for 2 min and quick-cool on ice. Then load 10 μL of each cleavage time-course reaction aliquot onto the gel and run at 1500–2000 V until the bromophenol blue dye has run two-thirds of the way to the bottom of a 40-cm gel, in order to separate the 5′-end-labeled cleavage product from the substrate.
9. The gel is then fixed in 40% methanol, 10% acetic acid, and 3% glycerol for 30–45 min at room temperature. The fixed gel is then dried, exposed to radioanalytic

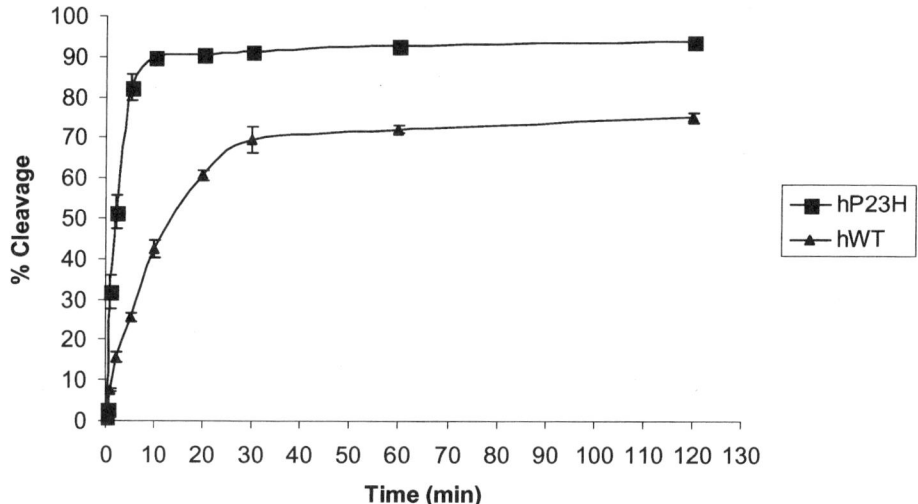

Fig. 2. Graphical representation of the cleavage time-course reaction of a hammerhead ribozyme designed to target the human P23H rhodopsin mRNA presented as percent (%) cleavage vs time.

phosphorescent screens, and analyzed using a Molecular Dynamics PhosphoImager system and ImageQuant software (Molecular Dynamics, Sunnyvale, CA). The percentage of substrate cleaved can then be determined from the ratio of radioactivity in the 5'-end-labeled cleavage product band (P) to the sum of the radioactivity in the 5'-end-labeled cleavage product band (P) and the substrate band (S): Percentage cleaved (%C) = 100 × [P/(P + S)].

10. Using Excel (Microsoft, Redmond, WA), the percentage of substrate cleaved is then averaged and plotted as a function of time to generate a graphical representation of the cleavage time-course (**Fig. 2**). The reaction velocity ($k_{obs}$) is calculated from the slope of this line during the linear phase of the reaction, and is reported as the concentration of substrate cleaved per min.

11. The process is then repeated using the wild-type substrate.

12. For each ribozyme, the Allelic Preference Quotient (APQ) is determined by calculating the $k_{obs}$ for each RNA substrate and taking the ratio of $k_{obs}$mutant to $k_{obs}$wild-type: APQ = $k_{obs}$mutant/$k_{obs}$wild-type. The APQ is a measure of the preference of a hammerhead ribozyme for its intended mutant substrate.

### 3.4. Cleavage Time-Course at Physiological Magnesium

For the initial cleavage time course described above in **Subheading 3.3.**, the final concentration of [$Mg^{2+}$] in solution was 20 m*M*. However, physiological magnesium concentrations are at least 10-fold less (0.5–2 m*M*). For this reason, ribozymes that successfully demonstrated cleavage at 20 mM $MgCl_2$ are further examined at 2 mM MgCl$_2$ by repeating the cleavage assay as described in **Subheading 3.3.**, but by adding 10 μL of 20 m*M* MgCl$_2$ in **step 4** rather than 10 μL 200 m*M* MgCl$_2$.

### 3.5. Competitive Cleavage Time-Course Analysis

This assay allows us to examine the ability of a given hammerhead ribozyme's to discriminate between its intended mutant substrate and the wild-type substrate. Competitive cleavage time course assays are performed using the protocol described in **Subheading 3.3.** with one notable difference: both alleles are present at a 1:1 molar ratio in the reaction solution.

1. In a sterile 1.5-mL microcentrifuge tube, combine the following reagents: 34 μL H$_2$O 4 μL (25 pmol/μL) unlabeled mutant substrate, 4 μL (25 pmol/μL) unlabeled wild-type substrate, and 2 μL (0.5 pmol/μL) [32]P-end-labeled mutant substrate.
2–9. These steps are identical to **steps 2–9** of **Subheading 3.3.**
10. The process is then repeated using 2 μL (0.5 pmol/μL) [32]P-end-labeled wild-type substrate rather than labeled mutant substrate.

### 3.6. Multiple-Turnover Kinetic Analysis

We perform multiple-turnover kinetic analysis (target excess) in order to determine kinetic parameters for the hammerhead ribozymes ($k_{cat}$, $K_M$, $V_{max}$). The hammerhead ribozyme catalytic reaction can be fitted to the Michalis-Menten equation as long as the following conditions are fulfilled: i) the substrate concentration must be in molar excess of the ribozyme concentration, ii) the formation of a ribozyme-substrate complex must be rapid and reversible, and iii) the rate-limiting step is the catalytic step *(28)*. To fulfill the first requirement, we perform the multiple-turnover kinetic analysis by holding the concentration of ribozyme constant at 15 n*M* and increasing the substrate concentration over a range from 150–1500 n*M*. Increasing ratios of ribozyme to target may be necessary to reach saturating conditions for the ribozyme. Multiple-turnover kinetic analysis should always be performed in triplicate, and replicates should yield reproducible cleavage levels. For more details, *see* Chapter 3. A minimal acceptable $k_{cat}$ value that justifies continued testing of a ribozyme in vivo is 0.1 min$^{-1}$.

1. Equilibrate the target stock solutions to 37°C.
2. Following **Table 1**, combine the indicated amounts of dH$_2$O, 400 m*M* Tris-HCl, and ribozyme in appropriately prelabeled sterile 1.5-mL microcentrifuge tubes.
3. Incubate the tubes at 65°C for 2 min to denature the ribozyme and allow the tubes to cool at room temperature for 10 min.
4. Then add the 200 m*M* MgCl$_2$ and RNasin, and equilibrate the tubes at 37°C for an additional 10 min.

**Table 1**
**Preparation of Multiple-Turnover Kinetic Reaction Tubes**

| Reaction | dH$_2$O | 400 mM Tris pH 7.4 | 0.3 pmol/µL Ribozyme | 1:10 RNasin: 0.1 M DTT | 200 mM MgCl$_2$ | 3 pmol/µL Substrate | 30 pmol/µL Substrate |
|---|---|---|---|---|---|---|---|
| 1,11,21 | 14 µL | 2 µL | 0 µL | 1 µL | 2 µL | 1 µL | |
| 2,12,22 | 10 µL | 2 µL | 1 µL | 1 µL | 2 µL | 4 µL | |
| 3,13,23 | 8 µL | 2 µL | 1 µL | 1 µL | 2 µL | 6 µL | |
| 4,14,24 | 6 µL | 2 µL | 1 µL | 1 µL | 2 µL | 8 µL | |
| 5,15,25 | 13 µL | 2 µL | 1 µL | 1 µL | 2 µL | | 1 µL |
| 6,16,26 | 12 µL | 2 µL | 1 µL | 1 µL | 2 µL | | 2 µL |
| 7,17,27 | 10 µL | 2 µL | 1 µL | 1 µL | 2 µL | | 4 µL |
| 8,18,28 | 8 µL | 2 µL | 1 µL | 1 µL | 2 µL | | 6 µL |
| 9,19,29 | 6 µL | 2 µL | 1 µL | 1 µL | 2 µL | | 8 µL |
| 10,20,30 | 4 µL | 2 µL | 1 µL | 1 µL | 2 µL | | 10 µL |

5. Stagger the addition of the target stocks to the ribozyme by 30–60 s and incubate reactions at 37°C for the time period estimated to reach 10–20% of full cleavage, as determined by the cleavage time-course reaction (*see* **Subheading 3.3.**).

6. Terminate the reactions after the appropriate time interval by adding 20 µL of ice-cold formamide RNA loading dye to the reaction tube. Place terminated reaction tubes on ice, and then store reactions at –20°C until the entire reaction set is ready to be analyzed by PAGE.

7. Denature the kinetic reactions at 85°C for 2 min and quick-cool on ice. Then load 10 µL of each reaction aliquot onto a 8 M urea, 10% acrylamide (w/v) denaturing gel and run in 1X TBE buffer at until the bromophenol blue dye has run two-thirds of the way to the bottom of a 40-cm gel in order to separate the 5'-end-labeled cleavage product from the substrate.

8. The gel is then fixed in 40% methanol, 10% acetic acid, and 3% glycerol for 30–45 min at room temperature.

9. The fixed gel is then dried and exposed to a radioanalytic phosphorescent screen along with the slot blot of the calibration curve (*see* **Subheading 3.7.**, step 3).

10. We use the equation determined from the calibration curve (*see* **Subheading 3.7.**) to convert the raw data obtained from the PhosphoImager to pmols of product per minute per reaction, and plot this against the mM values of target used per reaction. These data are used to generate an equation from a Lineweaver-Burke double reciprocal plot to calculate the kinetic parameters (**Fig. 3**).

### *3.7. Calibration Curve for Multiple-Turnover Kinetic Analysis*

1. **Table 2** shows the volumes of each target stock needed to create a calibration curve to facilitate the analysis of the multiple-turnover kinetic data.

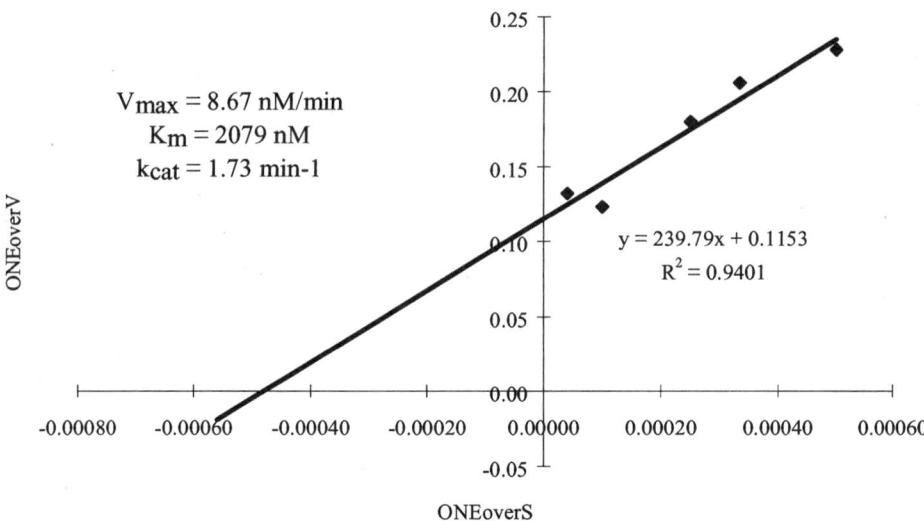

Fig. 3. Example of a Lineweaver-Burke plot produced from a multiple-turnover kinetic analysis on a hammerhead ribozyme designed to target human P23H rhodopsin.

**Table 2**
**Preparation of Stock Solutions Used to Generate the Calibration Curve**

| Tube # | dH$_2$O | 0.3 pmol/µL Substrate | 3 pmol/µL Substrate |
|---|---|---|---|
| 1,13 | 100 µL | 0 | |
| 2,14 | 99 µL | 1 | |
| 3,15 | 98 µL | 2 | |
| 4,16 | 96 µL | 4 | |
| 5,17 | 94 µL | 6 | |
| 6,18 | 92 µL | 8 | |
| 7,19 | 99 µL | | 1 |
| 8,20 | 98 µL | | 2 |
| 9,21 | 96 µL | | 4 |
| 10,22 | 94 µL | | 6 |
| 11,23 | 92 µL | | 8 |
| 12,24 | 90 µL | | 10 |

2. Using a slot blot apparatus; filter each target solution onto a HybondN$^+$ membrane that has been presoaked with dH$_2$O.

3. This blot must then be simultaneously exposed to a radioanalytic phosphorescent screen along with the dried gel of the multiple-turnover kinetic reaction, and analyzed in tandem using a Molecular Dynamics PhosphoImager system and ImageQuant software (Molecular Dynamics, Sunnyvale, CA).

4. After scanning the calibration slot blot, average the pixel-density values for each set of duplicate solutions and plot these values vs the known amount of 5'-end-labeled oligonucleotide substrate to generate the calibration curve.

5. Plot the raw numbers generated from a volume report of the PhosphoImager vs pmol of target and determine the equation for the best straight line through these data points. This equation is used to directly convert the raw data obtained for the amount of cleavage product formed into pmols.

## 3.8. Transient Cotransfection of HEK 293 Cells With Plasmid DNA Carrying Target and Ribozyme

1. Trypsinize and count HEK 293 cells the day before transfection so that they will give 70–90% of confluence on a day of transfection (usually 100% confluent growing cells on $100 \times 20$ mm plate are diluted 1:2, 1:3). Perform experiment in triplicate.

2. Carefully remove DMEM medium and add 5 mL of a fresh one without antibiotic.

3. For multiple dishes, make a "bulk mix" of plasmids carrying target and ribozymes (using molar ratio target to ribozyme as 1:6; 1:8, or 1:10) so that "control" variant would not have a plasmid expressing ribozyme. Dilute plasmids DNA in OPTI-MEM (10 µg in 500 µL for each dish).

4. Dilute LIPOFECTAMIN 2000 Reagent into OPTI-MEM medium (30 µL into 500 µL relatively for each dish). As a control, keep the ratios of LIPOFECTAMIN 2000 Reagent to plasmid DNA at 3:1 and LIPOFECTAMIN 2000 reagent to OPTI-MEM as 3 µL–50 µL relatively.

5. Once LIPOFECTAMIN 2000 reagent is diluted, combine it with the diluted plasmid DNA and incubate for 20 min at room temperature.

6. Add complex directly to each plate, gently rocking it.

7. Incubate plate with transfected cells at 37°C with 5% $CO_2$ for 24–72 h.

8. After incubation, tryptonize cells with 1 mL 1X trypsin and harvest them in a microcentrifuge at 4000$g$ for 5 min. Discard trypsin and wash cell pellet twice with 1X PBS.

## 3.9. Analysis of Ribozyme Activity by Quantitative RT-PCR

1. To assay activity of ribozyme targeted against a specific target RNA (e.g., rhodopsin) in HEK 293 cells, we prepare total RNA and compare the levels of target RNA to an unchanged cellular RNA. As a control for rhodopsin mRNA, we have used β-actin mRNA. We generally use one of several commercially available kits for RT-PCR. Set up the reverse transcriptase reaction as described in the protocol enclosed in a kit to synthesize first-strand cDNA. To avoid pipetting variation, the reverse transcription reaction for both genes (β-actin and rhodopsin) can be done in the same tube using sequence-specific antisense primers.

2. It is recommended that PCR for cDNAs be set up separately, since optimized conditions for genes amplification (primer sequence, annealing temperature, and extension time, as well as number of cycles) might be different.

3. For detection of PCR products (control and target of interest) add α-[$^{32}$P] dATP (10 µCi) to the amplification mixture.

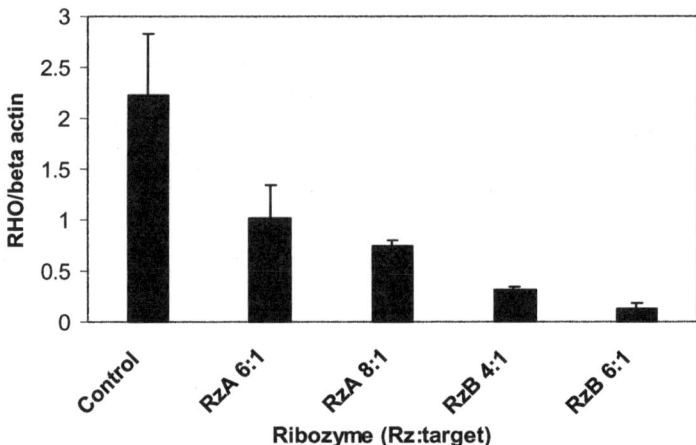

Fig. 4. Transient cotransfection of HEK 293 cells with plasmid DNA-encoding rhodopsin and/or rhodopsin-specific ribozymes (RzA and RzB). The ratio of ribozyme expressing plasmid to rhodopsin (target)-expressing plasmid is indicated. At 24 h posttransfection, the level of rhodopsin mRNA in the cells was assayed by RT-PCR. The relative reduction in the cellular pool of rhodopsin mRNA is given by the ratio of the absolute counts per min of the radiolabeled rhodopsin RT-PCR product to radiolabeled β-actin RT-PCR product (RHO/β-actin).

4. Load labeled PCR products in 1.0% agarose gel in TBE. Run the gel and dry it.
5. Perform detection of amplified cDNA using Molecular Dynamics PhosphoImager system and ImageQuant software (Molecular Dynamics).
6. Express relative target mRNA reduction by ratio of detected target product to β-actin product (**Fig. 4**).

## 4. Notes

1. The RNA synthesis process employed by Dharmacon has an average coupling yield of 99% per step, and thus the yield of full-length product is given by $(0.99)^{(\text{oligo length})} \times$ micromols ordered. The protected RNA oligonucleotides arrive as a dried pellet. We resuspend the entire stock in 50 μL dH$_2$O and aliquot it into $5 \times 10$ μL stocks that are then dried under a vacuum. Four of the five aliquots are stored at −80°C until needed. The remaining tube is deprotected as described.
2. Stocks of 5' end-labeled substrate RNAs should be stored in proper radioactive-material freezer boxes and can be used effectively as reagents for up to 2 wk.

## References

1. Adler, R. (1996) Mechanisms of photoreceptor death in retinal degenerations. From the cell biology of the 1990s to the ophthalmology of the 21st century? *Arch. Ophthalmol.* **114(1),** 79–83.

2. Phelan, J. K. and Bok, D. (2000) A brief review of retinitis pigmentosa and the identified retinitis pigmentosa genes. *Mol. Vision* **6**, 116–124.
3. Van Soest, S., Westerveld, A., Dejong, P. T. V. M., Bleeker-Wagemakers, E. M., and Bergen, A. A. (1999) Retinitis pigmentosa: defined from a molecular point of view. *Surv. Ophthalmol.* **43(4)**, 321–334.
4. Beelman, A. and Parker, R. (1995) Degradation of mRNA in eukaryotes. *Cell.* **81(2)**, 179–183
5. Mitchell, P. and Tollervy, D. (2000) mRNA stability in eukaryotes. *Curr. Opin. Gen. Dev.* **10(2)**, 193–198.
6. Symons, R. H. (1989) Self-cleavage of RNA in the replication of small pathogens of plants and animals. *Trends Biochem. Sci.* **14(11)**, 445–450.
7. Clouet-d' Orval, B. and Uhlenbeck, O.C. (1997) Hammerhead ribozymes with a faster cleavage rate. *Biochemistry* **36(30)**, 9087–9092.
8. Shimayama, T., Nishikawa, S., and Taira, K. (1995) Generality of the NUX rule: kinetic analysis of the results of systematic mutations in the trinucleotide at the cleavage site of hammerhead ribozymes. *Biochemistry* **34(11)**, 3649–3654.
9. Funato, T., Shitara, T., Tone, T., Jiao, L., Kashani-Sabet, M., and Scanlon, KJ.(1994) Suppression of H-ras-mediated transformation in NIH3T3 cells by a ras ribozyme. *Biochem. Pharmacol.* **48(7)**, 1471–1475.
10. Hormes, R., Homann, M., Oelze, I., Marschall, P., Tabler, M., Eckstein, F., et al. (1997) The subcellular localization and length of hammerhead ribozymes determine efficacy in human cells. *Nucleic Acids Res.* **25(4)**, 769–775.
11. Sijen, T., Fleenor, J., Simmer, F., Thijssen, K. L., Parrish, S., Timmons, L., et al. (2001) On the role of RNA amplification in dsRNA-triggered gene silencing. *Cell* **107(4)**, 465–476.
12. Werner, M. and Uhlenbeck, O. C. (1995) The effect of base mismatches in the substrate recognition helices of hammerhead ribozymes on binding and catalysis. *Nucleic Acids Res.* **23(12)**, 2092–2096.
13. Uhlenbeck, O. C., Pardi, A., and Feigon, J. (1997) RNA structure comes of age. *Cell* **90(5)**, 833–840.
14. Lieber, A. and Strauss, M. (1995) Selection of efficient cleavage sites in target RNAs by using a ribozyme expression library. *Mol. Cell Biol.* **15(1)**, 540–551.
15. Campbell, T. B., McDonald, C. K., and Hagen, M. (1997) The effect of structure in a long target RNA on ribozyme cleavage efficiency. *Nucleic Acids Res.* **25(24)**, 4985–4993.
16. Birikh, K. R., Berlin, Y. A., Soreq, H., and Eckstein, F. (1997) Probing accessible sites for ribozymes on human acetylcholinesterase RNA. *RNA* **3(4)**, 429–437.
17. Zuker, M. (1989) On finding all suboptimal foldings of an RNA molecule. *Science* **244(4900)**, 48–52.
18. Heidenreich, O., Kang, S. H., Brown, D. A., Xu, X., Swiderski, P., Rossi, J. J., et al. (1995) Ribozyme-mediated RNA degradation in nuclei suspension. *Nucleic Acids Res.* **23(12)**, 2223–2228.
19. Funato, T., Shitara, T., Tone, T., Jiao, L., Kashani-Sabet, M., and Scanlon, K. J. (1994) Suppression of H-ras-mediated transformation in NIH3T3 cells by a ras ribozyme. *Biochem. Pharmacol.* **48(7)**, 1471–1475.

20. Grassi, G., Forlina, A., and Marini, J. C. (1997) Cleavage of collagen RNA transcripts by hammerhead ribozymes in vitro is mutation-specific and shows competitive binding effects. *Nucleic Acid Res.* **25(17)**, 3451–3458.
21. Scherr, M., Grez, M., Ganser, A., and Engels, J. M. (1997) Specific hammerhead ribozyme-mediated cleavage of mutant N-ras mRNA in vitro and ex vivo. Oligoribonucleotides as therapeutic agents. *J. Biol. Chem.* **272(22)**, 14,304–14,313.
22. Drenser, K. A., Timmers, A. M., Hauswirth, W. W., and Lewin, A. S. (1998). Ribozyme-targeted destruction of RNA associated with autosomal-dominant retinitis pigmentosa. *Invest. Ophthalm. Visual Sci.* **39(5)**, 681–689.
23. Lewin, A. S., Drenser, K. A., Havswirth, W. W., Nishikawa, S., Yasumara, D., Flannery, J. G., La Vail, M. M., et al. (1998). Ribozyme rescue of photoreceptor cells in a transgenic rat model of autosomal dominant retinitis pigmentosa. *Nature Med.* **4(8)**, 967–971.
24. LaVail, M. M., et al. (2000) Ribozyme rescue of photoreceptor cells in P23H transgenic rats: long-term survival and late-stage therapy. *Proc. Natl. Acad. Sci. USA* **97(21)**, 11,488–11,493.
25. Shaw, L. C., Skold, A., Wong, F., Petters, R., Hauswirth, W. W., and Lewin, A. S. (2001) An allele-specific hammerhead ribozyme gene therapy for a porcine model of autosomal dominant retinitis pigmentosa. *Mol. Vision* **7**, 6–13.
26. Hertel, K. J., Herschlag, D., and Uhlenbeck, O. C. (1996) Specificity of hammerhead ribozyme cleavage. *EMBO J.* **15(14)**, 3751–3757.
27. Hendry, P. and McCall, M. (1996) Unexpected anisotropy in substrate cleavage rates by asymmetric hammerhead ribozymes. *Nucleic Acids Res.* **24(14)**, 2679–2684.
28. McConnell, T. S. (1997) Theoretical considerations in measuring reaction parameters, in *Methods in Molecular Biology, vol. 74, Ribozyme Protocols* (Turner, P.C., ed.), Humana Press, Totowa, NJ.
29. Hertel, K. J., Pardi, A., Uhlenbeck, O. C., Koizumi, M., Ohtsuka, P., Vesugi, S., et al. (1992) Numbering system for the hammerhead. *Nucleic Acids Res.* **20(12)**, 3252.

# 16

## Helicase-Attached Novel Hybrid Ribozymes

### Hiroaki Kawasaki, Masaki Warashina, Tomoko Kuwabara, and Kazunari Taira

### Summary

Ribozymes have potential as therapeutic agents and in functional studies of genes of interest. The activities of ribozymes in vivo depend on the accessibility of ribozymes to a cleavage site in the target RNA. At present, the selection of a target site for ribozymes is often based on a computer-aided structural analysis of the target RNA or trial-and-error experiments in which vast numbers of ribozymes are tested systematically. To overcome this problem, we have engineered intracellularly produced ribozymes with unwinding activity in vivo. We found that attachment to ribozymes (hybrid ribozymes) of an RNA motif with the ability to interact with intracellular RNA helicases, which create hybrid ribozymes, enhances ribozyme activity significantly in vivo. Thus, hybrid ribozymes can catalyze cleavage at the specified target site within an RNA in vivo almost independently of the secondary or tertiary structure of the target RNA around the cleavage site.

**Key Words:** Hammerhead ribozyme; RNA helicase; hybrid ribozyme; unwinding activity; poly(A); CTE.

## 1. Introduction

Hammerhead ribozymes are small RNA molecules that can catalyze cleavage of RNAs in a sequence-specific manner (1,2). The ribozyme can cleave oligoribonucleotides at specific sites (namely, after the sequence NUX, where N and X can be A, G, C, or U and A, C, or U, respectively (2). Thus, they have been used successfully to knock down the intracellular expression of a variety of specific viral and cellular targets (3,4).

Successful inactivation of a specific gene by the ribozyme in vivo depends largely on the appropriate design of the expression vector. The design can determine both the level of expression and the half-life of the expressed

From: *Methods in Molecular Biology, vol. 252: Ribozymes and siRNA Protocols, Second Edition*
Edited by: M. Sioud © Humana Press Inc., Totowa, NJ

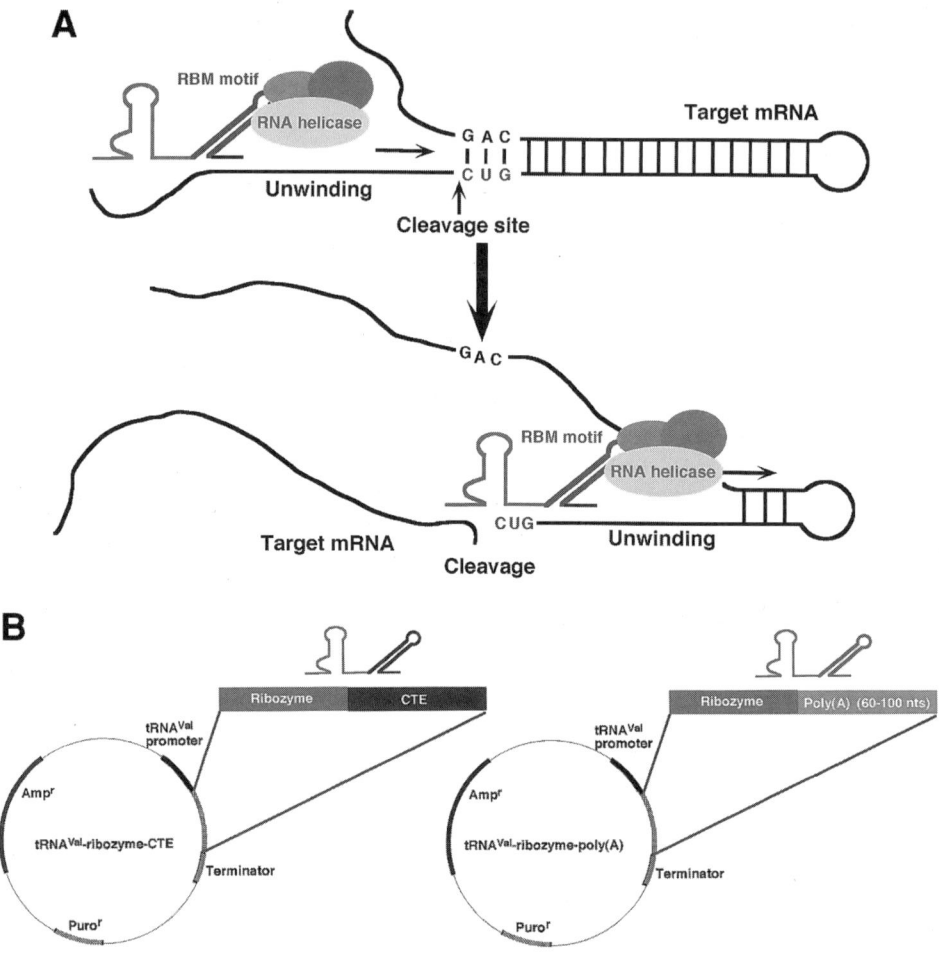

Fig. 1. (**A**) A schematic representation of the cleavage of a normally inaccessible target site by the hybrid ribozyme: the hybrid ribozyme connected to the sequence of RNA helicase binding motif (RBM) can recruit RNA helicase and can have the unwinding activity of the RNA helicase and the cleavage activity of the ribozyme. (**B**) Hybrid ribozyme expression plasmids [tRNA$^{Val}$-ribozyme-CTE and tRNA$^{Val}$-ribozyme-poly(A)]. Length of poly(A) sequence is 60–100 nucletiotides.

ribozyme. High-level expression under the control of the pol III promoter is clearly advantageous, since there is evidence that the association of a ribozyme with its target mRNA is the rate-limiting step in ribozyme-mediated reactions in cells *(5–8)*. Therefore, we have chosen to express ribozymes under the control of the promoter of a human gene for tRNA$^{Val}$. This system has previously been used successfully in the suppression of target genes by ribozymes *(9–11)*.

The efficiency of ribozyme-mediated cleavage in vivo is not always high in cells *(12)*. In cells, the activities of ribozymes depend on the level of their expression as well as their access to the cleavage site in the target RNA. To overcome this problem and to improve the efficiency of ribozymes in cells, we have created a hybrid ribozyme with the ability to access any target site and to cleave at a specific site *(5–8)*. The development of the hybrid ribozyme was accomplished by combining the cleavage activity of the ribozyme with the unwinding activity of the endogenous RNA helicase(s) (**Fig. 1A**). To connect the helicase to the ribozyme, we added an RNA motif, a poly(A) tail, or the constitutive transport element (CTE) sequence to the 3' end of the ribozyme. This poly (A) sequence interacts with the RNA helicase eIF4AI *(13)* via interactions with poly(A)-binding protein (PABP) and PABP-interacting protein-1 (PAIP) *(14)*. In addition, it is also known that CTE interacts with the RNA helicase A *(15)* and hDbp5 *(16–18)*. We have demonstrated that these hybrid ribozymes have strong cleavage activity as well as a substrate-unwinding activity *(5–8)*. Thus, they were able to cleave the target mRNA at any chosen site, regardless of the putative secondary or tertiary structure in the vicinity of the target site. In this chapter, we describe the construction of these hybrid ribozymes.

## 2. Materials
### 2.1. Reagents and Kits

1. TE buffer: 10 m$M$ Tris-HCl, pH 8.0, 1 m$M$ ethylenediamenetetraacetic acid (EDTA). Autoclave and store at room temperature.
2. 10X TBE buffer: 890 m$M$ Tris base, 890 m$M$ boric acid, 20 m$M$ EDTA, pH 8.0. Store at room temperature.
3. 50X TAE buffer: 2 $M$ Tris-acetate, 0.05 $M$ EDTA, pH 8.0. Store at room temperature.
4. DNA ligase (TaKaRa DNA Ligation Kit ver.2).
5. Calf intestine alkaline phosphatase (CIP, 10–30 U/≤L).
6. Restriction enzymes; *Kpn*I and *Csp*45I (TOYOBO) and reaction buffer; 10X L buffer: 100 m$M$ MgCl$_2$, 10 m$M$ dithiothreitol (DTT), 100 m$M$ Tris-HCl, pH 7.5.
7. Miniprep kit (QIAGEN).
8. Transfection reagent: Effectin (QIAGEN).
9. DMEM (Gibco-BRL).
10. FBS (Gibco-BRL).
11. PBS (TaKaRa).
12. 10X *Taq* polymerase buffer: 100 m$M$ Tris-HCl, pH 8.3, 15 m$M$ MgCl$_2$, 500 m$M$ KCl. Store at –20°C
13. 10X deoxyribonucleotide stock (deoxynucleotide 5' triphosphate [dNTP] mix): 2.5 m$M$ each dNTP in distilled water. Store at –20°C
14. 10X CIP buffer: 500 m$M$ Tris-HCl, pH 9.0, 10 m$M$ MgCl$_2$. Store at –20°C.

15. *Escherichia coli* (JM109)-competent cells.
16. 99.5% ethanol (special grade).
17. 70% ethanol (diluted with distilled water).
18. 3 *M* sodium acetate, pH 5.2 (RNase-free).
19. Mixture of phenol (TE-saturated, pH 8.0), chloroform, and isoamyl alcohol (IAA) (25:24:1, v/v).

## 2.2. Plasmids

1. pPURO-tRNA-CTE (Taira laboratory).
2. pPURO-tRNA-poly(A) (Taira laboratory).

# 3. Methods
## 3.1. Construction of Hybrid Ribozymes

To design hybrid ribozymes, we must first choose suitable target sites containing the NUX sequence that is required to be cleaved by a ribozyme (*see* **Note 1**). The number of bases at the target site is usually 8–12 nucleotides (*see* **Note 2**). After determination of the target site of ribozymes (*see* **Note 3**), construct hybrid ribozyme expression plasmids that are controlled by the tRNA$^{Val}$ promoter (*see* **Fig. 1B**).

1. Synthesis of a template DNA encoding a ribozyme sequence (5'-TCC CCG GTT CGA AAC CGG GCA CTA CAA AAA CCA ACT TCN NNN NNN NNC TGA TGA GGC CGA AAG GCC GAA NNN NNN NNN GGT ACC CCG GAT ATC TTT TTT T-3') by a DNA synthesizer. N means substrate-binding arms that are complementary to the target mRNA.
2. Amplify the template DNA with a specific forward primer (5'-TCC CCG GTT CGA AAC CGG GCA-3') and a reverse primer (5'-GCT TGC ATG CCT GCA GGT CGA CGC GAT AGA AAA AAA GAT ATC CGG GGT-3') using PCR. The reaction solution contains (final concentration): template DNA (0.02 µL), forward primer (0.5 µL), reverse primer (0.5 µL), *Taq* buffer, dNTP mix (0.25 m*M* each), and *Taq* DNA polymerase (10 U). Adjust the total volume to 100 µL. The thermal cycle is as follows: denaturing reaction (94°C, 1 min), annealing reaction (58°C, 1 min), polymerase reaction (73°C, 1 min 30 s), and repeat 25 cycles.
3. Remove and check 5 µL of polymerase chain reaction (PCR) products by electrophoresis with 1.5% agarose gels.
4. Purify the remaining 95-µL PCR products by using PCR purification kits or by phenol/chloroform extraction followed by ethanol-precipitation. PCR purification kits can remove enzymes, salts, and excess primers.
5. Digest 2 µg of the vector plasmid "pPUR-tRNA-CTE or pPUR- tRNA-poly(A)" and the PCR product with *Kpn*I and *Csp*45I in low-salt buffer for 2 h at 37C° in a volume of 50 µL.
6. Add 50 µL of a mixture of phenol (TE-saturated, pH 8.0), chloroform, and IAA (25:24:1, v/v) to each digested product. Extract restriction enzymes by mixing on

a vortex mixer and separate the phases by microcentrifugation for 5 min.

7. Transfer the aqueous phase to a new 1.5-mL microcentrifuge tube, add 5 μL of 3 $M$ sodium acetate and 30 μL of ethanol, mix, refrigerate for a few minuites, and collect the precipitate by microcentrifugation for 10 min at 4°C.

8. Remove the ethanol and save the pellet. Rinse the pellet with 1 mL of 70% ethanol, collect the rinsed precipitate by microcentrifugation for 5 min, remove the ethanol, and dry the pellet. Resuspend the pellet in 20 μL of TE buffer.

9. To prevent self-ligation of the vector, incubate the reaction product of pPUR-tRNA-CTE or pPUR-tRNA-poly(A) plasmid and 2 μL of a solution of CIP in 1X CIP buffer for 30 min at 37°C in a volume of 50 μL. Remove the enzyme by phenol-chloroform extraction and resuspend the pellet in 10 μL of TE buffer after ethanol-precipitation.

10. The ligation reaction is carried out by combining the following:
    a. 1 μL digested pPUR-tRNA-CTE or pPUR-tRNA-poly(A).
    b. 1 μL digested insert DNA.
    c. 2 μL DNA ligase solution 1.
    d. Incubate the reaction mixture for at least 30 min at 16°C.

11. Transform *E. coli* host cells with the ligation mixture and plate the cells on LB agar plates containing 100 μg/mL ampicillin. Recover plasmids from the positive colonies. Determine the nucleotide sequence of the hybrid ribozyme in plasmids to confirm the nature of the construct.

## 3.2. Effect of Hybrid Ribozymes on Expression of the Target mRNA

To confirm the effect of hybrid ribozymes, a transient assay is usually performed in mammalian cells. In this assay, hybrid ribozyme expression plasmids are introduced into mammalian cells. Then, the level of expression of target genes is examined using Western blot analysis.

1. Seed HeLa S3 cells at approx 25% confluence on 10-cm plate in Dulbecco's modified Eagle's medium (DMEM) with 10% fetal bovine serum (FBS). One should optimize the condition according to the cell types of interest.

2. After 24 h, the cells should be 40–60% confluent. Change to 7 mL fresh medium before transfection.

3. Add 200 μL buffer EC to 5 μg hybrid ribozyme expression plasmids in 5 μL TE buffer. Then add 16 μL eEnhancer solution to above mixture, and mix by vortexing for 1 s.

4. Incubate at room temperature for 5 min, and then spin down the mixture.

5. Add 50 μL Effectene transfection reagent and mix by vortexing for 10 s.

6. Incubate the mixture for 10 min at room temperature.

7. Add 1 mL DMEM medium to the mixture.

8. Mix by pipetting, and immediately drop onto the cells in each dish.

9. After 48 h, collect the cells, and extract proteins from the cells using lysis buffer.

10. Confirm the reduction of target protein using Western blot analysis with specific antibodies (*see* **Fig. 2**).

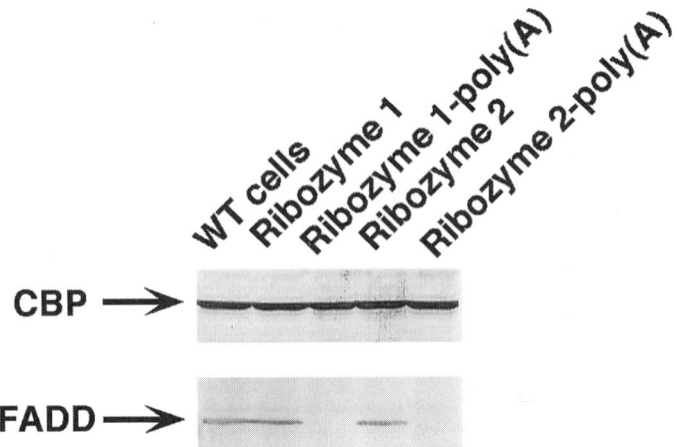

Fig. 2. The levels of FADD in cells that expressed the hybrid ribozyme-poly(A). Total proteins (each 20 μg) were resolved by sodium dodecyl sulfate polyacrylamide gel electrophoresis (SDS-PAGE) (10% polyacrylamide gel) and transferred to a polyvinylene difluoride (PVDF) membrane (Funakoshi Co., Tokyo, Japan) by electroblotting. Immune complexes were visualized with Amplified Immunblot kit (BIORAD, CA) using specific polyclonal antibodies against FADD (Santa Cruz, CA). CBP is an endogenous control.

## 4. Notes

1. One should choose GUC or other cleavable triplet as a target site because ribozymes require NUX (where N and X can be A, G, C, or U and A, C, or U, respectively) for effective cleavage in vitro and in vivo.
2. A short 7–10 sequence for binding of the ribozyme to its target is usually used. It has been shown that longer sequences reduce specificity and also inhibit the dissociation step of the ribozyme-catalyzed reaction, and that shorter sequences decrease the specificity for the target sequence.
3. Do not construct the ribozyme sequence that contains 5'-TTTT-3' sequence because this sequence acts as a terminator of pol III-dependent transcription.

## References

1. Uhlenbeck, O. C. (1987) A small catalytic oligoribonucleotide. *Nature* **328**, 596–600.
2. Zhou, D. M. and Taira, K. (1998) The hydrolysis of RNA: from theoretical calculations to the hammerhead ribozyme-mediated cleavage of RNA. *Chem. Rev.* **98**, 991–1026.
3. Rossi, J. J. and Couture, L. A. (1999) *Intracellular Ribozyme Applications.* Horizon Scientific Press, Norfolk, UK.
4. Krupp, G. and Gaur, R. K. (2000) *Ribozyme: Biochemistry and Biotechnology.* Eaton Publishing, Natick, MA.

5. Warashina, M., Kuwabara, T., Kato, Y., Sano, M., and Taira, K. (2001) RNA-protein hybrid ribozymes that efficientl cleave any mRNA independently of the structure of the target RNA. *Proc. Natl. Acad. Sci. USA* **98,** 5572–5577.

6. Kawasaki, H. and Taira, K. (2002) Identification of genes by hybrid-ribozymes that couple cleavage activity with the unwinding activity of an endogenous RNA helicase. *EMBO Rep.* **3,** 443–450.

7. Kawasaki, H., Ohnuki, R., Suyama, E., and Taira, K. (2002) Identification of genes that function in the TNF-α-mediated apoptotic pathway using randomized hybrid ribozyme libraries. *Nat. Biotechnol.* **20,** 376–380.

8. Kawasaki, H. and Taira, K. (2002) A functional gene discovery in the Fas-mediated pathway to apoptosis by analysis of transiently expressed randomized hybrid-ribozyme libraries. *Nucleic Acids Res.* **30,** 3609–3614.

9. Kawasaki, H., Ecker, R., Yao, T.-P., Taira, K., Chiu, R., Livingston, D. M., and Yokoyama, K. K. (1998) Distinct roles of the co-activators p300 and CBP in retinoic-acid-induced F9-cell differentiation. *Naure* **393,** 284–289.

10. Kuwabara, T., Warashina, M., Tanabe, T., Tani, K., Asano, S., and Taira, K. (1998) A novel allosterically *trans*-activated ribozyme, the maxizyme, with exceptional specificity *in vitro* and *in vivo. Mol. Cell* **2,** 617–627.

11. Koseki, S., Tanabe, T., Tani, K., Asano, S., Shioda, T., Nagai, Y., et al. (1999) Factors governing the activity *in vivo* of ribozymes transcribed by RNA polymerase III. *J. Virol.* **73,** 1868–1877.

12. Maddox, J. (1998) The great gene shears story. *Nature* **342,** 609–613.

13. Pause, A. and Sonenberg, N. (1992) Mutational analysis of a DEAD box RNA helicase: the mammalian translation initiation factor eIF-4A. *EMBO J.* **11,** 2643–2654.

14. Craig, A. W., Haghighat, A., Yu, A. T. and Sonenberg, N. (1998) Interaction of polyadenylate-binding protein with the eIF4G homologue PAIP enhances translation. *Nature* **392,** 520–523.

15. Tang, H., Gaietta, G. M., Fischer, W. H., Ellisman, M. H., and Wong-Staal, F., (1997) A cellular cofactor for the constitutive transport element of type D retrovirus. *Science* **276,** 1412–1415.

16. Braun, I. C., Rohrbach, E., Schmitt, C., and Izaurralde, E. (1999) TAP binds to the constitutive transport element (CTE) through a novel RNA-binding motif that is sufficient to promote CTE-dependent RNA export from the nucleus. *EMBO J.,* **18,** 1953–1965.

17. Schmitt, C., von Kobbe, C., Bachi, A., Pante, N., Rodrigues, J. P., Boscheron, C., et al. (1999) Dbp5, a DEAD box-protein required for mRNA export, is recruited to the cytoplasmic fibrils of Nuclear Pore Complex via a conserved interaction with CAN/Nup159p. *EMBO J.* **18,** 4332–4347.

18. Kang, Y. and Cullen, B. R. (1999) The human Tap protein is a nuclear mRNA export factor that contains novel RNA-binding and nucleocytoplasmic transport sequence. *Genes Dev.* **13,** 1126–1139.

# 17

# Functional Gene Discovery Using Hybrid Ribozyme Libraries

Yoshio Kato, Masaru Tsunemi, Makoto Miyagishi, Hiroaki Kawasaki, and Kazunari Taira

## Summary

Hybrid ribozymes that couple the cleavage activity of hammerhead ribozymes with the unwinding activity of RNA helicases are powerful tools in the study of cell genetics and pharmaceutical drug development. They are useful for targeting a specific gene as well as screening functional genes to show phenotypic alterations. By randomizing the binding arms within the ribozymes, we can create a library of ribozymes that are capable of cleaving any mRNA. After introducing the library into cells and recovering the sequence information of ribozymes that alter the cell phenotype of interest, genes that are responsible for the specific phenotype can be identified. This chapter describes a method known as the gene-discovery system to identify novel functional genes related to a specific phenotype. Our gene-discovery system should be a powerful tool for post-genome research.

**Key Words:** Ribozyme; library; randomization; hybrid ribozyme; screening system; recovery of plasmid; phenotype; tRNA promoter; RNA helicase.

## 1. Introduction

Ribozymes are catalytic RNA molecules that bind to defined RNA targets based on sequence complementarity and enzymatically cleave those RNA targets. To date, numerous studies directed toward the application of ribozymes in vivo have been performed, and many successful experiments aimed at the exploitation of ribozymes for the suppression of gene expression in various organisms have been reported (1,2).

Successful inactivation by ribozymes of a specific gene in cells depends very strongly on the appropriate design of the expression vector. High levels of

From: *Methods in Molecular Biology, vol. 252: Ribozymes and siRNA Protocols, Second Edition*
Edited by: M. Sioud © Humana Press Inc., Totowa, NJ

expression under the control of the promoter of a human gene for tRNA$^{Val}$ would obviously be advantageous for the exploitation of ribozymes in vivo. Ribozymes expressed under the control of the tRNA$^{Val}$ promoter with an appropriate linker are exported to the cytoplasm to colocalize with its target mRNA *(3–5)*. This expression system has been used successfully in the suppression of target genes by ribozymes *(6,7)*.

However, the efficiency of ribozyme-mediated cleavage in vivo is not always as high as anticipated or required. In vivo, the activities of ribozymes depend on access to their cleavage site in the target RNA, as there is evidence that the association of ribozymes with the target mRNA is the rate-limiting step in ribozyme-mediated reactions in cells *(4,8)*. To overcome this problem, efforts are being made that involve computer predictions of the secondary structure of the target RNA, or systematic trial-and-error experiments are being performed with large numbers of antisense molecules or ribozymes.

In order to solve this target problem and to improve the efficiency of ribozymes in vivo, we have created ribozymes with the ability to access any target site and to cleave at a specific site. This was accomplished by combining the cleavage activity of the hammerhead ribozymes with the unwinding activity of the endogenous RNA helicase. To connect the helicase to the ribozymes, we added a naturally occurring RNA motif to the 3' end of the ribozymes. We recently constructed two novel hybrid ribozymes, a poly(A)-connected ribozyme and a constitutive transport element (CTE)-connected ribozyme, and demonstrated that ribozymes of both types have strong cleavage activity and substrate-unwinding activity, regardless of the putative secondary or tertiary structure of the target in the vicinity of the actual target site *(8–10)*.

In addition to using these hybrid ribozymes to cleave specific known target mRNAs, they can also be used to identify genes associated with specific phenotypes in cells. This can be accomplished by creating ribozymes with randomized binding arms. The sequence of the human genome has become available, and it will be extremely valuable to have methods for the rapid identification of important genes. Since our hybrid ribozymes can attack structured sites, they can cleave mRNA with high-level efficiency. If libraries of hybrid ribozymes with randomized binding arms are introduced into cells, the genes associated with any changes in phenotypes—as a result of ribozyme-mediated cleavage of some mRNAs—can be readily identified by sequencing of the specific ribozyme clone. This gene-discovery system based on hybrid ribozyme libraries is indeed useful for the rapid identification of functional genes in the post-genome era.

## 2. Materials

### 2.1. Oligonucleotide

1. Primer A: 5'- CCG GTT CGA AAC CGG GCA C -3'.
2. Primer B: 5'- AAA AAA AGA TAT CCG GGG TAC CT -3'.
3. Template DNA: 5'- CCG GTT CGA AAC CGG GCA CTA CAA AAA CCA ACA **ANN NNN NN**C TGA TGA GGC CGA AAG GCC GAA **ANN NNN N**AG GT ACC CCG GAT ATC TTT TTT T -3'.

### 2.2. Reagents and Kits

1. *Taq* DNA polymerase for PCR (e.g., *Taq* DNA polymerase, TAKARA).
2. *E. coli* for transformation (e.g., DH5α strain).
3. Electroporation apparatus (e.g., Gene Pulser, Bio-Rad).
4. Ligase (e.g., TAKARA DNA ligation solution ver. 2).
5. Restriction enzymes; *Kpn*I and *Csp*45I (TOYOBO).
6. 10X L buffer: 100 m$M$ MgCl$_2$, 10 m$M$ dithiothreitol (DTT), 100 m$M$ Tris-HCl, pH 7.5.
7. Miniprep kit (QIAGEN).
8. PCR purification kits (e.g., QIAGEN or Promega).
9. TE buffer: 10 m$M$ Tris-HCl, pH 8.0, 1 m$M$ ethylenediaminetetraacetic acid (EDTA).
10. Transfection reagent: Lipofectamine 2000 (Invitrogen).
11. Dulbecco's modified Eagle's medium (DMEM) (Gibco-BRL).
12. Phosphate-buffered saline (PBS).
13. OPTI-MEMI (Gibco-BRL).

## 3. Methods

The gene-discovery system can be used in the genetic screening approach without prior sequence information of functional genes. This novel technology identifies all differentially regulated genes that directly or indirectly affect the phenotype of interest, even if the expression of the gene is extremely low for detection by other methods such as differential PCR. Moreover, after determining the sequence of target mRNA, the recovered ribozyme vector itself from the screening allows the analysis of the phenotype of interest.

### 3.1. Construction of the Plasmid Libraries Expressing the Ribozymes

The diversity of a ribozyme library is based on the randomization of the substrate-binding region of ribozymes. Although the way to construct the library of ribozymes is almost the same as that for the ribozyme against a specific target (*see* Chapter 16), there are several points that are important in order to maintain the diversity.

The library with randomized ribozyme sequences should be introduced into cells as the plasmid DNAs that express ribozymes constitutively. We have succeeded in establishing an effective ribozyme expression system by using the human tRNA$^{Val}$ promoter, which is appropriate to express small RNAs at a high level of transcription. Moreover, the attachment of an RNA helicase-binding motif to the ribozyme enhances the ability to bind a structured substrate.

Construction of the library consists of three steps (**Fig. 1**). At first, the double-stranded DNA fragments that encode the catalytic core of a ribozyme with both ends flanked by the randomized nucleotides should be chemically synthesized. After digestion of the synthesized fragments by appropriate restriction enzymes, they are ligated into the plasmid. Next, the plasmids are transformed into the *E. coli* cells.

1. Synthesize or purchase the DNA template of a ribozyme as follows, 5'- CCG GTT CGA AAC CGG GCA CTA CAA AAA CCA ACA A**NN NNN NN**C TGA TGA GGC CGA AAG GCC GAA A**NN NNN N**AG GTA CCC CGG ATA TCT TTT TTT -3', where the bold letters are the binding arms. N stands for any nucleotides (*see* **Note 1**).

2. Amplify the DNA templates by PCR. The reaction solution contains (final concentration): template DNA (0.02 μ*M*), primer A (0.5 μ*M*), primer B (0.5 μ*M*), *Taq* buffer, dNTP mix (0.25 m*M* each), and *Taq* DNA polymerase (10 U). The thermal cycle is as follows: denaturing (95°C, 30 s), annealing (55°C, 30 s), polymerization (74°C, 30 s), and repeat 6–8 cycles (*see* **Note 2**).

3. Purify the PCR products by using PCR purification kits or by phenol-chloroform extraction followed by ethanol-precipitation. PCR purification kits can remove enzymes and salts as well as excess amounts of primers.

4. Digest the PCR products with the restriction enzymes *Csp*45I and *Kpn*I. The reaction condition is as follows; L buffer, *Csp*45I and *Kpn*I, 37°C and >2 h. After the reaction, purify the DNA as noted previously.

5. Ligate 550 ng of the DNA with the 20-μg plasmids that had been digested in advance by *Csp*45I and *Kpn*I. Ligation reaction is carried out by using Ligation solution ver. 2 (Takara) at 16°C, >4 h. In order to check the quality of the vector, it is essential to ligate the vector by itself in the absence of the insert DNA, and it is important to confirmed that the transformed colonies are almost not found.

6. Transform the ligation products (1 mL) into *E. coli* cells (20 mL). Electroporation of plasmids transforms more efficiently than the heat-shock method using chemically

---

Fig. 1. *(opposite page)* (**A**) Schematic illustration of hybrid-ribozyme. tRNA$^{Val}$-driven hammerhead ribozyme with RNA helicase-binding motif (HBM) binds to its target mRNA. The cleavage site is indicated by an arrow. (**B**) The construction of a randomized ribozyme library and a map of the vector. Randomized DNA corresponding to the ribozyme is integrated into the plasmid that contains the human tRNA$^{Val}$ promoter, HBM, and the terminator sequence (T$_5$) for pol III transcription.

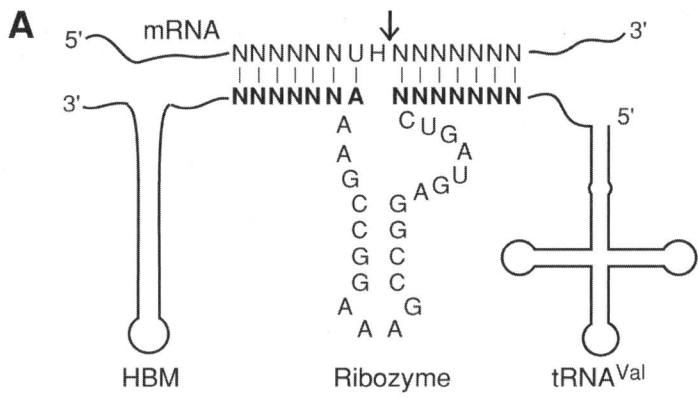

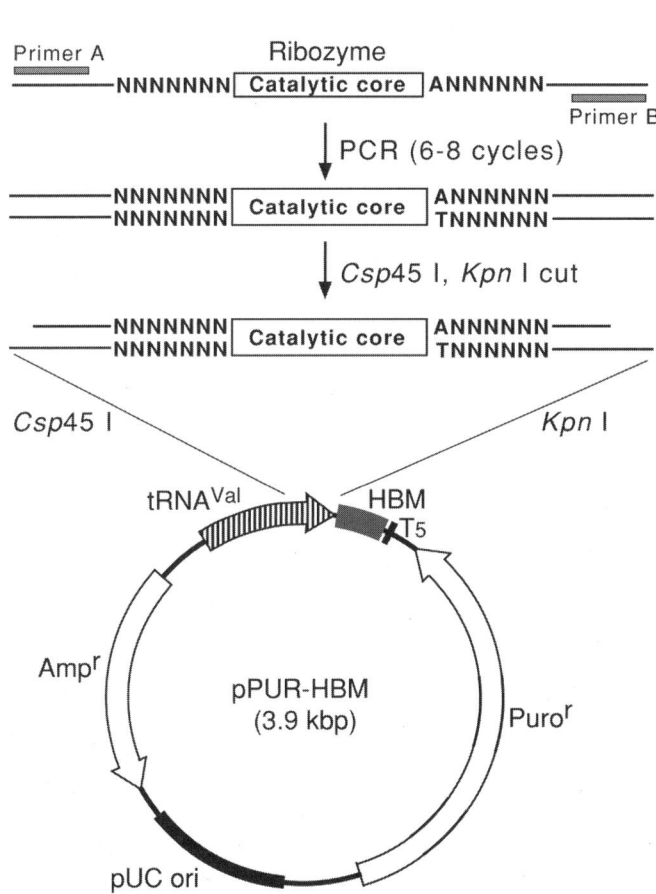

competent cells. However, the amount of competent cells in the electroporation method is too small to prepare with a sufficient amount of the library. Therefore, it is desirable to use the heat-shock method using chemically competent cells. After the stimulation, add SOC medium (180 mL) preheated at 37°C.

7. Incubate the *E. coli* solution at 37°C for 60 min. After the incubation, a part of *E. coli* solutions (10 μL and 100 μL) are plated to LB agarose plates containing ampicillin. Residual solution is subjected to the mini-preparation to extract the plasmid DNA.

8. After 12–14 h of incubation of LB plates, count the number of colonies of *E. coli* cells in order to check the diversity of libraries statistically. For example, if 1000 colonies are found on a plate to which 10 μL of 200 mL *E. coli* solution has been plated, the diversity of the library is estimated as $2 \times 10^7 = 1000$ (colonies)× 200 (mL)/10 (μL). Practically, about 50 independent clones can be picked and sequenced.

## 3.2. Transfection of Plasmids and Induction of Separable Pressure

In this system, the gene that is responsible for a specific phenotype of interest may be discovered. When the library is introduced into cells, one of the ribozymes of the library may cleave a message, and as a result, may change the phenotype of the cells. Identification of the target of the ribozyme leads directly to the identification of the gene that is responsible for the phenotype.

One of the most important steps in this procedure is to isolate cells with a phenotype that has changed by the introduction of ribozymes. The reported phenotypes for isolation of cells thus far are summarized in **Table 1**. This gene-discovery procedure may lead to identification of any genes of interest that are responsible for any phenotype. However, the application of the method is limited by the availability to separate the cells that had changed phenotypically from the remaining unchanged cells. Thus, one must fully consider the system for the procedure of isolation of cells corresponding to the phenotype of interest.

For this system, it is ideal that a single plasmid is introduced into a single cell. This is because when multiple plasmids are introduced into a cell and if one of the plasmids affects the phenotype, the other plasmids can also be rescued, and such plasmids may be regarded as the false-positives. Furthermore, we must carefully discriminate between the cells changed phenotypically and the cells unchanged, because the plasmids recovered from the unchanged cells constitute the false-positives.

In order to reduce such false-positives, one must optimize the separation conditions prior to the introduction of ribozyme library into cells. Although there are several separation systems, in this section we describe the conditions for identification of genes related to the induction apoptosis by ultraviolet (uv) irradiation, as an example. When cells are exposed to UV light, they die by apoptosis. Although several pathways lead to UV-induced apoptosis, the com-

**Table 1**
**Reported System to Separate From Phenotypic Converted Cells**

| Separation | Phenotype of interest | Candidate genes | Ref. |
|---|---|---|---|
| Survival | Apoptosis | Apoptosis-related genes | *10,11* |
| | Suicide gene as reporter | Viral translational factor | *12* |
| | Reagent sensitivity | Alzheimer related gene | *13* |
| Focus formation | Transformation | Telomerase | *14* |
| Attachment | Adherency | Adherence-related genes | *15* |
| FACS | GFP reporter | Translational factor | *16* |
| Permeability | Metastasis | Metastasis related genes | *17* |

plete cascade remains unknown. To identify genes that are related to UV-induced apoptosis, we can apply the ribozyme library. We can easily separate the cells converted by ribozymes from unchanged cells because the cells that were not affected by ribozymes die more easily and detach from the bottom of the culture dish, and the cells in which the apoptotic pathway was suppressed may remain attached to the dish. This protocol is written for the experiment of a 10-cm culture dish.

1. Optimize the condition of the separation of cells, depending on the presence or absence of a specific phenotype, as described previously. If the background is less than 1%, then proceed to the next step.
2. The day before transfection, trypsinize and count the number of cells, plating 90% confluent on a 10-cm dish on the day of transfection. Optimize the condition of confluency according to your research. The speed of growth of the cells depends on the kinds of cells.
3. Mix lipofectamine 2000 reagent (15 µg) and OPTI-MEM I (750 µL). The adequate transfection reagents should be used according to each cell type selected. This protocol describes a method of transfection for lipofectamine, as an example.
4. Mix the plasmid DNA (15 µg) and OPTI-MEM I (750 µL).
5. Combine the DNA in OPTI-MEM I with lipofectamine 2000 reagent.
6. Add the DNA-lipofectamine 2000 reagent complex to culture plate.
7. Incubate the cells at 37°C in a $CO_2$ incubator for a total of 24 h.
8. Induce the separable phenotype as shown in **Table 1**. (In the case of UV irradiation, after 24 h, irradiate UV to the culture plate. The strength and time of UV light exposure should be optimized in advance. The conditions to induce 99% of dead cells may be appropriate.)
9. Separate the selected cells. (In the case of UV irradiation, after 48 h, dead cells are detached from the bottom of dish, and the surviving cells remain attached. Attached cells are collected with treatment with trypsin.)
10. Extract plasmids (*see* **Subheading 3.3.**).

## 3.3. Recovery of Plasmids

Hirt's DNA isolation method is well-known as the protocol to isolate plasmid DNAs from mammalian cells. This method enables recovery of plasmid DNA by avoiding nicking. However, the method is time-consuming, and requires many steps to recover the plasmid just to be transformed into *E. coli* cells, since even nicked plasmid DNAs can be transformed into *E. coli* cells as efficiently as intact plasmids. Thus, we recommend a much easier approach, using such methods as a Miniprep kit (QIAGEN) for *E. coli* cells, as follows:

1. Collect cells into a 1.5-mL tube. In the case of adherent cells, detach the cells by 1 mL of trypsin solution, centrifuge at a maximum speed at 4°C and discard the supernatant.
2. Resuspend the cell pellet in buffer P1 (250 µL).
3. Add buffer P2 (250 µL) and invert the tube gently 4–6 times to mix.
4. Add cold buffer P3 (350 µL) and invert the tube immediately but gently 4–6 times.
5. Centrifuge for 10 min at a maximum speed in a centrifuge.
6. Apply the supernatant from **step 5** to the column by pipetting.
7. Centrifuge for 60 s. Discard the flowthrough.
8. Wash the column by adding buffer PE (500 µL) and centrifuge for 60 s.
9. Discard the flowthrough, and centrifuge for an additional 1 min to remove residual wash buffer.
10. Place the column in a clean 1.5-mL microcentrifuge tube. To elute DNA, add buffer EB (50 µL) twice to the top center of the column, let stand for 1 min, and centrifuge for 1 min.
11. Transform the resultant plasmids into *E. coli* cells as noted previously. If the number of colony of *E. coli* cells are less than—for example—100, then we can sequence the individual clones. If the colonies are extensive, we often recover the plasmid from the *E. coli* cells on the LB plate. *E. coli* cells on LB plate are directly added and suspended by SOC medium and followed by mini-preparation as described in **step 1** of this section. Resultant plasmids can be transfected as mentioned in **Subheading 3.2.**
12. Determine the sequences of plasmid DNAs by a sequencer.
13. Check the reproducibility of phenotypic conversion by using individual resultant plasmids. Note that false-positives should not change the phenotype, whereas positives should induce the expected phenotypic conversion.

## 3.4. Analysis of Sequences

By determining the sequence of the binding arms of a ribozyme, we can identify an mRNA that has been cleaved by the ribozyme. Since a ribozyme recognizes the substrate via Watson-Crick basepairing, the complementary sequence of the substrate-binding region of the recovered ribozyme is a part of the putative target mRNA. The sequence of the candidate mRNA estimated by the binding arms is approx 15 nt in length, with 1-nt gap at the center. The gap nucleotide, which is located at the 5' side of the cleavage site, cannot be identified as a unique nucleotide because it is not bound to the ribozyme.

The candidate genes can be determined by 5' or 3' RACE (rapid amplification of cDNA ends) using the obtained sequence as primers, by screening from cDNA libraries using the estimated sequence as probes, or by searching for homology using the entire sequence as query. The latter method easily allows for identification of the candidate. The BLAST program is well-known as a search engine to find homologies of sequences from the data base, and is suitable for our goals. The BLAST program is served on the website such as NCBI, and can be accessed at any time. Here, we describe the methods to find and list the candidate genes:

1. Access the web site; http://www.ncbi.nlm.nih.gov/blast/. Click "Search for short nearly exact matches" to open the format.
2. Fill in the sequence into the box of the search. The sequence may be 15 nt with one gap such as NNNNNNTHNNNNNN, where H can be filled in this format (*see* **Notes 4** and **5**).
3. Click the [BLAST!] button. The next page will be seen as follows, "Your request has been successfully submitted and put into the Blast Queue...". The results will be ready within 5 min.
4. Click the [Format!] button. The next page will appear.
5. Sequences and the names of genes with score (bits) and E value are listed at the middle of the page. A higher score and lower E value indicate higher homology.
6. If the gene of interest is found, confirm that the identified gene is the real target or not by targeting the same gene at different sites, or confirm by using another gene-suppression method such as RNAi (*see* Chapter 18). If one cannot find any homologous genes, try to carry out 5' or 3' RACE screening of cDNAs.

## 4. Notes

1. The length of the substrate-binding region must be weighed. We recommend the length of 7 and 7 at the 5' and 3' flanking sites of the ribozyme catalytic core. Shorter lengths of substrate-binding regions cause a decrease in the binding ability to a substrate and longer lengths require larger amounts of synthesis/reaction volumes in order to cover the entire population. In the scale of this protocol, about one molecule per one type of clone is included in a library.
2. The concentration of the template DNA should be in the range of 0.02–0.04 $\mu M$. After the 6–8 cycles of PCR, the concentration of the polymerase chain reaction (PCR) products may be approx 0.2 $\mu M$. More than eight cycles of PCR may bias the population of an individual ribozyme. Since the PCR amplifies the DNA exponentially, the difference in the concentration of individual DNA species increases with repeating cycles.
3. In order to avoid the replication of *E. coli* cells, do not incubate for more than 60 min since individual *E. coli* cells in solution replicate at different rates and the population of each cell may vary.

Randomized ribozyme library

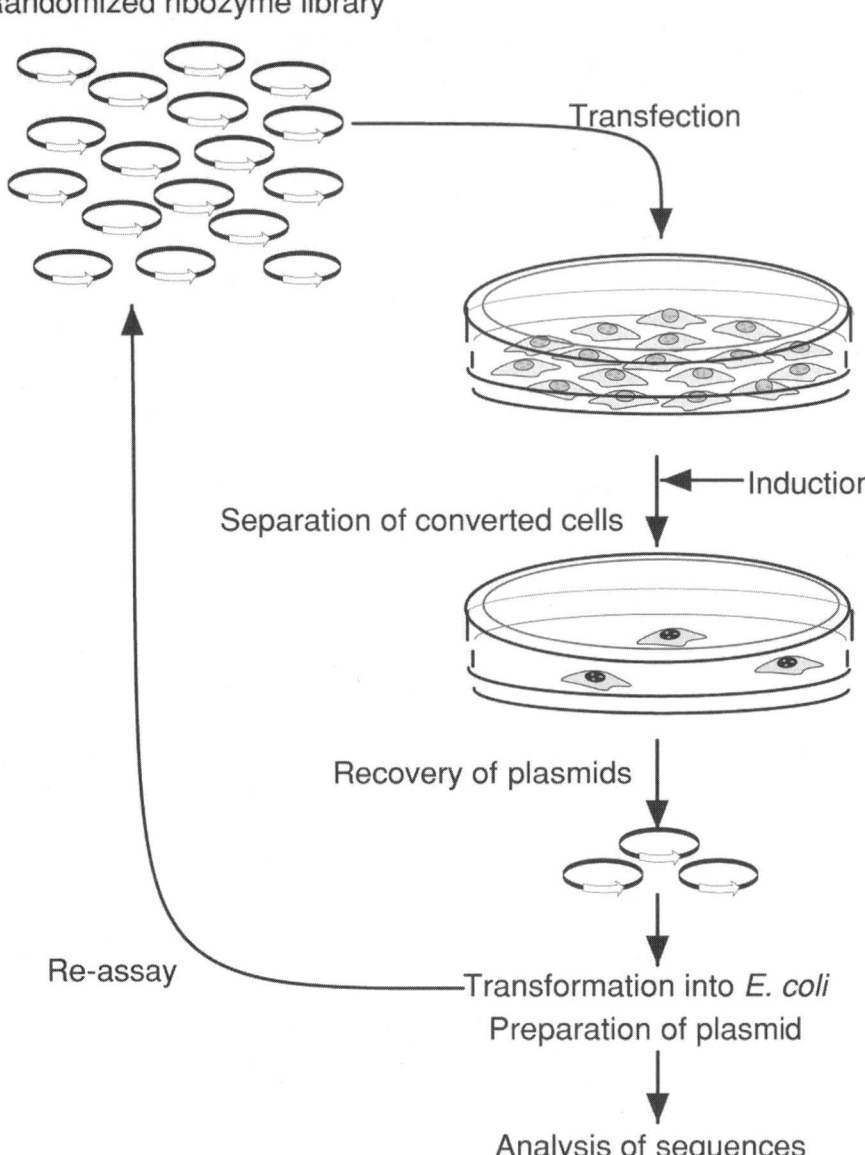

Fig. 2. Schematic diagram of the gene-discovery system. Cells are transfected by a random plasmid library and induced by selectable pressure. The cells that the phenotype of interest is converted by the effect of several ribozymes are isolated. After the recovery of the plasmids from the isolated cells, the sequences of the ribozymes that produce the phenotype of interest are analyzed, and the resultant plasmids are used for the second selection.

4. In the case of hammerhead ribozymes, a gap nucleotide, which does not form basepairing, is H (A, C, or U) nucleotide. Although we have adopted hammerhead ribozymes in this system, other ribozymes such as hairpin ribozymes can be used in a similar fashion. However, since the gap nucleotides of hairpin ribozymes are 4 nt in length, the identification of target genes may be more ambiguous than in the case of hammerhead ribozymes.
5. Optionally one can choose the species used in the study. These restrictions may decrease the hit of unrelated sequences and the searching time. Ribozymes often target EST (expression sequence tag), which means that the identified sequence is expressed to be transcribed.

## References

1. Rossi, J. J. and Couture, L. A. (eds.) (1999) *Intracellular Ribozyme Applications.* Horizon Scientific Press, Norfolk, UK.
2. Krupp, G. and Gaur, R. K. (eds.) (2000) *Ribozyme: Biochemistry and Biotechnology.* Eaton Publishing, Natick, MA.
3. Koseki, S., Tanabe, T., Tani, K., Asano, S., Shioda, T., Nagai, Y., et al. (1999) Factors governing the activity *in vivo* of ribozymes transcribed by RNA polymerase III. *J. Virol.* **73**, 1868–1877.
4. Kato, Y., Kuwabara, T., Warashina, M., Toda, H., and Taira, K. (2001) Relationships between the activities *in vitro* and *in vivo* of various kinds of ribozyme and their intracellular localization in mammalian cells. *J. Biol. Chem.* **276**, 15,378–15,385.
5. Kuwabara, T., Warashina, M., Sano, M., Tang, H., Wong-Staal, F., Munekata, E., and Taira, K. (2001) *Biomacromolecules* **2**, 1229–1242.
6. Kawasaki, H., Ecker, R., Yao, T.-P., Taira, K., Chiu, R., Livingston, D. M., and Yokoyama, K. K. (1998) Distinct roles of the co-activators p300 and CBP in retinoic-acid-induced F9-cell differentiation. *Nature* **393**, 284–289.
7. Tanabe, T., Kuwabara, T., Warashina, M., Tani, K., Taira, K., and Asano, S. (2000) Oncogene inactivation in a mouse model: tissue invasion by leukaemic cells is stalled by loading them with a designer ribozyme. *Nature* **406**, 473,474.
8. Warashina, M., Kuwabara, T., Kato, Y., Sano, M., and Taira, K. (2001) RNA-protein hybrid ribozymes that efficiently cleave any mRNA independently of the structure of the target RNA. *Proc. Natl. Acad. Sci. USA* **98**, 5572–5577.
9. Kawasaki, H. and Taira, K. (2002) Identification of genes by hybrid-ribozymes that couple cleavage activity with the unwinding activity of an endogenous RNA helicase. *EMBO Rep.* **3**, 443–450.
10. Kawasaki, H., Ohnuki, R., Suyama, E., and Taira, K. (2002) Identification of genes that function in the TNF-α-mediated apoptotic pathway using randomized hybrid ribozyme libraries. *Nat. Biotechnol.* **20**, 376–380.
11. Kawasaki, H. and Taira, K. (2002) A functional gene discovery in the Fas-mediated pathway to apoptosis by analysis of transiently expressed randomized hybrid-ribozyme libraries. *Nucleic Acids Res.* **30**, 3609–3614.
12. Krüger, M., Beger, C., Li, Q.-X., Welch, P. J., Tritz, R., Leavitt, M., et al. (2000) Identification of eIF2Bγ and eIF2γ as cofactors of hepatitis C virus internal

ribosome entry site-mediated translation using a functional genomics approach. *Proc. Natl. Acad. Sci. USA* **97,** 8566–8571.

13. Ohnuki, R., Kawasaki, H., and Taira, K. Submitted for publication.
14. Li, Q.-X., Robbins, J. M., Welch, P. J., Wong-Staal, F., and Barber, J. R. (2000) A novel functional genomics approach identifies mTERT as a suppressor of fibroblast transformation. *Nucleic Acids Res.* **28,** 2605–2612.
15. Welch, P. J., Marcusson, E. G., Li, Q.-X., Begar, C., Krüger, M., Zhou, C., et al. (2000) Identification and validation of a gene involved in anchorage-independent cell growth control using a library of randomized hairpin ribozymes. *Genomics* **66,** 274–283.
16. Beger, C., Pierce, L. N., Krüger, M., Marcusson, E. G., Robbins, J. M., Welcsh, P., et al. (2001) Identification of Id4 as a regulator of BRCA1 expression by using a ribozyme-library-based inverse genomics approach. *Proc. Natl. Acad. Sci. USA* **98,** 130–135.
17. Suyama, E., Kawasaki, H., Kasaoka T., and Taira, K. (2003) Identification of genes responsible for cell migration by a library of randomized ribozymes. *Cancer Res.* **63,** 119–124.

# 18

# Maxizyme Technology

## Mayu Iyo, Hiroaki Kawasaki, and Kazunari Taira

### Summary

Ribozymes are small and versatile nucleic acids that can cleave RNAs at specific sites. These molecules have great potential to be used as effective gene-therapeutic agents. However, because of the limitation for cleavable sequences within the target mRNA, in some cases conventional ribozymes have failed to exhibit precise cleavage specificity. A maxizyme is the dimer of minimized ribozymes (minizymes), which can specifically cleave two distinct target sites. The maxizyme also has an allosteric function in that it can form an active conformation and cleave the two target sites only when it recognizes two distinct target sites. We demonstrated previously that an allosterically controllable maxizyme was a powerful tool in the disruption of an abnormal chimeric RNA (*bcr-abl*) in cells and in mice. Furthermore, more than five custom-designed maxizymes have clearly demonstrated these allosteric functions in vitro and in vivo. Thus, maxizyme technology is not limited to one specific case, but may have broad general applicability in molecular biology and in molecular gene therapy.

**Key Words:** Hammerhead ribozyme; allosteric ribozyme; maxizyme; leukemia; gene therapy; *bcr-abl*; *Cis*-acting; sensor arm; *Trans*-maxizyme.

## 1. Introduction

Ribozymes, RNA enzymes, are RNA molecules that can cleave oligoribonucleotides at specific sites—namely, after the sequence UX, in which X can be A, C, or U *(1)*. Hammerhead ribozymes have been more commonly used because of their small size, only approx 30 nucleotides, and high specificity *(2–4)*. The ability of these ribozymes to cleave other RNA molecules at specific sites makes them useful as inhibitors of viral replication or of malignancy *(5)*. However, because of the limitation for cleavable sequences in the target mRNA, in some cases conventional ribozymes failed to possess precise cleavage specificity. It is possible that ribozymes may work in cancer cells as

From: *Methods in Molecular Biology, vol. 252: Ribozymes and siRNA Protocols, Second Edition*
Edited by: M. Sioud © Humana Press Inc., Totowa, NJ

well as normal cells, if the target genes of ribozymes are expressed both in normal cells and in cancer cells. In this case, they might cause toxicity in normal cells. For example, it is difficult to construct specific ribozymes that target an abnormal *bcr-abl* chimeric mRNA, because both *bcr* and *abl* sequences are expressed in normal cells.

A maxizyme invented by our group consists of two shortened ribozymes—minizymes—and can recognize and cleave two different target sites. A maxizyme results from deleting the stem-loop II region of a hammerhead ribozyme, and this minizyme is inactive as a monomer, but exhibits a strong catalytic activity as a dimer *(6,7)*. Furthermore, a heterodimeric system that is composed of two distinct monomers—maxizyme left (MzL) and maxizyme right (MzR)—was also designed *(8)*. This maxizyme also has an allosteric function that it can form an active conformation only when it specifically binds two target sites. We have previously shown that the *cis*-maxizyme targeting the *bcr-abl* chimeric mRNA, which causes chronic myelogenous leukemia (CML), is functional in vitro as well as a mouse model system *(9,10)*. In the case of targeting the *bcr-abl* chimeric mRNA, it is impossible to design a conventional hammerhead ribozyme that cleaves only the chimeric target without affecting each normal *bcr* or normal *abl* mRNA because there is no UX-cleavable site near the junction. The *cis*-acting maxizyme, which recognizes two distinct target sites in the target *bcr-abl* chimera mRNA, cleaved only the chimera mRNA and did not affect normal mRNAs, because the first binding region of the maxiyzme recognized the unique junction sequence and the second binding region was used to create a catalytic domain for an efficient cleavage.

In the case of targeting genes that are overexpressed in cancer cells but also needed in normal cells, decreasing the expression of these genes by ribozymes only in cancer cells may be impossible. For example, two distinct oncogenes—*hst-1* and *cyclin D1*—are overexpressed in breast-cancer cells, although these genes are also necessary for growth factor-dependent signal transduction and for cell cycling in normal cells *(11)*. Therefore, we have attempted to construct the *trans*-maxizyme (**Fig. 1**) havingwith an allosteric function, which can work only when it specifically recognizes the two target sites simultaneously in breast-cancer cells, in which the densities of target mRNAs are very high, and cannot readily form the active conformation in normal cells because the level of target mRNAs is low. The *trans*-maxizyme could be expected to cleave *hst-1* and *cyclin D1*, not in the normal cells but in breast-cancer cells.

In this chapter, we describe how to design and construct the *trans*-maxizyme expression vector. A simple method for in vitro analysis of the constructs is also described.

**A** Maxizyme

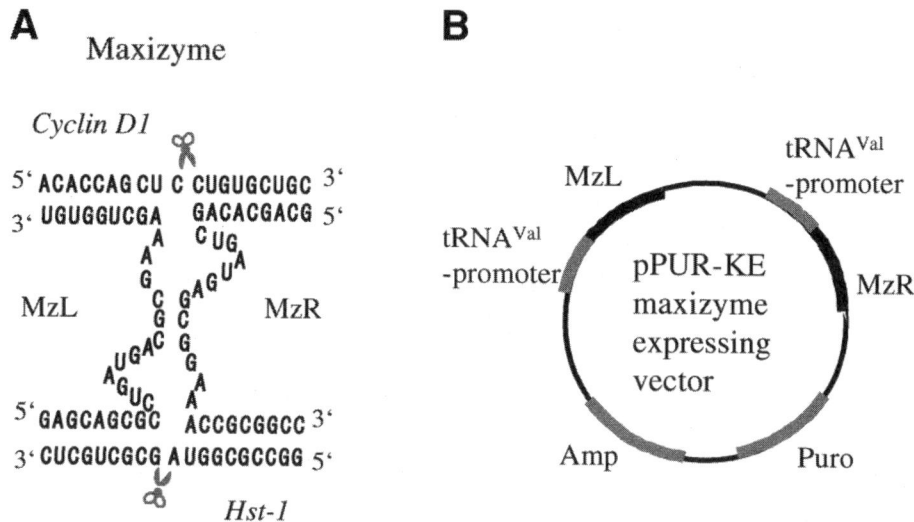

**B**

Fig. 1. **(A)** Predicted secondary structure, based on calculations by the mulfold program, of an active *trans*-maxizyme in the presence of both substrates. **(B)** A vector encoding tRNA$^{Val}$-MzL (maxizyme left) and tRNA$^{Val}$-MzR (maxizyme right).

## 2. Materials
### 2.1. Buffers

1. TE buffer: 10 m$M$ Tris-HCl, pH 8.0, 1 m$M$ ethylenediaminetetraacetic acid (EDTA). Use special-grade reagents and diethylpyrocarbonate-treated (DEPC-treated) distilled water. Autoclave and store at room temperature.
2. 10X TBE buffer: 890 m$M$ Tris base, 890 m$M$ boric acid, 20 mM EDTA, pH 8.0. Store at room temperature.
3. 50X TAE buffer: 2 $M$ Tris-acetate, 0.05 $M$ EDTA, pH 8.0. Store at room temperature.
4. 10X T7 RNA polymerase transcription buffer: 400 m$M$ Tris-HCl, pH 8.0, 80 m$M$ MgCl$_2$, and 20 m$M$ spermidine. Use special-grade reagents and DEPC-treated distilled water. Store at −20°C.
5. 2X RNA gel-loading buffer: 98% formamide (deionized), 10 m$M$ EDTA, pH 8.0, 0.1% xylene cyanol, and 0.1% bromophenol blue. Use special-grade reagents and DEPC-treated distilled water. Autoclave and store at room temperature.
6. 10X *Taq* polymerase buffer: 100 m$M$ Tris-HCl, pH 8.3, 15 m$M$ MgCl$_2$, and 500 m$M$ KCl. Store at −20°C.
7. 10X ribonucleotide stock (rNTP mix): 5 m$M$ each rNTP in DEPC-treated distilled water. Store at −20°C.
8. 10X deoxyribonucleotide stock (dNTP mix): 2.5 m$M$ each dNTP in distilled water. Store at −20°C.

9. DEPC-treated distilled water: Add 0.2% (v/v, final concentration) DEPC to distilled water. Shake vigorously and incubate at 37°C overnight. Autoclave for 20 min and store at room temperature.

10. 2X T4 DNA ligase buffer: 40 m$M$ Tris-HCl, pH 7.6, 10 m$M$ MgCl$_2$, 1 m$M$ adenosine 5' triphosphate (ATP), 10 m$M$ dithiothreitol (DTT), 30% (w/v) polyethylene, glycol 8000 (PEG 8000). Store at –20°C.

11. 10X CIAP buffer: 500 m$M$ Tris-HCl, pH 9.0, 10 m$M$ MgCl$_2$. Store at –20°C.

12. 10X K buffer: 200 mM Tris-HCl, pH 8.5, 100 mM MgCl$_2$, 10 mM DTT, 1000 m$M$ KCl. Store at -20∞C.

13. 10X Klenow fragment buffer: 100 m$M$ Tris-HCl, pH 7.5, 70 m$M$ MgCl$_2$, 1 m$M$ DTT. Store at –20°C.

14. 10X kinase buffer: 500 m$M$ Tris-HCl, pH 8.0, 100 m$M$ MgCl$_2$, 50 m$M$ DTT. Store at –20°C.

## 2.2. Other Reagents and Kits

1. 50 m$M$ DTT (RNase-free).
2. Ribonuclease inhibitor (30 U/mL).
3. T7 RNA polymerase (10–50 U/mL).
4. $^{32}$P-ATP (3,000 Ci/mmol, 10 mCi/mL).
5. 99.5% ethanol (special grade).
6. 70% ethanol (diluted with DEPC-treated distilled water).
7. 3 $M$ sodium acetate, pH 5.2 (RNase-free).
8. Mixture of phenol (TE-saturated, pH 8.0), chloroform, and isoamyl alcohol (25:24:1, v/v/v).
9. T4 DNA polymerase (5–10 U/mL)
10. *Taq* DNA polymerase (5 U/mL).
11. Restriction enzymes (*Kpn*I, *Csp*45I, *Eco*RI, *Bam*HI).
12. Calf intestine alkaline phosphatase (CIP), (10–30 U/mL).
13. Klenow fragment (4 U/mL).
14. T4 polynucleotide kinase (PNK) (from *E. coli* A19, 6 U/mL).
15. *E. coli* (HB101, DH-5, or XL-1Blue)-competent cells.
16. Kit for purification of DNA fragments from agarose gels (e.g., QIAquick Gel Extraction Kit; QIAGEN Inc.).

## 3. Methods

### 3.1. Design of the trans-Maxizyme and Plasmid Construction

#### 3.1.1. Design of the trans-Maxizyme and Construction of Each Maxizyme Unit Expression Vector

To design MzL and MzR of the *trans*-maxizyme that can form the active conformation only when it binds two distinct mRNAs, use the mulfold program (Biocomputing Office, Biology Department, Indiana University, Bloomington, IN) that can calculate a secondary structure of RNA as shown in

**Fig. 1.** Choose suitable target sites, which contain the NUX sequence that is necessary for the cleavage by the ribozyme or the maxizyme. The number of bases used to recognize the target site is usually 8–12 nucleotides (*see* **Note 1**). After determination of the sequences of MzL and MzR, construct the *trans*-maxizyme expression plasmid that is controlled by the tRNA$^{Val}$ promoter, as described in previous chapters. Synthesize MzL and MzR sequences connected to tRNA$^{Val}$ with a linker sequence (5'-GGG CAC TAC AAA AAC CAA CTT T-3') by a DNA synthesizer, and amplify with specific forward and reverse primers by polymerase chain reaction (PCR). Fractionate the products in 1.5% agarose gels, cut out each desired band from the gels, and purify the DNA fragments from gels using a kit. Then introduce each PCR product into pPUR-KE vector as follows:

1. In each 1.5-mL microcentrifuge tube, double-digest 2 mg of the vector plasmid pPUR-KE and each PCR product with *Kpn*I and *CSP*45I in low-salt buffer for 2 h at 37°C in a volume of 50 µL.

2. Add 50 µL of phenol (TE-saturated, pH 8.0), chloroform, and IAA (25:24:1, v/v) to each digested product. Vortex and separate the phases by microcentrifugation for 5 min at 12,000$g$ (this procedure is referred to as phenol-chloroform extraction).

3. Transfer the aqueous phase to a new 1.5-mL microcentrifuge tube, add 5 µL of 3 $M$ sodium acetate and 130 µL of ethanol, mix, and incubate at 44°C for few minutes. Collect the precipitate by microcentrifugation at top speed for 10 min at 4°C.

4. Remove the ethanol, wash the pellet with 1 µL of 70% ethanol, microcentrifuge for 5 min, remove the ethanol, and dry the pellet (procedures outlined in **steps 3** and **4** are referred to as ethanol-precipitation). Resuspend the pellet in 20 µL of TE buffer. Further purifications are not necessary for the insertion of maxizyme sequences.

5. To prevent self-ligation of the vector, incubate the pPUR-KE plasmid with CIPA (2 µL) in 1X CIP buffer for 30 min at 37°C (final volume 50 µL). Remove the enzyme by phenol-chloroform extraction and resuspend the pellet in 10 µL of TE buffer after ethanol precipitation.

6. The ligation reaction is carried our by combining the following reagents:
   a. 1 µL digested vector.
   b. 1 µL digested MzL or MzR.
   c. 2 µL distilled water.
   d. 5 µL 2X T4 DNA ligase buffer.
   e. 1 µL T4 ligase.

7. Incubate the reaction mixture for at least 30 min at 16°C.

8. Transform *E. coli* host cells with the ligation mixture and plate the cells on LB agar plates containing 100 µg/mL ampicillin. After colonies have become apparent, select desired recombinants by colony-PCR, and determine the nucleotide sequence of the positive clones to confirm the nature of the construct.

### 3.1.2. Ligation of MzR Unit to MzL Expression Vector

To generate MzL and MzR simultaneously from the same plasmid (**Fig. 1**), embed the tRNA$^{Val}$-connected MzR into the MzL cassette as follows:

1. Digest 2 μg of MzL-expressing vector with *Bam*HI and 2 mg of MzR-expressing vector with *Eco*RI and *Bam*HI in 1X K buffer for 2 h at 37°C in a volume of 50 μL (*see* **Note 2**).
2. Perform phenol-chloroform extraction and ethanol-precipitation, and then load the reaction mixtures onto a 1.5% agarose gel and subject to electrophoresis in 1X TAE buffer.
3. Excise gel pieces that contain the digested MzL-expressing vector (approx 3.7 kbp) or the MzR fragment (approx 200 bp), and purify them using a QIAquick Gel Extraction Kit.
4. To blunt-end each fragment, incubate 500 ng DNA, dNTP, and 1X Klenow fragment buffer for 1 h at 37°C in a volume of 20 μL. Remove the enzyme by phenol-chloroform extraction and perform ethanol-precipitation; then resuspend the pellet in 5 μL of TE buffer.
5. To prevent self-ligation of the vector, incubate the reaction product of MzL vector and 2 μL CIAP in 1X CIAP buffer for 30 min at 37°C in a volume of 50 μL. Remove the enzyme by phenol-chloroform extraction, perform ethanol-precipitation, and then resuspend the pellet in 5 μL of TE buffer.
6. Carry out the ligation by combining the followings reagents:
   a. 2 μL digested MzL vector.
   b. 2 μL digested MzR.
   c. 5 μL 2X T4 DNA ligase buffer.
   d. 1 μL T4 ligase.
7. Incubate the reaction mixture for at least 30 min at 16°C.
8. Transform *E. coli* host cells with the ligation mixture and plate the cells on LB agar plates containing 100 μg/mL ampicillin. After colonies have become apparent, select desired recombinants by colony-PCR, then determine the nucleotide sequence of the positive clones.

### 3.2. In Vitro Transcription of Maxizymes From the Expression Vector

To confirm whether the maxizyme shows an allosteric effect, an in vitro cleavage assay should be performed. In this assay, MzL and MzR are transcribed by T7 polymerase in vitro, and the substrates of the maxizyme are synthesized by a DNA/RNA synthesizer and then labeled at the 5'-end with $^{32}$P, as follows:

1. To prepare each maxizyme unit, amplify both regions of the MzL and MzR of the maxizyme vector by PCR with specific forward primers containing a T7 promoter and specific reverse primers. After incubation, remove the enzyme by phenol-chloroform extraction, perform ethanol-precipitation, and then resuspend the pellet in 10 μL of TE buffer.

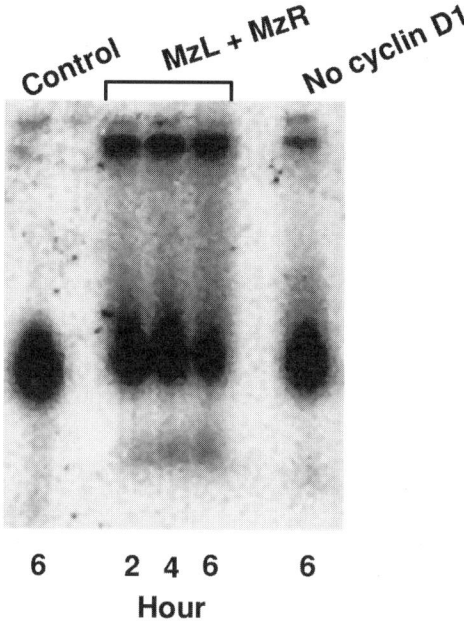

Fig. 2. Autoradiogram showing the substrates cleaved by the maxizyme in vitro. The substrate labeled by $^{32}$P was cleaved and reaction mixtures were fractionated by electrophoresis on a 15% polyacrylamide-8 $M$ urea gel. Lane 1 involved only he labeled substrate as a control and lane 5 did not involve effector substrate. Lane 2 to 4 reactions were incubated from 2 to 6 h.

2. Load the reaction mixtures onto a 1.5% agarose gel and subject to electrophoresis in 1X TAE buffer.
3. Excise gel pieces that contain the MzL or the MzR fragment, and purify them using a QIAquick Gel Extraction Kit.
4. Perform the in vitro transcription by mixing the following reagents:
   a. 1–2 μg of DNA.
   b. 5 μL 50 m$M$ DTT.
   c. 5 μL 10X T7 RNA polymerase transcription buffer.
   d. 1.5 μL ribonuclease inhibitor.
   e. 5 μL rNTP mix.
   f. 5 μL T7 RNA polymerase.
5. Complepe with DEPC-treated water to 50 μL final volume and incubate for at least 4 h at 37°C.
6. Remove the enzyme by phenol-chloroform extraction, perform ethanol-precipitation, resuspend the pellet in 10 μL TE buffer, add 10 μL of 2X RNA gel-loading buffer, and load the reaction mixtures onto a 10% polyacrylamide-8 $M$ urea gel. Electrophoresis was done in 1X TBE buffer.

7. Cut out the desired band, which can be seen as a black band by irradiating UV light, from the gel. Place the pieces of gel in a 1.5-mL microcentrifuge tube, and crush the gel with a disposable pipet tip. Add enough TE buffer to elute the RNA, and incubate overnight at 37°C with shaking.

8. Transfer the pieces of gel and the solution to a 0.45-μm membrane microcentrifuge tube (e.g., UFC3 OHV OS, Millipore Inc., Bedford, MA), and filter the solution by centrifugation at 5000*g* for 3 min at room temperature. Add 200 μL of TE buffer to the remaining pieces of gel, and filter the solution again by centrifugation.

9. After ethanol-precipitation, resuspend the pellet in 100 μL of TE buffer.

10. Label short substrates at 5'-end with $^{32}$P as follows: incubate 100 pmol substrate with 3 μL of $^{32}$P-ATP and 2 μL of T4 nucleotide kinase in 1X kinase buffer for 1 h at 37°C in a volume of 20 μL.

11. After phenol-chloroform extraction and ethanol-precipitation, load the reaction mixtures onto a 10% polyacrylamide-8 *M* urea gel and subject to electrophoresis in 1X TBE buffer.

12. Wrap the gel in plastic wrap and expose to X-ray film for 30 s.

13. Cut out the desired band by checking the gel against the result of X-ray film.

14. Purify RNA with TE as described previously, and carry out ethanol-precipitation, then resuspend the pellet in TE buffer to 3000 cpm/μL final concentration.

### 3.3. Cleavage Reaction In Vitro

1. Perform in vitro cleavage reaction by mixing the following reagents:
   a. X μL of 50 m*M* Tris-HCl, pH 8.0.
   b. 1 μL (3000 cpm) $^{32}$P-labeled first substrate.
   c. 1 μ*M* non-labeled effector (second) substrate.
   d. 100 n*M* MzR.
   e. 100 n*M* MzL.
   f. 10 m*M* MgCl$_2$.

2. Complete with DEPC-treated water to 10 μL final volume and incubate at 37°C for up to 10 h.

3. Remove aliquots of the reaction mixtures at appropriate intervals (e.g., 0, 20, 40, and 80 min) add an equal volume of 2X RNA gel-loading buffer, heat the samples for 1 min at 90°C, and store the sample on ice.

4. Fractionate the samples by electrophoresis on 10% polyacrylamide-8 *M* urea gels. Wrap the gel in plastic wrap and expose to an X-ray film for several hours at −80°C (*see* **Fig. 2**).

## 4. Notes

1. A short 7–10-nt sequence for binding of the maxizyme or ribozyme to its target is usually used. It has been shown sequences that are too long reduce specificity and also inhibit the dissociation step of the maxizyme-catalyzed reaction. However, sequences that are too short decrease the specificity for the target sequence.

2. Desired reaction temperature of *Bam*HI is 30°C. However, the activity of this enzyme incubated at 37°C is almost the same as at 30°C.

3. In treating RNA, great care must be taken to avoid contamination by RNases. Therefore, special-grade reagents (guaranteed RNase-free) and disposable sterile plasticware should be used. Moreover, disposable plastic gloves must be worn, since bare hands are a potential source of RNases.

## Reference

1. Shimayama, T., Nishikawa, S., and Taira, K. (1995) Generality of the NUX rule: kinetic analysis of the results of systematic mutations in the trinucleotide at the cleavage site of hammerhead ribozymes. *Biochemistry* **34**, 3649–3654.
2. Gesteland, R. F., Cech, T. R., and Atkins, J. F. (1999) *The RNA World, 2nd ed.* Cold Spring Harbor Laboratory Press, Cold Spring Harbor, NY.
3. Kawasaki, H., Eckner, R., Yao, T. P., Taira, K., Chiu, R., Livingston, D. M., et al. (1998) Distinct roles of the co-activators p300 and CBP in retinoic-acid-induced F9-cell differentiation. *Nature* **393**, 284–289.
4. Kawasaki, H., Suyama, E., Iyo, M., and Taira, K. (2003) siRNAs generated by recombinant human Dicer induce specific and significant but target site-independent gene silencing in human cells. *EMBO J.*, in press.
5. Krupp, G. and Gaur, R. K. (eds.) (2000) *Ribozyme, Biochemistry and Biotechnology.* Eaton Publishing Natick, MA.
6. Amontov, S. V. and Taira, K. (1996) Hammerhead minizymes with high cleavage activity: a dimeric structure as the active conformation of minizymes. *J. Am. Chem. Soc.* **118**, 1624–1628.
7. Kuwabara, T., Warashina, M., Orita, M., Koseki, S., Ohkawa, J., and Taira, K. (1998) Formation *in vitro* and in cells of a catalytically active dimer by tRNA$^{Val}$-driven short ribozymes. *Nature Biotechnol.* **16**, 961–965.
8. Kuwabara, T., Warashina, M., Nakayama, A., Ohkawa, J., and Taira, K. (1999) Novel tRNA$^{Val}$-heterodimeric maxizymes with high potential as gene-inactivating agents: Simultaneous cleavage at two sites in HIV-1 tat mRNA in cultured cells. *Proc. Natl. Acad. Sci. USA* **96**, 1886–1891.
9. Tanabe, T., Takata, I., Kuwabara, T., Warashina, M., Kawasaki, H., Tani, K., et al. (2000) Maxizymes, novel allosterically controllable ribozymes can be designed to cleave various substrates. *Biomacromology* **1**, 108–111.
10. Tanabe, T., Kuwabara, T., Warashina, M., Tani, K., Taira, K., and Asano, S. (2000) Oncogene inactivation in a mouse model: tissue invasion by leukaemic cells is stalled by loading them with a designer ribozyme. *Nature* **406**, 473,474.
11. Surekha, M., Zinged (2001) Cancer genes. *Curr. Sci.* **81**, 5.

# 19

## Target-Site Selection for the 10–23 DNAzyme

### Murray J. Cairns and Lun-Quan Sun

#### Summary

The 10–23 DNAzyme is capable of cleaving RNA with high sequence specificity at sites that contain purine-pyrimidine (R-Y) junctions. Although they are abundant in mRNA, many of these potentially cleavable junctions are protected from DNAzyme activity by secondary structure. To optimise the process of target-site selection in long RNA substrates, a multiplex assay was developed for simultaneous comparative analysis of 50 or more different DNAzymes in one reaction. Using this approach, the efficiency of 80 DNAzyme sites within the E6 component of a full-length HPV16 E6/E7 transcript was examined. The activity of molecules selected in this system was then compared in a conventional assay with DNAzymes of intermediate and low performance. This confirmed the results observed in the multiplex reactions, with 10% of DNAzymes inducing substantial cleavage of the long transcript. These DNAzyme-sensitive regions are potentially accessible to other RNA directed agents such as ribozymes or antisense oligonucleotides. Therefore, in addition to finding the most effective DNAzymes for a particular target mRNA, this method may also be applicable to locating accessible sites for other nucleic acid-based gene suppression strategies.

**Key Words:** DNAzymes; RNA accessibility.

## 1. Introduction

The enormous potential of catalytic DNA for both chemical and biological applications is exemplified by the 10–23 RNA-cleaving DNAzyme *(1)*. It has outstanding kinetic efficiency in catalyzing sequence-specific digestion of target RNA in vitro, and has been shown to be capable of achieving this under physiological conditions and against physiological substrates in vivo *(2)*. Indeed, the DNAzymes success in this environment is driving its development in biological applications as a general-purpose gene-suppression agent for gene function discovery, drug-target validation, and therapeutics *(3,4)*. The strength

From: *Methods in Molecular Biology, vol. 252: Ribozymes and siRNA Protocols, Second Edition*
Edited by: M. Sioud © Humana Press Inc., Totowa, NJ

of the 10–23 DNAzymes as a general-purpose RNA endonuclease is derived from the specificity and flexibility it enjoys as a consequence of employing Watson-Crick basepairing for targeting and binding RNA *(1)*. However, this feature has drawbacks, for although only one strand of RNA is transcribed from double-stranded DNA, the transcripts own intramolecular basepairing readily forms secondary structures which effectively compete with duplex-forming nucleic acid, such as antisense and DNAzymes, for their target sites. Efforts to predict RNA secondary structure and its influence on the binding of other nucleic acids using thermodynamic principles such as free-energy minimization are notoriously inaccurate, particularly for longer transcripts (*see* **Note 1**). This is further complicated in the intracellular context by the interactions of ribonuclear protein and other RNA-binding protein which can dramatically alter the configuration of intramolecular basepairing.

Despite these difficulties, most RNA targets do contain exposed single-stranded regions that are amenable to hybridization-based gene suppression. The challenge is to reliably identify these sites efficiently in a cost- and time-effective manner. In recent years, a number of methods for identifying accessible sites has been developed, particularly in the context of antisense oligonucleotides (ODNs). These range from the purely theoretical all the way to direct comparisons of in vivo activity. Obviously, the time, cost, and value of these approaches will vary significantly, with the theoretical approach being very quick, cheap, and probably of little value. At the other end of the scale, in vivo comparisons are slow, laborious, and expensive, and although they should be valuable, the analysis is usually far from comprehensive as the number of sites tested is rationalized in an attempt to contain costs. A number of assays have been developed to examine the accessibility of long RNA transcripts generated by in vitro transcription. These include the use of macro arrays containing oligonucleotide probes that represent every possible antisense of the target gene *(5)*. Probes, that correspond to accessible sites hybridize more efficiently to labeled transcript, allowing their identity to be revealed by autoradiography. Another approach relies on the activity of RNaseH to identify the location of accessible regions by inducing site-specific hydrolysis in which heteroduplexes are formed *(6–9)*. This methodology has one advantage—it encapsulates both the efficiency of binding and activation of RNaseH, which is also important to the intracellular mechanism of most antisense ODNs. Variations of this technique allow for the use of libraries of oligonucleotides with randomized and semi-randomized sequences, which allow for a universal approach to mapping accessibility in any target transcript *(10)*.

Target-site selection techniques based on the accessibility of transcripts produced in vitro, have also been a feature of catalytic nucleic acid-based gene-suppression strategies. In the case of the hammerhead ribozyme, hairpin ribozyme, and group I intron, a combinatorial approach using libraries of molecules containing randomized binding domains or guide sequences have been used against target transcripts to identify the most efficient cleavage sites *(11–13)*. Similarly combinatorial library of DNAzyme containing the 10–23 motif have been applied to target-site selection *(14)*. In each of these systems, the rare cleavage events produced by individual members of the library are detected by reverse transcription of the cleaved transcript from labeled primers, and mapped by alignment to sequencing fragments generated by extension from the same primers. In the methodology described in this chapter, we use a similar protocol for the analysis of DNAzyme cleavage sites, except one based on rationally designed DNAzymes that are tailored specifically to the target of interest. Although this is more expensive and less flexible than the irrational approaches previously described, we beleive that it is a far more robust and reliable methodology for target-site selection. We call this technique a "multiplex cleavage assay," and it is a very useful and effective method for screening cleavage sites along the entire length of target RNA—both in terms of accessibility and catalytic activity *(15)*.

## 2. Materials

### 2.1. DNAzyme Oligonucleotide Design

DNAzyme sequences for each target are assembled using the 10–23 catalytic motif [ggctagctacaacga] and hybridizing arms that are specific for each site along the target RNA transcript (**Fig. 1**). The length of each arm is usually fixed at nine bases; however, these can be shortened or lengthened depending (*see* **Note 1**) on their individual predicted hybridization free energy *(16)*. Each DNAzyme oligo is designed to target purine-uracil (RU). In most cases we ignore purine-cytosine sites as in our experience they are less reactive than RU sites particularly AC junctions (unpublished observation).

### 2.2. DNAzyme Synthesis, Purification, and Quality Control

DNAzyme oligonucleotides can be synthesized by conventional solid-phase methods or made to order by any commercial oligonucleotide supplier. We routinely check DNAzyme oligos for length and purity by denaturing polyacrylamide gel electrophoresis (PAGE). For this purpose we 5' end-label each oligo (prior to electrophoresis) in 60 m$M$ Tris-HCl, pH 7.5, 9 m$M$ MgCl$_2$, 10 m$M$ dithiothreitol (DTT), 1 U of polynucleotide kinase (PNK) (New England Biolabs), and 10 µCi of [$\gamma$-$^{32}$P] adenosine 5' triphosphate (ATP), and 1 m$M$ unlabeled ATP at 37°C for 10 min.

### 2.3. Transcription Template DNA Preparation

Whenever possible, the cDNA template used for transcription should contain as much of the target mRNA sequence as possible including the 5' and 3'

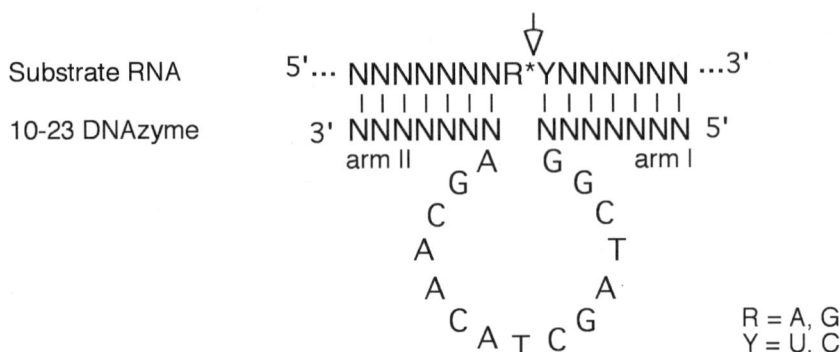

Fig. 1. The secondary structure of the 10–23 DNAzyme-substrate complex. shows the DNAzyme-substrate complex formed by Watson-Crick interactions between generic deoxyribonucleotides (N) in the arms of the DNAzyme (bottom) and the corresponding ribonucleotides (N) in the target (top). The defined 15-base sequence in the loop joining each arm represents the conserved catalytic motif and spans a single unpaired purine at the RNA target site.

untranslated region. Cloned cDNA containing a phage promoter sequence 5′ to the target sequence can be used for transcription, provided it is linearized by digestion near the 3′ end of the target insert to exclude extraneous vector sequence. However, we usually find that it is more convenient to prepare a tailored linear template sequence by polymerase chain reaction (PCR). This enables the promoter to be introduced immediately upstream of the target cDNA sequence at the site corresponding to the normal location of the 5′ terminus of the mRNA or cap site. This can be accomplished very easily by producing a forward sequence-specific primer that is appended at its 5′ end with the T7 phage-promoter sequence [TAATACGACTCACTATAGGGAGA]. PCR also allows precise control over the position of the 3′ terminus, and when possible, the reverse primer should contain a sequence complementary to the 3′ end of the 3′ untranslated region (UTR) (**Fig. 2**).

Linear template DNA for the production of a target transcript is prepared by PCR in a mixture containing 10 p$M$ plasmid clone containing the target cDNA, 1 μ$M$ of the forward T7 phage-promoter primer, and 1 μ$M$ of the reverse primer. The reaction is carried out over 25 temperature cycles at 95°C for 30 s, 60°C for 90 s, and 72°C for 60 s in a solution containing 16.6 m$M$ $(NH_4)_2SO_4$, 67 m$M$ Tris-HCl, pH 8.8, 6.7 m$M$ $MgCl_2$, 0.87 U of Ampli$Taq$ DNA polymerase (Perkin-Elmer), and 300 μ$M$ each of deoxyguanosine 5′ triphosphate (dGTP), deoxyadenosine 5′ triphosphate (dATP), deoxythymidine 5′ triphosphate (dTTP), deoxycytidine 5′ triphosphate (dCTP).

                                                        T7 Promoter

                                                        AGAACU
                        PE6T7
101  GCAAUGUUUC AGGACCCACA GGAGCGACCC AGAAAGUUAC CACAGUUAUG

151  CACAGAGCUG CAAACAACUA **UACAUGAUAU AAUAUUAGAA UGUGUGUACU**

201  GCAAGCAACA GUUACUGCGA CGUGAGG**UAU** AUGACUUUGC UUUUCGGGA**U**
     ▼PE6A

251  UU**AUGCAUAG UAUAUAGAGA UGGGAAUCCA UAUGCUGUAU GUGAUAAAUG**

301  **UUUAAAGUUU UAUUCUAAAA UUAGUGAGUA UAGACAUUAU UGUUAUAGUU**

351  UGU**AUGGAAC AACAUUAGAA CAGCAAUACA ACAAACCGUU GUGUGAUUUG**
                              ▼PE6B

401  **UUAAUUAGGU GUAUUAACUG UCAAAAGCCA CUGUGUCCUG AAGAAAAGCA**

451  **AAGACAUCUG GACAAAAAGC AAAGAUUCCA UAAUAUAAGG GGUCGGUGGA**

501  **CCGGUCGAUG UAUGUCUUGU UGCAGAUCAU CAAGAACACG UAGAGAAACC**

551  **CAGCUGUAAU CAUGCAUGGA GAUACACCUA CAUUGCAUGA AUAUAUGUUA**

601  GAUUUGCAAC CAGAGACAAC UGAUCUCUAC UGUUAUGAGC AAUUAAAUGA
            ▼PE7A

651  CAGCUCAGAG GAGGAGGAUG AAAUAGAUGG UCCAGCUGGA CAAGCAGAAC

701  CGGACAGAGC CCAUUACAAU AUUGUAACCU UUUGUUGCAA GUGUGACUCU

751  ACGCUUCGGU UGUGCGUACA AAGCACACAC GUAGACAUUC GUACUUUGGA

801  AGACCUGUUA AUGGGCACAC UAGGAAUUGU GUGCCCCAUC UGUUCUCAGA
                                              ▼PE7REV

851  AACCAUAAUC UA

Fig. 2. The HPV16 E6/E7 target sequence and primer arrangement used for multi-plex DNAzyme analysis. The RNA sequence of the HPV16 E6/E7 bicistronic tran-script showing potentially cleavable purine-uracil sites (bold U) in E6 targeted by multiplex DNAzyme pool. Arrows indicate the primer binding sites used for DNA template preparation and primer extension reactions. The underlined AUG motifs are at the E6 and E7 start codons, respectively.

## 2.4. DNA Template Purification

PCR products are electrophoresed in a 2% agarose gel alongside appropri-ate size/mass standards, and are visualized by ethidium fluorescence. The illu-minated amplicon bands are then excised from the agarose slab and extracted using Gene Clean (Bio 101). Template DNA is then redissolved in diethylpyrocarbonate (DEPC)-treated RNase-free water, and the concentration is adjusted to approx 100 ng/µL.

## 2.5. Target RNA Transcription

In vitro transcription was performed with 100 ng of purified template DNA and T7 RNA polymerase at 37°C for 3 h using an RNA transcription kit (Epicentre). To promote the effective synthesis of full length RNA transcripts, we employed the highest concentration of the four ribonucleotide triphosphates adenosine 5' triphosphate, cytidine 5' triphosphate, guanosine 5' triphosphate, and uridine 5' triphosphate (ATP, CTP, GTP, UTP) recommended for T7 RNA polymerase of 5 m$M$ in a final volume of 20 µL. After the 3-h incubation, the DNA template is degraded by the addition of 1 MBU of RNase free DNase I and an additional 15-min incubation at 37°C. This reaction is then terminated by adding an equal volume of phenol-chloroform-isoamyl alcohol. The RNA contained in the aqueous phase can be used directly in the multiplex cleavage reaction or precipitated in 0.3 $M$ sodium acetate and 2 vol of absolute ethanol. The direct use of unprecipitated RNA in cleavage reactions is intended to reduce the possibility of denaturing the secondary structured produced during RNA synthesis.

## 3. Methods

### 3.1. Multiplex Cleavage Reactions

Multiple DNAzyme oligos (0.5 n$M$–0.5 µ$M$) and synthetic RNA substrate (0.2 µ$M$) are pre-equilibrated separately for 10 min at 37°C in equal volumes of 50 m$M$ Tris-HCl, pH 7.5, 10 m$M$ MgCl$_2$, 150 m$M$ NaCl, and 0.01% sodium dodecyl sulfate (SDS). The reaction is then initiated by mixing the DNAzyme and substrate together. After 1 h, the reactions are stopped by emersion in ice and precipitation in 0.3 $M$ sodium acetate and 2 vol of ethanol.

### 3.2. Partial Ribonuclease Digestion

To support the association between strong DNAzyme cleavage sites and accessible regions of the target transcript, we find it valuable to have independent information about the transcript's secondary structure. The easiest way to produce empirical data concerning the disposition of RNA secondary structure is to employ a ribonuclease that preferentially digests single-stranded RNA. Ribonuclease T1 is ideal for this purpose, as it preferentially cleaves RNA at single-stranded G residues. Partial T1 digests of the target transcript can be analyzed in the same primer extension reaction alongside the G-specific sequencing fragments to produce a footprint of secondary-structure interaction.

For this purpose the target transcript (0.2 µ$M$) is subjected to partial digestion with ribonuclease T1 (0.2–1.0 U) in 50 m$M$ Tris-HCl, pH 7.5, 10 m$M$ MgCl$_2$, and 150 m$M$ NaCl for 10 min at 37°C. The reactions (20 µL) are stopped by extraction in 100 µL of phenol-chloroform and recovered by ethanol-precipitation.

### 3.3. Primer Extension Reaction

Each primer (1 µ$M$) is radiolabeled at the 5'-terminal in 60 m$M$ Tris-HCl, pH 7.5, 9 m$M$ MgCl$_2$, 10 m$M$ DTT, 1 U of PNK, and 10 µCi of [γ-$^{32}$P]ATP at 37°C for 30 min and 75°C for 5 min. Primer extension analysis is then performed on DNAzyme-cleaved and ribonuclease-digested RNA using either SuperScriptII reverse transcriptase (Life Technologies) or thermally stable Tth DNA polymerase (Promega) (*see* **Note 3**). In each reaction, 2 pmol of labeled primer is combined with 400 n$M$ of RNA and denatured at 90°C for 5 min. For extension with superscript reverse transcriptase, the primer is allowed to anneal slowly between 65°C and 45°C before adding the first-strand buffer, DTT, deoxyribonucleotides, and enzyme (according to the manufacturers instructions). This mixture (final vol 20 µL) is then incubated for 1 h at 45°C before being transferred to ice. Alternatively, after annealing at 65°C for 5 min, extension can be achieved at 72°C with Tth DNA polymerase over a period of 20 min. These reactions contained Tth polymerase buffer supplemented with MnCl$_2$ (in accordance with the manufacturer's instructions) for use with an RNA template. In this protocol the entire 20-µL reaction volume (including enzyme and dNTPs) was assembled on ice prior to the thermal denaturation step.

### 3.4. Sequencing Reactions

Sequencing fragments corresponding to each segment of the target are also generated by primer extension on the double-stranded DNA template in the presence of chain-terminating dideoxynucleotides (ddNTP). In these four reactions, the dNTP concentration is reduced to 2.5 µ$M$ while being supplemented by either 10 µ$M$ ddGTP, or 100 µ$M$ ddATP, or 200 µ$M$ ddTTP, or 100 µ$M$ ddCTP in 16.6 m$M$ (NH$_4$)$_2$SO$_4$, 67 m$M$ Tris-HCl, pH 8.8, 6.7 m$M$ MgCl$_2$, 0.87 U of *Taq* DNA polymerase, and 1.0 pmol of the $^{32}$P-labeled oligonucleotide. This reaction is performed as a linear amplification over 25 temperature cycles at 95°C for 30 s, 60°C for 90 s, and 72°C for 60 s *(17)*.

### 3.5. Electrophoresis and Densitometry

After primer extension, samples are combined with a equal volumes of stop buffer (formamide/EDTA/loading dye) before electrophoresis on a 6% denaturing polyacrylamide gel (**Fig. 3**). The corresponding image can then be

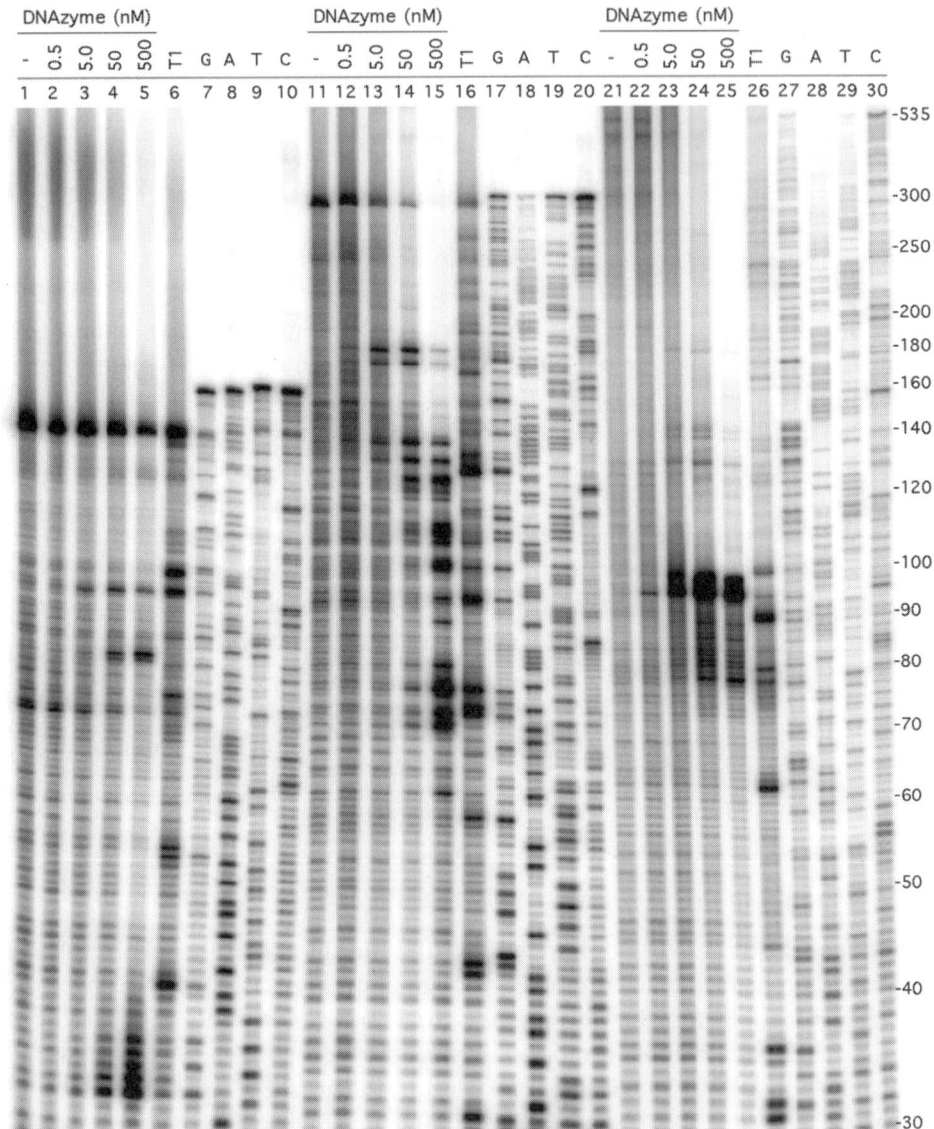

Fig. 3. Phosphorimage of a DNA sequencing gel containing the primer extension products of multiplex cleavage reactions in the E6/E7 transcript. Lanes 1–5 were derived from cleavage reactions with 0, 0.5, 5, 50, and 500-n$M$ concentrations of DNAzyme oligos directed to the 5' segment of the E6 gene and revealed by extension of primer PE6A. Lanes 6 was generated by extension of the same primer on substrate RNA that had been subjected to partial digestion with ribonuclease T1. Bands representing DNAzyme induced cleavage in lanes 2–5 were positioned by direct alignment

revealed and the band intensity quantified (*see* **Notes 4–6**) at each position with a Phosphorimager and ImageQuant software (Molecular Dynamics).

## 4. Notes

1. Prediction of hybridization free energy for individual DNAzyme arm-substrate interactions is highly recommended, as it can vary substantially from one sequence to another and it can provide some indication of a particular DNAzyme's cleavage potential. In our experience, DNAzymes containing a binding domain with an unusually high hybridization free-energy value is less likely to produce effective cleavage activity. Knowing this, the DNAzyme cleavage site can be ignored or the arm in question lengthened to compensate for its poor hybridization strength. Similarly, if the predicted hybridization strength is very high, the arm length can be reduced. This analysis can be performed on hundreds of DNAzyme simultaneously in a matter of seconds by doing it as a batch calculation on a spreadsheet.

2. DNAzymes, like any other single-stranded nucleic acid above a certain size, may be capable of forming their own stable intramolecular hairpin structures. In some cases, these oligonucleotides are incapable of achieving efficient *trans*-cleaving activity. To avoid this problem, the intramolecular stability of DNAzyme oligonucleotides should be predicted and used to exclude molecules with an unacceptably high propensity to form hairpins. There is a range of software available with the ability to carry out this analysis such as "Oligo 6.0" and a number of bioinformatic sites and oligo suppliers can provide this type of analysis online.

3. To enable simultaneous analysis of multiple RNA cleavage sites along an entire target transcript, primer extension reactions should be arranged so that the reverse primers are approx 300 bp apart (**Fig. 2**). This will enable basepair resolution of each extension product corresponding to a cleavage site, and allow for their direct identification by reference to the sequencing fragments run in parallel.

4. The relative intensity of each band determined by densitometry provides an indication of the individual DNAzymes efficacy. For example, a very intense banding down to very low DNAzyme concentrations is a good sign that there is little interference from secondary structure. In addition, the analysis of a ribonucleotide T1 partial digest can provided an independent indicator of the prevailing secondary structure to support this assertion.

---

Fig. 3. *(continued)* with the corresponding dideoxy sequencing reactions (lanes 7–10) also produced by the same primer. These extend slightly beyond the 5' end of the corresponding RNA (lanes 1–6), as the template contains the T7 promoter sequence. Lanes 11–20 and 21–30 contain the same order of samples as described for lanes 1–10, except that they were derived from reactions directed to the middle and 3' segments of E6, and revealed by extension of primers PE6B and PE7A, respectively. The scale down on the right-hand side of the gel indicates the mol wt (bp) of the primer-extension fragments.

5. In some target sequences, the primer extension reaction may be inhibited by secondary structure, resulting in nonspecific bands. In an attempt to eliminate this problem, we have also performed primer extension reactions at 72°C using the thermally stable reverse transcriptase activity of Tth DNA polymerase. However, although this seems to produce the same profile of multiplex cleavage, it does not prevent the appearance of nonspecific bands, and the signal is not as strong as that generated by conventional reverse transcriptases.

6. To achieve simultaneous analysis of 50 or more DNAzymes in a single primer extension, the DNAzyme:substrate ratio should be optimized. We have found that the most even distribution of DNAzyme cleavage is observed when the concentration of each oligo is 5 n$M$. At this concentration and below, only the most efficient DNAzyme in the mixed reaction are capable of producing a detectable signal. Under less stringent conditions (50 n$M$), the pattern and extent of DNAzyme activity on the complex substrate is more obvious. At even higher concentrations such as 0.5 $\mu M$ there is the emergence of multiple cleavage events per transcript occurs so that the cleavage pattern approaches total digestion rather than a distribution along the target RNA. This will be apparent by the overrepresentation of cleavage activity of DNAzymes proximal to the primer binding site at the expense of those that should be seen distal to the primer site.

# References

1. Santoro, S. W. and Joyce, G. F. (1997) A general purpose RNA-cleaving DNA enzyme. *Proc. Natl. Acad. Sci. USA* **94,** 4262–4266.
2. Santiago, F. S., Kavurma, M. M., Lowe, H. C., Chesterman, C. N., Baker, A., Atkins, D. G., and Khachigian, L. M. (1999) New DNA enzyme targeting Egr-1 mRNA inhibits vascular smooth muscle proliferation and regrowth after injury. *Nat. Med.* **5,** 1264–1269.
3. Cairns, M. J., Saravolac, E. G., and Sun, L. Q. (2002) Catalytic DNA: a novel tool for gene suppression. *Curr. Drug Targets.* **3,** 269–279.
4. Sun, L. Q, Cairns, M. J., Saravolac, E. G., Baker, A., and Gerlach, W.,L. (2000) Catalytic nucleic acids: from lab to applications. *Pharmacol. Rev.,* **52,** 325–347.
5. Milner, N., Mir, K. U., and Southern, E. M. (1997) Selecting effective antisense reagents on combinatorial oligonucleotide arrays. *Nature Biotechnol.* **15,** 537–541.
6. Ho, S. P., Britton, D. H, Stone, B. A., Behrens, D. L., Leffet, L. M., Hobbs, F. W., et al. (1996) Potent antisense oligonucleotides to the human multidrug resistance-1 mRNA are rationally selected by mapping RNA-accessible sites with oligonucleotide libraries. *Nucleic Acids Res.* **24,** 1901–1907.
7. Matveeva, O., Felden, B., Audlin, S., Gesteland, R. F., and Atkins, J. F. (1997) A rapid *in vitro* method for obtaining RNA accessibility patterns for complementary DNA probes: correlation with an intracellular pattern and known RNA structures. *Nucleic Acids Res.* **25,** 5010–5016.
8. Birikh, K. R., Berlin, Y. A, Soreq, H., and Eckstein, F. (1997) Probing accessible sites for ribozymes on human acetylcholinesterase RNA. *RNA* **3,** 429–437.
9. Lima, W. F., Brown-Driver, V., Fox, M., Hanecak, R., and Bruice, T. W. (1997)

Combinatorial screening and rational optimization for hybridization to folded hepatitis C virus RNA of oligonucleotides with biological antisense activity. *J Biol. Chem.* **272,** 626-638.

10. Ho, S. P., Bao, Y., Lesher, T., Malhotra, R., Ma, L. Y., Fluharty, S. J., et al. (1998) Mapping of RNA accessible sites for antisense experiments with oligonucleotide libraries. *Nature Biotechnol.* **16,** 59–63.

11. Bramlage, B., Luzi, E., and Eckstein F. (2000) HIV-1 LTR as a target for synthetic ribozyme-mediated inhibition of gene expression: site selection and inhibition in cell culture. *Nucleic Acids Res.* **28,** 4059–4067.

12. Yu, Q., Pecchia, D. B., Kingsley, S. L., Heckman, J. E., and Burke, J. M. (1998) Cleavage of highly structured viral RNA molecules by combinatorial libraries of hairpin ribozymes. The most effective ribozymes are not predicted by substrate selection rules. *J. Biol. Chem.* **273,** 23,524–23,533.

13. Campbell, T. B. and Cech, T. R. (1995) Identification of ribozymes within a ribozyme library that efficiently cleave a long substrate RNA. *RNA* **1,** 598–609.

14. Sriram, B. and Banerjea, A. C. (2000) In vitro-selected RNA cleaving DNA enzymes from a combinatorial library are potent inhibitors of HIV-1 gene expression. *Biochem. J.* **352,** 667–673.

15. Cairns, M. J., Hopkins, T. M., Witherington, C., Wang, L., and Sun, L. Q. (1999) Target site selection for an RNA-cleaving catalytic DNA. *Nature Biotechnol.* **17,** 480–486.

16. Sugimoto, N., Nakano, S., Katoh, M., Matsumura, A., Nakamuta, H., Ohmichi, T., et al. (1995) Thermodynamic parameters to predict stability of RNA/DNA hybrid duplexes. *Biochemistry* **34,** 11,211–11,216.

17. Murray, V. (1989) Improved double-stranded DNA sequencing using the linear polymerase chain reaction. *Nucleic Acids Res.* **17,** 8889.

# 20

## In Vitro Selected RNA-Cleaving DNA Enzymes From Combinatorial Libraries

### Samitabh Chakraborti, Bandi Sriram, and Akhil C. Banerjea

### Summary

The selective inactivation of a target gene by antisense mechanisms is an important biological tool in the delineation of gene functions. Ribozymes and RNA-cleaving DNA enzymes-mediated approaches are more attractive because of their ability to catalytically cleave the target RNA. DNA enzymes have recently gained great importance because they are short DNA molecules with simple structures that are expected to be stable to the nucleases present inside a mammalian cell. We have designed a strategy to identify accessible cleavage sites in human immunodeficiency virus-1 (HIV-1) gag RNA from a pool of random DNA enzymes, and for isolation of DNA enzymes. A pool of random sequences 29 nucleotides long that contained the previously identified 10–23 catalytic motif were tested for their ability to cleave the target RNA. When the pool of random DNA enzymes was targeted to cleave between A and U nucleotides, a DNA enzyme 1836 was identified. Although several DNA enzymes were identified using a pool of DNA enzyme that was completely randomized with respect to its substrate-binding properties, DNA enzyme-1810 was selected for further characterization. Both the DNA enzymes showed target-specific cleavage activities in the presence of $Mg^{2+}$ only. These strategies could be applied for the selection of desired target sites in any target RNA.

**Key Words:** DNA enzymes; HIV-1 gag; HIV-1 TAR; HIV-1 replication.

## 1. Introduction

Sequence-specific cleavage activities by short DNA molecules that possess two earlier identified catalytic motifs (10–23 and 8–17) (*1*) have recently been recognized as a powerful biological tool that can interfere with gene expression. These have been derived by in vitro selection from a combinatorial library of DNA sequences that are capable of cleaving 17-nucleotide-long synthetic RNA at simulated physiological conditions. These DNA enzymes—which are

From: *Methods in Molecular Biology, vol. 252: Ribozymes and siRNA Protocols, Second Edition*
Edited by: M. Sioud © Humana Press Inc., Totowa, NJ

expected to be more stable in contrast to short catalytic RNAs (ribozymes) that are inherently unstable—are short, and have simple secondary structures. They can potentially cleave any target RNA that contains a purine/pyrimidine junction, thus, allowing far greater flexibility in choosing the target sites. On the contrary, the target sites for hammerhead and hairpin motif containing ribozymes are limited. After their initial discovery of by Santoro and Joyce *(1)*, a number of investigators have used these DNA enzymes to selectively cleave and interfere with the function of the target genes *(2,3)*. Inhibition of infection of incoming HIV-1 was earlier reported by Zhang et al. *(4)*, by DNA enzymes that were targeted against the V3 loop of the envelope region. Most of these studies have used the DNA enzyme that possesses the 10–23 catalytic motif. We previously showed the cleavage of HIV-1 envelope RNA *(5)*, as well as the HIV-1 coreceptor- CCR5 *(6)* in a sequence-specific manner, and subsequently tested their ability to interfere with the function of the target gene when introduced into a mammalian cell. In recent studies, we reported a functional mono-DNA enzyme against the second most important HIV-1 coreceptor, CXCR-4, and when combined with CCR5 DNA enzyme in tandem, it had the ability to cleave both the target RNAs (CXCR-4 and CCR5) and to interfere with the respective coreceptor functions *(7)*. Despite all the success in constructing DNA enzymes, the selection and identification of DNA enzymes that might be effective in a complex cellular milieu under in vivo conditions has remained a problem, and must be determined empirically. Predicted secondary structures that are thermodynamically stable may help identify loop regions, but the physiological relevance of these structures inside a mammalian cell remains uncertain. It may still be possible to identify a good target site by targeting a small bulge region—in our experience, a relatively large loop region presents a poor cleavage site *(8)*. To overcome this problem, a "multiplex cleavage assay" was developed and shown to be very useful for screening cleavage sites along the entire length of target mRNA (*see* Chapter 19).

In this chapter, we present strategies to identify DNA enzyme target sites from a random pool of DNA enzymes, and demonstrate their sequence-specific cleavage activities. For this purpose, we amplified the HIV-1 gag *p24* gene and placed it downstream of the T7 promoter of the vector pGEM-T-Easy. It was linearized with *Sal*1 at the multiple cloning sites of the vector. A 740-base-long HIV-1 gag transcript was generated by in vitro transcription, and then subjected to cleavage by two types from a random pool of DNA enzymes—namely, random and AU-cleaving DNA enzymes.

## 2. Materials

Standard precautions should be taken when handling phenol, chloroform, polyacrylamide, ethidium bromide, formamide, and radioisotopes. Since RNA is extremely susceptible to degradation, diethylpyrocarbonate (DEPC)-treated water that is autoclaved should be used. Radioisotopes should be used, taking standard precautions. Most of the molecular biology related techniques were used according to the protocols developed by Promega Biotech. WI. Oligonucleotides were used directly for carrying out polymerase chain reaction (PCR) or for cleavage reaction (as in the case of the DNA enzymes). These should be dissolved in minimum amounts of 1X TE (10 m$M$ Tris-HCl, pH 8.0, and 1m$M$ ethylenediaminetetraacetic acid [EDTA]) and stored at –20°C. We have observed progressive loss of cleavage activity by DNA enzyme when stored for more than 1 mo. We recommend carrying out the cleavage reaction with the labeled target RNA (this also continues to be degraded, even when stored at –70°C) as soon as possible to achieve optimum results. Unlike ribozyme-mediated cleavage of the target RNA, DNA enzyme can cleave at 37°C with no denaturation of substrate and enzyme. When amplifying genes from human sources or pathogenic organisms, NIH safety guidelines should be followed.

## 2.1. Reagent Preparation

### 2.1.1. Cloning of Target Genes and In Vitro Transcription

1. Two types of targets were chosen—namely, the HIV-1 gag (amplified from pNL4-3 *[9]*) and the HIV-1 TAR element that encode the HIV-1 capsid and the characteristic stem-loop structure known as TAR, which is present at the 5'-end of all HIV-1 transcripts, respectively. Target genes against which the DNA enzymes need to be tested can be generated by carrying out PCR, using specific primers. We recommend that DNA enzymes be tested against short, synthetic target RNA first. This can be generated by simply designing two short (30–40 bases long) complementary oligonucleotides (which contain the target site) and annealing them. Thereafter, they can be cloned into a T-tailed vector. Alternatively, they can be PCR-amplified from the pathogen itself, using *Taq* polymerase, and then can subsequently be cloned into the same vector that contains either T7 or SP6 promoter (pGEM-T series of vector, Promega). Specific primers (*see* **Subheading 3.1.1.**) were designed and the PCR-amplified products were cloned into a T-tailed vector. These were then linearized by the use of restriction enzyme (located at the end of the cloned gene in multiple cloning sites of the vector) digestion and subjected to in vitro transcription using SP6 or T7 RNA polymerase (*see* **Subheading 3.1.1.**). Labeled transcripts ($^{32}$PUTP) (substrates) and DNA enzymes (unlabeled) are mixed in a 1:1 ratio (100 pmols each) and subjected to cleavage. Cleaved products are resolved by gel electrophoresis. Series of reverse primers are synthesized all along the target RNA, and a primer extension reaction is carried out to identify the cleavage site in the target RNA.
2. 10X annealing buffer: 100 m$M$ Tris-HCl, pH 8.3, 50 m$M$ MgCl$_2$.
3. Transcription buffer 5X: 400 m$M$ HEPES-KOH, pH 7.5, 120 m$M$ MgCl$_2$, 10 m$M$ spermidine, and 200 m$M$ dithiothreitol (DTT).

4. Acid (pH 4.5) phenol-chloroform-isoamyl alcohol (IAA) (25:24:1): Add equal parts of phenol and 50 m*M* sodium acetate, pH 4.5, until the pH of the aqueous phase is 4.5 (this may require overnight equilibration). Then mix the phenol phase with one part of chlorophorm-IAA (24:1).
5. T7 RNA polymerase.
6. Recombinant RNasin-ribonuclease inhibitor (Promega).
7. Nucleotide triphosphates (NTPs) (adenosine 5' triphosphate [ATP], guanosine 5' triphosphate [GTP], cytidine 5' triphosphate [CTP]—2.5 m*M* each). Prepared by mixing 1 vol of nuclease-free water with 1 vol of 10 m*M* NTP stock.
8. RQ1 RNase-free DNAse (Promega).
9. 100% and 80% ethanol.
10. Ligase 10X buffer: 300 m*M* Tris-HCl, pH 7.8, 100 m*M* MgCl$_2$, 100 m*M* DTT, and 10 m*M* ATP. This should be stored in small aliquots at –20°C to avoid degradation of ATP.
11. 10X TBE: 890 m*M* Tris-borate, pH 8.3, 20 m*M* EDTA.
12. Stop buffer: 95% formamide, 0.05% xylene cyanol FF, 0.05% bromophenol blue.
13. T7 Sequenase Quick-Denature plasmid sequencing kit (Amersham, Life Science).
14. Concert™ Matrix Gel extraction system (Gibco-BRL, Life Technologies, UK). PCR-amplified products were purified using this gel and then subjected for ligation.
15. Cleavage analyzing gel: 7 *M* urea-6% polyacrylamide gel (Mini-PROTEAN II System, Bio-Rad).
16. Qiagen purification kit: QIAprep Spin Miniprep Kit (Qiagen, Germany).
17. 10X kinase buffer: 500 m*M* Tris-HCl, pH 7.6, 100 m*M* MgCl$_2$, 50 m*M* DTT, 1 m*M* spermidine, 1 m*M* EDTA.
18. T4 polynucleotide kinase (PNK).

## 3. Methods

In order to find a cleavage site in any target RNA, labeled target RNA is subjected to cleavage in the presence of excess amounts (10- to 100-fold) of randomly synthesized DNA enzymes that contain the common 10–23 catalytic motif. Cleaved products (in presence and absence of Mg$^{2+}$) are analyzed by primer extension and a sequencing reaction is simultaneously performed to identify specific cleavage sites (**Fig. 1**). HIV-1 TAR element was subjected to DNA enzyme-mediated cleavage, and the specificity was confirmed by gel analyses.

———————

Fig. 1. *(opposite page)* Schematic representation of the experimental protocol. HIV-1 gag (*p24*) gene was amplified from the infectious HIV-1 DNA, pNL4-3 (*9*) and cloned into a T-tailed vector, pGEM-T-Easy (Promega Biotech.) and HIV-1 gag sequences were confirmed by sequencing. It was then linearized at the 3' end at the multiple cloning sites (MCS) to generate 740 bases long in vitro transcript using T7 RNA polymerase. This RNA (0.2 µm) was then mixed with the excess amount (20 µm)

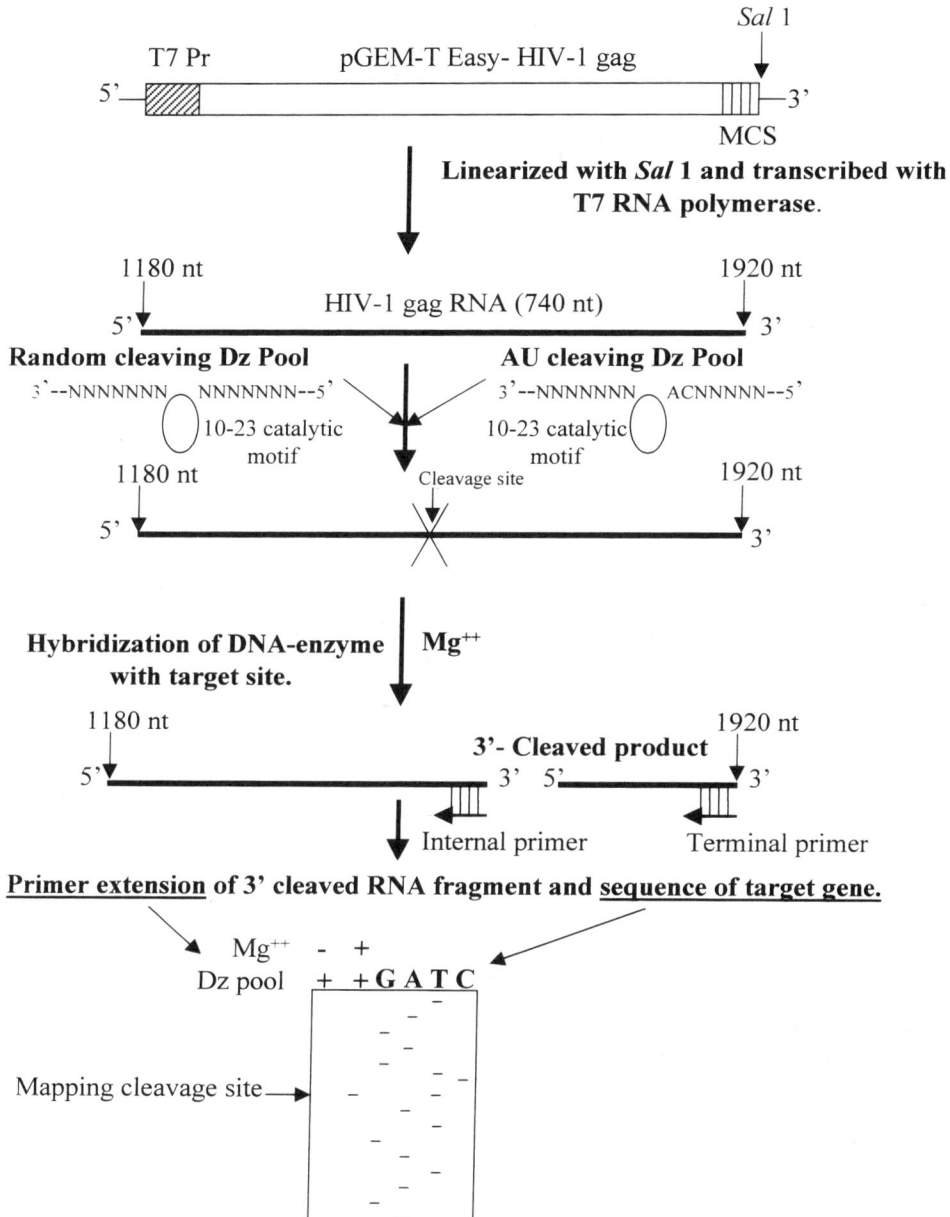

Fig. 1. *(continued)* of two different pools of DNA enzymes *(see* text) in presence or absence of MgCl₂. The cleaved products were subjected to primer extension using labeled primers. The primer-extended product was analyzed alongside the sequencing gel to map the cleavage site in the target RNA. Note that both the pools of DNA enzymes had a common 10–23 catalytic motif for carrying out sequence-specific cleavage of the target RNA in a catalytic manner.

## 3.1. Reagent Preparation

### 3.1.1. Cloning of Target Genes (HIV-1 gag Gene and HIV-1 TAR Element) and In Vitro Transcription

Specific primers were synthesized to amplify a 740-bp DNA fragment encoding the *p24* gene using the plasmid DNA pNL4-3 *(9)* and cloned into a T-tailed vector, pGEM-T-Easy (Promega Biotech, WI) to yield plasmid pGEM-gag. For ligation of the insert into pGEM-T-vector system, the PCR product is purified directly from agarose gel using the Wizard™ PCR Preps DNA purification system (Promega). The range of insert: vector is usually kept in the range of 3:1 (usually 30–60-ng T-tailed vector is used). We routinely use T4 DNA ligase supplied by Promega Biotech, using the ligase buffer (final volume 10 µL). We generally carry out ligation overnight at +4°C. Shorter duration (3–6 h) may result in less number of recombinant colonies. Multiple aliquots of T4 DNA ligase buffer should be stored at –20°C because the ATP is likely to become degraded with multiple freeze-thawing. High-frequency competent cells (JM 109) ($>1 \times 10^8$ colony-forming units/mg DNA) should be used for transformation. Primers possessed the following sequence:

> 5'-sense-CCCTATAGTGCAGAACCTCCA (1185–1205 nt)
> 5'antisense-CATTATGGTAGCTGGATTTGTTAC (1897–1920 nt)

Conditions are optimized for PCR by calculating the melting temperatures of the primers (50°C preferred), using the formula $T_m$ (°C) = 4(G + C) + 2(A + T). Recombinant clones should be confirmed by sequencing and correct clones are purified on a Qiagen column.

### 3.1.2. In Vitro Transcription

The cloned gene was linearized at the 3' end by *Sal*I (**Fig. 1**) digestion to obtain full-length, authentic HIV-1 gag RNA by in vitro transcription, using T7 RNA polymerase. The reaction mixture contained in a final volume of 20 µL of the following components:

1. Transcription buffer: 80 m*M* HEPES-KOH, pH 7.5, 25 m*M* MgCl$_2$.
2. 2 m*M* spermidine and 40 m*M* DTT.
3. 20 U ribonuclease inhibitor: RNasin.
4. Ribonucleotides 2.5 m*M* each.
5. 100 *M* of $^{32}$PUTP (400 Ci/mmol).
6. 1 µg linearized DNA.
7. 10–20 U of T7 RNA polymerase.

Transcription is usually carried out for 1 h at 37°C. The DNA template is removed by digestion with DNAse I (Promega's RQ1RNase-free-DNase (~1 U/µg of template DNA) for 15 min at 37°C, then, the RNA is recovered by standard acid (pH 4.5) phenol:chloroform:isoamylalcohol (25:24;1) extraction. Labeled RNA is resolved in a 6% polyacrylamide-7 $M$ urea gels.

### 3.1.3. Synthesis of Random Pool of DNA Enzymes

Two types of DNA enzyme pools were synthesized, which contained the previously identified 10–23 catalytic motif that had the following sequence: 5'-GGCTAGCTACAACGA-3'. In one case, the DNA enzyme was targeted to all the possible AUGs in the target RNA, and in second case, it was designed to cleave any potential DNA enzyme target site. This was possible by either totally randomizing the seven bases on either side of the catalytic motif (**Fig. 1**), or having CA immediately preceding the catalytic motif (for AUG cleaving). All the DNA enzymes that were synthesized chemically according to the standard procedures on an Applied Biosystems Oligonucleotide Synthesizer were 29 nucleotides long. Analysis on a 16% polyacrylamide gel electrophoresis (PAGE)-7 $M$ urea using 1X TBE buffer enabled a check of purification of the oligonucleotides.

### 3.1.5. Cleavage Reactions Using the Randomized Pool of DNA Enzymes

In vitro, synthesized and labeled HIV-1 gag transcript (740 nt long) that possessed the target sites was used for the cleavage reaction. This was obtained by linearizing the plasmid pGEM-gag with *Sal*I and subjecting it to in vitro transcription by T7 RNA polymerase. Unlabeled full-length (740-base-long HIV-1 gag RNA) substrate (0.1 $\mu M$) was incubated with a large excess (15 $\mu M$) of both types of randomized DNA enzyme pools in the presence and absence of 10 m$M$ MgCl$_2$ in a buffer containing 50 m$M$ Tris-HCl, pH 8.0 in a 10-μL volume reaction (*see* **Notes 1** and **2**). After 1 h at 37°C, cleaved RNA products were resolved by gel electrophoresis. Identical conditions were used for the cleavage of HIV-1 TAR RNA, using multiple DNA enzymes (**Fig. 2**).

### 3.1.6. Primer Extension and Identification of Cleavage Sites

Cleavage sites were mapped by primer extension of the cleavage products alongside a sequencing reaction generated using the reverse AMV transcriptase (*see* **Note 3**). The transcription buffer contains 1mM MgSO$_4$ and a 0.2 m$M$ concentration of all four dNTPs in a final volume of 50 μL. The primer (5'-antisense primer: 1897–1920, as described earlier and one internal primer-position 1792–1808: 5'-CCTGGTCCCAATGCTTT), were labeled at the 5' end with $\gamma^{32}$P ATP. The terminal and internal primer generated a sequence of ~200 nucleotides in length when analyzed on a 6%-PAGE-7 $M$ urea-sequencing gel. Primer extended product and the sequencing reactions were analyzed on a sequencing gel simultaneously and radioactive bands were detected by autoradiography (**Fig. 3A–D**; *see* **Notes 3–5**).

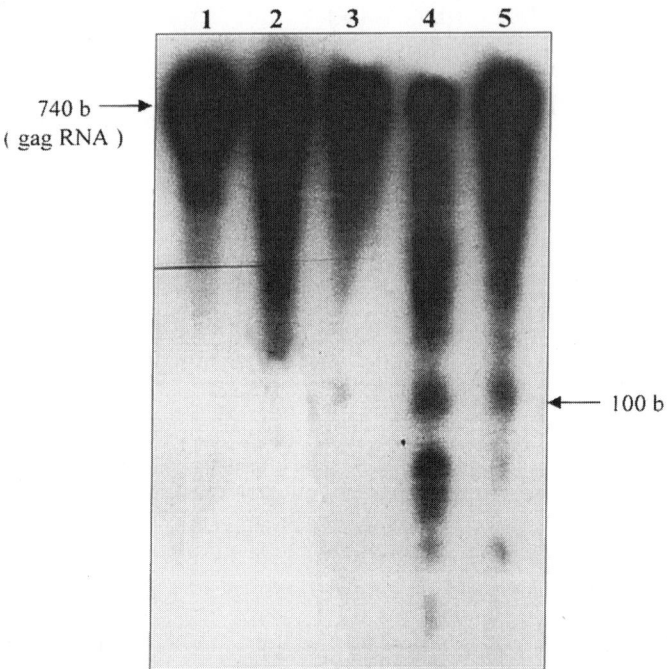

Fig. 2. Efficiency of cleavage of the target RNA by random pool of DNA enzymes. In vitro cleavage activity of the two pools of DNA enzymes. HIV-1 gag RNA (740 bases long) was synthesized by in vitro transcription as described in the text (lane 1). An excess amount of DNA enzyme was added and cleavage reaction was initiated in the presence (lanes 4 and 5) or in absence (lanes 2 and 3) of $MgCl_2$ (10 m$M$ final) and cleavage products were resolved by gel electrophoresis (7 $M$ urea-6% polyacrylamide gel) on a Mini-PROTEAN II gel apparatus (Bio-Rad). No cleavage was observed with Dz-1810 (lane 2) or Dz-1836 (lane 3) in the absence of $MgCl_2$. Multiple cleavage was observed with Dz-1810 (lane 4) and Dz-1836 (lane 5).

### 3.1.6. Construction of DNA Enzymes

Once the target site was identified, DNA enzyme with 10–23 catalytic motif was synthesized (*see* **Notes 6** and **7**). The strategy for synthesizing DNA enzymes as the same as that originally described by Santoro and Joyce *(1)*, and recently, by us *(5–7)*. Briefly, seven nucleotides on either side of the catalytic motif were made complementary to the target gene and the A (in case of AUG cleaving DNA enzyme) and G (in the case of the GC-cleaving DNA enzyme) was left unpaired. The cleavage is expected to take place after the A and G nucleotide, respectively (**Fig. 4A,B**: shown by arrows). The cleavage reaction with the labeled target RNA and library-selected DNA enzyme was carried out

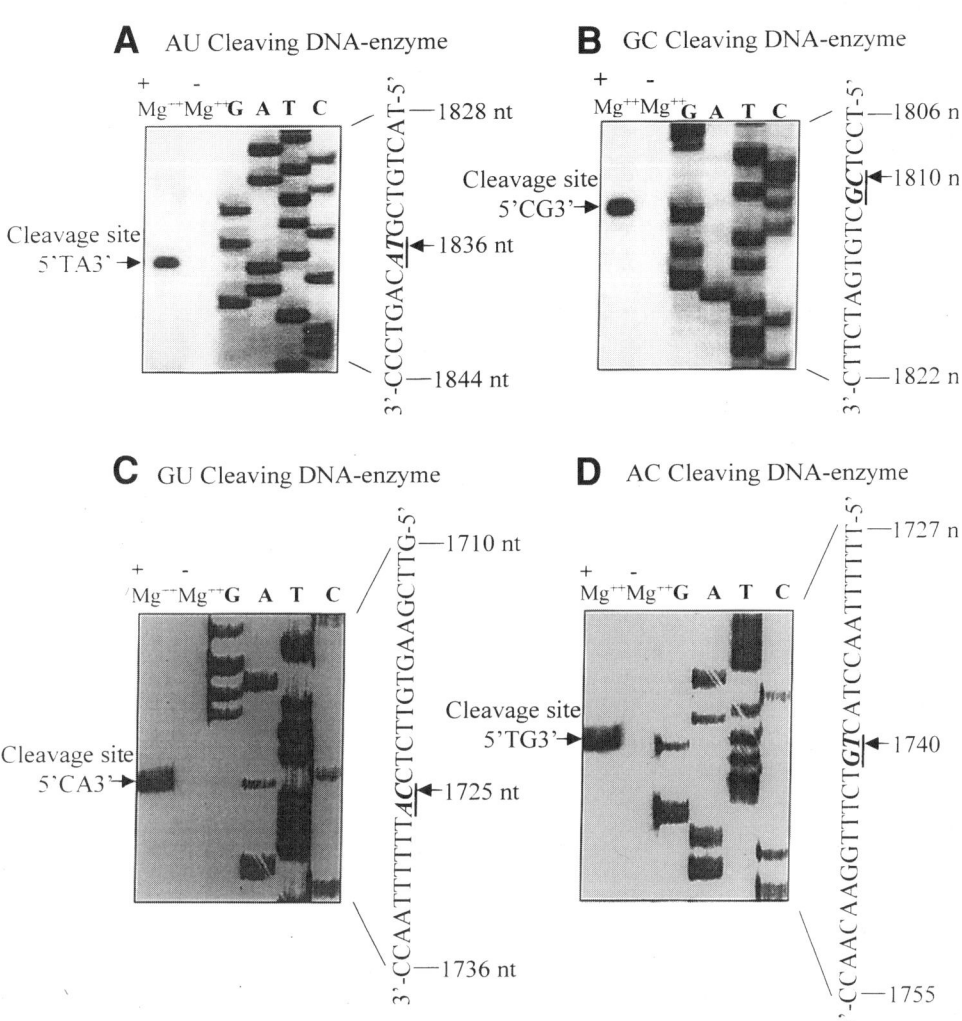

Fig. 3. Identification of DNA enzyme target sites. In vitro synthesized HIV-1 gag RNA (740 b) and the random pool of DNA enzymes were mixed as described in the text and the cleaved RNA products were subjected to primer extension as described before. Using A-U cleaving DNA enzyme, we identified a site at position 1836 in the target RNA (**A**, shown by an arrow) and using the pool of DNA enzyme that was totally randomized with respect to the target binding, a GC cleaving site was identified at position 1810 (**B**, shown by arrow). Note that the primer-extended products were observed only in presence of $Mg^{2+}$ in both the cases. Additional target sites were also found using a primer designed to hybridize internally (see text): at 1725 (GU cleaving) (**C**) and at 1740 position (AC cleaving) (**D**) by using the same kind of experimental protocol.

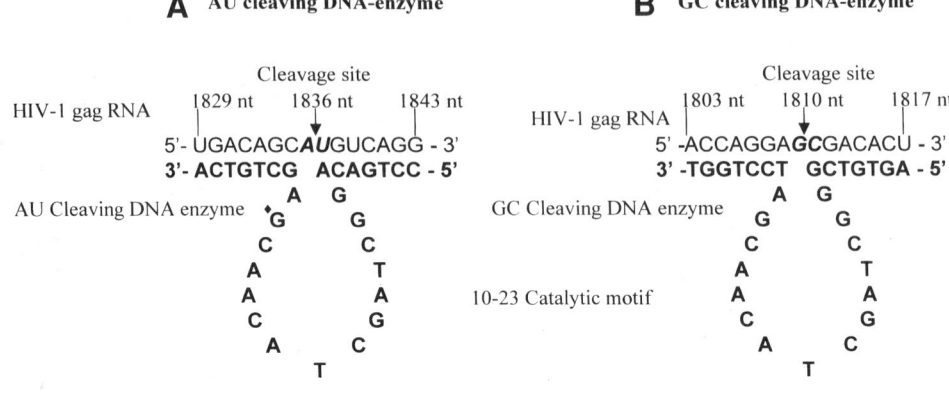

♦G to C mutation ( mut Dz1836 )

Fig. 4. Sequence of the target gene and DNA enzymes. Once the cleavage sites were identified in the target RNA as described previously, the DNA enzymes were synthesized chemically. The sequence of the target RNA for AU- and GC-cleaving DNA enzymes is shown in (**A**) and (**B**), respectively. They both contained the previously described 10–23 catalytic motif. Seven bases on either side of the cleavage site are synthesized that are complementary to the target sequence. In case of AU-cleaving DNA enzyme (**A**) the A is left unpaired whereas in GC cleaving DNA enzyme (**B**), G nucleotide is left unpaired. Both the DNA enzymes are 29 nucleotides long. A point mutation (G to C) was created in the catalytic motif of Dz-1836 to create mutant-Dz-1836 (panel A shown by ♦).

as described before. A point mutation (G to C) in the catalytic motif of Dz-1836 was also created that served as a disabled DNA enzyme control (**Fig. 4A**). The selected DNA enzymes against HIV-1 gag RNA exhibited significant cleavage activity in vitro.

## 4. Notes

1. A quick evaluation of whether the pool of DNA enzymes possess cleavage activity should be made by carrying out gel analysis of the cleaved product in the presence and absence of $Mg^{2+}$. For this purpose, we carried out in vitro cleavage of HIV-1 gag RNA in presence of *excess amounts* (10- to 100-fold) of random and AU-cleaving DNA enzyme pools in the presence or absence of $MgCl_2$, and the products were analyzed by gel electrophoresis, as described earlier.

2. Multiple cleavage of the target RNA by a random pool of DNA enzyme is greater, and can potentially hybridize to many sites in the target RNA. The critical control in these experiments is the difference in the cleavage pattern observed in the presence and absence of $MgCl_2$. A certain amount of nonspecific degradation of labeled RNA is always seen with prolonged storage at –70°C. We recommend

that the cleavage reaction be performed on a freshly synthesized $^{32}$P-labeled target RNA and the DNA enzyme oligonucleotide. Prolonged storage of DNA enzymes at –70°C also results in a loss of catalytic activity.

3. Before attempting to select a candidate DNA enzyme from a pool of DNA enzymes, it is important to confirm the presence of sequence-specific cleavage activities in the presence and absence of $Mg^{2+}$. This was obtained by carrying out a primer extension assay on the cleaved product using multiple reverse-sequencing primers, and carrying out a primer extension reaction. We identified several target sites in this way. Very often, more than one radioactive band shows up in the primer-extended products. Only those bands that appear in the presence of $Mg^{2+}$ are relevant.

4. The efficiency with which any DNA enzyme will cleave a target RNA depends largely on the secondary structure, and should be experimentally tested under identical cleavage conditions. Our experience with limited target RNA suggests that small, single-stranded regions are cleaved more efficiently than the DNA enzymes that are designed exclusively to hybridize with a large loop structure *(8)*. DNA enzymes may cleave the target RNA with varying efficiencies in presence of $Mg^{2+}$, and a range of concentrations should be used. DNA enzyme that retain appreciable sequence-specific cleavage activity in the range of 1–2 m$M$ $MgCl_2$ at 37°C (near physiological conditions) *(1)* may be more bio-efficacious *(8)*.

5. One important difference between ribozyme and DNA enzyme cleavage is that the latter does not usually require denaturation of substrate and DNA enzyme oligonucleotide to observe sequence-specific cleavage activities. Using a variety of target RNAs, we have found that ribozymes show significant enhancement (30–50%) at elevated temperatures (40–50°C), and this is not the case with DNA enzymes. We believe that the 10–23 catalytic motif containing DNA enzymes can cleave a target RNA more easily and efficiently as compared to hammerhead ribozymes, mainly because of the simple structure of the DNA enzyme that facilitates its docking with the target RNA. This must be experimentally determined, as the secondary structure of the target RNA may have a significant influence. In earlier studies, we found that DNA enzyme was more effective in cleaving the CCR5 (chemokine receptor-HIV-1 coreceptor) RNA than a hammerhead ribozyme *(6)*.

6. The length of the hybridizing arms of the DNA enzymes can also influence the cleavage efficiency. We find that 6–8 nucleotides on either side of the target RNA may be optimal for cleavage. In some instances, one extra nucleotide in the target-hybridizing arms improved its sequence-specific cleavage activities *(8)*.

7. By randomizing the target binding sites or introducing specific nucleotides in the design of the pool of DNA enzymes, it may be possible to select DNA enzymes that are targeted against a particular pair of purine or pyrimidine nucleotides. Our experience with several DNA enzymes against a variety of target RNAs suggests that almost all purine:pyrimidine combinations (AC; AU; GC; GU) work, and the efficiency of the cleavage will vary with each target RNA.

## Acknowledgments

This work was supported by the funding from the Department of Biotechnology, Government of India, to the National Institute of Immunology, New Delhi, India, and to the corresponding author, Akhil C. Banerjea.

## References

1. Santoro, S. W. and Joyce, G. F. (1997) A general purpose RNA-cleaving DNA enzyme. *Proc. Natl. Acad. Sci. USA* **94,** 4262–4266.
2. Santiago, F. S., Lowe, H. C., Kavurma, M. M., Chesterman, C. N., Baker, A., Atkins, D. G., and Khachigian, L. M. (1999) New DNA enzyme targeting Egr-1 mRNA inhibits vascular smooth muscle proliferation and regrowth after injury. *Nature Med.* **11,** 1264–1269.
3. Wu, Y., Yu, L., McMahon, R., Rossi, J. J., Forman, S. J., and Snyder, D. S. (1999) Inhibition of bcr-abl oncogene expression by novel deoxyribozymes (DNAzymes). *Human Gene Ther.* **10,** 2847-2857.
4. Zhang, X., Xu, Y., Ling, H., and Hattori, T. (1999) Inhibition of infection of incoming HIV-1 virus by RNA-cleaving DNA enzyme. *FEBS Lett.* **458,** 151–156.
5. Dash, B. C., Harikrishnan, T. A., Goila, R., Shahi, S., Unwalla, H., Husain, S., and Banerjea, A. C. (1998) Targeted cleavage of HIV-1 envelope gene by a DNA enzyme and inhibition of HIV-1 envelope-CD4 mediated cell fusion. *FEBS Lett.* **431,** 395–399.
6. Goila, R. and Banerjea, A. C. (1998) Sequence specific cleavage of the HIV-1 coreceptor-CCR5 gene by a hammer-head ribozyme and a DNA enzyme: inhibition of the coreceptor function by DNA enzyme. *FEBS Lett.* **436,** 233–238.
7. Basu, S., Sriram, B., Goila, R., and Banerjea, A. C. (2000) Targeted cleavage of HIV-1 coreceptor–CXCR-4 by RNA cleaving DNA enzyme: Inhibition of the coreceptor function by DNA enzyme. *Antiviral Res.* **46,** 125–134.
8. Unwalla, H. and Banerjea, A. C. (2001) Novel mono- and di-DNA enzymes targeted to cleave TAT and TAT-REV RNA inhibit HIV-1 gene expression. *Antiviral Res.* **51,** 127–139.
9. Adachi, A., Gendelman, H. E., Koenig, S., Folks, T., Willey, R., Rabson, A., and Martin, M. A. (1986) Production of acquired immunodeficiency syndrome associated retrovirus in human and nonhuman cells transfected with an infectious molecular clone. *J. Virol.* **59,** 284–291.
10. Goila, R. and Banerjea, A. C. (2001) Inhibition of hepatitis B virus X gene expression by novel DNA enzymes. *Biochem. J.* **353,** 701–708.

# 21

# Nucleic Acid Sequence Analysis Using DNAzymes

## Murray J. Cairns and Lun-Quan Sun

### Summary

   The sequence specificity of the "10–23" RNA-cleaving DNA enzyme can be utilized to discriminate between subtle differences in nucleic acid sequence. We examined this potential by comparing the cleavage activity of DNAzymes that target sequences derived from a relatively conserved segment of the L1 gene from different human papillomavirus (HPV) genotypes. DNAzyme activity was found to be highly sensitive to mismatches between its binding domain and substrate sequences containing polymorphisms. Type-specific DNAzyme-cleavable substrates can also be generated by genomic PCR using a chimeric primer containing three bases of RNA. The RNA component enables each amplicon to be cleavable in the presence of its matching DNAzyme. In this format, the specificity of DNAzyme cleavage is defined by Watson-Crick interactions between one substrate-binding domain (arm I) and the polymorphic sequence that is amplified during polymerase chain reaction (PCR). DNAzyme-mediated cleavage of amplicons generated by this method was used to examine the HPV status of genomic DNA derived from Caski cells, which are known to be positive for HPV16. This method is applicable to many types of nucleic acid sequence variation, including single-nucleotide polymorphisms (SNPs).

   **Key Words:** DNAzymes; genotyping; SNP; cycle cleavage.

## 1. Introduction

   With the potential to bind any RNA sequence and cleave purine-pyrimidine junctions, the 10–23 DNA enzyme has unprecedented target-site flexibility (*1*). Surprisingly, this enormous capacity for catalyzing different sequences appears to be matched by even greater specificity (*2*). This ability to discriminate makes the DNAzyme desirable for biological applications in which unwanted side reactions between the DNAzyme and some closely related or unrelated substrate could be detrimental to its use. This level of specificity also allows exquisite targeting of DNAzymes to RNA molecules containing disease-related

From: *Methods in Molecular Biology, vol. 252: Ribozymes and siRNA Protocols, Second Edition*
Edited by: M. Sioud © Humana Press Inc., Totowa, NJ

mutations. In the case of the *bcr-abl* fusion transcript, expressed in the CD34+ bone-marrow cells of patients with CML, the 10–23 DNAzyme demonstrated mutant-specific cleavage activity while leaving the wild-type *bcr* transcript intact *(3)*. This combination of flexibility, specificity, and activity has motivated the development of the DNAzyme as a gene-suppression agent for gene-function analysis, target validation, and potentially, therapeutics *(3–8)*. This activity has also been utilized in the laboratory as a mechanism for signal transduction in homogeneous real-time nucleic acid detection *(9)*.

Although the DNAzyme's ability to discriminate between SNPs is attractive, the extent will depend on its configuration in relation to its mismatched substrate—both in regard to the position and type of mismatch generated *(2)*. In general, mismatches that are induced close to the cleavage site will impair cleavage activity more than those on the periphery. The extent of impairment will also depend on the relative strength of pairing between the mismatch. Some mismatches introduce a considerable cost to the heteroduplex stability, whereas others such as wobble basepairs are relatively stable, and thus have less effect on DNAzyme cleavage activity. To heighten the DNAzyme's sensitivity to mismatches, particularly those distal to the target site or those consisting of wobble pairs, the DNAzyme's respective binding domain can be truncated without losing activity against the matched target *(10)*. Wu et al. *(3)* utilized this approach to enhance the specificity of their *bcr-able* transcript that cleaves DNAzyme. In the most clearly differentiated targets, the purine-pyrimidine composition of the cleavage site itself is formed through mutation from sites consisting of either purine-purine, pyrimidine-pyrimidine, or pyrimidine-purine, which are completely resistant to DNAzyme cleavage. These sequence variants affect the DNAzyme's mechanism of action directly, and therefore their influence is more dramatic than the more subtle disturbance of hybridization stability.

In this chapter, we further investigate the specificity of the 10–23 DNAzyme by observing its ability to discriminate between sequences that differ by as little as 1 SNP. For this purpose, reactions between DNAzyme and matching substrate sequences (derived from a polymorphic site in the L1 gene of six different clinically relevant HPV types) were compared with reactions in the unmatched substrates. In each case, only the perfectly matched type-specific DNAzymes were capable of achieving substantial cleavage of the corresponding substrate, despite the similarity between the different sequences.

To capitalize further on the high specificity and flexibility demonstrated by the 10–23 DNAzyme, we developed a system for generating DNAzyme-cleavable amplicons exclusively for the purpose of sequence analysis. In this system, a DNAzyme-cleavable ribonucleotide sequence is introduced to a polymorphic amplicon by using an RNA-containing primer. The polymorphic segment of the template DNA is thus transformed into a DNAzyme substrate by primer extension. This system, described as "substrate sequence amplification," should enable DNAzyme-based sequence analysis to be accessible to almost any nucleic acid, even for those in low abundance derived from biological material, provided they are capable of functioning as a template for amplification by PCR *(11)*.

## 2. Materials

1. Oligonucleotides: DNA/RNA chimeric oligonucleotides were synthesized and purified by Oligos Etc. The name, sequence, and origin of each oligonucleotide are provided in **Table 1**. All the oligonucleotides were checked for quality by 5'-end labeling using T4 DNA polynucleotide kinase (PNK).
2. Cellular DNA: The HPV16 positive Caski Cell line was cultured in Dulbecco's modified Eagle's medium (DMEM) supplemented with 10% fetal calf serum (FCS) at 37°C. Genomic DNA was extracted from the cells using a DNA Extraction Kit (Stratagene) according to the manufacturer's instructions.
3. Reaction buffer: 50 m$M$ Tris-Hcl, pH 7.5, 10 m$M$ Mg Cl$_2$.

## 3. Methods

### 3.1. Designing DNAzymes for Sequence Analysis

The DNAzymes that were used for genotyping are based on the "10–23" model. The design rules are generally similar to that utilized for gene inactivation in vitro and in vivo.

1. Perform sequence analysis to define the region(s) in which a significant homology exists among the different types of viruses or subtypes of genes of interest.
2. Scan the sequence for purine and pyrimidine junctions that are potentially cleavable by DNAzymes.
3. Define DNAzyme-binding sites that are affected by type-specific sequence variation, such as SNPs.
4. Adjust the length of the binding arms affected by substrate variants to attain the minimum heteroduplex required to achieve reasonable cleavage efficiency by the matched DNAzyme under the prevailing reaction conditions. The significance of minimizing binding-domain stability will depend on the type of mismatch generated by various sequences and their position with respect to the cleavage site (*see* **Notes 1–9**).
5. Design DNAzymes to match the wild-type sequence and each of the variants.
6. If the polymorphic sequence can be encompassed by more than one DNAzyme-binding region, a panel of DNAzymes can be used against each site. **Figure 1** shows the example in designing HPV-type-specific DNAzymes. Six chimeric substrates with sequences derived from a relatively conserved region of the L1 gene in different HPV types were each challenged with a perfectly matched DNAzyme, and five unmatched molecules were designed to cleave the alternative substrates. The sequence of arm I for each type-specific DNAzyme was slightly

**Table 1**
**Substrate, DNAzymes, and Primer Oligonucleotides**

| Name | Sequence | Target |
|---|---|---|
| Chimeric substrates[a] | | |
| DT148 | ACAGTAACA**AAU**AATTGATTA | HPV18 L1 |
| DT149 | ACAGTAACA**AAU**AGATGATTA | HPV11 L1 |
| DT150 | ACAGTAACA**AAU**AACTGATTG | HPV31 L1 |
| DT151 | ACAGTAACA**AAU**AGTTGATTA | HPV6 L1 |
| DT152 | ACAGTAACA**AAU**ACCTGATTG | HPV33 L1 |
| DT153 | ACAGTAACA**AAU**AGTTGGTTA | HPV16 L1 |
| DNAzymes[b] | | |
| DT160 | TAATCATCTAggctagctacaacgaTTGTTACTGT | HPV11 L1 |
| DT174 | TAATCAATTAggctagctacaacgaTTGTTACTGT | HPV18 L1 |
| DT175 | TAATCAACTAggctagctacaacgaTTGTTACTGT | HPV 6 L1 |
| DT176 | CAATCAGTTAggctagctacaacgaTTGTTACTGT | HPV31 L1 |
| DT177 | CAATCAGGTAggctagctacaacgaTTGTTACTGT | HPV33 L1 |
| DT178 | TAACCAACTAggctagctacaacgaTTGTTACTGT | HPV16 L1 |
| Primers[c] | | |
| DT184 | GTATCTACCACAGTAACAAAUA | HPV L1 |
| DT185 | AAYAATGGYATYTGYTGG | HPV L1 |
| DT280 | AATAATGGCATTTGTTGG | HPV16 L1 |

[a]The RNA component of the chimeric substrates are indicated by bold.
[b]The conserved 15-base motif of the 10–23 catalytic domain was denoted by lower case.
[c]Primer DT185 contained degenerate pyrimidines at four positions so that Y = T + C. A nondegenerate version of this primer, specific for HPV16, was also tested. All oligonucleotides were typed in the 5'–3' direction.

different, in order to correspond to the polymorphisms in each substrate. The remaining binding domain (arm II) and catalytic component of each DNAzyme were identical between HPV types.

### *3.2. Sequence-Specific DNAzyme Cleavage*

The specificity of DNAzyme-mediated cleavage can be examined by comparing the extent of cleavage achieved (during a 1-h incubation) with various matched and unmatched DNAzyme-substrate combinations (**Fig. 2**). The following protocol describes the procedure for measuring the extent of RNA cleavage reactions under single turnover conditions.

1. RNA cleavage reactions are usually performed under single-turnover conditions with a 10-fold excess of DNAzyme.
2. The RNA-containing substrate oligonucleotides (1 µ*M*) can be 5'-end labeled prior to the cleavage reaction with 1 U of PNK (New England Biolabs) in 60 m*M*

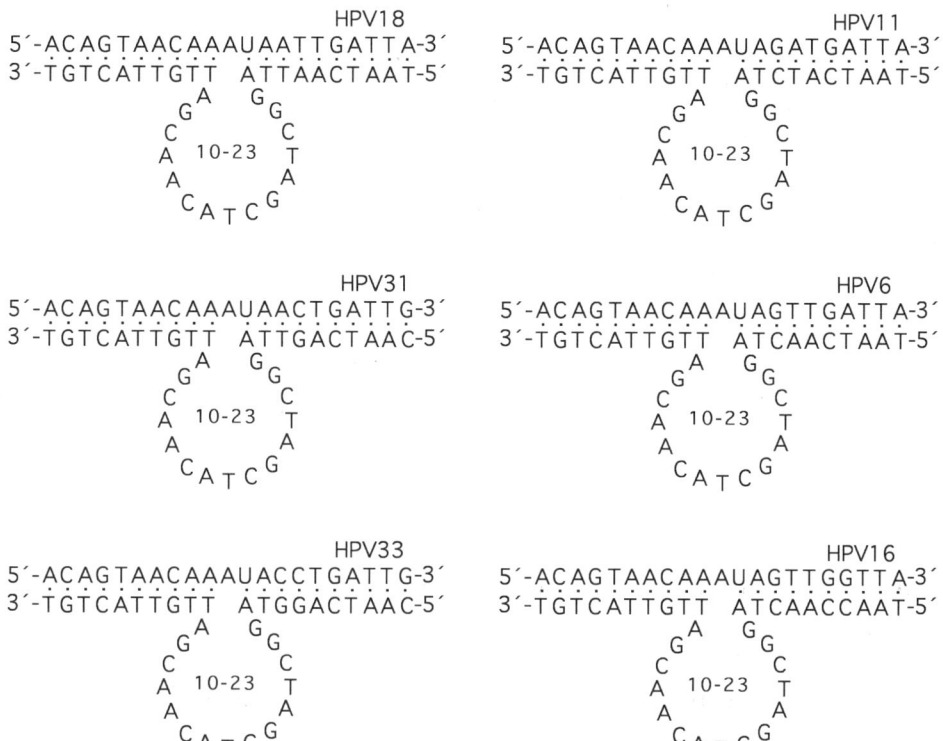

Fig. 1. HPV type-specific DNAzyme-substrate complexes. The secondary structure of six HPV type-specific DNAzyme with their corresponding substrate sequences. Each of the substrate sequences are derived from various HPV types (indicated above each complex) and differ from each other by polymorphisms contained within a small region corresponding to the 5' binding domain of the DNAzyme.

Tris-HCl, pH 7.5, 9 m$M$ MgCl$_2$, 10 m$M$ dithiothreitol (DTT), and 10 µCi of [γ-$^{32}$P] adenosine 5' triphosphate (ATP) (GeneWorks) at 37°C for 30 min and 75°C for 5 min.

3. For each cleavage reaction, the labeled substrate and DNAzyme are pre-equilibrated separately in the reaction buffer at 37°C for 5 min before being combined to a final concentration of 50 n$M$ and 500 n$M$, respectively.

4. After a suitable time interval (in the HPV L1 system we used 60 min), the reaction is stopped by mixing samples with an equal volume of ice-cold buffer containing 90% formamide, 20 m$M$ ethylenediaminetetraacetic acid (EDTA) and loading dye. After reactions, the uncleaved substrate and products were resolved by electrophoresis on a 10% denaturing polyacrylamide gel and analyzed using a phosphorimager (Molecular Dynamics).

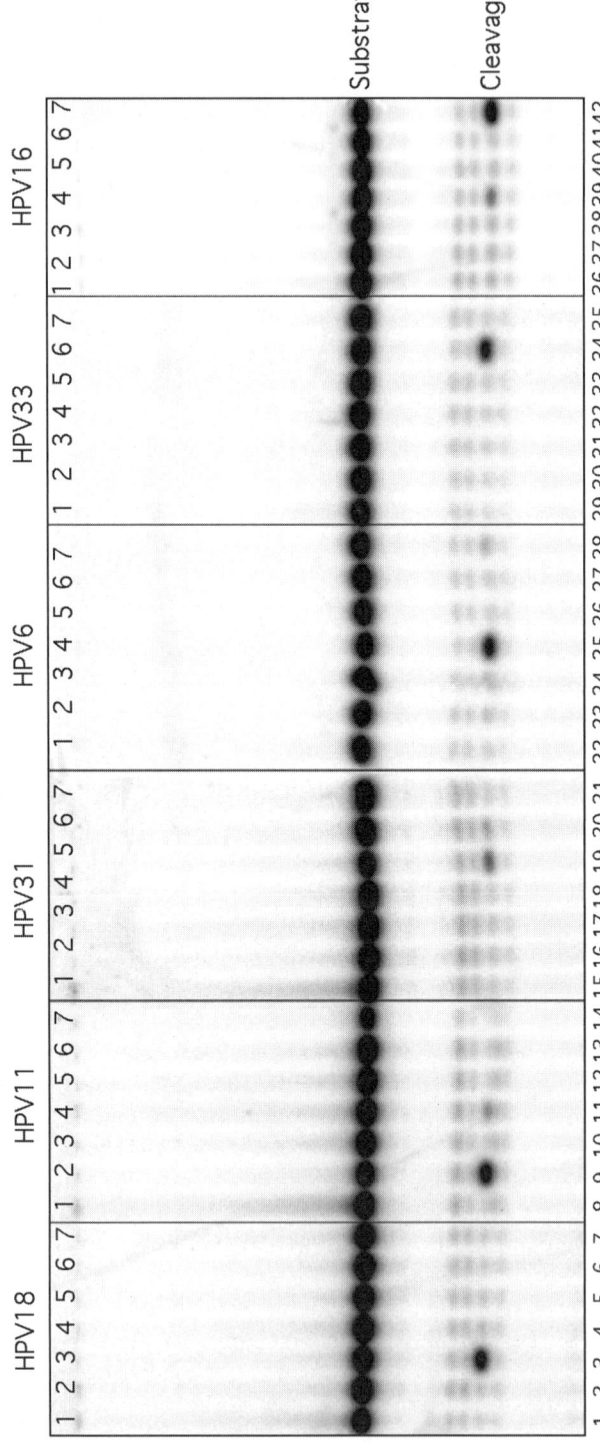

Fig. 2. DNAzyme cleavage based sequence analysis. Image contains a 16% polyacrylamide sequencing gel used to resolve end-labeled cleavage product from the uncleaved substrate. Each synthetic substrate derived from one of six HPV types was incubated with its matching DNAzyme and the unmatched counterparts in the presence of magnesium. The substrate-sequence origin for each divided reaction set is indicated at the top of the gel. The DNAzymes used in each set were numbered from 1–7 and code for i) no DNAzyme, ii) HPV11, iii) HPV18, iv) HPV6, v) HPV31, vi) HPV33, and vii) HPV16.

### 3.3. Strategy for Target Amplification and Cycle Cleavage

To effectively harness the capacity of 10–23 DNAzymes to discriminate between chimeric substrates with similar sequences, a DNA/RNA chimeric primer can be used to produce a DNAzyme-cleavable amplicon by PCR. The primer introduces the fixed RNA component, and extension produces the variable component as it traverses the template polymorphism (**Fig. 3**). We explored the potential of this system by generating a DNAzyme-cleavable amplicon from the L1 gene of HPV16-positive human cells, so that it contained the same chimeric HPV16 cleavage site examined using oligonucleotide-based substrates.

In this procedure, genomic DNA is thermally denatured and used as a template for amplification with generic HPV primers. These flank a polymorphic region, which enables the production of type-specific amplicons. As the reverse primer contains a 3-bp stretch of RNA (upper case), these amplicons are cleavable in the presence of a DNAzyme with an arm sequence complementary to the polymorphic region. In the later stages of this scheme, the substrate cleavage efficiency in this double-stranded format is enhanced by thermal cycling.

### 3.4. Substrate Amplification

A type-specific DNAzyme cleavable substrate can be generated directly at high copy number from small amounts of genomic DNA by PCR.

1. The RNA component of the cleavage site is incorporated into the amplicon by a chimeric primer containing three ribonucleotides.
2. The primer is then 5'-end labeled prior to amplification with PNK (as described for the previous substrates) and combined (2 pmol) of reverse primer and 10 ng of genomic DNA in a mixture consisting of 50 m$M$ KCl, 10 m$M$ Tris-HCl, pH 8.3, 2.5 m$M$ MgCl$_2$, 1.5 U of Ampli$Taq$ DNA polymerase (Perkin Elmer), 200 μ$M$ each of deoxyguanosine 5' triphosphate (dGTP), deoxyadenosine 5' triphosphate (dATP), deoxythymidine 5' triphosphate (dTTP), and deoxycytidine 5' triphosphate (dCTP).
3. After 25 temperature cycles at 95°C for 30 s, 50°C for 60 s, and 72°C for 90 s, the PCR product can be purified by 6% native polyacrylamide gel electrophoresis or used directly in cleavage reactions.

### 3.5. Cycle Cleavage Reaction

To maximize the cleavage extent in a double-stranded target, we used multiple cycles of binding, cleavage, and denaturation.

1. In this reaction, purified or unpurified PCR product containing the cleavable substrate is supplemented with a type-specific DNAzyme (1 μ$M$) and the MgCl$_2$ concentration made up to 10 m$M$.
2. The reaction is then carried out over 10 cycles of thermal denaturation at 80°C for 10 s followed by hybridization and cleavage at 37°C for 5 min.

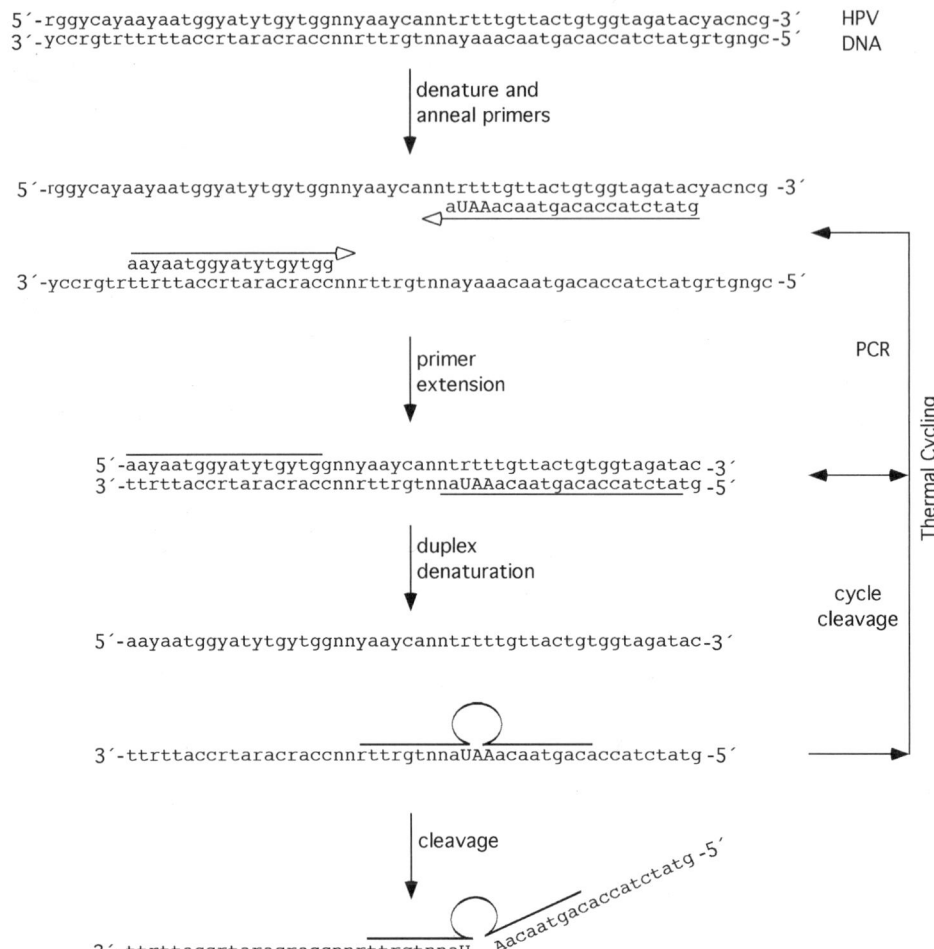

Fig. 3. Substrate sequence amplification and cycle cleavage. Schematic representation of a method for generating a DNAzyme-cleavable substrate from a small amount of genomic DNA using PCR and a cycle cleavage reaction. A 3-bp stretch of RNA is indicated as upper case for DNAzyme cleavage. Polymorphic purine and pyrimidine bases are denoted by r and y, respectively.

3. As a control of this cycle cleavage reaction, a static incubation at 37°C can also be set up for the same duration (approx 80 min) after an initial pre-denaturation at 95°C for 2 min.

4. After the cycle cleavage reaction, each sample is combined with an equal volume of formamide loading dye and electrophoresed on a 10% denaturing polyacrylamide gel. The respective amounts of cleaved and uncleaved substrate can be determined and analyzed using a phosphorimager.

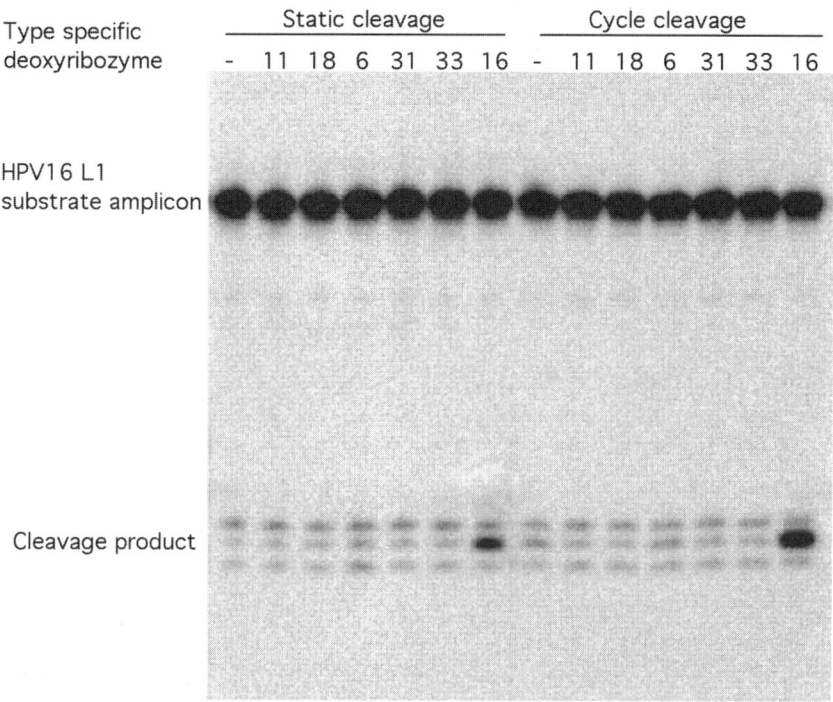

Fig. 4. Substrate amplification and cleavage. This image was derived from a 10% sequencing gel containing purified PCR products generated in the presence of HPV16-positive Caski-cell DNA followed by incubation with various HPV type-specific DNAzyme (indicated at the top of each lane). The relative positions of the cleaved and uncleaved substrate amplicon, which appear as intense low and high mol wt bands, are indicated down the left side of the gel. Each cleavage reaction was carried out both at a constant 37°C (static) and by thermal cycling between 80°C and 37°C min (cycle cleavage), as indicated on the top of the gel.

As shown in **Fig. 4**, the amplicon was challenged with the matching HPV16-specific DNAzyme and the other unmatched analogs (as in the oligonucleotide cleavage experiment) to ensure that the activity and specificity of the reaction was maintained with the substrate in this format. The results confirming this, as only the HPV16-specific DNAzyme was capable of producing cleavage activity. For this reaction, we found that the cleavage extent was enhanced by cycling between 37°C and 80°C. Under these reaction conditions, the thermal denaturation phases gave the DNAzyme multiple opportunities to compete with the template strand during hybridization with the substrate-containing strand. This dynamic reaction scheme was found to produce more cleavage product than a static incubation at 37°C.

## 4. Notes

1. To obtain the maximum target-sequence specificity and maintain efficient cleavage activity, the hybridization stability of the DNAzyme-substrate interactions should be optimized to suit the reaction system. If the binding-arm lengths are too long, the specificity of cleavage could be compromised; if they are too short, then the cleavage activity will be diminished. For DNAzyme applications in biological systems, this optimization must be tailored to physiological conditions. In contrast, mutation-detection analysis systems can take advantage of a broad variety of conditions, as there is the freedom to change reaction buffers and temperatures. For example, in the context of an amplification system, it may be desirable to increase the arm length of DNAzyme probes to achieve cleavage at higher temperatures. In this environment, specificity will be maintained if the temperature also remains relatively high.

2. In addition to arm length and reaction conditions, the hybridization stability of an individual binding domain will also depend on its sequence composition. The influence of sequence composition can be predicted to some extent using nearest neighbor analysis *(12)*. This will also change depending on the hybridization conditions and the RNA content of the substrate. The most reliable method for optimizing reaction schemes is to test them empirically.

3. The sensitivity of DNAzyme cleavage activity to mutations that give rise to base mismatch will vary according to the type and position of the mutation with respect to the cleavage site *(2)*. As a general rule, mismatches that are distal to the cleavage site will have less impact than those that are nearer. The effect of these proximal mutations is most profound if it changes the composition of the cleavage site itself from purine-pyrimidine to purine-purine, pyrimidine-purine, or pyrimidine-pyrimidine. Mutations that occur at a single base or a SNP will give rise to a single-base mismatch will have less impact on the cleavage activity of a paired DNAzyme than two or more mismatches or a deletion. However, the extent of tolerance of a SNP will depend on the type of mismatch it produces. For instance, a C to U mutation that induces a G-U "wobble" pair would not perturb the activity of the cognate DNAzyme nearly as much as a C-G mutation which produces a G-G mismatch. In the case of the L1 system used as an example, although some crossreactivity was observed between DNAzymes with substrates that differ by a SNP, the difference in cleavage intensity was great enough for clear discrimination between the two reactions. In particular, the HPV11 substrate with 20% cleavage from its matched DNAzyme also experienced 1% cleavage by the HPV6-specific DNAzyme (A-A mismatch), which represented 5% of the matched DNAzyme activity. A similar crossreaction was also observed in the HPV16 substrate with the same DNAzyme (HPV6) as a result of a distal T-G wobble pair. However, the other three substrates did not display any significant reaction with unmatched DNAzymes, including HPV31 and HPV 33, which only differ by a SNP.

4. The cleavage extent achieved by the HPV L1-specific DNAzymes, although sufficient to produce an unambiguous signal, were usually lower than expected. This is perhaps partly because of the majority DNA-DNA homoduplex composition of

the enzyme-substrate complex, which has already been shown to have the least stable hybrid and a lower cleavage efficiency compared to other duplex structures.

5. The catalytic efficiency for any given DNAzyme may be predicted based upon the hybridization stability *(12)*. However, the variation between different L1 substrates did not follow the predicted hybridization stability pattern closely. These differences in activity must therefore be the result of more subtle influences of the sequence polymorphism than gross helix stability.

6. Substrate-sequence amplification provides a convenient mechanism for generating a DNAzyme-cleavable substrate from very small amounts of genomic DNA and cDNA. However, this process adds an additional layer of complexity to the reaction system, which will invariably require some optimization. In most cases, the amplification will need to be arranged with an excess of the RNA-containing primer so that the complete substrate-containing strand is synthesized in greater numbers than its template strand. This reduces the DNAzyme's competition for the substrate stand, allowing more opportunities for cleavage to occur. Although this asymmetric PCR offers an advantage for the cleavage reaction, it comes at some expense to the amplification efficiency.

7. In some coupled reaction systems, the DNAzymes themselves may prime DNA synthesis or interfere with the amplification primers by binding or by forming primer dimers. In this event, the solution may be simply a matter of moving the primer binding site slightly upstream or downstream. It may also be advantageous to block the 3' end of the DNAzymes to prevent them from priming.

8. In addition to asymmetric amplification of the substrate sequence, the extent of cleavage can also be increased by thermocycling the cleavage reaction. This increases the DNAzyme's opportunity to hybridize and cleave substrate sequences that are in double-stranded amplicons.

9. In our pilot study, we employed DNA sequencing gels to visualize the products of DNAzyme cleavage reactions. In principle, these reactions could also be monitored in real time by measuring the increase in fluorescence produced as a result of cleavage of a substrate sequence that is acting as a linkage between a fluorophore and a quencher pair undergoing fluorescence energy transfer. This type of substrate arrangement could be introduced to an amplicon by substrate-sequence amplification and monitored as a closed system in a real-time PCR machine such as the ABI7700. In such systems, the DNAzyme would act as an agent for both mutation analysis and signal transduction.

# References

1. Santoro, S. W. and Joyce, G. F. (1997) A general purpose RNA-cleaving DNA enzyme. *Proc. Natl. Acad. Sci. USA* **94,** 4262–4266.
2. Santoro, S. W. and Joyce, G. F. (1998) Mechanism and utility of an RNA-cleaving DNA enzyme. *Biochemistry* **37,** 13,330–13,342.
3. Wu, Y., Yu, L., McMahon, R., Rossi, J. J., Forman, S. J., and Snyder, D. S. (1999) Inhibition of bcr-abl oncogene expression by novel deoxyribozymes (DNAzymes). *Hum. Gene Ther.* **10,** 2847–2857.

4. Santiago, F. S., Kavurma, M. M., Lowe, H. C., Chesterman, C. N., Baker, A., Atkins, D. G., and Khachigian, L. M. (1999) New DNA enzyme targeting Egr-1 mRNA inhibits vascular smooth muscle proliferation and regrowth after injury. *Nat Med.* **5,** 1264–1269.
5. Cairns, M. J., Saravolac, E. G., and Sun, L. Q. (2002) Catalytic DNA: a novel tool for gene suppression. *Curr. Drug Targets* **3,** 269–279.
6. Sun, L. Q, Cairns, M. J., Saravolac, E. G., Baker, A., and Gerlach,W. L. (2000) Catalytic nucleic acids: from lab to applications. *Pharmacol Rev.* **52,** 325–347.
7. Cairns, M. J., Hopkins, T. M., Witherington, C., Wang, L., and Sun, L. Q. (1999) Target site selection for an RNA-cleaving catalytic DNA. *Nature Biotechnol.* **17,** 480–486.
8. Sioud, M. and Leirdal, M. (2000) Design of nuclease resistant protein kinase calpha DNA enzymes with potential therapeutic application. *J. Mol. Biol.* **296,** 937–947.
9. Todd, A. V., Fuery, C. J., Impey, H. L., Applegate, T. L., and Haughton, M. A. (2000) DzyNA-PCR: use of DNAzymes to detect and quantify nucleic acid sequences in a real-time fluorescent format. *Clin Chem.* **46,** 625–630.
10. Cairns, M. J., Hopkins, T. M., Whitherington, C., and Sun, L. Q. (2000) The influence of arm length asymmetry and base substitution on the activity of the "10-23" DNA enzyme. *Antisense Nucleic Acid Drug. Dev.* **10,** 323–332.
11. Cairns, M. J., King, A., and Sun, L. Q. (2000) Nucleic acid mutation analysis using catalytic DNA. *Nucleic Acids Res.* **28,** e9.
12. Sugimoto, N., Nakano, S., Katoh, M., Matsumura, A., Nakamuta, H., Ohmichi, T., et al. (1995) Thermodynamic parameters to predict stability of RNA/DNA hybrid duplexes. *Biochemistry* **34,** 11,211–11,216.

# 22

## Crystallization of the Hairpin Ribozyme
*Illustrative Protocols*

## Peter B. Rupert and Adrian R. Ferré-D'Amaré

### Summary

Conditions and techniques that result in successful crystallization differ from RNA to RNA. However, there are some general principles that facilitate crystallization of most RNAs. Three procedures that were instrumental in obtaining well-ordered crystals of the hairpin ribozyme are described in this chapter. These are: i) the design of a series of candidate crystallization constructs; ii) the evaluation of conditions to obtain monodisperse RNA; and iii) the use of seeding techniques to separate nucleation and growth events during crystallization. These procedures can be usefully adapted for the crystallization of other RNAs.

**Key Words:** Catalytic RNA; crystallization; construct variation; RNA-binding protein; gel-shift assay; optimization; micro-seeding; macro-seeding.

## 1. Introduction

Preparation of well-ordered RNA crystals suitable for structure determination by X-ray crystallography proceeds through several steps. First, constructs of the RNA of interest are designed, synthesized, and purified to homogeneity. Second, the RNAs are characterized biochemically, to ensure that they adopt the relevant (or functional) conformation. Third, the RNA constructs are subjected to crystallization screens. Fourth, initial crystallization conditions usually must be optimized to obtain crystals large enough for X-ray analysis. Fifth, once large crystals are obtained, they can be examined by X-ray diffraction to establish whether their degree of internal order is sufficient to justify structure determination. Usually, the entire process must be iterated many times (*1*).

From: *Methods in Molecular Biology, vol. 252: Ribozymes and siRNA Protocols, Second Edition*
Edited by: M. Sioud © Humana Press Inc., Totowa, NJ

Procedures for the crystallization of an RNA are, by necessity, particular to that molecule. General protocols for the design of constructs to facilitate crystallization *(2,3)*, preparation of mg quantities of homogeneous RNA *(4–6)*, screening and optimization of crystallization conditions *(7,8)*, and X-ray characterization have been published elsewhere *(9)*. In this chapter, we describe three experimental approaches that were instrumental in the crystallization of the hairpin ribozyme *(10)*. **Subheading 3.1.** outlines the strategy used to generate a variety of candidate crystallization constructs of the hairpin ribozyme. In order to facilitate crystallization and structure determination, hairpin ribozymes were engineered to include a cognate site for an RNA-binding protein (the RNA-binding domain, or RBD, of the U1A protein *[3,11]*). **Subheading 3.2.** explains how the gel-mobility-shift assay was used to establish the RNA-protein ratio needed to produce a homogeneous complex for each ribozyme construct. Seeding is a powerful technique for optimizing crystal growth *(12)*. **Subheading 3.3.** describes how this technique was implemented in the case of the hairpin ribozyme. Although the details only apply to this catalytic RNA, the protocols serve to illustrate general approaches that can be adapted for use with other ribozymes.

## 2. Materials

### 2.1. Design and Production of RNA

1. Oligonucleotides for polymerase chain reaction (PCR) were produced by standard solid-phase synthesis.
2. *Taq* polymerase (Roche) was used to assemble the DNA templates by PCR using the supplier's recommended procedures.

### 2.2. Binding of RNA to U1A RBD

1. 40% polyacrylamide solution (29:1 acrylamide:*bis*-acrylamide), ammonium persulfate, *N,N,N',N'*-tetramethyl-ethylenediamine (TEMED).
2. 10X TBE buffer: 900 m$M$ Tris base, 900 mM boric acid, and 20 m$M$ ethylenediaminetetraacetic acid (EDTA).

### 2.3. Crystallization Screen and Seeding

All reagents should be analytical grade. All solutions should be made with diethylpyrocarbonate (DEPC)-treated water (*see* **Note 1**).

1. 24-well culture plate (e.g., Linbro plates from ICN Biomedicals or VDX plates from Hampton Research).
2. Glass cover slips (*see* **Note 2**).
3. 0.3-, 0.5-, and 0.7-mm Special Glass 14 capillary (Charles Supper or Hampton Research).
4. 9-well glass depression plate.

## 3. Methods

### 3.1. Design and Production of RNA

The choice of RNA sequence and length is critical for successful growth of crystals. One strategy for obtaining well-ordered crystals is to generate a collection of variant RNAs and subject them all to crystallization trials. The wild-type RNA may exist as one or several large strands in its natural form, which may not crystallize. However, the activity of interest may reside in a smaller segment within these larger RNA molecules. Our strategy involves removing all of the known dispensable RNA in order to arrive at a core RNA containing the native activity. Next, some or all of the strands that form the core are connected with short linkers, both to simplify and possibly to stabilize an initial prototype RNA. This prototype serves as a starting template for the generation of variants that contain different sequences in the formerly dispensable regions.

**Figure 1A** illustrates the steps used to construct a *trans*-cleaving prototype hairpin ribozyme. Four intervening regions were removed to produce a core ribozyme containing four blunt-ended helices (*see* **Note 3**). The distal ends of helical stems B and C were capped with GNRA tetraloops, thereby linking together the RNA into a single chain. Many variants were generated by inserting variable numbers of basepairs into the linker regions and by adding nucleotides onto the termini. **Figure 1B** shows a schematic representation of how the DNA template is made by PCR from multiple primers (*see* **Note 4**). The RNA for crystallization can be transcribed in vitro by T7 RNA polymerase from the DNA template (*6*; *see* **Note 5**). In our case, the substrate was synthesized as a separate, short piece of RNA (*see* **Note 6**). This allowed incorporation of site-specific chemical modifications.

### 3.2. Binding of RNA to U1A RBD

To help facilitate crystal growth, an U1A-binding site is grafted into a dispensable linker region as illustrated in **Fig. 1A** (*see* **Note 7**). Gel mobility-shift assays (on native polyacrylamide gels) are used to establish the stoichiometry of U1A:RNA binding (to account for inaccuracies in concentration estimates), and to determine whether the RNA-protein complex adopts a unique conformation (**Fig. 2**).

1. Incubate RNA construct (10 μg) with U1A RBD in a 0.5X TBE buffer at 25°C for 15 min. Prepare a series of samples with varying ratios of RNA to protein.
2. Resolve the free RNA from the RNA-protein complex on a 0.5X TBE, 10% native polyacrylamide gel. The gel should be run at a low voltage to avoid heating it above room temperature.
3. Stain with ethidium bromide, and visualize by *trans*-illumination (*see* **Note 8**).

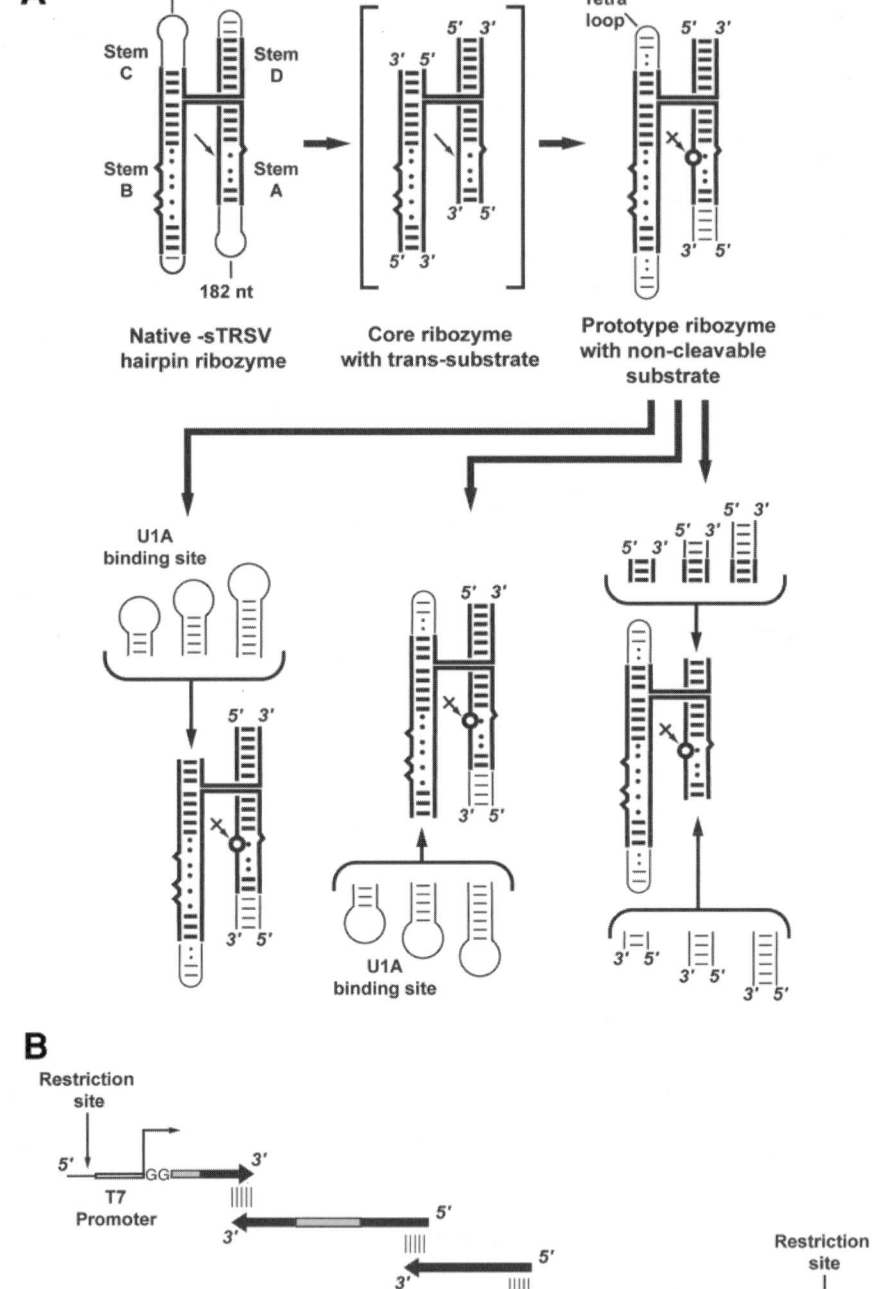

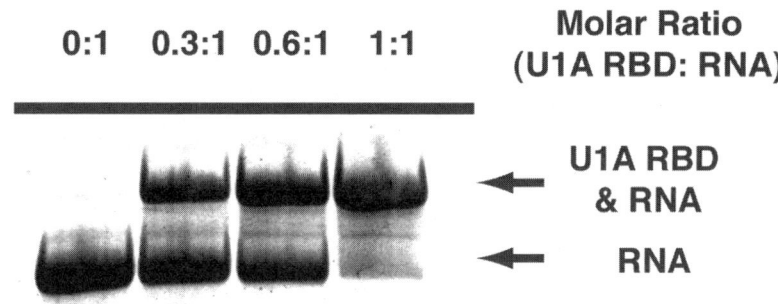

Fig. 2. Representative native polyacrylamide gel of mobility-shift assay for a crystallization construct and U1A RBD.

Only candidate RNA constructs that show a single band when incubated with equimolar U1A RBD are considered for crystallization. These ribozyme constructs are also tested for activity to ensure that modifications have not disrupted the core structure (*see* **Note 9**).

### 3.3. Crystallization Screen and Seeding

A comprehensive description of how to grow macromolecular crystals is beyond the scope of this chapter, and has been reviewed *(13)*. In general, all active RNA constructs are screened by the hanging drop vapor diffusion method against a standard, 48-condition sparse matrix at both 4 and 24°C (*see* **Note 10**). Conditions that produce crystalline precipitates or small crystals are further optimized by systematically varying the composition of the well solution and the drop, e.g., spermine and macromolecule concentration (*see* **Note 11**). The presence of intact RNA in the crystals should be confirmed by electrophoresis or chromatography *(1)*. Finally, crystals are mounted in glass capillary tubes, and initial diffraction images are acquired. Depending on the quality of the diffraction, further screening or optimization may be needed (*see* **Note 12**).

---

Fig. 1. Crystallization construct design. (**A**) Outline of the steps followed in the design of hairpin ribozyme constructs for crystallization. Thick lines represent regions of RNA that are essential for function, and thin lines expendable sequence. The substrate was made non-cleavable by replacing the 2'-OH with a methoxy group as indicated by the circle. A number of variant ribozymes were generated by systematically modifying the dispensable RNA regions. (**B**) Schematic representation of multiple PCR primers used to construct a ribozyme transcription template. Note that a particular dispensable region is spanned by a single primer.

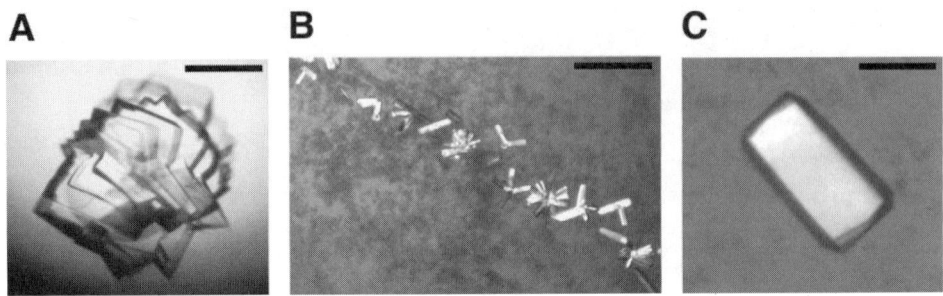

Fig. 3. Micro- and macro-seeding to optimize crystal size and morphology. (**A**) Typical crystal cluster that grew spontaneously in hanging-drop crystallization experiments. (**B**) Streak seeding experiments produced small, single crystals. (**C**) Crystals obtained after macro seeding with small, single crystals. Bars represent approx 100 μm.

One major difficulty in growing hairpin ribozyme crystals was poor crystal morphology. The construct that eventually proved to be successful crystallized initially as clusters (**Fig. 3A**). In order to grow larger single crystals that are suitable for diffraction experiments, micro- and macro-seeding were necessary. Eventually, through an iterative process of harvesting, washing, and reseeding, large, single crystals that diffracted X-rays to 2.4 Å resolution were grown.

### 3.3.1. Micro-Seeding

1. Set up a crystallization tray with solution compositions that are similar to those of a condition that produces crystalline clusters. For this tray, only one component of the well solution is varied. The other variable is the concentration of RNA-protein complex.
2. Allow the drop and well solutions to equilibrate. This may take a few days, depending on the precipitating agent in the well solution (*see* **Note 13**).
3. Transfer a large crystalline cluster into 1 mL of stabilizing solution (*see* **Note 14**).
4. Use a sharp instrument (e.g., fine stainless-steel acupuncture needle) to crush the crystal (*see* **Note 15**). Vortex the mixture.
5. Make serial dilutions of the crushed seed stock in stabilizing solution (1:10, 1:100, 1:1000, 1:10000).
6. Dip a cleaned hair (e.g., an eyelash, or a whisker shed by a cat; *see* **Note 15**) into the crushed crystal mixture, and streak across the equilibrated drops. Repeat streaking (after washing the hair) in adjacent equilibrated drops with lower dilutions of crushed crystal mixture. By optimizing the RNA-protein complex concentration and dilution of crushed seed stock, small, individual crystals can be grown (**Fig. 3B**).

## 3.3.2. Macro-Seeding

1. Prepare equilibrated tray to receive crystals by following **steps 1–3** as described in the previous section. The equilibrated drop should be 15 µL in size (*see* **Note 16**).
2. Fill three wells in a glass depression plate with 1 mL of stabilization solution.
3. Transfer selected candidate seed crystals into the first depression well (*see* **Note 17**).
4. Wash seed crystals by transferring successively into two additional wells (*see* **Note 18**).
5. Place seed crystal into equilibrated drops. After allowing the trays to equilibrate, the resulting larger crystal (**Fig. 3C**) can be used for subsequent rounds of macro-seeding, or capillary mounted for initial diffraction analysis.

## 4. Notes

1. DEPC water is prepared by mixing 1 mL of diethylpyrocarbonate into 4 L of deionized water, mixing well, and autoclaving to destroy the excess DEPC.
2. Siliconized cover slides can be purchased (e.g., Hampton Research). Alternatively, they can be siliconized by dipping in a approx 1% (v/v) solution of dimethyl-dichlorosilane in toluene, washed in ethanol, and dried individually.
3. Only stems A and B are required for hairpin ribozyme activity *(14)*. However, the ribozyme is more stable when it is part of a four-helix junction *(15)*.
4. Ideally, each dispensable region should be spanned by a single oligonucleotide, thus simplifying the production of different variants. A modest number of PCR primers can be combined to produce a large variety of constructs. We screened over 20 different hairpin ribozyme variants in order to find one that would produce diffraction quality crystals.
5. Pyrophosphate is a byproduct of transcription. As it accumulates, pyrophosphate chelates $Mg^{2+}$ and precipitates, thereby reducing the efficiency of transcription. By adding 1 U/mL of inorganic pyrophosphatase (IPP) from *E. coli* (EC 3.6.1.1, Sigma) to the transcription reaction, the pyrophosphate is converted to orthophosphate (which does not chelate $Mg^{2+}$). IPP is reconstituted in buffer containing 50% glycerol, and 10 m$M$ potassium phosphate pH 7.5, and stored at –20°C.
6. Chemically synthesized RNA oligonucleotides were purchased from Dharmacon Research.
7. The U1A RBD has been employed to aid crystal growth for both the hepatitis delta virus (HDV) and hairpin ribozymes *(3,11)*. In addition to helping in crystallization, chemically modified U1A protein can be used to introduce heavy atom sites needed for structure determination.
8. Toluidine blue O (methylene blue T50) staining is an alternative to ethidium bromide, and works well for visualizing RNAs less than 20 nucleotides in length. First stain the gel by soaking in a 1 g/L solution of Toluidine blue in 40% methanol, 1% glacial acetic acid. Destain the gel in a solution of 40% methanol and 1% glacial acetic acid.
9. An example of a hairpin ribozyme activity assay is shown in the supplementary information to *(10)*. (http://www.nature.com/nature/journal/v410/n6830/suppinfo/410780a0.html).

10. We used the 48-condition screen of Jancarik and Kim *(16)*, supplemented with 1 m$M$ spermine added to the RNA-protein complex.

11. In our experience, it is better to limit the amount of optimization to a small number of subsequent screens, and instead to screen a larger number of candidate constructs.

12. A crystal merits further optimization if it diffracts X-rays to 4–5 Å resolution after a reasonable exposure (e.g., 15 min on a home X-ray source).

13. An alternative to **step 2** is to prepare 2X concentration of well solution. This can be mixed with an equal volume of the RNA-protein solution, and diluted to 1X concentration in the wells. These crystallization drops can immediately be seeded.

14. Hairpin ribozyme/U1A RBD complex crystals were stabilized in a solution containing 20 m$M$ CaCl$_2$, 20 m$M$ spermine, 150 m$M$ NH$_4$Cl, and 30% MPD.

15. Instruments that contact solutions to be used with RNA are cleaned by placing in a 0.5 $N$ NaOH solution and then rinsing extensively with DEPC-treated water. This is to remove ribonuclease contamination.

16. Fifteen-well crystal plates from ICN Biomedicals are convenient for seeding experiments because the thick glass cover slips can be easily opened and resealed multiple times.

17. Crystals are transferred using a 0.5-mm-diameter glass capillary by aspiration, either by mouth pipeting with a rubber tube attached to the capillary, or by attaching a small syringe.

18. For the hairpin ribozyme, seed crystals were washed in stabilization solution. This differs from conventional macro-seeding technique, in which the seed crystal is dissolved slightly *(12)*.

## Acknowledgments

A. R. F. is a Rita Allen Foundation Scholar, and a W. M. Keck Foundation Distinguished Young Scholar in Medical Research, and P. B. R. is a trainee of the Chromosome Metabolism and Cancer training grant from the National Cancer Institute. This work was supported by grants from the National Institutes of Health.

## References

1. Ferré-D'Amaré, A. R. and Doudna, J. A. (2000) Methods to crystallize RNA. *Current Protocols in Nucleic Acid Chemistry* 7.6.1.–7.6.13.

2. Ferré-D'Amaré, A. R., Zhou, K., and Doudna, J. A. (1998) A general module for RNA crystallization. *J. Mol. Biol.* **279,** 621–631.

3. Ferré-D'Amaré, A. R. and Doudna, J. A. (2000) Crystallization and structure determination of a hepatitis delta virus ribozyme: use of the RNA-binding protein U1A as a crystallization module. *J. Mol. Biol.* **295,** 541–556.

4. Price, S. R., Ito, N., Oubridge, C., Avis, J. M., and Nagai, K. (1995) Crystallization of RNA-protein complexes I. Methods for the large-scale preparation of RNA suitable for crystallographic studies. *J. Mol. Biol.* **249,** 398–408.

5. Ferré-D'Amaré, A. R. and Doudna, J. A. (1996) Use of *cis*- and *trans*-ribozymes to remove 5' and 3' heterogeneities from milligrams of in vitro transcribed RNA. *Nucleic Acids Res.* **24,** 977,978.
6. Doudna, J. A. (1997) Preparation of homogeneous ribozyme RNA for crystallization. *Methods Mol. Biol.* **74,** 365–370.
7. Doudna, J. A., Grosshans, C., Gooding, A., and Kundrot, C. E. (1993) Crystallization of ribozymes and small RNA motifs by a sparse matrix approach. *Proc. Natl. Acad. Sci. USA* **90,** 7829–7833.
8. Scott, W. G., Finch, J. T., Grenfell, R., Fogg, J., Smith, T., Gait, M. J., and Klug, A. (1995) Rapid crystallization of chemically synthesized hammerhead RNAs using a double screening procedure. *J. Mol. Biol.* **250,** 327–332.
9. Drenth, J. (1999) *Principles of Protein X-Ray Crystallography*, Springer Verlag.
10. Rupert, P. B. and Ferré-D'Amaré, A. R. (2001) Crystal structure of a hairpin ribozyme-inhibitor complex with implications for catalysis. *Nature* **410,** 780–786.
11. Oubridge, C., Ito, N., Evans, P. R., Teo, C. H., and Nagai, K. (1994) Crystal structure at 1.92 Å resolution of the RNA-binding domain of the U1A spliceosomal protein complexed with an RNA hairpin. *Nature* **372,** 432–438.
12. Stura, E. A. and Wilson, I. A. (1990) Analytical and production seeding techniques. *Methods* **1,** 38–49.
13. McPherson, A. (1999) *Crystallization of Biological Macromolecules*, Cold Spring Harbor Laboratory Press, Cold Spring Harbor, NY.
14. Fedor, M. J. (2000) Structure and function of the hairpin ribozyme. *J. Mol. Biol.* **297,** 269–291.
15. Walter, F., Murchie, A. I., and Lilley, D. M. (1998) Folding of the four-way RNA junction of the hairpin ribozyme. *Biochemistry* **37,** 17,629–17,636.
16. Jancarik, J. and Kim, S. H. (1991) Sparse matrix sampling: a screening method for crystallization of proteins. *J. Appl. Cryst.* **24,** 409–411.

# 23

## An Experimental Method for Selecting Effective Target Sites and Designing Hairpin Ribozymes

Alicia Barroso-delJesus, Elena Puerta-Fernández,
Cristina Romero-López, and Alfredo Berzal-Herranz

### Summary

The proper selection of target sites and the correct design of specific ribozymes are decisive initial steps in any attempt to perform ribozyme-mediated gene silencing. Combinatorial methodologies can be used to improve ribozyme targeting and design. The in vitro selection strategy described in this chapter uses a combinatorial library of potentially self-cleaving RNA molecules. The hairpin ribozyme is attached to the target mRNA, and is adequately randomized to generate a population representing all possible substrate specificities. The selection procedure yields information on the best target sites, and provides information about optimal ribozyme sequences. Thus, this method helps in the rational design of efficient hairpin ribozymes for targeting purposes. and avoids trial-and-error assays usually associated with theoretical ribozyme design.

**Key Words:** Hairpin ribozyme; ribozyme targeting; in vitro selection; hairpin ribozyme design; gene silencing.

## 1. Introduction

Many factors determine the effectiveness of in vivo ribozyme use—e.g., the intracellular localization of the ribozyme and substrate, target accessibility, ribozyme half-life, catalytic activity in the cell environment, and specificity. However, some of these problems can be overcome through ribozyme design.

The first choice to be made is the selection of a target within the messenger RNA. Potential target sites can be occluded by the folded structure of the molecule. Computer-aided prediction of RNA structure can sometimes provide a rough estimate of the most inaccessible regions, but in order to be reliable, this requires previous structural knowledge based on available experimental data.

From: *Methods in Molecular Biology, vol. 252: Ribozymes and siRNA Protocols, Second Edition*
Edited by: M. Sioud © Humana Press Inc., Totowa, NJ

Even when available, the information obtained is often invalid or insufficient. Another strategy is to scan the messenger RNA by hybridization with small oligonucleotides and subsequent digestion with RNase H. However, for ribozymes, "target accessibility" involves not only hybridization with the target sequences, but also correct folding of the ribozyme-substrate complex into the catalytically active form. This process depends on specific sequences in both the target and ribozyme, as well as on possible interactions with neighboring sequences. The problems posed by this sequence interdependence can be better solved through the use of combinatorial methods (*see* Chapters 9, 19, and 20). In general, a ribozyme library that contains all possible substrate specificities processes an RNA of interest. Thus, sites cleaved are predominantly identified and chosen for targeting purposes. This strategy has been developed and successfully used with hammerhead, hairpin, and Group I intron ribozymes. However, none of these combinatorial methods provides information about the optimal ribozyme sequences that are responsible for cleavage. Substrate-recognition regions are derived theoretically assuming perfect Watson-Crick basepairing between the ribozyme and target. But this premise does not always have to be followed, especially when working with the hairpin motif (*1*).

## 1.1. Selection Procedure

The methodology described here can be considered an in vitro selection procedure (*2*). In addition to detecting the most accessible target sites within an RNA messenger, it also selects the best hairpin motif for cleaving those sites from the ribozyme library. The initial RNA library is randomized to contain more than $1.9 \times 10^5$ substrate specificities (*3–5*; **Fig. 1**). Although the method is based on a self-cleaving library, only the 5' half of the catalytic motif is covalently linked to the RNA of interest (**Fig. 2**). The use of a bimolecular hairpin motif (**Fig. 1**; *6*), allows the separation of the RNA synthesis from the self-cleavage reaction. Thus, a non-active RNA library can be synthesized, and the full-length population can be purified. The possibility of eliminating all the initial background derived from the transcription reaction is critical to the success of the selection. The RNA library is complemented by hybridization with the 3' half of the hairpin ribozyme, and undergoes self-cleavage. The chemistry of the cleavage reaction provides a free OH group at the 5' ends of the 3' cleavage products, which allows the possibility of selectively labeling and purifying these molecules. The 3' cleavage products carry all the information regarding both the cleavage site and the active ribozyme variant. Since their 5' end sequences are unknown, the molecules are circularized by an intramolecular RNA ligation reaction. This circular structure allows the retrotranscription and amplification of the full-length molecule, as well as the conservation of all sequence information.

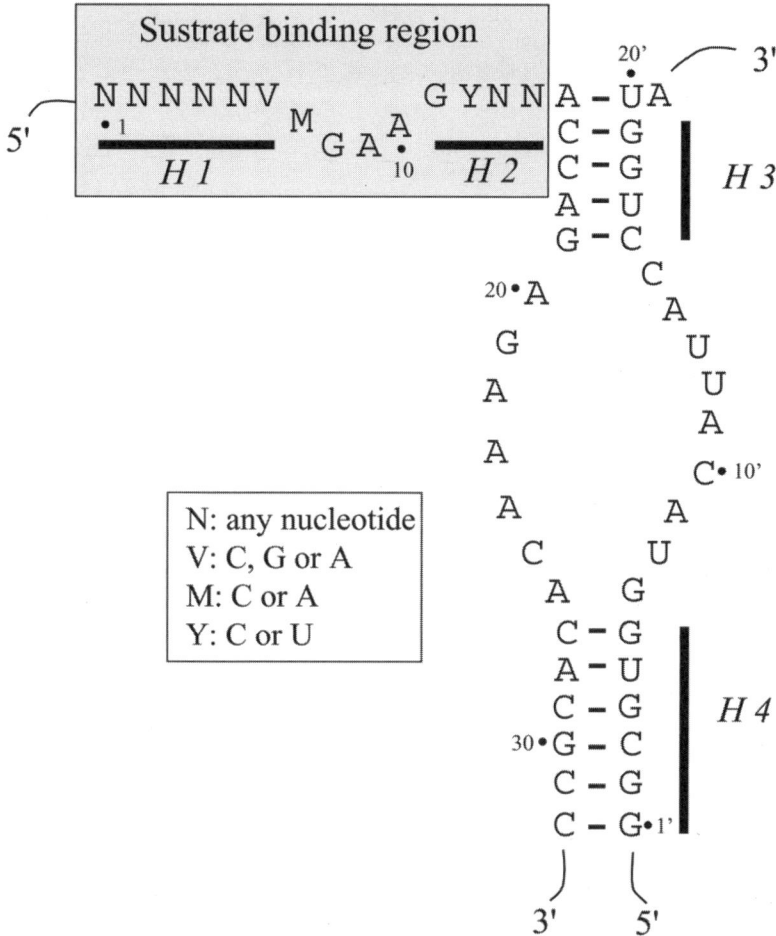

Fig. 1. Bimolecular hairpin ribozyme used in the selection procedure. H1-4 are ribozyme sequences involved in helical regions 1–4 in the ribozyme-substrate complex. Substrate interaction regions are shown with sequence variations corresponding to the initial RNA library. 5' half ribozyme sequences are numbered 1–32; 3' half ribozyme sequences are 1'–21'.

One round of selection identifies optimal target sites, but in order to obtain accurate information about active ribozymes, several rounds may be necessary. However, this strategy allows multiple, consecutive selection cycles. In addition, independence of the cleavage reaction from RNA synthesis allows important modifications of the cleavage reaction conditions, which determines the selection pressure.

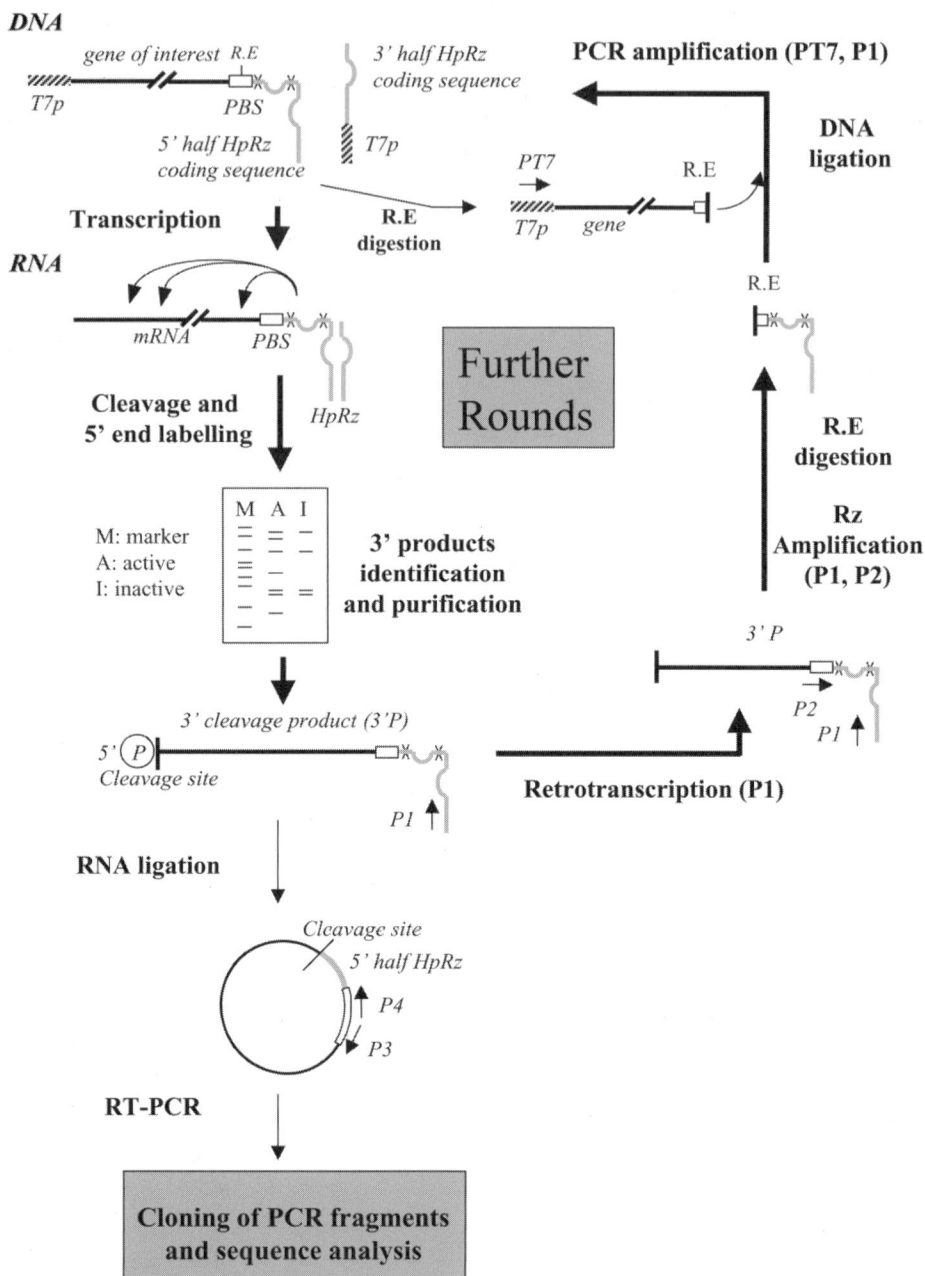

Fig. 2. Diagram of the in vitro selection procedure. Ribozyme sequences are shown in gray. X indicates degenerated sites within the substrate interaction regions of the ribozyme. P1-4 and PT7, primers used for cDNA synthesis and PCR amplification. T7p, T7 promoter. HpRz, hairpin ribozyme. PBS, primer binding site. R.E, recognition sequence for a DNA restriction enzyme.

## 2. Materials

An RNase-free environment is crucial when working with RNA. Indeed, in our in vitro selection protocol, RNA degradation creates a strong background of multiple-sized molecular species that obstructs the selection of the desired molecules. Thus, solutions and buffers should be prepared with diethylpyrocarbonate (DEPC)-treated $H_2O$ and autoclaved, in addition to all plasticware. Everything, including reagents, should be kept in RNase-free conditions.

### 2.1. PCR Amplifications

1. 10X deoxynucleotide 5' triphosphate (dNTP) mix stock solution: deoxyadenosine 5' triphosphate (dATP), deoxycytidine 5' triphosphate (dCTP), deoxyguanosine 5' triphosphate (dGTP), and deoxythymidine 5' triphosphate (dTTP), 2.5 m$M$ each.
2. 10X reaction buffer: 750 m$M$ Tris-HCl, pH 9.0, 20 m$M$ $MgCl_2$, 500 m$M$ KCl, 200 mM $(NH_4)_2SO_4$, 0.01% bovine serum albumin (BSA).
3. Thermostable DNA polymerase (e.g., Biotools).
4. The appropriate set of primers:
   a. SelT7 and Sel5'Rz for construction of the initial RNA library.
   b. P1 and P2 for amplification of the selected ribozyme population.
   c. P3 and P4 for full-length amplification of 3' cleavage products.
   d. T7P and P1 for amplification of the RNA library template entering further rounds of selection.

### 2.2. Nucleic Acid Purification

1. Phenol and chloroform:isoamyl alcohol (IAA) 24:1 (v/v) for nucleic acid extraction.
2. Absolute ethanol and 3 $M$ pH 5.2 sodium acetate for nucleic acid precipitation.

### 2.3. Transcription and Self-Cleavage Reactions

1. T7 RNA polymerase.
2. 10X NTP mixed stock solution: adenosine 5' triphosphate (ATP), cytidine 5' triphosphate (CTP), guanosine 5' triphosphate (GTP), and uridine 5' triphosphate (UTP) 10 m$M$ each.
3. RNase inhibitor (e.g., RNAguard™, Amershan Biosciences).
4. 10X transcription buffer: 400 m$M$ Tris-HCl, pH 8.0, 60 m$M$ $MgCl_2$, 10 m$M$ spermidine, 40 m$M$ NaCl, 100 m$M$ dithiothreitol (DTT), and 0.1% Triton X-100.
5. RNase-free DNase (e.g., RQ1 RNase-free DNase, Promega).
6. 10X ribozyme cleavage buffer: 500 m$M$ Tris-HCl, pH 7.0, 120 m$M$ $MgCl_2$.

### 2.4. 5'-End-Labeling Reaction

1. T4 polynucleotide kinase (PNK) (e.g., New England Biolabs).
2. 10X PNK buffer: 700 m$M$ Tris-HCl, pH 7.6, 100 m$M$ $MgCl_2$, 50 m$M$ DTT.
3. [$\gamma$-$^{32}$ATP] 3000–5000 mCi/mol.
4. Sephadex G-25.

## *2.5. Purification of 3' Cleavage Products*

1. 40% polyacrylamide stock solution (19:1 acrylamide:*bis*-acrylamide).
2. 10X TBE buffer: 0.89 $M$ Tris-base, 0.89 $M$ boric acid, 25 m$M$ ethylenediamine-tetraacetic acid (EDTA).
3. Ammonium persulfate 10% (w/v).
4. $N,N,N',N'$-tetramethyl-ethylenediamine (TEMED).
5. Formamide-loading buffer: 94% deionized formamide ultra-pure grade, 0.025% xylencyanol, 0.025% bromophenol blue, 17 m$M$ EDTA.
6. Sterile razor blades.
7. RNA elution buffer: 0.5 $M$ ammonium acetate, 0.1% sodium dodecyl sulfate (SDS), 1 m$M$ EDTA.

## *2.6. RNA Ligation Reaction*

1. T4 RNA ligase (e.g., USB).
2. 10X T4 RNA ligase buffer: 500 m$M$ Tris-HCl, pH 7.5, 50 m$M$ MgCl$_2$, 100 m$M$ DTT, 10 m$M$ ATP, 1 µg/mL BSA.

## *2.7. Reverse Transcription*

1. AMV reverse transcriptase (e.g., USB).
2. 5X reaction buffer: 250 m$M$ Tris-HCl pH 8.3, 40 m$M$ MgCl$_2$, 250 m$M$ NaCl, 5 m$M$ DTT.
3. RT primer (P1 or P3).

## *2.8. Restriction of DNA*

1. Appropriate restriction enzymes.
2. The corresponding manufacturer's 10X cleavage buffers.

## *2.9. DNA Ligation Reaction*

1. T4 DNA ligase (e.g., MBI Fermentas).
2. 10 m$M$ ATP.
3. 25% PEG.
4. 10X T4 DNA ligase buffer: 300 n$M$ Tris-HCl pH 7.8, 100 m$M$ MgCl$_2$, 100 m$M$ DTT, 10 m$M$ ATP.

## *2.10. Cloning and Sequence Analysis of Selected Molecules*

1. Appropriate plasmid vector for cloning selected sequences.
2. Sequencing facilities.

## 3. Methods

## *3.1. Preparation of the Combinatorial Self-Cleaving Library Templates*

As described in the Introduction, the selection method uses a bimolecular hairpin ribozyme to separate transcription and cleavage reactions. The self-cleaving library is assembled by hybridization of two independent RNA molecules: the RNA library that carries the messenger RNA and the 5' half of the ribozyme, and the so-called Sel3'Rz that consists of the 3' half of the hairpin ribozyme. This section describes how to generate the DNA templates for transcription of both RNA molecules.

1. The transcriptional template for the combinatorial RNA library is generated by PCR. The gene of interest serves as a DNA template during the amplification process. Oligonucleotides used as PCR primers are designed to hybridize at the 5' and 3' ends of the targeted gene, and to introduce the corresponding extra sequences. The annealing regions should be at least 17–20-nt long, and the general structure of primers should be as follows:

    Sense primer (SelT7): 5'-<u>TAATA CGACT CACTA TAGG</u> (X $_{17-20}$)-3'

    The T7 promoter sequence is underlined. (X$_{17-20}$) corresponds to the region of 17–20 nucleotides of the 5' end of the gene.

    Antisense primer (Sel5'Rz): 5'-GGCGT GTGTT TCTCT GG<u>TNN RC</u>TTC <u>KBNNN N</u> (PBS) (X'$_{17-20}$)-3'.

    Substrate-recognition regions are underlined, where N = any nucleotide, R = G or A, K = G or T, and B = G, C or T, (PBS) = primer binding sequence, and (X'$_{17-20}$) corresponds to the region of 17–20 nucleotides complementary to the 3' end of the gene.

    In 5' to 3' order, the cassette will finally contain (**Fig. 2**):
    a. A promoter for in vitro transcription. Our protocol uses the bacteriophage T7 RNA polymerase promoter followed by two 'Gs'.
    b. The gene of interest.
    c. A primer-binding site containing a unique restriction site.
    d. The 5' half of the bimolecular form of the hairpin ribozyme (**Fig. 1**), including the mutagenised substrate binding regions.
2. Design (*see* **Note 1**) and synthesize the required DNA oligonucleotides for PCR amplification. These can be purchased from commercial sources. Dissolve them in TE 1X or ddH$_2$O (10 pmol/mL).
3. Calculate the proper melting temperatures using appropriate software (e.g., OLIGO 4). Only those sequences that anneal with the gene are taken into consideration; all newly introduced sequences can be disregarded (*see* **Note 2**).
4. Set up the following PCR reaction in a final volume of 100 µL.

    | | |
    |---|---|
    | Water | 71 µL |
    | dNTPs (2.5 m$M$ each) | 7 µL |
    | 10X PCR buffer | 10 µL |
    | Sense primer | 5 µL |
    | Antisense primer | 5 µL |
    | DNA template (10 ng/µL) | 1 µL |
    | *Taq* polymerase (2 U/µL) | 1 µL |

5. Carry out a standard PCR protocol of at least 25 cycles consisting of: 30 s at 95°C, 30 s at the annealing temperature (the lowest melting temperature calculated

for the primers) and 30s at 72°C. The first cycle is preceded by an initial denaturing step of 2 min at 95°C, and the reaction is finished by a final elongation step of 7 min at 72°C and cooling to 4°C.

6. Purify the final PCR product by phenol-extraction and ethanol-precipitation. The pellet is washed with 70% ethanol and dissolved in 20 μL of DEPC-treated $H_2O$. Determine the final DNA concentration by ultraviolet (UV) spectrophotometry.

7. Mix oligos Sel3'Rz (5' TACCA GGTAA TATAC CACGC CATAT GTGAG TCGTA TTA 3') and PT7 (5' TAATA CGACT CACTA TA 3'), 500 pmol each, and bring the volume to 20 μL with $ddH_2O$. The sample is denatured and re-natured by heating for 2 min at 95°C followed by snap-cooling on ice plus further incubation on ice for another 15 min. The resulting partially dsDNA is used as a template for the synthesis of the 3' half ribozyme, since the T7 RNA polymerase only requires the 17-nt promoter region as dsDNA.

## 3.2. RNA Synthesis

Since the library is designed to contain all possible ribozyme specificities, the input for the first cycle of selection will include more than $1.9 \times 10^5$ variants (*see* **Note 3**). A preparative in vitro transcription reaction is performed in order to have sufficient RNA molecules of each sequence variant.

1. Set up a 250-μL reaction in 1X transcription buffer containing 1 m$M$ of each NTP, 0.5 U/ml RNAguard™, 40 ng/μL of library dsDNA, and 20 μg/mL of purified T7 RNA polymerase.

2. For the synthesis of the 3' half of the ribozyme, an identical reaction is prepared but in a final volume of 50 μL. The totality of the partially dsDNA obtained from the annealing reaction is used as a DNA template (*see* **Subheading 3.1., step 7**).

3. Incubation proceeds at 37°C for 2 h. Stop the transcription reactions by adding 0.02 U/μL of RNase-free DNase I. Incubate for a further 15 min at the same temperature.

4. Load the RNAs on preparative, standard denaturing-7 $M$ urea polyacrylamide gels (1.5-mm thickness). A 20% gel is used for the 3' half Rz RNA, and 4% gel for the RNA library. Prepare the gels in 1X TBE and use 0.5X TBE as running buffer.

5. Localize the full-size RNAs by UV shadowing (*see* **Note 4**) and excise the corresponding gel slices with a sterile razor blade. Cut them into smaller pieces and place them in a microfuge tube. Add 300 μL of RNA elution buffer and incubate the tubes overnight at 4°C.

6. Save the aqueous phase. Extract twice with phenol and once with chloroform:IAA (24:1). RNA is collected by ethanol-precipitation. Wash the pellet with 80% ethanol, dry briefly, and dissolve it in 20 μL of DEPC water.

7. Determine the RNA concentration by UV spectrophotometry.

## 3.3. Cleavage and Detection of Cleavage Products

To confer catalytic activity upon the self-cleaving library, both halves of the hairpin ribozyme must hybridize and reconstitute the catalytic motif.

1. Mix 30 pmol of the RNA library and 50 pmol of the 3' half Rz RNA. Heat at 65°C for 7 min, cool down on ice for 15 min, and then bring to reaction temperature by incubation at 37°C for an additional 15 min.

2. In parallel, prepare a control reaction that omits the 3' half Rz RNA.

3. To start the reaction, add the cleavage buffer at a final concentration of 12 m$M$ MgCl$_2$ and 50 m$M$ Tris-HCl. Incubate at 37°C for 90 min (*see* **Note 5**).

4. The newly generated 3' cleavage products have a 5' OH group that can be specifically radiolabeled. Collect the RNA by ethanol-precipitation. Dissolve the pellet in DEPC H$_2$O and set up the labeling reaction with 25 U of T4 PNK, 1 μL of 10X PNK buffer and 25 μCi of [γ-$^{32}$ATP] in a final volume of 10 μL. Incubate at 37°C for 45 min.

5. Non-incorporated radiolabeled nucleotides are removed by exclusion chromatography in a 1-mL G-25 Sephadex column. Pack the resin into a 1-mL standard syringe, and equilibrate the column twice with 100 μL of DEPC H$_2$O. Bring the volume of the labeling reaction to 100 μL with DEPC H$_2$O, and load the pre-equilibrated column with it. Recover the flowthrough by centrifugation and collect the RNA by ethanol-precipitation.

6. Add 1 vol of formamide-loading buffer to the control and to the sample before loading them onto a 4% (w/v) 7 $M$ urea-polyacrylamide denaturing gel (*see* **Note 6**). Before loading, samples should be denatured by heating for 2 min at 95°C followed by rapid cooling on ice.

7. After electrophoresis, place the gel on an exposure cassette and expose a sheet of X-ray film. To allow correct alignment of the film on the gel, use commercially available phosphorescent labels.

8. The autoradiography results are examined to determine which bands are specific for the problem lane in comparison to the control lane. These will be ribozyme-catalyzed 3' cleavage products (*see* **Note 7**).

9. With the help of the film, mark the exact position of each of the chosen RNA molecules on the gel. Excise the corresponding gel slices, elute and clean the RNA as previously described (*see* **Subheading 3.2.**, **steps 5** and **6**). Redissolve in DEPC-treated water.

## *3.4. Cleavage Site Mapping and Analysis of Ribozyme Sequences*

The 5' end sequence of the recovered cleavage products is unknown. One way to retrotranscribe and amplify the entire RNA molecule without information loss is to convert it into circular form (**Fig. 3**).

1. Each 3' cleavage product purified from the gel is fully used in the RNA ligation reaction. The reaction is performed at 4°C overnight with 24 U of T4 RNA ligase (USB) in the presence of 1X reaction buffer and 10 U of RNAguard™, in a total volume of 10 μL.

2. Design two divergent PCR primers to anneal on the PBS region as shown in **Fig. 2** (P3 and P4). The one that shows the opposite polarity and complementary sequence to the RNA strand (P3) is used as RT primer.

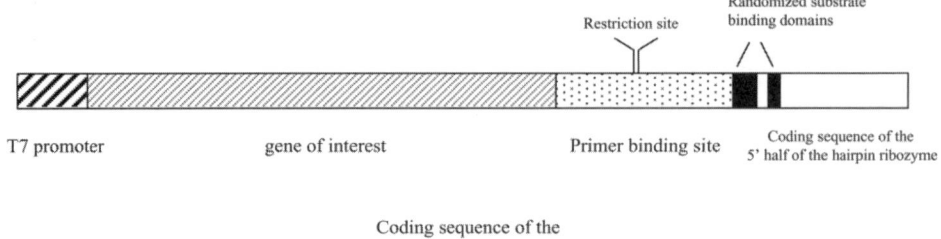

Fig. 3. Transcriptional cassettes for the two RNA molecules that make up the self-cleaving RNA library.

3. Set up the RT reaction with an aliquot of the circular RNA (for example, 2 μL from the 10 μL of the RNA ligation reaction) in the presence of 1X reaction buffer, 0.6 m*M* dNTPs, and 20 pmol of the RT primer. Heat the mixture for 2 min at 95°C and then cool it on ice for 15 min to allow annealing of the RNA and the RT primer. Then add 2 U/μL of AMV reverse transcriptase and incubate the reaction mix at 42°C for 30 min. The reaction is stopped by heat inactivation of the enzyme at 99°C for 5 min.

4. Full-length cDNA is PCR-amplified after addition of primer P4 and 0.02 U/μL of *Taq* polymerase.

5. Restriction sites that are suitable for cloning the final PCR product in a favorite cloning vector are included during the amplification step by primers P3 and P4 (*see* **Note 8**).

6. To obtain reliable conclusions (especially about proper ribozyme sequences for a specific site), at least 20 independent sequences for each 3' cleavage product should be analyzed.

7. Nucleotides juxtaposed at the original 3' end of the RNA library indicate the cleavage site, and sequences of the helix I and II regions of the ribozyme determine the substrate specificity of the active variant that processed the messenger at that particular site.

8. For definitive targeting experiments, the substrate-recognition regions of the hairpin ribozyme should be designed according to the most common nucleotides at each position, as determined from the population of sequences analyzed (*see* **Note 9**).

### *3.5. Further Rounds of Selection*

Thus far, we have described how to perform a complete, single cycle of selection. However, in an in vitro selection scheme, it is strongly recommended that more than one round be performed in order to achieve positive results. In our protocol, after the first cycle, cleavage sites are easily mapped, but ribozyme sequences are highly heterogeneous for each target (probably because of *trans*-cleavage), and no definitive conclusions about the best variant can be drawn.

1. To run an additional selection cycle, **Subheading 3.4.** is skipped.
2. Retrotranscribe each 3' cleavage product (*see* **Subheading 3.3., step 9**) using primer P1 (5' GGCGT GTGTT TCTCT GGT 3'). This primer hybridizes to the non-variable sequence located downstream to the mutagenized regions of the ribozyme (**Fig. 3**). Set up and perform the reaction as described in **Subheading 3.4., step 3**.
3. Amplify only the ribozyme portion by adding a second primer complementary to the primer-binding site region (P2; **Fig. 3**).
4. Digest the PCR product at the level of the unique restriction site included within the primer-binding site region.
5. Eliminate the restriction enzyme by extraction of the DNA with phenol and chloroform:IAA followed by ethanol-precipitation.
6. In parallel, digest the DNA template for the original RNA library with the same restriction enzyme. Purify the larger DNA fragment that contains the gene linked to the T7 promoter from an agarose gel.
7. Dissolve DNAs in 10 μL of ddH$_2$O each.
8. Set up the ligation reaction in 1X buffer containing 1 m$M$ ATP, 2.5% PEG, 5 μL of each DNA to be ligated, and 1 U of T4 DNA ligase in a final volume of 15 μL. Incubate overnight at 15°C.
9. Amplify the ligation product by PCR using oligos PT7 and P1 (use no more than 2 μL of the ligation reaction as DNA template for the amplification step). The PCR product will constitute the DNA template of the RNA library entering the next round of selection. The newly restricted population that emerges from a selection cycle has now substituted the initial population of ribozyme variants. This procedure can be repeated in as many rounds as desired (*see* **Note 10**) before performing the sequence analysis described in **Subheading 3.4.**

## 4. Notes

1. If the possibility of choosing the annealing regions for primers SelT7 and Sel5'Rz (at the 5' and 3' ends of the gene, respectively) exists, some simple rules should be followed that aid in the design of effective primers. The calculated T$_m$s should be balanced. Complementarity at the 3' ends of the primer pair should be avoided because it promotes the formation of primer-dimer artifacts and reduces the yield of the desired product. Finally, stretches of Cs or Gs at the 3' end, as well as palindromic sequences within primers, should also be avoided.
2. After several cycles of amplification, the extra sequences carried by primers SelT7 and Sel5'Rz have already been introduced into the DNA template. Consequently, the annealing region is extended and the T$_m$s of the primers increased substantially. If desired, the PCR protocol may include different annealing temperatures—the initial temperature (calculated as described in **Subheading 3.1.**) for the first 3–5 cycles, and a higher one for the remainder of the protocol.

3. Once the DNA template for the initial RNA library has been prepared, it may be useful to analyze the sequence variability of the ribozyme population in order to check that all nucleotides have been incorporated at the desired frequencies in every mutagenized position. This would guarantee that all possible variants are potentially present and equally represented in the initial pool. To do this, the ribozyme sequence from the starting library template is specifically amplified and cloned, and a sufficient number of independent clones are analyzed. If there is any bias for a specific nucleotide in any position, this should be taken into account during the analysis of the selection process.

4. Transcripts can be localized in the gel by UV shadowing using a silica gel 60 F254-coated plate and a UV epiluminiscence lamp (254 nm). Alternatively, poly-acrylamide gels may be stained with conventional dyes such as ethidium bromide or SYBR™ Green II (but with the resulting disadvantage of having to use an intercalating agent).

5. The specific conditions of cofactor concentration, reaction temperature, and reaction time are those recommended for earlier cycles. However, increasing the stringency of the reaction conditions may progressively strengthen selection pressure. This is strongly recommended for the last one or two cycles in order to recover and analyze only the most active variants. One possible strategy is to diminish reaction time to 10–15 min and $MgCl_2$ concentration to 4 m$M$.

6. The gel concentration will depend on the library RNA full-length measurements. Since more than 400 nt can be expected for most genes analyzed, a 4% gel is recommended. To adequately resolve the larger 3' cleavage products, it may be necessary to perform a second loading.

7. Since the T4 PNK incorporates only one label at the 5' end of each 3' cleavage product, the intensity of the signal for each independent band will depend on the quantity of RNA present, and not on product length. Therefore, every potential target site can be detected with the same sensitivity, regardless of its position in the messenger. Exposure time will determine, to some extent, the sensitivity of the selection procedure, since the most proficient sites are easily detected with shorter exposure times.

8. It is important to ensure that restriction sites used for cloning purposes, plus the one located in the PBS region, are not present within the messenger sequence.

9. When more than one ribozyme variant is selected with the same or similar frequency for a specific site, it is recommended that their in vitro *trans*-cleavage activity be compared to determine which is the most effective.

10. It is difficult to state *a priori* how many cycles of selection are necessary to obtain clear results, since this will depend on each particular assay as well as the stringency of conditions used during selection. A general selection scheme might be constituted by three initial permissive cycles plus one or two more restrictive cycles, making a total of four or five rounds. The final cycle must be performed as described in **Subheading 3.4.**, rather than as an additional round, in order to obtain combined information about ribozymes and target sites.

# References

1. Yu, Q., Pecchia, D. B., Kingsley, S. L., Heckman, J. E., and Burke, J. M. (1998) Cleavage of highly structured viral RNA molecules by combinatorial libraries of hairpin ribozymes. The most effective ribozymes are not predicted by substrate selection rules. *J. Biol. Chem.* **273(36)**, 23,524–23,533.
2. Barroso-delJesus, A. and Berzal-Herranz, A. (1999) Experimental strategies for hairpin ribozyme targeting of long RNAs. *Nucleic Acids Symp. Series* **41**, 63–65.
3. Berzal-Herranz, A., Joseph, S., Chowrira, B. M., Butcher, S. E., and Burke, J. M. (1993) Essential nucleotide sequences and secondary structure elements of the hairpin ribozyme. *EMBO J.* **12(6)**, 2567–2573.
4. Joseph, S., Berzal-Herranz, A., Chowrira, B. M., Butcher, S. E., and Burke, J. M. (1993) Substrate selection rules for the hairpin ribozyme determined by in vitro selection, mutation, and analysis of mismatched substrates. *Genes Dev.* **7(1)**, 130–138.
5. Shippy, R., Siwkowski, A., and Hampel, A. (1998) Mutational analysis of loops 1 and 5 of the hairpin ribozyme. *Biochemistry* **37(2)**, 564–570.
6. Chowrira, B. M. and Burke, J. M. (1992) Extensive phosphorothioate substitution yields highly active and nuclease-resistant hairpin ribozymes. *Nucleic Acids Res.* **20(11)**, 2835–2840.

# 24

# Design and Optimization of Sequence-Specific Hairpin Ribozymes

Cristina Romero-López, Alicia Barroso-delJesus,
Elena Puerta-Fernández, and Alfredo Berzal-Herranz

## Summary

The hairpin ribozyme belongs to a group of small catalytic RNAs that have been extensively used to *trans*-cleave RNA molecules. Many efforts have been made to elucidate its reaction mechanism, and there is great interest in designing hairpin ribozymes with improved catalytic activity for use in the development of agents that specifically inactivate RNA molecules. This chapter summarizes the general principles in the design of hairpin ribozymes for targeting purposes, and provides a brief overview of the well-characterized modifications of the ribozyme sequences and structural domains that are necessary for optimal activity. The main features of the target sequence are also examined and other procedures or modifications of interest are also discussed.

**Key Words:** Hairpin ribozyme; ribozyme targeting; specificity of the hairpin ribozyme; hairpin ribozyme design; improvement of the hairpin ribozyme; RNA catalysis.

## 1. Introduction

The hairpin ribozyme motif was first described as the RNA domain that is responsible for the self-cleavage and ligation of the negative strand of satellite RNA associated with tobacco ring spot virus [(–)sTRSV] *(1,2)*. The sequences required for this catalytic activity were identified by several techniques including insertions, deletions, and single mutagenesis. The catalytic domain was also engineered to support *trans*-cleavage reactions *(3,4)*. The resulting ribozyme belongs to the group of small catalytic RNAs, and catalyzes the reversible cleavage of a second RNA molecule (substrate) to yield products with 2', 3' cyclic phosphate and 5' hydroxyl termini. The most widely accepted construct and nomenclature is that proposed by Hampel and Tritz *(4)*. The minimal

From: *Methods in Molecular Biology, vol. 252: Ribozymes and siRNA Protocols, Second Edition*
Edited by: M. Sioud © Humana Press Inc., Totowa, NJ

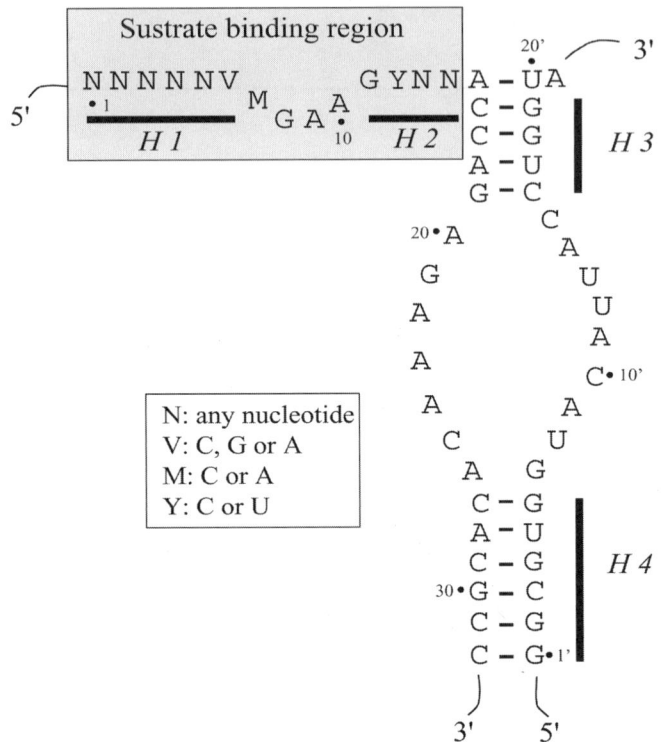

Fig. 1. Secondary structure of [–]sTRSV hairpin ribozyme:substrate complex. Ribozyme nucleotides are numbered 1–50, and substrate nucleotides (in lower case letters) are numbered –5 to +9. Helical regions are denoted as H1-H4, and junction regions are named as J1/2, J2/1, J3/4, J4/3. (**A** and **B**) represent the A- and B-defined domains of the ribozyme:substrate complex. Arrow indicates the cleavage site in the substrate.

hairpin ribozyme consists of a 50-nucleotide-long RNA that recognizes and cleaves an external 14-nucleotide-long RNA molecule.

Many studies have attempted to determine the sequence and structural requirements for optimal catalytic activities. These have included traditional mutagenesis *(5–9)* and in vitro selection strategies *(7,10–13)*, as well as chemical structure probing *(14,15)*. The hairpin ribozyme-substrate complex is comprised of two well-defined, independently folding domains, A and B. Each has two helical regions separated by an internal loop *(16;* **Fig. 1**). Domain A is formed by both the ribozyme and the substrate RNA molecules, and consists of the two intermolecular helices (1 and 2) and the internal loop A (**Fig. 1**). Domain B is entirely composed of the ribozyme sequences (helices 3 and 4 and loop B). Most of the nucleotides that are important for catalytic activity are contained within the impaired regions, yet there are almost no sequence restrictions for the bases involved in the formation of the helices as long as basepairing is achieved *(17,18)*. The active ribozyme can be assembled from the two separate domains, implying the existence of tertiary interactions *(16,19)*.

The specificity of the hairpin ribozyme resides in domain A (**Fig. 1**). Substrate recognition and binding by the ribozyme involves the formation of the two intermolecular helices—helix 1 and 2- (of 6 and 4 basepairs, respectively) *(6,7)*. These ribozyme sequences, which are responsible for substrate recognition, can tolerate almost any alteration. Therefore, the specificity of the ribozyme can be altered to target almost any RNA molecule of interest *(7,20)*. In addition to the formation of stable helices 1 and 2, some substrate-molecule sequence requirements must be met if cleavage by the catalytic domain is to be effective. Single-base substitutions as well as in vitro selection studies have determined the strong requirement of the hairpin ribozyme for a guanosine residue at the 3' of the scissile phosphodiester bond G + 1 *(11,21)*. Recently, an interdomain interaction involving G + 1 and C25 has been identified *(22)*. Similarly, in vitro selection studies, as well as site-directed mutagenesis, have identified the sequence preferences of the substrate RNA surrounding the cleavage site, and have given rise to the definition of two different consensuses concerning optimal cleavage *(7,8)*. Both have been used extensively for the design of ribozymes for targeting cellular and viral RNAs. Most recently, a detailed analysis of the region surrounding the cleavage site has been performed, and a redefinition of the requirements of this region has been made possible *(23)*.

Great efforts have been made to optimize the catalytic activity of the hairpin ribozyme by modifying its primary and secondary structure. Although most enhancing modifications have been demonstrated with the ribozyme derived from the ([–]sTRSV), these can be introduced into hairpin ribozymes of various specificities. Therefore, such modifications can be routinely included in hairpin ribozymes designed to inactivate specific cellular or viral RNAs *(7,8,11,20,24,25)*.

This chapter summarizes the various strategies used in the design of optimal activity hairpin ribozymes, and discusses the requirements of the substrate sequence as well as that of the catalytic domain itself.

## 2. Materials

This section describes the main strategies used to optimize the hairpin ribozyme for targeting purposes. All reactions for synthesizing RNA molecules and assaying the inhibitory activity of designed ribozymes should be performed in a ribonuclease-free environment (*see* **Note 1**).

1. RNAs: These can be obtained enzymatically by transcription using bacteriophage RNA polymerases. T7 RNA polymerase is the most commonly used, and is commercially available from many sources. Suitable plasmids are available that contain the T7 promoter and a multicloning site in which the specific template sequences can be included under the control of the T7 promoter. Alternatively, templates can be constructed using synthetic oligonucleotides (*see* **Note 2**).

2. Restriction enzymes for linearizing the plasmid templates to define the 3' end of the RNA.

3. 10X nucleotide triphosphates (NTP) mix stock solution: 10 m*M* adenosine 5' triphosphate (ATP), cytidine 5' triphosphate (CTP), guanosine 5' triphosphate (GTP), and uridine 5' triphosphate (UTP) (*see* **Note 3**).

4. 10X transcription buffer, made according to the manufacturer's instructions. A commonly used buffer is 1X 40 m*M* Tris-HCl, pH 8.0, 6 m*M* $MgCl_2$, 1 m*M* spermidine, 4 m*M* NaCl, 10 m*M* dithiothreitol (DTT), 0.01% Triton X-100 (*see* **Note 4**).

5. α-$^{32}$P-NTP. To be used within 14 d of the date of labeling to ensure high specific activity. This is a hazardous material and appropriate protection is required when handling.

6. RNase inhibitor (40 U/μL).

7. RNase-free DNase I (1–10 U/μL; e.g., RQ1 DNase from Promega).

8. 40% polyacrylamide stock solution (19:1 acrylamide:*bis*-acrylamide, *see* **Note 5**).

9. 10X TBE buffer: 0.89 *M* Tris base, 0.89 *M* boric acid, 25 m*M* ethylenediamine-tetraacetic acid (EDTA).

10. Ammonium persulfate 10% (w/v).

11. *N,N,N',N'*-tetramethyl-ethylemediamine (TEMED).

12. Formamide-loading buffer (97% formamide deionized ultra-pure grade, 17 m*M* EDTA, 0.025% xylene cyanol, 0.025% bromophenol blue).

13. RNA elution buffer: 0.5 *M* ammonium acetate, 0,1% sodium dodecyl sulfate (SDS), 1 m*M* EDTA.

14. Saturated phenol:chloroform:isopropyl alcohol 25:24:1 (v:v:v).

15. 3 *M* sodium acetate, pH 5.2 and 6.0.

16. Glycogen 20 mg/mL, to be used as a carrier for RNA precipitations.

17. Absolute ethanol.

18. 10X ribozyme cleavage buffer: 0.5 *M* Tris-HCl, pH. 7.5, 120 m*M* $MgCl_2$.

## 3. Methods

### 3.1. Synthesis and Purification of RNAs

1. An appropriate DNA template must be constructed for each required RNA molecule (ribozymes and substrates). The minimal requirement is a double-stranded T7 RNA promoter sequence upstream of the template of the desired RNA (*see* **Note 2**). The most common approach is to use suitable DNA plasmids carrying the T7 promoter upstream of a multicloning site in which the specific template sequences can be included. The various template sequences can easily be assembled with overlapping synthetic oligonucleotides.

2. Plasmid templates must be linearized with the appropriate restriction enzyme (*see* **Note 6**). After complete digestion, the DNA is cleaned by adding an equal volume of saturated phenol. It is then mixed well in a vortex to form an emulsion, and centrifuged for 2 min (>12,000$g$) in a microfuge. The upper phase is transferred to a new tube and an equal volume of chloroform:isoamyl alcohol (IAA) added (24:1). This is mixed to form an emulsion and centrifuged again. The upper phase is removed and kept. A 1/10 vol of 3 $M$ sodium acetate, pH 6.0, is then added, along with 3 vol of cold (−20°C) absolute ethanol. This is incubated at −80°C for 30 min and centrifuged for 15 min in a microfuge (>12,000$g$). The supernatant is discarded, and the pellet is washed with 500 µL of 70% ethanol. It is important to avoid disturbing the pellet. This is followed by an additional 5 min of centrifuging (>12,000$g$), after which the supernatant is discarded and dried briefly in a Speedvac. The pellet is then redissolved in RNase-free ddH$_2$O.
3. 50 µL transcription reactions are set up containing 1 µg of linearized and purified plasmid template in 40 m$M$ Tris-HCl, pH 8.0, 6 m$M$ MgCl$_2$, 1 m$M$ spermidine, 4 m$M$ NaCl, 10 m$M$ DTT, 0.01% Triton X-100, 1 m$M$ of each NTP, and 5 µCi of [α-$^{32}$P]UTP (3000 Ci/mmol), 0.5 U/µL RNAguard (Amersham Biosciences) and 20 µg/mL of purified T7 RNA polymerase. This is incubated at 37°C for 2 h before stopping the reaction by adding 1 U of RNase-free DNase and incubating at 37°C for 15 min.
4. An equal volume of formamide-loading buffer (50 µL) is added, and these samples are heated at 95°C in a dry bath before loading directly onto denaturing polyacrylamide-7 $M$ urea gels (*see* **Note 7**).
5. Electrophoresis should be performed under denaturing conditions. The gel is placed on an exposure cassette and a sheet of X-ray film exposed in a dark room (*see* **Note 8**). The acrylamide slice containing the full-length ribozyme or substrate is excised using a razor blade.
6. The gel slices are then soaked in 300 µL of RNA elution buffer and incubated overnight at 4°C. An equal volume of saturated phenol:chloroform:IAA is added, mixing well with a vortex, and then the sample is centrifuged for 2 min in a microfuge. The upper phase is removed and kept. Another volume of chloroform:IAA is then added before mixing and centrifuging again. The upper phase is again removed, and a 1/10 vol of 3 $M$ sodium acetate pH 5.2 is added, along with three volumes of cold (−20°C) absolute ethanol. Precipitation is allowed to occur overnight at −20°C (or 30 min at −80°C) before centrifuging for 30 min in a microfuge (>12,000$g$; *see* **Note 9**). The supernatant is discarded, and the pellet is washed with 500 µL of cold 80% ethanol. It is important to avoid disturbing the pellet. The sample is then centrifuged again for 5 min. The supernatant is discarded and dried briefly in a Speedvac. The pellet is then redissolved in RNase-free ddH$_2$O.
7. The product should be stored at −20°C until use.

## 3.2. Selection of Target Sites

1. Appropriate software should be used for searching specific sequences within the RNA molecule to be targeted (e.g., DNA Strider). A search should be made for

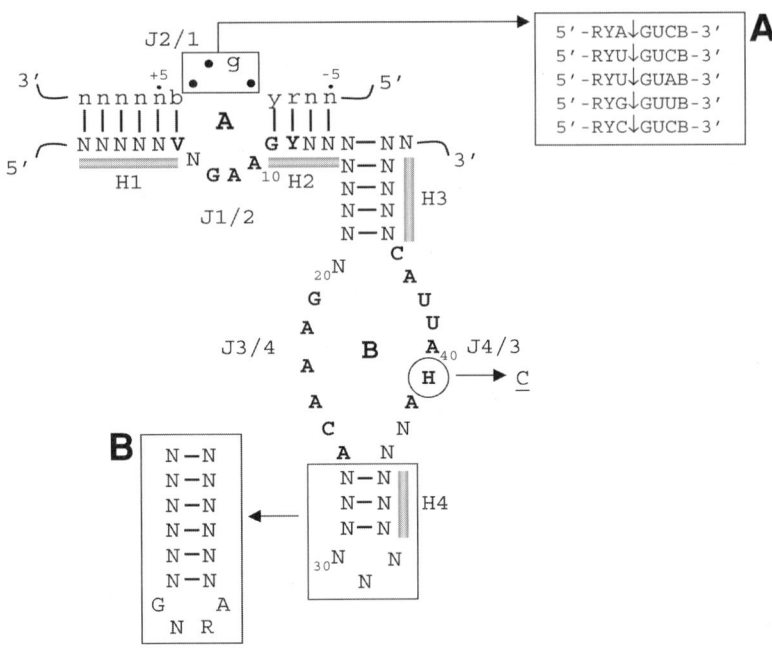

Fig. 2. Optimized hairpin ribozyme. N represents any nucleotide; Y represents C or U; R represents A or G; H represents A, C or U; and V represents A, C or G. The J2/1 region and helix 4 are boxed. The most effective J2/1 sequences for cleavage are shown in box (**A**). Box (**B**) contains proposed H4 extension for an optimal activity of the hairpin ribozyme. C indicates the substitution of $U_{39}$ to C.

any of the following combinations of nucleotides: 5'-RYA↓GUCB-3' > RYU↓GUCB > RYU↓GUAB > RYG↓GUUB > RYC↓GUCB (R:A or G; Y:C or U; B:C, G or U; ↓: cleavage site). These sequences, which corresponds to nucleotides from –3 to +4 (**Fig. 2**), represent the substrate RNAs that are cleaved more efficiently by the hairpin ribozyme. Other cleavable sequences and their cleavage rates are shown in **Fig. 3** (*see* also **Note 10**).

2. For cleavage to be effective, in addition to the sequence requirements of the substrate RNA, the target sequence must be accessible. RNA molecules may fold into more or less complex structures through intra- and/or intermolecular interactions. These interactions may impede the access of the ribozyme to its target sequence, and thus affect inhibitory efficiency. Methods for identifying less structured regions within RNA molecules are described elsewhere *(26–28)*.

### 3.3. Designing Ribozymes

1. Once the target site has been identified, the hairpin ribozyme that is specific for such a target must be designed. The ribozyme sequence should fulfill the requirements indicated in **Fig. 2**. Nucleotides involved in the interaction with the target

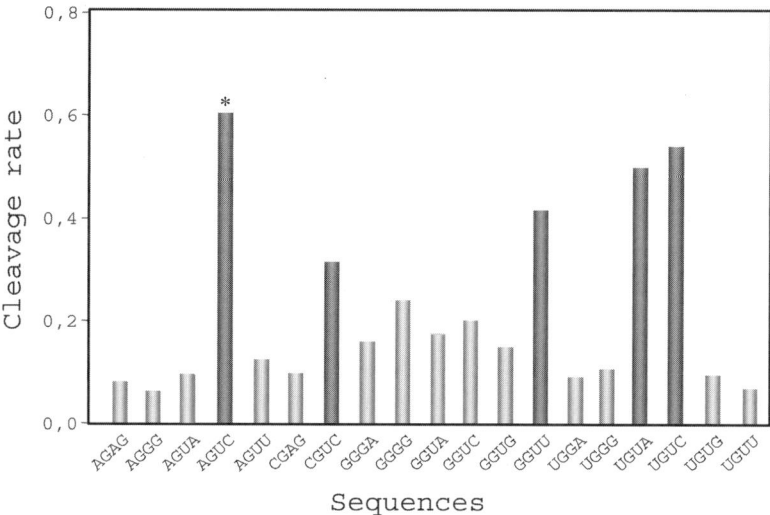

Fig. 3. Optimal cleavable substrates by the hairpin ribozyme. Bars represent the cleavage rates of various substrates, which are denoted by their J2/1 sequence. The wild-type substrate sequence is indicated with an asterisk.

sequence (helices 1 and 2) are susceptible to almost any change, and can therefore target any desired RNA molecule. In principle, ribozyme sequences involved in the interaction with the substrate molecule should form Watson-Crick basepairs with the corresponding substrate sequences, with the sole exception at position 11, where a G should always be included whether the substrate contains C or U at the −2 position (*20*; *see* also **Note 11**). The minimum length of helix 1 has been established as six nucleotides for optimum cleavage of the wild-type substrate (*3,4,7*). Helix 1 can be extended, but such an extension should be optimized for each target RNA (*see* **Note 12**).

2. Similarly, the remaining ribozyme sequences (domain B and J1/2) must meet the requirements indicated in **Fig. 2**. The use of these sequences from the wild-type hairpin ribozyme is strongly recommended (**Fig. 1**). Few modifications that increase the catalytic efficiency of the hairpin ribozyme have been well-characterized (**Fig. 2**). The substitution of U39 by C has been defined as a general upmutation (*7,11,20,25*). The same effect has been demonstrated for the stabilization of the helix 4 domain (*24,25*). The use of the extension 5'-GACGUAAGUC-3' (*29*; *see* also **Note 13**) is recommended.

## *3.4. Ribozyme Cleavage Assays*

1. Ribozyme excess can be used to quickly characterize the ability of the ribozyme to cleave the substrate molecule. Reactions can be performed with either a short substrate RNA molecule (17-nt-long), or a long RNA substrate containing the

target sequence. Short substrates provide information about potential problems arising from the ribozyme sequence; long substrates also provide information on substrate accessibility problems (*see* **Note 14**).

2. The ribozyme and substrate RNAs in 1X cleavage reaction buffer are heated in separate tubes for 2 min at 90°C. After snap-cooling on ice for 5 min, they are then incubated at 37°C for 15 min. Ribozyme is added to the substrate and incubated at 37°C. Aliquots are removed at defined times, and equal volumes of formamide-loading buffer are added to stop the reaction. Samples should be kept on ice until used.

3. Reaction products can be resolved in denaturing polyacrylamide gels. The samples should be heated for 2 min at 90°C, snap-cooled, and then loaded onto the gels.

4. Gels are dried and quantified using a betascan radioanalytic instrument or phosphorimager.

## 4. Notes

1. All reagents and solutions, as well as plastic and glassware, should be free of ribonucleases. Gloves should be used at all times, and kept as clean as possible. Solutions and buffers should be prepared in diehtylpyrocarbonat (DEPC)-treated distilled and deionized water to inhibit the activity of protein ribonucleases. DEPC should be added to yield a final concentration of 0.1% (v/v) and stirred vigorously for 10 min, and the solutions are autoclaved to destroy unreacted DEPC.

2. The T7 RNA polymerase efficiently transcribes partially double-stranded DNA templates. It requires only a double stranded promoter sequence and a single-stranded template. A 17-nt promoter followed by GGG or GCG is sufficient to achieve effective transcription initiation.

   5'-TAATACGACTCACTATA↓GGG
   3'-ATTATGCTGAGTGATAT CCCXXXX...................XXX-5'
   (Structure of the partially double-stranded DNA template; ↓ transcription initiation site.)

   There are several commercially available plasmid vectors that contain a multicloning site downstream of the promoter. This approach results in the incorporation of extra sequences at both ends of the transcript RNA derived from the plasmid vector. Alternatively, the T7 promoter can be added directly at the 5' end of the template sequence by polymerase chain reaction (PCR) to prevent the introduction of extra sequences at the 5' end of the resulting RNA product.

3. To achieve RNA molecules with high specific activity, the concentration of the cold nucleotide corresponding to the radioactive labeled counterpart used in transcription should be reduced. It is recommended that 0.4 m$M$ of this nucleotide is used while the other three are maintained at 1 m$M$.

4. DTT and Triton X-100 should not be included in the transcription buffer. Instead, they should be added directly to the transcription mix from 10X stock solutions (0.1 $M$ DTT and 0.1% Triton X-100.

5. The polyacrylamide solution should be stored at 4°C and protected from the light.

6. Complete digestion of the plasmid-template should be checked using an agarose gel. Since the restricted site will determine the transcription termination site, the presence of non-digested molecules in the reaction would result in high molecular wild-type RNA products that consume a large amount of nucleotides, resulting in a significant reduction in the yield of the desired molecule.

7. The gel concentration should be chosen according to the size of the RNA molecule to be purified: 20% is recommended for the 17-nt substrates and 15% for the ribozymes.

8. Gels should be left on top of an electrophoresis glass plate and covered with saran wrap. The saran wrap should be well fixed to the glass plate to avoid any slippage over the gel during manipulation. A sheet of X-ray film is placed on top. The position of the film on the gel (saran wrap cover) is recorded using a marker, with non-symmetric lines drawn across all edges of the film to ensure its correct repositioning on top of the gel after development. Alternatively, a variety of phosphorescent labels or markers available from various commercial sources can be used to guarantee the correct alignment of the film on the gel. A 2- to 5-min exposure should be sufficient for efficient transcription when using nucleotides with high specific activity.

9. To ensure effective precipitation of the RNA, glycogen can be added to the RNA-containing solution. It is recommended that up to 5 μg of glycogen be used per sample.

10. The hairpin ribozyme can cleave, to varying extents, 40 of the 64 sequences possible resulting from the randomization of positions −1, +2, and +3 of the substrate RNA. No cleavage is detected with the remaining 24 sequences. The presence of a C residue at position +2 is strongly discouraged. Other consensus for the J2/1 region requirements can be found in the literature but they are the result of incomplete analyses in which only a limited number of sequences were tested. For example, Hampel and colleagues proposed the use of N↓GUC *(8)*. Similarly, Burke and colleagues proposed the following: RYN↓GUCB based in in vitro selection as well as single mutagenesis studies *(7)*. But these authors only evaluated some of the possible sequence combinations of the J2/1 nucleotides. Therefore, although each of these two theories might be acceptable, they may lead to the under- or overestimation of the proficiency of certain sequences used as target sites for the hairpin ribozyme.

11. It has been proposed that a wobble interaction in the most distal basepair of helix 2 (nucleotides 14Rz/-5S; **Fig. 1**) may favor the necessary docking of the two domains of the hairpin ribozyme-substrate complex, and therefore may result in the optimization of the reaction *(30)*. This positive effect could be restricted to specific helix 2 sequences that result in a more rigid structure that may interfere with the docking process.

12. Helix 1 extension results in a more stable interaction between ribozyme and substrate, which challenges the turnover of the ribozyme. Dissociation of products would become the rate-limiting step, which may represent an important problem, depending on the conditions used. Furthermore, extra sequences at the 5' end of

the ribozyme sequence may cause undesirable intramolecular interactions and yield a non-active conformation of the ribozyme molecule. Therefore, it is important to test the cleavage efficiency of the ribozyme for each helix 1 extension.

13. The stabilization of helix 4 can be accomplished by the addition of RNA domains with defined activity—e.g., protein-binding domains *(24)*. This results in RNA molecules that are able to bind a protein and catalyze the cleavage of an RNA substrate, which may result in further improvement of inhibitory power. The protein may also help the correct subcellular compartmentalization of the ribozyme. It has also been proposed that the addition of a stable stem-loop extension at the 3' end of the ribozyme sequence can *per se* enhance the activity of the hairpin ribozyme *(31)*. In addition, such extension acts as an antisense domain that results in an improvement of the binding rates of the ribozyme to the target sequence when targeting long RNA molecules. This result has been only shown for the targeting of TAR-containing RNAs *(31)*.

14. In conditions of ribozyme excess or pre-steady-state conditions, the dissociation of the cleavage products do not affect the rates of reaction. The correct ribozyme/ substrate ratio must be determined for each pair of molecules in order to be assayed. The ribozyme should be in great excess: it is recommended that a start be made with at least a 10-fold excess of ribozyme over substrate—e.g., 1 $\mu M$ ribozyme for 0.1 $\mu M$ substrate. The observed rates should not vary significantly with a small increase in the ribozyme concentration, indicating that the ribozyme is in saturating conditions.

## Acknowledgments

This work was supported by the research grant BMC2000-1140 from the Spanish Ministry of Science and Technology.

## References

1. Buzayan, J. M., Gerlach, W. L., and Bruening, G. (1986) Satellite tobacco ringspot virus RNA: A subset of the RNA sequence is sufficient for autolytic processing. *Proc. Natl. Acad. Sci. USA* **83,** 8859–8862.

2. Prody, G. A., Bakos, J. T., Buzayan, J. M., Schneider, I. R., and Bruening, G. (1986) Autolytic processing of dimeric plant virus satellite RNA. *Science* **231,** 1577–1580.

3. Feldstein, P. A., Buzayan, J. M., and Bruening, G. (1989) Two sequences participating in the autolytic processing of satellite tobacco ringspot virus complementary RNA. *Gene* **82(1),** 53–61.

4. Hampel, A. and Tritz, R. (1989) RNA catalytic properties of the minimum (-)sTRSV sequence. *Biochemistry* **28(12),** 4929–4933.

5. Feldstein, P. A., Buzayan, J. M., van Tol, H., deBear, J., Gough, G. R., Gilham, P. T., and Bruening, G. (1990) Specific association between an endoribonucleolytic sequence from a satellite RNA and a substrate analogue containing a 2'-5' phosphodiester. *Proc. Natl. Acad. Sci. USA* **87(7),** 2623–2627.

6. Hampel, A., Tritz, R., Hicks, M., and Cruz, P. (1990) 'Hairpin' catalytic RNA model: evidence for helices and sequence requirement for substrate RNA. *Nucleic Acids Res.* **18(2)**, 299–304.

7. Joseph, S., Berzal-Herranz, A., Chowrira, B. M., Butcher, S. E., and Burke, J. M. (1993) Substrate selection rules for the hairpin ribozyme determined by in vitro selection, mutation, and analysis of mismatched substrates. *Genes Dev.* **7(1)**, 130–138.

8. Anderson, P., Monforte, J., Tritz, R., Nesbitt, S., Hearst, J., and Hampel, A. (1994) Mutagenesis of the hairpin ribozyme. *Nucleic Acids Res.* **22(6)**, 1096–1100.

9. Siwkowski, A., Shippy, R., and Hampel, A. (1997) Analysis of hairpin ribozyme base mutations in loops 2 and 4 and their effects on cis-cleavage in vitro. *Biochemistry* **36(13)**, 3930–3940.

10. Beaudry, A. A. and Joyce, G. F. (1990) Minimum secondary structure requirements for catalytic activity of a self-splicing group I intron. *Biochemistry* **29(27)**, 6534–6539.

11. Berzal-Herranz, A., Joseph, S., and Burke, J. M. (1992) In vitro selection of active hairpin ribozymes by sequential RNA- catalyzed cleavage and ligation reactions. *Genes Dev.* **6(1)**, 129–134.

12. Berzal-Herranz, A., Chowrira, B. M., Polsenberg, J. F., and Burke, J. M. (1993) 2'-Hydroxyl groups important for exon polymerization and reverse exon ligation reactions catalyzed by a group I ribozyme. *Biochemistry* **32(35)**, 8981–8986.

13. Sarguiel, B., McKenna, J., and Burke, J. M. (2000) Analysis of the functional role of a G.A sheared base pair by in vitro genetics. *J. Biol. Chem.* **275(41)**, 32,157–32,166.

14. Butcher, S. E. and Burke, J. M. (1994) Structure-mapping of the hairpin ribozyme. Magnesium-dependent folding and evidence for tertiary interactions within the ribozyme-substrate complex. *J. Mol. Biol.* **244(1)**, 52–63.

15. Butcher, S. E. and Burke, J. M. (1994) A photo-cross-linkable tertiary structure motif found in functionally distinct RNA molecules is essential for catalytic function of the hairpin ribozyme. *Biochemistry* **33(4)**, 992–999.

16. Butcher, S. E., Heckman, J. E., and Burke, J. M. (1995) Reconstitution of hairpin ribozyme activity following separation of functional domains. *J. Biol. Chem.* **270(50)**, 29,648–29,651.

17. Burke, J. M. (1994) The hairpin ribozyme. *Nucleic Acids Mol. Biol.* **8**, 105–118.

18. Burke, J. M. (1996) Hairpin ribozyme: current status and future prospects. *Biochem. Soc. Trans.* **24(3)**, 608–615.

19. Feldstein, P. A. and Bruening, G. (1993) Catalytically active geometry in the reversible circularization of 'mini-monomer' RNAs derived from the complementary strand of tobacco ringspot virus satellite RNA. *Nucleic Acids Res.* **21(8)**, 1991–1998.

20. Berzal-Herranz, A., Joseph, S., Chowrira, B. M., Butcher, S. E., and Burke, J. M. (1993) Essential nucleotide sequences and secondary structure elements of the hairpin ribozyme. *EMBO J.* **12(6)**, 2567–2573.

21. Chowrira, B. M., Berzal-Herranz, A., and Burke, J. M. (1991) Novel guanosine requirement for catalysis by the hairpin ribozyme. *Nature* **354(6351)**, 320–322.

22. Pinard, R., Lambert, D., Walter, N. G., Heckman, J. E., Major, F., and Burke, J. M. (1999) Structural basis for the guanosine requirement of the hairpin ribozyme. *Biochemistry* **38(49)**, 16035–16,039.

23. Perez-Ruiz, M., Barroso-DelJesus, A., and Berzal-Herranz, A. (1999) Specificity of the hairpin ribozyme. Sequence requirements surrounding the cleavage site. *J. Biol. Chem.* **274(41)**, 29,376–29,380.

24. Sargueil, B., Pecchia, D. B., and Burke, J. M. (1995) An improved version of the hairpin ribozyme functions as a ribonucleoprotein complex. *Biochemistry* **34(23)**, 7739–7748.

25. Barroso-delJesus, A., Tabler, M., and Berzal-Herranz, A. (1999) Comparative kinetic analysis of structural variants of the hairpin ribozyme reveals further potential to optimize its catalytic performance. *Antisense Nucleic Acid Drug Dev.* **9(5)**, 433–440.

26. Sczakiel, G. and Tabler, M. (1997) Computer-aided calculation of the local folding potential of target RNA and its use for ribozyme design, in *Ribozyme Protocols, Methods in Molecular Biology*, vol. 74, (Turner, P. C., ed.), Humana Press, Totowa, NJ, pp. 11–15.

27. James, W. and Cowe, E. (1997) Computational approaches to the identification of ribozyme target sites, in *Ribozyme Protocols, Methods in Molecular Biology*, vol. 74, (Turner, P. C., ed.), Humana Press, Totowa, NJ, pp. 17–26.

28. Frank, B. L. and Goodchild, J. (1997) Selection of accessible sites for ribozymes on large RNA transcripts in *Ribozyme Protocols, Methods in Molecular Biology*, vol. 74, (Turner, P. C., ed.), Humana Press, Totowa, NJ, pp. 37–43.

29. Esteban, J. A., Banerjee, A. R., and Burke, J. M. (1997) Kinetic mechanism of the hairpin ribozyme. Identification and characterization of two nonexchangeable conformations. *J Biol. Chem.* **272(21)**, 13,629–13,639.

30. Yu, Q., Pecchia, D. B., Kingsley, S. L., Heckman, J. E., and Burke, J. M. (1998) Cleavage of highly structured viral RNA molecules by combinatorial libraries of hairpin ribozymes. The most effective ribozymes are not predicted by substrate selection rules. *J. Biol. Chem.* **273(36)**, 23,524–23,533.

31. Puerta-Fernandez, E., Barroso-DelJesus, A., and Berzal-Herranz, A. (2002) Anchoring hairpin ribozymes to long target RNAs by loop-loop RNA interactions. *Antisense Nucleic Acid Drug Dev.* **12(1)**, 1–9.

# 25

## Design, Targeting, and Initial Screening of sTRSV-Derived Hairpin Ribozymes for Optimum Helix 1 Length and Catalytic Efficiency In Vitro

Max W. Richardson, Linda Hostalek, Michelle Dobson, Jason Hu, Richard Shippy, Andrew Siwkowski, Jonathan D. Marmur, Kamel Khalili, Paul E. Klotman, Arnold Hampel, and Jay Rappaport

### Summary

Hairpin ribozymes derived from the negative strand of satellite RNAs from the tobacco ringspot virus (sTRSV) can be engineered to target and cleave a variety of heterologous RNAs from both cellular and viral transcripts. Attention to design and targeting rules and optimization of helix 1 length and catalytic efficiency in vitro may increase the efficacy of hairpin ribozymes in reducing the expression of targeted transcripts. Here, principles for the design and targeting of sTRSV-derived hairpin ribozymes are described, as well as methods and materials for optimizing helix 1 length, and for conducting an initial screen of catalytic efficiency to identify promising candidates for further evaluation. Examples are provided for hairpin ribozymes that target human and mouse transforming growth-factor beta (TGF-$\beta$), as well as human polycystic kidney disease gene 1 (PKD1) and JC virus large T-antigen. The tetraloop modification of the sTRSV hairpin ribozyme is considered superior to designs based on the native sTRSV hairpin ribozyme, given its potential to yeild considerable improvements in stability and catalytic efficiency.

**Key Words:** Tetraloop; hairpin; ribozyme; sTRSV; TGF-$\beta$; HIV-1; PKD1; JCV; T-antigen.

## 1. Introduction

The development of recombinant hairpin ribozymes that are capable of targeting and cleaving heterologous RNA sequences in *trans* as tools for gene therapy for a variety of diseases has generated considerable excitement *(1,2)*. In particular, two catalytically efficient ribozymes were developed that targeted the 5' long terminal repeat (LTR) U5 leader sequence and the protease

From: *Methods in Molecular Biology, vol. 252: Ribozymes and siRNA Protocols, Second Edition*
Edited by: M. Sioud © Humana Press Inc., Totowa, NJ

gene of HIV-1, and initially showed very encouraging results in reducing or blocking viral replication in vitro in both transformed T-cell lines and primary cultures *(3–6)*. Unfortunately, despite preliminary evidence of the safety and stability in vivo of autologous CD8-depleted cells transduced ex vivo with a murine retroviral vector expressing an anti-HIV-1 ribozyme *(7)*, no clinical study to date has shown evidence of the utility of the hairpin ribozyme approach in vivo. This may reflect difficulty in maintaining long-term expression of ribozyme constructs, and the short half-life of ribozyme transcripts in vivo. Despite the lack of immediate success, hairpin ribozymes are still being studied as potential tools for gene therapy, and may show efficacy in the future as technology improves in terms of delivery and stability of expression, and the stability of the ribozymes themselves. It is also worth noting that although ribozymes are catalytically ineffective in vitro compared to conventional enzymes, they may be more efficient in vivo because of the presence of protein co-factors *(8)*, particularly if directed to the right subcellular environment.

The rules that govern the targeting and design of native and tetraloop hairpin ribozymes derived from the negative strand of the satellite RNA from tobacco ringspot virus (sTRSV) are well-defined, and were previously described in detail *(9)*. Caveats for the design of ribozymes derived from the satellite RNA of chicory yellow mottle virus (sCYMV1) have also been well-described *(10)*. Briefly, for sTRSV-derived hairpin ribozymes, targeting a heterologous RNA requires identification of the sequence BN*GUC at one or more positions in its transcript, in which B is any base but A, N is any base, "*" denotes the site of cleavage, and GUC are required. Particular attention should be given to identifying target sites in the 5' and 3' regions of transcripts, and splice donor sites and regions with considerable secondary structure in general should be avoided. In terms of substrate complementarity, helix 1 may be of variable length—generally from 6–10 bases—helix 2 is fixed at four bases. The tetraloop modification of the sTRSV ribozyme appears to have few drawbacks. It may lead to both increased stability and a considerable increase in catalytic efficiency relative to the native hairpin ribozyme, and is therefore recommended (*see* **Note 1**). This chapter provides specific examples from the development of sTRSV-derived native and tetraloop hairpin ribozymes that target both human transforming growth factor-β (hTGF-β) and murine TGF-β (mTGF-β), in the context of their overexpression in HIV-1-associated nephropathy (HIVAN) and in a transgenic mouse model with some characteristics of HIVAN *(11)*. TGF-β also has been implicated in HIV-induced central nervous system (CNS) disorders *(12,13)*. This chapter also describes the development of ribozymes targeting transcripts of the polycystic kidney disease gene 1 protein (PKD1) *(14)*, and the tumorigenic JC virus (JCV) large T-antigen *(15)*. The results are intended to provide an example of the application of targeting and design rules in an initial screen of ribozymes for catalytic efficiency to identify the most promising candidates for subsequent testing in vitro (*see* **Note 2**).

## 2. Materials

### 2.1. Identification of Potential Ribozyme Cleavage Sites in Targeted Transcripts

Sequences that are potentially amenable to targeting with hairpin ribozymes have been identified in the transcripts of targeted genes using the commercially available DNA sequence manipulation program MacVector (Accelrys Inc., San Diego, CA). Similarly, sequences of transcripts from targeted genes were obtained from publicly available sequence information (http://www.ncbi.nlm.nih.gov:80/entrez/query.fcgi?db=Nucleotide). Native hairpin and tetraloop ribozymes have been designed, as previously described *(9)*. A tetraloop design template is provided here (**Fig. 1**). T7 promoter regions must be added to the negative-strand template for in vitro transcription; restriction sites may be added and the plus strand must be synthesized for subsequent subcloning into appropriate expression vectors.

### 2.2. Oligonucleotide Templates for In Vitro Transcription of Ribozyme and Substrate RNA

Oligonucleotide templates for in vitro transcription using T7 RNA polymerase were ordered from commercial manufacturers (Midland Certified Reagent Company, Midland, Texas; Invitrogen, La Jolla, CA), or synthesized in house using an ABI 392 DNA/RNA synthesizer (Applied Biosystems, Foster City, CA). Commercial oligonucleotides were gel-purified. Oligonucleotides that were synthesized in-house were high-performance liquid chromatography (HPLC)-purified as previously described using an RP-300 Brownlee reverse-phase column and an acetonitrile-triethylammonium acetate, pH 7.0 gradient *(16,17)*. HPLC-purified oligonucleotides performed better in subsequent transcription reactions compared to those that were gel-purified. Oligonucleotide templates used for in vitro transcription of ribozymes and substrates, as well as the corresponding T7 promoter primers, are listed here; target regions of ribozymes are indicated in lower-case lettering. The mRNAs for human and murine TGF-β, PKD1, and JCV large T-antigen were targeted. The wild-type sTRSV ribozyme and substrate templates are also included, and were used as a control for kinetic analysis.

1. Human transforming growth factor beta (hTGF-β) ribozymes and substrates: 960R tetraloop (156–160) 5' TAC CAG GTA ATA TAC CAC GGA CCG AAG TCC GTG TGT TTC TCT GGT cca gtt ctt gag acg aCC CTA TAG TGA GTC GTA TTA 3'

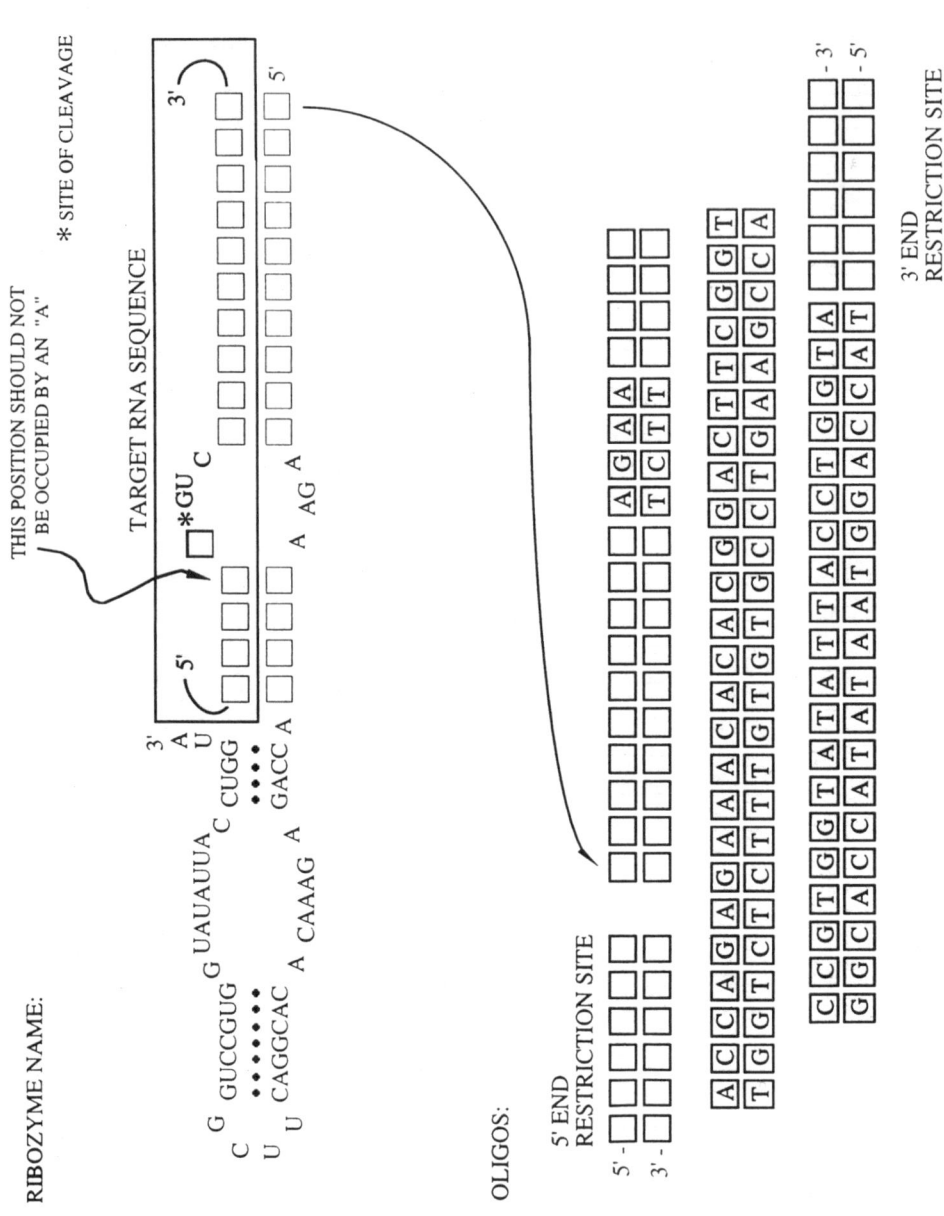

Fig. 1. sTRSV-derived tetraloop hairpin ribozyme design template.

760S substrate (156–160) 5' TCG TCT CAG ACT CTG GCG CTA TAG TGA GTC GTA TTA 3'

961R tetraloop (230–234) 5' TAC CAG GTA ATA TAC CAC GGA CCG AAG TCC GTG TGT TTC TCT GGT agc ttt ctg gga gaa gCC CTA TAG TGA GTC GTA TTA 3'

761S substrate (230–234) 5' CTT CTC CCG ACC AGC TCG CTA TAG TGA GTC GTA TTA 3'

962R tetraloop (400–404) 5' TAC CAG GTA ATA TAC CAC GGA CCG AAG TCC GTG TGT TTC TCT GGT acg gtt ctc ctc agg cCC CTA TAG TGA GTC GTA TTA 3'

762S substrate (400–404) 5' GCC TGA GGG ACG CCG TCG CTA TAG TGA GTC GTA TTA 3'

963R tetraloop (438–442) 5' TAC CAG GTA ATA TAC CAC GGA CCG AAG TCC GTG TGT TTC TCT GGT cgg gtt ctg ccg acc cCC CTA TAG TGA GTC GTA TTA 3'

763S substrate (438–442) 5' GGG TCG GCG ACT CCC GCG CTA TAG TGA GTC GTA TTA 3'

964R tetraloop (517–521) 5' TAC CAG GTA ATA TAC CAC GGA CCG AAG TCC GTG TGT TTC TCT GGT ggc ctt ctt cct gag cCC CTA TAG TGA GTC GTA TTA 3'

764S substrate (517–521) 5' GCT CAG GAG ACA GGC CCG CTA TAG TGA GTC GTA TTA 3'

966R tetraloop (2188–2192) 5' TAC CAG GTA ATA TAC CAC GGA CCG AAG TCC GTG TGT TTC TCT GGT ctg gtt ctt cca tcc cCC CTA TAG TGA GTC GTA TTA 3'

766S substrate (2188–2192) 5' GGG ATG GAG ACC CCA GCG CTA TAG TGA GTC GTA TTA 3'

967R tetraloop (2252–2256) 5' TAC CAG GTA ATA TAC CAC GGA CCG AAG TCC GTG TGT TTC TCT GGT tgc ctt ctt gca cta tCC CTA TAG TGA GTC GTA TTA 3'

767S substrate (2252–2256) 5' ATA GTG CAG ACA GGC ACG CTA TAG TGA GTC GTA TTA 3'

968R tetraloop (2377–2381) 5' TAC CAG GTA ATA TAC CAC GGA CCG AAG TCC GTG TGT TTC TCT GGT att ctt cta cca tag cCC CTA TAG TGA GTC GTA TTA 3'

768S substrate (2377–2381) 5' GCT ATG GTG ACT GAA TCG CTA TAG TGA GTC GTA TTA 3'

2. Murine transforming growth factor beta (mTGF-β) ribozymes and substrates:

835R native (23–27) 5' TAC CAG GTA ATA TAC CAC AAC GTG TGT TTC TCT GGT ctt gtt ctc ctc gca tcc CCC TAT AGT GAG TCG TAT TA 3'

838R tetraloop (23–27) 5' TAC CAG GTA ATA TAC CAC GGA CCG AAG TCC GTG TGT TTC TCT GGT ctt gtt ctc ctc gca tcc CCC TAT AGT GAG TCG TAT TA 3'

836S substrate (23–27) 5' GGA TGC GAG GGA CTC AAG CGC TAT AGT GAG TCG TAT TA 3'

842R native (532–536) 5' TAC CAG GTA ATA TAC CAC AAC GTG TGT TTC TCT GGT cgc ctt ctc ccc aag cca CCC TAT AGT GAG TCG TAT TA 3'
843R tetraloop (532–536) 5' TAC CAG GTA ATA TAC CAC GGA CCG AAG TCC GTG TGT TTC TCT GGT cgc ctt ctc ccc aag cca CCC TAT AGT GAG TCG TAT TA 3'
837S substrate (532–536) 5' TGG CTT GGG GGA CTG GCG CGC TAT AGT GAG TCG TAT TA 3'

3. Human polycystic kidney disease gene 1 protein (PKD1) ribozyme and substrate:
   PKDR native (13255–13259) 5' TAC CAG GTA ATA TAC CAC AAC GTG TGT TTC TCT GGT cat ctt ctt gtc tgt ggg CCC TAT AGT GAG TCG TAT TA 3'
   PKDS substrate (13255–13259) 5' CCC ACA GAC AGA CAG ATG CGC TAT AGT GAG TCG TAT TA 3'

4. JCV virus large T-antigen ribozyme and substrate.
   JCVR native (4931–4927) 5' TAC CAG GTA ATA TAC CAC AAC GTG TGT TTC TCT GT ttc ctt cta tga gaa aag ctC CCT ATA GTG AGT CGT ATT A 3'
   JCVS substrate (4931–4927) 5' AGC TTT TCT CAT GAC AGG AAC GCT ATA GTG AGT CGT ATT A 3'

5. sTRSV control ribozyme and substrate.
   sTRSVR native (control) 5' TAC CAG GTA ATA TAC CAC AAC GTG TGT TTC TCT GGT tga ctt ctc tgt ttC CCT ATA GTG AGT CGT ATT A 3'
   sTRSV substrate (control) 5' AAA CAG GAC TGT CAC GCT ATA GTG AGT CGT ATT A 3'

6. T7 promoter ribozyme and substrate oligonucleotides.
   773R (T7comp-R) 5' TAA TAC GAC TCA CTA TAG GG 3'
   772S (T7comp-S) 5' TAA TAC GAC TCA CTA TAG CG 3'

## 2.3. End-Labeling of DNA Mol-Wt Markers

Five micrograms each of oligonucleotides of 93, 71, 50, 47, and 22 nucleotides (nt) were pooled, lyophilized, and resuspended in 25 μL of dH$_2$O. The sequences of these oligonucleotides are as follows:

93 nt 5' CCC TGC AGA ATT CTT GGG AGA AGC GAA CCA GAG CGT CAC ACG GAC TTC GGT CCG TGG TAT ATT ACC TGG TAT TTT TTT AAG CTT ATC GAT CCC 3'
71 nt 5' AAT TCA CAC AAC AAG AAG GCA ACC AGA GAA ACA CAC GGA CTT CGG TCC GTG GTA TAT TAC CTG GTA CCC GG 3'
50 nt 5' GCT CGA GAC GTT GCA ACG TTG CAA CGT GGA TCC TCG ACG TGA GAG CTC GG 3'
47 nt 5' AAT TCT TTT TCT TCT AGA ATG TCT TGA TTG TTG AGG TAA GTG CTG CA 3'
22 nt 5' GGT CGA CTA CCA GGT AAT ATA C 3'

The following reagents are also necessary for end-labeling the oligonucleotide pool:

1. T4 polynucleotide kinase (PNK) and 10X kinase buffer (Composition), 10 m*M* spermidine.
2. $\gamma P^{32}$-adenosine 5' triphosphate (ATP) (4500 Ci/mmol); 660 $\mu M$ ATP.

## 2.4. In Vitro Transcription of Ribozyme and Substrate RNA

Reagents required for annealing of ribozyme and substrate templates to their respective T7-promoter primers, and for subsequent transcription and purification of transcription products, have been described previously in detail *(16)*. A brief summary of required reagents is provided here. The following reagents are required for annealing ribozyme and substrate templates with their respective T7 promoter primers, and for subsequent in vitro transcription of ribozyme and substrate RNAs.

1. Annealed ribozyme and substrate templates. Equimolar amounts of ribozyme or substrate template and the appropriate T7 promoter primer are required—approx 10.0 μg of ribozyme template with 2.5 μg T7 promoter 773, and 10.0 μg of substrate template with 5.5 μg of T7 promoter primer 772. The following formula may be used in a spreadsheet program such as Excel (Microsoft Inc., Bellevue, WA) to calculate equimolar amounts of T7 promoter primers. Amounts of T7 primers are calculated based on the use of 10 μg of ribozyme or substrate template, and equimolar amounts are calculated by adjusting for size in basepairs (bp) of primers:

$$\text{T7 } \mu L = [10 \times (\text{bp of T7 primer})/(\text{bp of ribo or sub primer})]/(\text{T7 primer conc.})$$

2. 1 *M* Tris-HCl, pH 7.5.
3. A source of fresh, RNase-free deionized water. DEPC treatment of deionized water is not necessary *(16)*.
4. An accurate, clean set of pipets such as Pipetman, preferably including a P10.
5. Latex or nitrile gloves to prevent RNase contamination.
6. Sterile pipet tips, preferably RNase- and DNase-free; 1.5-mL Eppendorf tubes.
7. A glass beaker or similar device to be heated to 90°C using either a microwave or ring-stand and Bunsen burner, a floating rack, and aluminum foil.
8. 2X transcription buffer mix: 80 m*M* Tris-HCl, pH 8.0 (1.6 mL 1 *M* Tris, pH 8.0), 12 m*M* MgCl$_2$ (240 μL 1 *M* MgCl$_2$), 10 m*M* dithiothreitol (DTT) (200 μL 1 *M* DTT), 2 m*M* Spermidine (400 μL 100 m*M* Spermidine), 8% polyethylene glycol (PEG) 3000 (4 mL 40% PEG 3000), 0.2% Triton X-100 (400 μL 10% Triton X-100), (13.16 mL dH$_2$O to 20 mL final). Aliquot and store frozen at –20°C or lower; resuspend completely before use to avoid precipitation.
9. 40 m*M* final nucleotide triphosphate (NTP) stock (10 m*M* each NTP). Aliquot and store frozen at –20°C or lower; avoid repeated freeze/thaw cycles.
10. 3000 Ci/mmol $\alpha P^{32}$-CTP (10 μCi/μL).
11. T7 RNA polymerase (20 U/μL).
12. RNasin or RNase inhibitor (40 U/μL).

13. RNase-free DNase I (2 U/μL).
14. Glycogen (20 mg/mL) from Roche Molecular Biochemicals (Indianapolis, IN).
15. 3 *M* sodium acetate, pH 5.4 (Sigma).
16. 100% reagent alcohol, HPLC-grade (Fisher Scientific, Pittsburgh, PA).
17. 70% reagent alcohol.
18. 2 m*M* ethylenediaminetetraacetic acid (EDTA).
19. 98% formamide dye mix—98% deionized formamide (49 mL) with 0.025% xylene cyanol (0.0125 g) and bromophenol blue (0.0125 g) as markers, and dH$_2$O (to 50 ml final). Aliquot and store at –20°C or lower.
20. 10% and 15% polyacrylamide, 7 *M* urea, 1X TBE gels. A 50l-mL solution for pouring a 10% gel consists of 12.5 mL of a 40% stock solution of 19:1 acrylamide:*bis*-acrylamide (0.22 μm filtered), 5 mL of 10X TBE, pH 8.0, solution, 21 g of urea, and dH$_2$O to 50 mL final; stir to mix into solution and filter through 0.22-μm filter unit prior to use. Immediately before pouring, add 350 μL of a 10% ammonium-persulfate solution, mix gently, and then add 75 μL of *N,N,N',N'*-tetramethyl-ethylenediamine (TEMED), mix again and pour. For a 15% gel, use 18.75 mL of the 40% stock solution.
21. Sterile razor blades and mini grinding pestles (Research Products International, Mt. Prospect, IL) designed for use in 1.5-mL Eppendorf tubes. Also, a standard Eppendorf microcentrifuge and a shaker/mixer (Fisher Scientific, Pittsburgh, PA).
22. Gel slice extraction buffer: 0.5 *M* ammonium acetate, 0.5 mg/mL sodium dodecyl sulfate (SDS), 2 m*M* EDTA.
23. Plastic wrap for covering gels; autoradiography cassettes, film, and development materials; scintillation vials capable of holding 1.5-mL Eppendorf tubes; a calibrated scintillation counter of known efficiency.
24. The following formulas for determining the concentration of gel purified ribozyme and substrate transcripts. These are easily incorporated in a spreadsheet format:
    a. dpm = Cerencov cpm/counting efficieny
    b. μCi in transcript = dpm/2,200,000; 1 μCi is 2,200,000 dpm
    c. μCi CTP label remaining = μCi added/decay factor; lose 3.5% per d past ref. date
    d. pmol C in transcript = (pmol cold CTP in rxn) × (B/C); 1 m*M* cold CTP is 50,000 pmol
    e. pmol RNA in transcript = D/(the number of C residues in the transcript)
    f. dH$_2$O μL for resuspension = E/(desired final concentration in μ*M*)

## 2.5. Helix 1 Length Optimization and Preliminary Analysis of Catalytic Efficiency

1. Gel-purified ribozyme and substrate RNA of known concentration; 0.25-mL Eppendorf tubes and a thermocycler with a heated lid may be preferable for cleavage reactions.
2. 4X cleavage buffer: 8 m*M* spermidine, 48 m*M* MgCl$_2$, 160 m*M* Tris-HCl, pH 7.5.
3. Scintillation fluid for quantifying radioactivity in gel slices containing uncleaved and cleaved substrate bands. Alternatively, cleavage may be quantified using a

phosphorimager, provided the signal is maintained within the linear range of the screen—i.e., the screen is not overexposed.

4. Software for calculating catalytic efficiency, preferably employing a nonlinear algorithm—for example, Tablecurve 2D (Jandel Scientific, San Rafael, CA) or more recent releases.

## 3. Methods

Methods for annealing ribozyme and substrate oligonucleotides with their respective T7 promoter primers—and for subsequent in vitro transcription, gel purification, quantification, helix 1 length optimization, and kinetic analysis—have been described in detail previously *(16)*. The following information is intended as a brief overview of relevant protocols.

### 3.1. Identification of Potential Ribozyme Cleavage Sites in Targeted Transcripts

Nucleotide sequences of transcripts encoding hTGF-β, mTGF-β, PKD1, and JCV large T-antigen were obtained from GenBank (*see* **Subheading 2.**), and imported into MacVector. Transcripts were then screened using MacVector for sites that could be potentially amenable to cleavage—e.g., the BN*GUC sequence, where B is any base but A, N is any base, and GUC are required. Particular attention was given to the 5' and 3' regions of transcripts, although rigorous analysis to identify highly conserved domains and regions of minimal secondary structure was not performed *(9,18,19)*. Once suitable sites were identified, hairpin and/or tetraloop ribozymes were designed to correspond to the targeted region using a convenient design template (**Fig. 1**), with a 9–10 basepair-complementary region in Helix 1. Substrate templates were designed accordingly, again with 9–10 complementary bases in Helix 1. The following sequences should be added to the 3' end of the minus strand of the ribozyme design template (**Fig. 1**), and to the 3' end of the complement of the substrate sequence—e.g., the substrate template for in vitro transcription:

Ribozyme: 5' C CCT ATA GTG AGT CGT ATT A 3'
Substrate: 5' C GCT ATA GTG AGT CGT ATT A 3'

Once ribozyme and substrate templates were designed, they were either synthesized commercially and gel-purified, or made in-house and HPLC-purified, which appeared to be preferable for subsequent in vitro transcriptions.

### 3.2. Annealing of Ribozyme and Substrate Templates With T7 Promoter Primers

1. In a 1.5-mL Eppendorf tube, add 10 μg of ribozyme or substrate oligonucleotide template, and an equimolar amount of appropriate T7 promoter primer as calculated using the formula from **Subheading 2.** Approximately 2.5 μg of T7

promoter primer 773 is used with ribozyme templates, and 5.0 μg of T7 promoter
primer 772 is used with substrate templates.

2. Add 1 μL of 1 *M* Tris, pH 7.5, and dH$_2$O to 100 μL final volume.
3. Heat to 95°C in a beaker of water, cover in aluminum foil, and cool slowly to
   room temperature. Store annealed templates at –20°C prior to use, or use immediately.

### 3.3. End-Labeling of DNA Mol-Wt Markers

1. Use 5 μL of the 25 μL oligo pool (1 mg/mL) described in **Subheading 2.** for the
   following reaction:
   a. 5 μL DNA pool
   b. 1 μL 10X kinase buffer
   c,. 1 μL γP$^{32}$-ATP (4500 Ci/mmol)
   d. 1 μL 10 m*M* spermidine
   e. 1 μL 660 μ*M* ATP
   f. 1 μL T4 PNK
2. Incubate at 37°C for 1 h.
3. Add 50 μL of formamide dye mix, and use 5 μL of the resulting solution per
   transcription gel. Alternatively, remove free nucleotide using an Eppendorf format G-25 spin column (5 Prime–3 Prime, Inc., Boulder, CO), and use a similar
   volume of the flowthrough per transcription gel. For overnight exposures, approx
   200,000–500,000 cpm is sufficient; it may be best to determine the exact amount
   empirically.

### 3.4. In Vitro Transcription of Ribozyme and Substrate RNA

1. Set up the transcription reaction (50 μL) as follows:
   a. 25 μL 2X transcription mix
   b. 5 μL 40 m*M* NTP mix (10 m*M* each NTP)
   c. 6 μL annealed template
   d. 2–4 μL αP$^{32}$-CTP 10 μCi/μL (Ribo = 2 μL; Sub = 4 μL)
   e. 1 μL RNasin (RNase inhibitor)
   f. 5–7 μL dH$_2$O (Sub = 5 μL; Ribo = 7 μL)
   g. 1 μL T7 RNA polymerase
2. Incubate at 37°C for 3 h; transcription yields may be improved by adding an
   additional 2.5 μL of 10 m*M* guanosine 5' triphosphate (GTP) to the reaction after
   1 h *(16)*.
3. After 3 h, add 1 μL of DNase I and incubate for an additional 30 min at 37°C.
4. Stop the reaction by adding 5 μL of 3 *M* sodium acetate, pH 5.4, 1 μL of glycogen, and 150 μL of 100% reagent alcohol. Freeze on dry ice or at –80°C for
   30 min, and precipitate RNA by centrifugation at full speed in a microcentrifuge
   (>10,000*g*) for 15 min. After removing the supernatant and air-drying briefly,
   resuspend the pellet in 8 μL of dH$_2$O, add 8 μL of formamide dye mix, heat to
   90°C for 3 min and cool rapidly on ice. The sample is now ready to be loaded on
   a denaturing polyacrylamide gel. Alternatively, reactions may be stopped by the
   addition of 50 μL of formamide dye mix, heated at 90°C for 3 min, and cooled

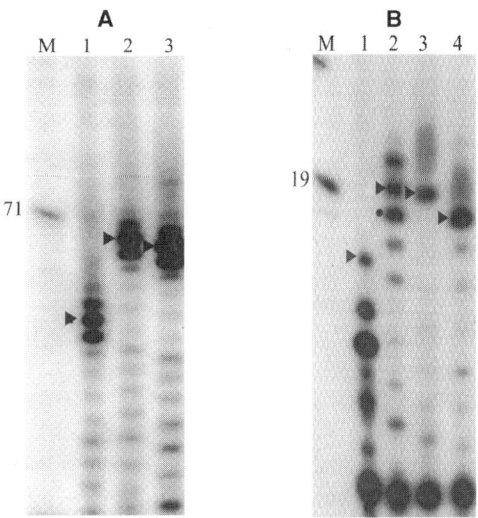

Fig. 2. Gel purification of ribozyme and substrate transcription products. (**A**) Ribozyme trancription products; the native sTRSV-derived ribozyme (lane 1) is somewhat smaller, and therefore migrates more quickly than the tetraloop sTRSV-derived ribozymes (lanes 2 and 3). (**B**) Substrate transcription products migrate somewhat variably, based on GC content. The transcript in lane 1 is shorter than it should be, and did not cleave efficiently; the other transcripts are of appropriate size and performed well in subsequent cleavage reactions. Full-length "n" transcripts are indicated with an arrow; one "n-1" transcript is marked with a circle.

    rapidly on ice, and loaded directly on a gel if it is of appropriate thickness and the wells are large enough to handle a sample volume of 100 µL.

5.  Load ribozyme transcripts on 10% polyacrylamide, 7 *M* urea gels; use 15% gels for purification of substrate transcripts. Run gels until the upper xylene cyanol dye front has migrated approx two-thirds of the way down, or approx 32 W, for 3 h.

6.  Disassemble gels, and transfer gel to some type of backing such as an older piece of exposed autoradiography film, wrap in plastic wrap, and expose for 2–10 min at room temperature, which should be sufficient if the transcription reaction has worked well. Place pieces of tape on the edges of the film, and mark them with a pen to orient the film properly after development for excision of transcript bands. Full-length ribozyme transcripts are relatively easy to identify. The lower 22-nt DNA marker migrates at approx 19 bp *(20)*, very close to many full-length substrates. Progressively smaller substrate bands (e.g., n-1) should also be excised for use in Helix 1 length optimization. Identification of the full-length substrate transcript can be somewhat difficult, and confirmation by direct RNA sequencing of purified substrate transcripts is an option *(20)*. An example of gel-purified ribozyme and substrate transcripts is provided (**Fig. 2**).

7. Excise bands with sterile razor blades, and grind in a 1.5-mL Eppendorf tube using 0.5 mL of gel-extraction buffer and a mini-grinding pestle. Shake homogenized gels slices vigorously in an Eppendorf mixer/shaker for 60 min, and then centrifuge full-speed (>10,000$g$) at room temperature in a microcentrifuge for 20 min. Transfer the upper phase to new 1.5-mL Eppendorf tubes, add 1 µL of glycogen and 1 mL of 100% reagent alcohol, and precipitate by freezing on dry ice or at −80°C for 30 min and then centrifuging at full speed in a microcentrifuge for 15 min at 4°C. Wash the pellet twice with 500 µL of ice-cold 70% reagent alcohol, and then dry pellets in a speed-vac, or simply by air-drying.

8. Count radioactivity in RNA pellets by Cerenkov counting—e.g., place the entire 1.5-mL Eppendorf containing the pellet in a scintillation vial (without scintillation fluid) and read it in a scintillation counter.

9. Calculate the pmoles of RNA of each ribozyme or substrate pellet using the formulas provided in **Subheading 2.** It is important to note that the number of Cs in a transcript can be determined by counting the number of Gs in the template oligonucleotide—excluding the T7 promoter region—but in the case of substrate templates including the additional G residue in the CGC sequence acted to enhance transcription initiation. Resuspend ribozymes at 80 n$M$, and substrates at 400 n$M$ in dH$_2$O initially. Using the formula given in **Subheading 2.**, the volume of dH$_2$O to add is calculated by dividing the pmols of RNA by the desired final concentration in µ$M$—e.g., 0.08 µ$M$ for ribozymes and 0.4 µ$M$ for substrates. If time allows and there is enough product, take wet counts in scintillation fluid using 1 µL of resuspended ribozyme or substrate, and adjust concentrations accordingly. Use ribozymes and substrates immediately or store at −80°C prior to use.

### 3.5. Helix 1 Length Optimization and Preliminary Analysis of Catalytic Efficiency

Formal kinetic analysis of individual hairpin ribozymes and their optimized substrates requires considerable effort, including repetition of analysis several times using material produced from various transcription reactions to demonstrate reproducibility. Using multiple-turnover conditions—albeit with less than 20% substrate cleavage—at least two time-points should be tested for each of several substrate concentrations around the apparent $K_M$. As an initial screen of several potential ribozymes that target the same gene, less rigorous methods may be used to narrow the field to a few promising candidates. The following protocols *(16)* are provided with the goal of a preliminary screen in mind.

1. Perform an initial cleavage reaction using a 1:5 ratio of ribozyme (0.08 µ$M$ stock) to substrate (0.4 µ$M$ stock) for 1 h at 37°C to determine whether turnover will occur. Run a native sTRMV ribozyme and substrate control reaction in parallel. Set up the reactions in a 0.25-mL Eppendorf tube suitable for use in a thermocycler with a heated lid as follows:

    a. 1 μL 4X cleavage buffer
    b. 1 μL 0.4 μ$M$ substrate stock
    c. 1 μL dH$_2$O
    d. 1 μL 0.08 μ$M$ ribozyme stock

2. Stop the reaction by adding 4 μL formamide dye, heat to 90°C, snap-cool on ice, and separate ribozyme, uncleaved substrate, and 3' (larger) and 5' substrate cleavage products on a 15% polyacrylamide, 7 $M$ urea gel; run gel for approx 1 h at 12 W. Expose to autoradiography film overnight at –80°C, or use a phosphorimager screen for an appropriate amount of time, as determined empirically.

3. Quantitate the percentage of cleavage by excising bands and counting in a scintillation counter, or by phosphorimager analysis. In order for turnover to occur, more than 20% of the substrate must be cleaved, considering the molar ratio of ribozyme to substrate. Percentage of cleavage and nM of substrate cleaved are calculated with the following formulas:

$$\% \text{ cleavage} = 100 \times [(5' \text{ product} + 3' \text{ product})/(\text{uncleaved} + 5' + 3')]$$

$$nM \text{ cleaved substrate} = (\% \text{ cleavage}) \times (nM \text{ substrate used in reaction})$$

4. If turnover did not occur in the initial cleavage reaction, it may be necessary to redesign the ribozyme with a shorter complementary region in helix 1. In either case, it is advisable to make use of the progressively shorter transcripts generated during substrate transcription to generate some information regarding the optimum length of helix 1. Smaller substrate transcripts are progressively shorter on the 3' end, and thus represent n-1, n-2, lengths of helix 1; an n + 1 substrate transcript is also typically present.

5. Set up reactions using equal volumes of 0.08 μ$M$ ribozyme and 0.4 μ$M$ substrate transcript stock solutions, including n + 1, n and n – 1, as previously with the initial cleavage reaction to determine the optimum helix 1 length. It may be necessary to incubate the reaction for a longer period—2 h instead of 1 h. Calculate the percentage of cleavage as in **step 3**. The optimum helix 1 length corresponds to the substrate transcript with which the highest percent cleavage was observed. An example of a helix optimization experiment is provided (**Fig. 3**). The ribozyme and substrate templates may be redesigned accordingly, or if there is enough of the appropriate substrate transcript available, further kinetic analysis may be performed immediately.

6. A time-course reaction should be performed next to confirm multiple turnover of the ribozyme, generate an estimate of the turnover rate, and refine the incubation time for subsequent kinetic analysis. Use a 1:5 molar ratio of ribozyme (0.08 μ$M$ stock) to substrate (0.4 μ$M$ stock). Lower concentrations of ribozyme should be used if >80% cleavage was observed in the initial reaction; lower concentrations of substrate may also be used. For reference, final ribozyme concentrations of 1–5 n$M$ are typically used later in formal kinetic analysis. Set up the time-course reaction as follows:

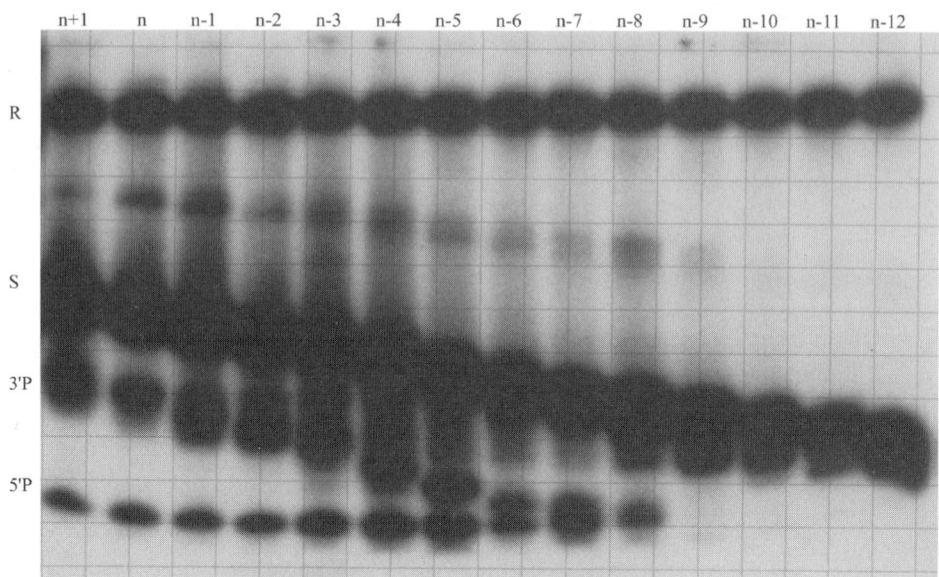

Fig. 3. Helix 1 length optimization. For this tetraloop ribozyme, helix 1 length was determined by phophorimager analysis to be optimized at n-5, corresponding to a helix 1 length of 5 nucleotides. No cleavage was observed past n-8, presumably because of the inability of the ribozyme to bind substrate efficiently with a helix 1 length of less than two bases.

     a. 8 μL 4X cleavage buffer
     b. 8 μL 0.4 μ$M$ substrate stock (or lower concentration)
     c. 8 μL dH$_2$O
     d. 8 μL 0.08 μ$M$ ribozyme stock (or lower concentration)
7. Withdraw 3-μL aliquots from the reaction at time 0, and every 15 min subsequently for approx 2 h. Stop reactions by adding an equal volume of formamide dye. Separate products by gel electrophoresis and determine the percentage of cleavage as in **step 3**. An example of a time-course experiment is provided (**Fig. 4**). The turnover rate is calculated with the following equation:

$$\text{turnover rate} = (\text{n}M \text{ cleaved substrate/n}M \text{ ribozyme in reaction})/(\text{time in min})$$

8. Set up multiple-turnover reactions with constant, limiting amounts of ribozyme and varying amounts of substrate to determine the rate constants $k_{cat}$ and $K_M$. Incubate reactions for the appropriate length of time based on the turnover rate. Keep the total percentage of cleavage of substrate below 20% to maintain linear conditions with regard to product vs time, yet maintain multiple-turnover conditions. Typical final ribozyme concentrations are from 1–5 n$M$. A range of substrate concentrations should be used, working down from the maximum volume

Fig. 4. Time-course experiment to verify multiple-turnover conditions, and obtain an estimate of the rate constant for subsequent kinetics experiments. The ribozyme, uncleaved substrate, and 3' and 5' substrate cleavage products are indicated. The minutes of incubation for each time-point are indicated. Although the amount of ribozyme appears lower at 30 and 60 min, all time-points were drawn from the same reaction, and variability was not observed with the substrate.

of concentrated substrate stock capable of being added to the reaction in 0.5-μL increments. High substrate concentrations are essential to approach $V_{max}$ experimentally, as evidenced by a plateau in a graph of velocity vs substrate concentration; accurate estimates of $k_{cat}$ and $K_M$ are dependent upon approximating $V_{max}$. The native sTRSV ribozyme and substrate should be run in parallel as a control. Separate products by gel electrophoresis and quantitate cleavage products as previously via scintillation counting or phosphorimager analysis. A typical gel from a successful kinetics reaction is provided for reference (**Fig. 5**). An example set of reactions follows:

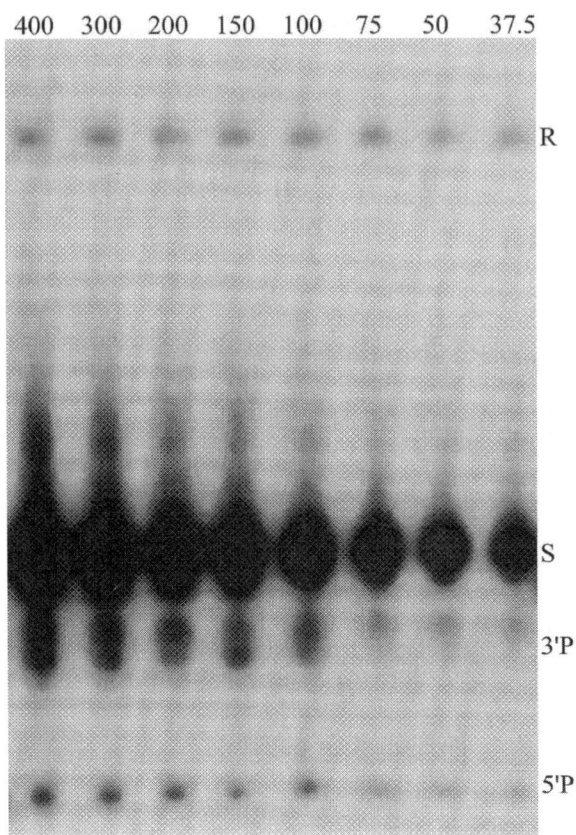

Fig. 5. Kinetics experiment with a constant and limiting amount of ribozyme, multiple-turnover conditions and varying amounts of substrate.

| substrate (400 n$M$) | ribozyme (5 n$M$) | dH$_2$O | 4X reaction buffer |
|---|---|---|---|
| 4.0 μL | 2 μL | 0 μL | 2 μL |
| 3.5 μL | 2 μL | 0.5 μL | 2 μL |
| 3.0 μL | 2 μL | 1.0 μL | 2 μL |
| 2.5 μL | 2 μL | 1.5 μL | 2 μL |
| 2.0 μL | 2 μL | 2.0 μL | 2 μL |
| 1.5 μL | 2 μL | 2.5 μL | 2 μL |
| 1.0 μL | 2 μL | 3.0 μL | 2 μL |
| 0.5 μL | 2 μL | 3.5 μL | 2 μL |

9. Calculate preliminary estimates of $k_{cat}$ and $K_M$ using Tablecurve 2D (Jandel Scientific, San Rafael, CA). Curves should be fit using nonlinear regression methods. The velocity of the reaction is determined first using the following equation:

$$\text{velocity} = \text{n}M \text{ cleaved substrate/min}$$

**Table 1**
**Summary of Results From the Initial Screen of Ribozymes Targeting hTGF-β, mTGF-β, PKD1, and JCV T-Antigen**

| Ribozyme/substrate | Gene | Site | Helix 1 | Turnover | $k_{cat}$ (min$^{-1}$) | $K_M$ (nM) | $k_{cat}/K_M$ (min$^{-1}$nM$^{-1}$) |
|---|---|---|---|---|---|---|---|
| 960R tetraloop/760S | hTGF-β | 157 | na | Yes | ND | ND | ND |
| 961R tetraloop/761S | hTGF-β | 231 | na | Yes | ND | ND | ND |
| 962R tetraloop/762S | hTGF-β | 401 | na | Yes | 0.093 | 298 | $3.12 \times 10^{-4}$ |
| 963R tetraloop/763Sb | hTGF-β | 439 | na | Yes | ND | ND | ND |
| 964R tetraloop/764S | hTGF-β | 518 | na | No | ND | ND | ND |
| 966R tetraloop/766S | hTGF-β | 2189 | na | ND | ND | ND | ND |
| 967R tetraloop/767S | hTGF-β | 2253 | na | Yes | 0.028 | 216 | $1.30 \times 10^{-4}$ |
| 968R tetraloop/768S | hTGF-β | 2378 | na | Yes | $1.86 \times 10^{13c}$ | $8.36 \times 10^{16c}$ | $2.22 \times 10^{-4c}$ |
| 835R native/836S | mTGF-β | 24 | n-2 | Yes | $0.013^{d}$ | $43^{d}$ | $3.02 \times 10^{-4d}$ |
| 838R tetraloop/836S | mTGF-β | 24 | n-2 | Yes | $0.05^{d}$ | $88^{d}$ | $5.68 \times 10^{-4d}$ |
| 842R native/837S | mTGF-β | 533 | n-4 | Yes | 0.038 | 192 | $1.98 \times 10^{-4}$ |
| 842R native /837S | mTGF-β | 533 | n | Yes | 0.025 | 168 | $1.49 \times 10^{-4}$ |
| 843R tetraloop/837S | mTGF-β | 533 | n-5 | Yes | 0.039 | 111 | $3.51 \times 10^{-4}$ |
| PKDR native/PKDS | PKD1 | 13256 | n-4 | Yes | 0.084 | 247 | $3.40 \times 10^{-4}$ |
| PKDR native/PKDS | PKD1 | 13256 | n | Yes | 0.021 | 87 | $2.41 \times 10^{-4}$ |
| JCVR native/JCVS | JCV-T$_{ag}$ | 4932 | n-1 | No | ND | ND | ND |
| sTRSVR/S control | Control | Control | n | Yes | 0.40 | 91 | $4.50 \times 10^{-4}$ |

ND, not determined.

[a]Helix 1 length optimization was not performed; preliminary kinetic analysis was performed with full-length substrate (n).

[b]Of the hTGF-β ribozymes for which preliminary kinetic analysis was not performed because of relatively inefficient substrate transcription, the 963R tetraloop ribozyme appeared to be most promising based on the initial time-course reaction using full-length substrate (n).

[c]Although the $k_{cat}/K_M$ value for this ribozyme is reasonable, the individual rate constants are unreliable because of the failure of this reaction to plateau in terms of a graph of velocity vs substrate concentration; higher substrate concentrations must be tested with this ribozyme.

[d]Although helix 1 was optimized at n-2 for these ribozymes, n was cleaved comparably, and was used in kinetic analysis because of the availability of the transcript in quantity.

The following equation is then used to calculate the $K_M$ and $V_{max}$:

$$y = A \times x/(B + x)$$

where y = velocity; x = initial $nM$ substrate; $A = V_{max}$; $B = K_M$.
The following equation is then used to determine $k_{cat}$:

$$k_{cat} = V_{max}/(nM \text{ ribozyme})$$

The overall catalytic efficiency of the ribozyme is expressed as $k_{cat}/K_M$.

10. Select the ribozymes with the highest catalytic efficiency for subsequent subcloning into the appropriate expression vectors for in vitro analysis of efficacy in reducing expression of the targeted transcripts (*see* **Notes 3** and **4**). If formal rate constants are desired, repeat the kinetic analysis with material from different transcriptions, using multiple time-points for each of several substrate concentrations around the apparent $K_M$.

## 4. Notes

1. Results of the initial screen of several ribozymes that target hTGF-β and mTGF-β, as well as PKD1 and the JCV large T-antigen are provided (**Table 1**). Several promising ribozymes directed against hTGF-β and mTGF-β have been identified. The ribozyme targeting PKD1 also appeared to be worthy of further consideration. In contrast, the ribozyme directed against JCV large T-antigen failed to turn-over, and thus must be redesigned with a shorter substrate or a different target region.

2. Generally, tetraloop ribozymes performed better than native varities. Therefore, it appears to be generally preferable to incorporate the tetraloop modification in the first place, and it may be of little value to also synthesize a native ribozyme for each target site in order to compare the two.

3. The studies described here illustrate methods to identify and optimize ribozyme targets in vitro. The extent to which catalytic efficiency ($k_{cat}/K_M$) is critical for ribozyme activity is unclear, since antisense effects may also be important in overall function within cells.

4. Additional considerations for the successful use of ribozymes for gene therapy should involve the utilization of genbank data on the human genome to identify potential RNA cleavage sites outside a particular target gene. As therapeutics, major barriers that must be addressed in the future include RNA trafficking and RNA-protein assembly, in which such processes may limit the diffusion, accessibility, and association of ribozymes with target RNAs within cells.

## Acknowledgments

This work was supported by NIH grants to J. R., P. E. K., and A. H.

## References

1. Hampel, A. (1998) The hairpin ribozyme: discovery, two-dimensional model, and development for gene therapy. *Prog. Nucleic Acid Res. Mol. Biol.* **58**, 1–39.

2. Shippy, R., Lockner, R., Farnsworth, M., and Hampel, A. (1999) The hairpin ribozyme. Discovery, mechanism, and development for gene therapy. *Mol. Biotechnol.* **12,** 117–129.

3. Ojwang, J., Hampel, A., Looney, D., Wong-Staal, F., and Rappaport, J. (1992). Inhibition of human immunodeficiency virus type-1 (HIV-1) expression by a hairpin ribozyme. *Proc. Natl. Acad. Sci. USA* **89,** 10,802–10,806.

4. Yu, M., Ojwang, J., Yamada, O., Hampel, A., Rappaport, J., Looney, D., and Wong-Staal, F. (1993) A hairpin ribozyme inhibits expression of diverse strains of human immunodeficiency virus type 1. *Proc. Natl. Acad. Sci. USA* **90,** 6340–6344.

5. Leavitt, M.C., Yu, M., Yamada, O., Kraus, G., Looney, D., Poeschla, E., and Wong-Staal, F. (1994) Transfer of an anti-HIV-1 ribozyme gene into primary human lymphocytes. *Hum. Gene Ther.* **5,** 1115–1120.

6. Yu, M., Poeschla, E., Yamada, O., Degrandis, P., Leavitt, M. C., Heusch, M., et al. (1995) In vitro and in vivo characterization of a second functional hairpin ribozyme against HIV-1. *Virology* **206,** 381–386.

7. Wong-Staal, F., Poeschla, E. M., and Looney, D. J. (1998) A controlled, Phase 1 clinical trial to evaluate the safety and effects in HIV-1 infected human of autologous lymphocytes transduced with a ribozyme that cleaves HIV-1 RNA. *Hum. Gene Ther.* **9,** 2407–2425.

8. Kruger, K., Grabowski, P. J., Zaug, A. J., Sands, J., Gottschling, D. E., and Cech, T. R. (1982) Self-splicing RNA: autoexcision and autocyclization of the ribosomal RNA intervening sequence of Tetrahymena. *Cell* **31,** 147–157.

9. Hampel, A., DeYoung, M. B., Galasinski, S., and Siwkowski, A. (1997) Design of the hairpin ribozyme for targeting specific RNA sequences, in *Methods in Molecular Biology, vol. 74,* (Turner, P., ed.), Humana Press, Totowa, NJ, pp. 171–177.

10. Lian, Y., DeYoung, M. B., Siwkowski, A., Hampel, A., and Rappaport, J. (1999) The sCYMV1 hairpin ribozyme: targeting rules and cleavage of heterologous RNA. *Gene Ther.* **6,** 1114–1119.

11. Rappaport J., Kopp, J. B., and Klotman, P. E. (1994) Host virus interactions and the molecular regulation of HIV-1: role in the pathogenesis of HIV-associated nephropathy. *Kidney Int.* **46,** 16–27.

12. Rappaport, J., Josesph, J., Croul, S., Alexander, G., Del Valle, L., Amini, S., and Khalili, K. (1999) Molecular pathway involved in HIV-1-induced CNS pathology: role of viral regulatory protein, Tat. *J. Leukoc. Biol.* **65,** 458–465.

13. Bonwetsch, R., Croul, S., Richardson, M.W., Lorenzana, C., Valle, L.D., Sverstiuk, A.E., et al. (1999) Role of HIV-1 Tat and CC chemokine MIP-1 alpha in the pathogenesis of HIV associated central nervous system disorders. *J. Neurovirol.* **5,** 685–694.

14. Arnaout, M. A. (2001) Molecular genetics and pathogenesis of autosomal dominant polycystic kidney disease. *Annu. Rev. Med.* **52,** 93–123.

15. Safak, M. and Khalili, K. (2001) Physical and functional interaction between viral and cellular proteins modulate JCV gene transcription. *J Neurovirol* **7,** 288–292.

16. DeYoung, M. B., Siwkowski, A., and Hampel, A. (1997) Determination of catalytic parameters for hairpin ribozymes, in *Methods in Molecular Biology, vol.* 74, (Turner, P., ed.), Humana Press, Totowa, NJ, pp. 209–220.
17. Siwkowski, A., Humphrey, M., DeYoung, M. B., and Hampel, A. (1997) Screening for important base identities in the hairpin ribozyme by in vitro selection for cleavage. *Biotechniques* **24,** 278–284.
18. DeYoung, M. B. and Hampel, A. (1997) Computer analysis of the conservation and uniqueness of ribozyme-targeted HIV sequences, in *Methods in Molecular Biology, vol.* 74, (Turner, P., ed.), Humana Press, Totowa, NJ, pp. 27–36.
19. Sczakiel, G. and Tabler, M. (1997) Computer-aided calculation of the local folding potential of target RNA and its use for ribozyme design, in *Methods in Molecular Biology, vol.* 74, (Turner, P., ed.), Humana Press, Totowa, NJ, pp. 11–15.
20. Siwkowski, A. (1997) T7 transcript length determination using enzymatic RNA sequencing, in *Methods in Molecular Biology, vol.* 74, (Turner, P., ed.), Humana Press, Totowa, NJ, pp. 91–97.
21. Cavusoglu, E., Chen, I., Rappaport, J., and Marmur, M. D. (2002) Inhibition of tissue factor gene induction and activity using a hairpin ribozyme. *Circulation* **105,** 2282–2287.

# 26

## Optimization and Application of the Group I Ribozyme *Trans*-Splicing Reaction

### Christer Einvik, Tonje Fiskaa, Eirik W. Lundblad, and Steinar Johansen

#### Summary

Group I ribozymes are naturally occurring catalytic RNAs that are able to excise themselves as introns (group I introns) from a precursor RNA, and to ligate the flanking exons. Group I ribozymes can be engineered to act in *trans* by recognizing a separate RNA molecule in a sequence specific manner, and to covalently link an RNA sequence to this separate RNA molecule. This ribozyme transesterification reaction has potential in molecular biology and in medicine as a new approach to gene therapy. Here we describe detailed optimized protocols where trans-splicing group I ribozymes are applied in mapping accessible sites in target messenger RNA, and in messenger RNA-repair by correcting mutations.

**Key Words:** Antisense; gene therapy; mRNA accessibility; ribozyme; RNA repair; *trans*-splicing; *trans*-tagging.

## 1. Introduction

Group I ribozymes are autocatalytic RNA molecules coded by group I introns. Group I ribozymes possess several catalytic activities, but most pronounced is the two-step transesterification reaction that leads to intron excision and exon ligation (*1*). A simplified view of the self-splicing reaction is presented in **Fig. 1A**. The first step of a regular group I intron *cis*-splicing is initiated by an exogenous guanosine attack at the 5' splice site (5'SS). This liberates the 5' exon and covalently links the guanosine to the 5' end of the intron. In the final step, the free 3' end of the 5' exon attacks the 3' splice site (3'SS) and results in exon ligation and intron excision.

Engineered group I ribozymes that lack the 5' exon are able to specifically *trans*-splice the 3' exon onto separate target RNA molecules both in vitro and

From: *Methods in Molecular Biology, vol. 252: Ribozymes and siRNA Protocols, Second Edition*
Edited by: M. Sioud © Humana Press Inc., Totowa, NJ

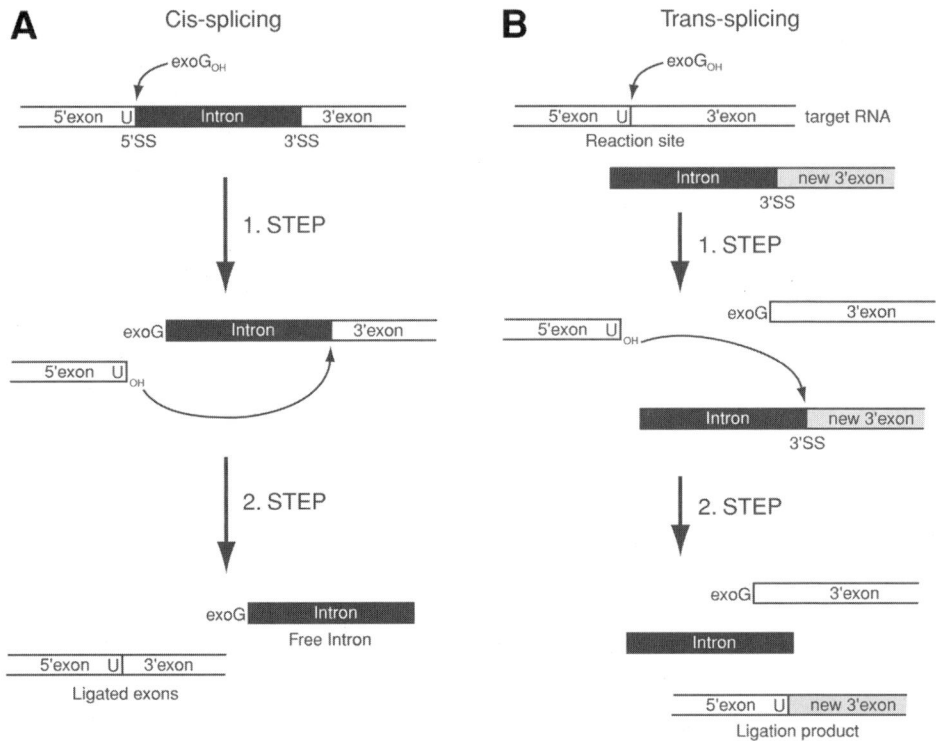

Fig. 1. Schematic view of the two-step group I intron splicing catalyzed by the group I ribozyme. (**A**) *Cis*-splicing of intron from a single precursor RNA resulting in ligated exon and excised intron. (**B**) *Trans*-splicing between a 5' exon (target RNA) and a new 3' exon located on a separate RNA. The *trans*-splicing reaction can be further developed into methods for mapping the accessibility of sites and correcting mutations in messenger RNAs. The only sequence requirement in target RNA is a uridine (U). 5' SS, 5' splice site; 3' SS, 3' splice site; exoG$_{OH}$, exogenous guanosine cofactor. See text for details.

in cell cultures (reviewed in **refs. *1–4***). Similar to the *cis*-splicing reaction, the substrate for the first transesterification reaction in *trans*-splicing is the phosphodiester bond located after a uridine in the target RNA (5'SS) (**Fig. 1B**) Here, the recognition at the 5'SS is accomplished by basepairing (P1) between the uridine and flanking exon nucleotides (nt) with a 4–9-nt internal guide sequence (IGS) located within the 5' region of the intron RNA. By changing the intronic IGS by mutagenesis, the ribozyme can be engineered to react with almost any RNA template *in trans*. The only sequence requirement is that the uridine at the 5'SS reaction site has to generate a wobble basepair to a gua-nosine in the IGS (*5*). Thus, a group I ribozyme can be designed to attach a 3' exon sequences onto almost any uridines positioned in a targeted RNA. By designing a ribozyme construct that recognizes a uridine upstream of a mutation in a messenger RNA (mRNA), group I ribozymes may replace a mutated region of an mRNA with a corrected sequence. Ribozyme-dependent RNA-repair has been demonstrated in several clinical relevant systems including, a trinucleotide repeat in myotonic dystrophy (*see* Chapter 27), a β-globin variant in sickle cell-anemia, and *p53* mutations in various cancer cells (*6–8*).

Successful application of group I ribozymes in RNA-repair approaches is dependent on optimized factors such as in vivo stability, target accessibility, cleavage efficiency, and specificity. One approach for improvement of the ribozyme tool is to evaluate and compare a number of different naturally occurring group I ribozymes. Thus, new and promising ribozymes can be selected for further molecular studies. However, one of the most important problems in ribozyme technology is that the majority of target sequences in a cellular mRNA are inaccessible to the ribozymes. This is the result of both a complex secondary and tertiary structure folding, and to the fact that all cellular RNAs are interacting with a great number of proteins in vivo (*9*). Several new approaches for mapping accessible sites on target RNAs have recently been described. These range from computer-assisted predictions, which are usually more or less unreliable, to complex hybrid ribozyme systems in which the cleavage activity of the ribozymes is coupled to the unwinding activity of RNA helicase (*10–13*).

Recent work in our lab has focused on the functional and structural characterization of a number of new group I ribozymes from various eukaryotic organisms. We have identified several interesting candidates for application in RNA-repair using mRNA targets that are relevant in molecular medicine, including AGU mRNA, α-mannosidosis mRNA, and *p53* mRNA (*14,15*). These studies have resulted in the development and improvement of a protocol for efficient mapping of accessible sites in a mutated mRNA in vitro based on engineered group I ribozymes. This method complements other available approaches for mapping the accessibility of specific target sequences in an RNA target, and serves as an integrated and important part of the RNA-repair approach.

## 2. Materials

All materials used should be of high-purity grade. To avoid degradation of RNA, we use 0.1% diethylpyrocarbonate (DEPC)-treated water when making all buffers and solutions. DEPC is a potential hazard and should be handled with care.

### 2.1. Synthesis of the GN5-Ribozyme Library

The oligonucleotide primers used to synthesize the template for making the GN5-ribozyme library contain both a T7 promoter and a unique arbitrary sequence used in the screening step.

1. Oligonucleotide primers used to make the polymerase chain reaction (PCR) product containing the GN5-ribozymes. Forward primer: 5'-<u>GGG AAT TAA TAC GAC TCA CTA TAG </u>GNN NNN $X_{20}$-3'. Reverse primer: 5'-*GCC CCG ATG CCG ACA GCA GAA TGG TTT CAC GAA CAA GAC G*TT TGG CAA AA$Y_{20}$-3'. Underlined: T7 RNA polymerase promoter, Italic: Inverted Dir.S956 ORF (unique arbitrary sequence), $X_{20}$: 20 nt identical to ribozyme 5' end, $Y_{20}$: 20 nt complementary to ribozyme 3' end.
2. T7 RNA polymerase and supplied reaction buffer (Stratagene).

## 2.2. In Vitro trans-Tagging of mRNA Isolated From Cells

1. Trizol-isolated total RNA from a mammalian cell line transfected with the mutated cDNA of interest.
2. 2 X standard low-salt self-splicing buffer: 80 m$M$ Tris-HCl, pH 7.5, 20 m$M$ MgCl$_2$, 400 m$M$ KCl, 4 m$M$ spermidine, 10 m$M$ dithiothreitol (DTT), 0.4 m$M$ guanosine 5' triphosphate (GTP).

## 2.3. Screening of Accessible Sites

1. Reverse oligonucleotide primer used in the reverse transcriptase (RT) reaction is 5'-GCC CGA TGC CGA CAG CA-3'.
2. Reverse transcriptase. We use the Moloney murine leukemia virus (M-MuLV) RT, which is included in the First-Strand cDNA Synthesis Kit (Amersham Biosciences).
3. PCR cloning kit. We use the pGEM-T and pGEM-T Easy Vector System (Promega).

## 2.4. Plasmid Cloning of the Repair-Ribozymes

1. Mammalian expression vector containing a T7 RNA polymerase promoter. We use the pcDNA3.1-vector (Invitrogen).

## 2.5. Analyzing the Repair-Ribozymes

1. STOP solution: 95% formamide, 50 m$M$ ethylenediaminetetraacetic acid (EDTA), 0.02% xylene cyanol, 0.05% bromophenol blue.
2. Lipofectamin 2000 (Invitrogen).
3. Trizol (Gibco-BRL).

## 3. Methods

## 3.1. Mapping Accessible Sites in Target RNA by trans-Splicing

To map accessible uridines in a template RNA, we use a GN5-ribozyme library with a tag sequence attached to its 3' end (GN5-ribozyme-tag). The tag sequence should be a unique arbitrary sequence. Here, we use an inverted sequence of the Dir.S956-1 intron ORF *(16)* in order to make the reverse transcriptase-polymerase chain reaction (RT-PCR) screening of accessible sites easier, but any unique sequence could be included (*see* **Notes 1–3**).

### 3.1.1. Synthesis of the GN5-Ribozyme-Tag Library

1. Standard PCR with a forward and reverse oligonucleotide primer that introduces the changes described here on a ribozyme plasmid template.
2. Isolation of the PCR product from a 2% agarose gel.
3. RNA synthesis from the PCR products using standard in vitro transcription with T7 RNA polymerase.
4. Purification of the GN5-ribozyme-tag RNA from a 5% denaturing polyacrylamide gel.

Ideally, this ribozyme pool should consist of all the 1024 different recognition sequences specified by the GN5 IGS sequence.

### 3.1.2. In Vitro trans-Tagging of mRNA Isolated From Cells

The source of target mRNA used in this assay is total RNA isolated from transient transfected COS-7 cells that overexpress a mutated form of the mRNA of interest. Expression plasmids were transfected into COS-7 cells by Lipofectamin 2000 (Invitrogen), with a transfection efficiency of approx 40%. RNA was isolated by the use of 2.5 mL Trizol solution per T-25 culture bottle from transfected cells, as described by the manufacturer.

1. 1 μg total cellular RNA and 2 m$M$ GN5-ribozyme-tag RNA are mixed with 2X standard low-salt self-splicing buffer in separate 1.5-mL eppendorf tubes. The RNA is allowed to fold correctly for 10 min at 50°C.
2. For *trans*-tagging to occur, refolded target RNA and ribozyme RNA are mixed and incubated for 3 h at 37°C.

A schematic view of the *trans*-tagging approach is shown in **Fig. 2**. In the *trans*-tagging reaction, the tag sequence is transferred from the GN5-ribozyme-tag RNA to accessible uridines in the template RNA.

### 3.1.3. Screening for Accessible Sites

Screening for accessible sites is performed by RT-PCR. *Trans*-tagging adds a unique sequence to the accessible uridines in the template RNA, and thus minimizes nonspecific PCR products (*see* **Notes 2** and **3**).

1. Reverse transcription of *trans*-tagged RNA with a reverse primer that is complementary to the unique tag sequence.
2. PCR with a nested reverse primer and a forward primer located in the target sequence upstream of the mutation. If a larger region of the template is analyzed for accessible sites, several forward primers at various locations in the template should be used in order not to favor smaller products in RT-PCR .
3. All products from the PCR reaction are cloned into a bacterial cloning vector and transformed into *E. coli* cells (Ultracompetent JM-109).
4. Positive selected clones are scored using PCR with the same set of primers as in **step 2**.
5. Plasmid inserts from the positive selected clones are confirmed by DNA sequencing.

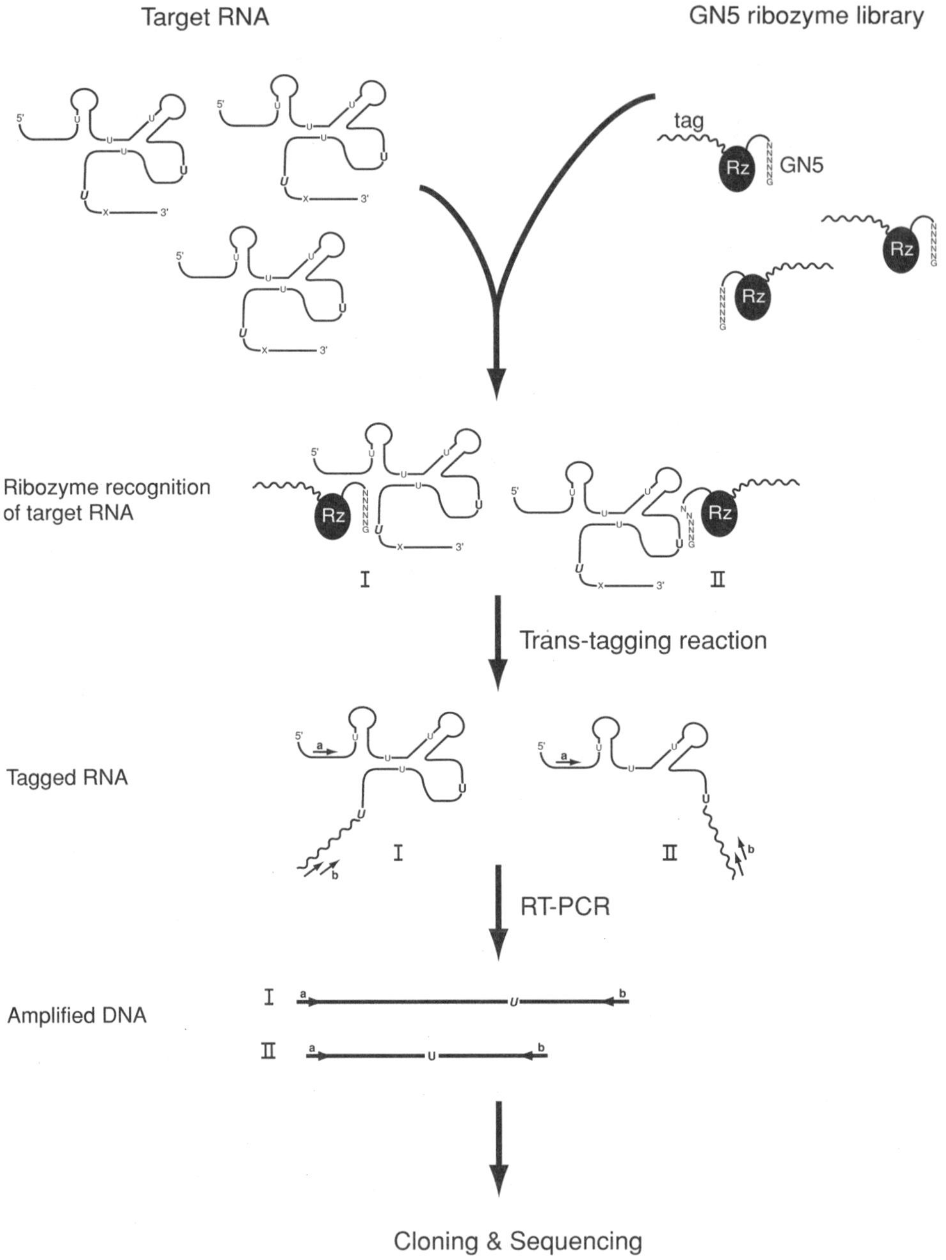

Fig. 2. Schematic representation of the accessible site-mapping approach based on *trans*-splicing group I ribozymes (Rz). Only a restricted number of sites are available in target RNA to a randomized ribozyme library (GN5 ribozyme library). Here, five positions in the internal guide sequence of the ribozyme are randomized. Two accessible sites (I and II) are exemplified after ribozyme recognition and binding. Subsequently, these sites become tagged by *trans*-splicing, and are subjected to RT-PCR

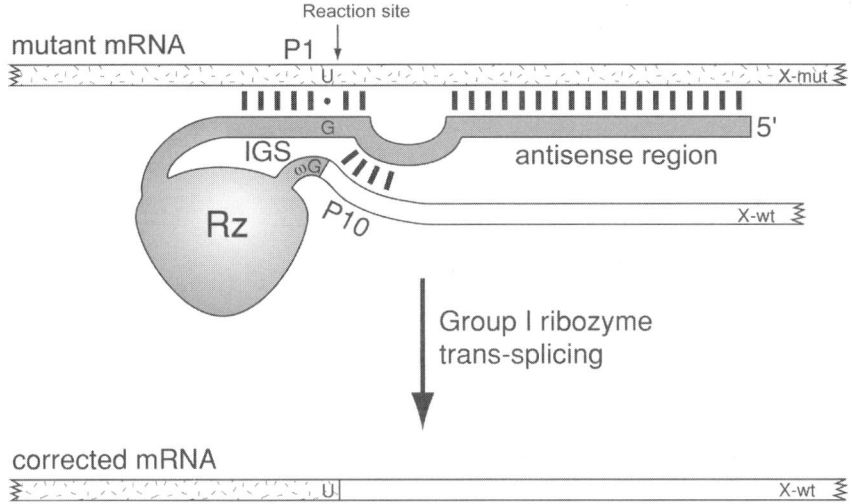

Fig. 3. Design of *trans*-splicing group I ribozyme for RNA repair, in which a mutant mRNA become converted into a corrected mRNA (see text for details). Important elements of ribozyme (Rz) design are the internal guide sequence (IGS), paired segment 10 (P10), and the antisense region. X-mut, mutant nucleotide position; X-wt, corrected nucleotide position corresponding to the wild-type sequence.

Application of this procedure to the α-mannosidosis mRNA identified three accessible sites within a 1000-nt region (unpublished results) (*see* **Note 4**).

### 3.2. Design of Repair-Ribozymes

#### 3.2.1. Target Site Recognition (P1-IGS)

The internal guide sequence (IGS) of *trans*-acting group I ribozymes is involved in the recognition of template RNA. The only sequence requirement in the IGS is a guanosine that interacts by wobble base pairing to the uridine (G:U basepair; **Fig. 3**) located at the reaction site.

1. IGS is designed with a guanosine followed by 5-nt complementary to the sequence immediately upstream of the detected accessible uridine.
2. In addition, 2–3 nucleotides complementary to the template are included upstream of the guanosine to resemble the P1 helix structure in natural group I intron ribozymes (*see* **Fig. 3**).

---

Fig. 2. *(continued)* amplification by the a/b primer set. Accessible sites are identified by DNA sequencing the plasmid cloned products.

### 3.2.2. Increasing the Specificity

Target recognition based on only a 6 nt IGS will cause the ribozyme to interact with numerous unintended RNAs in addition to the target RNA *(17)*. In a *trans*-tagging experiment using group I ribozymes in *E. coli*, Haseloff and colleagues *(18)* showed that a 40–200-nt sequence complementary to the template could be added to the 5' termini of the ribozyme, resulting in a 40- to 50-fold increase in activity.

1. To increase specificity of the repair-ribozymes, a 40-nt antisense sequence, complementary to target RNA, was added to the ribozyme at the 5' end (**Fig. 3**).
2. It is important to maintain a few single-stranded nucleotides between the IGS and the antisense region, corresponding to the P1 loop of natural group I intron ribozymes.

### 3.2.3. Design of the 3' Exon Region Carrying the Corrected Sequence

Several considerations must be included when designing the 3' exon region carrying the corrected sequence.

1. When correcting a mutation in an mRNA, reading frameshifts should be avoided at the junction between target RNA and the added 3' exon sequence. Thus, the 3' exon must start in-frame with the accessible uridine at the reaction site.
2. To avoid strong intermolecular basepairing to the antisense region, the first 30–40 nt of the 3' exon have to be degenerated by the use of alternative codons. The degenerated sequence makes it possible to construct a reverse primer that favors the corrected mRNA during RT-PCR analysis.
3. To improve the activity of the repair reaction, we include a 4-bp P10 structure in the repair-ribozymes (**Fig. 3**). Many natural group I intron ribozymes contain a P10 basepaired region (P10) between the IGS and the 3' exon. It has been reported that the maintenance of an artificial P10 of 4–6 basepair in a *trans*-tagging *Tetrahymena* ribozyme in *E. coli* yeilded a 50-fold increase in activity compared to a ribozyme that lacks P10 *(18)*.

### 3.2.4. Plasmid Cloning of the Repair-Ribozymes

Repair-ribozyme constructs are generated in a three-step cloning procedure. The ribozyme is inserted into a mammalian expression vector containing a T7 RNA polymerase promoter, allowing initial in vitro analysis of the ribozyme.

1. The antisense region is constructed by annealing two complementary deoxyoligonucleotides containing the entire antisense region. A unique restriction endonuclease site is added (in the loop region corresponding to P1) downstream of the antisense region. This restriction site corresponds to the upstream cloning site for the ribozyme-containing DNA. The ends of the annealed oligonucleotides contain restriction sites used to clone the insert into a mammalian expression vector. Annealing is performed by heating both deoxyoligonucleotides

(100 pmol each) in the same Eppendorf tube for 3 min at 95°C, followed by slow cooling to room-temperature.

2. The correct 3' exon sequence (wild-type sequence) is PCR-amplified from a template containing the wild-type cDNA sequence. A unique restriction site is included in the forward oligonucleotide as a downstream cloning site for the ribozyme-containing DNA in **step 3**. The PCR product is agarose gel-purified and ligated into the vector from **step 1** behind the antisense region.

3. The ribozyme is PCR-amplified with a forward and reverse oligonucleotide primer containing the same unique restriction sites used in **steps 1** and **2**, respectively. The forward primer also includes the correct IGS located just upstream of the nucleotides complementary to the ribozyme DNA template. The resulting PCR product is agarose gel-purified, digested by restriction endonucleases, and ligated into the vector between the antisense region and the corrected 3' exon sequence.

## 3.3. Analyzing the Repair-Ribozymes

### 3.3.1. In Vitro Analysis

For initial analysis of repair-ribozymes, we assay the reaction in vitro by denaturing polyacrylamide gel electrophoresis (PAGE). Possible repair product is then inspected by DNA sequencing to detect correct *trans*-splicing (**Fig. 4**).

1. In vitro transcription using T7 RNA polymerase on linearized vectors containing the repair-ribozyme construct and mutated cDNA template. A small amount of a radioactive labeled nucleotide (we use $\alpha$-35S cytidine 5' triphosphate [CTP]) is added to the reaction containing the template in order to visualize repair products on a denaturing polyacrylamide gel. All RNAs are polyacrylamide gel-purified.

2. Mutated template mRNA and an excess of ribozyme are mixed and incubated in a standard low-salt self-splicing buffer containing GTP (*see* **Subheading 2.2.**). The reaction is terminated at different time-points by addition of an equal volume of STOP-solution.

3. The repair reaction is analyzed on a 5% denaturing polyacrylamide gel (**Fig. 4A**)

4. The RNA species corresponding to the repair product is purified from the gel and subjected to RT-PCR analysis. The forward PCR primer is located ~200 nt upstream of the repair junction, and the nested reverse PCR and RT primers are located in the degenerated region of the wt 3' exon sequence (**Fig. 4B**).

5. The resulting PCR product is sequenced in order to verify that the repair reaction has proceeded correctly (**Fig. 4C**).

### 3.3.2. In Vivo Analysis

To analyze the repair-ribozyme in a cellular environment, the ribozyme containing plasmid from **Subheading 3.2.4.** is either co-transfected with plasmid containing the target sequence into CHO-K1 cells (Chinese hamster ovary), or transfected into a mammalian cell-line with an endogene expression of the target RNA.

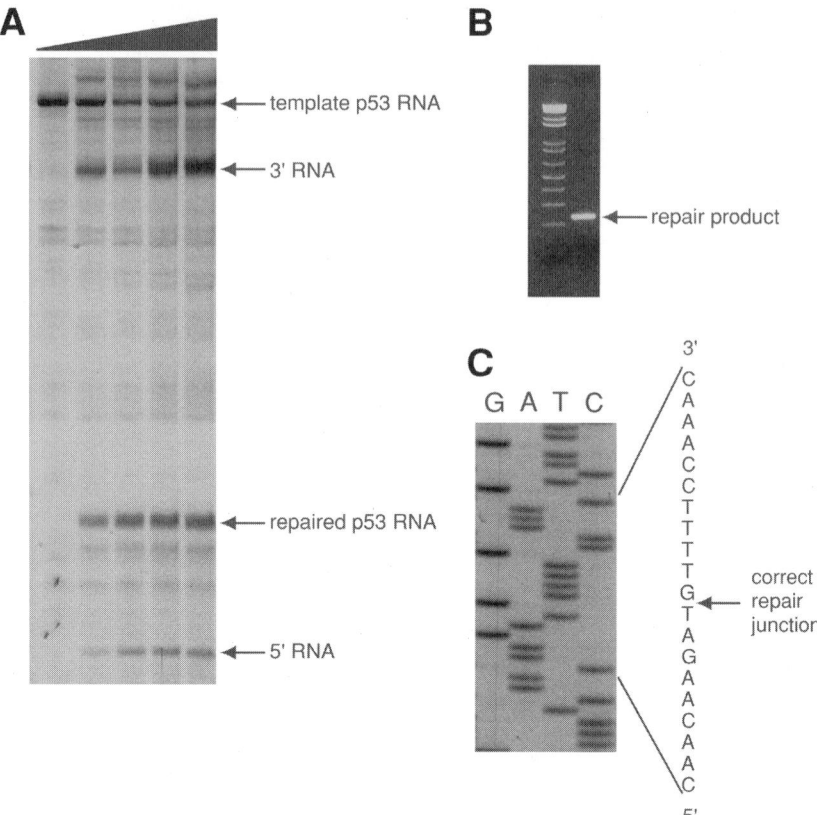

Fig. 4. Example of RNA-repair in vitro. (**A**) In vitro transcribed 35-S-labeled *p53* template RNA and unlabeled *Tetrahymena* repair-ribozyme RNA were mixed and incubated under standard low-salt conditions and analyzed in a time-course experiment (0, 5, 15, 30, and 60 min) on a 5% denaturing polyacrylamide gel. (**B**) Amplification of the expected repaired p53 RNA using RT-PCR. (**C**) Sequencing of the RT-PCR product shown in **B**. The sequencing gel shows an exact joining of the new 3' exon sequence derived from the ribozyme to the target *p53* RNA.

1. Plasmid containing the repair-ribozyme is transfected by Lipofectamine 2000 into a mammalian cell-line expressing the target RNA in standard 6-well plates.
2. Total RNA is isolated after 24 h using 0.5-mL Trizol (Gibco-BRL) per well. The isolated RNA is dissolved in 20 μL DEPC-treated water.
3. Isolated total RNA (50 ng–1 μg) is subjected to RT-PCR using the same primers as in **Subheading 3.3.1., step 4**.
4. The resulting PCR product is sequenced in order to verify that the repair reaction has occurred correctly in the cell (**Fig. 5**).

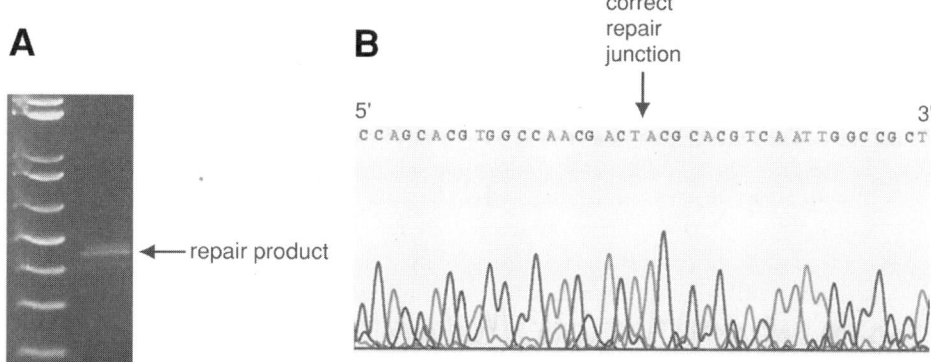

Fig. 5. Example of RNA-repair in vivo. (**A**) Amplification of the in vivo repaired product using RT-PCR. Total RNA isolated from fibroblasts expressing an endogenous α-mannosidosis target RNA is the template for the reverse transcription reaction. (**B**) Sequencing of the RT-PCR product shown in **A**. The sequence shows an exact joining of the new 3' exon sequence derived from the ribozyme to the target α-mannosidosis RNA. The correct repair junction is located after the uridine coded by T1312 in the α-mannosidosis cDNA.

## 4. Notes

1. We have described the experimental procedures for mapping accessible sites in mutated target RNAs for subsequent use in group I ribozyme mediated RNA-repair. Unlike most other available protocols, this method is based on a reaction (*trans*-tagging) similar to that of the reaction catalyzed by the repair-ribozymes. Thus, we score for available sites in the target RNA as well as most reactive sites catalyzed by the repair-ribozymes.
2. RNA molecules are known to adapt complex secondary and tertiary structures in vivo. RNA molecules also interact with proteins that may further reduce the accessibility of ribozyme to specific sites. Although the protocol described here is an in vitro accessibility mapping assay, several studies have shown that the most accessible sites found in vitro are also the most accessible in a cellular environment *(12,19,20)*. In a similar approach to our study, Sullenger and colleagues showed that the most efficient in vitro selected sites are the same as those found in vivo *(21)*.
3. We decided to optimize the in vitro-based protocol. Thus, because the *trans*-tagging reaction was expected to be more efficient in vitro on total RNA from transfected cells, because of an excess of ribozyme, compared to endogenous RNA within cells. Furthermore, during the screening process, we experienced several notable tagging products different from those expected. These were caused by ribozymes that had reacted with non-intended RNAs (usually ribosomal RNA), or were simply the result of primer ligation to random RNA molecules.

4. All previous reports on group I ribozyme-mediated RNA-repair and RNA *trans*-tagging are based on application of the *Tetrahymena* group I ribozyme (Tth.L1925). For current nomenclature of group I intron ribozymes, *see* **ref. *22***. In order to detect the most effective natural repair-ribozyme, we included several new group I ribozymes in our studies. Thus, three different group I ribozymes, in addition to the *Tetrahymena* ribozyme, were successfully included for the mapping of accessible sites on a mutated α-mannosidosis mRNA. These are Fse.L569 and Fse.L1898 from the myxomycete *Fuligo septica* and Dir.S956-1 from the myxomycete *Didymium iridis*. To compare the ribozymes against a particular target site, all ribozymes must have similar tolerance for nucleotide substitutions in the target recognition sequence (IGS). Interestingly, the experiments concluded that the most accessible site was the same for all ribozymes.

5. Hydrolysis at the 3'-splice site of a group I intron ribozyme is a competing reaction to the *trans*-splicing reaction. Thus, a significant fraction of the repair-ribozyme construct may cleave off the 3'-exon sequence and become unable to correct a mutated mRNA species. One solution to this problem is to select mutant versions of the *Tetrahymena* ribozyme, or other group I ribozymes, with a significant reduction in 3'SS hydrolysis compared to transesterification (repair reaction). We have obtained such mutants of Dir.S956-1 with substitutions and small deletions in the P9 peripheral domain that appear to be promising new candidates in RNA repair *(15)*.

## Acknowledgments

We appreciate the contributions by Ingrid Skjæveland, Jørn Henriksen, Eva Sjøttem, and all members of the RNA research group to the RNA-repair project. This work was supported by grants from the Norwegian Cancer Society, The Norwegian Research Council, and The Aakre Foundation for Cancer Research.

## References

1. Cech, T. R. (1990) Self-splicing of group I introns. *Ann. Rev. Biochem.* **59**, 543–568.
2. Johansen, S., Einvik, C., Elde, M., Haugen, P., Vader, A., and Haugli, H. (1997) Group I introns in biotechnology: prospects of application of ribozymes and rare-cutting homing endonucleases. *Biotechnol. Ann. Rev.* **3**, 111–150.
3. Watanabe, T. and Sullenger, B. A. (2000) RNA repair: a novel approach to gene therapy. *Adv. Drug Deliv. Rev.* **44**, 109–118.
4. Sullenger, B. A. and Gilboa, E. (2002) Emerging clinical applications of RNA. *Nature* **418**, 252–258.
5. Zaug, A. J., Been, M. D., and Cech, T. R. (1986) The *Tetrahymena* ribozyme acts like an RNA restriction endonuclease. *Nature* **324**, 429–433.
6. Phylactou, L. A., Darrah, C., and Wood, M. J. (1998) Ribozyme-mediated trans-splicing of a trinucleotide repeat. *Nature Genet.* **18**, 378–381.
7. Lan, N., Howrey, R. P., Lee, S. W., Smith, C. A., and Sullenger, B. A. (1998) Ribozyme-mediated repair of sickle beta-globin mRNAs in erythrocyte precursors. *Science* **280**, 1593–1596.

8. Watanabe, T. and Sullenger, B. A. (2000) Induction of wild-type p53 activity in human cancer cells by ribozymes that repair mutant p53 transcripts. *Proc. Natl. Acad. Sci. USA* **97,** 8490–8494.

9. Usman, N. and Stinchcomb, D. T. (1996) Design, synthesis, and function of therapeutic hammerhead ribozymes. *Nucleic Acids Mol. Biol.* **10,** 243–264.

10. Jaeger, J. A., Turner, D. H., and Zuker, M. (1989) Improved predictions of secondary structures for RNA. *Proc. Natl. Acad. Sci. USA* **86,** 7706–7710.

11. Sohail, M. and Southern, E. M. (2000) Hybridization of antisense reagents to RNA. *Curr. Opin. Mol. Ther.* **2,** 264–271.

12. Barroso-DelJesus, A. and Berzal-Herranz, A. (2001) Selection of targets and the most efficient hairpin ribozymes for inactivation of mRNAs using a self-cleaving RNA library. *EMBO Rep.* **2,** 1112–1118.

13. Warashina, M., Kuwabara, T., Kato, Y., Sano, M., and Taira, K. (2001) RNA-protein hybrid ribozymes that efficiently cleave any mRNA independently of the structure of the target RNA. *Proc. Natl. Acad. Sci. USA* **98,** 5572–5577.

14. Fiskaa, T. (2000) Characterization of two group I ribozyme catalysed reactions: full-length circle formation and trans-splicing. Master Thesis, University of Tromsø.

15. Haugen, P. (2001) Molecular characterization of complex self-splicing group I introns in nuclear ribosomal DNA. PhD Thesis, University of Tromsø.

16. Johansen, S. and Vogt, V. M. (1994) An intron in the nuclear ribosomal DNA of *Didymium iridis* codes for a group I ribozyme and a novel ribozyme that cooperate in self-splicing. *Cell* **76,** 725–734.

17. Jones, J. T., Lee, S. W., and Sullenger, B. A. (1996) Tagging ribozyme reaction sites to follow trans-splicing in mammalian cells. *Nature Med.* **2,** 643–648.

18. Kohler, U., Ayre, B. G., Goodman, H. M., and Haseloff, J. (1999) Trans-splicing ribozymes for targeted gene delivery. *J. Mol. Biol.* **285,** 1935–1950.

19. zu Putlitz, J., Yu, Q., Burke, J. M., and Wands, J. R. (1999) Combinatorial screening and intracellular antiviral activity of hairpin ribozymes directed against hepatitis B virus. *J. Virol.* **73,** 5381–5387.

20. Yu, Q., Pecchia, D. B., Kingsley, S. L., Heckman, J. E., and Burke, J. M. (1998) Cleavage of highly structured viral RNA molecules by combinatorial libraries of hairpin ribozymes. *J. Biol. Chem.* **273,** 23,524–23,533.

21. Lan, N., Rooney, B. L., Lee, S. W., Howrey, R. P., Smith, C. A., and Sullenger, B. A. (2000) Enhancing RNA repair efficiency by combining trans-splicing ribozymes that recognize different accessible sites on a target RNA. *Mol. Ther.* **2,** 245–255.

22. Johansen, S. and Haugen, P. (2001) A new nomenclature of group I introns in ribosomal DNA. *RNA* **7,** 935,936.

# 27

# Repair of Myotonic Dystrophy Protein Kinase (DMPK) Transcripts by *Trans*-Splicing Ribozymes

## Leonidas A. Phylactou

### Summary

The *Tetrahymena* group I intron ribozyme is an RNA molecule of approx 400 bases that is capable of base-specific RNA *trans*-splicing. These properties can be applied to repair mutations at the RNA level, and therefore can restore normal cellular functions in diseased cells. The purpose of this chaper is to present the methodology that can be used to repair mutations responsible for myotonic dystrophy. The design and construction of group I intron ribozymes and their transfection in myotonic dystrophy cells will be among the sections described in this chapter.

**Key Words:** Group I intron ribozymes; *trans*-splicing; myotonic dystrophy; DMPK.

## 1. Introduction

Since the original discovery of RNA enzymes (ribozymes), attention has been focused on the use of *trans*-cleaving ribozymes (e.g., hammerhead and hairpin motifs) to downregulate endogenous gene expression. Very little emphasis has been given to the original group I intron ribozyme as a means to modify cellular or viral gene expression. Our goal is to present a detailed analysis of how group I intron ribozymes can be used to repair mutant myotonic dystrophy protein kinase (DMPK) transcripts. Myotonic dystrophy (DM) is a common inherited neuromuscular disease and is caused by an expansion of the CTG trinucleotide repeat in the 3' untranslated region (UTR) of the DMPK gene.

The *Tetrahymena* group I intron ribozyme is an RNA molecule of approx 400 bases, which catalyzes its own intron splicing *(1)*. The *Tetrahymena* group I intron ribozyme is also capable of base-specific RNA *trans*-splicing. The reaction proceeds via two consecutive cleavage reactions in the presence of a divalent cation. The first of these is carried out in the target RNA (5' splice

From: *Methods in Molecular Biology, vol. 252: Ribozymes and siRNA Protocols, Second Edition*
Edited by: M. Sioud © Humana Press Inc., Totowa, NJ

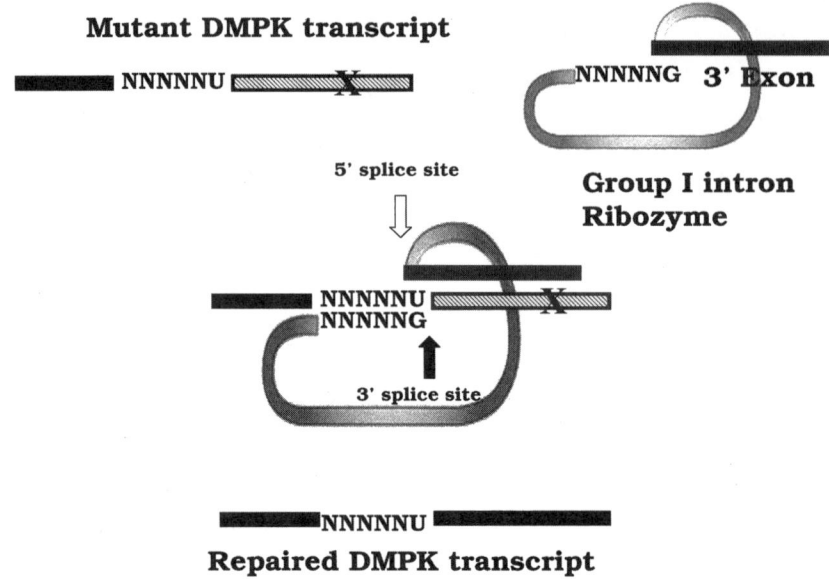

**Mutant DMPK transcript**

NNNNNU X

NNNNNG  3' Exon

5' splice site

**Group I intron Ribozyme**

NNNNNU
NNNNNG

3' splice site

NNNNNU
**Repaired DMPK transcript**

Fig. 1. Group I intron ribozymes are designed to repair mutant DMPK transcripts (X denotes the mutation). Ribozymes bind DMPK mRNA via a 6-nucleotide binding site GNNNNN where N is any nucleotide. Basepairing of ribozyme and target DMPK mRNA results in a two-step cleavage/splicing reaction (indicated by the arrows) and a replacement with the wild-type sequence (3' exon).

site), following binding of the ribozyme (**Fig. 1**). A uridine preceding the 5' splice site is the only sequence requirement. The second cleavage reaction releases the 3' exon attached to the ribozyme strand (3' splice site), which results in its ligation with the cleaved target RNA. This application, which has been demonstrated both outside cells with synthetic target RNAs or in cell culture *(2–5)*, has a wide application in RNA-mediated gene-therapy protocols.

This chapter describes the methodology on how to apply group I intron ribozyme technology in order to repair mutant DMPK transcripts (**Fig. 1**). The methodology section will be divided into:

1. Design and cloning of group I intron ribozymes.
2. In vitro testing of group I intron ribozyme activity.
3. Transfection of ribozyme constructs in DM fibroblasts.
4. Analysis of group I intron ribozyme cellular activity.

## 2. Materials
**Note:** All materials and chemicals that are described in this chapter should be considered hazardous and treated with caution. Particular care should be

taken during the handling of radioisotopes, phenol, polyacrylamide, diethylpyrocarbonate (DEPC), and ethidium bromide.

During RNA experimentation, all solutions used should be treated with 0.01% DEPC, and the work should be carried out in an RNase-free bench. Reaction buffers, nucleotides, oligonucleotide solutions, cloning vectors, and enzymes are stored at –20°C.

1. Restriction endonucleases and buffers.
2. Phenol-chloroform, pH 8.0.
3. T4 DNA ligase and buffer.
4. Cloning vectors: pL21 (*see* **Note 1**), pGEM-4Z (Promega), pRL-SV40 (Promega).
5. *E. coli* competent cells, DH5α (Invitrogen).
6. LB medium (per L): 10 g Bactotryptone, 5 g Bacto-yeast extract, and 5 g NaCl.
7. LB/ampicillin plates: 1 L Luria Bertani (LB) medium, 15 g agar, 100 µg/µL ampicillin.
8. Miniprep lysis buffers:
   a. Solution I: 25 m$M$ Tris-HCl, pH 8.0, 10 m$M$ ethylenediaminetetraacetic acid (EDTA), 50 m$M$ glucose.
   b. Solution II: 0.1 $N$ NaOH, 1% sodium dodecyl sulfate (SDS).
   c. Solution III (per 100 mL): 60 mL 5 $M$ potassium acetate, 11.5 mL glacial acetic acid, 28.5 mL H$_2$O.
9. In vitro transcription reagents (MAXIscript kit, AMBION).
10. RNA polymerase.
11. [α-$^{32}$P UTP] (800 Ci/mmol),
12. DNase I.
13. Polymerase chain reaction (PCR) reagents.
14. *Taq* DNA polymerase (Pharmacia).
15. RNA extraction kit (Perfect RNA kit—Eppendorf).
16. Gel extraction kit (PerfectPrep Gel Cleanup kit, Eppendorf).
17. cDNA synthesis reagents.
18. Moloney murine leukemia virus (MMuLV) reverse transcriptase (New England Biolabs).
19. Pfu DNA polymerase (Stratagene).
20. Superfect (Qiagen).

## *2.1. Preparation of Solutions*

### *2.1.1. Annealing of Oligonucleotides: IGS, IGS(N), IGS/AS*

1. Mix equimolar amounts (4 pmols) of both strands in 10 m$M$ Tris-HCl, 10 m$M$ MgCl$_2$, 10 m$M$ NaCl, and 50 m$M$ dithiothreitol (DTT) at pH 7.9.
2. Incubate at 95°C for 5 min, then immediately at 65°C for 10 min, followed by gradual cooling to room temperature.
3. Concentrate annealed oligonucleotides by ethanol-precipitation.
4. Perform a digest of the double-stranded oligonucleotides using the appropriate restriction enzymes.

5. Following digestion, extract the restriction endonucleases by phenol-chloroform and precipitate the digested double-stranded oligonucleotides by ethanol-precipitation.
6. Perform **steps 4** and **5** for the digestion of the cloning vectors.

### 2.1.2. PCR Amplification of 3' Exons

1. Extract total RNA from cells (e.g., normal human fibroblasts), expressing the DMPK gene (Perfect RNA kit—Eppendorf).
2. Carry out a reverse transcription with either an oligo(dT) primer or a primer that is specific for the DMPK RNA.
3. Set up a PCR amplification of the newly synthesized cDNA using as an upstream primer, a wild-type sequence that is complementary to the DMPK RNA, which starts immediately after the 5' splice site and as downstream primer, a wild-type sequence that is complementary to the 3' end of the DMPK RNA (**Fig. 2A**; *see* **Note 2**). The resulting PCR product is the 3' exon.
4. Ensure successful PCR reaction by checking a small sample on an agarose or polyacrylamide gel electrophoresis (PAGE) and visualize under ultraviolet (UV) after ethidium bromide staining.

### 2.1.3. Superfect: DNA Complexation

Superfect (Qiagen Inc.) and plasmid DNA are complexed according to the manufacturer's instructions. The optimum ratio of superfect to DNA can be accomplished by initial use of a plasmid, which contains a reporter gene (e.g., EGFP).

## 3. Methods

### 3.1. Design and Construction of Group I Intron Ribozymes and DMPK Target RNA

Group I intron ribozymes can *trans*-splice DMPK mRNA as follows: binding of the group I intron ribozyme to the DMPK target mRNA, can cause cleavage of the 3' end of the message, followed by the release of the wild-type DMPK sequence (3' exon) from the ribozyme strand and ligation to the 5' part of the cleaved mRNA (**Fig. 1**).

### 3.1.1. Theory of trans-Splicing Requirements

Group I intron ribozymes are designed to *trans*-splice the wild-type DMPK mRNA sequence to the mutant DMPK mRNAs. The design can be carried out in two ways. In the first, the most accessible sites for ribozyme binding on DMPK mRNA are chosen based on the calculations of RNA-predicting software. The second method is experimental, and allows the determination of optimal accessible RNA binding sites. *Trans*-splicing of wild-type DMPK RNA sequence is detected after incubating the ribozyme and target DMPK mRNA in a cell-free environment.

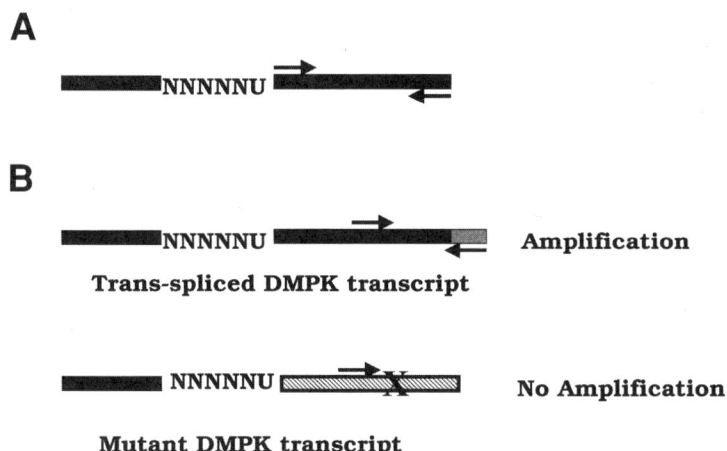

Fig. 2. PCR amplification techniques to construct 3' exon for ribozyme cloning (**A**) and detect *trans*-splicing products (**B**). (**A**)Primers are designed to amplify the wild-type sequence immediately downstream of the target cleavage/splice site (NNNNNU) and be used as a 3' exon. (**B**) Primers are designed to specifically amplify the *trans*-spliced DMPK product. As a 3' primer, a sequence is used that is complementary to the polyA$^+$ tail of the vector.

## 3.1.2. Theoretical Approach: RNA Accessibility Through Software Analysis of RNA Target

In order to determine the most accessible sites on the mutant DMPK mRNA, it is possible to use computer-aided software capable of predicting RNA folding. For example, one can use the Mfold software package to predict the structure and folding of the DMPK mRNA. The RNA sequence is inserted in the online software http://www.bioinfo.rpi.edu/applications/mfold/. The results of the survey show the sites where DMPK mRNA is more linear and therefore more accessible (**Note 3**). The experimental testing of the ability of the new ribozymes to *trans*-splice the DMPK RNA then follows the selection process.

### 3.1.2.1. CONSTRUCTION OF GROUP I INTRON RIBOZYME VECTOR

Following identification of possible *trans*-splicing sites, ribozyme-binding sites (internal guide sequences [IGS]) are designed and synthesized chemically. The IGS will replace the original insert in the pL21 plasmid, and will be in the form 5' GNNNNN 3', where N is any base complementary to the DMPK RNA target site and G is basepaired with uridine in the target DMPK RNA (**2**; **Note 4**). The new inserts will then be cloned in the vector, which contains the catalytic sequence (e.g., pL21). Inactive constructs should also be constructed by deleting part of the catalytic domain *(2)*.

1. Perform an overnight ligation of 0.1 pmol digested IGS double-stranded ribozyme oligonucleotides and the linearized pL21 vector by using 3–5 $M$ excess of the former in a total volume of 10 µL.
2. Transform 3 µL of the ligation mix into *E. coli*-competent cells and select recombinant clones on LB/ampicillin plates.
3. Extract plasmid DNA from individual colonies and screen for positive clones by restriction digestion and dideoxy sequencing.

### 3.1.2.2. Construction of DMPK cDNA Construct

DMPK RNA should be synthesized in a cell-free environment in order to determine the efficiency of the newly synthesized group I intron ribozymes.

1. Perform an overnight ligation of 0.1 pmol-digested DMPK cDNA and the linearized pGEM-4Z vector by using 3–5 $M$ excess of the former in a total volume of 10 µL.
2. Repeat **steps 2** and **3** in **Subheading 3.1.2.1.**

### 3.1.2.3. Synthesis of Group I Intron Ribozyme RNA

1. Linearize ribozyme-containing constructs by restriction digestion. Choose an appropriate restriction endonuclease so that the construct is linearized at the end of the cloned ribozyme sequence (*see* **Note 5**).
2. Mix 33 nmols of linearized plasmid, as template, with all four ribonucleotide triphosphates—adenosine 5' triphosphate (ATP), cytidine 5' triphosphate (CTP), uridine 5' triphosphate (UTP) and guanosine 5' triphosphate (GTP)—at a final concentration of 5 m$M$ in the presence of 15 m$M$ MgCl$_2$, 2 m$M$ spermidine, 50 m$M$ Tris-HCl, pH 7.5, 5 m$M$ dithiothreitol (DTT), 25 U of ribonuclease inhibitor, and 20 U of RNA polymerase in a total volume of 50 µL (MAXIscript kit, AMBION). All reagents should be added at room temperature to avoid precipitation of DNA.
3. Incubate the reaction at 37°C for 2–4 h and stop with the addition of 4 U of RNase-free DNase I by further incubating at 37°C for 15 min. The reaction is stopped by the addition of 0.5 $M$ EDTA.
4. Remove enzymes by phenol-chloroform extraction and recover the newly synthesized transcript by ethanol-precipitation.
5. Check the quality and amount of ribozymes either by reading their absorbance at 260 nm or by denaturing (7 $M$ urea) gel electrophoresis alongside markers of known concentration.

### 3.1.2.4. Synthesis of DMPK RNA Target

DMPK RNA target is then synthesized by in vitro transcription. Our standard protocol uses [α-$^{32}$P UTP] to incorporate labeled nucleotide. This method is similar to that used for ribozyme synthesis.

1. Mix 33 nmols of the linearized DMPK-containing construct product with ATP, CTP, and GTP at a final concentration of 0.5 m$M$, 50 µCi of [α-$^{32}$P UTP] (800 Ci/ mmol), 15 m$M$ MgCl$_2$, 2 m$M$ spermidine, 50 m$M$ Tris-HCl, pH 7.5, 5 m$M$ DTT, 12.5 U of ribonuclease inhibitor, and 10 U of RNA polymerase in a total volume of 20 µL. All reagents should be added at room temperature to avoid precipitation of DNA.
2. Incubate the reaction at 37°C for 60 min, halted with the addition of 4 U of RNase-free DNase I, and further incubated at 37°C for 15 min.
3. Remove all enzymes by phenol-chloroform extraction, and then recover the newly synthesized transcript by ethanol-precipitation.
4. Determine the specific activity of the labeled target by TCA precipitation followed by liquid scintillation counting, or by using a beta-counter.

### 3.1.3. Experimental Approach: Random Selection of Optimum Group I Intron Ribozymes

It is possible to randomly search for optimum active DMPK ribozymes (e.g., ribozymes that can access DMPK mRNA and possess enhanced repair efficiency) as described in **ref. 6**.

Briefly, a group I intron ribozyme, containing a degenerate target RNA-binding site (IGS[N]) in the positions of the five of the six nucleotides (e.g., a combination of all four nucleotides in every position, 5'-GNNNNN-3') is synthesized. Therefore, ribozyme and target are incubated in cell-free conditions. At this stage, the 3' exon can be of any sequence, since the goal is to identify effective *trans*-splicing. Detection of *trans*-splicing products is achieved by reverse transcription-polymerase chain reaction (RT-PCR), by using as a 5' primer, a sequence that is complementary to the 5' DMPK target cDNA and as a 3' primer, a sequence that is complementary to the 3' exon. This is followed by cloning of the *trans*-splicing PCR products and characterization by dideoxy sequencing *(6)*.

1. Perform an overnight ligation of 0.1 pmol-digested IGS(N) double-stranded ribozyme oligonucleotides and the linearized pL21 vector by using 3–5 $M$ excess of the former in a total volume of 10 µL.
2. Transform 3 µL of the ligation mix into *E. coli*-competent cells and select recombinant clones on LB/ampicillin plates.
3. Extract plasmid DNA from individual colonies, and screen for positive clones by restriction digestion and dideoxy sequencing.

In vitro transcription of ribozymes and target is carried out as described in **Subheading 3.1.2.**

### 3.2. Ribozyme trans-*Splicing Assays*

In vitro-synthesized DMPK ribozymes can now be tested for their ability to *trans*-splice the target RNA prior to cell-culture experiments. This can be done by incubating the ribozyme with the labeled DMPK mRNA target under optimum conditions. The 3' exon attached to the ribozyme should be a different size than the cleaved 3' end of the DMPK target RNA in order to distinguish between unspliced and *trans*-spliced products. If the experimental approach (*see* **Subheading 3.1.3.**) is followed rather than the theoretical (*see* **Subheading 3.1.2.**), the *trans*-splicing reaction is carried out with unlabeled target RNA and ribozyme and then subjected to RT-PCR, similarly to **Subheading 3.4.2.** in order to clone the *trans*-splicing products and determine ribozyme efficiency by counting the positive clones.

1. Preheat 100 n*M* of DMPK ribozyme in reaction buffer buffer (50 m*M* HEPES, pH 7.0, 150 m*M* NaCl, and 5 m*M* MgCl$_2$) in a total volume of 10 μL at 95°C for 2 min.
2. Preheat 10 n*M* labeled (or unlabeled) DMPK mRNA target at 95°C for 2 min, in the presence of 100 m*M* GTP in a total volume of 10 μL, and then add to the ribozymes and incubate at 37°C.
3. Remove aliquots at different time intervals between 30 min and 3 h and stop reactions by the addition of 20 m*M* EDTA.
4. Resolve target RNA and cleavage and *trans*-splicing products on a 4% denaturing (7 *M* urea) polyacrylamide gel electrophoresis (PAGE).

Following the cell-free *trans*-splicing assays effective ribozymes are identified based on their ability to *trans*-splice DMPK mRNA efficiently.

### 3.3. Cloning for Cell-Culture Experimentation

Following identification of optimum ribozyme-binding sites, new IGS and 3' exons are designed and cloned to construct full functional group I intron ribozymes for cellular DMPK *trans*-splicing.

### 3.3.1. Replacement of IGS

DMPK constructs with the highest efficiency identified during the cell-free experimentation are modified to accept a new IGS (IGS/AS) with an extended IGS and an antisense sequence (**Fig. 3**). This new version of group I intron ribozymes has been shown to result in an increased intracellular efficiency and specificity *(7)*. The original 6 nt of the IGS is followed by three extra nt (NNNNNGNNN) and a sequence of 40 nt, which should be complementary to the target DMPK RNA (**Note 6**).

1. Set a digest on ribozyme constructs to excise the original IGS.
2. Gel purify the digested plasmid without the insert using the PerfectPrep Gel Cleanup kit (Eppendorf).
3. Perform an overnight ligation of 0.1 pmol digested IGS/AS double-stranded ribozyme oligonucleotides and the linearized pL21 vector by using 3–5 *M* excess of the former in a total volume of 10 μL.

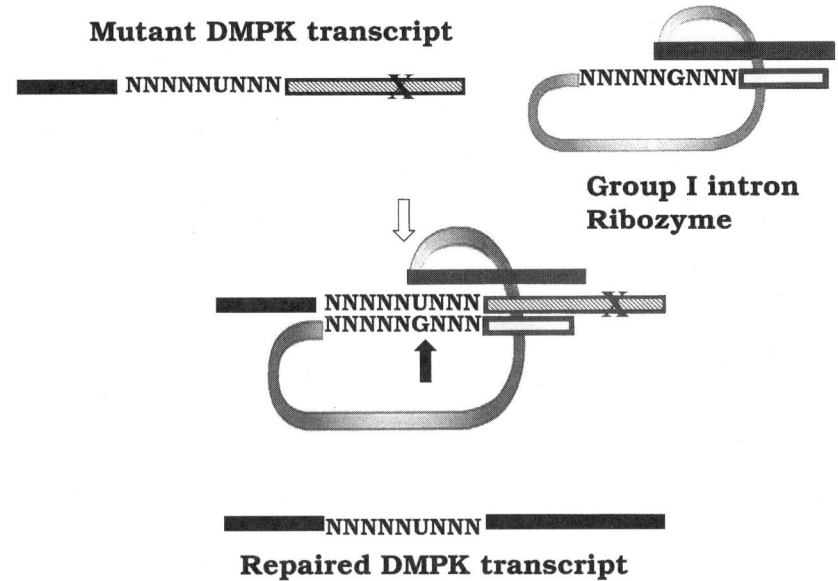

**Mutant DMPK transcript**

**Group I intron Ribozyme**

**Repaired DMPK transcript**

Fig. 3. A new version of group I intron ribozymes is designed for cell-culture experimentation. The ribozyme-binding site is extended to nine nucleotides (NNNGNNNNN, where N is any nucleotide) followed by an antisense region, indicated by the empty rectangle.

4. Transform 3 µL of the ligation mix into *E. coli*-competent cells and select recombinant clones on LB/ampicillin plates.
5. Extract plasmid DNA from individual colonies and screen for positive clones by restriction digestion and dideoxy sequencing.

### 3.3.2. Replacement of 3' Exon

1. Set digest to excise the 3' exon used during the cell-free experiments.
2. Gel purify the digested plasmid without the insert using the PerfectPrep Gel Cleanup kit (Eppendorf).
3. Perform an overnight ligation of 0.1 pmol 3' exons, prepared as PCR products (*see* **Subheading 2.1.2.**) and the linearized pL21 vector by using 3–5 $M$ excess of the former in a total volume of 10 µL.
4. Repeat **steps 4** and **5** from **Subheading 3.3.1.**

### 3.3.3. Cloning of New Generation of Ribozymes for Cell-Culture Delivery

1. Set up a digest to remove the ribozyme-exon insert from the original pL21 plasmid.
2. Perform an overnight ligation of 0.1 pmol-digested double-stranded ribozyme insert and the linearized pRLSV40 (Promega) vector by using 3–5 $M$ excess of the former in a total volume of 10 µL.
3. Repeat **steps 4** and **5** from **Subheading 3.3.1.**

### *3.4. Cell-Culture Investigations*

#### *3.4.1. Tranfections*

1. Seed $1.5 \times 10^6$ of DM fibroblast cells in 10% fetal calf serum (FCS) with Dulbecco's modified Eagle's medium (DMEM) in 60-mm plates and leave to attach for at least 12 h.
2. Remove medium, replace with Superfect:plasmid DNA mix, and leave incubating for 3 h.
3. Replace transfection mix with 3 mL fresh medium and leave cells to incubate at 37°C for 48 h.

### *3.5. RNA Analysis of Results*

1. Remove medium from cells and wash with PBS. Lyse cells and extract total RNA with the Perfect RNA kit (Eppendorf).
2. Carry out a reverse transcription with either an oligo(dT) primer or a primer that is specific for the RNA of interest.
3. Set up a PCR amplification in order to detect the *trans*-splicing DMPK RNA products. Choose as a 5' primer a sequence that will be complementary upstream of the target cleavage site. For a 3' primer, choose a sequence that will be complementary to the plasmid polyA$^+$ tail and therefore be specific only to *trans*-splicing products (**Fig. 2B**). Amplification will only occur when *trans*-splicing products are present in cell extracts.
4. Ensure a successful PCR reaction by checking a small sample on an agarose or PAGE.

## 4. Notes

1. pL21 is a plasmid that contains the group I intron catalytic sequence. It was kindly donated to our laboratory by Dr. Bruce Sullenger *(2)*.
2. It is best to use a high fidelity PCR DNA polymerase that does not create overhangs after PCR amplifications, so that cloning of 3' exon PCR products is accurate and efficient.
3. The output from the Mfold search shows the degree of basepairing of each nucleotide in the DMPK mRNA. It is best to select appropriate ribozyme-binding sites in areas in which there is a stretch of non-basepaired nucleotides.
4. Internal guide sequences for in vitro experimentation are composed of six nucleotides in the form of 5'-GNNNNN-3', where G will base pair with the uridine in the chosen 5' cleavage site in the DMPK mRNA. The following five nucleotides will be complementary to the five nucleotides preceding the target cleavage site (**Fig. 1**).
5. Restriction enzymes that create a 3' overhang should be avoided because the transcription will be inefficient.
6. This new version of group I intron ribozymes with extended IGS and antisense sequence has been recently described in **ref. 7**.

## Acknowledgments

This work was supported by grants from the A. G. Leventis Foundation, Muscular Dystrophy Campaign UK, and the Association Francaise Coytues Les Myopathies, France.

## References

1. Cech, T. R., Zaug, A. J., and Grabowski, P. J. (1981) *In vitro* splicing of the ribosomal RNA precursor of Tetrahymena: involvement of a guanosine nucleotide in the excision of the intervening sequence. *Cell* **27,** 487–496.
2. Sullenger, B. A. and Cech, T. R. (1994) Ribozyme-mediated repair of defective mRNA by targeted, *trans*-splicing. *Nature* **371,** 619–622.
3. Phylactou, L. A., Darrah, C., and Wood, M. J. A. (1998) Ribozyme-mediated trans-splicing of a trinucleotide repeat. *Nat. Genet.* **18,** 378–381.
4. Lan, N. Howery, R. P., Lee, S.-W., Smith, C. A., and Sullenger, B. A. (1998) Ribozyme-mediated repair of sickle beta-globin mRNAs in erythrocyte precursors. *Science* **280,** 1593–1596.
5. Watanabe, T. and Sullenger, B. A. (2000) Induction of wild-type p53 activity in human cancer cells by ribozymes that repair mutant p53 transcripts. *Proc. Natl. Acad. Sci. USA* **97,** 8490–8494.
6. Jones, J. T., Lee, S.-W., and Sullenger, B. A. (1996) Tagging ribozyme reaction sites to follow *trans*-splicing in mammalian cells. *Nat. Med.* **2,** 643–648.
7. Kohler, U., Ayre, B. G., Goodman, H. M., and Haseloff, J. (1999) Trans-splicing ribozymes for targeted gene delivery. *J. Mol. Biol.* **285,** 1935–1950.

# 28

# General Design and Construction of RNase P Ribozymes for Gene-Targeting Applications

## Hua Zou, Karen Chan, Phong Trang, and Fenyong Liu

### Summary

RNase P ribozyme, such as M1 RNA, the catalytic RNA subunit of RNase P from *Escherichia coli*, cleaves an RNA helix that resembles the acceptor stem and T-stem structure of its natural ptRNA substrate. When covalently linked with a guide sequence, the M1 ribozyme can function as a sequence-specific endonuclease, M1GS RNA, and cleave any target RNA sequences that basepair with the guide sequence. Using the mRNA coding for the major transcription regulatory protein ICP4 of herpes simplex virus 1 (HSV-1) as the model target, we describe in this chapter the general design and construction of M1GS ribozymes for gene-targeting applications. Specifically, methods are described in detail to determine ideal target regions of an mRNA for M1GS ribozymes and to construct highly active RNase P ribozymes that target these regions. Extensive protocols for in vitro synthesis of the ribozymes and for the cleavage assay of the ribozyme activity are also included. These methods are intended to provide general guidelines for the design and construction of M1GS ribozymes for gene-targeting applications.

Key Words: RNase P; ribozyme; gene targeting; gene therapy; M1 RNA; tRNA.

## 1. Introduction

### 1.1. RNA Enzyme As a Model System for Studying RNA Biology and As a Tool for Gene-Targeting Applications

Studies of RNA-RNA and RNA-protein interactions provide insight into numerous cellular events including RNA processing (e.g., splicing and editing) and translation that play significant roles in regulating gene expression in development as well as the progression of human cancer. The discovery of catalytic RNAs (ribozymes) has generated tremendous interest in understanding the structure and function of these RNA enzymes and their involvement in

From: *Methods in Molecular Biology, vol. 252: Ribozymes and siRNA Protocols, Second Edition*
Edited by: M. Sioud © Humana Press Inc., Totowa, NJ

biological catalysis *(1–3)*. One aspect of ribozyme research is an analysis of the general mechanisms of RNA catalysis and RNA-RNA and RNA-protein interactions by studying how ribozymes interact with their RNA substrates and protein cofactors and catalyze reactions *(1–3)*. These studies may provide insight into the general principles that govern the RNA-RNA and RNA-protein interactions, which play critical roles in regulating gene expression during human development and tumorigenesis. Another aspect of ribozyme research focuses on the development of RNA enzymes as gene-targeting tools for both in vitro and in vivo applications *(4–7)*. Antisense DNA and RNA *(8)* have been extensively used in basic research (e.g., studying gene functions during tumorigenesis), and are being developed for clinical applications (e.g., antiviral and anticancer therapy) *(9)*. RNA enzymes derived from hammerhead and hairpin ribozymes have also been shown to be promising gene-targeting agents to specifically cleave RNA sequences of choice *(7,10–13)*. These ribozymes contain a catalytic RNA domain that cleaves the target mRNA as well as a substrate-binding domain with an antisense sequence to the target mRNA sequence. Therefore, these gene-targeting ribozymes bind to the mRNA sequence through Watson-Crick interactions between the target sequence and the antisense sequence in the substrate-binding domain of the ribozyme. Compared to conventional antisense DNA and RNA, a ribozyme may have several unique features, as it can cleave its target irreversibly and a single ribozyme molecule can cleave multiple copies of its substrate. This chapter focuses on the general design and construction of ribonuclease P (RNase P) ribozymes that efficiently cleave an mRNA and abolish the expression of the target mRNA in tissue culture.

## 1.2. RNase P Catalytic RNA As a System for Ribozyme Studies

RNase P is a ribonucleoprotein complex responsible for the 5' maturation of tRNAs *(1,14)*. It catalyzes a hydrolysis reaction to remove a 5' leader sequence from tRNA precursors (pre-tRNA) and several small RNAs. In *E. coli*, RNase P consists of a catalytic RNA subunit (M1 RNA) of 377 nucleotides and a protein subunit (C5 protein) of 119 amino acids *(1,14)*. In the presence of a high concentration of salt, such as 100 m$M$ $Mg^{2+}$, M1 RNA acts as a catalyst and cleaves pre-tRNAs in vitro by itself *(15)*. The addition of C5 protein dramatically increases the rate of cleavage by M1 RNA in vitro and is required for RNase P activity and cell viability in vivo *(1,14)*. It has been proposed that C5 protein—an extremely basic protein—functions to stabilize the conformation of M1 RNA and enhance the interactions between the enzyme and the ptRNA substrate *(16–18)*. Extensive phylogenetic and biochemical analyses have established models for the secondary structure *(19–21)* and the three-dimensional structure of M1 RNA *(22,23)*. These models provide a framework

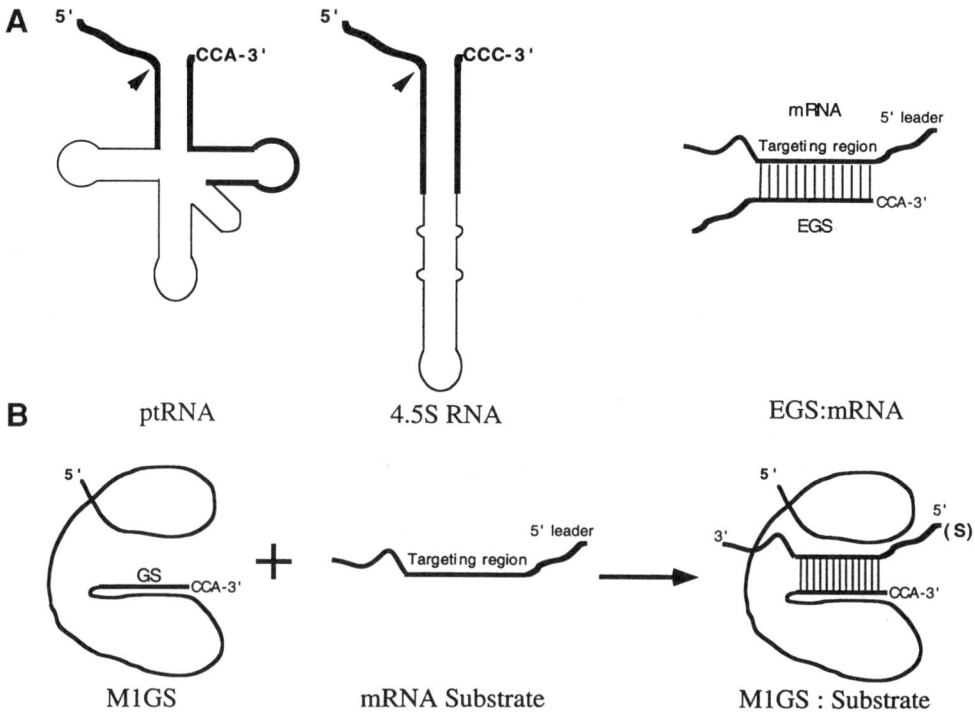

Fig. 1. (**A**) Representation of the natural substrates (ptRNA and 4.5S RNA) and a small model substrate (EGS:mRNA) for RNase P and M1 RNA. The site of cleavage is marked with a filled arrow. (**B**) Representation of an M1GS RNA construct to which a target RNA has hybridized.

for identifying the putative active site and substrate-binding site, and for the study of the catalytic mechanism of this ribozyme.

## 1.3. RNase P Ribozyme As a Tool for Gene-Targeting Applications

Studies on substrate recognition by M1 RNA and RNase P have led to the development of a general strategy in which M1 RNA and RNase P can be used as gene-targeting tools to cleave any specific mRNA sequences. The natural substrates for RNase P in *E. coli* include ptRNAs, the precursor to 4.5S RNA, and several small RNAs *(24–26)*. All these substrates can fold into a structure that is equivalent to the top portion of a ptRNA molecule (e.g., a 5' leader sequence, an acceptor-stem-like structure and a 3' CCA sequence) (**Fig. 1A**). Deletion analyses of a tRNA substrate have revealed that a small model substrate containing a structure equivalent to the acceptor stem and T stem, the 3' CCA sequence, and the 5' leader sequence of a pre-tRNA molecule can be cleaved effectively by M1 RNA (**Fig. 1A**; *27,28*). In this small model substrate, the 5' proximal sequence (the 5' leader and 5' proximal acceptor stem sequence) basepairs to the 3' proximal sequence (the 3' proximal acceptor stem sequence). This 3' proximal sequence is known as an external guide sequence (EGS) because it can basepair with the targeted sequence and guide M1 RNA to cleave the substrate (**Fig. 1A**). Thus, M1 RNA and RNase P can target any RNA sequence for cleavage, provided that an EGS is designed to hybridize with the target RNAs (**Fig. 1A**; *28,29*). M1 RNA can be converted into a sequence-specific ribozyme, M1GS RNA, by linking the ribozyme covalently to a guide sequence (GS) (**Fig. 1B**; *30,31*). An M1GS ribozyme, M1TK13, was designed to target the mRNA encoding the thymidine kinase (TK) of herpes simplex virus 1 (HSV-1). We have shown that M1TK13 is a sequence-specific endonuclease and cleaves substrates that basepair with the guide sequence. Furthermore, when M1TK13 was expressed in mammalian cells infected with HSV-1, a reduction of 80% in the expression levels of both TK mRNA and protein was observed *(31)*.

Targeted cleavage of mRNA by RNase P ribozymes provides a unique approach to inactivate any RNA of known sequence expressed in vivo. Thus, it can be used as a tool both in basic research—such as the regulation of gene expression during tumorigenesis and developmental processes—and in clinical applications, such as gene therapy. Using the mRNA coding for the HSV-1 major transcription regulatory protein ICP4 as the model target *(32)*, this chapter focuses on general design and generation of M1GS ribozymes that can efficiently cleave an mRNA sequence and effectively inhibit its expression in cultured cells.

## 2. Material

### 2.1. Buffer Solutions

1. 10X cleavage buffer A: 500 m$M$ Tris-HCl, pH 7.5, 1 $M$ NH$_4$Cl, 1 $M$ MgCl$_2$.
2. 2X binding buffer B: 100 m$M$ Tris-HCl, pH 7.5, 200 m$M$ NH$_4$Cl, 200 m$M$ CaCl$_2$, 6% glycerol, 0.2% xylene cyanol, 0.2% bromophenol blue.
3. 10X folding buffer C: 500 m$M$ Tris-HCl, pH 7.5, 1 $M$ NH$_4$Cl, 100 m$M$ MgCl$_2$.
4. Buffer A2: 50 m$M$ Tris-HCl, pH 7.5, 100 m$M$ NH$_4$Cl, 100 m$M$ MgCl$_2$.
5. Buffer C2: 50 m$M$ Tris-HCl, pH 7.5, 100 m$M$ NH$_4$Cl, 10 m$M$ MgCl$_2$.
6. Buffer E: 20 m$M$ sodium citrate, pH 5.0, 1 m$M$ ethylenediaminetetraacetic acid (EDTA), 7 $M$ urea.
7. Phosphate-buffered saline (PBS).
8. Lysis buffer: 150 m$M$ NaCl, 10 m$M$ Tris-HCl, pH 7.4, 1.5 m$M$ MgCl$_2$, 0.2% NP40.

### 2.2. PCR Reagents

1. 10X PCR buffer: 500 m$M$ KCl, 100 m$M$ Tris-HCl, pH 8.3.
2. 25 m$M$ MgCl$_2$.

3. Deoxynucleotide triphosphates (dNTPs). Each 10 m$M$ dNTP (deoxyadenosine 5' triphosphate [dATP], deoxycytidine 5' triphosphate [dCTP] deoxyguanosine 5' triphosphate [dGTP], deoxythymidine 5' triphosphate [dTTP]) is diluted from stock solution of 100 m$M$. The stock solution is dissolved from solid powder and adjusted to pH 7.0 by addition of 1 $M$ NaOH.
4. *Taq* DNA polymerase (Perkin).

### 2.3. Reverse Transcription

1. 5X MMTV reverse transcriptase (RT) buffer: 250 m$M$ Tris-HCl, pH 8.3, 375 m$M$ KCl, 15 m$M$ MgCl$_2$, 25 m$M$ dithiothreitol (DTT).
2. MMTV RT (Gibco-BRL).
3. 5X AMV RT buffer: 500 m$M$ Tris-HCl, pH 8.3, 50 m$M$ KCl, 30 m$M$ MgCl$_2$, 50 m$M$ DTT.
4. AMV RT (Boehringer Mannheim).

### 2.4. In Vitro Transcription

1. 5X buffer for in vitro transcription: 200 m$M$ Tris-HCl, pH 7.9, 30 m$M$ MgCl$_2$, and 50 m$M$ DTT, 10 m$M$ spermidine.
2. 100 m$M$ DTT.
3. 10 m$M$ ATP, 10 m$M$ GTP, 10 m$M$ CTP, 10 m$M$ uridine 5' triphosphate (UTP). Each NTP is diluted from stock solution of 100 m$M$. The stock solution is dissolved from solid powder and adjusted to pH 7.0 by addition of 1 $M$ NaOH.
4. RNasin RNase Inhibitor.
5. T7 RNA polymerase.

### 2.5. Radioisotope and Gel Electrophoresis

1. [$^{32}$P]-$\alpha$-GTP (Amersham).
2. 5% non-denaturing polyacrylamide gels.
3. 8% denaturing gels that contain 7 $M$ urea.
4. Restriction enzymes (New England Biolabs).

### 2.6. Cells and Plasmids

The Vero (African green monkey kidney) cells, PA317 cells, and ψCRE cells are maintained and propagated in Dulbecco's modified Eagle's medium (DMEM) supplemented with 10% fetal bovine serum (FBS). The propagation of HSV-1 (F) in these cells is carried out as described previously *(31)*. The anti-rabbit polyclonal antibody against HSV-1 (F) TK was a gift from Dr. Bill Summers *(33)*, and the anti-mouse monoclonal antibody (MAb) MCA406 against HSV-1 ICP35 protein *(34)* was purchased from Bioproduct for Sciences Inc. (Indianapolis, IN). The anti-mouse monoclonal antibodies c1101, c1113, and c1123, which react with the HSV-1 proteins ICP4, ICP27, and gB, respectively, were purchased from the Goodwin Institute for Cancer Research (Plantation, FL).

## 2.7. Major Equipment

1. Automated thermal cycle.
2. Molecular Dynamics phosphorimager (STORM840).
3. Oligonucleotide-synthesis facilities.
4. Tissue-culture facilities ($CO_2$ incubator, biosafety cabinet, microscope).

# 3. Methods

## 3.1. In Vivo Mapping of the Accessible Regions of ICP4 mRNA in HSV-1 Infected Cells

The RNA secondary structure and protein association can be mapped by two approaches to determining the accessible regions in RNA. First, RNA secondary structure and protein association can be mapped in living cells using dimethyl sulfate (DMS) *(31,35,36)*. Second, an in vitro-transcribed and end-labeled RNA is subjected to a partial digestion with RNase T1 to determine the sites that are accessible to enzymatic attack. The protocol described here is adopted from the procedures that have been used to map the targeted regions of the mRNAs that encode the TK of herpes simplex virus 1 (HSV-1) and protease of human cytomegalovirus (HCMV) *(31,37)*.

### 3.1.1. Dimethyl Sulfate Mapping of Secondary RNA Structure

DMS methylates N7 of guanine, N1 of adenine, and N3 of cytosine, of which the latter two can be detected by primer extension as stops in the transcript one base before the modified base. Therefore, determining the site of modification by primer extension will reveal the regions that are accessible to DMS, and presumably to M1GS binding.

1. Cells are grown in monolayer. In our experiments, Vero cells are used and grown in T25 flask and in DMEM supplemented with 5% FBS. Cells are infected with HSV-1 at a multiplicity of infection (MOI) of 5–10 for 4–8 h prior to the treatment with DMS.
2. Cells are washed with fresh media once, then incubated with 5 mL of fresh media that contain 1–2% of DMS for 5–10 min. We found that the DMS concentration and the length of incubation period are very critical to obtain reliable results (*see* **Note 1**).
3. After the incubation, the DMS media are immediately aspirated, and cells are washed three times with cold PBS that contains 1 m$M$ β-mercaptoethanol.
4. Cells are lysed by adding 0.5 mL of cold lysis buffer (*see* **Subheading 2.**) to the T25 flask.
5. Cell lysates are transferred to an Eppendorf tube and immediately spun in a microcentrifuge for 10 s at 4°C. The supernatant that contains the cellular lysate is transferred to another tube.
6. Total RNAs are isolated by phenol-chloroform extraction of the supernatant cellular lysate three times, followed by ethanol-precipitation.

7. Primer extension. Oligonucleotides of 17–25 nucleotides that are complementary to several regions of the targeted RNA are synthesized chemically in a 380B DNA synthesizer. These primers for the primer extension are 5'-labeled by T4 polynucleotide kinase (PNK) in the presence of $\gamma$-$[^{32}P]$-ATP.

8. 10 μg of total cellular RNA is mixed with 50,000 cpm of the primer in a volume of 8 μL, heated to 90°C for 2 min, and cooled down to allow them to be annealed to each other. Add the following sequentially:
    a. 4 μL 5X RT buffer
    b. 2 μL 10 m$M$ dATP
    c. 2 μL 10 m$M$ dTTP
    d. 2 μL 10 m$M$ dGTP
    e. 2 μL 10 m$M$ dCTP
    f. 0.5 μL RNasin
    g. 1 μL AMV RT
    The primer extension reactions are preceded for 2 h at 42°C.

9. The reaction products are extracted with phenol chloroform, than precipitated by cold ethanol, and finally are separated in 8% denaturing gels. Autoradiograph of the gels will reveal the sites that block the primer extension reaction by reverse transcription. These sites are the DMS modification site (*see* **Note 2**).

## 3.1.2. RNase T1 Mapping of Secondary RNA Structure

An end-labeled, in vitro-transcribed RNA is subjected to a partial digestion with RNase T1 to reveal regions that are vulnerable to enzymatic attack.

1. Synthesize a target RNA sequence by mixing the following:
    a. 4 μL (2 μg) plasmid DNA linearized
    b. 8 μL 5X transcription buffer
    c. 4 μL 100 m$M$ DTT
    d. 4 μL 10 m$M$ ATP
    e. 4 μL 10 m$M$ GTP
    f. 4 μL 10 m$M$ CTP
    g. 4 μL 10 m$M$ UTP
    h. 4 μL H$_2$O
    i. 1 μL RNasin (5 U/μL)
    j. 2 μL T7 RNA polymerase
    Incubate the reaction mixture at 37°C for 4 h or overnight.

2. Add 1 μL of DNase I (1 μg/μL) and incubate for 30 min at 37°C.

3. Extract the reaction mixture with an equal volume of phenol/chloroform/isoamyl alcohol (IAA) (25:24:1), and precipitate the RNA with ethanol.

4. Label the in vitro-transcribed RNA at the 3' end with 5'-$[^{32}P]$-pCp and RNA ligase or the 5' end with $\gamma$-$[^{32}P]$ATP and polynucleotide kinase. Purify the labeled RNA on a polyacrylamide-7M urea gel.

5. Dilute RNase T1 to $10^{-1}$, $10^{-2}$, $10^{-3}$ U/μL in water.

6. In nondenaturing conditions,1 μL of diluted RNase T1 is mixed with 9 μL of the end-labeled RNA (2000 cpm) in buffer A2 or buffer C2 (*see* **Subheading 2.**) that

has been supplemented with 5–10 μg bulk tRNA from *E. coli* and incubate the mixture at 37°C for 10 min. In denaturing condition, the RNase T1 digestion is carried out in buffer E (*see* **Subheading 2.**) at 37°C for 10 min.

7. Load the T1 digests on an 8% polyacrylamide-7 *M* urea gel and subject to autoradiography.

### 3.2. Construction of M1GS Ribozymes That Target HSV-1 ICP4 mRNA

The M1GS RNAs that target the ICP4 mRNA can be constructed by a method that is similar to the one described previously *(32)*. Briefly, in order to achieve effective cleavage and to target all strains of HSV-1, M1GSs should target a region that is accessible to ribozyme binding and is also highly conserved among all known HSV-1 strains and herpesviruses (e.g. HSV-2, HCMV, Epstein Barr virus [EBV]). A region of ICP4 mRNA (nucleotide positions 5–17, downstream from the ICP4 translation initiation site) was found to be most accessible to DMS methylation, and was used as the target sequence. Moreover, its flanking sequence exhibits several sequence features that must be present in order to interact with an M1GS ribozyme to achieve efficient cleavage. These features include the requirement for a guanosine and a pyrimidine to be the nucleotide 3' and 5' adjacent to the site of cleavage, respectively *(38)*. The interactions of these sequence elements with the M1GS ribozyme are critical for recognition and cleavage by the enzyme (*see* **Note 3**).

Plasmid pFL117 and pC102 contain the DNA sequence coding for M1 RNA and mutant C102 driven by the T7 RNA polymerase promoter, respectively *(31,39)*. Mutant ribozyme C102 contained several point mutations (e.g., $A_{347}C_{348} \rightarrow C_{347}U_{348}$, $C_{353}C_{354}C_{355}G_{356} \rightarrow G_{353}G_{354}A_{355}U_{356}$) at the catalytic domain (P4 helix) *(39)*. The DNA sequences that encode ribozymes M1-ICP4 and C-ICP4 were constructed by PCR using the DNA sequence of the ribozymes as the template and oligonucleotides AF25 (5'-GGAATTC<u>TAATACGACTCACTATAG</u>-3') and M1ICP4-2 (5'-**TGGT**<u>GCATCGGCGATGG</u>ACAGCTATGACCATG-3') as 5' and 3' primers, respectively. The underlined sequence in AF25 represents the promoter sequence for T7 RNA polymerase. The underlined sequence and the bold sequence in M1ICP4 correspond to the 3'ACCA sequence and the guide sequence, respectively (**Note 4**). The spacer connecting the 3' terminus of M1 RNA and the 5' end of the guide sequence is a 51 nucleotides-long sequence (5'-GAAGCTTGACCTGCAGGCATGCAAGCTTGGCGTAATCATGGTC ATAGCTGT-3') (*see* **Note 5**). The DNA sequence that encodes substrate icp32 was constructed by PCR using pGEM3zf(+) as template and oligonucleotide AF25 and sICP4 (5'-CGGGA**TCCGACGCCATCGCCGATGCGGG GCGATCC**<u>TATAGTGAG</u>-3') as 5' and 3' primers, respectively. The underlined sequence and the bold sequence in sICP4 correspond to the sequence that is complementary to a part of the promoter sequence for the RNA polymerase and the target ICP4 mRNA sequence, respectively. The construct pICP4 that codes for ICP4 DNA sequence is a gift from Dr. John Blaho (Mount Sinai School of Medicine) *(32)*.

### 3.3. In Vitro Cleavage and Binding Assay of M1GS Ribozyme

M1GS RNAs and the ICP4 mRNA substrates (e.g., icp32) can be synthesized in vitro by T7 RNA polymerase following the manufacturer's recommendations, and further purified on 8% urea/polyacrylamide gels (*see* **Note 6**). Subsequently, the M1GS RNAs (10 n$M$) can be mixed with the [$^{32}$P]-labeled mRNA substrate (10 n$M$) (*see* **Note 7**). The cleavage reactions are carried out at 37–50°C in a volume of 10 μL for 30 min in 1X buffer A (*see* **Subheading 2.**) *(40)*. Cleavage products are separated in denaturing gels and quantitated with a STORM840 phosphorImager.

The procedures to measure the equilibrium dissociation constants ($K_d$) of the M1GS-icp32 complexes can be modified from Pyle et al. *(41)*. In brief, various concentrations of ribozyme (0.05–50 n$M$) are preincubated in 1X buffer B (*see* **Subheading 2.**) for 10 min before mixing with an equal volume of 0.1–0.5 n$M$ of RNA substrate preheated under identical conditions. The samples are incubated for 15 min to allow binding, then loaded on a 5% polyacrylamide gel, and run at 10 W. The electrophoresis running buffer contains 100 m$M$ Tris-HEPES, pH 7.5 and 10 m$M$ MgCl$_2$ *(41)*. The amount of the bound complex is quantitated with a STORM840 Phosphorimager. The value of $K_d$ can then be extrapolated from a graph plotting the percentage of product bound vs ribozyme concentration (*see* **Note 8**).

Using these procedures, we have shown that the control ribozyme C-ICP4 does not cleave icp32 *(32)*. However, C-ICP4 binds to icp32 as well as the functional ribozyme M1-ICP4, suggesting that the mutations abolish the catalytic activity of C-ICP4, but do not significantly affect its binding affinity to icp32. Thus, C-ICP4 can be used as a control for the antisense effect of the ribozyme *(32)*.

### 3.4. Generation of Cell Lines Expressing M1-ICP4 Ribozyme

The ribozymes can be cloned into retroviral expression vector pLXSN-U6 that is derived from the original vector LXSN developed by Miller and Rosman *(42)*. Trang and Liu describe the protocols for the construction of M1GS-expressing cells using retroviral vectors in Chapter 32 in this book. In brief, amphotropic PA317 cells are transfected with retroviral vector DNAs (e.g., LXSN-U6-M1-ICP4 and LXSN-U6-C-ICP4) with the aid of a mammalian transfection kit purchased from Gibco-BRL (Grand Island, NY). The cells expressing these ribozymes are then selected and cloned. The expression of the

M1GS RNAs in these cells can be detected by Northern analyses, following the protocols described in Chapter 32.

### 3.5. Functional Analysis of M1GS RNA in Cells

To determine whether the M1GS RNA destroys the target RNA in the cells that express M1GS, various assays—including RNase protection, Northern and Western blot—are carried out to determine the expression level of the target gene. The procedures for Northern blot, Western analysis, and evaluating viral titers and growth *(43)* are described in Chapter 32. Our results indicate that a reduction of 80% in the ICP4 expression and a reduction of 1000 folds in HSV-1 growth were observed in cells that expressed M1-ICP4 *(32)*.

## 4. Notes

1. The authors found that the DMS concentration and the length of the incubation period are very critical in order to obtain reliable results. The authors usually carry out the experiments with several concentrations of DMS and several different periods of incubation in order to obtain specific and reproducible modification patterns.
2. To determine the modification sites accurately, it is useful to run a sequencing reaction on the same gel, which reveals the sequence of the targeted region. This sequencing reaction can be done by using a plasmid as the sequencing template that contains the targeted region and the same primer for the primer extension as the sequencing primer.
3. Although M1GS exhibits optimal activity in cleaving a mRNA sequence that contains a guanosine and a pyrimidine as the nucleotides 3' and 5' adjacent to the cleavage site, respectively, the ribozyme appears to be able to cleave any mRNA sequence that basepairs with the guide sequence *(38)*.
4. The 3' terminal sequence of M1GS ribozyme is important for its cleavage activity. Ribozymes with the exact 3' terminal CCA sequence are at least 500-fold more active (as indicated by the value of $k_{cat}/K_m$) than those with the deletion of the CCA sequence. Therefore, it is critical to include the 3' CCA sequence at the 3' terminal sequence of the guide sequence of the ribozyme *(38)*.
5. The length of the spacer connecting the 3' terminus of M1 RNA and the 5' end of the GS appears to be important for the activity of M1GS, presumably because the spacer affects the proper docking of the mRNA-guide sequence complex into the active site of the ribozyme. Our unpublished results suggest that M1GS ribozymes with a spacer of 28–65 nucleotides achieve efficient cleavage activity in vitro.
6. The ribozymes that are in vitro synthesized by T7 RNA polymerase should be purified using denaturing gels containing 8 $M$ urea. The gel slices that contain the ribozyme fractions are crushed in an Eppendorf tube with a minihomogenizer bar and then soaked in RNase-free water for 20 min in ice. Precautions should be taken to avoid degradation of RNA as a result of RNase contamination. Since most RNases do not require divalent cations, it is not recommended to soak the

gel crush with TE buffer. The ribozyme fractions in the soaked tube are separated from the crushed gel slices by micro-centrifugation. The ribozyme-containing supernatants are extracted with phenol-chloroform solutions twice, and then the ribozymes are precipitated in the presence of ethanol.
7. The ribozymes are resuspended in the 1X folding buffer C (50 m$M$ Tris, pH 7.5, 100 m$M$ NH$_4$Cl, and 10 m$M$ MgCl$_2$), incubated at 65°C for 5 min, and then allowed to fold by gradually lowering the temperature. This folding treatment will allow most of the ribozymes to fold into their active conformations, and in our experience, has significantly increase the activity of the ribozymes.
8. The values for $K_d$ are usually the average of three experiments.

## Acknowledgments

Gratitude goes to Ahmed Kilani and Joe Kim for sharing their research experiences on RNase P ribozymes. P. T. is a recipient of the American Heart Association Predoctoral Fellowship (Western States Affiliate). K. C. is supported by a NIH predoctoral training grant (AI07620). F. L. is a Pew scholar in Biomedical Sciences, a Scholar of Lymphoma and Leukemia Society of America, and an Established Investigator of American Heart Association. F. L. also acknowledges support from a Hellman Family Faculty Fellowship and a Regent's Junior Faculty Fellowship (University of California). The research has been supported by grants from NIH (AI41927 and GM54815).

## References

1. Altman, S. and Kirsebom, L. A. (1999) Ribonuclease P, in *The RNA World* (Gesteland, R. F., Cech, T. R., and Atkins, J. F., eds.), 2nd ed., Cold Spring Harbor Laboratory Press, Cold Spring Harbor, NY, pp. 351–380.
2. Cech, T. R. and Golden, B. L. (1999) Building a catalytic active site using only RNA, in *The RNA World* (Gesteland, R. F., Cech, T. R., and Atkins, J. F., eds.), 2nd ed., Cold Spring Harbor Laboratory Press, Cold Spring Harbor, NY, pp. 321–350.
3. McKay, D. B. and Wedekind, J. E. (1999) Small ribozymes, in *The RNA World* (Gesteland, R. F., Cech, T. R., and Atkins, J. F., eds.), 2nd ed., Cold Spring Harbor Laboratory Press, Cold Spring Harbor, NY, pp. 265–286.
4. Kruger, M., Beger, C., and Wong-Staal, F. (1999) Use of ribozymes to inhibit gene expression. *Methods Enzymol.* **306,** 207–225.
5. Guo, H., Karberg, M., Long, M., Jones, J. P., 3rd, Sullenger, B., and Lambowitz, A. M. (2000) Group II introns designed to insert into therapeutically relevant DNA target sites in human cells. *Science* **289(5478),** 452–457.
6. Sullenger, B. A. (1999) RNA repair as a novel approach to genetic therapy. *Gene Ther.* **6(4),** 461,462.
7. Rossi, J. J. (1999) Ribozymes, genomics and therapeutics. *Chem. Biol.* **6(2),** R33–37.
8. Zamecnik, P. C. and Stephenson, M. L. (1978) Inhibition of Rous sarcoma virus replication and cell transformation by a specific oligodeoxynucleotide. *Proc. Natl. Acad. Sci. USA* **75,** 280–284.

9. Stein, C. A. and Cheng, Y. C. (1993) Antisense oligonucleotides as therapeutic agents—is the bullet really magical? *Science* **261,** 1004–1012.

10. Yu, M., Ojwang, J., Yamada, O., Hampel, A., Rapapport, J., Looney, D., and Wong-Staal, F. (1993) A hairpin ribozyme inhibits expression of diverse strains of human immunodeficiency virus type 1. *Proc. Natl. Acad. Sci. USA* **90,** 6340–6344.

11. Sarver, N., Cantin, E. M., Chang, P. S., Zaia, J. A., Ladne, P. A., Stephens, D. A., and Rossi, J. J. (1990). Ribozymes as potential anti-HIV-1 therapeutic agents. *Science* **247,** 1222–1225.

12. Poeschla, E. and Wong-Staal, F. (1994). Antiviral and anticancer ribozymes. *Curr. Opin. Oncol.* **6,** 601–606.

13. Castanotto, D., Rossi, J. J., and Sarver, N. (1994). Antisense catalytic RNAs as therapeutic agents. *Adv. Pharmacol.* **25,** 289–317.

14. Frank, D. N. and Pace, N. R. (1998). Ribonuclease P: unity and diversity in a tRNA processing ribozyme. *Ann. Rev. Biochem.* **67,** 153–180.

15. Guerrier-Takada, C., Gardiner, K., Marsh, T., Pace, N., and Altman, S. (1983) The RNA moiety of ribonuclease P is the catalytic subunit of the enzyme. *Cell* **35,** 849–857.

16. Gopalan, V., Talbot, S. J. and Altman, S. (1995) RNA-protein interactions in RNase P, *RNA-Protein Interactions in RNase P*. RNA-protein interactions. (Nagai, K. and Mattaj, I. W., eds.), Oxford University Press, Oxford, pp. 103-126.

17. Niranjanakumari, S., Stams, T., Crary, S. M., Christianson, D. W., and Fierke, C. A. (1998) Protein component of the ribozyme ribonuclease P alters substrate recognition by directly contacting precursor tRNA. *Proc. Natl. Acad. Sci. USA* **95,** 15,212–15,217.

18. Reich, C., Olsen, G. J., Pace, B., and Pace, N. R. (1988) Role of the protein moiety of RNase P, a ribonucleoprotein enzyme. *Science* **239,** 178–181.

19. Haas, E. S., Brown, J. W., Pitulle, C., and Pace, N. R. (1994) Further perspective on the catalytic core and secondary structure of ribonuclease P RNA. *Proc. Natl. Acad. Sci. USA* **91,** 2527–2531.

20. Haas, E. S. and Brown, J. W. (1998) Evolutionary variation in bacterial RNase P RNAs. *Nucleic Acids Res.* **26(18),** 4093-9.

21. Haas, E. S., Armbruster, D. W., Vucson, B. M., Daniels, C. J., and Brown, J. W. (1996) Comparative analysis of ribonuclease P RNA structure in Archaea. *Nucleic Acids Res.* **24,** 1252-9.

22. Chen, J. L., Nolan, J. M., Harris, M. E., and Pace, N. R. (1998) Comparative photocross-linking analysis of the tertiary structures of Escherichia coli and Bacillus subtilis RNase P RNAs. *EMBO J.* **17,** 1515-25.

23. Massire, C., Jaeger, L., and Westhof, E. (1998) Derivation of the three-dimensional architecture of bacterial ribonuclease P RNAs from comparative sequence analysis. *J. Mol. Biol.* **279,** 773–793.

24. Bothwell, A. L., Garber, R. L., and Altman, S. (1976) Nucleotide sequence and in vitro processing of a precursor molecule to Escherichia coli 4.5 S RNA. *J. Biol. Chem.* **251,** 7709–7716.

25. Alifano, P., Rivellini, F., Piscitelli, C., Arraiano, C. M., Bruni, C. B., and Carlomagno, M. S. (1994) Ribonuclease E provides substrates for ribonuclease P-dependent processing of a polycistronic mRNA. *Genes Dev.* **8,** 3021–3031.
26. Komine, Y., Kitabatake, M., Yokogawa, T., Nishikawa, K., and Inokuchi, H. (1994) A tRNA-like structure is present in 10Sa RNA, a small stable RNA from *Escherichia coli. Proc. Natl. Acad. Sci. USA* **91,** 9223–9227.
27. McClain, W. H., Guerrier-Takada, C., and Altman, S. (1987) Model substrates for an RNA enzyme. *Science* **238,** 527–530.
28. Forster, A. C. and Altman, S. (1990) External guide sequences for an RNA enzyme. *Science* **249,** 783–786.
29. Yuan, Y., Hwang, E., and Altman, S. (1992) Targeted cleavage of mRNA by human RNase P. *Proc. Natl. Acad. Sci. USA* **89,** 8006–8010.
30. Frank, D., Harris, M., and Pace, N. R. (1994) Rational design of self-cleaving pre-tRNA-Ribonuclease P RNA conjugates. *Biochemistry* **33,** 10,800–10,808.
31. Liu, F. and Altman, S. (1995) Inhibition of viral gene expression by the catalytic RNA subunit of RNase P from *Escherichia coli. Genes Dev.* **9,** 471–480.
32. Trang, P., Kilani, A. F., Kim, J., and Liu, F. (2000) A ribozyme derived from the catalytic subunit of RNase P from *Escherichia coli* is highly effective in inhibiting replication of herpes simplex virus 1. *J. Mol. Biol.* **301,** 817–826.
33. Liu, Q.-Y. and Summers, W. C. (1988) Site-directed mutagenesis of a nucleotide-binding domain in HSV-1 thymidine kinase: effects on catalytic activity. *Virology* **163,** 638–642.
34. Liu, F. and Roizman, B. (1993) Characterization of the protease and other products of amino-terminus-proximal cleavage of the herpes simplex virus 1 UL26 protein. *J. Virol.* **67,** 1300–139.
35. Ares, M. and Igel, A. H. (1990) Lethal and temperature-sensitive mutations and their suppressors identify an essential structural element in U2 small nuclear RNA. *Genes Dev.* **4,** 2132–2145.
36. Zaug, A. J. and Cech, T. R. (1995) Analysis of the structure of Tetrahymena nuclear RNAs in vivo: telomerase RNA, the self-splicing rRNA intron, and U2 snRNA. *RNA* **1,** 363–374.
37. Trang, P., Lee, M., Nepomuceno, E., Kim, J., Zhu, H., and Liu, F. (2000) Effective inhibition of human cytomegalovirus gene expression and replication by a RNase P ribozyme. *Proc. Natl. Acad. Sci. USA* **97,** 5812–5817.
38. Liu, F. and Altman, S. (1996) Requirements for cleavage by a modified RNase P of a small model substrate. *Nucleic Acids Res.* **24,** 2690–2696.
39. Kim, J. J., Kilani, A. F., Zhan, X., Altman, S., and Liu, F. (1997) The protein cofactor allows the sequence of an RNase P ribozyme to diversify by maintaining the catalytically active structure of the enzyme. *RNA* **3,** 613–623.
40. Kilani, A. F., Trang, P., Jo, S., Hsu, A., Kim, J., Nepomuceno, E., Liou, K., and Liu, F. (2000) RNase P ribozymes selected in vitro to cleave a viral mRNA effectively inhibit its expression in cell culture. *J. Biol. Chem.* **275,** 10,611–10,622.

41. Pyle, A. M., McSwiggen, J. A., and Cech, T. R. (1990) Direct measurement of oligonucleotide substrate binding to wild-type and mutant ribozymes from Tetrahymena. *Proc. Natl. Acad. Sci. USA* **87,** 8187–8191.
42. Miller, A. D. and Rosman, G. J. (1989) Improved retroviral vectors for gene transfer and expression. *BioTechniques* **7,** 980–990.
43. Liu, F. and Roizman, B. (1991) The herpes simplex virus 1 gene encoding a protease also contains within its coding domain the gene encoding the more abundant substrate. *J. Virol.* **65,** 5149–5156.

# 29

## In Vitro Selection of RNase P Ribozymes That Efficiently Cleave a Target mRNA

### Kihoon Kim and Fenyong Liu

#### Summary

An in vitro selection procedure for identifying highly efficient RNase P ribozyme (M1GS RNA) variants is presented as a model system for engineering ribozymes to improve their catalytic efficiency. Detailed protocols as well as the rationale for setting up such a system are included, using the mRNA sequence that encodes the thymidine kinase (TK) of herpes simplex virus 1 (HSV 1) as a target substrate of choice. Using the selection system, we have successfully generated M1GS RNA variants that more efficiently cleave the TK mRNA in vitro and more effectively inhibit the TK expression in cultured cells than the ribozyme derived from the wild-type RNase P ribozyme sequence. The in vitro selection system represents a novel and effective approach for engineering highly active RNase P ribozymes that can be used in both basic research and clinical therapeutic settings.

**Key Words:** RNase P; ribozyme; in vitro selection; in vitro evolution; herpes simplex virus 1 (HSV 1); thymidine kinase (TK); gene therapy; M1 RNA.

### 1. Introduction

RNase P is a ribonucleoprotein complex that is responsible for processing transfer RNA (tRNA) by catalyzing a hydrolysis reaction to remove a 5' leader sequence from the precursor tRNA (ptRNA) to form a mature tRNA molecule *(1,2)*. Its presence is well-documented in all types of cells—bacterial, archaebacterial, and eukaryotic. In *Escherichia coli*, RNase P consists of a catalytic RNA subunit (M1 RNA) and a protein subunit (C5 protein) *(3)*. Extensive studies of the interaction between M1 RNA and its natural substrate have revealed that the structures of the substrates themselves, rather than specific sequences or nucleotides, dictate recognition and cleavage by the enzyme *(4,5)*. These findings suggest that for any given sequence that meets the structural

From: *Methods in Molecular Biology, vol. 252: Ribozymes and siRNA Protocols, Second Edition*
Edited by: M. Sioud © Humana Press Inc., Totowa, NJ

requirements, M1 RNA can be used to target against the sequence for a specific and irreversible cleavage. For example, M1 RNA can be covalently linked to a guide sequence (GS) that is complementary to a target mRNA sequence *(4,6)*. The resulting ribozyme, M1GS RNA, can bind the mRNA substrate through Watson-Crick basepairing interaction between the guide sequence and the mRNA *(7,8)*. Thus, the intrinsic catalytic activity of M1GS, in principle, can be used to target virtually any known mRNA sequence as long as the structural requirements are met.

### 1.1. M1 RNA in Therapeutic Applications

The M1GS strategy of gene silencing instantly opened up a vast number of opportunities in the both basic research and therapeutic applications. For example, downregulation of the expression of a certain gene to investigate its role in cellular processes can be readily achieved using the M1GS-based as well as other ribozyme-based technologies, bypassing the shortcomings of the conventional antisense oligodeoxynucleotides or antisense RNA, such as cytotoxicity and nonspecific inactivation of gene functions *(9)*. In therapeutic applications of M1GS technology, targeted cleavage of viral mRNA and irreversible destruction of cellular oncogenes are being actively explored to evaluate the potentials of the technology in clinical settings *(10–14)*. We previously reported the use of the M1GS ribozyme to target the mRNA sequence encoding the thymidine kinase (TK) of human herpes simplex virus 1 (HSV-1). We showed that M1GS ribozyme effectively cleaved the TK mRNA in vitro *(13)*. In addition, a reduction of 80% in the expression of both viral TK mRNA and protein was observed when the ribozyme was expressed in mammalian cells infected with HSV-1 *(13)*. More recently, we have reported that M1GS ribozyme could effectively target the mRNA that codes for IE1/IE2 of human cytomegalovirus (HCMV), whose gene products are major viral regulatory proteins required for HCMV gene expression and growth *(14)*. This chapter focuses on generating highly active M1GS ribozyme variants using in vitro selection. Related topics regarding general design of RNase P ribozymes are discussed in other chapters in this volume.

### 1.2. In Vitro Selection of M1GS Ribozymes

To understand the enzyme kinetics and related parameters of the M1GS ribozyme, further studies are needed to determine the rate-limiting steps of the cleavage reaction. Many studies have been carried out to address the topic of in vitro cleavage by M1GS. However, little is currently known about the rate-limiting steps of M1GS targeting in cultured cells. Previous studies on hammerhead and hairpin ribozymes have suggested that binding of the ribozyme to its target RNA appears to be the rate-limiting step in vivo *(15–17)*. In addition, studies on these ribozymes to colocalize with the substrates within a particular cellular compartment and on designing ribozymes to target the regions in mRNA that are readily accessible to binding have led to significant improvements of the efficacies of hammerhead and hairpin ribozymes in cellular environments *(15–20)*.

In our studies, M1GS RNA was targeted to a region of TK mRNA that is accessible to modification by dimethyl sulfate (DMSO) in cell culture and accessible to ribozyme binding *(4,21)*. Moreover, the ribozyme was expressed primarily in the nuclei by using the promoter of the small nuclear U6 RNA *(4,19,21,22)*. This design would increase the probability for the constructed RNase P ribozyme to locate and bind to its target mRNA sequence. Under such conditions, it is possible that the efficiency of the RNase P ribozyme cleavage in cultured cells is dictated by the overall cleavage rate ($k_{cat}/K_m$) of the ribozyme. Indeed, our recent results are consistent with this notion, and indicate that increasing the catalytic efficiency of the ribozyme leads to more effective inhibition of the target mRNA expression in cultured cells *(23)*.

In order to improve the catalytic efficiency of the M1GS ribozyme, an in vitro selection procedure can be used to select for more catalytically active variants of ribozymes from a pool of M1 RNA that contains randomized mutations. Using this approach, we have previously shown that ribozymes selected in vitro are at least 10× more active in cleaving TK mRNA sequence than the wild-type M1GS ribozyme *(23)*. Furthermore, the selected ribozymes exhibited up to 30× higher efficiency in cleaving the HSV-1 ICP4 mRNA and the HCMV IE1/IE2 mRNAs than the M1GS ribozyme derived from the wild-type M1 sequence *(12,23)*. Taken together, these findings suggest that the active domains of the selected ribozymes can be used to construct effective ribozymes to target any specific mRNA.

### *1.3. Strategies for Engineering M1GS Ribozymes*

In vitro selection has been widely used to generate either new RNA catalysts or more effective variants from known ribozyme molecules. Here, we report the use of in vitro selection/evolution to engineer M1GS ribozymes targeting HSV-1 TK mRNA. Mutations were introduced into three discrete regions of M1 RNA to generate a pool of M1 RNA variant molecules. These regions include the nucleotide positions that have been found to be conserved among all known RNase P catalytic RNAs and play an important role in catalysis and substrate binding (**Fig. 1**; *24–26*). The pool of variant ribozymes was then constructed with a 10% mutation rate from the wild type M1 RNA sequence (e.g., 90% G, 3.3% A, 3.3% C, and 3.3% U at a G position of M1 RNA) in these regions.

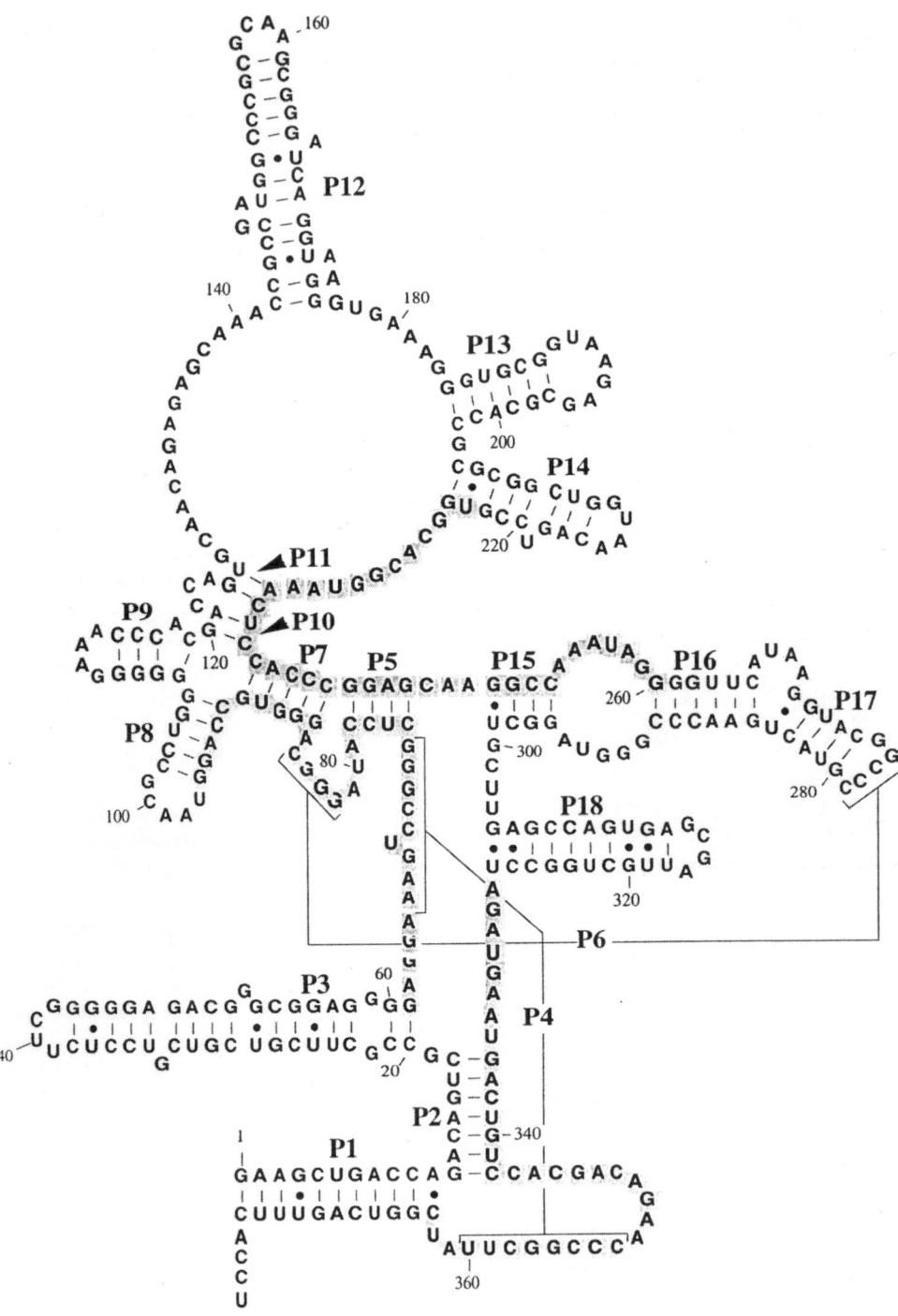

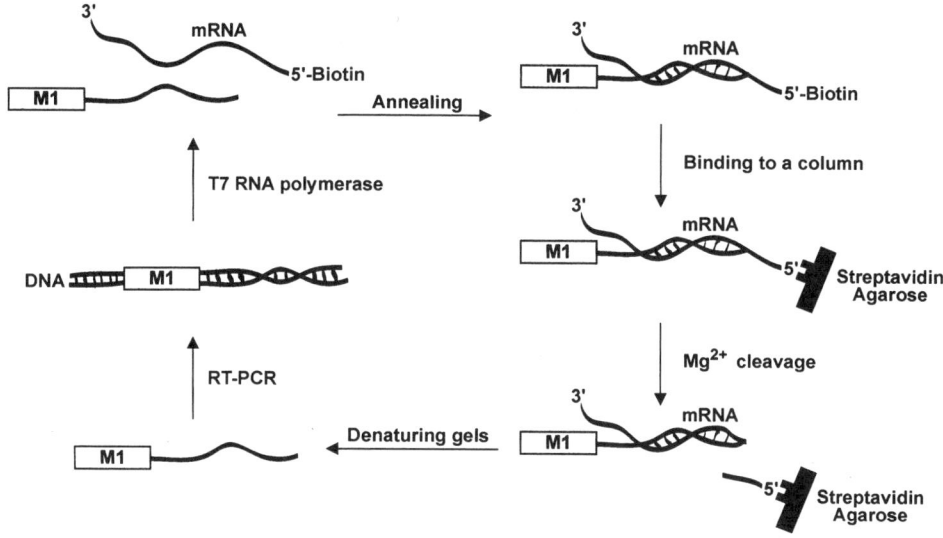

Fig. 2. The evolution in vitro procedure to select M1GS variants to cleave the TK mRNA sequence.

The mutant ribozyme pool was subjected to an in vitro selection procedure to generate highly active variants. In this procedure (**Fig. 2**), the ribozyme molecules were annealed to 5'-biotinylated RNA substrate tk46 in annealing buffer (50 m$M$ Tris-HCl, pH 7.5, 100 m$M$ NH$_4$Cl). This substrate corresponds to the portion of the mRNA sequence that is accessible to M1 RNA binding in mammalian-cell culture. The annealed complexes of the ribozyme and 5'-biotinylated tk46 substrate were allowed to bind to a streptavidin column in the absence of divalent ions such as Mg$^{2+}$, which is essential for RNase P ribozyme catalysis *(3)*, and all unbound ribozymes were washed away (**Fig. 2**). Cleavage buffer (buffer A: 50 m$M$ Tris-HCl, pH 7.5, 100 m$M$ NH$_4$Cl, 100 m$M$ MgCl$_2$) was added to the column to allow cleavage reaction to occur. The ribozymes that cleaved their substrates were released from the column and loaded on a denaturing gel, and these active ribozymes were then recovered from the gel. The cDNA copies of these RNA molecules were synthesized and amplified by reverse transcriptase-polymerase chain reaction

Fig. 1. *(opposite page)* The proposed secondary structure of the RNA subunit (M1 RNA) of RNase P from *E. coli*. The shaded sequences represent the three regions (positions 61–96, 223–260, and 328–363), which were randomized at the 10% level in our studies *(23)*. Most of the sequences of these regions are highly conserved in all known RNase P RNA subunits from eubacteria and archaea *(25,27)*.

(RT-PCR) to create DNA templates for synthesis of ribozyme molecules for the next round of selection. The sequences isolated after rounds of selection can be cloned and sequenced. Representative sequences from each round of selection can be analyzed for their catalytic efficiency to evaluate the progress of the selection process. Using this selection procedure, we have successfully generated RNase P ribozyme variants that efficiently cleave a mRNA sequence in vitro and effectively inhibit the expression of the target mRNA in cultured cells *(23)*.

## 2. Materials

### 2.1. Chemicals and Stock Solutions

1. 1 *M* Tris-HCl, pH 8.0.
2. 1 *M* Tris-HCl, pH, 7.5.
3. 1 *M* Tris-acetate, pH 7.8.
4. 0.5 *M* ethylenediaminetetraacetic acid (EDTA).
5. 5 *M* NaCl.
6. 1 *M* $MgCl_2$.
7. 40% Acrylamide/*N,N′* methylene *bis*-acrylamide (29:1).
8. 10 *M* urea.
9. 10X TBE: 0.89 *M* Tris-borate, 10 m*M* EDTA.
10. DEPC-treated $H_2O$: double-distilled water is mixed with 0.1% diethylpyrocarbonate (DEPC) and stirred overnight. The DEPC is inactivated by autoclaving for 20 min.
11. 10 mg/mL ethidium bromide.
12. Phenol saturated with 10 m*M* Tris-HCl, pH 7.5, 1 m*M* EDTA.
13. Chloroform/isoamyl alcohol (IAA) (24:1).
14. 1 *M* dithiothreitol (DTT).
15. [$^{32}$P]-labeled nucleotides (Amersham, Arlington Heights, IL).

### 2.2. Solutions and Buffers

1. 10X cleavage buffer A: 500 m*M* Tris-HCl, pH 7.5, 1 *M* $NH_4Cl$, 1 *M* $MgCl_2$.
2. 10X buffer B: 166 m*M* PIPES, 400 m*M* Tris-HCl, pH 6.0, 1 *M* NaCl, 1 *M* $CaCl_2$.
3. 10X annealing buffer: 500 m*M* Tris-HCl, pH 7.5, 1 *M* $NH_4Cl$.
4. 10X buffer C: 166 m*M* PIPES, 400 m*M* Tris-HCl, pH 6.0, 1 *M* NaCl, 1 *M* $MgCl_2$.
5. 2X RNA dye solution: 9 *M* urea, 20 m*M* EDTA, 0.25 mg/mL bromophenol blue, 0.25 mg/mL xylene cyanol.
6. 1X washing buffer: 25 m*M* Tris-HCl, pH 7.5, 50 m*M* $NH_4Cl$.
7. Buffer A: 50 m*M* Tris-HCl, pH 7.5, 100 m*M* $NH_4Cl$, 100 m*M* $MgCl_2$.

### 2.3. Enzymes and Reagents

1. Restriction endonucleases and 10X reaction buffers (New England Biolabs).
2. Exonuclease III (US Biochemical).

3. T4 DNA ligase and 10X ligation buffer: 500 m*M* Tris-HCl, pH 7.8, 100 m*M* MgCl$_2$, 100 m*M* DTT, 10 m*M* adenosine 5' triphosphate (ATP), 250 mg/mL bovine serum albumin (BSA).
4. T4 polynucleotide kinase (PNK) and 10X kinase buffer: 700 m*M* Tris-HCl, pH 7.6, 100 m*M* MgCl$_2$, 50 m*M* DTT.
5. In vitro transcription system (Promega) including: T7 RNA polymerase, 5X transcription buffer (200 m*M* Tris-HCl, pH 7.5, 30 m*M* MgCl$_2$, 10 m*M* spermidine, 50 m*M* NaCl), 100 m*M* DTT, 10 m*M* each nucleotide 5' triphosphate (NTP) (ATP, guanosine 5' triphosphate [GTP], cytidine 5' triphosphate [CTP], uridine 5' triphosphate [UTP]), and RNasin ribonuclease inhibitor.
6. AMV reverse transcriptase with 5X RT buffer: 250 m*M* Tris-HCl, pH 8.5, 40 m*M* MgCl$_2$, 150 m*M* KCl, 5 m*M* DTT.
7. PCR system including 10X PCR buffer (200 m*M* Tris-HCl, pH 8.4, 500 m*M* KCl, 1 mg/mL BSA), 25 m*M* MgCl$_2$, four 10 m*M* dNTPs, *Taq* DNA polymerase (Perkin-Elmer Cetus).
8. Dideoxynucleotide sequencing kit (Amersham Life Sciences, Cleveland, OH).

## *2.4. Miscellaneous Materials*

1. AffiniTip streptavidin columns (Genosys Biotechnologies).
2. G-50 Sephadex gel-filtration columns (Roche Molecular Biochemicals).

## 3. Methods
## *3.1. Construction of Ribozymes and Substrates*

1. The biotinylated RNA substrate tk46 (5'-biotin-GACCCCUGCCAU CAACACGC**G21TCTGCGTTCGAC33**CAGGCTGCGCGGU-3') is synthesized using a DNA oligonucleotide synthesizer. The portion of the substrate in boldface (positions 21–33) basepairs with the guide sequence of M1GS RNAs. The sequences upstream and downstream from the targeting sequence are the 5' leader sequence of 20 nucleotides and 3' tail sequence of 13 nucleotides, respectively. The substrate tk46 can be further purified by treatment of the oligonucleotide with tetrabutylammonium fluoride followed by purification in 15% polyacrylamide gels that contain 8 *M* urea.
2. Plasmids pFL120 and pTyr contain the DNA sequences that code for the RNA substrate tk46, and ptRNA$^{Tyr}$, respectively.
3. Uniformly radiolabeled RNA molecules can be synthesized in vitro by T7 RNA polymerase in the presence of $\alpha$-[$^{32}$P]-GTP using the following reaction mixture:
   a. 4 μL 5X transcription buffer
   b. 2 μL 100 m*M* DTT
   c. 6 μL H$_2$O
   d. 1 μL 10 m*M* ATP
   e. 1 μL 10 m*M* CTP
   f. 1 μL 10 m*M* UTP
   g. 1 μL 1 m*M* GTP

h.  2 µL template DNA
i.  50 µCi α-[$^{32}$P]-GTP
j.  0.5 µL RNasin ribonuclease inhibitor
k.  2 µL T7 RNA polymerase

Incubate the reaction mixture at 37°C for 4–16 h. Next, add 1 µL of DNase I (1 µg/µL) and incubate for 30 min. Extract the reaction mixture with an equal volume of phenol:chloroform:IAA (25:24:1) and precipitate the RNA with ethanol. Purify the labeled RNA on a denaturing polyacrylamide gel with 8 *M* urea.

4.  The DNA sequences coding for ribozymes M1-TK, R43, 107, 29, R29, and R6 were obtained from the wild-type M1 sequence *(4,5)* or the selected variants *(23)*. In vitro synthesis of these M1GS ribozymes can be carried out following the protocol similar to **Subheading 3.1., step 3**, except that no α-[$^{32}$P]-GTP is used, and 10 m*M* GTP is used instead.

## *3.2. Construction of M1 RNA Variant Sequences*

1.  Plasmid pFL117 is linearized with restriction enzyme *Afl*III and further digested with exonuclease III from *E. coli.*
2.  The linearized *Exo*III-digested plasmid is annealed to three oligonucleotides FL111, FL112, FL113. FL111 (5'-TTCCCCCCAGGCGTTAcctggcaccctgcc ctatggagcccggacttcctcCCCTCCGCCCGTC-3'), FL112 (5'-GCCGTACCTT ATGAACCcctatttggccttgctccgggtggagtttaccgTGTTACGGACTGTTAC-3'), and FL113 (5'-AAGCTTCAGGTGAAACTGACCgacaagccgggttctgtcgtgacagtc attcatctAGGCCAGCAATCGCT-3') contain 10% mutations in the regions corresponding to positions 61–96, 223–270, and 328–363 of M1 RNA, respectively (**Fig. 1**). Mix 5 µg of FL111, FL112, and FL113 each in 50 µL of annealing buffer (50 m*M* Tris, pH 7.5, 100 m*M* NH$_4$Cl) that contains 20 µg *Exo*III-digested pFL117 DNA. Heat the mixture at 95°C for 2 min and allow it to cool down slowly (>30 min) to room temperature.
3.  T4 DNA polymerase and ligase are added to the mixture to extend and ligate the three oligonucleotides in order to generate the full-length DNA sequences that code for M1 RNA.
4.  The full-length DNA template of M1 RNA is amplified by PCR with Oligo101 (5'-GTGGTGTCTGCGTTCGACTATGACCATG-3') and OligoTK31(5'-GTGGTGTCTGCGTTCGACTATGACCATG-3') as 5' and 3' primers, respectively. Oligo101 contains the 5' sequence of M1 RNA and the promoter sequence for the bacteriophage T7 RNA polymerase. OligoTK31 contains the guide sequence (GS) that is complementary to the TK mRNA sequence.

## *3.3. Evolution In Vitro*

1.  [$^{32}$P]-labeled ribozyme molecules are synthesized from the full-length DNA templates of M1 RNA that contain randomized sequences (**Subheading 3.1., step 3**) (*see* **Note 1**).
2.  Mix 100 pmol of ribozymes containing randomized mutations with the same amount of biotinylated tk46 in 1X annealing buffer (*see* **Note 2**). Heat the reac-

tion mixture at 95°C for 2 min and let it cool down slowly (>30 min) to room temperature. The divalent $Mg^{2+}$ ions are omitted from the annealing buffer to prevent cleavage reaction.

3. Run the annealed complexes of the ribozyme and the 5'-biotinylated tk46 substrate through the AffiniTip streptavidin columns to allow them to bind. Add the 1X washing buffer to wash away all unbound ribozymes (see Note 3).

4. Add 1X cleavage buffer and allow the cleavage reaction to proceed at 37°C for the following incubation times for each cycle of selection: 1st and 2nd cycle, 720 min, 3rd and 4th, 120 min, 5th and 6th, 30 min, 7th, 8th, and 9th, 5 min (*see* **Note 4**).

5. Load the effluent of the column onto 4% denaturing polyacrylamide gels containing 8 *M* urea. Expose the gels to a Phosphorscreen and scan the screen using a PhosphorImager (STORM840). The RNA bands are located and excised from the gels. Extract the RNA molecules from the excised gel slice by the crush-soak method using DEPC-treated water (*see* **Note 1**).

6. Synthesize the cDNA of the recovered RNA molecules by reverse transcription in the presence of OligoTK31. Add the following reagents sequentially:
   a. 5 μL 5X reverse transcription buffer
   b. 5 μL 1 mM dATP
   c. 5 μL 1 mM dGTP
   d. 5 μL 1 mM dCTP
   e. 5 μL 1 mM dTTP
   f. 1 μL RNasin ribonuclease inhibitor
   g. 1 μL (20 U) AMV reverse transcriptase
   Incubate the reaction mixture at 42°C for 2 h.

7. Amplify the cDNA template by PCR with Oligo101 and OligoTK31 as 5' and 3' primers, respectively. Use the following temperature cycle:
   a. Denature for 2 min at 94°C
   b. 30 cycles of:
   c. Denature for 2 min at 94°C
   d. Anneal for1 min at 47°C
   e. Extend for 1 min at 72°C
   f. Final extension 10 min at 72°C

8. The PCR products from the previous step are used again for the next round of selection. This procedure can be repeated many times until no further improvement of the cleavage rate of the ribozyme population is observed after a short period of incubation (e.g., 5 min) (*see* **Note 5**).

9. The cDNAs of the recovered ribozymes are cloned into pUC19 and sequenced.

## *3.4. Kinetic Analyses*

### *3.4.1. Assaying of the Overall Cleavage Rate $(k_{cat}/K_m)_{app}$ of the Selected Ribozymes*

In our selection procedure (**Fig. 2**), ribozyme-substrate complexes are bound to a streptavidin column, and the ribozymes are specifically selected for activity to cleave the biotinylated substrate under single-turnover conditions. Accordingly, kinetic analyses of the cleavage reactions catalyzed by the selected ribozymes are performed under single-turnover conditions in order to determine their catalytic activity. In our study, the cleavage reactions of the mRNA substrate (e.g., tk46) by the in vitro-selected ribozymes are carried out in buffer A, which is used for the selection (*see* **Subheading 2.2., item 7**) The kinetic analysis should be performed in single-turnover conditions to determine the values of $(k_{cat}/K_m)_{app}$. Using this assay, we showed that the selected ribozymes are up to 20× more active $[(k_{cat}/K_m)_{app}]$ in cleaving tk46 in vitro than the ribozyme derived from the wild type M1 RNA *(23)*.

1. A trace amount of radiolabeled substrate ($<0.1$ n$M$) and an excess amount of ribozymes (from 0.5–100 n$M$) are incubated in cleavage buffer A (*see* **Note 6**).
2. Aliquots are withdrawn from reaction mixtures at regular intervals (from 0–160 min), and the reactions are stopped by adding an equal volume of 2X RNA dye solution (9 $M$ urea, 20 m$M$ EDTA, 0.25 mg/mL bromphenol blue, 0.25 mg/mL xylene cyanol). The cleavage products are separated on 15% polyacrylamide denaturing gels, autoradiographed, and quantitated with a STORM 840 PhosphorImager (Molecular Dynamics, Sunnyvale, CA).
3. The pseudo-first-order rate constants ($k_{obs}$) of cleavage are calculated by observing the slope of the line for the plot of $\ln[(F_t - F_e)/(1 - F_e)]$ vs time using Kaleidagraph (Synergy Software, Reading, PA). $F_t$ and $F_e$ represent the fraction of the substrate quantified at time $t$ and the end point ($>10$ h) of the experiments, respectively.
4. The values of overall cleavage rate $(k_{cat}/K_m)_{app}$ are determined by calculating the slope of the least-squares linear-regression line for the values of $k_{obs}$ vs the concentrations of the ribozymes. The values obtained are the average of three experiments.

### 3.4.2. Assaying the Rate of Chemical Cleavage of the Selected Ribozymes

The values of $(k_{cat}/K_m)_{app}$ obtained under single-turnover conditions reflect the rates of substrate binding and chemical cleavage of the phosphodiester linkage *(23)*. Accordingly, experiments can be further carried out with these selected variants to determine whether a change in the rates of these steps contributes to the increased $(k_{cat}/K_m)_{app}$ values. The protocols for assaying substrate binding of RNase P ribozymes (e.g., equilibrium dissociation constant $K_d$) are described in the other chapters on RNase P ribozymes in this volume. This section focuses on the procedure for assaying the rate of chemical cleavage ($k_{app}$). To determine the values of $k_{app}$, the ribozyme-substrate complexes are first allowed to form active complexes in the absence of MgCl$_2$, then diluted in different concentrations, and finally incubated in the presence of 100 m$M$ MgCl$_2$ to allow cleavage. Our results showed that the rates of chemical cleavage for some of the selected ribozymes are up to fivefold higher than that of the ribozyme derived from the wild-type M1 RNA *(23)*.

1. Equimolar amounts of substrates and ribozymes are incubated in buffer B (16.6 m*M* PIPES, 40 m*M* Tris-HCl, pH 6.0, 100 m*M* NaCl, 100 m*M* CaCl$_2$) at 37°C for 10 min to allow binding. In order to reduce the rate of cleavage while allowing proper folding of the ribozymes and substrates and preserving the interactions between the ribozyme and the substrate in an active ribozyme-substrate complex, CaCl$_2$ is used instead of MgCl$_2$ as the source of divalent ions. In the presence of CaCl$_2$, the rate of cleavage of substrate tk46 is at least 100-fold slower than the rate observed in the presence of MgCl$_2$ *(23)*. Kinetic and structural analyses suggest that interactions between the ribozymes and the substrates in the presence of Ca$^{2+}$ ions are similar to those found in the presence of Mg$^{2+}$ ions *(23)*.
2. The unbound substrates are separated using G-50 Sephadex gel-filtration columns (Roche Molecular Biochemicals, Indianapolis, IN). The presence of ribozyme-substrate complexes is detected by using 5% non-denaturing polyacrylamide gel and quantitation with STORM 840 PhosphorImager (Molecular Dynamics, Sunnyvale, CA).
3. Ribozyme-substrate complexes are diluted into different concentrations (2–70 n*M*) and incubated in 1X buffer C (**Subheading 2.2., item 4**). Aliquots are drawn from reaction mixtures at regular time intervals (from 0–160 min). The cleavage products are separated on 15% denaturing polyacrylamide gels, and quantitated.
4. To calculate the apparent rate constant $k_{app}$, the slope of the line for the plot of ln $(S_o/S_t)$ vs time is calculated. $S_o$ equals the initial substrate concentration and $S_t$ equals the substrate concentration at a given time-point.

## 4. Notes

1. The in vitro-synthesized ribozymes and substrates should be purified using denaturing gels containing 8 *M* urea. The gel slices that contain the ribozyme fractions are crushed and then soaked in RNase-free water for 20 min. The ribozyme fractions are separated from the crushed gel slices by microcentrifugation, extracted with phenol-chloroform solutions twice, and recovered by ethanol-precipitation.
2. In order to ensure sufficient complexity of the ribozyme variants in the initial cycle of selection, we usually use 100 pmol of the initial RNA pools. To increase the stringency of the selection, the concentration of the substrates is gradually reduced to select those ribozymes that exhibit better cleavage efficiency and binding affinity to the mRNA substrate.
3. The streptavidin column (AffiniTip) (Genosys Biotechnologies, Inc. Woodlands, TX) is usually prewashed with the annealing buffer three times according to manufacturer's recommendations, prior to the use for the selection.
4. The rate for folding of the proper secondary and tertiary structure of M1GS RNA may be a critical factor in determining the catalytic efficiency of the ribozyme. Our selection procedure discriminates against those species that fold slowly.

Moreover, our selection protocol can be used to select ribozymes that exhibit optimal cleavage activity under different ionic strengths (e.g., physiological buffer condition), and in the presence of cellular proteins by including different buffers and cellular protein extracts in the cleavage step.

5. One of the most common approaches to select for ribozymes that are more robust is to successively decrease the time of incubation. The shorter the time of incubation, the more active ribozymes will be selected by the procedure, since only highly active ribozymes will cleave the 5'-biotinylated tk46 and release from the column after a short period of incubation.

6. The variations in the amount of substrate do not affect the observed cleavage rate ($k_{obs}$), and the reaction follows pseudo-first-order kinetics.

## Acknowledgment

We thank Ahmed Kilani and Stephen Raj for helpful discussions and for sharing their research experience on RNase P ribozymes. K. K. is a recipient of the summer student fellowship from University of California at Berkeley. F. L. is a Pew scholar in Biomedical Sciences, a Scholar of the Lymphoma and Leukemia Society of America, and an Established Investigator of American Heart Association. The research has been supported by grants from State of California Universitywide AIDS research program, March of Dimes National Birth Defects Foundation, and NIH (GM54815 and AI41927).

## References

1. Altman, S. and Kirsebom, L. A. (1999) in *The RNA World* (Gesteland, R. F., Cech, T. R., and Atkins, J. F., eds.) 2nd ed., Cold Spring Harbor Laboratory Press, Cold Spring Harbor, NY, pp. 351–380.
2. Frank, D. N and Pace, N. R. (1998) Ribonuclease P: unity and diversity in a tRNA processing ribozyme. *Annu. Rev. Biochem.* **67,** 153–180.
3. Guerrier-Takada, C., Gardiner, K., Marsh, T., Pace, N., and Altman, S. (1983) The RNA moiety of ribonuclease P is the catalytic subunit of the enzyme. *Cell.* **35,** 849–857.
4. Liu, F. and Altman, S. (1995) Inhibition of viral gene expression by the catalytic RNA subunit of RNase P from *Escherichia coli. Genes Dev.* **9,** 471–480.
5. Liu, F. and Altman, S. (1996) Requirements for cleavage by a modified RNase P of a small model substrate. *Nucleic Acids Res.* **24,** 2690–2696.
6. Frank, D., Harris, M., and Pace, N. R. (1994) Rational design of self-cleaving pre-tRNA-Ribonuclease P RNA conjugates. *Biochemistry* **33,** 10,800–10,808.
7. McClain, W. H., Guerrier-Takada, C., and Altman, S. (1987) Model substrates for an RNA enzyme. *Science* **238,** 527–530.
8. Forster, A. C. and Altman, S. (1990) External guide sequence for an RNA enzyme. *Science* **249,** 783–786.
9. Curcio, L. D., Bouffard, D. Y., and Scanlon, K. J. (1997) Oligonucleotides as modulators of cancer gene expression. *Pharmacol Ther* **74(3),** 317–332.

10. Cobaleda, C., and Sanchez-Garcia, I. (2000) *In vivo* inhibition by a site-specific catalytic RNA subunit of RNase P designed against BCR-ABL oncogenic products: a novel approach for cancer treatment. *Blood* **95**, 731–737.
11. Trang, P., Lee, J., Kilani, A. F., Kim, J. and Liu, F. (2001) Effective inhibition of herpes simplex virus 1 gene expression and growth by engineered RNase P ribozyme. *Nucleic Acid Res.* **29**, 5071–5078.
12. Trang, P., Hsu, A., Zhou, T., Lee, J., Kilani, A. F., Nepomuceno, E., and Liu, F. (2002). Engineered RNase P ribozymes inhibit gene expression and growth of cytomegalovirus by increasing rate of cleavage and substrate binding. *J. Mol. Biol.* **315**, 573–586.
13. Trang, P., Kilani A. F., Kim, J., and Liu, F. (2000) A ribozyme derived from the catalytic subunit of RNase P from *Escherichia coli* is highly effective in inhibiting replication of herpes simplex virus I. *J. Mol. Biol.* **301**, 817–826.
14. Trang, P., Lee, M., Nepomuceno, E., Kim, J., Zhu, H., and Liu, F. (2000) Effective inhibition of human cytomegalovirus gene expression and replication by a ribozyme derived from the catalytic RNA subunit of RNase P from *Escherichia coli. Proc. Natl. Acad. Sci. USA* **97**, 5812–5617.
15. Pal, B. K., Scherer, L., Zelby, L., Bertrand, E., and Rossi, J.(1998). Monitoring retroviral RNA dimerziation in vivo via hammerhead ribozyme cleavage. *J. Virol.* **72**, 8349–8353.
16. zu Putlitz, J., Yu, Q., Burke, J. M., and Wands, J. R. (1999) Combinatorial screening and intracellular antiviral activity of hairpin ribozymes directed against hepatitis B virus. *J. Virol.* **73**, 5381–5387.
17. Lee, N. S., Bertrand, E., and Rossi, J. (1999) mRNA localization signals can enhance the intracellular effectiveness of hammerhead ribozymes. *RNA* **5**, 1200–1209.
18. Sullenger, B. A. and Cech, T. R. (1993) Tethering ribozymes to a retroviral packaging signal for destruction of viral RNA. *Science* **262**, 1566–1569.
19. Bertrand, E., Castanotto, D., Zhou, C., Carbonelle, C., Lee, N. S., Good, P., et al. (1997) The expression cassette determines the functional activity of ribozymes in mammalian cells by controlling their intracellular localization. *RNA* **3**, 75–88.
20. Yu, Q., Pecchia, D. B., Kingsley, S. L., Heckman, J. E., and Burke, J. M. (1998) Cleavage of highly structured viral RNA molecules by combinatorial libraries of hairpin ribozymes. The most effective ribozymes are not predicted by substrate selection rules. *J. Biol. Chem.* **273**, 23,524–23,533.
21. Kawa, D., Wang, J., Yuan, Y., and Liu, F. (1998) Inhibition of viral gene expression by human ribonuclease P. *RNA* **4**, 1397–1406.
22. Yuan, Y., Hwang, E., and Altman, S. (1992) Targeted cleavage of mRNA by human RNase P. *Proc. Natl. Acad. Sci. USA* **89**, 8006–8010.
23. Kilani, A. F., Trang, P., Jo, S., Hsu, A., Kim, J., Nepomuceno, E., Liou, K., and Liu, F. (2000) RNase P ribozymes selected *in vitro* to cleave a viral mRNA effectively inhibit its expression in cell culture. *J. Biol. Chem.* **275**, 10,611–10,622.
24. Chen, J. L., Nolan, J. M., Harris, M. E., and Pace, N. R. (1998) Comparative photocross-linking analysis of the tertiary structures of *Escherichia coli* and *Bacilus subtilis* RNase P RNAs. *EMBO J.* **17**, 1515–1525.

25. Haas, E. S., Brown, J. W., Pitulle, C., and Pace, N. R. (1994) Future perspective on the catalytic core and secondary structure of ribonuclease P RNA. *Proc. Natl. Acad. Sci. USA* **91,** 2527–2531.
26. Massire, C., Jaeger, L., and Westhof, E. (1998) Derivation of the three-dimensional architecture of bacterial ribonuclease P RNAs from comparative sequence analysis. *J. Mol. Biol.* **279,** 773–793.
27. Haas, E. S., Armbruster, D. W., Vucson, B. M., Daniels C. J., and Brown, J. W. (1996) Comparative analysis of ribonuclease P RNA structure in Archaea. *Nucleic Acids Res.* **24,** 1252–1259.

# 30

## In Vitro Selection of External Guide Sequences for Directing Human RNase P to Cleave a Target mRNA

### Stephen Raj and Fenyong Liu

#### Summary

External guide sequences (EGSs) are oligonucleotides that consist of a sequence that is complementary to a target mRNA and recruit intracellular RNase P for specific degradation of the target RNA. Recent studies indicate that increasing the targeting activity of EGSs in directing human RNase P to cleave an mRNA in vitro can lead to better efficacies of the EGSs in inducing RNase P-mediated inhibition of the expression of the target mRNA in cultured cells. This chapter will describe the procedure for the generation of highly functional EGSs by in vitro selection. We also describe protocols for in vitro evaluation of the activity of the EGSs. These methods should provide general guidelines for using in vitro selection for generating highly active EGSs for gene-targeting applications.

**Key Words:** RNase P; external guide sequence; in vitro selection; herpes simplex virus 1 (HSV 1); in vitro evolution; gene targeting.

## 1. Introduction
### 1.1. In Vitro Selection: A Tool for EGS Design

In vitro selection has been widely used to generate novel classes of ribozymes and ligands with superior catalytic activity and target specificity (*1–3*). This technique can be exploited to identify and optimize the external guide sequences (EGSs) that direct cellular RNase P to irreversibly cleave the target mRNA. Based on the single-stranded regions of target mRNA, the EGSs can be designed to anneal with the mRNA to form a bimolecular complex. This EGS-mRNA complex may resemble a precursor tRNA (ptRNA)-like structure, and may in turn become a substrate for RNase P (*4,5*). Many studies have been carried out to study how EGSs direct RNase P to cleave a target mRNA in vitro. However, little is currently known about the rate-limiting steps of the

From: *Methods in Molecular Biology, vol. 252: Ribozymes and siRNA Protocols, Second Edition*
Edited by: M. Sioud © Humana Press Inc., Totowa, NJ

EGS-directed RNase P cleavage in cultured cells. Recent studies on ribozymes and antisense molecules suggest that binding of these molecules to their target RNAs appears to be a rate-limiting step for depleting the target message in vivo *(6–9)*. It is generally believed that the optimal annealing of the EGSs to the target mRNA, the cleavage efficiency, and the colocalization of the EGS and the target in the relevant subcellular compartment are considered major determinants in governing the success of the EGSs in vivo. Indeed, our recent results indicate that the EGS efficacy in culture cells is dictated by the overall efficiency ($k_{cat}/K_m$) of the EGS-induced RNase P cleavage, and that increasing the targeting activity of the EGS leads to more effective inhibition of target mRNA expression in cultured cells *(10)*. Moreover, these observations demonstrate that highly efficient EGSs can be generated using in vitro selection, and are also highly effective in directing RNase P for inhibition of gene expression in cultured cells *(10)*. Thus, in vitro selection represents a promising approach for generating highly effective EGSs for gene-targeting applications.

EGSs can either be directly transfected *(11)* or stably expressed by manipulation of the cells *(12,13)*. Each of these techniques has its own unique advantages. For example, the nuclease-resistant synthetic EGSs can be incorporated with phophorothioates and 2' modifications to enhance their stability against other cellular endonucleases *(14)*. These synthetic EGSs can directly be transfected into the cells by encapsulating with liposomes, as described in Chapter 31. However, stable intracellular expression of EGSs may be a desirable strategy in which EGSs can be expressed in the nucleus and become instantly available for binding with the target RNA. However, in direct transfection, EGSs must passage across the cytoplasm and the nuclear envelope to initiate target depletion. Because RNase P is known to reside within the nuclei *(5,15)*, it is necessary to co-localize the enzyme and EGS within the same subcellular compartment.

## 1.2. Outline of Overall Strategy

This chapter focuses on various experimental methodologies involved in the in vitro selection of efficient EGS against thymidine kinase (TK) mRNA of herpes simplex virus I (HSV-I). The in vitro-selected EGSs can be transfected into human 143tk⁻ cells and stably expressed to elicit sustained effect on the expression of the target mRNA *(10)*. The series of experiments undertaken to downregulate the expression of the TK message by EGS are i) identifying the potential single-stranded regions of target mRNA for EGS binding, ii) optimal design of EGS for target-specific binding and effective suppression of targeted message, iii) in vitro selection to develop EGSs that direct RNase P for rapid degradation of target mRNA, iv) determination of the efficiency of RNase P-mediated degradation of the target mRNA in the presence of the selected

EGSs by in vitro cleavage assay, v) intracellular expression of EGS using a RNA polymerase III promoter, and vi) evaluation of the levels of suppression of the target message and its putative gene product formation in EGS-expressing cells using molecular techniques such as Northern and Western blot analysis.

Experimental protocols on the identification of single-stranded accessible regions on a target mRNA, purification of human RNase P *(16)*, in vitro cleavage assays, in vivo expression, and evaluation of the levels of suppression of the target mRNA expression are described in Chapters 28 and 29. Thus, this chapter primarily describes the procedures on in vitro selection for efficient EGS and in vitro characterization of the targeting activity of the selected EGSs.

### *1.3. In Vitro Selection Strategy*

#### *1.3.1. Principles Underlying EGS Design*

The RNA substrate tk46 used in the selection experiment contains 46 nucleotides derived from the sequences adjacent to the initiator ATG of TK mRNA. These nucleotides are selected on the basis of their susceptibility to modifications by dimethyl sulfide (DMS) within the cellular environment *(13)*. We have previously shown that EGSs derived from ptRNA$^{Ser}$ could direct human RNase P to cleave tk46 in vitro and inhibit HSV-1 TK expression in cultured cells *(13)*. However, the cleavage reaction is relatively inefficient compared to the cleavage of a natural tRNA substrate (e.g., ptRNA$^{Ser}$). The acceptor stem, D-stem, and D-loop sequences within a natural tRNA molecule are believed to be involved in tertiary interactions among the various domains of tRNA (e.g., variable, T-stem, and loop regions) (**Fig. 1A**). This enables the folding of the tRNA molecule into a stable tertiary conformation to facilitate the accessibility of its domains for interactions with RNase P *(17,18)*. The reduced RNase P cleavage of the tk46-EGS complex, compared to a natural tRNA, is probably a result of the nucleotide substitution at the tk46 sequence equivalent to the acceptor and D-stem regions. These nucleotide substitutions may cause perturbations within interacting nucleotides, and eventually disrupt some of the tertiary interactions important for maintaining the proper tRNA-like conformation that is essential for recognition by human RNase P. Restoration of these interactions or the introduction of additional sequence manipulation in other parts of the EGS sequence, such as those that resemble the variable region, T-stem, and T-loop, may increase the susceptibility of the mRNA-EGS complex to be cleaved by RNase P *(10)*.

#### *1.3.2. In Vitro Selection of EGS-tk46 Hybrid Molecule*

To generate EGSs that are highly active in directing RNase P to cleave tk46, a pool of chimeric, covalently linked tk46-EGS substrates that contain partially

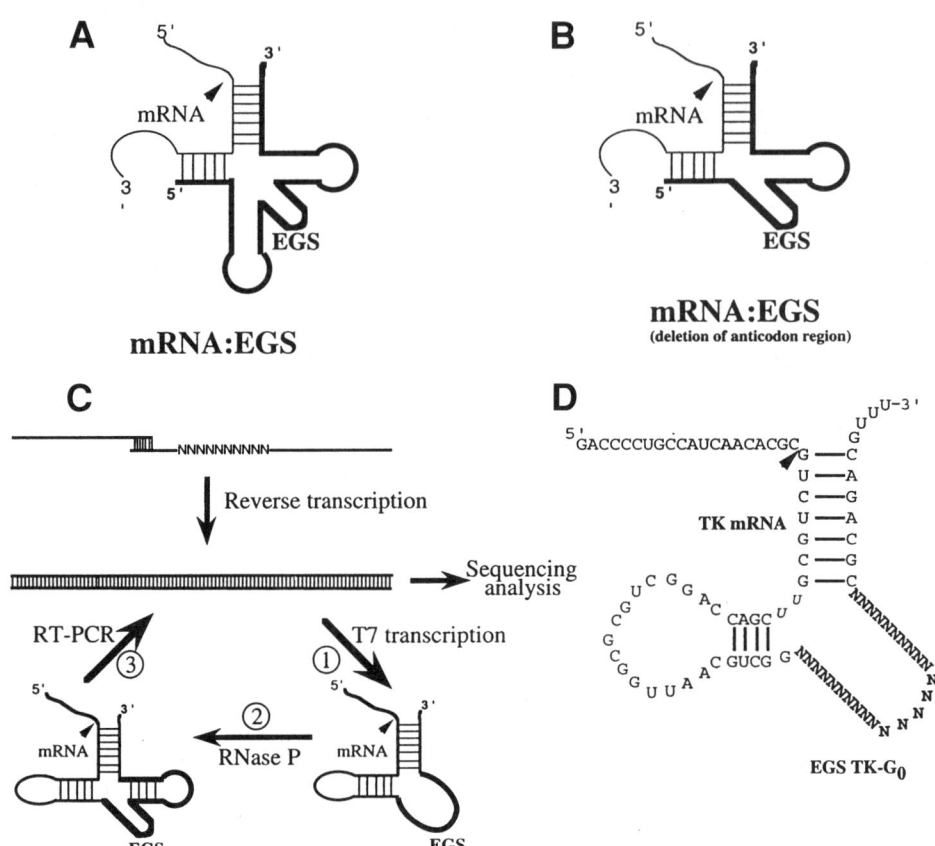

Fig. 1. (**A**) A hybridized complex of a target RNA (e.g., mRNA) and an EGS that resembles the structure of a tRNA and can be cleaved by RNase P. (**B**) results from (**A**) achieved deleting the anticodon domain of the EGS, which is indispensable for EGS-targeting activity *(20)*. (**C**) Schematic representation of in vitro selection experiments. (**D**) The substrate used for selection. The 5' part of the sequence shown, including the large loop, consists of the portion of the sequence of the mRNA that encodes HSV-1 TK. The rest of the sequence was designed to hybridize to the TK mRNA-encoding sequence, or was randomized. The 25 nucleotides that were randomized are each indicated by N. They are shown in bold type in (**D**) and as a bold line in (**C**). The site of cleavage by RNase P is marked by an arrowhead.

randomized sequences is constructed and selected based on their ability to act as a substrate for human RNase P (**Fig. 1**). The chimeric RNA integrating the sequences of both tk46 and the EGS contains a randomized sequence of 25 nucleotides at the positions corresponding to the regions resembling the variable stem, T-stem, and T-loop (**Fig. 1D**), whereas the 5' proximal sequence that spans the D-stem, D-loop, acceptor stem, and leader sequence contains the sequence of tk46. The anti-codon region, which has been shown to be dispensable for RNase P activity *(19)*, is not included in the chimeric substrate (**Fig. 1B,D**). The pool of tk46-EGS chimeric substrates is synthesized in vitro by T7 RNA polymerase. In each round of selection, the RNA variants are digested with human RNase P and the 3' cleavage products are isolated in denaturing gels (**Fig. 1C**). The cDNA molecules are synthesized from these RNA molecules by reverse transcription and amplified by polymerase chain reaction (PCR). The resulting PCR products are used as templates for the synthesis of EGS RNA molecules for the next round of selection (**Fig. 1C**). The 5' primer for the PCR reaction contains the leader sequence and the promoter sequence for the T7 RNA polymerase, and thus enables the restoration of these sequences in the PCR products that will be used for the next cycle of selection. The stringency of the selection is increased at each cycle by reducing the amount of human RNase P and the time allowed for the cleavage reaction, such that only those EGS-mRNA chimeric substrates that exhibit better susceptibility to be cleaved by RNase P under suboptimal conditions will be selected. The efficacy of the selected EGSs can be evaluated by in vitro cleavage assays using the purified human RNase P. The relative efficiency of the RNase P-mediated cleavage directed by the in vitro selected EGSs is estimated using Michaelis-Menton kinetics to identify EGS molecules that confer superior targeting activity.

## 2. Materials

### 2.1. Reagents

1. DEAE Sepharose (Sigma).
2. Leupeptin (Sigma).
3. Deoxynucleotide 5' triphosphates (dNTPs) and nucleotide 5' triphosphates (NTPs) (10 m$M$ each).
4. β-mercaptoethanol.
5. Buffer A: 50 m$M$ Tris-HCl, pH 7.5, 10 m$M$ MgCl$_2$, 100 m$M$ NH$_4$Cl.
6. 5$X$ MMTV reverse transcription buffer: 500 m$M$ Tris-HCl, pH 8.3, 50 m$M$ KCl, 30 m$M$ MgCl$_2$, 30 m$M$ dithiothreitol (DTT).
7. 5X binding buffer: 250 m$M$ Tris-HCl, pH 7.5, 0.5 m$M$ ethylenediaminetetraacetic acid (EDTA), 50 m$M$ NaCl, 50 m$M$ MgCl$_2$, 15% glycerol, 0.1% xylene cyanol.
8. RNasin ribonuclease inhibitor (Promega).
9. γ-[$^{32}$P]-adenosine 5' triphosphate (ATP) and α-[$^{32}$P]-ATP (Amersham).
10. Loading dye: 9 $M$ urea, 0.05% (w/v) xylene cyanol, and 0.05% (w/v) bromophenol blue.
11. Denaturing dye: 1 m$M$ Tris-HCl, pH 7.0, 8 $M$ urea, 0.01% xylene cyanol, and 0.01 bromophenol blue.
12. Annealing buffer: 10 m$M$ Tris-HCl, pH 7.5, 10 m$M$ KCl.

## 2.2. Enzymes and Kits

1. RNase T1 (Amersham Pharmacia Biotech).
2. T4 polynucleotide kinase (PNK) (New England Biolabs).
3. MMTV reverse transcriptase (Roche Molecular Biochemicals).
4. Avian myeloblastosis virus reverse transcriptase (Roche Molecular Biochemicals).
5. RNase A (Sigma).
6. Taq polymerase (Perkin-Elmer).
7. T7 RNA polymerase (Promega).
8. DNase-free RNase (Roche Biochemicals).
9. PNK (New England Biolab).
10. Dideoxynucleotide sequencing kit (Amersham Biosciences).
11. PCR kit (Perkin Elmer).

## 3. Methods

### 3.1. Construction of DNA Sequences Coding for the EGSs

Oligonucleotide TK21 (5'GGAATTCTAATACGACTCACTATAGA CCCCTGCCATCAACACGCGTCTGCGTTCGACCAGGCTGCGCGGTTAACGTCG-3') and oligonucleotide TK22 (5'-AAACGTCTGCGNNNNNNNN NNNNNNNNNNNNNNNNNNCCG-ACGTTAA-3') are used to synthesize the double-stranded DNA templates using MMTV reverse transcriptase. Oligo-nucleotide TK21 contains the leader sequence and T7 promoter sequence, and TK22 contains a randomized sequence indicated as $(N)_{25}$ (**Fig. 1**). Equimolar amounts of A, C, G, and T are incorporated in the positions represented as N during the chemical synthesis of the oligomer. The overall strategy for the design of these oligonucleotides and the selection procedure are in part, based on similar method described by Yuan and Altman *(20)*.

1. Resuspend an equal amounts (5 nmol) of oligonucleotides TK21 and TK22 into 500 μL of annealing buffer, heat at 95°C for 2 min, and then allow to anneal by dropping the temperature gradually to 37°C.
2. The annealed oligonucleotides are filled in using the MMTV reverse transcriptase. Add to the mixture 8 μL of 5X MMTV reverse transcription buffer, 4 μL each of 10 m*M* deoxyadenosine 5; triphosphate (dATP), deoxycytidine 5' triphosphate (dCTP), deoxyguanosine 5' triphosphate (dGTP), deoxythymidine 5' triphosphate (dTTP) and 2 μL MMTV reverse transcriptase, and then incubate at 37°C for 3–4 h (*see* **Note 1**).
3. Load the reaction on 5% nondenaturing polyacrylamide gels and cut out the full-length DNA products
4. Place the fragments into a dialysis bag (2,000-Dalton cut off) with 500 μL water electroelute with TBE buffer for 20 min.
5. Filtrate the solution containing the DNA sample through a microcentrifuge filter, extract with phenol-chloroform, and then precipitate with ethanol (*see* **Note 2**).

6. It is important to preamplify the DNA products prior to in vitro transcription of the RNA library to increase the copy number of each individual DNA products (*see* **Note 3**). In our experience, an amplification of five- to 10-fold is usually adequate.

7. For a 100-µL PCR reaction, mix the following:
   a. 10 µL of 10X PCR buffer.
   b. 6 µL of 25 m*M* MgCl$_2$.
   c. 2 µL each of 10 m*M* dATP, dGTP, dCTP, dTTP.
   d. 20 pmol of RT-generated DNA templates.
   e. 400 pmol each of TK21 and TK23 oligomer.
   f. 2.5 U *Taq* DNA polymerase.

8. Overlaid the PCR reaction mixture with mineral oil and cycle at 94°C for 1 min, 47°C for 1 min, and 72°C for 1 min.

9. Load the PCR products on 5% non-denaturing polyacrylamide gel and purify the full-length DNA products as described previously. To carry out the selection process, we would generate 1–5 nmol of the DNA templates coding for the EGS RNAs that contain randomized sequences *(10)*.

### 3.2. Generation of EGS RNA Library for In Vitro Selection by In Vitro Transcription

1. The gel-purified PCR DNA products are used as a template for the in vitro transcription of EGSs. A 100 µL in vitro transcription reaction mix is prepared by adding the following:
   a. 20 µL of 5X in vitro transcription buffer.
   b. 10 µL of 100 m*M* DTT.
   c. 5 µL each of 10 mM ATP, UTP, GTP, CTP.
   d. 7 µL of PCR-generated DNA templates (about 10 µg DNA).
   e. 2-3 µL (20–30 µCi) α-[$^{32}$P]-GTP.
   f. 1.5 µL of RNasin RNase inhibitor.
   g. 1.5 µL of T7 RNA polymerase.

2. After incubation at 37°C for at least 16 h, stop the reaction by adding 100 µL of denaturing dye and load on 8% denaturing gels containing 7 *M* urea.

3. Visualize the full-length RNA products by autoradiography, excise them from the gel, and purify by extraction followed by ethanol precipitation (*see* **Note 4**).

4. The yield of the in vitro RNA synthesis is determined either by measuring the concentration of the RNA using a spectrophotometer or quantitation of the total radioactivity using a STORM 840 PhosphorImager. We have routinely obtained 50 pmol of RNA transcripts from a 100-µL reaction.

### 3.3. In Vitro RNase P Cleavage Assay for the Selection of Effective EGSs

The selection procedures are illustrated in **Fig. 1C**, and can be summarized as follows:

1. In the first (initial) selection cycle, 20 nmol of the pool of in vitro synthesized RNA substrates possessing random sequences are digested by human RNase P in

a volume of 2 mL reaction mixture. The RNA substrate is incubated at 37°C with human RNase P in buffer A (*see* **Subheading 2.** and **Note 5**). The cleavage reaction is stopped by adding the loading dye solution (*see* **Subheading 2.**), and reaction mixtures are loaded in denaturing 8% polyacrylamide gels.

2. The 3' proximal cleavage product is separated in denaturing 8% polyacrylamide gels, visualized by autoradiography, excised from the gel, and purified by extraction followed by ethanol-precipitation (*see* **Note 6**).

3. The 3' proximal sequence that lacks leader sequence is used as a template for reverse transcription to generate cDNA using 20 m*M* oligoTK23 (5'-AAACGTCTGCG-3'), and avian myeloblastosis virus reverse transcriptase. The mixture is incubated for 2 h at 42°C.

4. The cDNA is subsequently amplified by PCR with oligodeoxynucleotide primers TK21 and TK23, and is used to generate RNA substrates for the next round of selection. TK21, the 5' primer for PCR, allows restoration of the T7 promoter sequence and the leader sequence of the RNA substrates.

5. Initially, a pool of 20-nmol RNA substrates with the randomized sequence is synthesized in vitro and digested by human RNase P in a volume of 2 mL. In the subsequent cycles of selection, 5 pmol of substrate is used and digested with human RNase P in a volume of 20 μL. During the first four cycles of selection, substrates are incubated for 120 min at 37°C with 100 U of human RNase P. During the final five cycles of selection, a higher stringency is applied to the selected substrate population by reducing the incubation time to 10 min and the amount of human RNase P by 200-fold.

6. The selection process can be repeated for many cycles (*see* **Note 7**). The cleavage of chimeric RNA for the each population (after different cycles) can be monitored by human RNase P reaction followed by gel electrophoresis. A progressive increase in the ability of the chimeric RNAs to be cleaved by RNase P with each successive cycle of selection is expected.

7. After several cycles of selection, the pool of selected RNA molecules is cloned into pUC19 and sequenced. We usually sequence 20–30 individual clones by dideoxynucleotide chain-termination method using M13 reverse primer (*10*).

### 3.4. In Vitro Evaluation of the Selected EGSs

Using the sequences of the selected chimeric substrates, individual EGS RNAs are prepared and evaluated for their ability to direct human RNase P to cleave the target mRNA.

### 3.4.1. Template Preparation

1. Prepare DNA templates for EGSs by PCR, using the sequenced chimeric clones as templates and primers: TK23 (5'-AAACGTCTGCG-3') and TK24EGS (5'-GGAATTCTAATACGACTCACTATAGGTTAACGTC-3'), which contains a promoter sequence for T7 RNA polymerase.

2. Synthesize EGS RNAs by in vitro transcription by T7 RNA polymerase from the PCR-generated DNA templates, as described in **Subheading 3.2.**

### 3.4.2. Determination of Cleavage Efficiency of the RNase P Reaction With Selected EGSs

Each of the individual EGS RNAs is mixed with radiolabeled target RNA, and then the mixtures are incubated with human RNase P.

1. Mix 3000 cpm of target RNA with 5 ng of EGS RNA in 8.5 µL of $H_2O$.
2. Heat at 80°C for 3 min.
3. Add 1 µL of buffer A.
4. Add 0.25 µL of RNasin and 0.5 µL (0.25 U) of human RNase P.
5. Incubate the reaction mixture at 37°C for 10 min.
6. Add 9 µL of 2X RNA dye solution and 1 µL of phenol to each reaction.
7. Load onto an 8% polyacrylamide 7 $M$ urea gel.
8. Expose the gel to a phosphor screen, which will be analyzed with a PhosphorImager. A successful selection will result in new EGSs that direct efficient cleavage by human RNase P.

### 3.4.3. Measurement of the Equilibrium Dissociation Constant ($K_d$) for the Target RNA by the Selected EGSs Using a Band-Shift Method

The in vitro-selected EGSs may have mediated enhanced rates of cleavage by RNase P in part by increasing the stability of the EGS-mRNA complex. Therefore, the stability of such complexes should be determined.

1. Prepare a 5% polyacrylamide gel with a 30:1 weight ratio of acrylamide to *N,N'*-methylene *bis*-acrylamide in 36 m$M$ Tris base/64 m$M$ HEPES, pH 7.5, 10 m$M$ $MgCl_2$, and 0.1 m$M$ EDTA. Prerun the gel in the same buffer at constant power to raise, and maintain the temperature of the gel at 37°C (*see* **Note 8**).
2. Dilute EGS to varying concentrations (ranging from 0.1–1000 n$M$). Mix 10 µL EGS dilutions at 2X final concentration with 6 µL [32]P-labeled target RNA (1000 cpm and 0.1 n$M$). Heat at 80°C for 3 min and add 4 µL of 5X binding buffer
3. Incubate the samples at 37°C for 10 min (to reach an equilibrium). Load immediately onto the prewarmed gel and run at a constant temperature of 37°C.
4. Dry the gel and quantitate free target RNA and bound RNA at each concentration of EGS on a PhosphorImager. The dissociation constant of an EGS is determined by inspection of the gel midpoint, where RNA[$P_{free}$] = [$P_{bound}$]. $K_d$ = [$E_{total\ at\ midpoint}$ ]–1/2[$P_{total}$] (here [P] and [E] symbolize the target RNA and the EGS, respectively).

### 3.4.4. Kinetic Analyses of the Cleavage Reactions With Selected EGSs

These kinetics are performed to determine the Michaelis constant ($K_m$) and the maximum velocity ($V_{max}$) of the enzymatic reactions *(10)*.

1. Mix varying amounts of [32]P-labeled in vitro-transcribed target RNA with an excess amount of EGS RNA in buffer A, and calculate the actual concentration of target RNA-EGS complex in each mixture based on the dissociation constants ($K_d$).

2. Perform standard RNase P reactions for 0.5, 1, 2, 5, and 10 min, and extrapolate an initial rate of RNase P reaction for each concentration of RNA-EGS complex.
3. Determine the values of $K_m$ and $V_{max}$ using Lineweaver-Burk plots (double reciprocal plots) *(10)*.

## 4. Notes

1. Various methods (including using *E. coli* DNA polymerase I Klenow fragment and PCR with Taq DNA polymerase) are used to generate the double-stranded DNA templates from single-stranded oligomers. Our results indicate that reverse transcription by MMTV reverse transcriptase yields the most efficient synthesis (about 60%) from the double-stranded DNA templates.
2. Although there are several methods available for oligonucleotide purification, gel purification offers unique advantages over other methods. First, this method could eliminate the partially extended cDNA products that are generated as a result of the premature termination of reverse transcription. Second, the unused primers (TK21 and TK22) are removed in this purification process.
3. There is probably only a single copy for each specific sequence in our RT-generated DNA pool. Some of these sequences might be lost during the selection procedures in which complete recovery of nucleic acid molecules might not be possible (e.g., gel purification or ethanol-precipitation). This amplification step would increase the probability that the sequences efficiently cleaved by the enzyme in the original pool would not be lost during the selection procedure.
4. Gel purification of the RNA molecules is necessary in order to obtain the full-length substrates prior to the cleavage reaction.
5. The stringency of the selection is increased after each cycle of selection by reducing the amount of the enzyme and by shortening the incubation time to ensure that less than 5% of total population of the chimeric RNAs are cleaved during each cycle.
6. Warming up the RNA samples in 37°C before loading on the gel for at least 5 min, and flushing the wells to avoid urea precipitation ahead of loading the samples help to produce sharp bands and acquire high-quality data.
7. To generate EGSs that are highly active in a cellular environment, the selection can be carried out in a buffer that resembles the physiological ionic and pH conditions, and in the presence of protein extracts isolated from various cellular compartments (e.g., nuclear protein extracts).
8. Prerun the gel in the same buffer at constant power to raise and maintain temperature of the gel at 37°C is critical to obtain accurate values of $K_d$.

## Acknowledgments

We would like to thank Yan Yuan, Tianhong Zhou, and Ahmed Kilani for helpful discussion and sharing their unpublished results. F. L. is a Pew Scholar in Biomedical Sciences, a Scholar of the Leukemia and Lymphoma Society, and a recipient of Established Investigator Grant award from the American Heart Association. This research has been supported by UC-Berkeley

(Chancellor's Special Initiative Award) and NIH (RO1-AI43250 and RO1-DE14842).

## References

1. Joyce, G. F. (1998) Nucleic acid enzymes: playing with a fuller deck. *Proc. Natl. Acad. Sci. USA* **95,** 5845–5847.
2. Szostak, J. W. (1992) In vitro genetics. *Trends Biochem. Sci.* **17,** 89–93.
3. Gold, L., Polisky, B., Uhlenbeck, O., and Yarus, M. (1995) Diversity of oligonucleotide functions. *Annu. Rev. Biochem.* **64,** 763–797.
4. Forster, A. C. and Altman, S. (1990) External guide sequences for an RNA enzyme. *Science* **249,** 783–786.
5. Yuan, Y., Hwang, E. S., and Altman, S. (1992) Targeted cleavage of mRNA by human RNase P. *Proc. Natl. Acad. Sci. USA* **89,** 8006–8010.
6. Lee, N. S., Bertrand, E., and Rossi, J. (1999) mRNA localization signals can enhance the intracellular effectiveness of hammerhead ribozymes. *RNA* **5,** 1200–1209.
7. Pal, B. K., Scherer, L., Zelby, L., Bertrand, E., and Rossi, J. J. (1998) Monitoring retroviral RNA dimerization in vivo via hammerhead ribozyme cleavage. *J. Virol.* **72,** 8349–8353.
8. Sullenger, B. A. and Cech, T. R. (1993) Tethering ribozymes to a retroviral packaging signal for destruction of viral RNA. *Science* **262,** 1566–1569.
9. zu Putlitz, J., Yu, Q., Burke, J. M., and Wands, J. R. (1999) Combinatorial screening and intracellular antiviral activity of hairpin ribozymes directed against hepatitis B virus. *J. Virol.* **73,** 5381–5387.
10. Zhou, T., Kim, J., Kilani, A. F., Kim, K., Dunn, W., Jo, S., Nepomuceno, E., and Liu, F. (2002) In vitro selection of external guide sequences for directing RNase P-mediated inhibition of viral gene expression. *J. Biol. Chem.* **277,** 30,112–30,120.
11. Dunn, W., Trang, P., Khan, U., Zhu, J., and Liu, F. (2001) RNase P-mediated inhibition of cytomegalovirus protease expression and viral DNA encapsidation by oligonucleotide external guide sequences. *Proc. Natl. Acad. Sci. USA* **98,** 14,831–14,836.
12. Plehn-Dujowich, D. and Altman, S. (1998) Effective inhibition of influenza virus production in cultured cells by external guide sequences and ribonuclease P. *Proc. Natl. Acad. Sci. USA* **95,** 7327–7332.
13. Kawa, D., Wang, J., Yuan, Y., and Liu, F. (1998) Inhibition of viral gene expression by human ribonuclease P. *RNA* **4,** 1397–1406.
14. Ma, M. Y., Jacob-Samuel, B., Dignam, J. C., Pace, U., Goldberg, A. R., and George, S. T. (1998) Nuclease-resistant external guide sequence-induced cleavage of target RNA by human ribonuclease P. *Antisense Nucleic Acid Drug Dev.* **8,** 415–426.
15. Bertrand, E., Houser-Scott, F., Kendall, A., Singer, R. H., and Engelke, D. R. (1998) Nucleolar localization of early tRNA processing. *Genes Dev.* **12,** 2463–2468.
16. Bartkiewicz, M., Gold, H., and Altman, S. (1989) Identification and characterization of an RNA molecule that copurifies with RNase P activity from HeLa cells. *Genes Dev.* **3,** 488–499.
17. Frank, D. N. and Pace, N. R. (1998) Ribonuclease P: unity and diversity in a tRNA processing ribozyme. *Annu. Rev. Biochem.* **67,** 153–180.

18. Altman, S. and Kirsebom, L. (1999) in *The RNA World* (Gesteland, R.F., Cech, T.R., and Atkins, J.F., eds.), 2nd ed., Cold Spring Harbor Laboratory Press, Cold Spring Harbor, NY, pp. 351–380.
19. Yuan, Y. and Altman, S. (1995) Substrate recognition by human RNase P: identification of small, model substrates for the enzyme. *EMBO J*. **14,** 159–168.
20. Yuan, Y. and Altman, S. (1994) Selection of guide sequences that direct efficient cleavage of mRNA by human ribonuclease P. *Science* **263,** 1269–1273.

# 31

## RNase P-Mediated Inhibition of Viral Growth by Exogenous Administration of Short Oligonucleotide External Guide Sequence

### Walter Dunn and Fenyong Liu

### Summary

The use of external guide sequence (EGS) in directing endogenous ribonuclease P (RNase P) for inhibition of viral propagation is described in this chapter, with an emphasis on chemically modified EGSs and their extracellular delivery. Targeting of the mRNA-encoding human cytomegalovirus (HCMV) protease by DNA-based EGSs is presented as an example of how to design chemically modified EGSs for antiviral applications. General information about the EGS-based technology is included, followed by detailed protocols for EGS design, human RNase P purification, in vitro assay of EGS activity, liposome-mediated delivery of chemically modified EGSs and detection of their distribution in cells, and an assay of EGS activity for blocking growth of HCMV in cultured cells.

**Key Words**: RNase P; external guide sequence (EGS); human cytomegalovirus (HCMV); antisense; gene therapy; HCMV protease; liposome.

## 1. Introduction

External guide sequence (EGS) technology is based on the principle that a short nucleic acid sequence hybridized with a targeted mRNA can adopt a pre-tRNA-like conformation that is recognized by endogenous RNase P. RNase P identifies this bipartite complex as its natural pre-tRNA substrate and irreversibly cleaves the mRNA component *(1,2)*. The use of EGSs to combat viruses is especially relevant because pathogenesis is contingent on viral gene activity (as opposed to disease-causing toxins or chemical agents), and this therapy is designed specifically to block gene expression at the mRNA level *(3,5)*. Several groups have demonstrated the use of EGSs for antiviral applications in tissue culture against pathogens such as herpes simplex virus-1 (HSV-1),

From: *Methods in Molecular Biology, vol. 252: Ribozymes and siRNA Protocols, Second Edition*
Edited by: M. Sioud © Humana Press Inc., Totowa, NJ

influenza, human immunodeficiency virus (HIV), human cytomegalovirus (HCMV), and hepatitis B *(3,4,6–8)*. As previously described, RNA-based EGSs are typically expressed endogenously, and therefore must be delivered by a gene-expression vector *(4)*. Gene delivery alone presents a myriad of complexities that must be resolved for effective therapeutic activity to be realized. Therefore, in order to develop the EGS technology for the clinical setting, exogenous modes of delivery also must be developed and explored. In this chapter, we describe the application of the EGS-based technology with nucleic acid modifications that allow extracellular delivery of the EGS. In order to achieve effective antiviral therapy, the mRNA that the EGS directs RNase P to degrade must code for an essential protein(s) of the virus. The target for gene inactivation will also be discussed in the context of HCMV infection.

### 1.1. Human RNase P

RNase P is responsible for the 5' maturation of precursor tRNAs (ptRNAs) in all living organisms (**Fig. 1A**). The eubacterial RNase P consists of a catalytic RNA subunit (M1 RNA in *E. coli*) and a protein subunit (C5 protein in *E. coli*). The M1 RNA alone is catalytically active in elevated salt concentrations (100 m*M* MgCl$_2$) *(10)*. By contrast, the eukaryotic RNase P is more complex in its subunit composition, containing one RNA and several protein subunits. Unlike the prokaryotic RNase P, the protein components are essential for RNase P activity of the eukaryotic enzyme, because the RNA subunit alone does not exhibit catalytic activity in vitro *(11,12)*. Since this chapter focuses on the value of RNase P as a gene-inactivation tool in human cells, readers can refer to articles by van Eenennaam et al., Gopalan et al., and Jarrous et al. to gain insight into the basic biochemistry of human RNase P *(13–15)*. For information on the use of RNase P ribozymes for gene-targeting application, please refer to Chapters 28, 29, and 32.

### 1.2. General EGS Design

Generally, an ideal three-quarter EGS (**Fig. 1B**) imitates the structural features of a ptRNA (**Fig. 1A**). However, studies on the systematic deletion of various domains of ptRNA have revealed that the anticodon stem and loop are dispensable for RNase P cleavage *(16)*. This observation helps to design EGSs with minimal features, facilitating more efficient transfection into mammalian cells. As is the case with many of the natural tRNAs, the unpaired acceptor stem and D-stem of the EGS contains seven and at least three nucleotides, respectively, to facilitate effective basepairing with the complementary regions of target mRNA. It is important to have a spacer of two nucleotides between EGS-binding sequence segments (seven and three nucleotides) on the target mRNA, as found in the natural ptRNA substrates. To rule out the possibility of

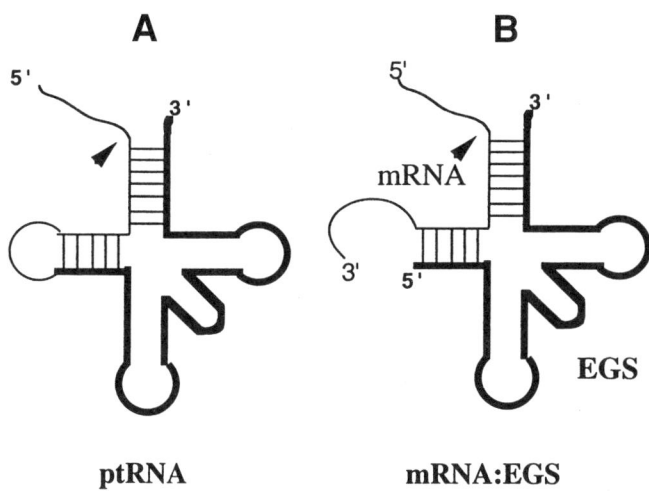

Fig. 1. Schematic representation of substrates for RNase P. (**A**) A natural substrate (ptRNA). (**B**) A hybridized complex of a target RNA (e.g., mRNA) and an EGS that resembles the structure of a tRNA.

antisense-mediated degradation of target mRNA, the positions of an EGS equivalent to the conserved nucleotides in the T-loop, which are essential for RNase P recognition and cleavage, can be substituted or reversed so that target mRNA interacting with such a mutant EGS is no longer cleaved by endogenous RNase P (**Fig. 2B**). These mutant EGSs would help to determine whether the cleavage of target mRNA is mediated by RNase P or conventional antisense-induced RNase H cleavage. The targeting efficacy of EGSs in directing RNase P to cleave a mRNA can be evaluated by in vitro cleavage assays using the purified human RNase P.

### 1.3. Modified EGSs

Unlike RNA EGSs delivered by gene-expression vectors and expressed endogenously, extracellular delivery of EGSs requires special modifications of the RNA molecules in order to avoid nuclease activity that degrades the EGS RNAs. Several types of modifications are available, and have proven to be compatible with the EGS-based technology. The primary modification strategy is to synthesize EGS oligonucleotides with the 2'-hydroxyl group removed or replaced. It is this functional group that makes RNA highly susceptible to nuclease degradation. The most straightforward modification is the replacement of the 2'-hydroxyl group with a hydrogen molecule (*3*). The resulting EGS is simply a DNA EGS. The advantage of this approach is that chemical synthesis of DNA oligonucleotides is widely available and relatively inexpensive.

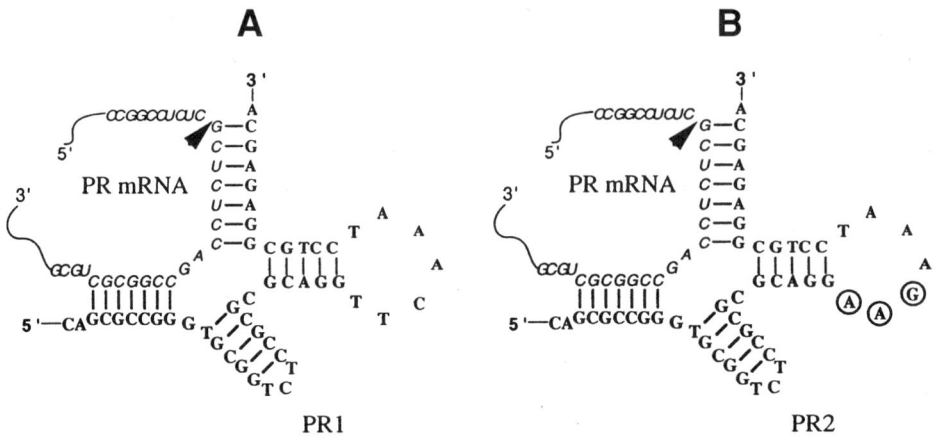

**PR mRNA:EGS PR1**                    **PR mRNA:EGS PR2**

Fig. 2. Complexes between PR mRNA sequence and EGS PR1 (**A**) and PR2 (**B**), respectively. The sequence of these EGSs equivalent to the tRNA sequence was derived from tRNA[ser] and resembles the T-stem, loop, and variable region of the tRNA molecule.

A possible drawback is that RNA and DNA molecules have dissimilar non-canonical basepairing potentials. Thus, the tertiary conformation of the complex of a DNA based EGS and its targeted mRNA may not resemble an authentic pre-tRNA molecule, thereby obviating RNase P recognition. Some modification strategies replace the 2'-hydroxyl group, but attempt to preserve the interaction potential at that position. 2'-*O*-methyl, phosphorothioate, and 2'-fluoro group modification are other possible modifications that are compatible with EGS technology (*8*). The disadvantage of these group substitutions is that they are relatively specialized and costly to incorporate into an oligonucleotide. However, these modifications have been shown to substantially increase the half-life of the EGS molecule to approx 18 h in fetal calf serum (FCS) (*8*), as opposed to <5 min for an all-RNA EGS (*17*). These EGSs, which are adapted for extracellular delivery, can be introduced by such means as direct injection or liposome-mediated delivery.

### 1.4. EGS As a Therapeutic for HCMV Infection

To demonstrate the use of EGSs against viral agents, treating HCMV infection with a chemically modified EGS is outlined here. One of the most important issues is the delivery of the therapeutic EGS. In order for the EGS-based technology to be a viable antiviral strategy, the EGS will be modified to allow exogenous administration of the oligonucleotide. A hydrogen group will replace the 2'-hydroxyl group of the EGS bases. The resulting DNA EGS will be chemically synthesized and delivered via liposomes.

The gene targeted for inactivation is the protease gene of HCMV. The protease is highly conserved throughout the herpesviridae family, and has been shown to be essential for the replication of HSV 1 *(18)*. It functions as a structural virion component, and also, possesses enzymatic activity necessary for the proper assembly of the viral capsid *(19)*. Because the protease is involved in several aspects of viral maturation, this enzyme serves as a good target for shutting down viral propagation. The mRNA of the protease gene will be examined as previously described, to determine which sequences are accessible to EGS binding *(3)*. The following sections describe the necessary protocols for treatment of HCMV infection with a modified EGS. Procedures for human RNase P purification, anti-HCMV protease EGS design, and an in vitro cleavage assay are first outlined. The subsequent sections cover liposome-mediated delivery and detection of the EGS molecule in the cells. Finally, an assay on the effectiveness of the EGS treatment in blocking HCMV replication in human cells is described.

## 2. Materials
### 2.1. Purification of RNase P From HeLa Cells

Human RNase P is prepared from HeLa cells through a DEAE Sepharose chromatography and a glycerol gradient *(20)*.

### 2.2. Solutions and Buffers

1. 10X buffer A: 500 m$M$ Tris-HCl, pH 8.0, 100 m$M$ MgCl$_2$. Autoclaved.
2. 15% glycerol solution: 1X buffer A, 100 m$M$ KCl, 15% (v/v) glycerol.
3. 30% glycerol solution: 1X buffer A, 100 m$M$ KCl, 30% (v/v) glycerol.
4. 10X buffer B (RNase P reaction buffer):
   a. Type I: 500 m$M$ Tris-HCl, pH 7.5, 1 $M$ NH$_4$Cl, 20 m$M$ MgCl$_2$.
   b. Type II: 500 m$M$ Tris-HCl, pH 7.5, 1 $M$ NH$_4$Cl, 100 m$M$ MgCl$_2$.
   c. Type III: 500 m$M$ Tris-HCl, pH 7.5, 1 $M$ NH$_4$Cl, 250 m$M$ MgCl$_2$.
5. 2X RNA dye solution: 9 $M$ urea, 20 m$M$ ethylenediaminetetraacetic acid (EDTA), 0.25 mg/mL bromophenol blue, 0.25 mg/mL xylene cyanol FF.

## 3. Methods
### 3.1. Preparation of S100 Extract From HeLa Cells

1. Suspend 10 g of HeLa cells in 5 vol (50 mL) of lysis buffer (0.2X buffer A, 10 m$M$ KCl, 2 m$M$ dithiothreitol [DTT], 0.5 µg/mL leupeptin, and 0.2 m$M$ PMSF [add right before using]) and place on ice for 10 min. Homogenize the cells using a Dounce homogenizer for 10 strokes.

2. Centrifuge the lysate in a Sorvall centrifuge SS-34 rotor at 5000 rpm at 4°C for 10 min.
3. Save the supernatant and spin it in a 60Ti rotor at 40,000 rpm (100,000*g*) for 60 min. The supernatant (S100) can be frozen at –70°C until ready to use for further purification.

## 3.2. DEAE-Sepharose Chromatography

1. Pack a 50–100-mL column with DEAE-Sepharose and wash with 1 L 1X buffer A, 50 m*M* KCl.
2. Load S100 onto the column. Wash the column with 1 L 1X buffer A, 100 m*M* KCl, 0.2 m*M* PMSF, and 1 m*M* DTT.
3. Elute the enzyme with a linear gradient formed in a gradient maker using 200 mL buffer 100 m*M* KCl, 1 m*M* DTT, 0.2 m*M* PMSF, and 200 mL from 1X buffer A, 500 m*M* KCl, 1 m*M* DTT, 0.2 m*M* PMSF. Collect 4-mL fractions using a fraction collector.
4. Assay 5 µL of each fraction for protein concentration using the Bio-Rad Protein Assay (Bradford Assay). Protein peak should be in the fractions eluted, approx 200 m*M* KCl. Assay 20 fractions beginning at the protein peak for RNase P activity using 1 µL of fraction. The assay for RNase P activity has been described by Guerrier-Takada et al. (*10*; *see* **Note 1**). The enzymatic peak should be in the fractions eluted from 200–250 m*M* KCl.
5. Pool the fractions that exhibit RNase P activity and concentrate using Amicon Centriprep-30 concentrators. Store the sample on ice.

## 3.3. Glycerol Gradient

1. Prepare gradients in 12-mL ultraclear tubes using 5.5 mL each of buffer A, 100 mL KCl, 1 m*M* DTT, 15% glycerol, and 1X buffer A, 100 mL KCl, 1 m*M* DTT, and 30% glycerol.
2. Load 1 mL of sample per tube, and spin at 30,000*g* at a SW41 rotor for 22 h at 0°C.
3. Collect 400-µL fractions and assay 1 µL of each fraction for the enzymatic activity for each fraction.
4. Pool the fractions that exhibit RNase P activity and concentrate using Amicon Centricon-30 concentrators. Dialyze against solution containing 1X buffer A, 100 mL KCl, 1 m*M* DTT, and 50% glycerol. Such a two-step purification (DEAE-Sepharose chromatography and velocity sedimentation through glycerol gradients) provides partially purified human RNase P (~ 200-fold purification), which is sufficiently clean material for in vitro cleavage assay and kinetic analysis. The enzyme is stable at –20°C for several years (*see* **Note 2**).

## 3.4. Design of EGS Against HCMV Protease mRNA

1. The accessible region of the protease gene mRNA can be determined by the mapping procedure described in Chapter 28.
2. Two 2'deoxynucleotide EGSs can be constructed. The functional EGS PR1 (5'-CAGCGCCGGGTGCGGTCTCCGCGCGCAGGTTCAAATCCTGCGGAGAGCA-

3') is designed based on the T-stem and loop, and variable region of the tRNA[Ser] and basepairs to the mRNA encoding the HCMV protease (**Fig. 2A**; *3*). The control EGS PR2 (5'-CAGCGCCGGGTGCGGTCTCCGCGCGCAGGAAGA AATCCTGCGGAGAGCA-3') is derived from PR1, except that the 5'-TTC-3' sequence in the T-loop is changed to AAG (**Fig. 2B**). PR2 is constructed as a control to account for the antisense activity of PR1. Altering these three bases essentially abolishes the activity of the EGS to direct RNase P because these three nucleotides in the T-loop are essential for recognition by RNase P *(20,21)*. Both PR1 and PR2 are chemically synthesized using a DNA oligonucleotide synthesizer. Moreover, PR1 and PR2 that contain a 5' fluoresceinisothiocyanate (FITC) label (Glen Research, Sterling, VA) can also be chemically synthesized from the DNA oligonucleotide synthesizer.

### 3.5. Assay of EGS Activity In Vitro

1. To study the EGSs targeting the HCMV protease gene, the DNA template coding for the substrate was generated by annealing oligonucleotides OliT7 (5'-TAATACGACTCACTATAG-3') and sAP1 (5'-CGGGATCCGCAGCGC CGGCTGGAGAGCGAGAGGCCGGCCTATAGTGAGTCGTATTA-3'), and then transcription was performed using the following mixture:
   a. 4 µL (2 µg) DNA template
   b. 8 µL 5X transcription buffer
   c. 4 µL 100 m*M* DTT
   d. 4 µL 10 m*M* adenosine 5' triphosphate (ATP)
   e. 4 µL α-[32P]-guanosine 5' triphosphate (GTP)
   f. 2 µL 10 m*M* GTP
   g. 4 µL 10 m*M* cytidine 5' triphosphate (CTP)
   h. 4 µL 10 m*M* uridine 5' triphosphate (UTP)
   i. 4 µL $H_2O$
   j. 1 µL RNasin (5 U/uL)
   k. 2 µL T7 RNA polymerase
   Incubate the reaction mixture at 37°C for 4 h or overnight. Add 1 µL of DNase (1 µg/µL) and incubate for 30 min. Extract the reaction mixture with an equal volume of phenol/chloroform/isoamyl alcohol (IAA) (25:24:1) and precipitate the RNA with ethanol.
2. Add an equal volume of 2X dye solution and load onto an 8% polyacrylamide-7-*M* urea gel.
3. Expose the gels to a phosphor screen and scan the screen using a PhosphorImager (e.g., STORM840). The RNA bands are located and excised from the gels. Extract the RNA molecules from the excised gel slice by the crush-soak method using diethylpyrocarbonate (DEPC)-treated water.

### 3.6. EGS-Directed Cleavage of RNA In Vitro

1. Mix 1000 cpm of target RNA with 1, 5, and 50 ng of EGS DNA, respectively, in 8.5 µL of $H_2O$.

2. Heat at 80°C for 3 min and then place on ice.
3. Add 1 μL of 10X buffer B (Type II), 0.25 μL of RNasin and 0.5 μL (0.25 U) of human RNase P.
4. Incubate the reaction mixture at 37°C for 30 min.
5. Terminate the reaction by adding 10 μL of 2X RNA dye solution and 1 μL of phenol. Load onto a 5% polyacrylamide-7-*M* urea gel.
6. Dry the gel and expose it to a phosphor screen. The results will be analyzed with a STORM840 PhosphorImager.

## *3.7. Preparation of Cells for Liposome-Mediated Delivery of EGS*

1. Using forceps, place a single 18-mm round glass cover into each of the wells of the 6-well tissue-culture plate (*see* **Note 3**).
2. Count human foreskin fibroblasts (HFFs) (Clonetics, San Diego, CA) and plate directly onto the glass covers in the wells of the 6-well plate. Plate cells at a density so that cells will be 80% confluent the following day (*see* **Note 4**).
3. The cells should be evenly distributed throughout each well in order to achieve optimal transfection efficiency.

## *3.8. Preparing EGS Liposome Formulation for Delivery and EGS Visualization*

1. Dilute Lipofectamine 2000 reagent (Gibco-BRL) into 100 μL of Opti-MEM medium with EGS to yeild a final concentration of 10 μg/mL lipid and 100 n*M* EGS (*see* **Note 5**).
2. The volume of the EGS liposome formulation should be kept at a minimum (~100 μL) during formulation, since the volume will affect the efficiency of EGS liposome-complex formation.
3. Allow the EGS-liposome complexes to form for 30 min. After this, an additional 900 μL of Opti-Mem media can be added to the EGS liposome formulation (*see* **Note 6**).
4. Aspirate the media from the 6-well plate that contains the cells growing on top of the glass cover slips. Wash the cells twice with phosphate-buffered saline (PBS) to remove any residual FBS.
5. Add 1 μL of EGS-liposome formulation for each of the wells. Place the plate back inside the incubator at 37°C for 7 h (*see* **Note 7**).
6. After 7 h, aspirate the liposome formulation and wash twice with PBS.
7. Fix cells with 4% paraformaldehyde for 20 min at 4°C. Wash cells twice with PBS.
8. Remove glass cover from wells and place on dry towel, with cells facing up. Add 1 U of Texas Red phalloidin diluted in 100 μL of PBS directly onto the cover slip and allow to sit at room temperature for 5 min (*see* **Note 8**).
9. Wash glass cover with PBS by handling cover slips with forceps and dunking in 50-mL conical tube of PBS several times.
10. Mount cover slip, cells face down, on microscope glass slide and seal edges with clear nail polish. Allow nail polish to dry, and view under confocal microscope.

11. The Texas red phalloidin (Molecular Probe, Portland, OR) will stain the actin in the cell, thereby defining the cell cytoplasm. Texas Red phalloidin will not stain the nucleus, and therefore, the nucleus should appear as a hollow circle in the cell.

12. View with the confocal microscope (e.g., Nikon-PCM2000), using lasers to excite the Texas Red phalloidin and FITC labels. If the FITC-conjugated EGS oligonucleotides concentrate to the nucleus then the FITC and Texas Red signals should not overlap or co-localize. Our results indicate that the FITC-labeled EGS oligonucleotides are primarily localized within the nuclei *(3)*.

## 3.9. Assaying the Antiviral Activity of EGS With Single-Step Growth Curve

1. Count human foreskin fibroblast and plate into 6-well plates so that they will reach 60% confluency the following day.

2. Formulate EGS-liposome complexes and apply them to the cells for 7 h in the same manner described in **Subheading 3.8.**

3. Appropriate controls should also be transfected in parallel to the functional EGS. Controls that may be used include: i) liposomes complexed with an EGS with mutations in the T-loop (control for the antisense effect), ii) liposomes complexed with an EGS that targets another unrelated mRNA (e.g., EGS against the mRNA coding for thymidine kinase [TK] of HSV 1), and iii) liposomes alone without an EGS.

4. Aspirate the EGS liposome complexes and wash twice with PBS, then replenish with Dulbecco's modified Eagle's medium (DMEM) containing 10% FBS.

5. Wait 12 h after washing EGS liposome complexes, and then infect the cells at a multiplicity of infection (MOI) of 5 with HCMV AD169 diluted in 1 mL of DMEM for 1–2 h (*see* **Note 9**). The viral inoculum is removed, the cells are washed with PBS, and DMEM supplemented with 10% FBS is added back to the cells.

6. The cells are then harvested at 1, 2, 3, 4, 5, 6, and 7 d post-infection. Viral stocks are prepared by adding an equal volume of 10% skim milk, followed by sonication.

7. Infecting $2 \times 10^5$ HFF cells in 6-well plates and counting the number of plaques 7–10 d post-infection will determine the titers. The titers for each day of the growth curve are done in triplicate.

## 4. Notes

1. A gene for the precursor to tRNA$^{Tyr}$ (su3) from *E. coli* adjacent to an upstream T7 promoter sequence, obtained from Dr. Sidney Altman, is used to prepare pre-tRNA substrate for RNase P assay *(10)*. In vitro-transcribed and $^{32}$P-labeled pre-tRNA$^{Tyr}$ is incubated with 1 μL of fraction sample in buffer B (50 m*M* Tris-HCl, pH 7.5, 100 m*M* NH$_4$Cl, 10 m*M* MgCl$_2$) containing 0.5 μL of RNasin at 37°C for 30 min. The reactions are assayed on an 8% polyacrylamide-7-*M* urea gel.

2. RNase P should be concentrated immediately after glycerol gradient. Purified human RNase P is unstable in a diluted condition.

3. If the glass cover slips are not previously sterile before placing into the 6-well plate, expose the plate containing the cover slips to the ultraviolet (UV) light source in a biosafety cabinet for 1 h.

4. The cells should be evenly distributed throughout each well so that some cells will grow on the cover slip and others will adhere to the well surface.
5. The concentration of the lipid and the EGS can be varied depending on the types of the cells used and the conditions of the experiment. For the delivery of EGS PR1 and PR2 to the HFF cells, 10 μg/mL lipid concentrations and 100 n$M$ EGS was determined to be optimal.
6. The total volume of the EGS liposome formulation should be ~1 mL, which is the minimum volume required to cover the entire surface of a single well in a 6-well plate. This volume is ideal because the EGS liposome mixture must cover the cells growing on top of the glass cover slip at a maximal EGS-Liposome complex concentration. Therefore, increasing the formulation volume to more than 1 mL would adversely affect the transfection efficiency.
7. The transfection/incubation time can be varied, depending on the EGS liposome formulation used. Extended transfection/incubation periods may result in changes in cell morphology, and will not necessarily increased transfection/delivery efficiency.
8. Add carefully; the dye and PBS will remain on the glass cover, resembling a large bead of liquid.
9. Although the manufacturer may claim that the liposome treatment is nontoxic, cell stress is noticeable after extended treatment with the EGS liposome complexes. Therefore, the cells are incubated for 12 h in fresh DMEM after transfection and allowed to recover before being subjected to viral infection. Furthermore, the authors believe that despite extensive washing of the cells after liposome treatment, a small amount of liposome-EGS complexes may still attach to the surface of the cells, and may interfere with viral attachment and entry. Therefore, the 12-h recovery period permits these residual complexes to be either internalized or degraded so that they will not impede viral attachment and entry during infection.

## Acknowledgment

We thank Jiaming Zhu and Phong Trang for technical assistance. W. D. is partially supported by a Block Grant Predoctoral Fellowship (UC-Berkeley). F. L. is a Pew Scholar in Biomedical Sciences and a Scholar of Leukemia and Lymphoma Society. This research has been supported by UC-Berkeley (Chancellor's Special Initiative Award) and NIH (RO1-AI43250 and RO1-DE14842).

## References

1. Forster, A. C. and Altman, S. (1990) External guide sequences for an RNA enzyme. *Science* **249,** 783–786.
2. Yuan, Y., Hwang, E. S., and Altman, S. (1992) Targeted cleavage of mRNA by human RNase P. *Proc. Natl. Acad. Sci. USA* **89,** 8006–8010.
3. Dunn, W., Trang, P., Khan, U., Zhu, J. and Liu, F. (2001) RNase P-mediated inhibition of cytomegalovirus protease expression and viral DNA encapsidation

by oligonucleotide external guide sequences. *Proc. Natl. Acad. Sci. USA* **98**, 14,831–14,836.

4. Kawa, D., Wang, J., Yuan, Y., and Liu, F. (1998) Inhibition of viral gene expression by human ribonuclease P. *RNA* **4**, 1397–1406.

5. Kraus, G., Geffin, R., Spruill, G., Young, A. K., Seivright, R., Cardona, D., et al. (2002) Cross-clade inhibition of HIV-1 replication and cytopathology by using RNase P-associated external guide sequences. *Proc. Natl. Acad. Sci. USA* **99**, 3406–3411.

6. Plehn-Dujowich, D. and Altman, S. (1998) Effective inhibition of influenza virus production in cultured cells by external guide sequences and ribonuclease P. *Proc. Natl. Acad. Sci. USA* **95**, 7327–7332.

7. Hnatyszyn, H., Spruill, G., Young, A., Seivright, R., and Kraus, G. (2001) Long-term RNase P-mediated inhibition of HIV-1 replication and pathogenesis. *Gene Ther.* **8**, 1863–1871.

8. Ma, M. Y., Jacob-Samuel, B., Dignam, J. C., Pace, U., Goldberg, A. R., and George, S. T. (1998) Nuclease-resistant external guide sequence-induced cleavage of target RNA by human ribonuclease P. *Antisense Nucleic Acid Drug Dev.* **8**, 415–426.

9. Altman, S., Kirsebom, L., and Talbot, S. (1993) Recent studies of ribonuclease P. *FASEB J.* **7**, 7–14.

10. Guerrier-Takada, C., Gardiner, K., Marsh, T., Pace, N., and Altman, S. (1983) The RNA moiety of ribonuclease P is the catalytic subunit of the enzyme. *Cell* **35**, 849–857.

11. Bartkiewicz, M., Gold, H., and Altman, S. (1989) Identification and characterization of an RNA molecule that copurifies with RNase P activity from HeLa cells. *Genes Dev.* **3**, 488–499.

12. Lee, J. Y. and Engelke, D. R. (1989) Partial characterization of an RNA component that copurifies with Saccharomyces cerevisiae RNase P. *Mol. Cell Biol.* **9**, 2536–2543.

13. van Eenennaam, H., van der Heijden, A., Janssen, R. J., van Venrooij, W. J., and Pruijn, G. J. (2001) Basic domains target protein subunits of the RNase MRP complex to the nucleolus independently of complex association. *Mol. Biol Cell.* **12**, 3680–3689.

14. Gopalan, V., Vioque, A., and Altman, S. (2002) RNase P: variations and uses. *J. Biol. Chem.* **277**, 6759–6762.

15. Jarrous, N. (2002) Human ribonuclease P: subunits, function, and intranuclear localization. *RNA* **8**, 1–7.

16. Yuan, Y. and Altman, S. (1995) Substrate recognition by human RNase P: identification of small, model substrates for the enzyme. *EMBO J.* **14**, 159–168.

17. Yuan, Y. and Liu, F. (1998) Targeted cleavage of RNA using external guide sequences and eukaryotic RNase P, in *Therapeutic Applications of Ribozymes* (Scanlon, J. K., ed.), Humana Press, Totowa, NJ, pp. 397–413.

18. Liu, F. Y. and Roizman, B. (1991) The herpes simplex virus 1 gene encoding a protease also contains within its coding domain the gene encoding the more abundant substrate. *J. Virol.* **65**, 5149–5156.

19. Welch, A. R., Woods, A. S., McNally, L. M., Cotter, R. J., and Gibson, W. (1991) A herpesvirus maturational proteinase, assemblin: identification of its gene, putative active site domain, and cleavage site. *Proc. Natl. Acad. Sci. USA* **88,** 10,792–10,796.
20. Sprinzl, M., Dank, N., Nock, S., and Schon, A. (1991) Compilation of tRNA sequences and sequences of tRNA genes. *Nucleic Acids Res.* **19,** 2127–2171.
21. Frank, D. N. and Pace, N. R. (1998) Ribonuclease P: unity and diversity in a tRNA processing ribozyme. *Annu. Rev. Biochem.* **67,** 153–180.

# 32

## RNase P Ribozyme As an Antiviral Agent Against Human Cytomegalovirus

### Phong Trang and Fenyong Liu

### Summary

Human cytomegalovirus (HCMV) represents one of the most medically important human viruses and causes a wide spectrum of human diseases, including birth defects and mental retardation in newborns, common opportunistic infections in aquired immunodeficiency syndrom (AIDS) patients (e.g., CMV-associated retinitis and pneumonia), and possibly cardiovascular diseases such as atherosclerosis. This chapter describes the utilization of RNase P ribozyme—specifically, M1GS ribozyme, as a gene-targeting agent for blocking HCMV gene expression and growth. The target for the RNase P ribozyme is the overlapping region of the mRNAs that code for HCMV major transcription factors IE1 and IE2, which are essential for viral gene expression and replication. The methods described in this chapter focus primarily on i) construction of the retroviral vector for expression of M1GS ribozymes in cultured cells, ii) generation of stable cell lines expressing ribozymes, iii) determination of the expression of M1GS RNAs in human cells, and iv) evaluation of the efficacy of ribozymes in inhibiting HCMV IE1/IE2 expression and viral growth. Using these methods, we successfully constructed M1GS RNAs against the IE1/IE2 mRNA sequence and recently showed that a reduction of up to 150- to 3000-fold in HCMV growth is found in cells that express the ribozymes.

**Key Words:** RNase P; M1 RNA; ribozyme; human cytomegalovirus (HCMV); gene therapy; antiviral; retroviral vector.

## 1. Introduction

RNase P is a ribonucleoprotein complex that is responsible for the 5' maturation of tRNA (1). It catalyzes a hydrolysis reaction to remove a 5' leader sequence from tRNA precursors (ptRNA) and several other small RNAs. In *E. coli*, RNase P consists of a catalytic RNA subunit (M1 RNA) and a protein subunit (C5 protein) (1). In vitro, RNase P ribozyme can cleave its pre-tRNA

From: *Methods in Molecular Biology, vol. 252: Ribozymes and siRNA Protocols, Second Edition*
Edited by: M. Sioud © Humana Press Inc., Totowa, NJ

substrate at high divalent ion concentrations (e.g., 100 m$M$ Mg$^{2+}$) in the absence of C5 protein. M1 RNA can cleave an mRNA sequence effectively if an additional small RNA (known as external guide sequence [EGS]), which contains a sequence complementary to the substrate and a 3' proximal CCA, is present *(2)*. EGSs are antisense oligoribonucleotides that have been used to diminish gene expression in mammalian cells *(3,4)* with the help of either RNase P or its catalytic subunit (e.g., M1 RNA). The EGS-based technology takes advantage of RNase P or M1 RNA to cleave a targeted mRNA when the EGS hybridizes to the target RNA, and forms a structure resembling a portion of the natural tRNA substrates of the enzymes *(4)*. Recent studies have shown that expression of EGSs in tissue culture inhibits the gene expression of herpes simplex virus (HSV) and influenza virus, and also abolishes the replication of influenza virus *(5,6)*. To increase the targeting efficiency, the EGS can be covalently linked to M1 RNA (e.g., to the 3' end) to generate a sequence-specific ribozyme, M1GS RNA *(3)*. Thus, in principle, any RNA could be targeted by a custom-designed M1GS for specific cleavage.

### 1.1. RNase P As a Unique Class of Gene-Targeting Ribozymes for Therapeutic Applications

Compared to hammerhead and hairpin ribozymes, RNase P ribozymes possess several unique features as gene-targeting tools. First, RNase P ribozymes can fold into a defined active conformation in the absence of their substrates *(1,7)*. Second, RNase P ribozymes can interact with specific cellular factors, which may include the protein subunits of RNase P *(1,3)*. Their activities have been shown to be stimulated significantly in the presence of cellular proteins. Therefore, RNase P ribozymes may be less susceptible to degradation by intracellular RNases, and may function more efficiently in the presence of cellular proteins. Moreover, the target sites for both hammerhead and hairpin ribozymes are limited by the requirement of specific sequence consensus in the mRNA targets in order for the ribozymes to achieve optimal cleavage *(8)*. RNase P ribozyme has no such sequence requirement at the cleavage site *(1,8)*. Targeted cleavage of mRNA by RNase P ribozymes provides a unique approach to inactivate any RNA of known sequence expressed in vivo. Thus, it can be used as a tool both in basic research—such as the regulation of gene expression during developmental processes—and in clinical applications, such as gene therapy.

### 1.2. Expressing M1GS Ribozymes in Cells

To evaluate the therapeutic potential of M1GS against a human virus, cell lines that express M1GS are generated. The ribozyme is cloned into a retroviral vector, pLXSN-U6 (**Fig. 1**), which is then transfected into amphotrophic packaging cells (e.g., PA317) to generate replication-defective retroviruses. The

**A**

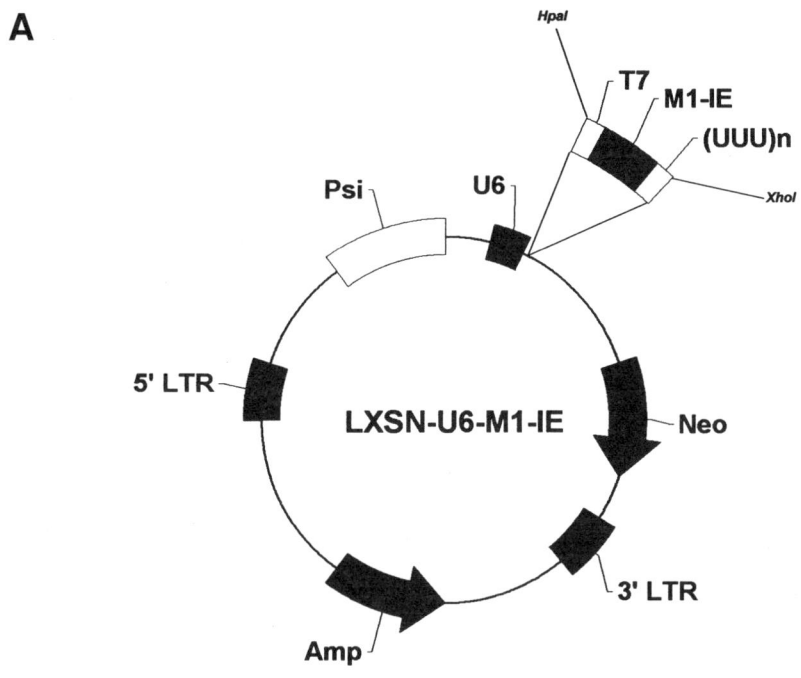

**B**

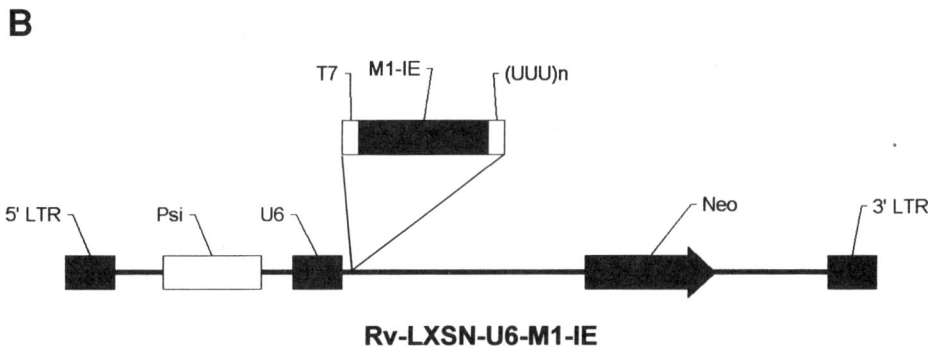

Fig. 1. Schematic presentation of the retroviral vector pLXSN-U6-M1-IE (**A**) and the genome of the replication-defective retrovirus that contains the DNA sequence coding for M1-IE (**B**). pLXSN-U6-M1-IE is derived from pLXSN-U6 by inserting the DNA sequence coding for M1-IE and the transcription terminal signal for RNA polymerase III [(UUU)$_n$] between the *Hpa*I and *Xho*I sites of construct pLXSN-U6, which is derived from retroviral vector construct pLXSN and contains the promoter sequence for small nuclear U6 RNA (*3,19*).

retroviruses are then used to infect human cells (e.g., U373MG) and stable cell clones expressing M1GS are generated.

In order to produce M1GS RNAs endogenously, we have used the mouse U6 small nuclear RNA promoter to express the ribozymes in culture cells *(3,9,10)*. The reasons for choosing this promoter are that: i) U6 snRNA is a very strong polymerase III promoter, which supports the synthesis of more than $10^6$ U6 sn RNAs in every mammalian cell; ii) RNA transcripts synthesized from this promoter remain primarily in the nucleus where RNase P is located; and iii) this promoter has been successfully used to express M1GS RNAs and other functional RNAs in the nucleus at a high level *(4,6,11–13)*. The U6 promoter with the 5' sequence that is required for capping will also be used for the expression of the M1GSs in order to increase their stability and steady-state expression levels *(11–13)*. In our study, M1GS sequences are inserted downstream from the promoter, and are followed by a signal for termination of transcription. Thus, when such a synthetic gene for M1GS is introduced into cells through the retroviral vector, the M1GS sequence is expressed by RNA polymerase III. In contrast, pol II-transcribed RNAs are expected to be transported to the cytoplasm.

Yu et al. *(14)* have compared various gene-expression strategies for the expression of hairpin ribozymes and their efficacy. Their results showed that the expression of the ribozyme driven by a promoter for tRNA$^{Val}$, which is driven by RNA polymerase III, is 88% higher than the RNA polymerase II-controlled expression by human β-actin promoter, and the ribozyme directed by RNA polymerase III yielded 15–25% more inhibition than RNA polymerase II-directed ribozyme *(14)*. These results strongly suggest that the small size, high rate of transcription, and ubiquitous expression of RNA polymerase III-transcription units make them a good choice for expressing ribozymes in cultured cells.

### 1.3. Therapeutic Potential of M1GS As an Antiviral Agent

The M1GS-based gene-targeting method has been used in a variety of applications, from targeting oncogenes to viral genes *(3,15)*. One study has shown that M1GS RNA efficiently cleaves a mRNA substrate that basepairs with the guide sequence in vitro, and effectively inhibits the expression of the target mRNA in mammalian cells *(3)*. For M1GS to be effective as an antiviral agent, the viral gene targeted by the ribozyme must be essential for viral replication and growth. One such target is the overlapping region of the mRNAs (IE1/2 mRNA) coding for human cytomegalovirus (HCMV) immediate early proteins IE1 and IE2, which are the major viral transcriptional factors responsible for activation of viral gene expression *(16)*. Shutting down the expression of these proteins would severely limit the ability of HCMV to replicate and grow. We have recently shown that M1GS RNAs with guide sequences (GS) complementary to the IE1/2 mRNA region reduce more than 80% of the expression levels of IE1 and IE2, and inhibit viral growth by about 150-fold in human cells *(9)*.

HCMV is an important opportunistic pathogen that affects individuals with immune functions that are compromised or immature *(16)*. For example, HCMV is a leading cause of retinitis-associated blindness and other debilitating conditions such as pneumonia and enteritis among AIDS patients. This virus also causes mental and behavioral dysfunctions in children who are infected in utero (16). The continued development of effective antiviral compounds and approaches is central in controlling HCMV infections and preventing HCMV-associated complications. Our recent finding that M1GS ribozymes are effective in inhibiting the growth of HCMV as well as herpes simplex virus 1 (HSV-1) clearly demonstrate the value of M1GS RNAs as a new class of ribozymes for antiviral applications *(9,10,17)*.

This chapter focuses on the expression of M1GS through the generation of stable M1GS-expressing cell lines and characterization of the ribozyme activity in cultured cells by Northern and Western analysis using IE1/2 mRNA as a model target for antiviral therapy. The construction and engineering of M1GS ribozymes are described in Chapters 28 and 29.

## 2. Materials
### 2.1. Reagents and Stock Solutions

1. 1 $M$ Tris-HCl, pH 8.0.
2. 1 $M$ Tris-HCl, pH 7.5.
3. 1 $M$ Tris-HCl, pH 7.0.
4. 0.5 $M$ ethylenediaminetetraacetic acid (EDTA).
5. 5 $M$ NaCl.
6. 1 $M$ MgCl$_2$.
7. 5 $M$ NaOH.
8. 20% sodium dodecyl sulfate (SDS).
9. 30% acrylamide/0.8% N,N' methylene *bis*-acrylamide.
10. 8 $M$ urea.
11. 10X TBE: 0.89 $M$ Tris-borate, 10 m$M$ EDTA.
12. 20X standard saline citrate (SSC): 3 $M$ NaCl, 0.3 $M$ trisodium citrate.
13. Diethylpyrocarbonate (DEPC)-treated H$_2$O: double-distilled water is mixed with 0.1% DEPC and stirred overnight. The DEPC is inactivated by autoclaving for 20 min.
14. 10 mg/mL ethidium bromide.
15. Phenol.
16. Chloroform/isoamyl alcohol (IAA) (24:1, v/v).
17. 1 $M$ dithiothreitol (DTT).

18. β-mercaptoethanol.
19. Dimethyl sulfate (DMS).
20. Neomycin (Gibco-BRL).
21. Trizol reagent (total RNA isolation reagent) (Gibco-BRL).
22. ECF Western blotting kit (Amersham).
23. [$^{32}$P]-labeled nucleotides (Amersham).

## 2.2. Solutions and Buffers

1. 10X buffer A (cleavage assay buffer): 500 m$M$ Tris-HCl, pH 7.5, 1 $M$ NH$_4$Cl, 1 $M$ MgCl$_2$ (autoclaved).
2. 2X RNA dye solution: 8 $M$ urea, 20 m$M$ EDTA, 0.25 mg/mL bromophenol blue, 0.25 mg/mL xylene cyanol FF.
3. Phosphate-buffered saline (PBS).
4. Prehybridization buffer: 6X SSC, 0.05% sodium pyrophosphate, and 2X Denhardt's solution.
5. 1X Denhardt's solution: 0.1% bovine serum albumin (BSA), 0.1% polyvinylpyr-rolidone, 0.1% Ficoll), 0.1% SDS, and 200 μg/mL of denatured salmon sperm.
6. TNT: 10 m$M$ Tris-HCl, pH 7.0, 150 m$M$ NaCl, 0.05% Tween-20.

## 2.3. Enzymes and Reaction Buffers

1. T4 DNA ligase and 10X ligation buffer (New England Biolabs).
2. Polynucleotide kinase (PNK) and 10X kinase buffer (New England Biolabs).
3. T7 in vitro transcription system and 5X transcription buffer (Promega).
4. Polymerase chain reaction (PCR) system including 10X PCR buffer, 25 m$M$ MgCl$_2$, four 10 m$M$ deoxynucleotide 5' triphosphate (dNTP), $Taq$ DNA polymerase (Promega).

## 2.4. Virus, Cells, and Plasmids

1. HCMV (strain AD169).
2. Human foreskin fibroblasts (HFF).
3. Human U373MG cells (American Type Culture Collection [ATCC]).
4. PA317 cells (amphotropic retrovirus packaging cell line).
5. M1 RNA clone (plasmid FL117).

## 3. Methods

## 3.1. Construction of M1GS Ribozyme

M1GS can be designed to target any specific mRNA sequence of interest because the target sequence contains accessible regions for the guide sequence (GS) to bind (18). The guidelines and protocols for mapping the accessible regions of an mRNA and designing a M1GS that targets the mRNA region are described in Chapters 28 and 29. Following these guidelines, we designed M1GS RNAs (e.g., M1-IE) that target a previously mapped accessible region of HCMV IE1/2 mRNA *(9)*. The following protocols are for the generation of the DNA templates coding for the ribozymes and for the in vitro transcription synthesis of the ribozymes.

1. The DNA template for the generation of M1GS ribozyme (M1-IE) targeting HCMV IE1/2 mRNA is generated by PCR amplification from plasmid FL117 (the DNA sequence encoding the wild-type M1 RNA sequence cloned into pUC19 vector) with the following oligonucleotides:

   5' primer (OliT7) 5'-TAATACGACTCACTATAG-3' (contains T7 promoter)
   3' primer (M1EI12) 5'- CCGCTCGAGAAAAAATGGTGCAACGAGAACC CTGTGGAATTG-3'

   Add the following sequentially:
   a. 10 μL 10X PCR buffer (Mg$^{2+}$ free)
   b. 2 μL 10 m$M$ deoxyadenosine 5' triphosphate (dATP)
   c. 2 μL 10 m$M$ deoxythymidine 5' triphosphate (dTTP)
   d. 2 μL 10 m$M$ deoxyguanosine 5' triphosphate (dGTP)
   e. 2 μL 10 m$M$ deoxycytidine 5' triphosphate (dCTP)
   f. 8 μL MgCl$_2$
   g. 10 pmol 5' primer (OliT7)
   h. 100 pmol 3' primer (M1IE12)
   i. 1 μg *Pvu*II-digested FL117 plasmid
   j. 1 μL *Taq* DNA polymerase
   k. DEPC-treated water final reaction volume to 100 μL
   PCR amplification of the DNA template using the following amplification program
   Denaturing for 2 min at 94°C
   30 cycles of:
   Denaturingfor 2 min at 94°C
   Annealing for 1 min at 47°C
   Extension for 1 min at 72°C
   Final Extension for 10 min at 72°C
   The PCR DNA products are separated in 5% polyacrylamide gels under non-denaturing conditions, and are purified and used as the template for the in vitro transcription synthesis of the ribozymes.
2. The M1-IE ribozyme is synthesized by an in vitro transcription using the follow-ing mixture:
   a. 4 μL (2 μg) M1-IE plasmid DNA from PCR reaction
   b. 8 μL 5X transcription buffer
   c. 4 μL 100 m$M$ DTT
   d. 4 μL 10 m$M$ ATP
   e. 4 μL 10 m$M$ GTP
   f. 4 μL 10 m$M$ CTP
   g. 4 μL 10 m$M$ UTP

    h.  4 μL H$_2$O
    i.  1 μL RNasin (5 U/μL)
    j.  2 μL T7 RNA polymerase

3.  Incubate the reaction mixture at 37°C for 4 h to overnight. Add an equal volume of 2X RNA dye solution and load onto an 8% polyacrylamide-7-*M* urea gels. Place the gel on a TLC plate (silica gel UV$_{256}$). Visualize RNA bands by briefly shadowing with a shortwave UV lamp. Extract RNA from the excised gel slice by the crush-soak method using DEPC-treated water.

4.  To make a DNA template for a sample substrate (ie37) representing the targeted sequence of HCMV IE1/2 mRNA, two oligonucleotides are synthesized using a DNA synthesizer.
(OliT7) 5'-TAATACGACTCACTATAG-3' (contain T7 promoter)
(sIE1) 5'-CGGGATCCTTTCTCGGGGTTCTCGTTGCGATTCCCGGTCCT
ATAGTGAGTCGTATTA-3'

    The two oligonucleotides are then use in a PCR reaction to generate the DNA template. Uniformly labeled sample substrate is synthesized by in vitro transcription of the DNA template as described in **Subheading 3.1.2.**, except including 30 μCi α-[$^{32}$P]-GTP and using 1 m*M* GTP stock instead of 10 m*M* in the reaction.

### 3.2. M1-IE Ribozyme Cleavage of Substrate ie37 In Vitro

1.  Mix 10 n*M* of M1-IE and 10 n*M* of $^{32}$P-labeled ie37 mRNA, 1 μL of 10X buffer A, and H$_2$O in a final volume of 10 μL.
2.  Incubate the reaction mixture at 37°–50°C for 30 min.
3.  Terminate the reaction by adding 10 μL of 2X RNA dye solution and 1 μL of phenol. Load onto an 8% polyacrylamide-7-*M* urea gel.
4.  Expose the gel to a phosphorscreen and scan with a PhosphorImager.

### 3.3. Construction of a Retroviral Vector Containing M1-IE Ribozyme Expressed by RNA Polymerase III In Vivo

1.  The PCR-generated DNA sequences coding for the M1-IE ribozymes are digested with XhoI restriction enzyme.
2.  The sequences are then subcloned into retroviral vector pLXSN-U6 at the *Xho*I/ *Hpa*I (blunt-end) site and placed under the control of the U6 RNA promoter (*3,19*; **Fig. 1**). The pLXSN-U6 vector is derived from LXSN, and contains the promoter sequence for small nuclear U6 RNA. Because extra flanking sequences may affect the folding of the ribozyme, the ribozyme is expressed with minimal flanking sequences between the 3' terminal CCA sequence of the guide sequence (GS) and the terminus of the transcript. To achieve this, the transcription termination signal of RNA polymerase III (5'-TTTTTT-3') is included in the 3' PCR primer and is placed immediately downstream from the 3' terminal CCA of the guide sequence (*see* **Note 1**).

## 3.4. Expression of M1-IE Ribozyme in Tissue-Culture Cells

The retroviral vector carrying M1-IE ribozyme (e.g., pLXSN-U6-M1-IE) (**Fig. 1A**) is used to transfect an amphotrophic packaging cell line, PA317. The transfection can be performed with a variety of methods, such as calcium phosphate precipitation or cationic liposomes.

1. Count PA317 cells and plate onto 6-well plate. The cells should be 80% confluent the day of transfection (*see* **Note 2**).
2. 12–16 h after cell plating, transfect the cells with 10–20 μg of plasmid LXSN-U6-M1-IE DNA using a calcium phosphate transfection kit (Stratagene, San Diego, CA).
3. 48 h after transfection, collect culture supernatants (about 3 mL). These supernatants are used to infect human U373MG cells.
4. One day before retroviral infection, plate U373MG cells onto a 6-well plate, so that they are 80% confluent the day of infection with the culture supernatant containing retroviruses.
5. Aspirate out the media and add 1.5 mL of the collected retroviral stock (*see* **step 3**) to the U373MG cell culture, and incubate for 4–12 h (*see* **Note 3**). Shake occasionally. Replace the inoculum with fresh DMEM supplemented with 10% fetal bovine serum (FBS).
6. 48–72 h after infection, add neomycin (Gibco-BRL) to the culture medium at a final concentration of 600 μg/mL.
7. Cells are selected subsequently in the presence of neomycin for 2 wk, and neomycin-resistant cells are cloned (*see* **Note 4**).
8. Aliquots of these cells are frozen for long-term storage in liquid nitrogen or used for further studies.

## 3.5. Expression of M1-IE Ribozyme is Analyzed by a Northern Blot (see Note 5)

### 3.5.1. Isolation of Total RNAs From Cells Expressing the Ribozymes

1. $1-5 \times 10^7$ cells are used to isolate total RNA with Trizol reagent.
2. Wash cells twice with PBS.
3. Add 7.5 mL of Trizol, lyse cells by repetitive pipeting and incubate for 5 min at room temperature.
4. Add 1.5 mL of chloroform, shake vigorously for 15 s, and incubate for 3 min at room temperature.
5. Centrifuge the mixture at no more than 12,000$g$ for 30 min at 4°C.
6. Collect the aqueous layer, mix with 3.75 mL isopropanol, and incubate at room temperature for 10 min.
7. Centrifuge the mixture at no more than 12,000$g$ for 10 min at 4°C.
8. Wash the pellet with 3.75 mL of ethanol and mix by vortexing.
9. Centrifuge at no more than 7500$g$ for 5 min at 4°C.
10. Resuspend the pellet in RNase-free water.

## 3.5.2. Northern Blot Analysis

1. Load 10 μg of total RNA prepared *above* onto a 1–2.5% formaldehyde agarose gel containing 45 mL of formaldehyde, 50 mL of 5X Northern buffer (0.2 *M* MOPS, pH 7.0, 50 m*M* NaOAc, 5 m*M* EDTA [*see* **Note 6**]). Separate the RNAs by running the gel at a constant voltage.
2. Wash the gel three times for 10 min each in $H_2O$.
3. Transfer the RNAs from the agarose gel onto a nitrocellulose membrane.
4. Place a sponge slightly larger than the gel in a glass dish. Fill the dish with enough 20X SSC to leave the soaked sponge about half-submerged in buffer.
5. Cut three pieces of Whatman 3MM paper to the same size as the sponge. Place them on the sponge and wet with 20X SSC.
6. Place the gel on the filter paper and squeeze out air bubbles by rolling a glass pipet over the surface.
7. Cut four strips of plastic wrap and place over the edges of the gel to prevent buffer from "short-circuiting" around the gel rather than passing through it.
8. Cut a piece of nitrocellulose membrane just large enough to cover the exposed surface of the gel. Wet the membrane with RNase-free water.
9. Place the wetted membrane on the surface of the gel. Remove any air bubbles under the membrane by carefully rolling a glass pipet over the surface.
10. Flood the surface of the membrane with 20X SSC. Cut five sheets of Whatman 3MM paper to the same size as the membrane and place on top of the membrane. Place paper towels on top of the Whatman 3MM paper to a height of ~4 cm.
11. Lay a glass plate on top of the structure and add a weight (1 kg) to hold everything in place. Leave overnight.
12. Rinse the membrane with deionized water three times for 10 min each.
13. Bake the nitrocellulose membrane at 80°C for 1.5 h in a vacuum oven.
14. The nitrocellulose membrane is pre-hybridized for 4 h and hybridized for 16 h with labeled oligonucleotide complementary to M1-IE RNA at 65°C in 6X SSC, 0.05% sodium pyrophosphate, 2X Denhardt's solution, 0.1% SDS, and 200 μg/mL of salmon sperm. Wash the filter with 2X SSC, 1X SSC, and 0.5X SSC containing 0.1% SDS, each for 15 min at 42°C.
15. Expose the membrane to a phosphorscreen and scan with a Phosphorimager.

## 3.6. Functional Analysis of M1-IE in Cells

Northern blot, Western blot and viral titering are performed to determine the efficacy of M1-IE ribozyme in reducing the expression of target viral mRNA and protein, thereby inhibiting viral replication (*see* **Note 7**).

## 3.6.1. Viral Infection and Preparation of RNA

1. $1 \times 10^6$ human U373MG cells containing the M1-IE ribozymes are infected with HCMV at a multiplicity of infection (MOI) of 1 in an inoculum of 1.5 mL of Dulbecco's modified Eagle's medium (DMEM) supplemented with 1% FCS and incubated for 2 h (*see* **Note 8**).

2. Replace the inoculum with DMEM supplement with 10% FBS and incubate for 24–48 h (mRNA extract) or 48–72 h (protein extract).
3. Trizol reagent is used to isolate mRNA as described in **Subheading 3.4.2.** The level of viral IE1/2 mRNA inhibition is determined by Northern blot as described in **Subheading 3.2.**

## 3.6.2. Isolation of Protein Extracts From Viral Infected Cells

1. Aspirate out the culture media and rinse the cells twice with 5 mL of PBS each.
2. Scrape the cells off the flask, and spin down at 3000$g$ for 5 min at 4°C.
3. Resuspend cell pellets into 50–100 µL of cold PBS, then add the same volume of 2X disruption buffer (4% SDS, 100 m$M$ β-mercaptoethanol, 100 m$M$ Tris-HCl, pH 7.0, 6% sucrose, and trace amount of bromophenol blue and xylene dye), and mix by vortexing for 1 minute.
4. Sonicate the mixture for 20–30 s for three times on ice.
5. Boil the sample for 1 min before loading on SDS-polyacrylamide gel electrophoresis (PAGE) gel. The level of viral IE1/2 protein is determined by Western blot.
6. Load 50 µg of proteins onto 9% SDS-polyacrylamide gels crosslinked with $N,N''$–methylenebisacrylamide with a stacking layer of 4.5% acrylamide/*bis*-acrylamide. Separate the proteins by running the gels at a constant power setting.
7. Transfer the proteins from the polyacrylamide gel onto a nitrocellulose membrane, using an electrophoretic transfer apparatus with a constant current of 150 mA for 2 h.
8. Air-dry the membrane.
9. Incubated (pre-blocking step) the nitrocellulose membrane with TNT buffer (10 m$M$ Tris, pH 7.0, 150 m$M$ NaCl, 0.05% Tween-20) plus 2.5% nonfat dry milk (NFDM) for 1 h in an orbital shaker.
10. After blocking, incubate for 1 h with primary antibody (anti-IE1/2) at a dilution of 1:500 in TNT supplemented with 1.25% of NFDM.
11. Wash the membrane with TNT buffer three times (5 min each), and then incubate with fluorescein-linked anti-species antibody (secondary antibody) diluted at 1:1000 dilution. Incubate for 1 h in an orbital shaker (ECF Western blotting kit (Amersham).
12. Wash the membranes with TNT three times (5 min each), and then incubate with anti-fluorescein alkaline phosphatase conjugate dilute 1:2500 dilutions for 1 h in an orbital shaker.
13. Wash the membrane with TNT three times and then incubate with ECF substrate for 20 min at room temperature.
14. Air-dry the membrane for 20 min and scan the membrane using a STORM840 PhosphorImager.

## 3.6.3. M1-IE Ribozyme Activity In Vivo

The level of viral growth inhibition by M1-IE ribozyme is determined by assaying the viral titers in tissue-culture cells.

1. The level of viral growth inhibition by M1-IE ribozyme is determined by assaying the viral titers in tissue culture cells.
2. Infect $5 \times 10^5$ ribozyme-expressing cells with HCMV at a MOI of 1 in 0.5 mL of DMEM for 2 h.
3. Wash the cells with PBS, then add 0.5 mL of DMEM supplemented with 10% FBS, and incubate for an interval of 1–7 d.
4. Harvest cells at 1-d intervals throughout the 7 d after infection. Prepare viral stock by adding 0.5 mL of 10% FBS/10% nonfat milk solution. Scrape the cells and sonicate them three times.
5. Prepare 10-fold serial dilution of the viral stock in 1 mL of DMEM for each dilution.
6. Infect $1 \times 10^5$ cells of human foreskin fibroblasts with 1 mL of viral dilution and incubate for 2 h.
7. Wash the cells with DMEM, then overlay with 1% agarose and DMEM containing 10% FBS in a 1:1 ratio.
8. Count the number of viral plaques 10–14 d after infection. Plaque-forming unit (PFU) is determined by the highest viral dilution that yields plaque.

## 4. Notes

1. The extra sequence 3' downstream from the 3' CCA sequence significantly affects the cleavage activity of the ribozyme. Our unpublished results suggest that the shorter the extra sequence 3' downstream from the 3' CCA sequence is, the more active ribozyme will be. Therefore, in our retroviral expression construct for the ribozyme, the transcriptional terminal signal is placed immediately downstream from the 3' terminal CCA sequence of the M1GS ribozyme so that the ribozyme expressed from the U6 promoter terminates immediately after the 3'CCA sequence.
2. The cells should be evenly distributed throughout each well in order to achieve optimal transfection efficiency.
3. To increase the efficiency of retroviral infection to target cells, add polybrene to a final concentration of 8 μg/mL. The efficiency of infection can also be improved by carrying out multiple rounds of infection and using a high level of multiplicity of infection (MOI). The retroviral stock obtained from our experiments is usually about $1 \times 10^3$ PFU/mL.
4. Cells infected with retrovirus are subsequently splitted sparsely over 10 cultured flasks and placed under neomycin to select for cloned ribozyme-expressing cell lines. In our experiments, up to six clonal cell lines from each ribozyme construct are selected, and most of the selected lines are found to express a high level of ribozymes.
5. To determine if the M1GS RNAs expressed by the U6 promoter is primarily localized in the nucleus, cellular RNAs can be isolated from the nuclear and cytoplasmic fractions separately. The presence of M1GS in these fractions can be determined using Northern analysis. Our results indicate that M1-IE is expressed primarily in the nucleus (*9*).

6. This gel-running step should be completely RNase-free, since mRNA is extremely sensitive to degradation. It is recommended that DEPC-treated $H_2O$ is used for the entire gel-running process from making the gel to running buffer.

7. In addition to the functional ribozyme M1-IE, a control ribozyme should also be constructed. In our study (*9*), mutant ribozyme C102-IE contains several point mutations (e.g., $A_{347}C_{348} \rightarrow C_{347}U_{348}$, $C_{353}C_{354}C_{355}G_{356} \rightarrow G_{353}G_{354}A_{355}U_{356}$) at the catalytic domain (P4 helix), which render it nonfunctional (*20*). However, the mutant ribozyme contains the exact same guide sequence as M1-IE, which can bind complementarily to the target mRNA. Indeed, C102-IE binds to the IE1/2 mRNA sequence in vitro as well as M1-IE (*9*). These observations suggest that C102-IE ribozyme may be a good control for the antisense effect.

8. Among the selected M1GS-expressing cell lines that are generated in the previous steps, only the cells that express similar levels of M1-IE and control ribozymes are subsequently used to determine the inhibition level of viral gene expression and viral load.

## Acknowledgments

P. T. is a recipient of the American Heart Association Predoctoral Fellowship (Western States Affiliate). F. L. is a Pew scholar in Biomedical Sciences, a Scholar of Lymphoma and Leukemia Society of America, and an Established Investigator of American Heart Association. The research has been supported by grants from the March of Dimes National Birth Defects Foundation, and NIH (RO1-AI41927 and R29-GM54815).

## References

1. Altman, S. and Kirsebom, L. A. (1999) RNase P, in *The RNA World* (Gesteland, R. F., Cech T. R., and Atkins J. F., ed.), Cold Spring Harbor Laboratory Press, Cold Spring Harbor, NY, pp. 351–380.
2. Forster, A. C. and Altman, S. (1990) External guide sequences for an RNA enzyme. *Science* **249,** 783–786.
3. Liu, F. and Altman, S. (1995) Inhibition of viral gene expression by the catalytic RNA subunit of RNase P from Escherichia coli. *Genes Devel.* **9,** 471–480.
4. Yuan, Y., Hwang, E. S. and Altman, S. (1992) Targeted cleavage of mRNA by human RNase P. *Proc. Natl. Acad. Sci. USA* **89,** 8006–8010.
5. Plehn-Dujowich, D. and Altman, S. (1998) Effective inhibition of influenza virus production in cultured cells by external guide sequences and ribonuclease P. *Proc. Natl. Acad. Sci. USA* **95,** 7327–7332.
6. Kawa, D., Wang, J., Yuan, Y. and Liu, F. (1998) Inhibition of viral gene expression by human ribonuclease P. *RNA* **4,** 1397–1406.
7. Frank, D. N. and Pace, N. R. (1998) Ribonuclease P: unity and diversity in a tRNA processing ribozyme. *Annu. Rev. Biochem.* **67,** 153–180.
8. Cobaleda, C. and Sanchez-Garcia, I. (2001) RNase P: from biological function to biotechnological applications. *Trends Biotechnol.* **19,** 406–411.

9. Trang, P., Lee, M., Nepomuceno, E., Kim, J., Zhu, H., and Liu, F. (2000) Effective inhibition of human cytomegalovirus gene expression and replication by a ribozyme derived from the catalytic RNA subunit of RNase P from Escherichia coli. *Proc. Natl. Acad. Sci. USA* **97,** 5812–5817.

10. Trang, P., Lee, J., Kilani, A. F., Kim, J., and Liu, F. (2001) Effective inhibition of herpes simplex virus 1 gene expression and growth by engineered RNase P ribozyme. *Nucleic Acids Res.* **29,** 5071–5078.

11. Das, G., Henning, D., and Reddy, R. (1987) Structure, organization, and transcription of Drosophila U6 small nuclear RNA genes. *J. Biol. Chem.* **262,** 1187–1193.

12. Bertrand, E., Castanotto, D., Zhou, C., Carbonnelle, C., Lee, N. S., Good, P., et al. (1997) The expression cassette determines the functional activity of ribozymes in mammalian cells by controlling their intracellular localization. *RNA* **3,** 75–88.

13. Good, P. D., Krikos, A. J., Li, S. X., Bertrand, E., Lee, N. S., Giver, L., et al. (1997) Expression of small, therapeutic RNAs in human cell nuclei. *Gene Ther.* **4,** 45–54.

14. Yu, M., Ojwang, J., Yamada, O., Hampel, A., Rapapport, J., Looney, D., and Wong-Staal, F. (1993) A hairpin ribozyme inhibits expression of diverse strains of human immunodeficiency virus type 1. *Proc. Natl. Acad. Sci. USA* **90,** 6340–6344.

15. Cobaleda, C. and Sanchez-Garcia, I. (2000) In vivo inhibition by a site-specific catalytic RNA subunit of RNase P designed against the BCR-ABL oncogenic products: a novel approach for cancer treatment. *Blood* **95,** 731–737.

16. Mocarski, E. S. and Courcelle, C. T. (2001) Cytomegaloviruses and their replication, in *Fields Virology,* 3rd ed., (Knipe, D. M. a. H., P.M., ed.), Lippincott-Williams & Wilkins, Philadelphia, PA, pp. 2629–2673.

17. Trang, P., Kilani, A., Lee, J., Hsu, A., Liou, K., Kim, J., et al. (2002) RNase P ribozymes for the studies and treatment of human cytomegalovirus infections. *J. Clin. Virol.* **25,** 63–74.

18. Yuan, Y. and Liu, F. (1998) Targeted cleavage of RNA using external guide sequences and eukaryotic RNase P, in *Therapeutic Applications of Ribozymes* (Scanlon, J. K., ed.), Humana Press, Totowa, NJ, pp. 397–413.

19. Miller, A. D. and Rosman, G. J. (1989) Improved retroviral vectors for gene transfer and expression. *Biotechniques* **7,** 989,990.

20. Kim, J. J., Kilani, A. F., Zhan, X., Altman, S., and Liu, F. (1997) The protein cofactor allows the sequence of an RNase P ribozyme to diversify by maintaining the catalytically active structure of the enzyme. *RNA* **3,** 613–623.

# 33

# Inhibition of Gene Expression by Nucleic Acid Enzymes in Rodent Models of Human Disease

## Per Ole Iversen and Mouldy Sioud

## Summary

Nucleic acid enzymes have emerged as a versatile technique for sequence-specific gene silencing in a wide range of cells. However, the question remains as to whether, for example, DNA enzymes and ribozymes are functional in animals. In this chapter, we describe two different rodent models of human diseases—namely, leukemia and chronic heart failure. We specifically reduced *Raf-1* expression in leukemic mice using an anti-*Raf-1* DNA enzyme. A continuous supply of this catalytic molecule led to a substantial reduction in leukemic-cell burden and survival. Rats with postinfarction heart failure were treated with a DNA enzyme targeting TNFa, and this led to a substantial improvement of cardiac function concomitant with a restoration of the hemodynamic status of the animals. The described protocols should facilitate the *in vivo* evaluation of other oligonucleotide-based therapy such as small interferering RNAs (siRNAs).

**Key Words:** Cardiac failure; DNA enzyme; ribozyme leukemia; osmotic minipump.

## 1. Introduction

In order to elucidate the molecular mechanism of most physiological and pathological processes, it is essential to understand the function and regulation of single gene products. This is often a major challenge, and is not always possible, in the intact organism. The advent of targeted disruption of genes in rodents has offered new insights into, and promises of, future mechanism-based treatment modalities. However, the engineering of gene-knockout models is a strenuous exercise with high costs. In addition, uncertainty remains as to how the missing, single gene may adversely affect the animal as a whole during its development.

From: *Methods in Molecular Biology, vol. 252: Ribozymes and siRNA Protocols, Second Edition*
Edited by: M. Sioud © Humana Press Inc., Totowa, NJ

The introduction of nucleic acid enzymes (ribozymes and DNA enzymes) that target mRNAs offers a method of circumventing some of these obstacles in the adult animal. Thus, it is being increasingly used in various models of human disease such as septicemia and inflammation (1), neoplasia (2), and circulatory disorders (3). A major challenge has been to determine that these molecules reach their target cells in sufficient quantities, and are not degraded by endonucleases, once given to the living animal (4,5).

We have developed DNA enzymes to study the effect of inhibiting the gene expression of the pleiotropic cytokine tumor necrosis factor α (TNFα) and Raf-1 kinase in rodent models of leukemia and heart failure (2,3,6). Considerable efforts have been made to develop efficient delivery systems for these compounds in vivo, and design them to be nuclease-resistant.

## 2. Materials

### 2.1. Catalytic Nucleic Acids

DNA enzymes that target TNF-α and *Raf-1* kinase have the following sequences: 5'-GTGCTCA**GGCTAGCTACAACGA**GGTGTCCTTT-3'; 5'-TATGTGCTCCA**GGCTAGCTACAACCGA**TGATGCA-3'. Bold letters correspond to the DNA enzyme catalytic domain. To increase stability, the hydroxyl group of the phosphate backbone within the DNA antisense arms were replaced with sulphur atoms to make them phosphorothioate-modeled DNA enzymes. The inactive DNA enzymes contain reversed antisense arms. The molecules were chemically synthesized by Eurogentec (Belgium).

### 2.2. Animals and Transplantable Cells

#### 2.2.1. Mice

In the leukemia studies we have used nucleotide-binding oligomerization/ severe combined immunodeficient (NOD/SCID)-mice of both genders. These mice are severely immunocompromised, and lack sufficient numbers of functional phagocytes. Upon sublethal irradiation, the lymphocytes are also severely impaired, rendering these animals useful for xenograft transplantation (*see* **Note 1**). This transplantation model has been extensively validated (7,8).

#### 2.2.2. Juvenile Myelomonocytic Leukemia

We used primary malignant cells from the bone marrow of untreated patients with established juvenile myelomonocytic leukemia (JMML). This rare childhood leukemia has a very poor prognosis with few treatment options, and cure can only be offered with bone-marrow transplantation (9). It has been well-documented that the JMML cells are dependent on cytokines, particularly

TNFα and granulocyte-macrophage colony-stimulating factor (GM-CSF), for their growth and survival *(9)*.

Following density-gradient centrifugation, JMML cells were isolated from the mononuclear-cell suspension using monoclonal antibodies (MAbs) that recognize the heterodimeric GM-CSF receptor and the TNFα receptors *p75/p55*, and a magnetic sorting system (*see* **Note 2**).

## 3. Methods
### 3.1. Preparation of Animal Disease-Models
#### 3.1.1. Juvenile Myelomonocytic Leukemia: Mice

We injected 20–30 million JMML cells iv. Before injection, the mouse had been lightly anesthetized with barbiturate (25 mg/mL, ip). In addition, we made a laparatomy on the day of transplantation or 4 wk after. An osmotic mini-pump (Alzet, Palo Alto, CA) was then carefully placed into the peritoneal cavity (*see* **Note 3**).

#### 3.1.2. Post-Infarction Heart Failure: Rats

The rats received a surgical plane of anesthesia (barbiturate 50 mg/kg, ip). A midline incision was made in the chest and the left ventricle exposed before the left coronary artery (LCA) was identified. The LCA was ligated, and the immediate bleeching of the left ventricular myocardium was taken to signify cessation of its blood supply resulting from an acute myocardial infarction. Sham-operated animals underwent similar surgery, but without ligation of their LCAs (*see* **Note 4**). For other hemodynamic measurements, catheters were placed in carotid and femoral arteries and secured (**Fig. 1**).

#### 3.1.3. In Vivo Detection of Catalytic Activity

An essential part of the experimental set-up involved validating that the catalytic enzyme activities were maintained following delivery to the animals. Therefore, we sampled tissue specimen from the target organs, homogenized them, and measured their catalytic activity on artificial substrates (*see* **Note 5**).

### 3.2. Delivery of the DNA Enzymes In Vivo

DNA enzymes were encapsulated with cationic liposomes at 25 µg/mL; 1,2,-*bis*-(oleoxyloxy)-3-trimethylammanium propane (DOTAP), Boehringer Mannheim, Mannheim, Germany) using the following procedure:

1. In polystyrene tubes, mix 25 µL of DOTAP and 75 µL of transfection buffer (20 m*M* HEPES, 150 m*M* NaCl, pH 7.4) (*see* **Note 6**).

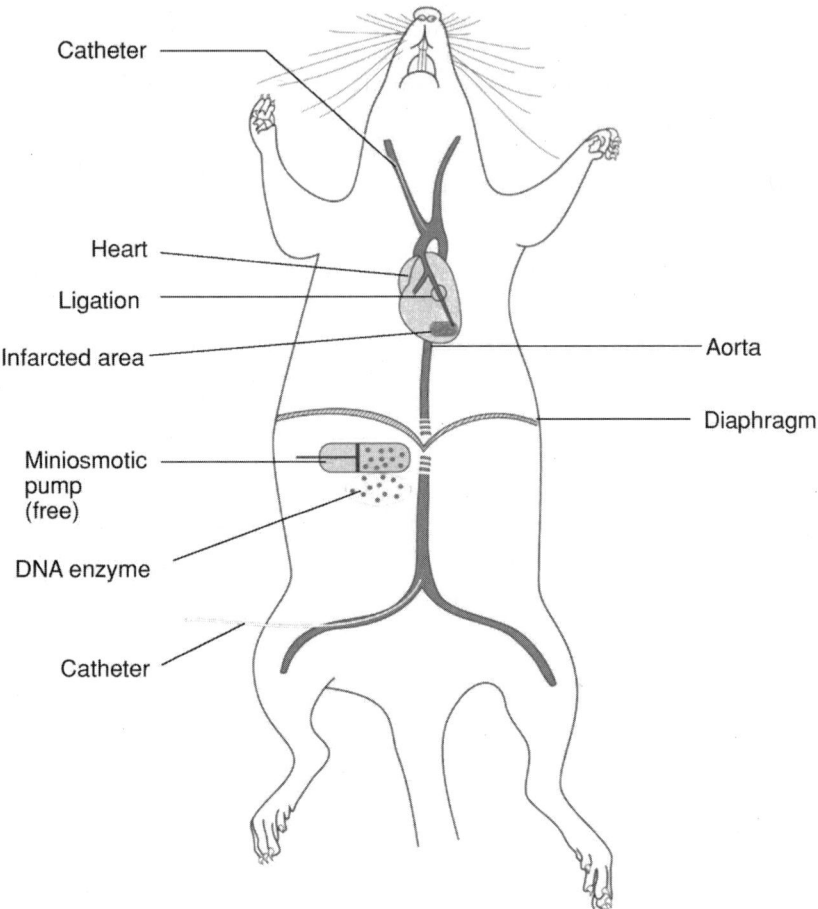

Fig. 1. A schematic illustration of the experimental setup for continuous hemody-namic measurements and delivery of test molecules intraperitoneally to rats with myo-cardial infarction.

2. In an Eppendorf tube, mix 100–300 μg (10–30 μL) of DNA enzyme and 50 μL of transfection buffer.
3. Mix solution II, and I vortex gently, and incubate at room temperature for at least 15 min. Adjust to the desired volume with NaCl (0.9 g/L). The TNF-α DNA enzyme was delivered at 1 μg/kg/h. NOD/SCID mice received 10 μg of anti-*Raf* DNA enzyme/d.
4. The osmotic minipump was filled with either the mixture (DOTAP and test mol-ecules) or the vehicle (DOTAP only) before it was inserted into the peritoneal cavity of the animal under anesthesia.

## 4. Notes

1. In our experience, the JMMl cells engrafted to the highest numbers when the animals had received transplantation twice on separate days. In order to minimize the mortality rate resulting from infection of these immunodeficient animals, it is essential to keep them in specific pathogen-free conditions.
2. We usually suspended the sorted JMML cells in RPMI-1640 medium supplemented with glutamine and antibiotics, and added bicarbonate to ensure a physiological pH.
3. The pumps were filled with the compounds to be tested. It is essential that the pump delivers the calculated amount of test molecules, and this should be tested in vitro. It is also advisable to determine plasma concentrations (e.g., using appropriate enzyme-linked immunosorbent assay [ELISA] or radioimmunoassays) of the test molecules to ensure that a steady-state concentration has been achieved.
4. To establish that these animals did develop a post-infarction heart failure, we routinely examined histological sections of the left ventricular myocardium for fibrosis. We also looked for enlargement of the left ventricular cavity, thinning of the intraventricular septum, and pleural effusion. Immediately before sacrificing the animal, we measured the left ventricular end-diastolic pressure (LVEDP) using the indwelling carotid catheter. LVEDP values that exceeded 15 mmHg were an indication heart failure, as this cut-off value correlates excellently with ultrasound examinations in the rat model *(10)*.
5. It is advisable to limit the size of the pump to the required test period. In our experience, a 4-wk treatment will require a rather large pump that will occupy most of the peritoneal cavity. In our models, the test molecules would freely diffuse into the peritoneal cavity to later be absorbed by the peritoneal epithelium before entering the circulation. Alternatively, one could place a small catheter so that one end is placed at the tip of the pump and the other can be positioned close to the target organ; however, this is sometimes rather difficult, particularly in small animals such as the mouse.
6. Similar to synthetic DNA oligonucleoties, naked DNA enzymes are *in vivo* taken up by cells. However, to increase the intracellular uptake and stability, we have included a small amount of cationic liposomes.

## Acknowledgments

Financial support was obtained by the Norwegian Cancer Society, the Norwegian Research Council and Throne Holst Foundation.

## References

1. Sioud, M. (1996) Ribozyme modulation of lipopolysaccharide-induced tumor necrosis factor-alpha production by peritoneal cells in vitro and in vivo. *Eur. J. Immunol.* **26**, 1026–1031.
2. Iversen, P. O., Lewis, I. D., Turczynowicz, S., Hasle, H., Niemeyer, C., Schmiegelow, K., et al. (1997) Inhibition of granulocyte-macrophage colony-stimulating factor

prevents dissemination and induces remission of juvenile myelomonocytic leukemia in engrafted immunodeficient mice. *Blood* **90,** 4910–4917.

3. Iversen, P. O., Nicolaysen, G., and Sioud, M. (2001) DNA enzyme targeting TNF-α mRNA improves hemodynamic performance in rats with postinfarction heart failure. *Am. J. Physiol. Heart. Circ. Physiol.* **281,** H2211–H2217.

4. Stull, R. A. and Szoka, Jr., F. C. (1995) Antigene, ribozyme and aptamer nucleic acid drugs: progress and prospects. *Pharm. Res.* **12,** 465–483.

5. Sioud, M. (1999) Application of preformed hammerhead ribozymes in the gene therapy of cancer. *Int. J. Mol. Med.* **3,** 381–384.

6. Iversen, P. O., Emanuel, P. D., and Sioud M (2002) Targeting *Raf-1* gene expression by a DNA enzyme inhibits juvenile myelomonocytic leukaemia cell growth. *Blood* **99,** 4147–4153.

7. Uckun, F. M. (1996) Severe combined immunodificient mouse models of human leukemia. *Blood* **88,** 1135–1146.

8. Kamel-Reid, S. and Dick, J. E. (1988) Engraftment of immune-deficient mice with human hematopoietic stem cells. *Science* **242,** 1706–1709.

9. Arico, M. and Biondi, A. (1997) Juvenile myelomonocytic leukemia. *Blood* **90,** 479–488.

10. Sjaastad, I., Sejersted, O. M., Ilebekk, A., and Bjørnerheim, R. (2000) Echocardiographic criteria for detection of post-infarction congestive heart failure in rats. *J. Appl. Physiol.* **89,** 1445–1454.

# 34

## Potential Design Rules and Enzymatic Synthesis of siRNAs

### Mouldy Sioud and Marianne Leirdal

#### Summary

Small interfering RNAs (siRNAs) have emerged as a powerful technique for sequence-specific gene silencing in a wide variety of organisms. However, the base composition of the siRNA sequence is not the only determinant of efficacy. Intrinsic factors related to mRNA structures are likely to be crucial determinants for siRNA activity. Indeed, placing the recognition site of an active siRNA into a structured mRNA region has abrogated the siRNA activity. Therefore, a successful gene-targeting project may require the design of many distinct siRNAs at a high cost. Here, potential design rules, cost-effective strategies for producing siRNAs by T7 RNA polymerase, and expression cassettes for in vivo testing are described.

**Key Words:** RNA interference; small interfering RNAs; antisense RNA; H1 promoter.

## 1. Introduction

The inactivation of gene function by reverse genetics is important in the study gene function, and may also have an impact on the treatment of diseases initiated and/or perpetuated by aberrant gene expression. Among the strategies, antisenses and catalytic nucleic acids have been extensively applied *(1)*. Recently, the discovery of RNA interference (RNAi) has offered an additional way to inhibit the expression of virtually any gene in living cells. RNAi is a powerful technique for sequence-specific gene silencing in a wide variety of organisms *(2–5)*. Potentially, this process has major implications for target validation and drug development. Although RNA interference with chemically made siRNA provides a promising tool for studying gene functions *(5–8)*, the approach is limited by its high cost, since it requires the chemical synthesis of siRNAs without a guarantee that the designed

From: *Methods in Molecular Biology, vol. 252: Ribozymes and siRNA Protocols, Second Edition*
Edited by: M. Sioud © Humana Press Inc., Totowa, NJ

molecules will be effective in silencing. Notably, selection of small interfering RNAs (siRNAs) is still an empirical process. A successful gene-targeting project may require the design of many distinct siRNAs at a high cost. To overcome some of these problems, a set of potential design rules and transcription strategies are presented.

## 2. Materials

1. Synthetic DNA oligonucleotide templates containing T7 promoter bottom (−) strand sequence and siRNA antisense or sense sequence.
2. A synthetic DNA oligonucleotide containing T7 promoter top (+) strand sequence. This sequence can generally be used for transcribing all siRNAs.
3. Synthetic oligonucleotides containing the H1 promoter sequence.
4. Synthetic oligonucleotides containing the desired hairpin siRNA sequences. Oligonucleotides can be synthesized in-house, or ordered from any commercial oligonucleotide supplier.
5. Optimized 5X transcription buffer: 200 m$M$ Tris-HCl, pH 7.9, 30 m$M$ MgCl$_2$, 10 m$M$ spermidine, and 50 m$M$ NaCl.
6. 10 X Klenow buffer: 100 m$M$ Tris-HCl, pH 7.5, 500 m$M$ NaCl.
7. 10 m$M$ nucleotide triphosphate (NTP) mixture: adenosine 5' triphosphate (ATP), cytidine 5' triphosphate (CTP), guanosine 5' triphospahte (GTP), and uridine 5' triphosphate (UTP) each at 10 m$M$. Store at −20°C in aliquots.
8. 10 mM deoxynucleotide 5' triphosphate (dNTP) mixture: deoxyadenosine 5' triphsophate (dATP), deoxythymidine 5' triphosphate (dTTP), deoxycytidine 5' triphosphate (dCTP), and deoxyguanosine 5' triphosphate (dGTP) each at 10 m$M$. Store at −20°C in aliquots.
9. 100 m$M$ dithiothreitol (DTT).
10. T7 RNA polymerase (17 U/μL).
11. Recombinant RNasin (40 U/μL).
12. RNase-free DNase I and ribonuclease T1.
13. Klenow DNA polymerase
14. 100% and 75% ethanol.
15. 3 $M$ sodium acetate (Na-Ac), pH 7.0.
16. Acrylamide/*bis*-acrylamide (39:1), *N,N,N',N'*-tetramethyl-ethylemediame (TEMED) (Bio-Rad) and 10% ammonium phosphate.
17. CHROMA SPIN-10 column (Clontech).
18. 10X electrophoresis buffer (TBE): 890 m$M$ Tris-borate, pH 8.3, 20 m$M$ ehtylenediaminetetraacetic acid (EDTA).
19. Standard laboratory microcentrifuge.
20. Polyacrylamide or agarose gel electrophoresis equipment.
21. Diethylpyrocarbonate (DEPC)-treated water (RNase-free water).
22. Phenol/chloroform (1:1, v/v) and chloroform/isoamyl alcohol (IAA) (24:1, v/v).

## 3. Methods

### 3.1. General Design Rules

RNA molecules can specifically recognize other RNAs by sequence, structure, or a combination of both. Watson-Crick basepairing allows one RNA to bind another RNA on the basis of the sequence only. In mammalian cells, like antisense and ribozymes, siRNAs recognize target mRNA via Watson-Crick basepairing, and induce site-specific cleavage of RNA by specific nucleases *(5)*. siRNAs must overcome the same problems as RNase H-dependent antisense oligonucleotides and ribozymes—mainly, target accessibility, intracellular localization, and delivery. Thus, the design of effective siRNAs may depend on parameters that are impossible to predict with any certainty, such as target accessibility. The effects of ribozymes and antisense oligonucleotides are dependent on target position *(9,10)*.

To identify accessible sites in mRNA one could perform experimental analyses *(11; see*, for example, Chapters 9, 19, 20, and 23). However, computer programs can identify putative accessible sites that do not exhibit extensive basepairing *(12)*. However, the relevance of such analysis is still uncertain, since the method does not consider either tertiary structure or RNA-protein interactions. Despite these limitations, we have found that computational approaches to the identification of ribozymes or siRNA target sites can be helpful *(1,6)*. Based on the published data, the following criteria should facilitate the design of effective siRNAs.

1. In principle, any site of mRNA can be targeted. However, avoid AUG start and the termination codon, since these regions are believed to be sites for intracellular proteins.
2. Choose a site that is balanced in (GC) content. Ideally, the GC content is 50%, but it can be higher or lower.
3. Avoid repetitive nucleotides (more than three)—particularly long stretches of G's because they form G-quartet structures.
4. Start your siRNA target sequence with a double AA or IG leader (*see* **Note 1**). However, this has been shown to not be required for activity.
5. Perform a Blast search with the selected sequence against sequence databases, to ensure that the designed siRNA is specific (*see* **Note 2**).
6. For each target, we suggest the design of 2–4 siRNAs to ensure that one is working. Notably, not all sites along a given mRNA are equally sensitive to siRNA-mediated gene silencing *(6,13,14)*.
7. Once an active site is identified, shifting 2–3 nucleotides at 5' or 3' end can further optimize it.

## 3.2. siRNA Synthesis

Once the site is selected chemically, sense and antisense siRNAs can be ordered from any commercial RNA oligonucleotide supplier (e.g., Eurogentec, Integrated DNA Technologies). To illustrate the design of siRNA for in vitro and in vivo expression, the following target-site sequence is used as an example:

5'-GCACGACUUCUUCAAGUCC-3'

1. Select a target site and design sense and antisense RNA oligonucleotides. For the example given here, this could be the following sequences:
   Sense strand          5'-GCACGACUUCUUCAAGUCC-3'
   Antisense strand      5'-GGACUUGAAGAAGUCGUGC-3'
2. Processed siRNAs contain two nucleotide-overhangs at their 3' ends, a characteristic of RNase III activity. In the original paper by Elbastir et al. (5), the 2-nt overhangs were replaced with DNA thymidines (dTdT), which provide more reliable synthesis and stability than ribonucleotides. However, siRNAs with 2'-ribonucleotide overhangs exhibited the same silencing effect when compared with the same siRNA with 2'-deoxythymidine overhangs. Whatever the downstream and upstream leader sequences of the siRNA site (see Note 1), synthesize the sense and the antisense siRNA as 5'-(N19)TT and 5'-(N<<19)TT, respectively. Order 0.1–0.2 μM for evaluation.
   Sense siRNA          5'-GCACGACUUCUUCAAGUCCdTdT-3'
   Antisense siRNA      5'-GGACUUGAAGAAGUCGUGCdTdT-3'
3. When working with single-stranded RNA molecules, an annealing step is necessary. Handling should be conducted under RNase-free conditions.
4. Dissolve each RNA oligo in RNase-free water at 2–5 μg/μL.
5. To prepare the working stock (20–40 μM), mix an aliquot of each RNA oligo in 10 mM HEPES, pH 7.4, and 150 mM NaCl (final volume 100–200 μL). Heat for 1 min at 80°C, spin down, and then incubate at 37°C for 1 h (see Note 3).
6. After annealing, the mixture can be stored frozen at –70°C and freeze-thawed several times. However, we recommend making several small aliquots and freeze-thawing only a few times (<6).
7. Based on the previous example, the siRNA duplex should have the following sequence:
   5'-GCACGACUUCUUCAAGUCCdTdT-3'
   3'-TdTdCGUGCUGAAGAAGUUCAGG-5'

The annealed duplex can be applied to cells after cationic liposome encapsulation and/or standard transfection procedures such as electroporation (see Chapters 35, 40, and 43). As for ribozyme and antisense molecules, the choice of the appropriate controls is important for siRNA studies. In this respect, siRNAs with four consecutive nucleotide mismatches exhibited no RNA interference activity; thus, they represent an excellent control (14,15). Alternatively, a scrambled siRNA or an inactive siRNA that target the same mRNA can be used (see Note 4).

### 3.3. T7 siRNAs

As mentioned previously, the siRNA efficacy is highly dependent upon target position. The secondary structure of the target was found to have an equally important effect on siRNA and RNase H-dependent oligonucleotide activity. In this respect, placing the recognition site of an active siRNA into a highly structured RNA region abrogated the siRNA activity *(16)*. A successful siRNA research project may require the design of various distinct siRNAs at a high cost. To overcome this problem, siRNAs can be synthesized in vitro by T7 RNA polymerase *(14)*. The T7 promoter has the requirement of a guanosine (G) in the +1 position. Therefore, in vitro transcribed siRNA must begin with a least one G. Based upon these roles, the siRNA site requirement is **G-N17-C** (site length is 19 nt). This sequence is frequently present in mRNAs. The T7 RNA polymerase functions best if the first three nucleotides of the RNA are G-purine-purine, so the full use of the enzyme activity may require addition of one or more nucleotides to the 5'-end of the siRNA. By adding a GGG sequence at the 5'end of both sense and antisense strands, any site sequence in mRNAs can be targeted using T7 siRNAs (M.S., unpublished results; *see* **Note 5**). Notably, the production of RNA that does not start with G can be produced using T7 RNA polymerase (*see* Chapter 2).

Once the siRNA site is selected, design two oligonucleotides that code for the sense and the antisense strands under the control of T7 promoter. A schematic representation and detailed illustration of the method are shown in **Fig. 1A** and **B**, respectively.

1. Combine equimolar amounts of the DNA oligonucleotide with the T7 promoter top strand oligonucleotide (2 nmol from each oligo). Heat to 90°C, and then allow to equilibrate at room temperature. The T7 promoter primer is common for all siRNAs to be synthesized. This template can be used for in vitro transcription. However, we recommend the use of a double-stranded template that can be generated via primer extension reaction (*see* **Notes 6–8**).

Jacque et al. *(15)* showed that cotransfection of permissive cells with a proviral clone—T7-DNA template expressing a hairpin siRNAs against, for example, *Vif* sequences and recombinant T7 RNA polymerase—significantly suppressed virus production. Thus, double-stranded DNA templates harboring desired siRNA under the T7 promoter could be directly tested in vivo prior cloning into eukaryotic expression vectors.

### 3.4. Generation of Double-Stranded DNA Template for In Vitro Transcription of siRNAs

1. Mix the following components in an Eppendorf tube:
   a. 20 nmol DNA oligonucleotide

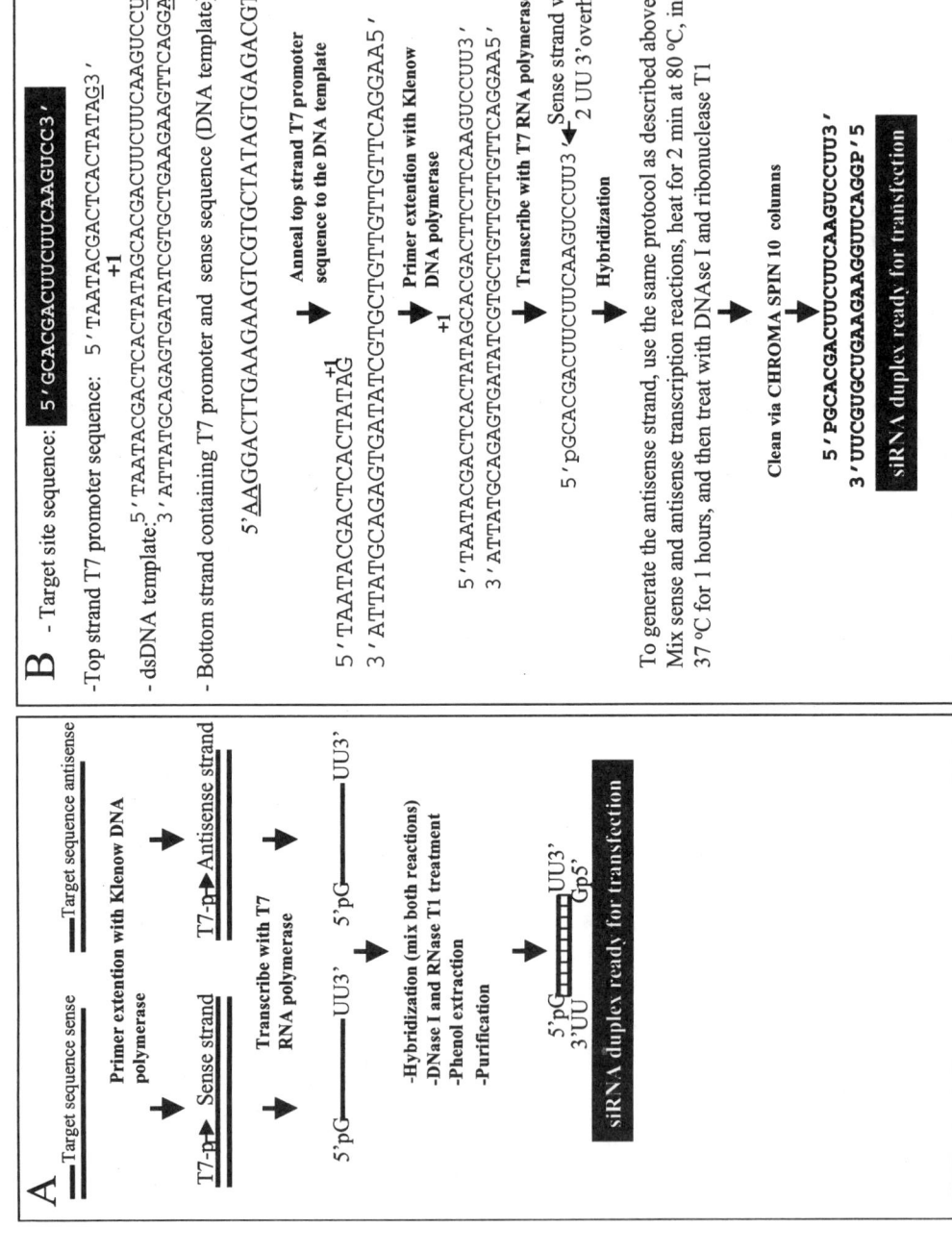

    b. 20 nmol T7 promoter top-strand sequence

    c. 20 μL Klenow buffer

    d. 5 μL dNTPs mixture

2. Complete to 198 μL with water. We find it valuable to set up the extension mixture for several transcription reactions.
3. Heat the reaction mixture to 90°C for 2 min, and cool slowly to anneal the T7 promoter primer.
4. Spin down briefly, add 2 μL Klenow enzyme (1–5 U), and then incubate at 37°C for 30 min.
5. Add 100 μL of phenol/chloroform (1:1), vortex for 15 s, and microcentrifuge for 1 min at 10,000–12,000$g$.
6. Pipet out the upper phase and extract it with 50 μL chloroform/IAA (24:1)
7. Transfer the supernatant to a new tube, add 0.1 vol of 3 $M$ Na-Ac and precipitate DNA with 2.5 vol of absolute ethanol. Incubate at –20°C for 1 h, or –70°C for 15 min.
8. Pellet the DNA in a microcentrifuge at >10,000$g$ for 10 min, and then wash the pellet with 70% ethanol.
9. Dry briefly and dissolve in 50 μL of DEPC-treated water. This double-stranded template is now ready for transcription.

## 3.5. Synthesis of Large Amounts of siRNA

1. For in vitro transcription, combine the following components:

    a. DNA template coding for the sense or antisense strand 0.1–1 μg (1–10 μL)

    b. 5 μL NTP mix (each 10 m$M$)

    c. 20 μL 5X transcription buffer

    d. 1 μL RNasin

    e. 2 μL T7 RNA polymerase

    f. 10 μL DTT (100 m$M$)

2. Complete with RNase-free water to 100 μL and incubate at 37°C for 2–3 h.
3. Check transcription by analyzing a 5-μL aliquot with a 15% polyacrylamide gel.
4. To hybridize sense and antisense siRNAs, mix sense and antisense transcription reactions, heat at 80°C for 2 min, and then incubate at 37°C for 1 h.
5. To eliminate DNA template and single-stranded RNAs, treat the sample with DNAse I (1 U) and ribonuclease T1 (1 U) for 15 min at 37°C. Ribonuclease T1 preferentially cleaves RNA at single stranded G residues. Notably, extra non-paired

---

Fig. 1. Experimental procedures. (**A**) Schematic DNA templates for in vitro synthesis of siRNAs using T7 RNA polymerase. First, an oligonucleotide is synthesized, which contains the (–) strand of the T7 RNA polymerase promoter followed by a sense or antisense sequence. The oligonucleotide is then annealed to the top (+) strand of the T7 promoter. This results in a double-stranded T7 promoter region, which is required for T7 RNA polymerase transcription. (**B**) Experimental design of a siRNA duplex directed against the 5'GCACGACUUCUUCAAGUCC3' site sequence. +1, indicates the transcription initiation site. The two uridine (U) 2' overhangs are underlined.

Gs at the 5'end of the siRNA duplex are cleaved away. Therefore, any target site can be in vitro transcribed (*see* **Note 5**).

6. Extract the mixture with 0.5 vol phenol-chloroform and then with chloroform/IAA.
7. Transfer the supernatant to a new tube and then remove incorporated NTPs and degraded materials via CHROMA SPIN-10 column as described by the manufacturer.
8. Adjust to 0.3 *M* Na-Ac, and then precipitate the RNA by adding 3 vol of absolute ethanol.
9. Wash the RNA pellet with 70% ethanol, dry, and dissolve in 50 µL of RNase-free water.
10. Determine the concentration by measuring the optimum density (OD) ($1OD_{260}$ U = 33 µg/mL). Always use gloves when handling RNA.

**Note:** the entire procedure requires less than 24 h, and multiple siRNAs can be synthesized simultaneously.

## 3.6. Production of Hairpin siRNAs

The synthesis of one siRNA usually requires the production of two independent sense and antisense RNA. However, it is possible to synthesize siRNA duplexes as one single short transcript (*17–20*).

1. Design a DNA oligonucleotide containing the T7 promoter sequence followed by 19-nt sense strand, a loop sequence (*see* **Note 10**), and the 19-nt antisense strand. Add two extra uridines at the 3'end. A schematic representation of the method is shown in **Fig. 2**.
2. Once the oligonucleotide is designed, order the corresponding bottom-strand sequence (DNA template).
3. Anneal the T7 top strand sequence to the template. This results in a double-stranded T7 promoter region, which is required for T7 RNA polymerase transcription (*see* **Subheading 3.4.**).
4. Proceed to primer extension reaction (*see* **Subheading 3.4.**).
5. Proceed to in vitro transcription (*see* **Subheading 3.5.**).
6. After transcription, treat samples with DNase I to remove the DNA template and purify hairpin siRNA via CHROMA SPIN-10 columns as described by the manufacturer.
7. The hairpin RNA preparations can be used without further precipitation.
8. Determine the concentration by measuring the OD at 260 nm.

The designed hairpin RNA can be in vitro or in vivo synthesized using T7 RNA polymerase.

## 3.7. Expression of the Hairpin siRNAs in Mammalian Cells

The in vivo expression of siRNA using U6 and H1 promoters is described in Chapters 36, 3, and 41. However, we find that the compact H1 promoter (100-nt) is more convenient for preparing DNA template using overlapping

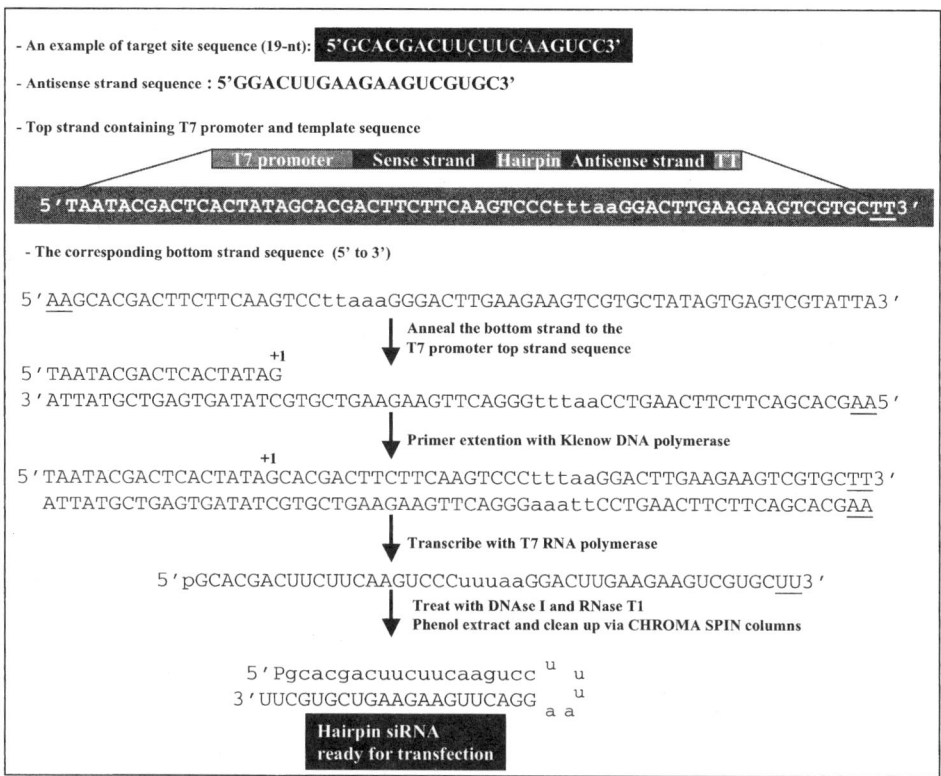

Fig. 2. An example of hairpin siRNA design. The siRNA was designed to target the GFP sequence: 5'GCACGACUUCUUCAAGUCC3'.

synthetic oligonucleotides. This enables the promoter to be introduced immediately before the siRNA, and can be a time-saving process. This is accomplished easily by synthezing two overlapping oligonucleotides containing the H1 promoter sequence, followed by an 19-nt sense strand, loop sequence (*see* **Note 10**), 19 antisense strand, and a termination signal of five thymidines. In this system, the cleavage of the transcript takes place after the second uridine. This would generate a perfect hairpin siRNA with 2'-uridines overhangs. In contrast to the U6 promoter, no specific nucleotides are required at position +1 (Agami, R., personal communication); thus, any target site sequence can be expressed. A schematic illustration of the method is shown in **Fig. 3**.

1. Design two overlapping oligonucleotides coding for the hairpin siRNA under the human or mouse H1 promoter. The 5' and the 3' ends should contain restriction sites that are suitable for cloning.

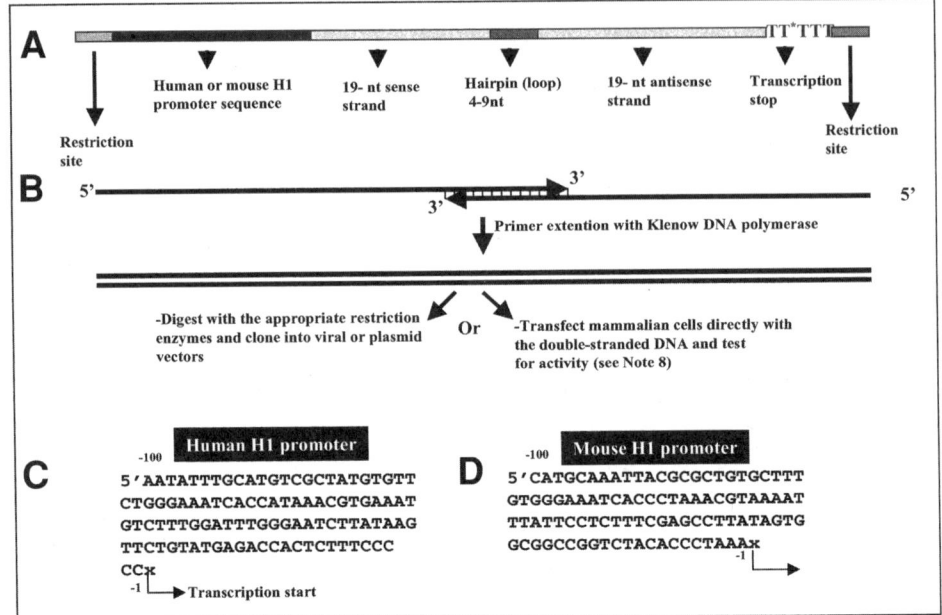

Fig. 3. Schematic illustration of the overlapping oligonucleotide strategy to construct a synthetic siRNA minigene. (**A**) The location of H1 promoter and hairpin siRNA sequences. \*, Indicates the site of the RNA cleavage. (**B**) Two overlapping oligonucleotides containing promoter and template sequences. The human (**C**) and mouse (**D**) H1 promoter sequences. The H1 promoter sequence can be PCR amplified from genomic DNA using specific primers (*see* Chapter 38). +1, indicates the transcription initiation site. X = G, A, T, or C.

2. Anneal both oligonucleotides and perform primer extension with Klenow DNA polymerase as described in **Subheading 3.4.** In principle it is preferable to choose a target sequence with low thermodynamic stability, in particular at the ends. This would facilitate the melting of the siRNA duplex and therefore increasing the incorporation of the single stranded siRNA antisense strand into the active complex. Notably, pyridmidine rich target sites were found to be suitable for ribozyme targeting (*23*).

3. After primer extension, phenol-extract the sample and collect the DNA by ethanol-precipitation as described in **Subheading 3.4.**

4. Digest about 0.5 µg of the DNA with the two restriction enzymes whose recognition sequences were engineered into the 5' and 3' ends, and then polyacrylamide gel electrophoresis (PAGE) or agarose gel-purify the digested DNA.

5. Clone the digested DNA into the appropriate vector (*see* Chapter 36 and **Note 11**). For stable expression in mammalian cells, we have used the pCDNA3 mammalian expression plasmid (Clontech) after removing the parental CMV promoter (unpublished observations). Since not all sites along a given mRNA are equally sensitive to siRNA, it may be important to test various siRNA cassettes in vivo (*21*; *see* Chapter 39).

## 4. Notes

1. Although the rules that govern the design of effective siRNAs are unknown, targeting the following sequence of mRNA motifs may enhance the probability of selecting effective molecules: 5'AA(**N19**)TT, 5'AA(**N19**)NN, and 5'NA(**N19**)NN, where N is any nucleotide in the mRNA sequence.

2. Standard computer programs for choosing sequences for PCR primer or DNA probes can be used to select the optimal target for siRNAs, which must be examined for analogies with other sequences in genome databanks.

3. If the transfection buffer is not needed, use the annealing buffer provided with the siRNA molecules or prepared in-house (200 m*M* potassium acetate, 4 m*M* $MgCl_2$, 60 m*M* HEPES-KOH, pH 7.4).

4. Measuring the target RNA and protein levels in treated cells vs controls can demonstrate siRNA-specific gene inhibition. Several control sequences can be used. siRNA with four consecutive nucleotide mismatches seems to be an ideal control. However, they may exhibit some activity, because siRNAs can function as microRNA *(22)*. Therefore, the best control sequences are mismatched siRNA, which maintain the same base composition as the siRNA.

5. To improve the siRNA yield, a GGG sequence can be added to the 5'end of the siRNA sequences. After transcription, these unpaired Gs are removed by treatment with ribonuclease T1. This process increases the number of sites that can be targeted with T7-siRNAs. Thus, all sequence motifs can be in vitro transcribed using T7 RNA polymerase. So, search for sequences 5'-(**N19**)-3' in the mRNA sequence (e.g., 5'-UACCGGAUGCUUAACCAUA-3-), add a GGG sequence at the 5'end of both sense and antisense strands and then design DNA oligonucleotide for in vitro transcription (*see* **Subheading 3.3.**). After transcription and annealing, the designed siRNA for the previous site should have the following sequence:

   5'-<u>GGG</u>UACCGGAUGCUUAACCAUAUU-3'
   3'-UUAUGGCCUACGAAUUGGUAU<u>GGG</u>-5'

   After ribonuclease T1 treatment that removes the extra G, a perfect siRNA duplex with 2'-UU overhangs would be generated:

      5'-**UACCGGAUGCUUAACCAUAUU**-3'
   3'-**UUAUGGCCUACGAAUUGGUAU**-5'

6. If the template is partially single-stranded synthetic DNA, and only the 18-mer-promoter part is double-stranded, this occasionally results in poor transcript yields. This problem can be overcome by the primer extension reaction, which would generate double-stranded DNA template. Co-transfection of such DNA template with recombinant T7 RNA polymerase mediated gene silencing *(15)*.

7. Although a filled and precipitated DNA template is fine for in vitro transcription, we have found that it is important to PAGE purify the DNA template.

8. T7 RNA polymerase functions best if the first three nucleotides of the RNA are G-purine-purine.

9. If the siRNA is going to be expressed in vivo, during the in vitro selection step use hairpin siRNAs.

10. If hairpin siRNA is made, the selection of the loop sequence may be critical for the siRNA activity. In principle, longer loop sequences are more likely to be cleaved by Dicer than short sequences. Notably, most microRNA precursors have longer loop sequences. The following loop sequences were tested and found to be effective in mammalian cells: 5'-UUCG-3', 5'-UUAA3-3', 5'-UUUGCC-3', 5'-UUUGUGUAG-3', 5'-UUCAAGAGA-3', and 5'-UUUGCUGUAG-3' (*see* Chapters 36, 38, 39, and 41).

11. For longer silencing effect, a vector-based approach is used. Generation of plasmid and viral vectors are described in Chapters 36, 38, and 41. Because there is no guarantee that the designed siRNA will work in vivo, prior cloning into vector siRNA cassette(s) under the U6, H1, or T7 promoter could be tested in cells. The 5' overlapping primer is expected to contain most of the H1 promoter sequence. This primer can be combined with a variety of oligonucleotides that contain different siRNAs. This makes it possible to create and test a variety of siRNA cassettes to identify the one that works most effectively in vivo. Once an interesting siRNA is identified, it can be cloned into the expression vector of choice (*see* Chapter 39).

12. The development of gene-specific knockdown siRNA vectors should facilitate the construction of RNAi vector libraries that should permit large-scale loss-of function genetic screens in mammalian cells. Because of the problem of target accessibility, these libraries must contain multiple RNAi vectors (2- to 3/gene). Technical advances in oligonucleotide synthesis may facilitate the design of RNAi libraries with high degree of complexity.

## References

1. Sioud, M. (2001) Nucleic acid enzymes as a novel generation of anti-gene agents. *Curr. Mol. Med.* **1**, 575–588.
2. Sharp, P. A. (2001) RNA interference. *Genes Dev.* **15**, 188–200.
3. Fire, A., Xu, S., Montgomery, M. K., Kostas, S. A., Driver, S. E., and Mello, C. C. (1998) Potent and specific genetic interference by double-stranded RNA in *Caenorhabditis elegans. Nature* **391**, 806–811.
4. Hamilton, A. J. and Baulcombe, D. C. (1999) A species of small antisense RNA in posttranscriptional gene silencing in plants. *Science* **286**, 950–952.
5. Elbashir, S. M., Harborth, J., Lendeckel, W., Yalcin, A., Weber, K., and Tuschl, T. (2001) Duplexes of 21-nucleotide RNAs mediate RNA interference in cultured mammalian cells. *Nature* **411**, 494–498.
6. Sorrensen, D. R., Leirdal, M., and Sioud, M. (2003) Gene silencing by systemic delivery of synthetic siRNAs in adult mice. *J. Mol. Biol.* **327**, 761–766
7. Lewis, D. L., Hagstom, J. E., Loomis, A. G., Wolff, J. A., and Herweijer, H. (2002) Efficient delivery of siRNA for inhibition of gene expression in postnatal mice. *Nature Genet.* **32**, 107,108.
8. McManus, M. T. and Sharp, P. A. (2002) Gene silencing in mammalians by siRNAs. *Nat. Genet. Rev.* **3**, 737–740.

9. Bennett, C. F., Chiang, M.-Y., Wilson-Lingardo, L., and Wyatt, J. R. (1994) Sequence specific inhibition of human type II phospholipase A2 enzyme activity by phosphorothioate oligonucleotides. *Nucleic Acids Res.* **22,** 3202–3209.

10. Sioud, M. (1997) Effects of variations in length of hammerhead ribozyme antisense arms upon the cleavage of longer RNA substrates. *Nucleic Acids Res.* **25,** 333–338.

11. Lee, N. S., Dohjima, T., Bauer, G., Li, H., Li, M.J., Ehsani, A., et al. (2002) Expression of small interfering RNAs targeted against HIV-1 *rev* transcripts in human cells. *Nat. Biotech.* **20,** 500–505.

12. Sczakiel, G., Homann, M., and Rittner, K. (1993) Compute-aided search for effective antisense RNA target sequences of the human immunodeficiency virus type 1. *Antisense Res. Dev.* **3,** 45–52.

13. Holen, T., Amarzguioui, M., Wiiger, M. T., Babaie, E., and Prydz, H. (2002) Positional effects of short interfering RNAs targeting the human coagulation trigger Tissue Factor. *Nucleic Acids Res.* **30,** 1757–1766.

14. Leirdal, M. and Sioud, M. (2002). Gene silencing in mammalian cells by preformed small RNA duplexes. *Biochem. Biophys. Res. Commun.* **295,** 744–748.

15. Jacque, J. M., Triques, K., and Stevenson, M. (2002) Modulation of HIV-1 replication by RNA interference. *Nature* **896,** –4.

16. Vickers, T. A., Koo, S., Bennet, C. F., Crooke, S. T., Dean, N. M., and Baker, B. F. (2003) Efficient reduction of target RNAs by small interfering RNA and RNase H-dependent antisense agents. *J. Biol. Chem.* **278,** 7108–7118.

17. Brummelkamp, T. R., Bernards, R., and Agami, R. (2002) A system for stable expression of short interfering RNAs in mammalian cells. *Science* **296,** 550–553.

18. Paul, C. P., Green, P. D., Winer, I., and Engelke, D. R. (2002) Effective expression of small interfering RNA in human cells. *Nat. Biotechnol.* **19,** 505–508.

19. Miyagishi, M. and Taira, K. (2002) U6 promoter-driven siRNA with four uridines 3'overhangs efficiently suppress targeted gene expression in mammalian cells. *Nat. Biotechnol.* **19,** 497–500.

20. Hannon, G. J. (2002) RNA interference. *Nature* **418,** 244–251.

21. Castanotto, D., Li, H., and Rossi, J. J. (2002) Functional siRNA expression from transfected PCR products *RNA*, **8,** 1454–1460.

22. Doech, J. G., Petersen, C. P., and Sharp, P. A. (2003) siRNA can function as miRNAs. *Genes Devel.* **17,** 438–442.

23. Sioud, M. and Jespersen, L. (1996) Enhancement of hammerhead ribozyme catalysis by glyreraldehyde-3-phophate dehydrogenase. *J. Mol. Biol.* **257,** 775–789.

# 35

## RNA Interference (RNAi) With RNase III-Prepared siRNAs

### Dun Yang, Andrei Goga, and J. Michael Bishop

### Summary:

Small interfering RNA (siRNA) has become a powerful tool for selectively silencing gene expression in cultured mammalian cells. Because different siRNAs of the same gene have varying silencing capacities, several different siRNAS typically must be screened to obtain a region that will effectively silence the gene of interest. However, RNA interference with synthetic siRNA is inefficient and cost-intensive, especially for large, functional genomic studies. Here, we describe the use of *E. coli* endoribonuclease III to cleave double-stranded RNA (dsRNA) into esiRNA (endoribonuclease-prepared siRNA) that can target multiple sites within an mRNA. EsiRNA mediates effective RNA interference with no apparent nonspecific effects in cultured mammalian cells. Since the whole gene can be used at once, screening for an active siRNA for an individual gene is eliminated. Because of its simplicity and potency, this approach is useful for large-scale analysis of mammalian gene function.

**Key Words:** RNA interference; RNAi; dsRNA; RNase III; Dicer; siRNA; esiRNA.

## 1. Introduction

RNAi has become a powerful method to specifically inhibit gene expression in both genetically tractable and intractable organisms *(1,2)*. Although RNAi has also been observed in vertebrate systems such as mouse oocytes and pre-implantation mouse embryos, large dsRNA triggers nonspecific gene silencing in adult vertebrates and commonly used vertebrate cell lines *(3–5)*. Large dsRNA has profound negative effects on cell proliferation in vertebrate cells by inducing synthesis of γ-interferon (IFN) *(5)*. The IFN-responsive pathway, activated by dsRNA molecules longer than 30 basepairs (bp), triggers activation of dsRNA-dependent protein kinase (PKR) and 2'-5'oligoadenylate synthetase (2'-5'-AS). The activated PKR globally inhibits translation by

From: *Methods in Molecular Biology, vol. 252: Ribozymes and siRNA Protocols, Second Edition*
Edited by: M. Sioud © Humana Press Inc., Totowa, NJ

phosphorylating the translation factor eIF-2α, and activated 2'-5'-AS causes widespread mRNA degradation through activation of RNase L. Neither of these processes is sequence-specific to the large trigger dsRNA *(5)*.

Other methods of RNA-mediated gene silencing include: small interfering RNAs (siRNAs), and small hairpin RNAs (shRNAs). In contrast to the nonspecific effects of a large dsRNA, synthetic siRNA mediates selective gene silencing in cultured vertebrate cell lines by eliciting destruction of target mRNAs *(6–9)*. siRNAs are 21 to 27 nucleotides in length, and are double-stranded. By virtue of its small size, siRNA avoids activation of the IFN response. SiRNA has the signature termini structure of RNase III digestion: 5' phosphate, 3' hydroxyl, and 2- to 3-nucleotide 3' overhangs *(10–12)*. These end structures of siRNA are reported to be important for RNAi activity *(13)*. shRNA made in vitro or expressed in vivo has also been reported to trigger selective gene silencing *(8,14,15)*. This is probably the result of conversion of shRNA into siRNA in vivo. Application of siRNA or shRNA made in vitro causes transient inhibition of its cognate gene, and expression of a shRNA in vivo continuously from a stably integrated plasmid DNA vector or retroviral vector presumably allows permanent gene silencing *(14–16)*.

Although RNA interference with siRNA or shRNA provides a promising tool for studying gene functions in cultured mammalian cells, the approach is limited by its expense and inefficiency. siRNAs to different sequences within a gene have dramatically varied inhibitory ability, and it is difficult to predict an active siRNA for a particular gene *(8,17)*. Therefore, each mRNA must be screened for an efficient siRNA. This is an ineffective and costly process, especially for functional genomic studies.

We have solved this problem by using siRNA hydrolyzed from a large dsRNA with a dsRNA-specific endoribonuclease *(8,18)*. We have called these small dsRNA esiRNA—for endoribonuclease-prepared siRNA. We used *E. coli* RNase III to prepare siRNA in vitro because it can digest dsRNA very efficiently into short pieces with the same end structures of siRNA. Digestion of a large dsRNA with RNase III generates a large variety of siRNAs that are capable of interacting with multiple sites within target mRNAs. Thus, the requirement to screen for an active siRNA for each individual gene is eliminated. Other advantages of simultaneous targeting of multiple sites of a specific RNA by esiRNA include: decreasing the likelihood of mutations that would render siRNA resistance if RNA is used as an anti-retroviral therapy; and allowing systematic high-throughput genomic studies. The power of RNAi as a genetic tool can be greatly enhanced by using esiRNA.

The protocol summarized in **Fig. 1**. has three major steps:

1. Prepare dsRNA by in vitro transcription of PCR-amplified DNA templates;
2. Digest dsRNA with RNase III and purify esiRNA;
3. Transfect esiRNA into cells and evaluate the selectivity of gene silencing.

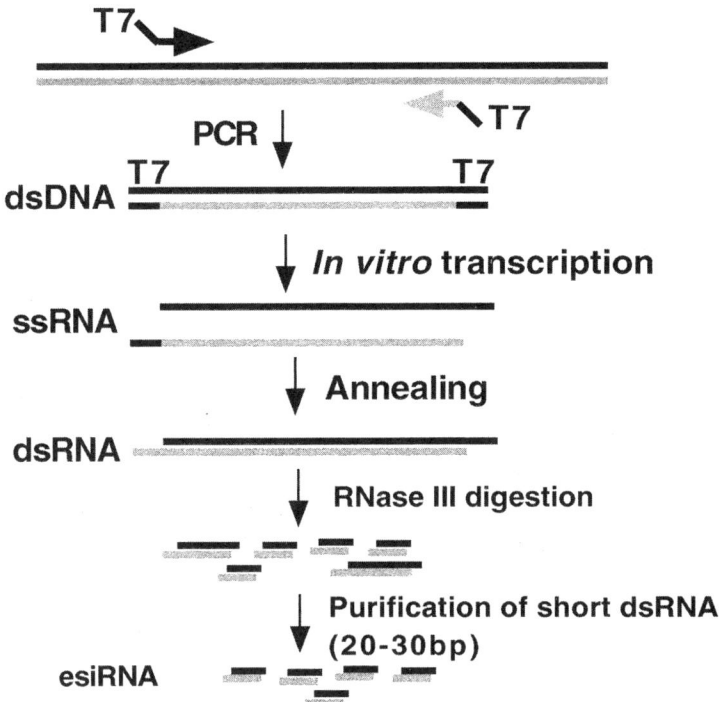

Fig. 1. Schematic of in vitro esiRNA production. T7 RNA polymerase promoter sequence is added to the 5' end of each PCR primer for amplification: 5'TAATACGACTCACTATAGGG......3'. The cDNA of interest will thus have T7 sites at both ends, to allow simultaneous production of in vitro transcribed RNA from both directions. In vitro transcribed dsRNA is cleaved by GST-RNase III.

For each gene of interest, a fragment corresponding to the coding sequence or untranslated region (UTR) was polymerase chain reaction (PCR)-amplified (*see* **Notes 1** and **2**). Primers were designed to incorporate T7 promoter sequences at each end (**Fig. 1**). In vitro transcription was performed on PCR-amplified DNA to make sense and anti-sense RNA in one reaction. Single-stranded RNA was made double-stranded by allowing annealing of the two strands. After digestion of dsRNA with glutathione *S*-transferse (GST)-RNase III, 20- to 30-bp digestion products were purified and introduced into cells to trigger selective gene silencing. The potency and specificity of esiRNA-mediated gene silencing could be evaluated by Northern blotting, Western blotting, or a functional assay.

## 2. Materials

**Note:** All materials and chemicals used in the experiment should be considered to be a potential health hazard. Particular care should be taken during handling of phenol, chloroform, ethidium bromide, polyacrylamide, and diethylpyrocarbonate (DEPC). All reagents are prepared using double-distilled water that has been treated with DEPC and autoclaved. Reaction buffers, nucleotide, dsDNA, dsRNA, esiRNA, and enzymes are stored at –20°C. There was no significant loss of activity after 6 mo of storage of esiRNA and GST-RNase III fusion protein at –20°C.

## 2.1. Expression and Purification of E. coli Endonuclease RNase III As a GST Fusion

1. Luria broth (LB) medium.
2. LBA: LB medium with 100 µg/mL ampicillin.
3. IPTG (isopropylthio-β-D-galactoside): Dissolve IPTG in sterile water to make a 1-$M$ stock solution. Dispense as 1-mL aliquots and store at –20°C.
4. 100 m$M$ dithiothreitol (DTT).
5. Phosphate-buffered saline (PBS).
6. Solublization buffer: 5 $M$ NaCl, 2% Triton X-100, 20 m$M$ DTT.
7. 50% (v/v) glutathione agarose beads (Sigma).
8. RNase-free water: Add 1 mL of DEPC per L of H$_2$O, stir well, and incubate for a period of several hours to overnight at 37°C. Autoclave for 45 min.
9. Falcon tube (50 mL).
10. Washing buffer: PBS with 500 m$M$ NaCl.
11. Disposable column (Pharmacia Biotech).
12. 100 m$M$ reduced glutathione: Dissolve reduced glutathione (Sigma) in RNase-free water to make a 100-m$M$ stock, and store at –20°C until needed. Avoid more than five freeze-thaw cycles.
13. Glutathione elution buffer: 5 m$M$ reduced glutathion in 200 m$M$ Tris-HCl, pH 7.9.
14. Glycerol.
15. Dialysis buffer: 20 m$M$ Tris-HCl, pH 7.9, 0.5 m$M$ ethylenediaminetetraacetic acid (EDTA), 5 m$M$ MgCl$_2$, 1 m$M$ DTT, 140 m$M$ NaCl, 2.7 m$M$ KCl, 30% glycerol in RNase-free water.

## 2.2. RNase III/esiRNA Procedure

### 2.2.1. Preparation of dsRNAs

1. Deoxynucleotide triphosphates (dNTPs), deoxyadenosine 5' triphosphate (dATP), deoxycytidine 5' triphosphate (dCTP), deoxyguanosine 5' triphosphate (dGTP), and deoxythymidine 5' triphosphate (dTTP) dissolved at 100 m$M$ in dH$_2$O, equilibrated to pH 7.0 with 50 m$M$ Tris-HCl, pH 7.0. The stock solution is equally mixed to yeild a working solution containing all four dNTPs at a concentration of 25 m$M$ each.
2. *Taq* DNA polymerase.
3. 10X *Taq* buffer.
4. Forward and reverse primers: TAATACGACTCACTATAGGGN$_{18-24}$, the minimal promoter sequence for efficient transcription by T7 phage polymerase was

added upstream of a gene-specific primer ($N_{18-24}$). Three Gs (in bold) that will form the first three bases of the transcribed RNA were added to increase the yields of transcription product.

5. 10X transcription buffer.
6. Ribonucleotide triphosphates: adenosine 5' triphosphate (ATP), guanosine 5' triphosphate (GTP), cytidine 5' triphosphate (CTP), and uridine 5' triphosphate (UTP), 75 m$M$ for each NTP.
7. T7 RNA polymerase enzyme mix. **Items 5–7** are from MEGAscript™ in vitro transcription kit (Ambion, Austin, TX).
8. Phosphate-buffered saline (PBS).
9. RNase-free DNase I digestion solution. Add 1 mL of DNase I into 20 mL of 10 m$M$ Tris-HCl buffer, pH 7.9, containing 10 m$M$ MgCl$_2$ and 10 m$M$ CaCl$_2$.
10. 3 $M$ Na acetate, pH 5.6.
11. 100% and 75% ethanol.
12. TE: 10 m$M$ Tris-HCl, pH 7.9 and 1 m$M$ EDTA.
13. Phenol:chloroform (1:1).
14. Ethidium bromide.

## 2.2.2. esiRNA Preparation

1. DsRNA digestion buffer: Same as dialysis buffer, except for 5% glycerol.
2. 0.5 $M$ EDTA.
3. 4% agarose in TAE.
4. Preparative 8% polyacrylamide gel (19:1) in 1X TBE buffer. The gel is 3-mm thick and 20-cm long with a 5-cm-wide well.
5. 10-bp DNA ladder.
6. TE: 10 m$M$ Tris-HCl, pH 7.9 and 1 m$M$ EDTA.
7. 3 $M$ Na acetate, pH 5.6.
8. 100% and 75% ethanol.
9. PD-10 column (Pharmacia Biotech).
10. Phenol:chloroform (1:1).
11. Ethidium bromide.

## 2.2.3. Selective Gene Silencing by Introduction of esiRNA Into Cells

1. Lipofectamine™2000 (Invitrogen Lifetechnologies).
2. Serum-free cell-culture medium, such as Dulbecco's modified Eagle's medium (DMEM).
3. PBS.
4. Fetal bovine serum (FBS).

## 3. Methods

RNase III/esiRNA is divided into three steps that are schematized in **Fig. 1**. These steps can be conveniently performed with the following schedule. Experiments in **step 1**—including preparation of in vitro transcription templates by PCR, large-scale synthesis of single-stranded RNA, and preparation of dsRNA by annealing—could be performed within d 1. Hydrolysis of dsRNA with GST-RNase III and separation of digestion product in a polyacrylamide gel in **step 2** is usually performed on d 2. Elution of esiRNA from gel slices is allowed to proceed overnight, and is stopped at the beginning of d 3. After further purification, esiRNA could be transfected into cultured cells on d 3.

### 3.1. Expression and Purification of E. coli RNase III As a GST Fusion

1. Use a single colony of *E. coli* BL21 cells containing the recombinant pGEX-RNase III plasmid to inoculate 10 mL of LBA medium.
2. Shake at 250 rpm for 12–16 h at 37°C.
3. Dilute the culture 1:100 into fresh LBA medium, distribute it into four 1-L flasks, and grow at 37°C with shaking until the $OD_{600}$ reaches 0.3–0.4.
4. Add IPTG to a final concentration of 1.0 m$M$ and continue shaking for an additional 2 h.
5. Transfer the culture to centrifuge bottles and sediment the cells by centrifugation at 5000$g$ for 5 min.
6. Discard the supernatant and place the pellet on ice.
7. Suspend the cell pellet in 40 µL of ice-cold 1X PBS (40 mL of PBS per mL of culture).
8. Disrupt cells by sonication on ice in short bursts. Cell disruption may be checked by microscopy examination.
9. Add 5 mL of ice-cold solubilization solution and incubate on ice for 30 min to aid in solubilization of the fusion protein.
10. Pellet cell debris by centrifugation at 12,000$g$ for 10 min at 4°C and transfer the supernatant to a fresh 50-mL tube.
11. Add 2 mL of the 50 % slurry of Glutathione agarose beads pre-equilibrated with 1X PBS.
12. Incubate with gentle agitation at 4°C for 30 min.
13. Centrifuge at 500$g$ for 5 min and discard the supernatant.
14. Wash the pellet with 50 mL of washing buffer.
15. Repeat **steps 13** and **14** two more times.
16. Resuspend glutathione beads into 10 mL of washing buffer and transfer them to a disposable column.
17. Wash the matrix by adding 10 mL of 1X PBS. Allow the column to drain.
18. Elute the fusion protein by the addition of 4 mL of glutathione elution buffer per mL of bed volume.
19. Collect the eluate and dialyze it against 500 mL of dialysis buffer at 4°C for 10–12 h.
20. Store purified protein at –20°C (30% glycerol should keep it from freezing).

### 3.2. RNase III/esiRNA Procedure

#### 3.2.1. Step 1: Preparation of dsRNA

1. Mix the following reagents in a PCR tube:
   a. 10 µL of 10X PCR buffer
   b. 1 µL of forward primer (1 µg/µL)

     c. 1 µL of backward primer (1 µg/µL)

     d. 1 µL of plasmid DNA (1–10 ng)

     e. 1 µL of 10 m$M$ each dNTP

     f. 85 µL of $H_2O$

     g. 1 µL of DNA polymerase 1 U/µL

2. Perform 30 cycles of PCR, 94°C 30", 57°C 30", and 72°C 1 min/1 kb.
3. Run an aliquot on 1% agrose gel and ethidium bromide staining to check purity and concentration.
4. Prepare a 20-µL reaction mix at room temperature as follows (*see* **Note 3**):

     a. 2.0 µL of 10X transcription buffer

     b. 2.0 µL of each of 75 m$M$ ATP, GTP, CTP, and UTP

     c. 5.0 µL of DNA template from a PCR reaction (0.2 mg)

     d. 3.0 µL of ddH2O

     e. 2.0 µL of T7 RNA polymerase enzyme mix

5. Incubate the reaction at 37°C for 3–5 h.
6. Add 20 µL of RNase-free DNase I digestion solution and incubate at 37°C for 15 min (*see* **Note 4**).
7. Add 160 mL of PBS buffer.
8. Heat the sample at 90°C for 5 min in a beaker with 300 mL of $H_2O$, and then let it cool down to room temperature during a period of a few hours.
9. Add 200 µL of phenol/chloroform (1:1), mix vigorously, and quick-spin.
10. Transfer the upper aqueous phase to a new tube and extract with 200 µL of phenol/chloroform (1:1) once more.
11. Transfer the upper phase to a new tube and add 1/10 vol of 3 $M$ NaOAC and 2.5 vol of –20°C 100% ethanol.
12. Mix vigorously and centrifuge at 14,000$g$ for 10 min.
13. Discard the supernatant and wash the pellet with 1 mL of cold 75% ethanol.
14. Dry the pellet and suspend it in 100 µL of TE buffer.
15. Check the dsRNA on 1% agarose gel and RNA can be quantified by determining $OD_{260}$ with 1 µL of sample.

### 3.2.2. Step 2: Preparation of esiRNA (see **Notes 5–7**)

1. Mix 100 µg of dsRNA with 1 µg of GST-RNase III protein in a 200-µL reaction.
2. Incubate at 37°C for 15–60 min.
3. Terminate the reaction by adding EDTA to 20 m$M$.
4. Run an aliquot on 4% agarose gel along with a 10-bp DNA ladder to check appropriateness of digestion (*see* **Fig. 2**).
5. Prepare a 8% native acrylamide gel (19:1) in 1X TBE. Use a 3.0-mm-thick gel with well-forming combs.
6. Dissolve 100 µg of dsRNA in PBS and mix with loading buffer.
7. Flush wells before loading dsRNA and 10-bp molecular ladder and electrophoresis at 200 V.
8. Place gel in a glass container and stain with 0.5 µg/mL of ethidium bromide in 1X TBE for 10 min.

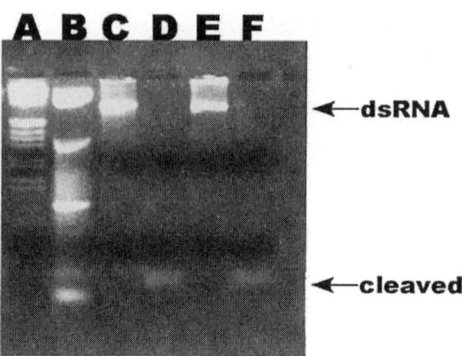

Fig. 2. In vitro digestion of large dsRNAs with GST-RNase III. 4% agarose gel stained with ethidium bromide is shown. Lanes: (**A**) 1-kb DNA ladder, (**B**) 10-bp DNA ladder, (**C**) in vitro transcribed cyclin-dependent kinase 1 (CDK1) dsRNA, (**D**) cleaved product of Cdk1 dsRNA (~20-bp esiRNA), (**E**) full-length Cdk2 dsRNA, (**F**) cleaved product of Cdk2 dsRNA (~20-bp esiRNA).

9. Examine gel with ultraviolet (UV) light and excise the areas of the gel containing dsRNA from 20–30 basepairs.
10. Crush the gel slice and elute the dsRNA into 2 mL of PBS overnight with shaking.
11. Transfer the eluted solution to a fresh tube, wash the gel slice with 0.5 mL of PBS, and combine with the eluted solution.
12. After washing a PD-10 column with 12 mL of autoclaved water, add the eluted dsRNA to the column, and discard the flowthrough.
13. Elute the dsRNA by adding 2 mL of TE with 50 m$M$ NaCl to the column and collect this fraction in a sterile tube.
14. Dry the dsRNA under vacuum and dissolve it in 100 µL of TE.
15. Add 100 µL of phenol-chloroform and vortex thoroughly.
16. Spin at room temperature for 5 min at 14,000$g$. Remove 90 µL of upper aqueous layer and transfer it to a fresh tube.
17. Add 10 µL of 3.0 $M$ NaOAc and 350 µL absolute ethanol. Vortex to mix, store on ice for 10 min, and spin for 10 min at 14,000$g$ at 4°C.
18. Remove supernatant and wash with 70% ethanol.
19. Dry esiRNA and dissolve it in 60 µL of TE. Verify by OD$_{260}$ that the esiRNA concentration is approx 0.5 µg/µL.

### 3.2.3. Step 3: Selective Gene Silencing by Introduction of esiRNA Into Cells

1. Plate cells so that they will be 80% confluent on the day of transfection.
2. Change medium 4 h before transfection.
3. Dilute 0.1–0.5 µg of esiRNA with 250 µL of serum-free medium.
4. Dilute 10 µL of lipofectamine™ 2000 with 250 µL of serum-free medium.

5. Add the RNA solution from **step 3** to the diluted lipofectinamine™ 2000 from **step 4**. Incubate at room temperature for 20 min.
6. Add esiRNA/lipofectamine mix to cells growing in serum-containing medium. Incubate at 37°C for 3–24 h.
7. Add fresh medium to cells as needed, passage cells into fresh medium. RNA interference can be detected within 24 h post-transfection. RNAi usually lasts at least for 6 d after one transfection. Evaluation of RNAi activity by Western blotting and functional assays should be performed to evaluate the effectiveness of protein-expression suppression and selectivity (e.g., compare to an endogenous non-targeted control protein) (*see* **Notes 8** and **9**).

## 4. Notes

1. In the selection of a target region for RNAi, sequences anywhere within the mRNA can be used. We had success with either translated regions or 3' UTR sequences. No obvious difference in effectiveness between open reading frame (ORF) and 3' UTR sequences has been observed, although we cannot rule out a subtle quantitative effect of position. We recommend segments between 0.5 and 2.0 kb for ease of RNA synthesis. We had success with regions shorter than 0.5 kb. Regions longer than 2.0 kb can be used, but RNA yields may be lower, with early termination products more common.
2. The most striking feature of RNAi is sequence-specificity of inhibition. EsiRNA depletes the expression of the gene from which the esiRNA is derived, without affecting the function of genes that are unrelated in sequence. In a mammalian system, we reported that esiRNA from *clathrin light chain a* (*LCa*) specifically silenced *LCa* expression without affecting *clathrin light chain b* (*LCb*) expression although these two genes have 66% homology in DNA sequence *(8)*. Recently, we observed esiRNA from human c-*myc* gene had no cross-silencing of mouse c-*myc*, although they share more than 80% homology in the region used to prepare esiRNA (Yang, D. unpublished data). Sequence-specificity of RNAi has also been previously highlighted in *Drosophila* by the observation that dsRNA generated from the sequence encoding the bHLH domain of one member in the *achaete-scute* complex only affected expression of that bHLH protein but no other member of the complex, despite their closely related bHLH domain *(19)*. However, there are also reports that dsRNA generated from one gene may inactivate more than one gene if their sequences are too closely related to each other *(20)*. Thus far, there are no reports about the minimal homology required for a cross-silencing. We suspect that cross-silencing is only a problem for genes that are very closely related at the DNA sequence level. If a homologous gene presents a potential unintended target, we suggest using different (non-overlapping) segments of the intended target gene for RNAi. siRNA or small hairpin RNA could be alternatively used.
3. We routinely prepare our templates for in vitro transcription by PCR-amplifying a segment from cDNA, although linearized plasmid DNA with appropriate phage promoters flanking the gene of interest could also be used as templates for in

vitro transcription. We generally add amplified DNA directly to transcription reactions with no further purification after the PCR. Typically, 5 µL of a 100-µL PCR reaction, corresponding to about 0.2 µg of double-stranded DNA, is used as a template in a standard transcription reaction. If the PCR-generated DNA is not concentrated enough, it can be precipitated with 2 vol of ethanol in the presence of 0.5 $M$ ammonium acetate after first extracting with phenol-chloroform. Although dsRNA could be formed spontaneously from sense and antisense RNA made in the one reaction, an annealing step is strongly recommended to maximize the yield of dsRNA. Hybrid RNA molecules with a double-stranded region linked to single-stranded tails can be digested by *E. coli* RNase III effectively. These single-stranded tails do not contribute to interference. DsRNA or esiRNA can be easily quantified by reading the $A_{260}$. The intensity of ethidium bromide staining of dsRNA or esiRNA in an agrose gel alongside an aliquot of RNA of known concentration can also be used to obtain a rough estimation of the RNA yield.

4. You generally do not need to remove DNA from in vitro transcription reactions because single-stranded RNAs are in excess amounts to the template DNA. You can always get rid of DNA templates when you purify esiRNA. If you use unpurified RNase III end-digestion products to perform RNAi, you may consider removing DNA templates by DNase I treatment (that has been certified RNase-free).

5. We found that it is difficult to obtain dsRNA digestion products ranging from 20–30 bp with the wild-type *E. coli* RNase III. Instead, it produced end-digestion products 12–14 bp that are too short to trigger RNA interference. The difference observed between GST-RNase III fusion and RNase III alone is probably the result of oligomerization properties of GST portion in the GST-RNase III. Aside from the *E. coli*, RNase III fusion could be used to prepare siRNA, human Dicer has also been successfully used to make small interfering RNA in vitro *(21)*.

6. For each batch of enzyme, you must determine the optimal digestion time. We usually pause the reaction at 4°C and run an aliquot on a 4% agarose gel to check the digestion status, and the digestion reaction could be resumed by incubating at 37°C if necessary. Avoid using excess recombinant GST-RNase III enzyme, as it may decrease the amount of esiRNA that can be generated from the digestion reaction. Aside from the commercially available dsDNA and dsRNA molecular markers, two synthetic DNA oligos of 20 and 30 bp in size can be used to estimate migration of esiRNA.

7. Detection of esiRNA in gel by ethidium bromide staining. Because short dsRNA does not stain well by ethidium bromide, to stain esiRNA in agarose gel, we usually added ethidium bromide into both gel and running buffer. In addition, you should always dilute esiRNA into a buffer with 100 m$M$ NaCl, not pure water.

8. To control for the specificity of RNAi, unrelated esiRNA targeting for example: luciferase, green fluorescent protein (GFP) or LacZ could be used as negative controls. Phenotypes should be reproduced by esiRNAs corresponding to different regions within the gene of interest. The strictest way to confirm that RNAi works specifically is to conduct a rescue experiment. Investigators may design

esiRNA to target either 5' UTR or 3' UTR, and then introduce expression cDNA to complement the defect caused by RNAi.

9. Although esiRNA is effective for many genetic disruption applications, it is not a guaranteed means to obtain null phenotypes. The effectiveness of esiRNA will depend on several variables: the target gene, the amount of RNA delivered, and the magnitude of inhibition required for a phenotypic effect. In preparing to interpret RNAi results, investigators should also consider the possibility that homeostatic mechanisms for a particular gene might allow cells to compensate for RNAi through auto-regulation or post-translational modulation.

## Acknowledgments

This work was supported by NIH grant CA44338 to J. M. Bishop, and the G. W. Hooper Research Foundation. D. Yang is supported by a Postdoctoral Fellowship from the Susan G. Komen Breast Cancer Fund. A. Goga is supported by a Postdoctoral Research Fellowship for Physicians from the Howard Hughes Medical Institute. We thank N. L'Etoile for critical reading of the manuscript.

## References

1. Sharp, P. A. (2001) RNA interference—2001. *Genes Dev.* **15,** 485–490.
2. Bosher, J. M. and Labouesse, M. (2000) RNA interference: genetic wand and genetic watchdog. *Nature Cell Biol.* **2,** E31–36.
3. Paddison, P. J., Caudy, A. A., and Hannon, G. J. (2002) Stable suppression of gene expression by RNAi in mammalian cells. *Proc. Natl. Acad. Sci. USA* **99,** 1443–1448.
4. Wianny, F. and Zernicka-Goetz, M. (2000) Specific interference with gene function by double-stranded RNA in early mouse development. *Nature Cell Biol.* **2,** 70–75.
5. Stark, G. R., Kerr, I. M., Williams, B. R., Silverman, R. H., and Schreiber, R. D. (1998) How cells respond to interferons. *Annu. Rev. Biochem.* **67,** 227–264.
6. Elbashir, S.M., Harborth, J., Lendeckel, W., Yalcin, A., Weber, K., and Tuschl, T. (2001) Duplexes of 21-nucleotide RNAs mediate RNA interference in cultured mammalian cells. *Nature* **411,** 494–498.
7. Caplen, N. J., Parrish, S., Imani, F., Fire, A., and Morgan, R. A. (2001) Specific inhibition of gene expression by small double-stranded RNAs in invertebrate and vertebrate systems. *Proc. Natl. Acad. Sci. USA* **98,** 9742–9747.
8. Yang, D., Buchholz, F., Huang, Z., Goga, A., Chen, C. Y., Brodsky, F. M., and Bishop, J. M. (2002) Short RNA duplexes produced by hydrolysis with *Escherichia coli* RNase III mediate effective RNA interference in mammalian cells. *Proc. Natl. Acad. Sci. USA* **99,** 9942–9947.
9. Schwarz, D. S., Hutvagner, G., Haley, B., and Zamore, P. D. (2002) Evidence that siRNAs function as guides, not primers, in the *Drosophila* and human RNAi pathways. *Mol. Cell* **10,** 537–548.
10. Amarasinghe, A. K., Calin-Jageman, I., Harmouch, A., Sun, W., and Nicholson, A. W. (2001) *Escherichia coli* ribonuclease III: affinity purification of

hexahistidine-tagged enzyme and assays for substrate binding and cleavage. *Methods Enzymol.* 2001. **342**, 143–158.

11. Bass, B. L. (2000) Double-stranded RNA as a template for gene silencing. *Cell* **101**, 235–238.

12. Elbashir, S. M., Lendeckel, W., and Tuschl, T. (2001) RNA interference is mediated by 21- and 22-nucleotide RNAs. *Genes Dev.* **15**, 188–200.

13. Elbashir, S. M., Martinez, J., Patkaniowska, A., Lendeckel, W., and Tuschl, T. (2001) Functional anatomy of siRNAs for mediating efficient RNAi in Drosophila melanogaster embryo lysate. *EMBO J.* **20**, 6877–6888.

14. Paddison, P. J., Caudy, A. A., Bernstein, E., Hannon, G. J., and Conklin, D. S. (2002) Short hairpin RNAs ( shRNAs ) induce sequence-specific silencing in mammalian cells. *Genes Dev.* **16**, 948–958.

15. Brummelkamp, T. R., Bernards, R., and Agami, R. (2002) A system for stable expression of short interfering RNAs in mammalian cells. *Science* **296**, 550–553.

16. Barton, G. M. and Medzhitov, R. (2002) Retroviral delivery of small interfering RNA into primary cells. *Proc. Natl. Acad. Sci. USA* **99**, 14,943–14,945.

17. Holen, T., Amarzguioui, M., Wiiger, M. T., Babaie, E., and Prydz, H. (2002) Positional effects of short interfering RNAs targeting the human coagulatinon trigger Tissue Factor. *Nucleic Acids Res.* **30**, 1757–1766.

18. Calegari, F., Haubensak, W., Yang, D., Huttner, W. B., and Buchholz, F. (2002) Tissue-specific RNA interference in postimplantation mouse embryos with endoribonuclease-prepared short interfering RNA. *Proc. Natl. Acad. Sci. USA* **99**, 14,236–14,240.

19. Misquitta, L. and Paterson, B. M. (1999) Targeted disruption of gene function in Drosophila by RNA interference (RNA-i): a role for nautilus in embryonic somatic muscle formation. *Proc. Natl. Acad. Sci. USA* **96**, 1451–1456.

20. Fire, A., Xu, S., Montgomery, M. K., Kostas, S. A., Driver, S. E., and Mello, C. C. (1998) Potent and specific genetic interference by double-stranded RNA in *Caenorhabditis elegans*. *Nature* **391**, 806–811.

21. Kawasaki, H., Suyama, E., Iyo, M., and Taira, K. (2003) siRNAs generated by recombinant human Dicer induce specific and significant but target site-independent gene silencing in human cells. *Nucleic Acids Res.* **31**, 981–987.

# 36

## RNAi Expression Vectors in Mammalian Cells

### Makoto Miyagishi and Kazunari Taira

#### Summary

RNA interference (RNAi) is a recently developed technique for gene silencing by introducing dsRNA into cells, and it is shown to work in mammalian cells when siRNAs are used. Several groups have developed vector-based siRNA expression systems that can induce RNAi in living cells. These vector systems use a pol III promoter, such as U6 or H1, and are classified into two groups based on the form of expressed RNAs: tandem-type and hairpin-type. Here, we describe how to generate these siRNA expression vectors and outline the experimental procedure for suppressing the expression of a reporter gene by transient transfection of a siRNA expression vector.

**Key Words:** RNAi; RNA interference; gene silencing; polymerase III; U6 promoter.

## 1. Introduction

RNA interference is a phenomenon in which expression of an individual gene can be specifically silenced by introducing a double-stranded RNA (dsRNA), that is homologous to the gene into cells, and it is receiving attention as the most powerful tool for reverse genetics in the post-genome era *(1)*.

Following the discovery of RNAi in *C. elegans (2)*, it was also observed in various organisms including plants, *Drosophila*, and a protozoan *(3–6)*. Although the mechanism of RNAi is not fully understood, recent genetic and biochemical studies have revealed some details at the molecular level. Inside the cell, dsRNAs are digested into fragments of 21–23 nucleotides (nt) with a two-nt 3'-overhang by an RNase III-related enzyme, which was identified as Dicer in *Drosophila (7)* and recently in humans *(8,9)*. Subsequently, these small fragments, known as small interfering RNAs (siRNAs), are incorporated into protein/RNA complexes *(10)* or the RISC (RNA-induced silencing complex) *(11)*. Then, they appear to be unwound to yield single-stranded RNAs *(10)* that

From: *Methods in Molecular Biology, vol. 252: Ribozymes and siRNA Protocols, Second Edition*
Edited by: M. Sioud © Humana Press Inc., Totowa, NJ

act as guide sequences (GS). The active complexes containing the guide RNA recognize and cleave the target RNA.

Although RNAi works well in various organisms, the silencing of specific genes by RNAi has proven to be difficult to detect in mammalian systems because of the dsRNA-dependent nonspecific inhibition of protein synthesis, which is part of the host's system for defense against viral infections. However, recent studies by Tuschl and colleagues demonstrated that 21- or 22-nt RNAs with two-nt 3'-overhangs could induce gene silencing without the non-specific inhibition of translation in cultured mammalian cells *(12–14)*.

Various groups, including our own, have developed systems for vector-mediated specific RNAi in mammalian cells *(15–20*; reviewed in **refs. *14*** and **22**). These vector systems use a pol III promoter, such as the U6, or H1, and the systems have been classified into two groups—depending on whether the expressed RNAs are tandem-type or hairpin-type (**Figs. 1, 2**). In this chapter, we describe how to construct these vectors, and show an example of a suppressive experiment targeted to reporter genes.

## 2. Materials

### 2.1. Buffers

1. *Bsp*MI (NewEngland Biolabs Inc., Beverly, MA).
2. 10X BAP buffer (Toyobo).
3. 10X pyrobest™ DNA polymerase buffer (Takara).
4. 10X deoxynucleotides (dNTPs) stock mixture: 2.5 m$M$ each dNTP.
5. TE buffer: 10 m$M$ Tris-HCl, pH 8.0, 1 m$M$ ethylenediaminetetraacetic acid (EDTA).
6. Buffer EC, enhancer (supplied in Effectene transfection reagent, Qiagen).

### 2.2. Other Reagents and Kits

1. Bacterial alkaline phosphatase (BAP).
2. MinElute Gel purification kit (QIAGEN).
3. Pyrobest™ DNA polymerase (TAKARA).
4. Ligation high (TOYOBO).
5. HiSpeed midi purification kit (QIAGEN).
6. DsRed protein expression plasmid (e.g., pDsRed2, BD Biosciences Clontech).
7. Hygromycin/GFP fusion protein expression plasmid (e.g., pHygEGFP, BD Biosciences Clontech).
8. Effectene transfection reagent (Qiagen).

## 3. Methods

### 3.1. Preparation of Vector DNA

1. Digest 3–5 μg of pU6icassette plasmid (**Fig. 2A**) with *Bsp*MI in a reaction volume of 100 μL.
2. Load the reaction mixture onto a 0.8% agarose gel and electrophoresis.

**A**
**Tandem type**

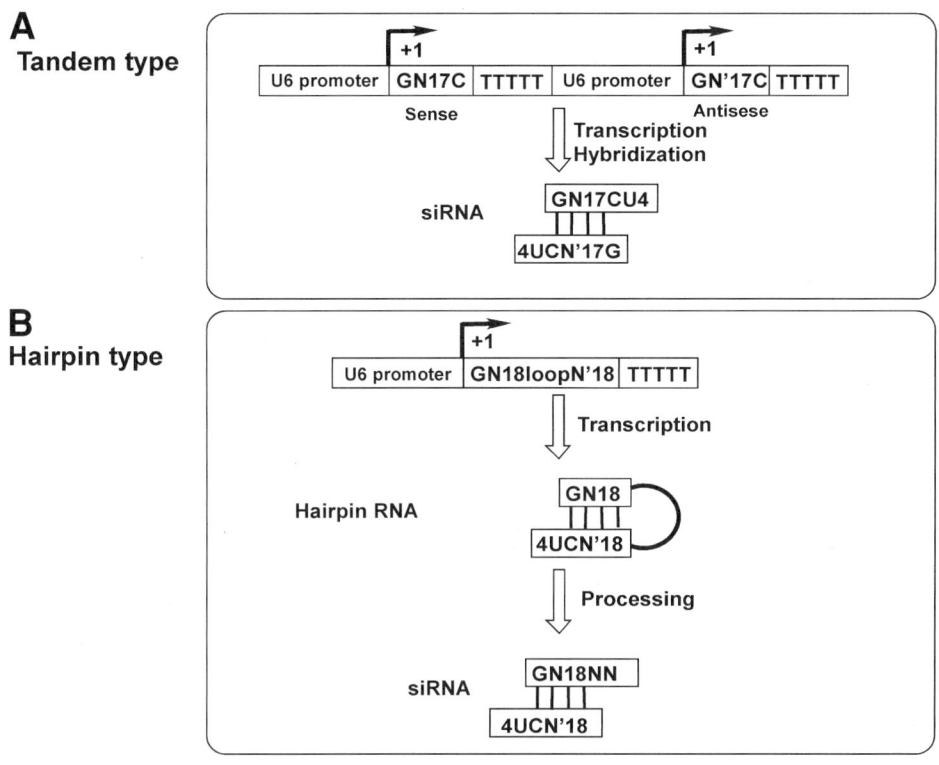

Fig. 1. Schematic representation of tandem-type and hairpin-type siRNA-expression vectors. In the tandem type, the U6 promoter drives both sense and antisense strands transcription. Subsequently, sense and antisense RNAs anneal and form siRNA duplexes with a 4-nt overhang at each 3'-end. In the hairpin type, sense and antisense nucleotides are expressed as a single unit. The transcribed RNA forms a hairpin, with a stem structure that is processed as indicated.

3. Excise gel pieces that contain the DNA fragments and purify via—for example— MinElute Gel purification kit.
4. In the case of construction of tandem-type siRNA expression vector, the digested vector should be dephosphorylated to reduce no insert background. The digested vector must not be phosphorylated for construction of hairpin-type siRNA expression vector. Combining the following produces a dephosphorylation reaction:
   a.  10 µL digested vector.
   b.  10 µL 10X BAP buffer.
   c.  3 µL BAP.
   d.  77 µL distilled water.
   Incubate the reaction for at least 1 h at 65°C.

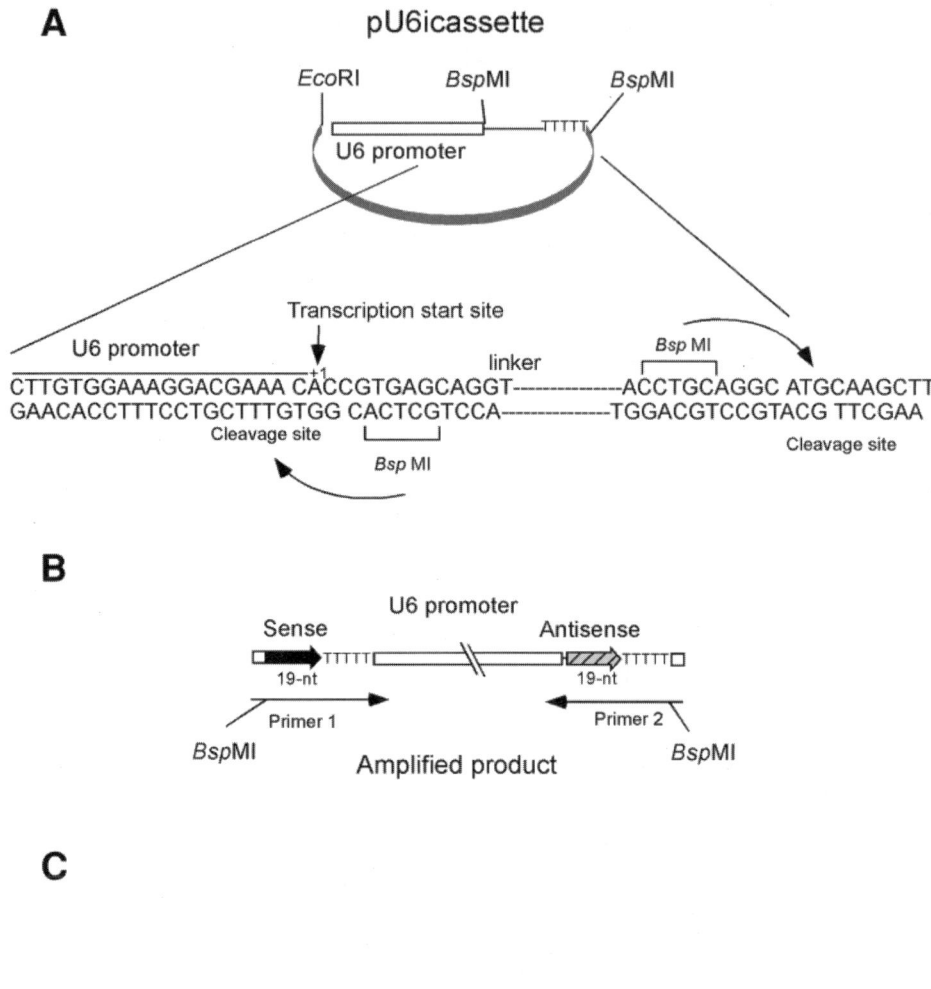

Fig. 2. siRNA expression vector. (**A**) Cloning site of pU6icassette vector. (**B**) DNA insert for a tandem-type siRNA expression vector. (**C**) The sequence of DNA insert for a hairpin-type siRNA expression vector.

5. After BAP treatment, the vector is purified by MinElute PCR purification kit or MinElute Reaction Cleanup Kit. Final elution volume is 30 μL, and 1 μL (approx 100 ng) is used in each ligation reaction.

## 3.2. Preparation of Insert DNA

### 3.2.1. Construction of Tandem-Type siRNA Expression Vector

1. Order or synthesize the following primer pairs:
   a. 5'-ggc tct aga ACC TGC cgg cca ccN NNN NNN NNN NNN NNN NNN ttt ttc aat tca agg tcg ggc ag-3'
   b. 5'-ggc tct aga ACC TGC tag cgc ata aaa aNN NNN NNN NNN NNN NNN NNg gtg ttt cgt cct ttc cac aag-3'
2. Replace N19 with sense or antisense target sequence. When 5'-GTG CGC TGC TGG TGC CAA C-3' is used as a target sequence. For example, the paired primers have the following sequence:
   a. 5'-ggc tct aga ACC TGC cgg cca ccG TGC GCT GCT GGT GCC AAC ttt ttc aat tca agg tcg ggc ag-3'
   b. 5'-ggc tct aga ACC TGC tag cgc ata aaa aGT GCG CTG CTG GTG CCA ACg gtg ttt cgt cct ttc cac aag-3'
3. Polymerase chain reactions are carried out under the following conditions: Reaction mixture:
   a. 10 μL 10X pyrobest polymerase buffer.
   b. 0.5 μL pyrobest polymerase.
   c. 1 μL (100 m$M$) each primer.
   d. 8 μL (2 m$M$ each) dNTP mix.
   e. 79.5 μL distilled water.
   Samples were incubated at 95°C for 1 min, and then amplification was performed for 30 cycles (10 s at 98°C, 1 min at 55°C; 1 min at 72°C), followed by 4°C incubation.
4. PCR products (**Fig. 2B**) are purified by MinElute PCR purification kit and then digested with *Bsp*MI for at least 2 h in 100 μL reaction volume.
5. After digestion, samples are separated by agarose gel electrophoresis and purified by MinElute Gel purification kit. Final elution volume is 10 μL.

### 3.2.2. Construction of Hairpin-Type siRNA Expression Vector

1. In the case of the target site sequence given in the previous section, order or synthesize the following oligonucleotides (**Fig. 2C**; *see* **Notes 1** and **2**).
   a. 5'-cac cGT GGC TGC TGG TGC CCA ACC Cgg aca gca cac GGG TTG GGC ACC AGC AGC GCA Ctt ttt-3'
   b. 5'-gca taa aaa GTG CGC TGC TGG TGC CCA ACC Cgt gtg ctg tcc GGG TTG GGC ACC AGC AGC GCA C-3'
2. For annealing, combine the following materials:
   a. 5 μL (100 m$M$) each oligonucleotide.
   b. 1 μL 1 $M$ NaCl.
3. After mixing, perform the following profile using thermal cycler:
   a. Heat to 95°C and remain at 95°C for 2 min.
   b. Rapid cool to 72°C.
   c. Ramp cool to 4°C over a period of 2 h.

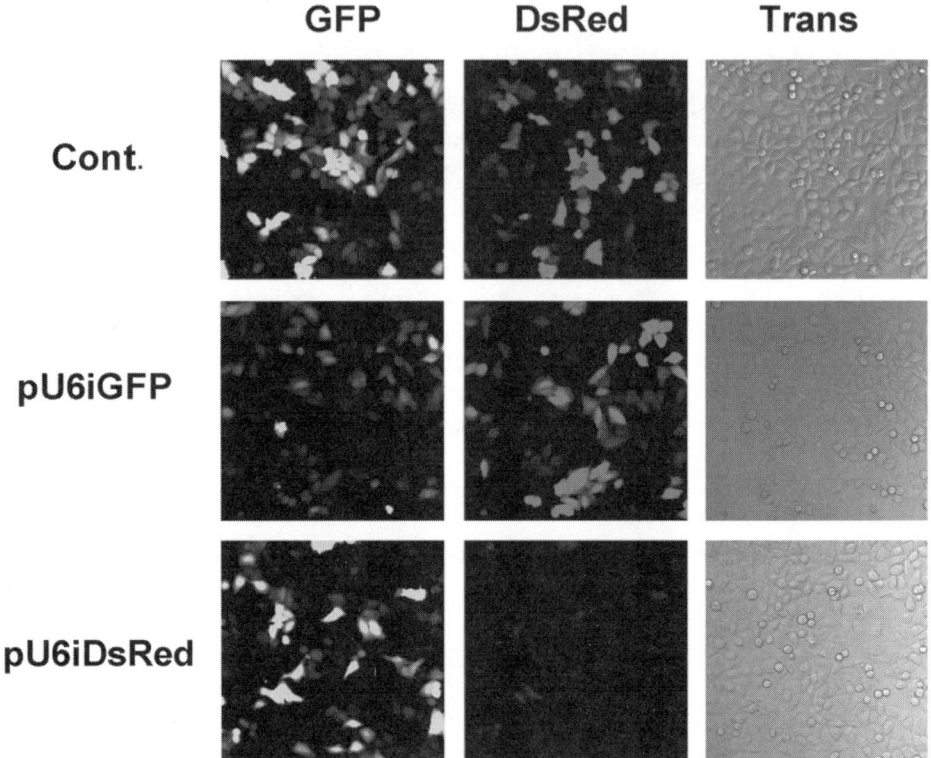

Fig. 3. The effect of hairpin-type siRNA expression vector targeted to Hygromycin/ GFP fusion or DsRed gene. The target sequence of each gene is as follows: Hygromycin/GFP, 5'-GGC TAC GTC CAG GAG CGC ACC-3'; DsRed, 5'-GTG GGA GCG CGT GAT GAA CTT-3'.

4. After incubation at 4°C, briefly spin the tubes in a microfuge, and dilute the annealed oligonucleotide 200-fold with TE buffer. Use 1 μL of for ligation described in the following section.

### 3.2.3. Ligation of the Insert DNA to the pU6icassette Vector and Transformation Into E. coli

1. The ligation reaction is carried out by combining the following components:
   a. 1 μL annealed oligonucleotide.
   b. 1 μL gel-purified vector.
   c. 1 μL ligation high.
2. Incubate the reaction mixture for 30 min at 16°C.
3. Transform *E. coli* host cells by adding the ligation mixture and plate the cells on LB agar plates containing ampicillin at 100 μg/mL. We use DH5α cells as the host. More than 90% of colonies should be positive (**Fig. 3**; *see* **Note 3**)

## 3.3. Suppression of the Expression of an Exogenously Introduced Genes (e.g., Hygromycin/GFP, DsRed)

### 3.3.1. Preparation of Plasmid for Transfection

Plasmids used in transfection experiments are prepared by the HiSpeed midi purification kit. After preparation, plasmids were precipitation by ethanol, washed, and then dissolved in TE buffer at 1 μg/μL. At this stage, it is important to check the purity and quality of the plasmids by agarose gel electrophoresis.

### 3.3.2. Transient Transfection of DsRed Expression Vector and siRNA Expression Vector Into HeLa S3 Cells

1. Seed HeLa S3 cells in 6-well plate. Approximately $1 \times 10^5$ cells/well/1 m: complete medium (Dulbecco's modified Eagle's medium [DMEM] +10% fetal bovine serum [FBS]) and incubate the cells at 37°C in a humidified 5% incubator.
2. Change the medium before transfection, when the cell reach 40–60% confluency (usually 1 d after seeding).
3. Prepare the following plasmids in a microcentrifuge tube:
   a. 50 ng pDsRed2.
   b. 500 ng pHygEGFP.
   c. 500 ng siRNA expression vector (targeted to the Hygromycin/GFP or DsRed gene).
4. Mix the mixture and adjust to a total volume of 10 μL by adding TE buffer.
5. Add 140 μL buffer EC, and subsequently, 8 μL Enhancer, vortex for 1 s.
6. Incubate at room temperature for 5 min and then spin down the mixture.
7. Add 25 μL Effectene transfection reagent and vortex for 10 s.
8. Incubate the sample for 10 min at room temperature.
9. Add 1 mL growth medium to the sample.
10. Mix by pipetting, and immediately added to one well containing cells.
11. 2 d after transfection, cells can be analyzed by fluorescence microscopy (*see* **Note 4**).

## 4. Notes

1. In this protocol, we use a microRNA-derived 11-nt sequence, as the loop sequence in hairpin type siRNA expression vector.
2. When constructing hairpin-type siRNA-expression vectors, some technical problems related to sequencing can be observed. Such problems can be reduced by introducing mutations in the sense strand of hairpinRNA. Typically we introduce three to four C>T or A>G mutations in the sense strand.
3. If stable transfection is required, subclone the siRNA expression cassette into an appropriate vector.
4. Comparative analysis of tandem-type and hairpin-type siRNA expression vectors reveals that the hairpin-type siRNA expression vector has a higher suppressive activity than the tandem-type siRNA expression vector at low concentrations of plasmid. When a retrovirus or lentivirus vector was used, a hairpin-type siRNA expression cassette seemed to be favorable.

## References

1. Couzin, J. (2002) Breakthrough of the year: small RNAs make big splash. *Science* **298,** 2296.
2. Fire, A., Xu, S., Montgomery, M. K., Kostas, S. A., Driver, S. E., and Mello, C. C. (1998) Potent and specific genetic interference by double-stranded RNA in *Caenorhabditis elegans. Nature* **391,** 806–811.
3. Fire, A. (1999) RNA-triggered gene silencing. *Trends Genet.* **15,** 358–363.
4. Hammond, S. M., Caudy, A. A., and Hannon, G. J. (2001) Post-transcriptional gene silencing by double-stranded RNA. *Nat. Rev. Genet.* **2,** 110–119.
5. Sharp, P. A. (2001) RNA interference 2001. *Genes Dev.* **15,** 485–490.
6. Zamore, P. D. (2001) RNA interference: listening to the sound of silence. *Nature Struct. Biol.* **8,** 746–750.
7. Bernstein, E., Caudy, A. A., Hammond, S. M., and Hannon, G. J. (2001) Role for a bidentate ribonuclease in the initiation step of RNA interference. *Nature* **409,** 363–366.
8. Provost, P., Dishart, D., Doucet, J., Frendewey, D., Samuelsson, B., and Radmark, O. (2002) Ribonuclease activity and RNA binding of recombinant human Dicer. *EMBO J.* **21,** 5864–5874.
9. Zhang, H., Kolb, F. A., Brondani, V., Billy, E., and Filipowicz, W. (2002) Human Dicer preferentially cleaves dsRNAs at their termini without a requirement for ATP. *EMBO J.* **21,** 5875–5885.
10. Nykanen, A., Haley, B., and Zamore, P. D. (2001) ATP requirements and small interfering RNA structure in the RNA interference pathway. *Cell* **107,** 309–321.
11. Hammond, S. M., Boettcher, S., Caudy, A. A., Kobayashi, R., and Hannon, G. J. (2001b) Argonaute2, a link between genetic and biochemical analyses of RNAi. *Science* **293,** 1146–1150.
12. Tuschl, T. (2002) Expanding small RNA interference. *Nature Biotechnol.* **20,** 446–448.
13. Caplen, N. J., Parrish, S., Imani, F., Fire, A., and Morgan, R. A. (2001) Specific inhibition of gene expression by small double-stranded RNAs in invertebrate and vertebrate systems. *Proc. Natl. Acad. Sci. USA* **98,** 9742–9747.
14. McManus, M. T. and Sharp, P. A. (2002) Gene silencing in mammals by small interfering RNAs. *Nat. Rev. Genet.* **3,** 737–747.
15. Brummelkamp, T. R., Bernards, R., and Agami, R. (2002a) A system for stable expression of short interfering RNAs in mammalian cells. *Science* **296,** 550–553.
16. Lee, N. S., Dohjima, T., Bauer, G., Li, H., Li, M.J., Ehsani, A., et al. (2002) Expression of small interfering RNAs targeted against HIV-1 rev transcripts in human cells. *Nat. Biotechnol.* **20,** 500–505.
17. Miyagishi, M. and Taira, K. (2002) U6 promoter-driven siRNAs with four-uridine 3' overhangs efficiently suppress targeted gene expression in mammalian cells. *Nat. Biotechnol.* **20,** 497–500.
18. Paddison, P. J., Caudy, A. A., Bernstein, E., Hannon, G. J., and Conklin, D. S. (2002) Short hairpin RNAs (shRNAs) induce sequence-specific silencing in mammalian cells. *Gens Dev.* **16,** 948–958.

19. Paul, C. P., Good, P. D., Winer, I., and Engelke, D. R. (2002) Effective expression of small interfering RNA in human cells. *Nat. Biotechnol.* **20,** 505–508.
20. Sui, G., Soohoo, C., Affar, el B., Gay, F., Shi, Y., Forrester, W. C., and Shi, Y. (2002) A DNA vector-based RNAi technology to suppress gene expression in mammalian cells. *Proc. Natl. Acad. Sci. USA* **99**, 5515–5520.
21. Yu, J. Y., DeRuiter, S. L., and Turner, D. L. (2002) RNA interference by expression of short-interfering RNAs and hairpin RNAs in mammalian cells. *Proc. Natl. Acad. Sci. USA* **99,** 6047–6052.
22. Tuschl, T. (2002) Expanding small RNA interference. *Nat. Biotech.* **20,** 446–448.

# 37

## Gene-Array Analysis of Glioma Cells After Treatment With an Anti-PKCα siRNA

*A General Protocol*

### Marianne Leirdal and Mouldy Sioud

### Summary

Post-transcriptional gene silencing is a powerful tool to reveal gene function. In this chapter, we have used gene-array technology to analyze changes in gene expression after specific targeting of the protein kinase Cα isoform with small interfering RNAs (siRNAs). Expression profiles were investigated on a human gene array of 588 key genes involved in DNA synthesis, apoptosis, cell–cell communication, intracellular signal-transduction pathways, cell-surface molecules and transcription factors. The data indicate that siRNAs are an ideal tool for target validation, and show that RNA-dependent RNA polymerase is not required for RNAi in mammalian cells.

**Key Words:** DNA enzymes; RNA interference; small interfering RNAs; expression array; microarray.

## 1. Introduction

Cancer is a genetic disease that arises from an accumulation of mutations that promote clonal selection of cells with increasingly aggressive behavior. Most of the mutations described in cancer are somatic, and are found only in an individual's cancer cells. These somatic mutations frequently alter the functions of cell-signaling pathways (*1–7*).

The ability to design effective treatments for cancer depends on our understanding of the mechanisms surrounding cell proliferation and differentiation. The most valuable information would be obtained if the function of the gene of interest could be blocked in specific cells. One technique for turning off a single gene is the use of ribozymes, DNA enzymes, and siRNAs (*8–11*). Because of the specificity of Watson-Crick basepairing, these agents have the ability to

From: *Methods in Molecular Biology, vol. 252: Ribozymes and siRNA Protocols, Second Edition*
Edited by: M. Sioud © Humana Press Inc., Totowa, NJ

block expression of individual genes that are structurally similar, such as isoenzymes. Here, we have used gene-array technology to identify the molecular targets of protein kinase Cα that are known to be involved in glioma-cell proliferation *(12,13)*. The experimental protocol is described in **Fig. 1**. A strong interaction between the PKCα and some key cellular compounds such as *c*-jun, Her-3, integrins, and p21[WAF1/CIP1] were found. Despite the high degree of homology between the various human PKC isoforms, only the PKCα was downregulated by the designed siRNA. This observation would indicate that 5' nor 3' spreading of the silencing signal from siRNA do not occur in mammalian cells.

## 2. Materials

### 2.1. Anti-PKCα siRNA

The siRNA targeting the protein kinase Cα has the following sequence: 5'GGCUGAGGUUGCUGAUGAA-3'. Only the sense strand is shown. The target site has a 5' double AA leader (*see* Chapter 34).

### 2.2. Glioma Cells

The human glioma-cell line U87-MG was obtained from American Type Culture Collection (ATCC) and grown in Dubelcco's modified Eagle's medium (DMEM) from Gibco-BRL supplemented with 10% fetal bovine serum (FBS) and antibiotics (penicillin/streptavidin).

### 2.3. Transfection Reagent

1. Cationic liposomes 1,2-bis-(oleoyloxy)-3-trimethylammonium) propane (DOTAP).
2. Transfection buffer: 20 m$M$ HEPES, 150 m$M$ NaCl, pH 7.4.

### 2.4. Isolation of Total RNA

1. RNA working solution: 4.2 $M$ guanidine thiocyanate, 26.4 m$M$ sodium citrate, pH 7.0, 9 m$M$ sarcosyl. Add 0.360 mL β-mercaptoethanol per 50 mL of stock solution.
2. 2 $M$ NaAc, pH 4.0.
3. Phenol, chloroform/isoamyl alcohol (IAA) (24:1).
4. Isopropanol, 75% ethanol, and sterile water.

### 2.5. DNase I Treatment of Total RNA

1. RNase-free DNase I: 10 U/µL.
2. 10X DNase I buffer: 400 m$M$ Tris-HCl, pH 7.5, 100 m$M$ NaCl, 60 m$M$ MgCl$_2$.
3. DNase I solution: 1 U/µL diluted 1/10 in 1X DNase I buffer.
4. 10X termination mix: 0.1 $M$ EDTA, pH 8.0, 1 mg/mL glycogen.
5. Phenol/chloroform/IAA (25:24:1) equilibrated with 100 m$M$ sodium citrate, pH 4.5, 1 m$M$ ethylenediaminetetraacetic acid (EDTA).
6. 95% ethanol and 2 $M$ NaAc, pH 4.5.

---

## Tranfect the cells with siRNAs

↓

**After 48 h transfection time, prepare total RNA from siRNA- transfected and control cells**

↓

**Prepare cDNA probes**

↓

**Hybridyze to atlas arrays**

↓

**Analyze the data**

Fig. 1. Schematic presentation of the experimental protocol.

### 2.6. cDNA Probe Synthesis

1. 50 m$M$ dinucleotide triphosphate (dNTP) mix: 5 m$M$ each deoxycytidine 5' triphosphate (dCTP), deoxyguanosine 5' triphosphate (dGTP), deoxythymidine (dTTP).
2. 5X reaction buffer: 250 m$M$ Tris-HCl, pH 8.3, 375 m$M$ KCl, 15 m$M$ MgCl$_2$.
3. Moloney murine leukemia virus (MMLV) reverse transcriptase.
4. 100 m$M$ dithiothreitol (DTT).
5. α-32P dATP: 10 µCi/µL; 3000 Ci/mmol.
6. CDS primer mix supplied with the array.
7. 10X termination mix: 0.1 $M$ EDTA, pH 8.0, 1 mg/mL glycogen.

### 2.7. Probe Purification

After labeling, cDNA probes were purified using column chromatography, as described by the manufacturer (Clontech).

### 2.8. Hybridization

1. 10 mg/mL sheared salmon testes DNA "ExpressHyb solution" supplied with the array.
2. 10X denaturing solution: 1 $M$ NaOH, 10 m$M$ EDTA.
3. 2X neutralizing solution: 1 $M$ NaH$_2$PO$_4$, pH 7.0.
4. C$_0$t-1 DNA (1 mg/mL) supplied with the array. Deionized H$_2$O.
5. Atlas cDNA expression array membranes (Clontech) (*see* **Note 1**).

### 2.9. Washing

1. 20X SSC: 3 $M$ NaCl, 0.3 $M$ sodium citrate, pH 7.0.
2. Washing solution I: 2X standard saline citrate (SSC), 1% sodium dodecyl sulfate (SDS).

3. Washing solution II: 0.1X SSC, 0.5% SDS.
4. PhosphorImager.
5. Orientation grid supplied with the array.

## 3. Methods
### 3.1. Transfection

1. The day before transfection, trypsinize and plate the cells (U87-MG) in a 75-cm$^3$ flask, so they are 30% confluent the day of transfection. Incubate the cells at 37°C in a humidified 5% $CO_2$.
2. The day of transfection, replace the medium with 6 mL fresh medium.
3. Mix in a 1.5-mL microcentrifuge tube, 6 μL of a stock siRNA solution (100 μ*M*), and 44 μL transfection buffer to yeild a 100 n*M* final concentration (based on 6 mL total volume of cells).
4. Mix 50 μL of DOTAP and 100 μL of transfection buffer in a polystyrene tube.
5. Mix both solutions and incubate at room temperature for 15–30 min.
6. Add the mixed solutions to cells and rock the culture flask to ensure uniform distribution of the transfection mixture.
7. Transfect control cells with an inactive or irrelevant siRNA.
8. Isolate total RNA at 24–48 h posttransfection time.

### 3.2. Isolation of Total RNA

1. Rinse the cells with phosphate-buffered saline (PBS), add 5 mL working solution and rock the culture flask for 15–30 s to ensure cell lysis.
2. Transfer the cell lysate into 25–50-mL Falcon tube.
3. Add 1 mL of 2 *M* NaAc, pH 4.0, 5 mL of phenol and 2 mL of chloroform/IAA (24:1).
4. Mix well by inverting the tube and incubate for 15 min on ice.
5. Spin the mixture for 20 min at 2500*g* at 4°C.
6. Transfer the upper phase to a new Falcon tube, add 5 mL of isopropanol and precipitate overnight at –70°C.
7. Pellet the RNA by centrifugation for 30 min at 3000*g* at 4°C.
8. Discard the supernatant and wash the RNA pellet with ice-cold 75% ethanol.
9. Air-dry the RNA pellet and dissolve in 50 μLof diethylpyrocarbonate (DEPC)-treated water.

### 3.3. DNase I Treatment of Total RNA

1. For each sample mix in an eppendorf tube the following components:
   a. 50 μL total RNA (approx 0.1 μg/μL)
   b. 40 μL 10X DNase I buffer
   c. 20 μL DNase I solution
   d. 290 μL deionized $H_2O$
2. Incubate for 30 min at 37°C.
3. Add 100 μL of 10X termination mix. Mix by pipetting.

4. Add 400 μL of phenol/chloroform/IAA (25:24:1; pH 4.5) and vortex thoroughly.
5. Centrifuge for 10 min at 12,000$g$ at 4°C.
6. Carefully transfer the upper phase to a new Eppendorf tube. Add 400 μL of chloroform and vortex thoroughly.
7. Centrifuge for 10 min at 12,000$g$ at 4°C. Transfer the upper phase to a new microcentrifuge tube.
8. Add 1/10 vol (40 μL) of 2 $M$ NaAc, pH 4.5, and 2.5 vol of 95% ethanol. Vortex thoroughly and incubate on ice for 10 min.
9. Centrifuge at 12,000$g$ for 15 min at 4°C, wash the pellet with 500 μL of 80% ethanol, air-dry, and dissolve RNA pellet in 10 μL of DEPC-treated water.
10. At this stage, the quality of the RNA should be analyzed by running 1 μL of the RNA onto a 1% agarose gel.

### 3.4. cDNA Probe Synthesis

1. For each cDNA probe mix in a 0.5-mL polymerase chain reaction (PCR) tube the following components:
   a. 2 μL 5X reaction buffer
   b. 1 μL 50 m$M$ dNTP mix (for dATP label)
   c. 3.3 μL α$^{32}$P dATP
   d. 0.5 μL 100 m$M$ DTT
   e. (Prepare a mater mix for all probes)
2. Preheat a PCR thermal cycler to 70°C.
3. For each reaction, mix the following components in a 0.5-mL PCR tube:
   a. 1–2 μL RNA (2–5 μg)
   b. 0.8 μL CDS primer mix
   c. Water to a final volume of 3 μL
4. Incubate tubes in the preheated PCR thermal cycler at 70°C for 2 min. Reduce the temperature of the thermal cycler to 50°C, and further incubate for 2 min.
5. During this incubation, add 1 μL of MMLV reverse transcriptase per reaction to the master mixture (step 1) mix, and keep at room temperature.
6. After 2 min incubation at 50°C, add 8 μL of master mix to each reaction tube. Mix by pipeting, immediately return the tubes to the thermal cycler at 50°C, and incubate for 1 h.
7. Stop the reaction by adding 1 μL of 10X termination mix, and then purify the marked cDNA (probe) by column chromatography (*see* **Note 2**).

### 3.5. Column Chromatography

Column chromatography is performed according to the array instructions. After probe purification, check the radioactivity of the probe by scintillation counting. In this respect, add 2 μL probe to 5 mL of scintillation fluid and count $^{32}$P- or $^{33}$P-labeled samples on the $^{32}$P channel. Calculate the total number of counts in each sample. Probes synthesized using the procedure described above (**Subheading 3.4.**) should have a total of 2–10 × 10$^6$ cpm. However, 1 × 10$^6$ cpm will yeild good hybridization signals (*see* **Fig. 2A,B**).

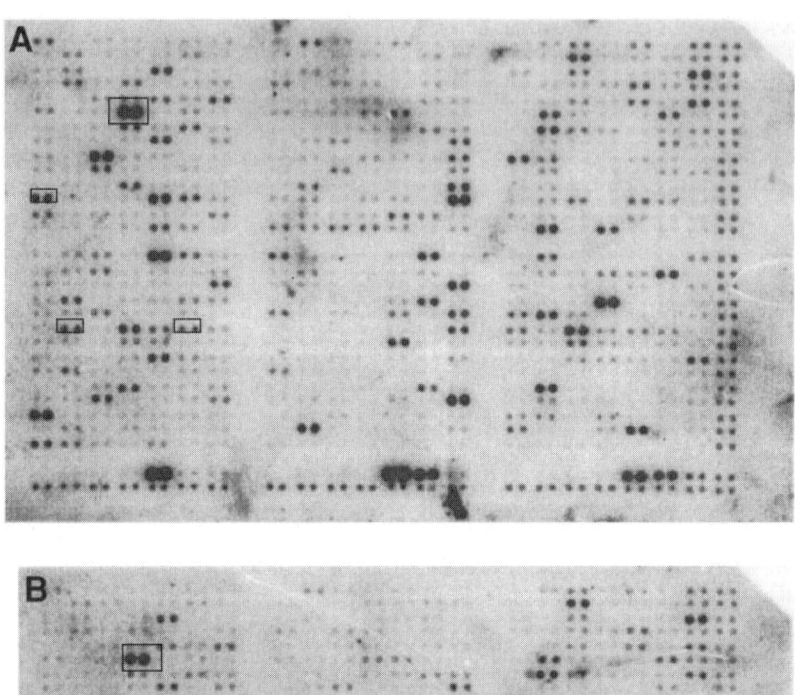

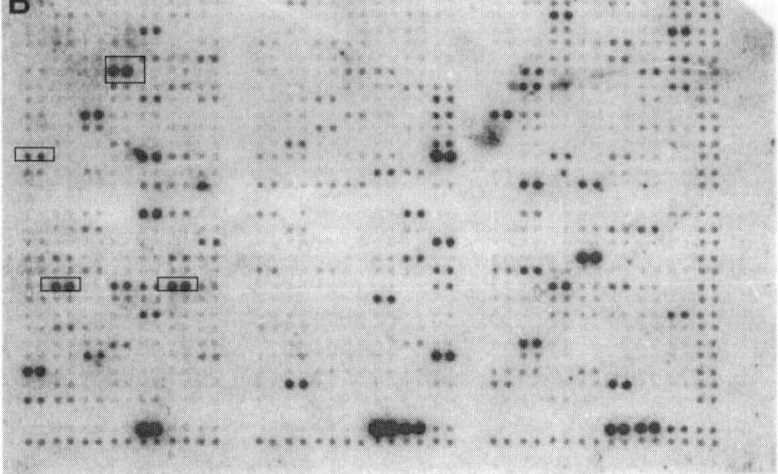

Fig. 2. A representative example of expression array analysis. Total RNA was prepared from anti-PKCα-treated cells (**A**) and untreated U87MG cells (**B**). The autoradiograph images are shown. Among the genes that are downregulated by anti-PKCα
siRNA treatment is the *p21*$^{WAF1/CIP1}$.

## 3.6. Hybridizing cDNA Probes to the Atlas Array

1.  Prewarm 5 mL of ExpressHyb solution at 68°C. At the same time, heat 0.5 mg of
    the sheared salmon testes DNA at 95–100°C for 5 min and chill on ice. Mix heat-
    denatured sheared salmon testes DNA with prewarmed ExpressHyb solution.

2. Wet the Atlas Array by placing it in a dish of deionized water, roll the membrane into a mesh, and then place the membrane and mesh into a hybridization bottle. Fill the hybridization bottle 1/4 with deionized water. Be sure that the membrane adheres to the inside walls of the container without creating air pockets, and then pour off all the water from the bottle (*see* **Note 1**).

3. Add 5 mL of the ExpressHyb solution containing salmon testes. Ensure that the solution is distributed over the membrane and prehybridize for 30 min with continuous agitation at 68°C.

4. Mix 100 μL of labeled probe, 11 μL of 10X denaturing solution and incubate at 68°C for 20 min.

5. Add 5 μL of $C_0t$-1 DNA and 115 μL of 2X neutralizing solution and incubate at 68°C for 10 min

6. Add the mixture (**steps 4** and **5**) directly to the array and the prehybridization solution, and hybridize overnight with continuous agitation at 68°C.

## *3.7. Washing*

1. Prewarm washing solutions II and I at 68°C.

2. Remove the hybridization solution. Replace with 200 mL of prewarmed washing solution I and incubate for 30 min with continuous agitation at 68°C. Repeat this step two to three times. Check with a Geiger hand counter between the washes.

3. Wash for 30 min with 200 mL of prewarmed solution II at 68°C.

4. Remove the atlas array from the container and shake off excess wash solution. Do not allow the membrane to dry. Immediately seal the atlas array in plastic and expose it to a PhosphorImaging screen for 3–72 h (**Fig. 1A,B**; *see* **Note 3**). For conservation and re-use of the membranes, *see* **Notes 4** and **5**.

## 4. Notes

1. Always use forceps to handle the membranes and grip the membranes by the edges only. Never allow the membranes to dry, even slightly. After the final wash, shake off excess solution with forceps, and immediately wrap the membrane completely with plastic wrap. If the membrane dries even partially, subsequent removal of the probe (stripping) from the atlas array will be difficult.

2. The quality of each probe should be checked before hybridization to Atlas Array Membranes. Use a control (blank) nylon membrane and follow the procedure for hybridizing cDNA probes to the atlas array, except that only 1/5 to 1/10 of your total pool of probe will be used. This will allow you to estimate the level of nonspecific background.

3. We usually confirm the array data by Northern and Western blot analysis. Specific downregulation of the PKCα by the designed siRNA did not affected the expression of other PKC isoforms.

4. After you have finished exposing the membrane to PhosphorImager, immediately strip the membrane, wrap it in plastic, and place it in a –20°C freezer.

5. For effective stripping of a cDNA probe from the membranes, it is important that the membranes are stored at –20°C when not in use. Stripping of the membranes.

Heat 500 mL 0.5% SDS solution to boiling. Remove the plastic wrap from the atlas array and immediately place the membrane into the boiling solution for 5–10 min. Remove the solution from the heat and allow to cool for 10 min. Rinse the atlas array in washing solution and immediately seal the atlas array in plastic. Check the efficiency of stripping with a Geiger hand counter and by phosphorimaging.

## References

1. Brazma, A. and Vilo, J. (2000) Gene expression data analysis. *FEBS.* **480,** 17–24.
2. Celis, J. E., Kruhøffer, M., Gromova, I., Frederiksen, C., Østergaard, M., Thykjaer T., Gromov, P., et al. (2000) Gene expression profiling: monitoring transcription and translation products using DNA microarrays and proteomics. *FEBS* **480,** 2–16.
3. Kinzler, K. W. and Vogelstein, B. (1996) Lessons from hereditary colorectal cancer. *Cell* **87,** 159–170.
4. Campbell, S. L., Khosravi-Far, R., Rossman, K. L., Clark, G. J., and Der, C. J. (1998) Increasing complexity of Ras signalling. *Oncogene* **17,** 1395–1413.
5. Robinson, M.J., and Cobb, M.H., (1997) Mitogen-activated protein kinase pathways. *Curr. Opin. Cell Biol.* **9,** 180–186.
6. Hug, H. and Sarre, T. F. (1993) Protein kinase C isoenzymes: divergence in signal transduction? *Biochem. J.* **291,** 329–343.
7. Dhanasekaran, N. (1998) Cell signaling: an overview. *Oncogene* **17,** 1329–1330.
8. Sioud, M. (2001) Nucleic acid enzymes as a novel generation of anti-gene agents. *Curr. Mol. Med.* **1,** 575–588.
9. Joyce, G. F. (1996) Building the RNA world. Ribozymes. *Curr. Biol.* **6,** 965–967.
10. Elbashir, S. M., Harborth, J., Lendeckel, W., Yalcin, A., Weber, K., and Tuschl T. (2001) Duplexes of 21-nucleotide RNAs mediate RNA interference in cultured mammalian cells. *Nature* **411,** 494–498.
11. Caplan, N. J., Parrish, S., Imani, F., Fire, A., and Morgan, R. A. (2001) Specific inhibition of gene expression by small double-stranded RNAs in invertebrate and vertebrate systems. *Proc. Natl. Acad. Sci. USA* **98,** 9742–9747.
12. Sioud, M. and Sørensen, D. R. (1998) A nuclease-resistant protein kinase Ca ribozyme blocks glioma cell growth. *Nat. Biotech.* **16,** 556–561.
13. Couldwell, W. T., Uhm, J. H., Antel, J. P., and Young, V. W. (1991) Enhanced protein kinase C activity with the growth rate of malignant gliomas in vitro. *Neurosurgery* **29,** 880–887.
14. Sioud, M. and Leirdal, M. (2000) Design of nuclease resistant protein kinase Ca DNA enzyme with potential therapeutic application. *J. Mol. Biol.* **296,** 937–947.
15. Leirdal, M. and Sioud, M. (2002) Gene silencing in mammalian cells by preformed small RNA duplexes. *Biochem. Biophys. Res. Comm.* **295,** 744–748.

# 38

# RNAi in Living Mice

## Hidetoshi Hasuwa and Masaru Okabe

## Summary

By introducing double-stranded RNAs (dsRNAs), it was shown that mRNAs that share sequences are destroyed, and that the translation step is severely downregulated (RNAi). This technique was demonstrated to be a very powerful tool for reverse genetics in *Caenorhabditis elegans*. However, studies have shown that RNAi can be achieved in living mice. In this chapter, we used "green mice" and "green rats" as model animals to demonstrate silencing of green fluorescent protein (GFP) expression by RNAi. In order to silence the gene, we produced transgenic mouse lines that produce dsRNA that is driven by the H1 promoter. It was demonstrated that the transgenically expressed, double stranded RNA could silence the GFP throughout the body of the mouse or rat and throughout the lifetime.

**Key Words:** RNAi; siRNA; gene silencing; RNA polymerase III promoter; transgenic animal.

## 1. Introduction

Double-stranded RNA (dsRNA)-based gene silencing or RNA interference (RNAi) is considered to be an ancient and evolutionarily conserved mechanism for sequence-specific post-transcriptional gene silencing among species from various kingdoms (reviewed in **refs. *1*** and ***2***). Until recently, the use of RNAi to silence mammalian genes was not applicable, because the introduction of dsRNAs longer than 30 nt elicits a viral response that is sequence-nonspecific *(3)*. Recently, the introduction of short duplexes of synthetic 21–23-nt RNAs (siRNA) into mammalian cells was found to have a gene-specific silencing function *(4)*. However, the transfected synthetic siRNA works for only a few days in mammalian cells. If the siRNAs were produced from the vector that is integrated into the genome, the RNAi would continue throughout the cellular lifespan.

From: *Methods in Molecular Biology, vol. 252: Ribozymes and siRNA Protocols, Second Edition*
Edited by: M. Sioud © Humana Press Inc., Totowa, NJ

We chose the green fluorescent protein (GFP) as a model gene to be silenced and used the "green mice" and "green rats" that express GFP all over their bodies *(5,6)*. Here, we describe the silencing of the ubiquitously expressed GFP by injecting vectors that consisted of RNA polymerase III (pol III) promoter (H1 or U6) that is designed to produce 21-nt siRNA specific to GFP *(7)*.

The success of the transgene-based RNAi system suggests that the method can be used as an alternative method for gene disruption (or gene "knockdown"). Moreover, the silencing in the rat *(7)* indicates that the method may be applicable to animals in which homologous recombination is not possible because of the lack of ES cell lines.

## 2. Materials

### 2.1. Preparation of RNAi Vector

#### 2.1.1. pH1 Plasmid (pH1)

5'-CCGCTCGAGAAGCTTCTAGACTGCAGTGGTCTCATACAG AACTTATAAGATTCCC-3' and 5'-GATATCGGATCCGCGGAATTCGA ACGCTGACGTCATCAACCC-3' were used to amplify the H1 promoter *(8)* from human 293 cell-genomic DNA. The PCR product digested with *Eco*RV and *Xho*I, was cloned into the *Xho*I and blunted SacI site of pBluescript II SK+.

#### 2.1.2. DNA Oligos

DNA oligos that encode the siRNA of target gene and termination signal were chemically synthesized (C-18 cartridge-purified grade). The sequence that encodes EGFP siRNA is shown in **Fig. 2**.

#### 2.1.3. pRed Gene

The HcRed1 gene was amplified from pHcRed1-N1 (Clontech) by PCR using primers including *Not*I or *Xho*I sites at their 5' ends. The PCR product was digested with *Not*I and *Xho*I, and was ligated to *Not*I and *Xho*I sites introduced between the CAG promoter and the rabbit beta-globin polyA region of pCX/EGFP *(5)*.

#### 2.1.4. E. coli Strains

JM109, SURE2 were purchased from Stratagene.

### 2.2. Purification of the Transgene

1. SeaKem Gold Agarose (TaKaRa 50150).
2. QIAEX II gel extraction kit (Qiagen 20021).
3. Phenol/chloroform/isoamyl alcohol (IAA) (25:24:1).
4. SUPREC-01 (TaKaRa 9040).

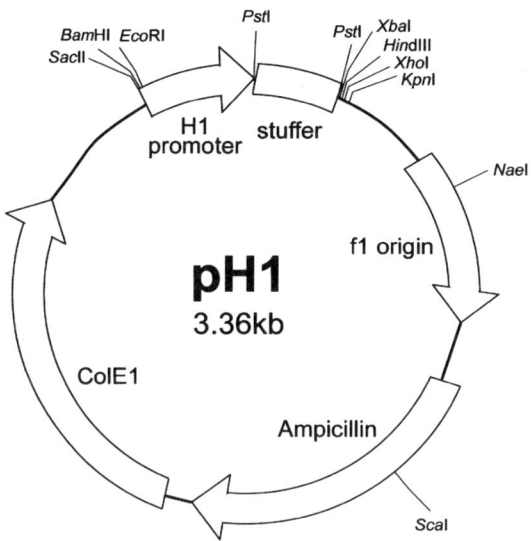

Fig. 1. Map of pH1 vector. The map is drawn approximately to scale. The annealed oligos are inserted into the *Pst*I site, and *Xba*I. *Hind*III, *Xho*I, and *Kpn*I restriction sites can also be used.

```
cccAACCACTACCTGAGCACCCAGttcaagagaCTGGGTGCTCAGGTAGTGGTTtttttggaaa
acgtgggTTGGTGATGGACTCGTGGGTCaagttctctGACCCACGAGTCCATCACCAAaaaaacctttgatc
        sense                          antisense           termination
                                                            signal
```

Fig. 2. Sequence of oligos encoding the siRNA and termination signal. Sense and antisense sequences are shown in bold upper case, and the termination signal is shown in bold lower case.

## 2.3. Preparations of Eggs

1. Dissection microscope.
2. Inverted type microscope with micromanipulator.
3. $CO_2$ incubator.
4. Injector (Eppendorf, FEMT-JET 5247).
5. Capillary puller (Sutter, P-97/IVF) with capillaries B100-75-10 and BF100-78-10 for egg holding and injection, respectively.
6. Microforge (Technical Products International, MF1).
7. The "green" mouse and rat. C57BL/6TgN(act-EGFP)OsbC14-Y01-FM131, SD TgN(act-EGFP)Osb4, are available to the scientific community at http://kumikae01.gen-info.osaka-u.ac.jp/tg/greenmouse.cfm.
8. Culture medium (FHM and Modified kSOM).
9. Egg-handling pipet (finely drawn capillary tube [FUNAKOSHI 1-40-7500]) with mouth pipet (Sigma).

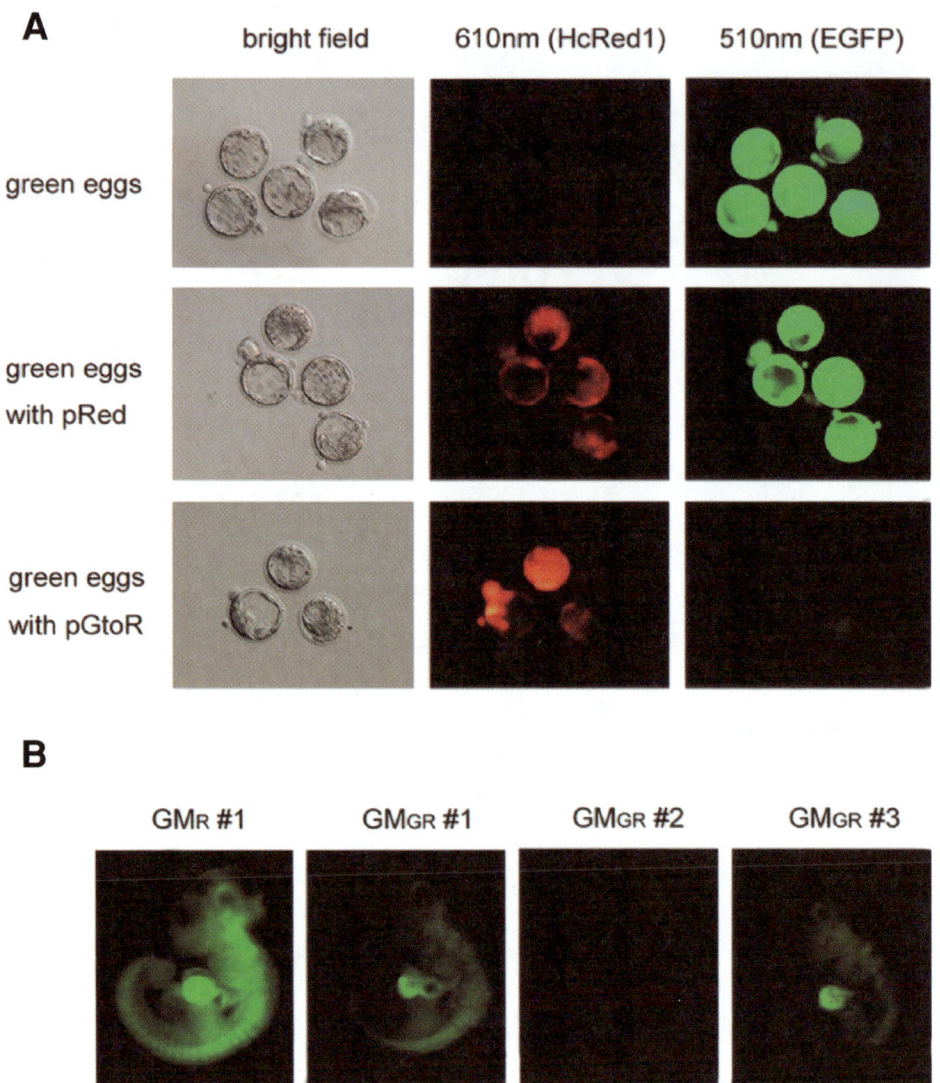

Fig. 3. Experimental illustrations. (**A**) Green eggs, obtained from wild-type females mated with homozygous "green" males, were injected with pGtoR, cultured in kSOM for 3 d, and imaged for HcRed (middle column) and EGFP (right column) expression by fluorescence microscopy. "Green" blastocysts with no additional transgene (top panel), "green" blastocysts with the pRed transgene (middle panel), and "silenced" blastocysts with the pGtoR transgene (bottom panel). (**B**) 10.5 d later embryos were recovered and analyzed for GFP gene expression. (from the left, "green" embryo with pRed $GM_R$ #1, with pGtoR $GM_{GR}$ #1, 2 and 3, respectively).

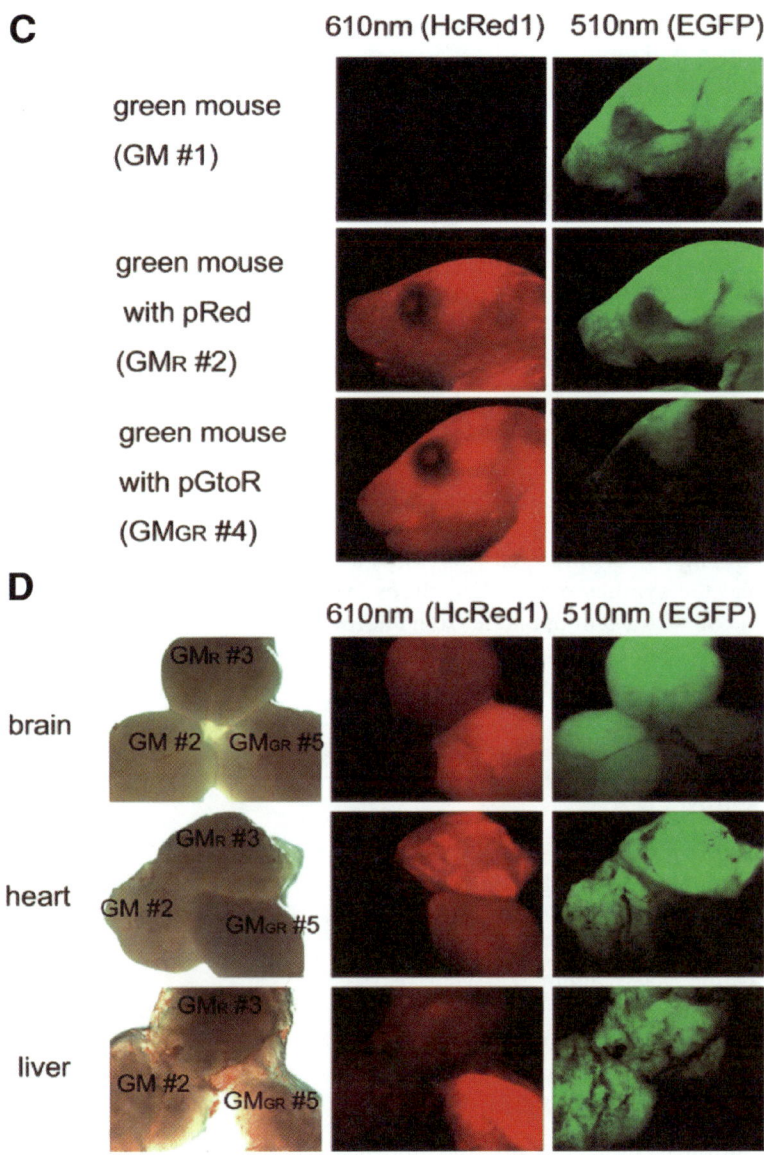

Fig. 3. **(C)** A newborn "green" mouse without an additional transgene (GM #1), a "green" mouse containing the pRed transgene (GM$_R$ #2), and a "silenced green" mouse with the pGtoR transgene (GM$_{GR}$ #4). **(D)** Various organs were removed from newborn founder mice (GM #2, GM$_R$ #3, GM$_{GR}$ #5) containing the genetic backgrounds indicated above and photographed under normal lighting (left), 510 nm (right), and 610 nm (center) band pass filter. The exposure time for each panel was adjusted to obtain optimal images. Reprinted by permission of **ref. 7**.

10. Pregnant mare serum gonadotropin and human chorionic gonadotropin (HCG) for super ovulation.
11. Hyaluronidase type IV-S (Sigma).

### 2.4. Embryo Transfer

1. D 0.5 pseudopregnant females (mated with vasectomized males).

### 2.5. Observation of Fluorescence

1. Fluorescent dissection microscope.
2. Fluorescent inverted microscope.

## 3. Methods

### 3.1. Preparation of RNAi Vector

#### 3.1.1. pH1/siRNA$_{EGFP}$

1. Digest 5 μg of pH1 vector with *Xba*I and *Pst*I.
2. Electrophorese the digested vector and extract 3.1-kb vector from the gel.
3. Anneal the synthesized siRNA oligos. Mix 10 nM sense and antisense oligos, heat at 95°C for 5 min, and let stand at room temperature overnight.
4. Ligate the annealed oligos and digested pH1 vector.
5. Transform into the *E. coli* strain JM109 or SURE2 by manufacturer's protocol.
6. Prep the plasmid from the transformed colonies and sequence the insert oligos (*see* **Note 1**).

#### 3.1.2. pGtoR

1. Digest the pH1/siRNA$_{EGFP}$ and pRed with *Bam*HI and *Hin*dIII.
2. Electrophorese both digested vectors and gel-extract the 300-bp fragment (RNAi part: H1 promoter + annealed oligos) and 5-kb fragment (marker gene part: CAG promoter + HcRed1 + polyA signal + pBlueScript).
3. Ligate the RNAi and marker-gene part.
4. Transform into the *E. coli* strain JM109 or SURE2 by manufacturer's protocol.
5. Prep the plasmid (*see* **Note 1**).

### 3.2. Purification of Transgene

1. Digest transgene with restriction enzymes (*Bam*HI, *Sal*I) (*see* **Note 2**).
2. Apply 30–50 μg of the DNA on agarose gel (*see* **Note 3**).
3. Recover the band and purify with QIAEX II kit.
4. Dissolve in 400 μL of 10 m*M* Tris-HCl, pH 8.5.
5. Extract with phenol-chloroform and ethanol-precipitate.
6. Dissolve in TE buffer (10 m*M* Tris-HCl, 0.1 m*M* EDTA, pH 8.0).
7. Filter with SUPREC-01 and measure DNA concentration.
8. Adjust concentration to 0.54 mg × mW (kb) of DNA. If the fragment is greater than 5 kb, use 2.7 μg/mL as an injection concentration.

### 3.3. Microinjection Into Eggs

### 3.3.1. Preparation of Eggs

1. Inject 5 IU of pregnant mare serum gonadotropin and 5 IU of HCG (ip) in 48-h interval to 8-wk old female mice and mate with males
2. Sacrifice the mice at 22–24 h after HCG injection.
3. Collect eggs from the ampullar portion of the oviduct into FHM.
4. Treat eggs with hyaluronidase (final concentration 350 U/mL) for 5 min, wash 4× with FHM, and cultivate in kSOM.

### 3.3.2. Microinjection

1. Place 20 eggs in 20 μL of FHM drops prepared on 60-mm dish and covered with paraffin oil (use lid of the dish for easier manipulation).
2. Set under the micromanipulator.
3. Hold the eggs with holding pipet by suctioning.
4. Focus the rim of male pronuclei, and inject about 1–2 pl DNA into pronuclei (*see* **Note 4**).
5. Transfer eggs to kSOM after injection and cultivate until transplant into oviducts of pseudopregnant mothers.

### 3.3.3. Egg Transfer

Egg transfer was performed according to **ref. 9**.

### 3.4. Observation of GFP Fluorescence

The GFP and HcRed fluorescence can be visualized using fluorescence microscopy using filters designed for fluorescein isothiocyanate (FITC) and rhodamine, respectively. It is also possible to observe the expression of GFP with fluorescence-activated cell sorting (FACS). The GFP and HcRed localize throughout the cytoplasm and nucleus, and are best observed when the cells are alive. Once the cells are dead, the GFP and HcRed diffuse out through the cell membrane. Therefore, a quick fixation of the cells is important when sectioning of the organ is planned. If necessary, we recommend 4% paraformaldehyde in phosphate-buffered saline (PBS) as a fixative, and the organs should be cut in a small block to allow the fixative to penetrate the organ as quickly as possible. After the fixation, the samples can be embedded in Tissue-Tek OCT compound (Miles Inc., Elkhard, IN) and quickly frozen in liquid nitrogen or embedded in glycol methacrylate (Technovit 8100; Heraeus Kulzer GmbH) for sectioning.

## 4. Notes

1. The double-stranded sequences are not stable in *E. coli*. They are sometimes eliminated during the cultivation. It is recommended to sequence the construct at each step. However, in some cases, the sequencing is not possible. In our case,

we were able to sequence the dsRNA region. It is presently unknown which part of the mRNA should be targeted for efficient silencing, and unfortunately, this remains a matter of trial and error.

2. When preparing the transgene, choose a restriction enzyme that can eliminate the plasmid backbone sequence.
3. The agar used for DNA purification should be a high-quality product, and should be used in a low concentration to minimize agar contamination.
4. The size of pipet tip is critical for survival of injected eggs. We break the tip by making contact with the holding pipet. The size can be estimated by injection pressure. We use the pipet that works best with pressures from 700–1000 psi.

## Acknowledgments

We thank Ann O. Sperry for critically reading the manuscript.

## References

1. Hannon, G. J. (2002) RNA interference. *Nature* **418,** 244–251.
2. Hutvagner, G., and Zamore, P. D. (2002) RNAi: nature abhors a double-strand. *Curr. Opin. Genet. Dev.* **12,** 225–232.
3. Gil, J. and Esteban, M. (2000) Induction of apoptosis by the dsRNA-dependent protein kinase (PKR): mechanism of action. *Apoptosis* **5,** 107–114.
4. Elbashir, S. M., Harborth, J., Lendeckel, W., Yalcin, A., Weber, K., and Tuschl, T. (2001) Duplexes of 21-nucleotide RNAs mediate RNA interference in cultured mammalian cells. *Nature* **411,** 494–498.
5. Okabe, M., Ikawa, M., Kominami, K., Nakanishi, T., and Nishimune, Y. (1997) 'Green mice' as a source of ubiquitous green cells. *FEBS Lett.* **407,** 313–319.
6. Ito, T., Suzuki, A., Imai, E., Okabe, M., and Hori, M. (2001) Bone marrow is a reservoir of repopulating mesangial cells during glomerular remodeling. *J. Am. Soc. Nephrol.* **12,** 2625–2635.
7. Hasuwa, H., Kaseda, K., Einarsdottir, T., and Okabe, M. (2002) Small interfering RNA and gene silencing in transgenic mice and rats. *FEBS Lett.* **532,** 227–230.
8. Baer, M., Nilsen, T. W., Costigan, C., and Altman, S. (1990) Structure and transcription of a human gene for H1 RNA, the RNA component of human RNase P. *Nucleic Acids Res.* **18,** 97–103.
9. Hogan, B., Beddington, R., Costantini, F., and Lacy, E. (1994) *Manipulating the Mouse Embryo: A Laboratory Manual*, 2nd Ed., Cold Spring Harbor Laboratory Press, Cold Spring Harbor, NY.

# 39

## Construction and Transfection of PCR Products Expressing siRNAs or shRNAs in Mammalian Cells

### Daniela Castanotto and John J. Rossi

#### Summary

In mammalian cells, the RNA interence (RNAi) effect has been observed through expression of 21–23 base transcripts capable of forming duplexes, or via expression of short hairpin RNAs. Here, we describe a facile polymerase chain reaction (PCR)-based strategy for rapid synthesis and evaluation of small interfering RNAs (siRNA) expression units in mammalian cells. The siRNA expression constructs are constructed by PCR, and the PCR products are directly transfected into mammalian cells for functional testing. This method is fast and inexpensive, allowing several different siRNA gene candidates to be rapidly screened for efficacy.

**Key Words:** Pol III expression; short hairpin RNAs (shRNAs); small interfering RNAs (siRNAs); RNA interference.

## 1. Introduction

RNA interference (RNAi) is perhaps the most powerful target specific knock-down approach available in mammalian cell biology *(1,2)*. The most popular methods for using the active component of RNAi, the small interfering RNAs (siRNAs), are transfection of chemically synthesized siRNA duplexes or intracellular expression of siRNAs using Pol III and Pol II cassettes *(1,3–5)*. Aside from the great popularity of RNAi in mammalian cell biology, there are problems in identifying optimal target sites for siRNA function *(6,7)*. One possible way to overcome this limitation is to have a facile screening procedure that allows the testing of several different sites along a messenger RNA for sensitivity to siRNA. The polymerase chain reaction (PCR)-based method previously described *(8)* and discussed in this chapter has proven to be both facile and robust. The method involves the creation of Pol III transcription units by

From: *Methods in Molecular Biology, vol. 252: Ribozymes and siRNA Protocols, Second Edition*
Edited by: M. Sioud © Humana Press Inc., Totowa, NJ

PCR, and direct transfection and testing of these products for siRNA function in cell culture.

## 2. Materials

1. Plasmid containing the U6 promoter.
2. 5' oligonucleotide (5' U6 universal primer): 5' ATCGCAGATCT **GGATCCAAGGTCGGGCAGGAAGAGGGCCT-3'**. This oligo is complementary to 29 nt at the end of the U6 promoter (bold), and therefore is used for all PCR.
3. 10 m$M$ deoxynucleotides (dNTPs).
4. 10 m$M$ adenosine 5' triphosphate (ATP).
5. *Taq* polymerase.
6. T4 DNA kinase.
7. QIAquik PCR purification Kit (Qiagen).
8. PCR apparatus.

## 3. Methods

The transfection-PCR methodology can be used to rapidly test siRNA targeting and function in cells. One critical element in the design of effective siRNAs is the selection of siRNA/target sequence combinations that yield the best inhibitory activity. Although it can be accomplished using siRNAs and transfection procedures, this can be a costly and time-consuming step. By utilizing the PCR strategy, several siRNA genes can be simultaneously tested in a single transfection experiment.

1. The procedure for the PCR approach employs a universal primer complementary to the 5' end of the U6 promoter (or possibly other Pol III and possibly Pol II promoters) along with a primer complementary to the 3' end of the promoter that harbors appended sequences, which are complementary to the sense and the antisense siRNA gene (**Fig. 1**). The sense and the antisense sequences are separated by a 9-nucleotide loop (UUUGUGUAG), and followed by a stretch of six deoxyadenosines (Ter) and by a short additional "stuffer-tag" sequence that includes a restriction site for possible cloning at a later stage. When using the UUUGUGUAG loop sequence (*see* **Note 1**), it is important that the siRNA sense strand does not contain a U at its 3' terminus, since this would create a stretch of four uridines that could serve as a Pol III terminator element. Similarly, the sense and antisense strands should not contain more than three Ts in a row. Finally,

---

Fig. 1. *(opposite page)* Schematic representation of the PCR strategy used to generate U6 transcription cassettes that express siRNAs. The 5' PCR primer is complementary to the 5' end of the U6 promoter, and is standard for all PCR reactions. (**A**) The first 3' PCR primer is complementary to sequences at the 3' end of the U6 promoter followed by the sense sequences and a 9 nt loop. The second 3' PCR primer contains a sequence complementary to the 9-nucleotide loop followed by the antisense sequences,

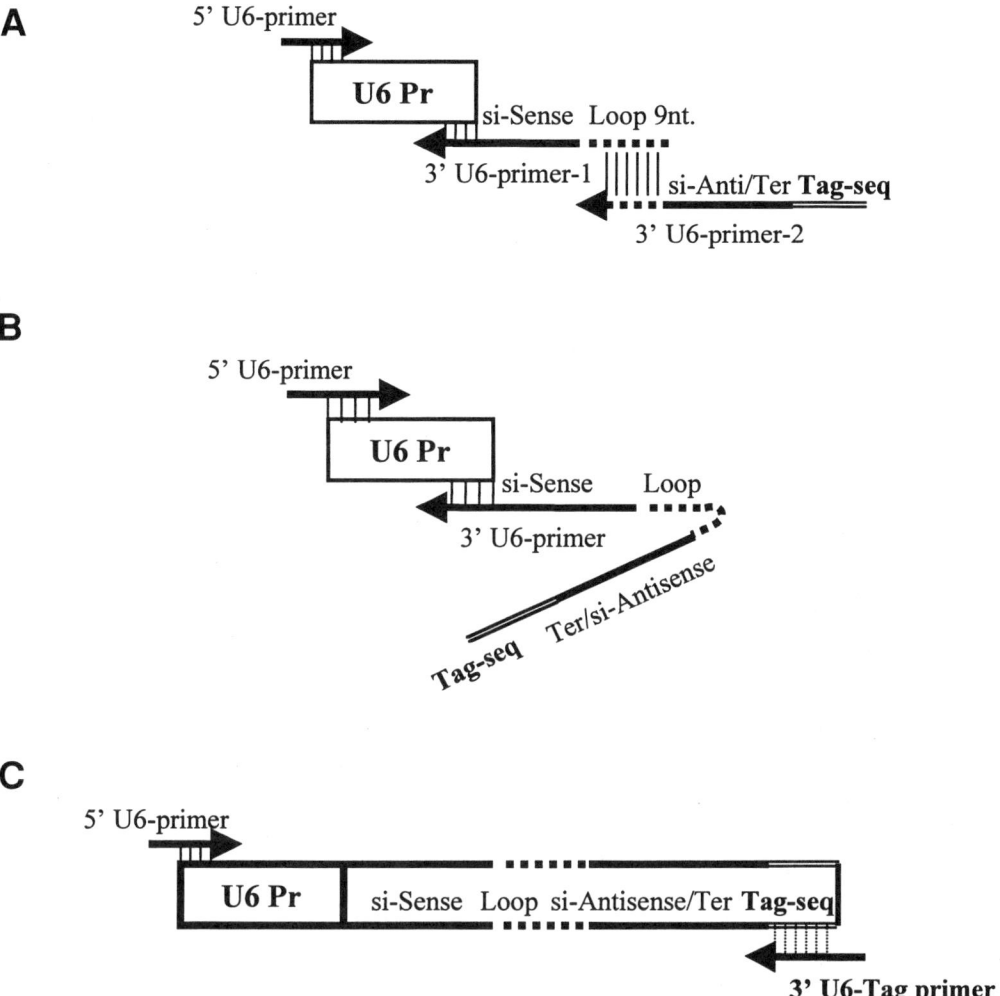

Fig. 1. *(continued)* a stretch of 5–6 deoxyadenosines (Ter) and an additional "stuffer-Tag" sequence. The adenosines code for the Us that serve as the Pol III termination signal. Therefore, any sequence added after this signal will not be transcribed by the Pol III polymerase and will not be part of the siRNA. The sense and antisense sequences are inserted in the cassette by a two-step PCR reaction (*see* **Suhbeading 3.2.**). (**B**) The sense and antisense sequences linked by a 9-nucleotide loop and followed by the stretch of adenosines and by the stuffer-Tag sequences are included in a single 3' primer. (**C**) Complete PCR expression cassette obtained by the PCR reaction. To amplify and identify functional siRNAs from the transfected cells, or to increase the yield of the PCR product shown in **B**, a nested PCR can be performed using the universal 5' U6 primer and a 3' primer complementary to the Tag sequence, as indicated.

since U6 initiates transcription with a G, this will be the first base of the sense strand. Optimally, the target sequence should also begin with a G so that the sense/antisense constructs are fully basepaired, but this is not always necessary.

2.  The resulting PCR product includes the U6 promoter and the sense and antisense siRNAs in the form of a stem-loop, the terminator sequence, and the tag sequence. To construct this cassette, two 3' primers or a single 3' primer can be used. When two 3' primers are used, the first PCR reaction employs the 5' U6 universal primer and a 3' primer complementary to 25 nucleotides of the U6 promoter, followed by sequences complementary to the sense and the 9-nt loop (**Fig. 1**). One µL of this first reaction is re-amplified in a second PCR reaction that employs the same 5' U6 primer and a 3' primer that harbors sequences complementary to the 9-nt loop appended to the antisense strand, Ter, and "stuffer" tag sequence (**Fig. 1**). The resulting PCR products include the U6 promoter, the sense and antisense coding sequences followed by the Pol III terminator sequence and the stuffer-tag sequence. If the H1 promoter is used, this can be facilitated by the annealing and extension of two oligonucleotides containing the hairpin and promoter sequences (*see* Chapter 34).

3.  When a single 3' primer is used, the procedure consists of a one-step PCR reaction with a 3' primer that harbors the sense, loop antisense, Ter, and stuffer-tag sequences, as depicted in **Fig. 1**. This second approach employs a considerably long and structured 3' PCR primer that may cause difficulties in the amplification reaction with a few sequences. However, this strategy eliminates the possibility of inserting any polymerase-induced mutations in the siRNA sequence during the amplification reaction.

4.  The PCR conditions are relatively standard for all siRNA genes, since the regions complementary to the U6 promoter do not change. PCR reactions are performed using 5–10 ng of a plasmid containing the human U6 promoter as template. A PCR fragment of the U6 promoter will also work. The 5' oligonucleotide (5' U6 universal primer) is complementary to 29 nucleotides at the 5' end of the U6 promoter, and is used for all PCR steps. A restriction site should be included at the 5' end of this oligo for possible subsequent cloning procedures. The last 25 nucleotides at the 3' end of all 3' PCR primers are complementary to the last 25 nucleotides of the U6 promoter. All PCR reactions are carried out as follows: 1 min at 94°C, 1 min at 55°C, and 1 min at 72°C for 30 cycles. The use of Vent Polymerase is recommended for the amplifications.

5.  For direct transfections and testing of the PCR amplified siRNA genes, the 5' termini of the PCR primers must be phosphorylated using DNA polynucleotide kinase (PNK) and non-radioactive ATP. This modification results in enhanced expression of the PCR products, perhaps by stabilizing them intracellularly or promoting ligation (multimerization) of the PCR cassettes. Thus, the PCR primers should be kinased for 30 min with non-radioactive ATP using T4DNA ligase buffer and 1 µL of kinase enzyme (New England Biolabs). The kinased oligos should be purified on G50 columns (Amersham-Pharmacia Biotech), as suggested by the manufacturer, prior to using them in the PCR reactions. The concentration

of the primers should be verified after the column purification. The PCR products are also purified using the QIAquick PCR purification Kit (Qiagen) or gel purified.

6. Once the PCR reaction is completed and the products are column-purified from the primers, they can be applied to cells using cationic liposomes, calcium phosphate, or electroporation, depending upon what the best transfection agent or condition is for the cells in question. 250 ng of the target plasmid can be co-transfected with 25–100 ng of the PCR cassette expressing the siRNA. As little as 25 ng of the PCR product can be effective in producing siRNAs. However, it is advisable to transfect at least 50 ng of PCR product.

7. If co-transfections of the target gene on a plasmid and the PCR products are to be used for the experiments, the ratio between the PCR products and the plasmid expressing the target should be calculated based on the size of the plasmid and the corresponding number of moles for the PCR cassettes. Various ratios from 1:1 to 1:5 may be tested. To facilitate transfection of small amounts of PCR-amplified DNA, 400 ng–1 µg of Bluescript plasmid should be added to each reaction to serve as a carrier. We have had poor results using chromosomal DNAs as carriers, and these should be avoided. It is important that the total amount of DNA is the same for each transfection. The Bluescript plasmid can be used in each case to achieve the desired amount of DNA recommended for each tranfection procedure. For readily transfectable cell lines, transfections can be performed in 6-well plates using Lipofectamine Plus™ (Life Technologies, Gibco-BRL) or other transfection reagents, as described by the manufacturer.

8. Strong and specific downregulation of the target gene by the siRNAs should be detected 36–48 h post-transfection, although for some targets, the effect lasts at least 1 wk. We have not examined knock-downs for longer than 6 d, but longer effects are clearly possible. A non-functional mutant siRNA, or a target with silent codon changes in the region of the siRNA basepairing, should be used as controls for nonspecific effects. Four consecutive nucleotide mismatches should work for the majority of siRNA/target combinations.

9. One of the most useful aspects of this technique is that once the PCR product that works best for a given target is identified, it can be cloned into a plasmid or viral vector for transfection/transduction into primary cells and long-term studies of knock-down. Finally, multiplexing of the PCR products can be used in conjunction with targets that can be analyzed using fluorescence-activated cell sorting (FACS). The most effective PCR cassettes will sort with the cells with the most potent knock-downs. The PCR cassettes can be reamplified from DNA preparations of these cells using the stuffer-Tag specific primer and the upstream U6 primer *(8)*.

## 4. Notes

1. The selected length and sequence of the nine-base loop (UUUGUGUAG) used is based on comparisons of loops found in several micro-RNA precursors. We find that this nine-base loop is more effective than other lengths and sequences.

2. The method described here can be used for screening siRNA gene libraries.

## References

1. Tuschl, T. (2001) RNA interference and small interfering RNAs. *Chembiochemistry* **2,** 239–245.
2. Hannon, G. J. (2002) RNA interference. *Nature* **418,** 244–251.
3. Elbashir S. M., Lendeckel W., Tuschl T. (2001) RNA interference is mediated by 21- and 22-nucleotide RNAs. *Genes Dev.* **15,** 188–200.
4. Elbashir, S. M., Harborth J., Lendeckel W., Yalcin, A., Weber, K., and Tuschl, T. (2001) Duplexes of 21-nucleotide RNAs mediate RNA interference in cultured mammalian cells. *Nature* **411,** 494–498.
5. Tuschl, T. (2002) Expanding small RNA interference. *Nat. Biotechnol.* **20,** 446–448.
6. Holen T., Amarzguioui M., Wiiger M. T., Babaie E., and Prydz H. (2002) Positional effects of short interfering RNAs targeting the human coagulation trigger Tissue Factor. *Nucleic Acids Res.* **30,** 1757–1766.
7. Lee, N. S., Dohjima, T., Bauer G., et al. (2002) Expression of small interfering RNAs targeted against HIV-1 rev transcripts in human cells. *Nat. Biotechnol.* **20,** 500–505.
8. Castanotto D., Li, H., and Rossi, J. J. (2002) Functional siRNA expression from transfected PCR products. *RNA* **8,** 1454–1460.

# 40

# Systemic Delivery of Synthetic siRNAs

## Mouldy Sioud and Dag R. Sørensen

### Summary

The ability to regulate endogenous gene expression by cleaving mRNA is important in basic and applied biological research. This is now driven predominantly by small interfering RNAs (siRNAs) as they induce sequence-specific gene silencing. However, the major obstacle to the use of siRNAs as therapeutics is the difficulty involved in effective in vivo delivery. This chapter describes the liposomal delivery of siRNAs into adult mice.

**Key Words:** RNA interference; siRNA; antisense RNA; fluorescein isothiocyanate (FITC).

## 1. Introduction

Novel tools for evaluating gene function in vivo such as ribozymes and RNA interference (RNAi) are emerging as the most effective strategies *(1,2)*. RNAi is a sequence-specific post-transcriptional gene silencing that is triggered by double-stranded RNA (dsRNA). This process, in which dsRNA mediates the degradation of homologous transcript, was first described in the nematode worm *Caenorhabditis elegans (3)*. Long dsRNAs are cleaved into small interfering RNAs (siRNAs) with two-nucleotide 3' overhangs and 5'-phosphate termini by Dicer (for review, *see* **ref. *2***). Thereafter, the processed siRNAs are incorporated into a multicomponent nuclease complex, known as RNA-induced silencing complex (RISC) that is responsible for the destruction of cognate mRNA (*see* Chapter 1; Fig. 3).

The detection of RNAi in mammalian cells by long dsRNA has been hampered by the activation of a nonspecific pathway that represents a host response to viral infection. In this process, both the dsRNA-dependent protein kinase (PKR) and 2', 5' oligoadenylate synthetase are activated, leading to a general

From: *Methods in Molecular Biology, vol. 252: Ribozymes and siRNA Protocols, Second Edition*
Edited by: M. Sioud © Humana Press Inc., Totowa, NJ

inhibition of protein synthesis and nonspecific degradation of mRNA by RNase L. Although it seemed for some time that the use of RNAi in mammalian systems would not be feasible, the first demonstration that the technology might work came when RNAi was indication in early mouse embryos *(4)*. Indeed, mouse embryos microinjected with long dsRNA exhibited specific silencing of the cognate gene. Thus, the RNAi pathway specifically functions in embryonic mammalian cells.

To overcome the nonspecific response to long dsRNA observed with mammalian somatic cells, Tuschl and colleagues showed that the siRNA that mimic Dicer products are incorporated into RISC, and induce gene-specific silencing in a variety of mammalian cell lines *(5)*. Similar results were reported by Caplen et al. *(6)*. In contrast to long dsRNA, siRNAs can bypass the activation of the nonspecific pathway, leading to translation inhibition. In contrast to worms and plants, mammalian cells apparently lack the mechanisms that amplify silencing *(7)*.

Although siRNAs are a powerful tool to silence gene expression post-transcriptionally, their therapeutic potential remains unproven. Therapeutic application of siRNAs requires a delivery agent that can be administered efficiently, safely, and repeatedly. Cationic liposomes represent one of the few examples that can meet these requirements *(8)*. In vivo gene targeting by means of systemic delivery of cationic liposomes and siRNAs indicate that cationic lipids such as 1,2-*bis*-(oleoyloxy)-3-(trimethylammonium) propane (DOTAP) mediate efficient delivery of siRNA into peritoneal cells *(9)*. Additionally, systemic liposomal delivery of plasmid-expressing ribozymes targeting the NF-κB inhibited NF-κB in vivo *(10)*. Thus, it is possible to use these carrier systems for the delivery of synthetic and plasmid-expressing siRNAs in vivo. This chapter describes liposomal delivery of siRNA into adult mice.

## 2. Materials

### 2.1. Equipment

1. Coulter cell counter.
2. Fluorescence microscope with appropriate fluorescence filters.
3. Flow cytometer.
4. Tissue-culture incubator.
5. Electrosquare porator ECM 830 (BTX) and appropriate electrode gag cuvets.
6. Eppendorf centrifuge.
7. Falcon tube centrifuge.

### 2.2. Mice

Balb/C mice, 6–8 wk of age and weighting 20–25 g.

## 2.3. Reagents

1. Chemically made siRNAs.
2. Cationic liposomes DOTAP.
3. Transfection buffer 5X: 100 m$M$ HEPES, 750 m$M$ NaCl, pH 7.4.
4. RPMI culture medium supplemented with 10% fetal bovine serum (FBS) and antibiotics.
5. Pasteur pipets.
6. Sterile syringes 1, 5, and 10 mL.
7. Glass tubes (5 and 10 mL).
8. Polystyrene and Eppendorf tubes.
9. Tissue-culture dishes (25 and 75 cm$^3$).
10. 96-well cell-culture plates.

## 3. Methods

### 3.1. Peritoneral Lavage

To increase the number of residual peritoneal cells, 1 d before cell preparation, the mice receive (ip) a 200 µL of phosphate-buffered saline (PBS) containing 10 µg of DOTAP (*see* **Note 1**).

1. Sacrifice the mouse by cervical dislocation and place the animal into 95% ethanol.
2. Restrain the mouse in the supine position, make an incision in the abdominal wall, and gently lift the abdominal wall.
3. With a syringe, inject 5–8 mL of saline to the abdominal cavity, and massage the abdomen for 1 min.
4. Make a small hole to admit Paster pipet by cutting through the skin and the muscle (*see* **Note 2**).
5. Aspirate the fluid by Pasteur pipet, while holding the muscle up by a forceps. Place the fluid into a 5-mL glass tube.
6. Spin down and wash the cells with RPMI medium.

### 3.2. Preparation of Adherent Peritoneal Cells

1. Plate peritoneal cells at $10 \times 10^6/20$ mL complete medium in a 75-cm$^3$ culture dish and allow adhering for 90 min at 37°C.
2. Gently aspirate non-adherent cells, add 20 mL complete medium and continue incubation at 37°C for 2 h or more until use.
3. Harvest the adherent cells by gentle scraping.

### 3.3. Transfection of Adherent Peritoneal Cells

There are several techniques available to introduce siRNA into cells—e.g., cationic liposomes and electroporation (*see* Chapter 43). In contrast to primary non-adherent cells, a significant fraction of adherent peritoneal cells can be

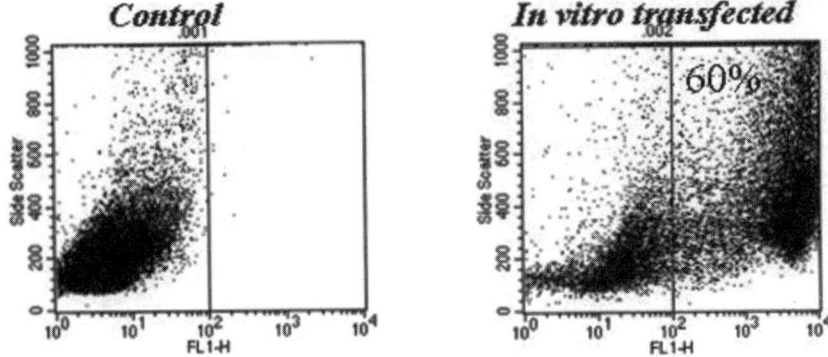

Fig. 1. Dot plots of flow cytometric analysis of murine peritoneal cells. Adherent peritoneal cells were prepared as described in **Subheading 3.2.**, and after 24 h were tranfected with DOTAP (20 µg/m:) and FITC-labeled siRNA (300 n*M*). 16 h post-transfection time, cells were washed three times with PBS and then analyzed with flow cytometry. Nearly 60% of the cells are positive for FITC-labeled siRNA.

transfected with cationic lipids such as DOTAP (*see* **Note 3**). We also found that electroporation routinely worked well for transfection of non-adherent and adherent cells. Transfection of adherent peritoneal cells with cationic liposomes is performed as described in Chapter 37. **Figure 1** shows in vitro transfection of adherent peritoneal cells with siRNA/DOTAP complexes.

### 3.4. In Vivo Delivery of siRNAs

The prospect of application of ribozymes to inhibit gene expression for therapeutic purposes is discussed in several accompanying chapters, and these also apply to siRNAs. We have been exploring the use of liposomes for delivering synthetic ribozymes and siRNAs to animals, and eventually to human patients. Notably, a variety of cationic liposomes have been used and shown to deliver DNA and RNA oligonucleotides into cells (*see* Chapter 43). Although we will not provide an in-depth discussion of all the technical issues involved in the in vivo experiments, we will provide our protocol for in vivo delivery of siRNA into adult mice. However, whatever the type cationic liposomes used, the formulated complexes should have a net positive charge. Therefore, it is desirable to test a range of cationic liposome concentrations and siRNA ratio.

### 3.5. Intraperitoneal (ip) Delivery

1. In a sterile microcentrifuge tube, mix 100 µg of siRNA (up to 20 µL) and 80 µL of 1X transfection buffer.

2. In a separate polystyrene tube, mix 100–200 μL (1 μg/μL) of DOTAP and 300–600 μL of transfection buffer. The ratio of siRNAs and DOTAP is 1:1 and 2:1, respectively.
3. Transfer the siRNA mixture to the polystyrene tube containing the DOTAP.
4. Mix gently by pipeting several times, and incubate at room temperature for 30 min. The final volume should be approx 500 or 800 μL.
5. Complete to 1 mL with 1X transfection buffer, mix gently, and then inject ip into mice.
6. After the desired time of post-injection (12–48 h), investigate the in vivo biological effect of the designed siRNAs (*see* **Notes 4–7**).

### *3.6. Intravenous (iv) Delivery*

1. In a sterile microcentrifuge tube, mix 50–100 μg siRNA (up to 20 μL) and 40 μL of transfection buffer.
2. In a separate polystyrene tube, mix 50–100 μL (1 μg/μL) of DOTAP and 90 μL transfection buffer. Charge ratio of siRNAs and DOTAP is 1:1 (*see* **Note 8**).
3. Transfer the siRNA mixture to the polystyrene tube containing the DOTAP.
4. Mix gently by pipetting several times, and incubate at room temperature for 30 min. The final volume should be approx 200 μL.
5. Inject the mixture into the tail vein.

### *3.7. In Vivo Uptake of siRNA by Peritoneal Cells*

The most important aspect of in vitro and in vivo delivery of siRNA is the careful optimization of transfection conditions. In the case of siRNA delivery, cationic lipids have many advantages in terms of enhancing the binding to cells and blood circulation time. The analysis of siRNA uptake should provide us with the principle factors that govern the distribution of siRNA in vivo. For therapeutic purposes, it is important to identify the organs in which the siRNA are taken easily or naturally (without delivery agents).

Using the ip delivery protocol described in **Subheading 3.5.**, a fluorescein isothiocyanate (FITC)-labeled siRNA (100 μg) was delivered to peritoneal cells. The siRNA and DOTAP ratio is 1:1. **Figure 2A** and **B** show the uptake of the siRNA by peritoneal cells 20 h after ip injection. Flow cytometry analysis indicates that nearly 30% of peritoneal cells were transfected (**Fig. 2C**). Most of the transfected cells are adherent cells. This observation is of interest from the point of view of observing a significant fraction of adherent cells are in vivo transfected. Thus, in vivo functional genomic studies using siRNAs can be performed. However, some nonspecific effects were observed in human monocytes (*see* **Note 9**).

## 4. Notes

1. In a normal situation, the peritoneum contains some residual cells. However, they are not sufficient for in vitro studies. Liposomes such as DOTAP can recruit cells into the peritoneal cavity. During in vivo experiments, the number of recruited

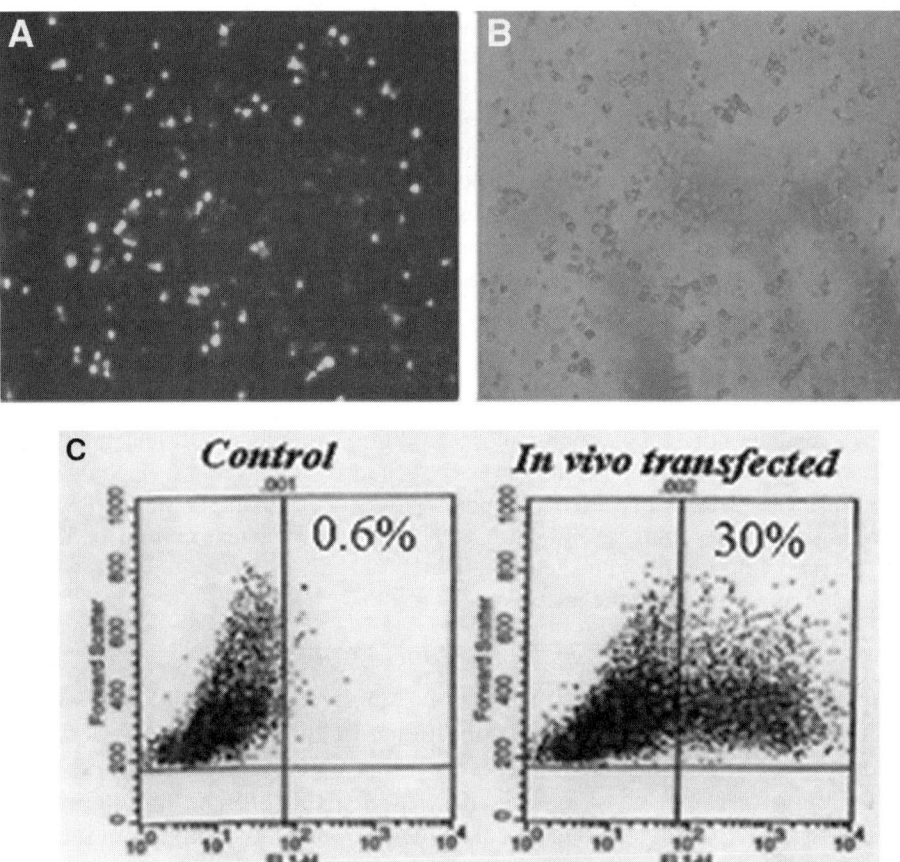

Fig. 2. Analysis of murine peritoneal cells by an epifluorescence microscope after ip delivery of FITC-labeled siRNA. Cells within peritoneal lavage were washed with PBS and examined by an epifluorescence microscope. Fluorescence (**A**) and light (**B**) images of the same field are shown. Dot plots of flow cytometric analysis of the same cells are shown in **C**. Nearly 30% of the cells show siRNA uptake.

cells can vary between animals. Thus, in vivo siRNA effects should be adjusted to the number of recruited cells.
2. A syringe with 25-gage needle can aspirate peritoneal fluids. In this case, there is no need to make a small hole within the abdominal cavity.
3. Unlike many transfection reagents, DOTAP does not need to be washed from the cells after transfection. Prolonged exposure does not induce cytoxicity in most cells tested.

4. Before analyzing siRNA activity in peritoneal cells or other organs, transfection efficiency should be analyzed. For this purpose, it will desirable to use 3'-FITC-labeled siRNAs or include a small amount of FITC-labeled DNA oligonucleotide.

5. The maximum period available for observation of the effects of suppression of a protein after a single treatment depends on the stability of the protein after elimination of mRNA. Therefore, it is important to determine the half-life of the target proteins after a single injection. If external triggers (e.g., lipopolysaccharide [LPS]) are used to induce gene expression, the effects of siRNAs would depend on the concentration of the trigger.

6. Changes in gene expression can be directly monitored from extract of peritoneal cells by Western and/or Northern blot analysis. If desired, cytokine contents can be measured in the peritoneal lavage fluids using commercially available ELISA.

7. The difference between in vivo and in vitro experiments, in vivo cells are mostly non-dividing, so there is no siRNA dilution with cells division. Thus, the in vivo effect of a single injection of siRNA could be longer than that seen in tissue culture.

8. The condensation that occurs during complex formation is progressive, and within minutes may result in the precipitation of large aggregates that are not suitable for iv delivery. Small particles are desirable. In this respect, cationic polyspermine could be a good carrier system for iv delivery.

9. During our studies with siRNAs, we have noted that some molecules do activate the nonspecific pathway in human freshly isolated monocytes, leading to TNF-$\alpha$ and IL-6. This effect seems to be cell and sequence dependent are more likely to be mediate via the activation of NF-k$\beta$ and Ap-1 transcription factors.

## References

1. Sioud, M. (2001) Nucleic acid enzymes as a novel generation of anti-gene agents. *Curr. Mol. Med.* **1**, 575–588.
2. Hannon, G. J. (2002) RNA interference. *Nature* **418**, 244–251.
3. Fire, A., Xu, S., Montgomery, M. K., Kostas, S. A., Driver, S. E., and Mello, C. C. (1998) Potent and specific genetic interference by double-stranded RNA in *Caenorhabditis elegans. Nature* **391**, 806–811.
4. Wianny, F. and Zernicka-Goetz, M. (2000) Specific interference with gene function by double-stranded RNA in early mouse development. *Nature Cell Biol.* **2**, 70–75.
5. Elbashir, S. M., Harborth, J., Lendeckel, W., Yalcin, A., Weber, K., and Tuschl, T. (2001) Duplexes of 21-nucleotide RNAs mediate RNA interference in cultured mammalian cells. *Nature* **411**, 494–498.
6. Caplen, N. J., Parrish, S., Imani, F., Fire, A., and Morgen, R. A. (2001) Specific inhibition of gene expression by small double-stranded RNAs in invertebrate and vertebrate systems. *Proc. Natl. Acad. Sci. USA* **98**, 9742–9747.
7. Sijen, T., Fleenor, J., Simmer, F., Thijssen, K. L., Parrish, S., Timmons, L., et al. (2001). On the role of RNA amplification in dsRNA-triggered gene silencing. *Cell* **107**, 465–476.
8. Templeton, S. N. (2002) Liposomal delivery of nucleic acids in vivo. *DNA Cell Biol.* **21**, 859–867.

9.  Sørensen, D. R., Leirdal, M., and Sioud, M. (2003) Gene silencing by systemic delivery of synthetic siRNAs in adult mice. *J. Mol. Biol.* **327,** 761–766

10. Kashani-Sabet, M., Liu, Y., Fong, S., Desprez, P.-Y., Liu, S., Tu, G., et al. (2002) Identification of gene function and functional pathways by systemic plasmid-based ribozyme targeting in adult mice. *Proc. Natl. Acad. Sci. USA* **99,** 3878–3883.

# 41

## Adenovirus-Delivered siRNA

### Changxian Shen and Sven N. Reske

### Summary

RNA interference is the process that double-stranded RNA (dsRNA) induces the homology-dependent degradation of cognate mRNA mediated by 21- to 23-nt small interfering RNA (siRNA). Successful application of RNAi in functional genomics and proteomics, cancer gene therapy, and virus protection depends on the efficient delivery of siRNA into mammalian cells. The availability of high virus titer, infection of a broad spectrum of cell types, and independence on active cell division makes adenovirus the vector of choice for siRNA delivery. To this end, we developed a new adenovirus shuttle vector designated as pShuttle-H1 to host H1-RNA promoter and unique *Bgl*II and *Hin*dIII sites for insertion of oligos for expression of siRNA. In this chapter, we describe an adenovirus system that uses a commercially available adenovirus system and pShuttle-H1 to deliver siRNA-expressing cassette into cells to silence a specific gene in mammalian cells.

**Key Words:** RNAi; siRNA; p53; adenovirus; functional genomics.

## 1. Introduction

RNA interference (RNAi) or RNA silencing is the process by which double-stranded RNA (dsRNA) induces the homology-dependent degradation of cognate mRNA. This is an ancient and ubiquitous antiviral system used by organisms to maintain the integrity of the genome, to defend cells against viral infection, and to regulate the expression of cellular genes. In experimental biology, RNAi has been widely used in the identification and characterization of genes and inhibition of viruses *(1)*. In functional genomics, a large number of genes that control cell division and metabolism have been identified by screening with RNAi in the chromosome I and III of *C. elegans (1)*.

In mammalian cells, dsRNA larger than 30 bp induces general nonspecific suppression of gene expression by activating the antiviral interferon response

From: *Methods in Molecular Biology, vol. 252: Ribozymes and siRNA Protocols, Second Edition*
Edited by: M. Sioud © Humana Press Inc., Totowa, NJ

*(1)*. This obstacle was overcome by the discovery that the effector in RNAi is a small or short interfering RNA (siRNA) which is 21- to 23-nt long with a 2-nt 3' overhang and 5' phosphate. Since synthetic 21–23 nt siRNA was shown to induce efficient RNAi in mammalian cells *(2)*, siRNA has been routinely used in gene silencing by transfection of chemically synthesized siRNA. To circumvent the high cost of synthetic siRNA, and in order to establish stable gene knock-down cell lines by siRNA, several plasmid vector systems were designed to produce siRNA inside cells driven by RNA polymerase III-dependent promoters such as U6 and H1-RNA gene promoters *(3,4)*. With these plasmid vectors, the phenotypes of gene silencing could be observed by stable transfection of cells *(3)*. Nevertheless, transient siRNA expression and low and variable transfection efficiency remains a problem for chemically synthesized and vector-derived siRNA. For this purpose, several virus vectors have been developed for effective delivery of siRNA into mammalian cells *(5–7)*. Retroviral vectors were designed to produce siRNA driven by either U6 or H1-RNA promoter for efficient, uniform delivery and immediate selection of stable knockdown cells *(5,6)*. Meanwhile, an adenovirus vector using RNA polymerase II CMV promoter was also developed and demonstrated to mediate gene silencing both in vitro and in vivo *(7)*. More recently, several groups developed lentiviral vectors using U6 or H1-RNA promoter to deliver siRNA into mammalian cells *(8,9)*. With the completion of whole-genome sequencing of several organisms and extensive studies of functional genomics and proteomics, more and more genes will be validated for gene therapy.

It has been demonstrated that a vector derived from pol III-dependent H1-RNA gene promoter can produce siRNA and cause effective and specific downregulation of gene expression, resulting in functional inactivation of the targeted genes *(3)*. Almost all the elements of the H1 RNA promoter are located upstream of the transcribed region, and it is ideally suited to the expression of approx 21-nt siRNA or ~50-nt RNA stem-loops. The stem-loop precursor transcript is generated and processed to functional siRNA by cellular enzymes *(3)*. The small size of siRNA prevents activation of the dsRNA-inducible interferon system that is present in mammalian cells and avoids the nonspecific phenotypes normally produced by dsRNA larger than 30 bp in somatic cells.

The availability of high virus titer, infection of a broad spectrum of cell types, and independence on active cell division makes adenovirus the vector of choice for siRNA delivery. The human adenovirus serotype 5 was used in our system. This is a replication-defective adenovirus with deletion of E1 and E3 genes, which render this virus incapable of replication itself *(10)*. Moreover, this system makes use of the efficient homologous recombination machinery in *E. coli (10)*, to produce recombinant adenovirus by a double-recombination event between cotransformed adenoviral backbone plasmid pAdEasy-1 and a shuttle vector pShuttle-H1. Here, we describe a new adenovirus shuttle vector to express functional hairpin siRNAs directed against the *p53* gene.

## 2. Materials

1. pSUPER plasmid containing the H1 promoter (kindly provided by Reuven Agami, The Netherlands Cancer Institute, Amsterdam, The Netherlands).
2. High-performance liquid chromatography (HPLC)-purified 64-mer oligonucleotides.
3. Annealing buffer:100 m*M* potassium acetate, 30 m*M* HEPES-KOH, pH 7.4, 2 m*M* Mg-acetate.
4. T4 DNA ligase (1 U/μL).
5. T4 polynucleotide kinase (PNK), 10 U/μL.
6. Calf intestinal phosphatase (CIP), 10 U/μL.
7. Restriction endonucleases *Eco*RI, *Bgl*II, *Hin*dIII, *Pme*I, *Pac*I, and *Bam*H.
8. DOSPER liposome.
9. Elution buffer (EB): Tris-HCl, pH 8.0.
10. Plasmid miniprep kit.
11. Plasmid maxiprep kit.
12. Nucleotide removal kit.
13. Gel extraction kit.
14. 50X TAE buffer.
15. BJ5183 AD Easy-1 (BJ5183 bacterium transformed with adenoviral back AD Easy-1, maintained by ampicillin selection) from Stratagene (*see* **Note 1**).
16. AD-293 packaging-cell line from Stratagene.
17. Dulbecco's modified Eagle's medium (DMEM) with 10% fetal bovine serum (FBS) supplemented with 100 U/mL penicillin and 100 μg/mL streptomycin.
18. Kanamycin and ampicillin antibiotics.
19. 100-bp DNA marker and 1-kb DNA marker.

## 3. Methods

### 3.1. Construction of pShuttle-H1 Plasmid

H1-RNA promoter was cloned into the promoterless shuttle vector pShuttle, which can drive the expression of siRNA in recombinant adenovirus. For convenient cloning and confirmation, the *Bgl*II site of the original promoterless pShuttle vector was converted to *Eco*RI site by site-directed mutagenesis (Stratagene) with the following primers: forward 5'-GCTTGTCGACTCGAATTCCTGGGCGTGGTTAAGGG–3' and reverse 5'-CCCTTAACCACGCCCAGGAATTCGAGTCGACAAGC-3'. H1-RNA promoter was cut from pSUPER (kindly provided by Reuven Agami, The Netherlands Cancer Institute, Amsterdam, The Netherlands) with *Xba*I and *Hin*dIII, cloned into the same restriction sites of the promoterless mutagenized pShuttle (*Bgl*II to *Eco*RI) and confirmed by DNA sequencing (**Fig. 1**).

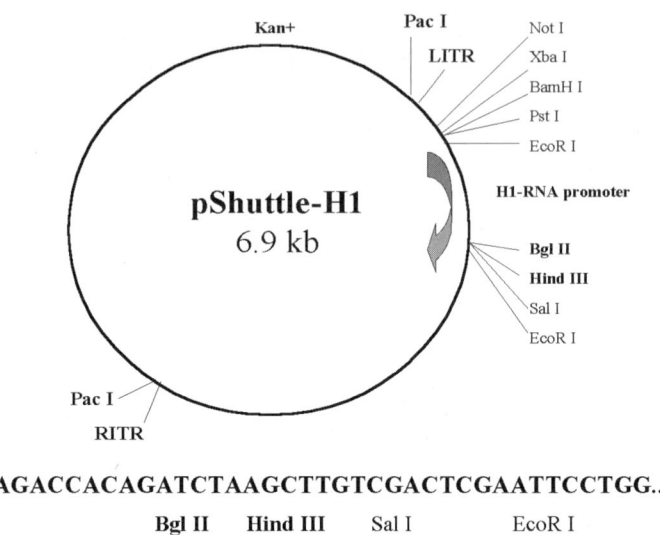

....GAGACCACAGATCTAAGCTTGTCGACTCGAATTCCTGG....

Fig. 1. Schematic diagram of siRNA-expressing adenovirus shuttle vector pShuttle-H1. This vector was derived from the promoterless adenovirus shuttle vector pShuttle (Stratagene). The *Bgl*II site of the original promoterless pShuttle was converted to *Eco*RI by site-directed mutagenesis. Stem-loop-producing oligonucleotides are cloned into the unique *Bgl*II and *Hin*dIII sites. Upon ligation, the *Bgl*II site is destroyed. The insert is confirmed with *Eco*RI digestion by distinguishing the 360-bp band of a positive clone from the 300-bp band of an empty vector. *Pol*III dependent H1-RNA promoter drives the expression of the hairpin siRNA, which is processed into functional siRNA by cellular enzymes.

### 3.2. Selection of Target Sequences

Selection of target sequences is similar to selection of synthetic siRNAs (**Fig. 2**). Preferably, the 19-nt target sequence should be flanked in the mRNA, with AA at the 5' and TT at the 3'. Target sequences with flanking 5' AA can be also selected. However, these design rules are not required (*see* Chapter 34). Regions at the mRNA to select the 19-nt from are preferably in the coding region: 100 bp from start and termination of translation. Any selected sequences should be scanned by BLAST to make sure that they are unique to the gene to be silenced (*see* **Note 2**).

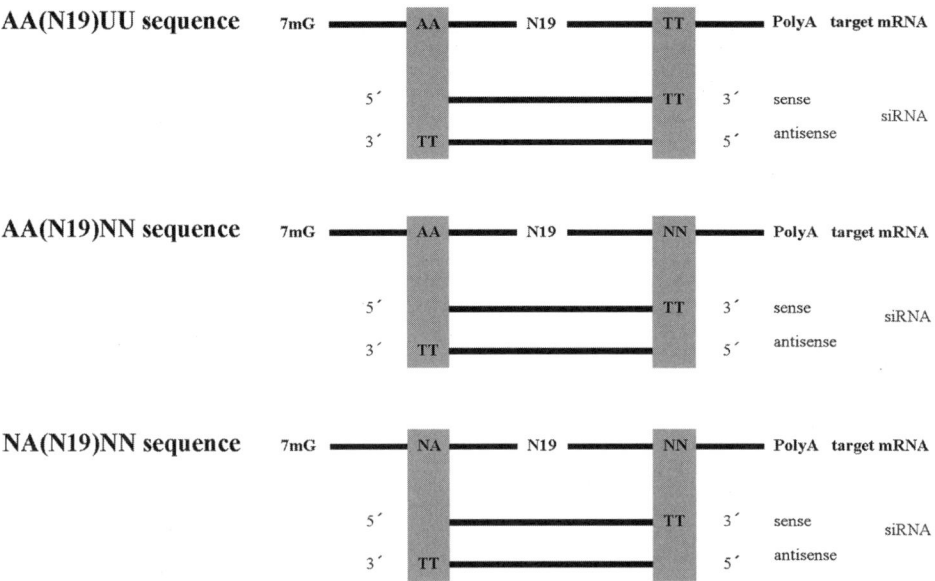

Fig. 2. Selection of target sequences.

## 3.3. Design of Oligos

For convenient cloning into pShuttle-H1, oligos should be 64 nt. After annealing, the double-stranded oligo should generate cohensive *Bgl*II and *Hind*III sites at 5' and 3', respectively. An example:

Forward oligo
        Sense orientation        antisense orientation

GATCCCC**gactccagtggtaatctac**ttcaagaga**gtagattaccactggagtc**TTTTTGGAAA

Reverse oligo
        Sense orientation        antisense orientation

*AGCTTTTCCAAAAA***GACTCCAGTGGTAATCTAC***tctcttgaa***GTAGATTACCACTGGAGTC***CGGG*

To design your oligonucleotides, just replace the 19-nt bold sequence with any target gene sequence. After annealing, a dsDNA with cohesive *Bgl*II and *Hind*III sites is generated and ready for cloning into *Bgl*II and *Hind*III-cleaved pShuttle-H1 (**Fig. 3**).

## 3.4. Annealing of Oligos (see Note 3)

1. Dissolve oligos in $H_2O$ to 100 pmol/µL (100 µ*M*).
2. Take 1.5 µL from each oligo (forward + reverse).
3. Add 47 µL annealing buffer.

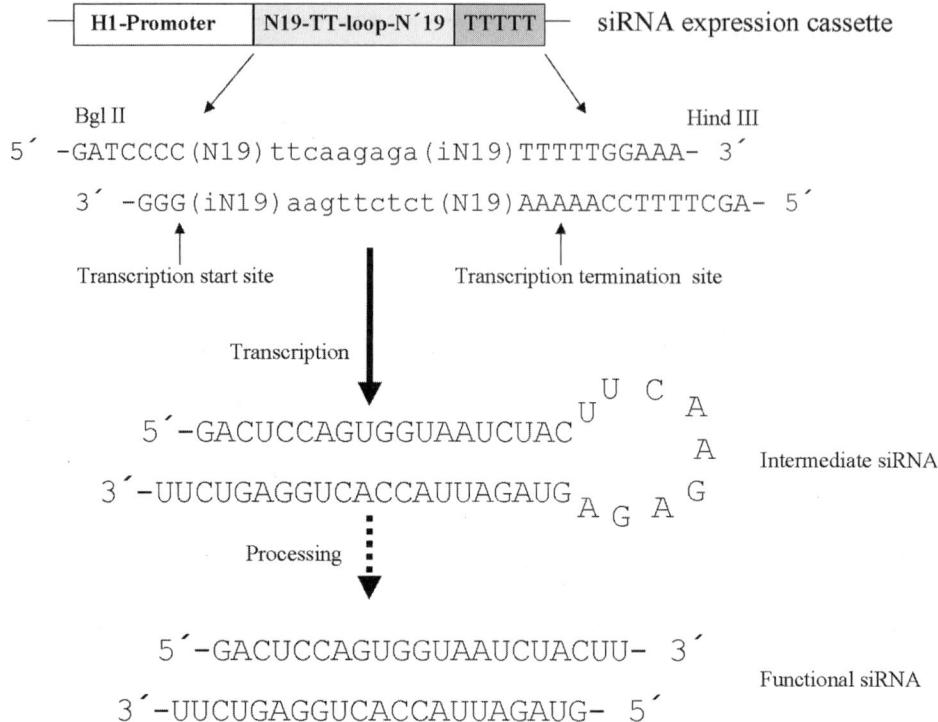

Fig. 3. siRNA produced by H1-RNA promoter. The siRNA is directed against *p53*.

4. Incubate for 4 min at 94°C and then for 10 min at 70°C.
5. Slowly cool down the annealed oligos to 4°C (or 10°C), Cooled samples are stored at –20°C until use.

### 3.5. Phosphorylation of Oligos (Final Concentration is 3 µM)

1. Take 2 µL of the annealed oligos.
2. Add 1 µL T4 PNK buffer.
3. Add 1 µL 1 m*M* ATP.
4. Add 1 µL T4 PNK.
5. add 5 µL $H_2O$ and incubate for 30 min at 37°C.
6. To inactivate PNK, incubate reaction for 10 min at 70°C.

### 3.6. Digestion of pShuttle-H1 With Bgl*II* and Hind*III*

1. Take 5 µg of pShuttle-H1 plasmid.
2. Add 10 µL NEB buffer 3.
3. Add 10 µL *Bgl*II.
4. Supplement with $H_2O$ to 100 µL.
5. Incubate at 37°C for 2 h.

6. Purify DNA with nucleotide removal kit and elute in 45 µL $H_2O$.
7. Take 40 µL of purified DNA and add 10 µL NEB buffer 2.
8. Add 10 µL *Hin*dIII.
9. Supplement with $H_2O$ to 100 µL.
10. Incubate at 37°C for 2 h.
11. Purify DNA with nucleotide removal kit and elute in 35 µL $H_2O$.

## 3.7. Dephosphorylation of Digested Vector DNA

1. Take 30 µL of *Bgl*II and *Hin*dIII cleaved pShuttle-H1 DNA.
2. Add 5 µL CIP buffer.
3. Add 5 µL CIP.
4. Supplement with $H_2O$ to 50 µL.
5. Incubate at 37°C for 2 h.
6. Purify DNA with nucleotide removal kit and elute in 30 µL $H_2O$.

## 3.8. Ligation With pShuttle-H1

1. Take 2 µL of the annealed phosphorylated oligos (0.6 µ*M* in ligation reaction).
2. Add 1 µL ligase buffer.
3. Add 1 µL dephosphorylated pShuttle-H1 plasmid.
4. Add 5 µL $H_2O$.
5. Add 1 µL ligase and incubate for 1–2 h at room temperature.

## 3.9. Preparation of Competent Bacterium DH5α

1. Pick one clone from DH5α plate (no older than 2 wk) into 2-mL LB and shake (100–150*g*) overnight at 37°C.
2. The next day, take 0.5–1.0 mL of this overnight culture into 50-mL LB medium and culture for 3–4 h.
3. Cool the culture on ice for 10 min.
4. Centrifuge at 4000*g* for 5 min.
5. Discard supernatant completely and let it stand upside down for 1 min.
6. Resuspend bacterium in 10 mL precooled 0.1 *M* $CaCl_2$ and place in ice for 10–30 min.
7. Centrifuge at 4000*g* for 5 min.
8. Discard supernatant completely and let it stand upside down for 1 min.
9. Resuspend bacterium in 2–5 mL precooled 0.1 *M* $CaCl_2$ and aliquot it to 100–200 µL.

## 3.10. Transformation

1. Apply 1–5 µL DNA (should be less than 5 µL) to 200-µL competent DH5a bacterium, mix gently, and place on ice for 30 min.
2. Place the sample in 42°C (prewarmed) for exactly 90 s.
3. Rapidly take it out and place it on ice for 2 min.
4. Add 800 µL LB medium to each tube, incubate in 37°C water bath for 5 min, transfer to 37°C shaker, and incubate at 37°C with shaking (150*g*) for 45 min.
5. Plate around 200 µL on LB plates with 50 µg/mL ampicillin.
6. Incubate the LB plates at 37°C overnight (12–16 h).

### 3.11. Clone Characterization

1. The next day, pick 5–10 clones of each ligation and grow overnight in 2 mL LB with 50 µg/mL ampicillin.
2. Extract plasmid DNA from each clone by plasmid miniprep kit and elute the plasmid DNA in 50 µL elution buffer.
3. Digest plasmid DNA (10–20 µL DNA in 50-µL volume) with *Eco*RI and run 1.5% agarose gel. A positive clone should have a second band of approx 360 bp, and an empty vector has a second 300-bp band.
4. Confirm the inserts by DNA sequencing.

### 3.12. Production of Recombinant Adenoviral Plasmid

1. Prepare shuttle plasmid DNA (e.g., pShuttle-H1-*p53*) in sufficient quantity.
2. Linearize shuttle plasmid DNA by Pme I restriction enzyme, and confirm complete digestion by agarose gel electrophoresis.
3. Remove the enzyme and buffer by nucleotide removal kit.
4. Treat the purified DNA with CIP for 30 min at 37°C and purify by agarose gel electrophoresis.
5. Elute the DNA in 30 µL of EB buffer.
6. Prepare competent BJ5183-AD-1 cells as described in **Subheading 3.8.** Ampicillin should be used for selection.
7. Pipet 1–5 µL (10–50 ng) of linearized, dephosphorylated shuttle plasmid DNA to 200 µL competent BJ5183-AD-1 cells, mix gently, and keep on ice.
8. After follow the transformation steps described in **Subheading 3.9.** Use kanamycin for selection.
9. Pick about 10 smaller, well-isolated colonies each from the recombination plate into 2 mL LB with kanamycin.
10. Incubate at 37°C overnight with shaking at 100–150*g*.
11. Prepare miniprep DNA and elute in 30 µL elution buffer.
12. Cut 10 µL of the miniprep DNA with *Pac*I restriction enzyme and run the entire digest on a 0.8% agarose gel next to 10 µL of uncut miniprep DNA. Positive recombinant adenoviral plasmid should have a band of approx 30 kb and a second band of either 3.0 kb or 4.5 kb.
13. Plasmid DNA from correct clone is transformed into competent DH5α cells with kanamycin LB broth and reconfirmed (miniprep and *Pac*I digestion).
14. Prepare sufficient recombinant adenoviral plasmid DNA by maxiprep kit.
15. Linearize the adenoviral DNA with *Pac*I digestion and purify by ethanol-precipitation.

### 3.13. Packaging of Adenovirus in AD-293 Cells

1. Plate $1.5 \times 10^6$ AD-293 cells in 25 cm$^2$ flask in 6 mL DMEM the day before transfection.
2. Dilute 6.0 µg of *Pac*I linearized, purified recombinant adenoviral plasmid (e.g., AdH1-*p53*) DNA in 1.5 mL serum-free DMEM medium.

3. Dilute 24 µL liposome in 1.5 mL serum-free DMEM medium.
4. Combine these two solutions and mix gently.
5. Incubate at room temperature for 20–30 min to form DNA/liposome complex.
6. Immediately before transfection, wash cells with PBS once.
7. Add the DNA/liposome complexes (3 mL) onto cells, distributing it around the well. Swirl the wells to ensure even dispersal.
8. Return the cells to incubator for 5 h.
9. Add 3 mL fresh DMEM medium containing 20% FBS and return to incubator.
10. The next day, replace with 6 mL fresh DMEM medium containing 10% FBS and continue incubation for additional 8–10 d. Change medium every 2–3 d.

### *3.14. Harvest Primary Virus*

1. Carefully remove medium and wash once with PBS.
2. Add 1.5 mL PBS to each flask, harvest cells by pipeting up and down and transfer the cell suspension to a 2-mL screw-cap microcentrifuge tube.
3. Release virus by 3–4 rounds of freeze-thaw by alternating the tube between nitrogen bath (30 s–1 min) and 37°C water bath (5 min), vortexing briefly after each thaw.
4. Transfer the virus solution to another 2-mL screw-cap microcentrifuge tube and collect cell debris by 12,000$g$ for 10 min at room temperature.
5. Aliquot the supernatant (primary virus stock) to a new 2-mL screw-cap microcentrifuge tube and store it under –80°C. The primary virus stock is ready for amplification, titering, and infection of cells for silencing the targeted gene.

## 4. Notes

1. BJ5183 AD Easy-1 competent cells can also be prepared in-house. In this case transform pAdEasy-1 plasmid into BJ5183 cells with ampicilin selection.
2. The shosen target sequence (19 nt) must not contain a stretch of four or more adenines or thymidines, since this will give premature termination of the transcript.
3. T4 DNA ligation buffer containing 1 m$M$ ATP.

## References

1. Hannon, S. J. (2002) RNA interference. *Nature* **418,** 244–251.
2. Elbashir, S. M., Harborth, J., Lendeckel, W., Yalcin, A., Weber, K., and Tuschl, T. (2001) Duplexes of 21-nucleotide RNAs mediate RNA interference in cultured mammalian cells. *Nature* **411,** 494–498.
3. Brummelkamp, T. R., Bernards, R., and Agami, R. (2002) A system for stable expression of short interfering RNAs in mammalian cells. *Science* **296,** 550–553.
4. Paddison, P. J., Caudy, A. A., Bernstein, E., Hannon, G. J., and Conklin, D. S. (2002) Short hairpin RNAs (shRNAs) induce sequence-specific silencing in mammalian cells. *Genes Dev.* **16,** 948–958.
5. Eevroe, E. and Silver, P. A. (2002) Retroviral-delivered siRNA. *BMC Biotechol.* **2,** 15–19.
6. Barton, G. M. and Medzhitov, R. (2002) Retroviral delivery of small interfering RNA into primary cells. *Proc. Natl. Acad. Sci. USA.* **99,** 14,943–14,945.

7. Xia, H., Mao, Q., Paulson, H. L., and Davidson, B. L. (2002) siRNA-mediated gene silencing in vitro and in vivo. *Nat. Biotechnol.* **20,** 1006–1010.

8. Qin, X. F., An, D. S., Chen, I. S., and Baltimore, D. (2003) Inhibiting HIV-1 infection in human T cells by lentiviral-mediated delivery of small interfering RNA against CCR5. *Proc. Natl. Acad. Sci. USA.* **100,** 183–188.

9. Tiscornia, G., Singer, O., Ikawa, M., and Verma, I. M. (2003) A general method for gene knockdown in mice by using lentiviral vectors expressing small interfering RNA. *Proc. Natl. Acad. Sci. USA* **100,** 1844–1848.

10. He, T.-C., Zhou, S., Da Costa, L. T., Yu, J., Kinzler, K. W., and Vogelstein, B. (1998) A simplified system for generating recombinant adenoviruses. *Proc. Natl. Acad. Sci. USA* **95,** 2509–2514.

11. Shen, C., Buck, A., Liu, X., Winkler, M., and Reske, S. N. (2003) Gene silencing by adenovirus-delivered siRNA. *FEBS Lett.* **539,** 111–114.

# 42

# RNAi Expression Vectors in Plant Cells

## Hideo Akashi, Makoto Miyagishi, and Kazunari Taira

### Summary

Suppression by double-stranded RNA (dsRNA) of expression of a target gene is known as RNA interference (RNAi). Tobacco BY-2 cell suspension has been used as a model cultured plant cell, because it is possible to produce populations of tobacco BY-2 cell suspensions that are uniform and divide synchronously for functional gene analysis. Here, we describe a method to induce RNAi by introducing a hairpin-type dsRNA expression vector into BY-2 cells via electroporation. This methodology should facilitate the analysis of individual gene function in plant cells.

**Key Words:** BY-2; electroporation; plant cells; RNAi; tobacco.

## 1. Introduction

Generally, plant cells have many sets of multiple copies of a gene or homologous genes with related functions that can compensate for each other. Many genes have been identified successfully by gene-disruption methods with *Agrobacterium* T-DNA inserts, but only one copy of a T-DNA insert can be inserted into a genome, and such an insert can thus only disrupt the function of a single gene *(1,2)*. The situation is more severe in the case of cultured plant cells, because they become polyploid during a short period of culturing. For this reason, cultured plant cells have played only a limited role in the analysis of individual gene function. Since multiple copies of the target gene or related genes would produce mRNAs with homologous sequences to that of the target mRNA, it would be advantageous to target these mRNAs as well.

The phenomenon known as RNA interference (RNAi) triggered by double-stranded RNA (dsRNA) has been described in the nematode *Caenorhabditis elegans (3)*. Further studies indicate that RNAi operate by a mechanism that is widely conserved among species that include *C. elegans, Trypanosoma brucei,*

From: *Methods in Molecular Biology, vol. 252: Ribozymes and siRNA Protocols, Second Edition*
Edited by: M. Sioud © Humana Press Inc., Totowa, NJ

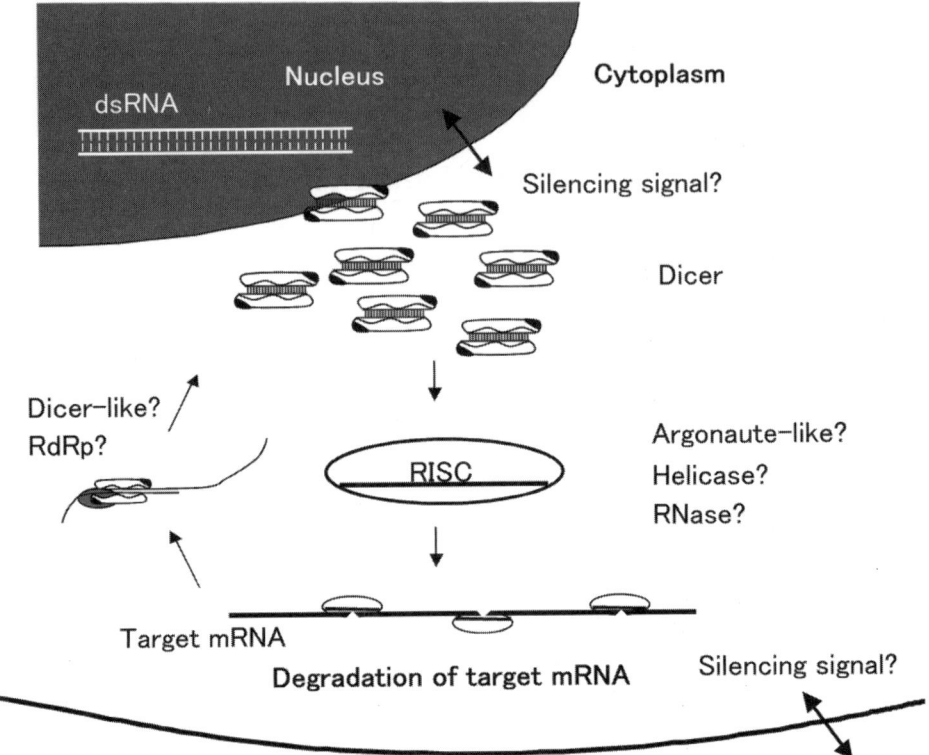

Fig. 1. A model for the mechanism of RNAi. The dsRNA-processing proteins—
which belong to a family of RNase III—bind to the dsRNA. The RNase III cleaves the
dsRNA into siRNAs that are incorporated into multicomponent nuclease complexes
(RISC). The target mRNA recognized by the RISC is cleaved in the center of the
region that is complementary to the siRNA.

the cnidarian *Hydra magnipapillata*, the planarian *Schmidtea mediterranea*,
the fungus *Neurospora crassa*, the fruit fly *Drosophila melanogaster*, and various mammalian and plant cells (**Fig. 1**) *(4)*. The dsRNA is degraded by a member of the RNase III family (Dicer) into small RNAs of 21–23 nucleotides (nt)
in length, which are designated as short interfering RNAs (siRNA) *(5)*. Then,
each siRNA molecule forms multicomponent nuclease complexes (RISC) that
destroy mRNA that is complementary to the siRNA *(6)*. However, the exact
biochemical mechanism has not been clarified. In plants, siRNAs that are characteristic of RNAi have also been detected when post-transcriptional gene
silencing occurs as a result of invasion by viruses or stress caused by chemical
treatment.  RNAi can also be induced via sense cosuppression, and even

antisense suppression. Whatever the nature of the trigger, the siRNA species is common, and is assumed to play a central protective role.

Tobacco BY-2 cultured cells have been used as a model plant cultured cell, because of their high rate of propagation, early establishment of methods for transient or stable cells, highly homogenous and synchronous character, and the significant size of the intracellular organs *(7)*. A conventional knockout strategy could not be applied to plant cultured cells because of the redundancy of genes. However, RNAi has made it possible to analyze gene function by using cultured plant cells. Here, we describe how the RNAi effect caused by a dsRNA expression plasmid through electroporation would provide an effective knockout strategy.

## 2. Materials

### 2.1. Buffers and Media

1. 1 *M* MES buffer: 1 *M* 2-morpholinoethanesulfonic acid, pH to 5.8. The pH should be adjusted with 5 *N* KOH. Sterilize by filtration and store at room temperature in darkness.
2. 0.4 *M* mannitol: Sterilize with filtration and store at 4°C.
3. Enzyme solution: 5 m*M* MES buffer, 0.4 *M* mannitol, 1% cellulase Y-C, 0.1% Pectolyase Y-23. Sterile by filtration and store at –20°C.
4. EP buffer: 5 m*M* MES buffer, 0.3 *M* mannitol, 70 m*M* KCl. Sterilize by filtration and store at 4°C.
5. PEG solution: 10% PEG6000 in EP buffer. Sterilize by filtration and store at 4°C.
6. MSD medium: full-strength MS salts, 3% sucrose, 0.2 g/L KH$_2$PO$_4$, 0.01% myo-inositol, 1 µg/mL thiamine-HCl, 0.2 µg/mL 2,4-dichlorophenoxyacetic acid. Adjust the pH to 5.8, using 1 *N* KOH. Sterilize by autoclaving and store at room temperature in darkness.
7. LS medium for protoplasts: full-strength MS salts, 0.2 g/L KH$_2$PO$_4$, 0.01% myo-inositol, 1 µg/mL thiamine-HCl, 0.2 µg/mL 2,4-dichlorophenoxyacetic acid, 0.4 *M* mannitol. Adjust the pH to 5.8 using 1 *N* KOH. Sterilize by autoclaving and store at room temperature in darkness.

### 2.2. Other Reagents and Kits

1. *E. coli* alkaline phosphatase.
2. T4 DNA ligase.
3. Kit for purification of DNA fragments from agarose gels (e.g., Qiaquick Gel Extraction Kit; Qiagen GmbH, Hilden, Germany).
4. *Taq* DNA polymerase.
5. Mixture of phenol (TE-saturated, pH 8.0), chloroform, and isoamyl alcohol (IAA) (25:24:1, v/v).
6. Kit for preparation of plasmid DNA (e.g., Hispeed Plasmid Midi Kit; Qiagen GmbH).
7. Dual-Luciferase Reporter Assay System (Promega Corporation, Madison, WI).

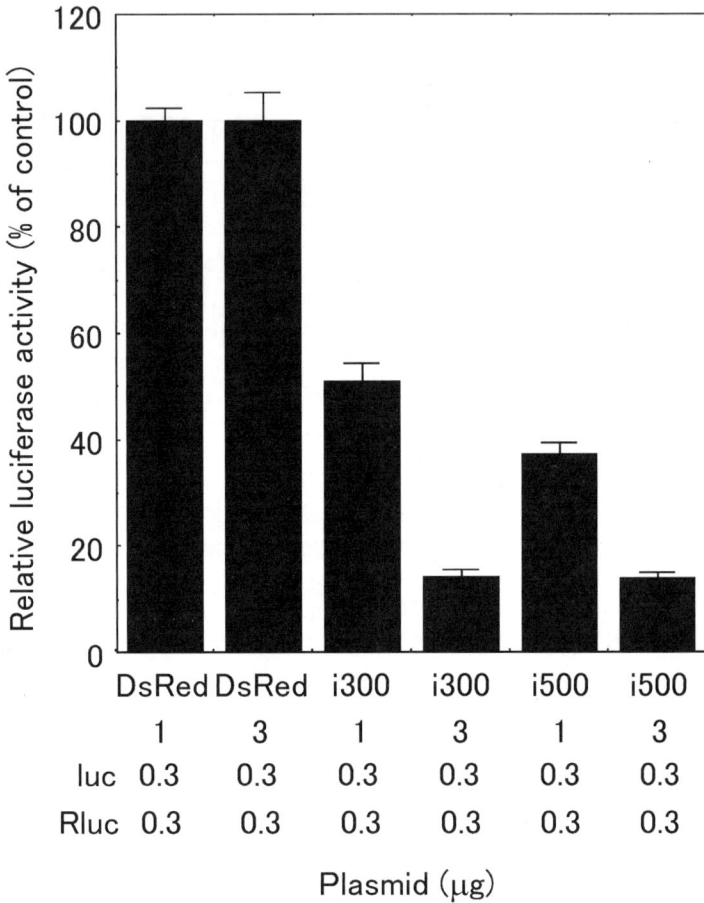

Fig. 2. The dependence of the RNAi effect on firefly luciferase activity in BY-2 cells on the amount of various dsRNA expression plasmids. Tobacco BY-2 cells were converted to protoplasts, and were cotransformed by electroporation with three plasmids—namely, p35Sluc and p35SRluc (0.3 μg each) and one dsRNA expression plasmid. The amounts of dsRNA expression plasmids are indicated below the histogram. Experiments were performed in triplicates. Columns and bars show mean results and SD, respectively.

## 3. Methods

### 3.1. Design of Hairpin-Type dsRNA Expression Vector

#### 3.1.1. Stem Region

The length of the stem region of a hairpin-type dsRNA is usually longer than 500 bp. This construct has been shown to cause about 90% reduction of expression of the target gene *(8)*. A shorter dsRNA expression vector with a 300-bp or 100-bp stem region is also effective. However, the shorter stem region exhibited lower RNAi activity, particularly at a low dose (**Fig. 2**). Notably, plasmids with shorter stem regions are more stable than plasmid with long stems. In addition, we found that plasmids with a long stem region often produce low yields and a high rate of dimmer formation. The target region is usually translated region or 5' untranslated region (UTR). Using a BLAST search, one should confirm that the selected target sequence has no homology with other sequences of the genome.

### 3.1.2. Loop Region

It is known that a long, perfect palindromic sequence without a loop cannot survive in a common strain of *E. coli* because of frequent recombination (*see* **Note 1**). Therefore, it is advisable to use *E. coli* in which the palindromic sequences are more stable than in other strains (e.g., SURE II; Stratagene). The stability of the plasmid is highly improved in the presence of the loop region. We used approx 80 bp sequence of a 3' downstream part following the sense region as the loop sequence. However, the length of 80 bp was found to be inadequate for constructing the dsRNA expression vector with a 500-bp stem region. In fact, even with the spacer (loop) region, dsRNA expression plasmids were somewhat difficult to construct, and unusual products with—for example—duplicated plasmids were often detected (**Fig. 3**). The loop with several hundred basepairs in length appears to be more favorable. Optimization of the length of the loop sequence has not been attempted. The use of an intron sequence as a loop showed strong silencing effects *(9)*.

### 3.1.3. Primers

We constructed two dsRNA expression plasmids under the control of the 35S promoter of CaMV, targeted to the firefly luciferase gene (**Fig. 4**). The length of the antisense sequence, which is defined as the length of the double-stranded region, is approx 500 bp and 300 bp. The sense strand in each dsRNA expression plasmid is the region that corresponds to nucleotides +1 to +586 relative to the site for initiation of translation of the luciferase gene (*see* **Note 2**). The spacer region between the sense and the antisense strand is designed to form a loop within a putative hairpin structure. In both cases, the spacer is the same 86-bp region, and corresponds to nt +500 to +586 from the start codon of the firefly luciferase gene. The following primer sequences were used:

1. Antisense strand: 5'-AAACCATGGtgatcaACGTGTACATCGACTGAAATCCC-3'
   *Bcl*I luc + 500-475
2. Sense strand for 500 bp of dsRNA:
   5'-GAAgcggccgcATGGAAGACGCCAAAAACATAAAGAAAG-3'
   *Not*I luc + 1-28

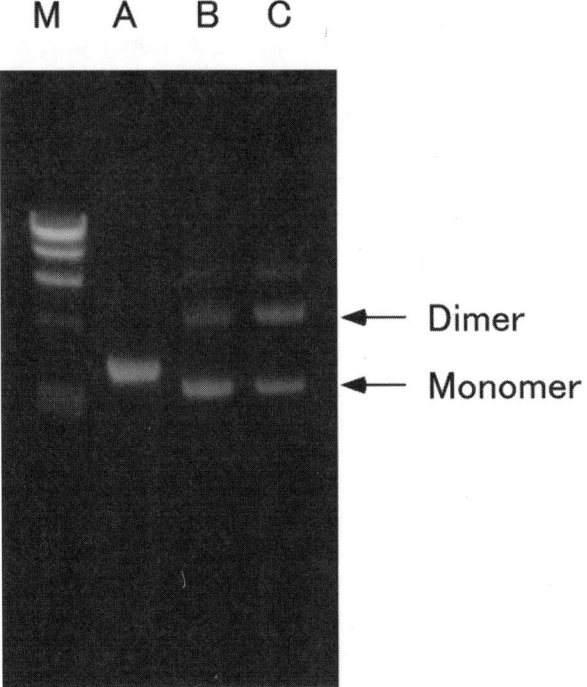

Fig. 3. Agarose gel electrophoresis of a firefly luciferase expression plasmid and dsRNA expression plasmids. **(A)** *p35*Sluc+, **(B)**, **(C)**, dsRNA expression plasmids derived from different bacterial colonies, **(M)** λ *Hin*dIII marker.

3. Sense strand for 300 bp of dsRNA:
   5'-CTAAGTCGGGgcggccgcTATGAAACGATATGGGCTGAATC-3'
   *Not*I luc + 195-209

Lowercase letters correspond to restriction enzyme sites. Numbers correspond to the nucleotide position from the luc start codon.

### 3.2. Preparation of the Vector DNA (see Notes 3–6)

1. In a 1.5-mL microcentrifuge tube, digest 3–5 µg of plasmid vector p35Sluc+ (**Fig. 4**) with 20-30 U of *Bcl*I for 1–2 h at 37°C (final volume 50 µL).
2. Add 7.5 µL of 10X high-salt buffer and 20–30 U of *Not*I, complete to 100 µL and incubate for 1–2 h at 37°C.
3. Add 10 µL of 3 *M* sodium acetate (pH 5.2) and 50 µL of phenol/chloroform/IAA (25:24:1, v/v), Vortex well, and then separate the phases by centrifugation at 12,000*g* for 2 min (hereafter, this procedure is referred to as PCI extraction).
4. Transfer the aqueous phase to a new 1.5-mL tube, add an equal volume of 2-propanol, mix by inverting the tube, and incubate at room temperature for 10 min.

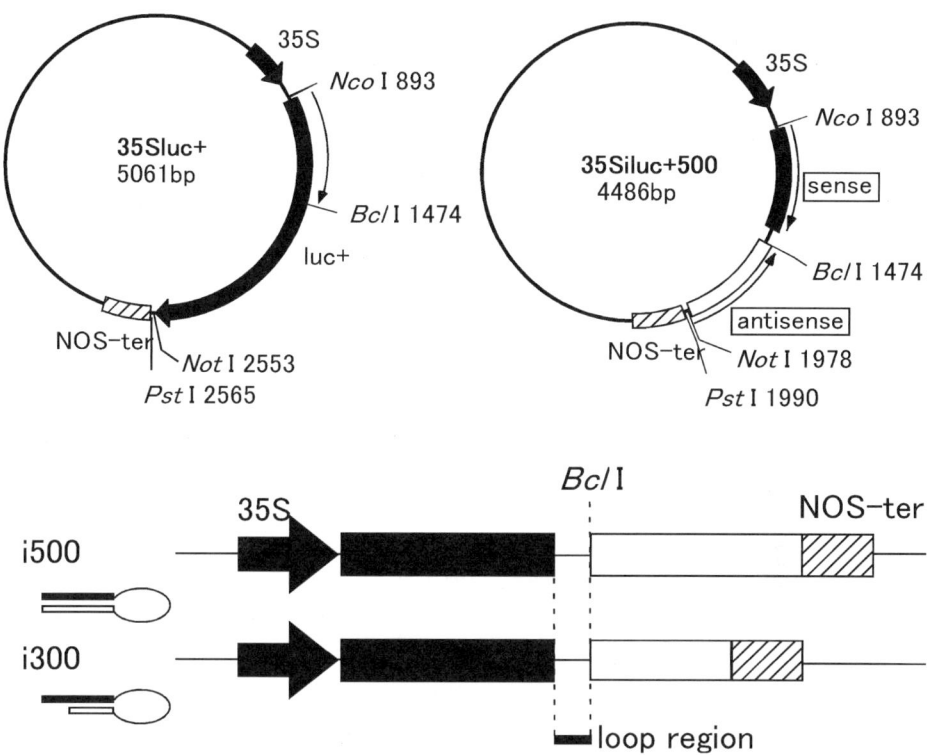

Fig. 4. Schematic representation of firefly luciferase expression plasmid, and dsRNA expression plasmid targeted to the firefly luciferase gene. The numbers after the letters indicate the approximate length of the region of the dsRNA. The thin arrows show the orientation and the length of the dsRNA region. The thick arrows show promoter sequences The spacer (loop) region between sense and antisense sequences is also indicated. 35S, 35S promoter of CaMV; NOS-ter, nopaline synthase terminator.

5. Centrifuge at 12,000g for 10–20 min and discard the supernatant. Rinse the pellet with 500 μL of 70% ethanol. Centrifuge at 12,000g for 10 min and discard the supernatant.

6. Dry the pellet at 65°C for 1–2 min in an oven (hereafter, this procedure is referred to as 2-propanol precipitation).

7. Dissolve the pellet in 20 μL of distilled water. Add 3 μL of *E. coli* alkaline phosphatase (BAP) in a 100 μL reaction volume in appropriate buffer and incubate at 65°C for 1 h.

8. Extract with PCI and precipitate with 2-propanol.

9. Dissolve the pellet in 20 μL TE buffer, and subject to 0.8% agarose gel electrophoresis.

10. Excise the appropriate band from the gel, and extract the DNA fragment from the agarose using appropriate methods.

### 3.3. Preparation of DNA Insert for Cloning

1. Amplify the antisense strand by PCR with primers described in **Subheading 3.1.3.**
2. Extract with PCI and precipitate with 2-propanol.
3. Digest polymerase chain reaction (PCR) products with *Bcl*I and *Not*I.
4. Extract with PCI and precipitate with 2-propanol.
5. Purify PCR products by agarose gel electrophoresis.

### 3.4. Ligation and Transformation

1. Ligate the vector DNA (*see* **Subheading 3.2.**) with antisense fragments (*see* **Subheading 3.3.**), using T4 DNA ligase at 16°C overnight (final volume 10 μL).
2. Add 10 μL of the ligation mixture to 100 μL competent cells (SURE II), and incubate for 30 min on ice.
3. Incubate the tube for 45 s at 42°C, and then place on ice for 2 min. Add 900 μL SOC medium and incubate at 37°C for 1 h.
4. Collect bacteria by centrifugation and resuspend in 200 μL LB medium, plate on a LB plate containing 100 μg/mL of ampicillin, and then incubate the plate at 37°C overnight.
5. Pick eight colonies, add 3 mL LB medium containing 100 μg/mL ampicillin to each sample and culture overnight at 37°C.
6. Store a small portion of the bacterial medium at 4°C. Recover the plasmids from the remainder by alkali-sodium dodecyl sulfate (SDS) methods.
7. Check plasmids by cutting appropriate restriction enzymes. Also check the level of dimmer formation by electrophoresis of non-cutting plasmids (**Fig. 3**).
8. Culture the saved portion of the bacterial medium overnight in 100–200 mL Luria broth (LB) medium containing 100 μg/mL of ampicillin.
9. Prepare plasmids by using a kit for plasmid DNA preparation.
10. Measure the absorption of the DNA solution at 260 nm to determine the concentration.
11. Precipitate with 2-propanol and wash with 70% ethanol.
12. Dissolve the plasmid in distilled water at 1 μg/μL.

### 3.5. Preparation of Protoplasts

1. Subculture tobacco BY-2 cells at 7-d intervals with 1-mL inoculum/95-mL MSD medium in a 300-mL Erlenmeyer flask.
2. Collect the BY-2 cells from two flasks at 3-d culture, which is at the initial stage of logarithmic growth phase, into two 50-mL Falcon tubes. Hereafter, the volume of each reagent is described per 50-mL Falcon tubes.
3. Centrifuge at 700*g* for 2 min, and remove the supernatant.
4. Wash cells with 25 mL of 0.4 *M* mannitol with pipetting. Centrifuge at 700*g* for 2 min, and remove the supernatant.

5. Again, wash cells as previously described.
6. Add 25 mL of enzyme solution, mix by pipetting, and incubate the tube at 30°C for 1 h. During the incubation, mix the solution with a 25-mL pipet about ten times at every 20-min interval (*see* **Note 7**).
7. Centrifuge at 300*g* for 2 min, and remove the supernatant.
8. Wash cells with 0.4 *M* mannitol twice with pipetting. Before removing 0.4 *M* mannitol, determine the number of cells by hemocytometer. Approximately $2 \times 10^7$ cells are recovered from 100 mL medium.
9. Resuspend cells in ice-cold EP buffer to a concentration of around $1.5 \times 10^6$ cells per 700 µL and place the tube on ice for 10 min.

### *3.6. Electroporation*

1. Mix appropriate amounts of firefly and *Renilla* luciferase expression plasmids and the dsRNA expression plasmid in a 1.5-mL tube. Add 80 µL of ice-cold PEG solution to the plasmid mix.
2. Transfer protoplasts into the plasmid mix and incubate for 10 min on ice.
3. Quickly mix the solution with pipetting, transfer the solution into an ice-cold 0.4-cm gapped cuvet, and perform electroporation at 300 V and 125 µF *(10)*. Time const. is about 5.5–6.0 ms.
4. Quickly transfer the protoplasts from the cuvet to the former ice-cold Eppendorf tube by decantation, and place the tube for 10 min on ice.
5. Mix with pipetting. Transfer 200 µL of protoplasts into a well of a 12-well plate containing 2.8 mL LS medium for protoplasts. Incubate at 28°C for 24 h (*see* **Note 8**).

### *3.7. Measurement of Luciferase Activity*

1. Discard 2 mL of medium. Transfer the remaining medium containing protoplasts to a 1.5-mL Eppendorf tube, centrifuge at 700*g* for 2 min, and discard the supernatant.
2. Add 100 µL of the passive lysis buffer (Promega), mix by tapping, and incubate the tube for 15 min at room temperature.
3. Add 50 µL of the luciferin solution to the 10 µL of the solution in passive lysis buffer, and measure the luminescence derived from firefly luciferase by luminometer. Add 50 µL of the Stop & Glo solution, and measure the luminescence derived from *Renilla* luciferase by luminometer (*see* **Notes 8–10**).

## 4. Notes

1. Plasmids with a palindromic sequence usually undergo recombination and easily form a dimeric structure. Therefore, we chose plasmids that had a lower level of dimeric form checked by agarose gel electrophoresis, and proceeded with a large-scale preparation of plasmids. We could not detect the insert by colony PCR, because uncertain bands or no bands were observed. Similarly, it is very difficult to analyze the sequence corresponding to the dsRNA region. Therefore, we neglected the mutation introduced by PCR and only checked constructed plasmids

by digestion with restriction enzymes. One method that overcomes these problems is the use of a loop sequence of sufficient length. An alternative method is to design dsRNA that contains mismatches by introducing some C-to-T conversion or insertion/deletion of bases in the sense region of dsRNA at every 7–10 nucleotide interval, to form G:U mismatches or bulge structures, respectively. In these cases, it maybe possible to improve the stability of the dsRNA expression plasmid in bacterial and plant cells without interfering with the RNAi effect, to perform colony PCR, and also to read the whole sequence corresponding to the dsRNA region.

2. We used the sense sequence of the luciferase gene, including the start codon. However, it may be advisable to make sense sequence by amplification with PCR so that it might not contain the start codon.

3. The yield of the plasmid containing a palindromic sequence without a loop sequence is as low as 50–100 µg/L, and can be improved by the existence of a loop sequence.

4. Because the *Bcl*I cannot cut methylated sites, a special host (e.g., SCS110; Stratagene) is required.

5. After cutting the vector with a restriction enzyme and the BAP treatment, the DNA fragment should be excised from the gel. This is because denatured plasmids that cannot be cut with a restriction enzyme must be removed to decrease the background. Using this treatment, almost all plasmids contain the insert sequence.

6. If no plasmid DNA was recovered, the wavelength of UV transilluminator used in excising the band may be inappropriate. Ligation reaction cannot continue after a few seconds of irradiation by some of the UV transilluminator.

7. The level of removal of cell walls in preparing protoplasts can be checked by staining with Calcofluor White ST (American Cyanamid Co., Wayne, NJ) or Bioglo (Calbiochem Co.). This procedure is not always required.

8. The RNAi effects can be observed 6–48 h after the introduction of plasmids.

9. One of the major reasons for the inefficiency of gene expression by electroporation is the inappropriate state of the cells. The cells from solid medium must be maintained in a liquid medium for several months until cells rapidly propagate. We usually maintain BY-2 cells only in liquid medium.

10. To compensate for the variation of transfection efficiency, *Renilla* luciferase expression vector was co-introduced, and the luminescence was measured by a dual luciferase kit (Promega). No treatment for destruction of cell walls, such as sonication, is required when the PLB solution supplied in the kit is used. Other reporter proteins, such as green fluorescent protein (GFP) and DsRed can be extracted and measured in the PLB buffer.

## Acknowledgments

We thank Professor Toshiyuki Nagata (Department of Biological Sciences, School of Science, The University of Tokyo, Japan) for helpful advice and comments.

## References

1. Bouche, N. and Bouchez, D. (2001) *Arabidopsis* gene knockout: phenotypes wanted. *Curr. Opin. Plant Biol.* **4,** 111–117.
2. Parinov, S. and Sundaresan, V. (2000) Functional genomics in *Arabidopsis*: large-scale insertional mutagenesis complements the genome sequencing project. *Curr. Opin. Plant Biol.* **11,** 157–161.
3. Fire, A., Xu, S., Montgomery, M. K., Kostas, S. A., Driver, S. E., and Mello, C. C. (1998) Potent and specific genetic interference by double-stranded RNA in *Caenorhabditis elegans. Nature* **391,** 806–811.
4. Cogoni, C. and Macino, G. (2000) Post-transcriptional gene silencing across kingdoms. *Curr. Opin. Genet. Dev.* **10,** 638–643.
5. Elbashir, S. M., Harborth, J., Lendeckel, W., Yalcin, A., Weber, K., and Tuschl, T. (2001) Duplexes of 21-nucleotide RNAs mediate RNA interference in cultured mammalian cells. *Nature* **411,** 494–498.
6. Hammond, S. M., Bernstein, E., Beach, D., and Hannon, G. J. (2000) An RNA-directed nuclease mediates post-transcriptional gene silencing in Drosophila cells. *Nature* **404,** 293–296.
7. Nagata, T. and Kumagai, F. (1999) Plant cell biology through the window of the highly synchronized tobacco BY-2 cell line. *Methods Cell Sci.* **21,** 123–127.
8. Akashi, H., Miyagishi, M., and Taira, K. (2001) Suppression of gene expression by RNA interference in cultured plant cells. *Antisense Nucleic Acid Drug Dev.* **11,** 359–367.
9. Wesley, S. V., Helliwell, C. A., Smith, N. A., Wang, M. B., Rouse, D. T., Liu, Q., et al. (2001) Construct design for efficient, effective and high-throughput gene silencing in plants. *Plant J.* **27,** 581–590.
10. Okada, K., Nagata, T., and Takebe, I. (1986) Introduction of functional RNA into plant protoplasts by electroporation. *Plant Cell Physiol.* **27,** 619–626.

# 43

## Delivery Agents for Oligonucleotides

### Olivier Seksek and Jacques Bolard

### Summary

This chapter provides a basic overview of most of the oligonucleotide delivery systems available for an in vitro use. Two major classes are described: systems that act through an endocytosis process (e.g., lipid-based vectors, nanoparticles, and polycations) and systems that by-pass this endocytosis process (e.g., peptides and pore-forming agents). Each technique is briefly described to allow a critical choice of the best delivery systems suitable for specific purposes in cultured cells.

**Key Words:** Oligonucleotides; antisense; RNA; delivery systems; cationic lipids; polyamines; nanoparticles.

## 1. Introduction

In the past decade, oligonucleotide strategies such as antisense, triplex-forming *(1)*, chimeric RNA/DNA *(2)*, small interfering RNA (siRNA) *(3)*, ribozymes, and DNA enzymes *(4–6)* have been among the most exciting advances in targeting gene expression. They all rely on the ability of the oligonucleotide (of any type) to interact directly with a specific intracellular target. Unfortunately, in order to successfully accomplish their purposes, the stability and delivery of these agents must be improved *(7)*. Indeed, nuclease degradation before activity and the inability to penetrate cellular membranes are the major drawbacks of such technologies. Chemical modification of the oligonucleotides have led to some effective solutions. However, these chemical alterations, although improving some characteristics, may also negatively alter other ones (lack of selectivity and/or efficiency). Thus, further development of delivery systems is critical to enhance the cellular oligonucleotide uptake.

This chapter provides the reader with an overview of the problems and solutions of effective delivery of oligonucleotides in cell culture. Since it is not

From: *Methods in Molecular Biology, vol. 252: Ribozymes and siRNA Protocols, Second Edition*
Edited by: M. Sioud © Humana Press Inc., Totowa, NJ

feasible to comprehensively describe all of the literature on this topic, we have limited ourselves to a short description of the systems proposed, indicating only the most recent reviews covering each of them. To avoid redundancy, only chronologically posterior data are included. Moreover, most of the delivery systems developed for nucleic acids are initially tested for gene transfection. It is only in a second step that they are applied to oligonucleotides. For this reason, several new approaches studied for plasmid or large DNA, which could be *a priori* of interest for oligonucleotides, are described succinctly or not at all.

The design of the delivery systems should represent a compromise between: i) an easy preparation consisting of the simple mixture of oligonucleotides/ vector with the oligonucleotides adsorbed onto the surface of the particles but rapidly released in the presence of protein-rich biological medium; and ii) an oligonucleotide encapsulation in the particle, more stable in serum but possibly not releasing oligonucleotide or being not internalized by the target cells. Since only in vitro experiments are described here, a number of issues directly related to the bioavailability, biodistribution, and pharmacokinetics of oligonucleotides are not included.

## 2. Oligonucleotide Cell Uptake Using Delivery Systems

Because phospholipid bilayers are strong barriers for the diffusion of oligonucleotides, this process is of limited importance in cell uptake. Internalization is believed to involve a membrane protein and an endocytotic mechanism. However, the involvement of a protein-independent endocytic, or potocytotic, mechanism has also been considered, as well as protein-dependent but not endocytotic mechanisms, such as carrier or pore-mediated uptake. These discrepancies may indicate complex transport mechanisms involving more than one uptake pathway *(8)*. However, for in vitro studies, the level of uptake is relatively low, variable, and inefficient, and, as already stated, has led to the development of efficient vector systems. The internalization of naked oligonucleotides is not covered here.

Except by functional assays, which are the ultimate evidence of cell uptake, the internalization of oligonucleotides delivered into cells in culture can be followed by a few methods that are widely described in the literature, often with contradictory results *(9)*. Indeed, they rely mostly on the use of fluorophores either bound to the 5' or 3' end or on one of the bases present in the oligonucleotide, with general agreement that the label does not perturb the uptake. Furthermore, private companies now provide various fluorescent labels on a standard basis. Thus, fluorescence experiments can be performed in many ways, from spectrofluorometry to confocal microscopy or fluorescence imaging. However, depending on the mode of detection of uptake, the distinction between total uptake and membrane binding may be unclear, and differentiation between vesicular, cytoplasmic, nuclear, or perinuclear localizations is ambiguous. As an alternative method, the use of radioactive labeling (e.g., $^{32}$P) may yield quantitative results, but less trivial to carry out.

Microinjection experiments of naked oligonucleotides in the cytoplasm indicate that the ultimate destination of oligonucleotide is the nucleus *(10)*. However, from many experiments in literature, it must be pointed out that it is quite difficult to get a clear view on the real cellular end point of delivered oligonucleotides, either cytoplasmic or nuclear. There are also several discrepancies in the literature regarding the intracellular localization of the target on which the nucleic acids would be very efficient *(11,12)*; there is no general rule for vectorized oligonucleotide-cell uptake and localization, and it would depend on many parameters: the delivery system that is used and the cell cycle would influence the cell distribution of the vectorized antisense oligonucleotides *(13)*, as well as the fixative protocol *(14)*.

Certain recent studies may illustrate the type of results that can be obtained. In cultured keratinocytes, antisense oligonucleotides were tested with various lipid-based delivery systems to point out the optimal conditions *(15)*. These experiments were done in conjunction with the cell confluence status, and this latter appeared to influence the nuclear localization *(16)*. In the same manner, cationic liposomal vectorization of antisense oligonucleotides on mouse peritoneal macrophages was evaluated *(17)*: fluoresceinated-oligonucleotide mixed with different systems yielded the same punctuate pattern visualized by confocal microscopy. It was then determined that these liposomal vectors would not be optimal carriers to promote the release of ODNs from the endosomal compartments in this cell type. With 5' fluoresceinated-chimeric RNA/DNA oligonucleotides *(18)*, data showed that in human airway epithelial cells, this type of oligonucleotide would have to be used with PEI or Cytofectin instead of Gene Porter transfection agents; all three commercially available agents gave 100% uptake, but only the first two facilitated nuclear localization of the oligonucleotide, where the oligonucleotide effect is assumed to take place. With triplex-forming oligonucleotides (TFO), effective triplexes were formed inside epithelial cells when PEI-adenovirus complexes *(19)* delivered oligonucleotides. In this case, TFOs were followed by incorporation of an Auger-emitting radionucleotide. Based on the published data, three types of delivery systems can found: systems that act by endocytotic process; systems that would bypass this process; and the physical methods.

## 3. Oligonucleotide Delivery Systems Acting by Endocytosis

### 3.1. Lipid-Based Nucleic Acid Delivery Systems

Lipid-based delivery systems for nucleic acids are probably the most extensively studied systems, and many review articles have thoroughly described

**Table 1**
**Lipid-Based Commercially Available Nucleic Acid Delivery Systems**

| Brand name | Company | Lipid type | Cell type | Nucleic acid type |
|---|---|---|---|---|
| Cellfectin | Invitrogen | TMTPS[1]/DOPE[2] (1:1.5 w/w) | Most effective on insect cells | Any |
| Cytofectene | Bio-Rad | Cationic lipid/DOPE (1:1) | Mammalian cells | Any |
| Cytofectin GSV | Gilead Sciences | GS 3815[3]/DOPE (2:1) | Mammalian cells | Any |
| CLONfectin | Clontech | Cationic lipid[4] | Mammalian cells | Any |
| DMRIE-C | Invitrogen | DMRIE[5]/cholesterol (1:1) | Mammalian suspension cells | Recommended for RNA |
| DOSPER | Roche Applied Sciences | DOSPER[6] liposomes | Mammalian cells | Any |
| DOTAP | Roche Applied Sciences | DOTAP[7] liposomes | Mammalian and insect cells | Any |
| Effectene | Qiagen | Non-liposomal lipid + "DNA condensing enhancer" | Mammalian cells | Any |
| Escort series | Sigma-Aldrich | Polycationic lipids/DOPE | Mammalian cells | Any |
| Eufectin | Novagen | Cationic lipid | Insect cells | Any |
| Fluorofectin | Qbiogene | DOTAP/cholesterol + fluorescent tag | Mammalian cells | Any |
| Fugene 6 | Roche Applied Sciences | Non-liposomal lipids | Mammalian cells | Any |
| GeneLimo | CPG Inc. | Polycationic lipids | Mammalian and insect cells | Any |
| GenePorter 1 & 2 | Gene Therapy Systems Inc | Cationic lipids /DOPE | Mammalian cells | Any |
| GeneSHUTTLE | Qbiogene | Polycationic lipids | Mammalian and insect cells | Any |
| Genetransfer | Wako | TMAG[8]/DLPC[9]/DOPE | Mammalian cells | Any |
| Insectogene | Biontex | Cationic lipids | Insect cells | Any |
| Lipofectamine /2000 | Invitrogen | Cationic lipids | Mammalian | Any |

| Product | Company | Composition | Cells | Nucleic acid |
|---|---|---|---|---|
| Lipofectamine Plus | Invitrogen | Cationic lipids + "precomplexing DNA agent" | Mammalian | Any |
| Lipofectin | Invitrogen | DOTMA[10]/DOPE (1:1) | Mammalian cells | Any, proteins |
| LipoTAXI | Stratagene | Liposome formulation | Mammalian cells | Any |
| Metafectene | Biontex | Cationic lipids | Mammalian cells | Any |
| NeuroPorter | Gene Therapy Systems Inc | Cationic lipid | Neuronal cells | Any |
| Oligofectamine | Invitrogen | Cationic lipids | Mammalian cells | Oligonucleotides |
| Perfectin | Gene Therapy Systems Inc | Cationic lipid / DOPE | Mammalian cells | Any |
| SiPORT lipid | Ambion | Cationic lipids | Mammalian cells | siRNA |
| Tfx series | Promega | Cationic lipid[11]/DOPE (1:n) | Mammalian and insect cells | Any |
| TransFast | Promega | Cationic lipid[12]/DOPE | Mammalian and insect cells | Any |
| Transfectam | Promega | DOGS[13] | Mammalian cells | Any |
| TransIT-Insecta | Mirus | Cationic lipid/neutral lipid | Insect cells | Any |
| TransMessenger | Qiagen | Lipid + "RNA-condensing enhancer" | Mammalian cells | RNA |
| X-tremeGENE Q2 | Roche Applied Sciences | Lipid film | Mammalian cells | Any |

[1] $N,N^I,N^{II},N^{III}$-tetramethyl- $N,N^I,N^{II},N^{III}$-tetrapalmitylspermine

[2] Dioleoyl phosphatidylethanolamine.

[3] Dimyristylamidoglycyl-N-isopropoxycarbonyl-arginine dihydrochloride.

[4] N-t-butyl-N'-tetradecyl-3-tetradecyl-aminopropion-amidine.

[5] dimyristoyl Rosenthal inhibitor ether.

[6] 1,3-di-oleoyloxy-2-(6carboxy-spermyl)-propylamid.

[7] N-[1-(2,3-dioleyloxy)propyl]-N,N,N-trimethylammonium methyl-sulfate.

[8] N-(alpha-trimethylammonioacetyl)-didodecyl-D-glutamate chloride.

[9] Dilauroyl phosphatidylcholine.

[10] N-[1-(2,3-dioleyloxy)propyl]-n,n,n-trimethylammonium chloride.

[11] [N,N,N',N'-tetramethyl-N,N'-bis2-hydroxy-ethyl)-2,3-dioleoyloxy-1,4-butanediammonium iodide.

[12] N,N bis(2-hydroxyethyl)-N-methyl-N-[2,3di(tetradecanoyloxy)propyl] ammonium iodide.

[13] Dioctadecylamidoglycyl spermine.

**Table 2**
**Non-Lipid-Based Commercially Available Nucleic Acid Delivery Systems**

| Brand name | Company | Molecule type | Cell type | Nucleic acid type |
|---|---|---|---|---|
| DuoFect | Qbiogene | PEI[1]/transferrin complex | Mammalian cells | DNA |
| ExGen 500 | Fermentas | Linear PEI | Mammalian cells | Any |
| GeneJammer | Stratagene | Polyamines | Mammalian cells | Any |
| GeneJuice | Novagen | Protein/polyamine | Mammalian cells | Any |
| JetPEI | Qbiogene | PEI | Mammalian cells | Any |
| JetSI | Qbiogene | Proprietary formulation | Mammalian cells | RNA |
| Penetratin 1 | Qbiogene | Antennapedia-derived peptide | Mammalian cells | Any, peptides |
| Polyfect | Qiagen | Activated dendrimer | Mammalian cells | Any |
| Ribojuice | Novagen | Proprietary formulation | Mammalian cells | RNA |
| SiPORT Amine | Ambion | Polyamine | Mammalian cells | RNA |
| Superfect | Qiagen | Activated dendrimer | Mammalian cells | Any |
| TransIT series | Mirus | Polyamine reagents | Mammalian and insect cells | Any (specific reagents) |

[1]Polyethyleneimine.

them *(9,20–28)*. Furthermore, they outnumber the non-lipid based formulations in the commercially available delivery systems set (**Tables 1** and **2**). To get the chemical formula of the main molecules, the reader is referred to Garcia-Chaumont et al. *(9)*.

Two main categories of lipid-based delivery systems can be described. The first involves chemical conjugation between oligonucleotides and lipid molecules such as cholesterol, phospholipids, or other alkyl chains *(29)*. The justification for such a modification would be to increase the lipid solubility of nucleic acids in order to successfully allow them to cross the plasma membrane and thus increase the cell uptake; they are consequently poorly soluble in aqueous solution. Nevertheless, even if the intracellular delivery is greatly improved, less protection—if any—against extracellular or serum nucleases is conferred by the chemical modification in vivo; this issue could be minimized in the case of an in vitro utilization. Another strategy for increasing oligonucleotide uptake and selectivity consists of conjugation to a ligand recognized by membrane proteins and receptors or to membrane-active effectors; several studies have demonstrated its relative utility *(9)*.

The non-covalent oligonucleotide/lipid complexes form the second and most successful set of lipid-based oligonucleotide delivery systems. Two subcategories can be distinguished: the systems in which oligonucleotides are encapsulated into the aqueous core of liposomes (mainly anionic) and the systems where complexes are directly formed between lipids (mainly cationic) and oligonucleotides. It must be emphasized that this category of colloidal lipid-based delivery systems is very effective for an in vitro purposes, but although it is used for systemic delivery, it is often toxic and eliminated from the blood by the reticuloendothelial system, with a relative immunostimulatory effect *(30)*.

True liposomal encapsulations of oligonucleotides are essentially formed by anionic *(31,32)* and neutral or zwitterionic lipids *(33,34)*. First, it must be noted that their poor encapsulation efficiency resulting from the negative charge borne by the nucleic acid and its high mol wt, has prevented their widespread use; although many studies have investigated their use for oligonucleotide delivery. A few methods could be used to improve the encapsulation process *(26)*: for example, oligonucleotide chemically modified (cholesterol-bound) oligonucleotides generally have greater encapsulation efficiency. Also, a high lipid-to-oligonucleotide ratio could be used, although it often gives a high percentage of large multilamellar vesicles with a reduced oligonucleotide-containing-aqueous core. Freeze-thawing and freeze-drying techniques must be used when oligonucleotides are added to preformed lipid films. Indeed, breaking and reforming the vesicles could rectify the imbalance in the distribution of oligonucleotide between external and internal vesicular spaces. External oligonucleotide could be then removed. With anionic lipids, 10–70%

encapsulation efficiency may be obtained with these methods and others such as reverse-phase evaporation. Typical lipids used in these formulations are PG (phosphatidylglycerol, negatively charged lipid) and PC (phosphatidylcholine, neutral lipid) often mixed with cholesterol. Practically, the liposomes that are made in these conditions can be extruded to achieve any desired size.

The second type of lipid-based delivery systems, in which nucleic acids—negatively charged at physiological pH—interact electrostatically with cationic liposomes, is the most successfully used delivery system in vitro *(35)*, and offers the most commercially available systems, often with proprietary synthetic formulations (**Table 1**). The tedious step of encapsulating oligonucleotide into delivery systems is avoided: the system is simply obtained by mixing the components, and allows the use of preformed formulations. It is called "lipoplex" or "cytofectin" or "transfectin"; however, many variations in the terminology can be found in the literature—for example, "cationic liposomes" are lipoplex formulations.

Unlike neutral and anionic lipids, a significant proportion of oligonucleotides can associate with the external surface of the liposome, although this fraction of external oligonucleotide may dissociate when used with serum. The addition of oligonucleotide to cationic liposomes induces their aggregation and fusion. In the presence of dioleolyl phosphatidyl ethanolamine (DOPE), hexagonal lipid tubule formation occurs *(9)*. The mechanism of the cell membrane destabilization by cationic lipids is not fully understood, but it is acknowledged that the lipid/oligonucleotide charge ratio should be positive, allowing a high affinity for most cell membranes, and the intracellular uptake is achieved by an adsorptive endocytic process. Apparently, cationic lipids can destabilize lipid bilayers by promoting the formation of nonbilayer lipid structures in combination with the anionic phospholipids present in the bilayer *(36)*. With this type of lipids, classical liposome formulation techniques such as dry lipid hydration can be used, as well as reverse-phase hydration or encapsulation by detergent dialysis.

Typical cationic lipids that can be used in oligonucleotide strategies can be displayed in three groups *(9)*. The first contains a single quaternary ammonium salt with long aliphatic chains (for example, $N$-[1-(2,3-dioleoxy)propyl]-$n,n,n$-trimethylammonium chloride [DOTMA]; dioctadecyl dimethylammonium bromide [DODAB]; 1,2-*bis*-(oleoyloxy)-3-(trimethylammonium) propane [DOTAP], dimyristooxypropyl dimethyl hydroxyethyl ammonium bromide [DMRIE]). The second group contains DC-Chol ($3\beta$-($N$-($N$-(dimethylaminoethane)carbamoyl) cholesterol) or TMAE-Chol (trimethyl aminoethane carbamoyl cholesterol) *(37)* as cholesterol derivatives. The last group bears multivalent headgroups such as DOGS (dioctadecylamidoglycylspermine) or DOSPA (2,3-dioleoyloxy-$N$-[2(sperminecarboxamido)ethyl]-$N,N$-dimethyl-1-propanaminium triflmuoroacetate).

All these compounds are cationic amphiphiles with hydrophobic chains that will adjust the molecule in the lipid bilayer with the cationic head group at the surface. For all these cationic lipid formulations, in order to successfully transfect in vitro cells, an experimental optimization is necessary in terms of lipid/oligonucleotide ratio, cationic lipid concentration, complex formation, and oligonucleotide length; the intracellular bioavailability would then be increased and the potential toxicity would be minimized.

Because most liposome formulations are internalized by endocytosis, during which they would fuse with lysosomes, which could be an end point, pH-sensitive and fusogenic formulations have been developed in order to address this issue. Indeed, these particular liposomes get to the low-pH endosomal compartment, where they fuse with the endosomal membrane and deliver their load to the cytoplasmic compartment. They are often composed of a non-bilayer-forming lipid (e.g., DOPE, which is an inverted cone-shaped lipid), and a biodegradable pH-sensitive surfactant *(38)* or a titratable amphiphile, such as OA (oleic acid) *(39)* or CHEMS (cholesteryl hemisuccinate) *(40,41)*. These latter molecules act by retaining the lipid mixture in a bilayer form at pH 7.0 and at lower pH, they induce membrane fusion by a non-bilayer phase formation. It has been shown that such formulation would greatly improve oligonucleotide transfer from endosomal to cytoplasmic compartments *(42)*.

In order to improve the targeting ability of liposomes, other formulations have been investigated, such as glycoliposomes *(43)*, antibody-associated liposomes *(44)*, or the combination of a viral vector (for example: Hemagglutinating virus of Japan, Sendai virus, adenovirus, Influenza type A) and a liposome ("virosome") *(45–48)*. Lactosylated low-density lipoprotein has also been used as a targeting factor to specific cells *(29)*. In all cases, oligonucleotide delivery was shown to be greatly improved.

In vivo polyethylene glycol (PEG)-lipid derivatives are often employed to stabilize or prolong the circulation lifetime of liposome formulations ("stealth" liposomes). In this case, PEG acts as an exchangeable lipid, and its loss will allow the liposome to recover its fusogenic ability, yielding programmable fusogenic vesicles *(49,50)*. Moreover, it has recently been shown *(51)* that high spontaneous entrapment efficiency could be achieved when this type of PEG-coated cationic liposomes interacted with oligonucleotide in the presence of ethanol.

### 3.2. Nanoparticles

The term "nanoparticle" is often used in general sense to encompass solid colloidal particles ranging in size from 10–1000 nm *(52)*. According to the process used for the preparation, nanospheres or nanocapsules can be obtained. Nanocapsules are vesicular systems in which the drug is confined to a cavity (an oily or aqueous core) surrounded by a unique membrane; nanospheres are matrix systems in which the drug is dispersed throughout the particles *(53)*. When the size of the sphere exceeds 1 μm, it is commonly called "microsphere." Liposomes could be considered to be nanocapsules, but the term is generally used for nanoparticles composed of polymers. Microemulsions are composed of an oily or non-polar phase, an aqueous phase, a surfactant, and possibly a cosurfactant *(54)*. They are smaller than 0.2 μm and they are transparent to translucent; no energy is required for their formation, and they are thermodynamically stable *(55)*.

Nanospheres, which exhibits an internal structure consisting of a matrix made up of a dense polymeric network, represent the most common nanoparticulate systems used for oligonucleotide delivery *(53)*. Two types of methods are developed for their preparation. One is based on the polymerization of monomers, and has mainly been applied to an polyalkylcyanoacrylate (PACA) *(56)*, approach that represented the first trial with promising results. The second one uses preformed polymers instead or monomers, with or without emulsification. It has been applied to polylactic acid (PLA), which has the major advantage of being biocompatible and is well-tolerated. PLA nanoparticles were shown to efficiently deliver oligonucleotides against lymphocytic cells *(57)*. In order to render the surface potential of the nanospheres negative and to facilitate the condensation of the negatively charged oligonucleotides, a cationic hydrophobic detergent (CTAB, DEAE) was combined with the PACA polymer. With the same purpose, a positive charge was generated in polystyrene nanoparticles by performing the emulsion polymerization in the presence of cationic initiators *(58)*.

More recently, spongelike alginate and poly-lysine nanospheres were prepared *(59)*. In this case, a diffusion process achieves the loading of oligonucleotide. Another solution consists of covalently linking the oligonucleotides to a hydrophobic molecule, allowing a hydrophobic interaction with the polymer core. This approach was developed together with a double-emulsion process, coupled with the addition of oligonucleotides associated to CTAB, leading to the high entrapment of the oligonucleotides in the core of the PLA matrix *(60)*. However, these approaches result in a rather low rate of oligonucleotide uptake and protection, although nanospheres prepared by the solvent evaporation method presented increased encapsulation *(61)*. Therefore, efforts were developed to prepare nanocapsules with an aqueous core instead of nanospheres *(62)*. In this system, oligonucleotides were indeed, efficiently protected from degradation by nucleases.

Diblock copolymer nanoparticles offer additional opportunities for the development of new oligonucleotide delivery systems *(52,63)*. The term "block" denotes the linear architecture of the polymer in which the end of one segment is covalently joined to the head of the other segment to yield a diblock AB or multiple-block (AB)$_n$-type copolymers. By playing on the hydrophilic/ hydrophobic balance of the segments as well as on their electrostatic charge, it is possible to modulate the structure of the particles. "Graft" copolymers have a comb-like structure with hydrophilic segments attached on the side of the cationic segments. Hydrophobic blocks include the poly (L amino-acid), PLA, and PACA. When present, the hydrophilic block is usually poly (ethylene glycol) (PEG) (PEG is the term used for polyoxyethylene polymers with low mol wt, and the term poly (ethyleneoxide) (PEO) is preferred to designate polyoxyethylenes with high mol wt. When the mol wt of the hydrophilic block exceeds the mol wt of the hydrophobic block, a micelle-like structure (20–40 nm in diameter and fairly monodisperse) is likely to result from self-assembly. In the reverse case, larger colloids may form, usually nanospheres (approx 100– 200 nm in diameter). The separation of hydrophilic and hydrophobic blocks into entirely distinct domains is uncertain for nanospheres, although with PEG, their surface may be enriched with this segment.

The polyesters made by polycondensation of L-lactide or glycolide have a long history of use as biodegradable biomaterials. They have been used for the delivery of oligonucleotides *(7,64,65)*. The chemical conjugation of oligonucleotides to poly D, L-lactic-co-glycolic acid has led to copolymer micelles efficiently transported within cells *(66)*. Recently, it has been shown that when oligonucleotides were loaded in these microparticles associated with PEI, the intercellular penetration of the released oligonucleotide was improved *(67)*.

The systems described here with PACA or PLA, in which oligonucleotides are absorbed at the surface of the nanoparticles, are promising in vitro, but there is a lack of information about their efficacy in vivo: clearance of the targeted oligonucleotide from the bloodstream—for instance, resulting in the uptake by Kuppfer cells—may be rapid. Incorporation of PEG into the design of the vector may improve its biocompatibility: it is well-known that strong hydration and high conformational flexibility provide PEG with the steric stabilization effect. Therefore, PEO-polyspermine and PEG-pLL as well as PEG-tertiary amine methacrylate polymers *(68)* have been studied as delivery systems. It should be noted that PEG-pLL/oligonucleotide systems dissociate upon dilution *(69)*, but this dissociation can be suppressed by glutathione–sensitive stabilization *(70)*.

Pluronic® block copolymers (also known as Poloxamer or Synperonic) were recently proposed for the delivery of oligonucleotides *(64,71)*. They consist of ethyleneoxide (EO) and propyleneoxide (PO) blocks arranged in a triblock structure EO$_x$–PO$_y$–EO$_x$. They self–assemble into micelles, and the inner core of these is constituted of the hydrophilic PO blocks and is covered by the hydrophilic corona from EO blocks. The interest of poly ethylene-co-vinylacetate has also been demonstrated in an in vitro study *(72)*. A positively charged emulsion based on lecithin, triglycerides (the oil core), poloxamer 188, and stearylamine has recently been proposed as a delivery system for oligonucleotide *(54)*. This system exhibited prolonged stability in serum because of steric stabilization caused by Poloxamer molecules.

Monomethylaminomethylmethacrylate (MMAEMA) methylmethacrylate copolymers nanoparticules, upon which oligonucleotides were absorbed by condensation upon the cationic MMAEMA, demonstrated a promising activity of carrier *(73)*. Recently a new family of nanoscale materials was proposed, on the basis of dispersed networks of crosslinked ionic (PEI) and non-ionic (PEO) hydrophilic polymers. Interaction of oligonucleotides with these structures results in the formation of nanocomposite materials leading to the collapse of the dispersed gel particles. Efficient cellular uptake and intracellular release were achieved and antisense activity was detected *(74)*.

It should be noted that a new class of surfactants has recently appeared. They are made up of two amphiphilic moieties connected at the level of the head groups, or very close to the head groups by a spacer group. Such surfactants have been referred to as cationic detergents, *bis*-quaternary ammonium surfactants, or gemimi surfactants. Although at the present time only applied to gene delivery *(75)*, they appear to be promising candidates for oligonucleotide delivery.

### 3.3. Polycations

#### 3.3.1. Polyaminoacids

Polyelectrolytes such as polylysine (PLL or PLK) or polyornithine (POR) have been used for a long time as carriers for transportation of bioactive agents. As a result, PLL-oligonucleotide conjugates were one of the first delivery systems proposed for oligonucleotide delivery. Later on, the conjugate PLL-oligonucleotide was addressed to specific cell types or specific cell receptor by covalently binding ligands such as glycoproteins, transferrin, insulin, antibodies, epidermal growth factor (EGF), lectins, mannose, polymeric immunoglobulins, signal peptides, and folic acid *(9,64,76)*. The use of these vectors allowed cell-specific targeting and increased cellular uptake and antisense effect. However, PLL is cytotoxic in high doses and ineffective in some cell lines. Nevertheless, the use of PLL is still advocated, and a large range of synthetic PLL products is available commercially. Recently, data obtained with H5WYG, a histidine-rich, uncharged (at neutral pH) derivative of E5 (*see* the previous description of endosome destabilization) has led to substitute PLL with histidyl residues. Highly substituted histidylated oligolysines seemed to be suitable vectors for oligonucleotide transfer after simple complex formation of both components, and appeared to result in a 10-fold increase of antisense activity *(77)*.

Protamine is a relatively small polycationic peptide (mol wt 4000–4500) with a high content (two-thirds) of arginine residues. By simple mixing with oligonucleotides, solid particles with initial diameter of 90–150 nm are formed, which facilitate the oligonucleotides internalization in the cytoplasm and the nucleus *(78)*. The same effect was observed with the complex formed by condensation of oligonucleotide on protamine sulfate and coated with a negatively charged liposomal formulation *(79)*. A bifunctional peptide composed of a protamine fragment and a nuclear localization signal, facilitated the intracellular oligonucleotide uptake and produced an excellent downregulation of the expression of complementary mRNAs. However, under certain conditions, incubation of the cells with chloroquine was required to produce antisense activity *(80)*. With the exception of Benimetskaya's *(80)* Astriab-Fisher's *(81)*, and Pichon's *(82)* studies, in almost all the examples of peptide-mediated delivery, it is unclear whether any sequence-specific downregulation of the expression of a target mRNA is observed.

### 3.3.2. Polyamines

Starburst polyamidoamine (PAMAM) dendrimers are spherical, highly ordered, dendritic polymers with positively charged primary amino groups on the surface at physiological pH *(83)*. They form stable complexes with oligonucleotides, with limited cell toxicity and are effective in oligonucleotide delivery and development of antisense activity at the cell-culture level *(9)*. The charge ratio of oligonucleotide:dendrimer and the size (generation) of the dendrimers appeared to be critical variables for the antisense effect. Sixth-generation PAMAM are used either in intact (Polyfect) or fractured (Superfect) form. For gene delivery, the fractured dendrimers are much more efficient than the intact ones. Lower generation dendrimers have relatively small molecular masses, and may form relatively small complexes. Compared to other types of delivery agents, PAMAM dendrimers were more effective in delivering oligonucleotides into the nucleus of cells in the presence of serum proteins *(84)*.

Polyethylenimine (PEI) is the focus of more and more interest for the study of gene delivery. It is the compound with the highest charge density and a high intrinsic endosomolytic activity because of a strong buffer capacity at virtually any pH ("proton sponge" mechanism). Polymers with different mol wts and degrees of branching have been synthesized and used for oligonucleotide delivery, and highly branched polymers such as the 25-kDa PEI and the 800-kDa PEI are most frequently used. However, the huge amount of positive charges results in a rather high toxicity, which is one of the major limitations for its use *(85)*. Linear PEI (ExGen 500) has been synthesized with a low toxicity, but studies have been limited to gene transfer, until now. Its activity as a phosphorothioates oligonucleotide vector was demonstrated in a variety of cell lines *(9)*. Furthermore it is also efficient in improving the antisense activity of 3'-capped phospho*diester* oligonucleotides, which cannot be achieved by other delivery systems—for instance, lipofection as a result of insufficient protection from nuclease degradation *(86)*. The very low cost of PEI compared to cytofectins, and the increased affinity for target mRNA and decreased affinity for proteins of phosphodiesters compared to phosphorothioates, makes its use very attractive.

The polyamines spermine and spermidine have been used as vectors for gene transfection, and their cationic charge allows the condensation of DNA and the formation of condensed structures. Even in deionized water, only a very small amount of particles were found, and nearly all oligonucleotides remained unbound *(87)*. In addition, the solutions of spermine significantly reduced the cell viability. However, agents termed "molecular umbrellas," containing two or more cholic acid moieties covalently coupled to spermine and/or spermidine, provide a cationic "tail" or "handle" to which anionic oligonucleotides can bind. They have shown substantial delivery activity, even in the presence of high concentrations of serum *(88)*.

### 3.4. Oligodeoxyribonucleotide Delivery Systems That Bypass the Endosomal Pathway

All the oligonucleotide-delivery systems considered until now are assumed to be internalized into cells by endocytosis. This pathway brings the oligonucleotide into the endosomal vesicles, which represent an end point for them, unless the chosen vector allows their escape. Short peptides, that present fusogenic or permeabilizing properties at the acidic pH encountered in the lysosomes can fulfill this requirement. However, applying these properties of membrane fusion or transient permeability to the plasma membrane reorganization may represent a simpler method, allowing the direct oligonucleotide transfer into the cell and bypass the endocytosis step. With or without endocytosis, delivery across cellular membranes is required.

Two types of approaches may be used. Ionic complexes may be formed between the oligonucleotide and the peptide, but the peptide can also be covalently linked to the oligonucleotides, which gives a more consistent product in which the desired properties can be carefully controlled by structural variations *(89)*. However, the latter approach implies a chemical synthesis step.

### 3.4.1. Permeabilization of Endosome Membranes

There are two known types of pH-sensitive peptides for oligonucleotide delivery. They include fusogenic peptides based on sequences derived from viral proteins such as the influenza virus hemagglutinin (the N-terminal segment of its HA-2 subunit) and amphipathic helix peptides with the prototype being the "GALA" peptide *(76,90)*. The amphiphilic anionic peptides of the INF family, of the E5 family such as E5CA, GALA, are almost inactive in the presence of serum because they have the disadvantage of binding to serum proteins. However, they can deliver oligonucleotide to the nucleus, in vitro, if the pH of the medium is lowered *(91)*. Conversely, amphiphilic basic peptides such as melittin, K5 the cationic counterpart of E5 and KALA the lysil counterpart of GALA, exhibit more efficient membrane fusion and permeabilization activities at both neutral and acidic pH than anionic peptides, but they interact mainly with the plasma membrane and are more cytotoxic. A new basic membrane-destabilizing peptide has not yet been studied with oligonucleotides, but shows promising properties for nucleic acid delivery *(92)*. Finally H5WYG— a histidine-rich, uncharged (at neutral pH) derivative of E5—was designed, but the amount of oligonucleotide it could deliver in the cytosol from the endosomes was too weak to provide an efficient biological activity *(77)*.

In contrast, the peptide KDEL is recognized by receptors located in various subcellular compartments, and it shuttles between them. It was hypothesized that oligonucleotides bearing a KDEL motif would more easily enter the cytosol via translocation throughout one of these membranes. Actually, this was not the case, but oligonucleotide antisense activity was increased *(93)*.

### 3.4.2. Direct Transfer Through the Plasma Membrane (**42,94**)

#### 3.4.2.1. MPG Peptide

This peptide is derived from the fusion sequence of human immunodeficiency virus (HIV) gp41 active at neutral pH coupled to a hydrophilic domain, derived from the nuclear localization sequence of Simian virus 40T-antigen. Its complex with oligonucleotides was shown to be effectively delivered into the cells. However, other peptides containing the hydrophobic region of a signal peptide sequence and a nuclear localization sequence, also allowed an efficient intracellular penetration, but failed to show any serious antisense effect of conjugated oligonucleotides *(95)*. On the other hand, we have seen that fusogenic liposomes may be prepared with inactivated Sendai virus, which acts at neutral pH *(96)*.

#### 3.4.2.2. Protein Transduction Domains

Recently, several small regions of proteins known as protein transduction domains (PTD) have been identified that possess the ability to traverse biological membranes efficiently in a process termed protein transduction, in a receptor- and transporter-independent fashion that appears to target the lipid bilayer directly. They include the Tat protein from the HIV virus, the Drosophilia homeotic transcription factor ANTP (encoded by the *antennapedia* gene), and the herpes simplex virus type 1 (HSV-1) VP22 transcription factor. Therefore, a peptide from Tat *(49–57)* was conjugated to an antisense oligonucleotide, which resulted in an increased internalization of the oligonucleotide as compared to that of free oligonucleotide *(20)*. Similar results were obtained with ANTP *(97)*. Antisense inhibition of P-glycoprotein expression could be obtained with these conjugates *(81)*. Cellular delivery using these cell-penetrating peptides has the advantage (or the inconvenience) of not being cell-specific *(98)*. A α-helical amphipathic 18-mer model peptide (and several of its derivatives), which was shown to possess analogous cell-penetrating properties, was bound to oligonucleotides by complex formation or conjugation. It appeared that the uptake differences between naked oligonucleotides and their respective peptides complexes or conjugates were generally confined to one order of magnitude for a particular cell type, independently of the peptide structural properties. These results questioned the common belief that these peptides act by improved membrane translocation *(99)*.

### 3.4.2.3. PORE-FORMING AGENTS

Another approach takes advantage of the transmembrane pore-forming ability of the bacterial toxin streptolysin-O, which is used for controlled permeabilization of the cell membrane. Despite the absence of information on possible formation of a complex between oligonucleotides and streptolysin-O, and the need for membrane resealing after treatment, this approach was used successfully in several studies *(9,42)*. Similarly, a cationic derivative of the polyene antibiotic amphotericin B, AMA (amphotericin B 3-dimethylamino-propyl amide), has been shown to interact with antisense oligonucleotides as well as plasmid DNA, and allows their vectorization into mammalian cells. The interaction between AMA and nucleic acids, which has been well-characterized by spectroscopic techniques, has led to complexes in which nucleic acids are protected from serum degradation or DNase I attack, which is a fundamental advantage for clinical use. Biological effects of antisense oligonucleotides vectorized by AMA were investigated with a 20-mer oligonucleotide directed against the human MDR1 mRNA of the multidrug resistance phenotype *(100)*. The advantages of this novel vector involves the fact that i) it is a single-component vector; ii) it bears an amide bond that is chemically stable and biodegradable; iii) it is a derivative of a clinically widely used drug; and iv) it can interact directly with plasma membrane lipids, which creates a novel route of entry for nucleic acids into cells through membrane pores. We are currently evaluating the most innovative contribution of cationic derivatives of polyene antibiotics to the oligonucleotide delivery approach in targeting to fungal cells.

## 3.5. Physical Methods

### 3.5.1. Shock Waves

Shock waves are generated by an electrohydraulic lithotripter, and offer a new method for oligonucleotide cytoplasmic delivery, effective for the uptake as well as for the biological activity of antisense oligonucleotide *(101)*. Pressure-mediated transfection has been applied to the oligonucleotide delivery to human venous endothelium ex vivo and to the non-vascular cells of rat myocardium *(102)*.

### 3.5.2. Electroporation

This method involves the use of an electric field to open transitory pores into the plasma membrane. During the time the pores are open, oligonucleotides can enter the cell directly. This process can be used with virtually any cell type *(42)*. However, a large number of cells are required because efficient electroporation results in 20–50% cell death. It was also recently demonstrated that square-wave electroporation allows the delivery of a more defined and regulated electrical pulse than the conventional method (e.g., exponential decay) and is associated with high transfection efficiencies in a variety of systems *(103)*. Antisense strategy was shown to be relevant to this delivery approach *(104)*.

### 3.5.3. Microinjection

Oligonucleotides can be directly injected into the cytoplasm or nucleus. This procedure is highly efficient, but requires specialized equipment, and can be only performed on a limited amount of cells *(10,42)*.

## 3.6. Conclusion

Over the years, various delivery systems have been developed for oligonucleotide internalization into cultured cells, and many of them have been patented and are commercially available. According to the manufacturer's handbooks, they are all highly effective and all "give the best internalization rate ever"; however, caution must be taken and an evaluation must be thoroughly conducted for each cell type, each oligonucleotide type and sequence, and each delivery system to get the optimal delivery conditions. Furthermore, it must be kept in mind that the presence of serum often represents a serious impediment in the in vitro delivery of oligonucleotides. Although lipid-based systems appear to be the most numerous, other non-lipid carriers give a very good result; some of them tend to achieve a non-endocytic process in order to avoid an endosomal oligonucleotide storage, and they might be very promising in terms of internalized oligonucleotide amount.

# References

1. Faria, M., Wood, C. D., White, M. R., Helene, C., and Giovannangeli, C. (2001) Transcription inhibition induced by modified triple helix-forming oligonucleotides: a quantitative assay for evaluation in cells. *J. Mol. Biol.* **306,** 15–24.
2. Lai, L. W. and Lien, Y. H. (2002) Chimeric RNA/DNA oligonucleotide-based gene therapy. *Kidney Int* **61 Suppl 1,** 47–51.
3. Paul, C. P., Good, P. D., Winer, I., and Engelke, D. R. (2002) Effective expression of small interfering RNA in human cells. *Nature Biotechnol.* **20,** 505–508.
4. Sun, L. Q., Cairns, M. J., Saravolac, E. G., Baker, A., and Gerlach, W. L. (2000) Catalytic nucleic acids: from lab to applications. *Pharmacol. Rev.* **52,** 325–347.
5. Jen, K. Y. and Gewirtz, A. M. (2000) Suppression of gene expression by targeted disruption of messenger RNA: available options and current strategies. *Stem Cells* **18,** 307–319.
6. Sioud, M. (2001) Nucleic acid enzymes as a novel generation of anti-gene agents. *Curr. Mol. Med.* **1,** 575–588.
7. Akhtar, S., Hughes, M. D., Khan, A., Bibby, M., Hussain, M., Nawaz, Q., et al. (2000) The delivery of antisense therapeutics. *Adv. Drug Deliv. Rev.* **44,** 3–21.
8. Wu-Pong, S. (2000) Alternative interpretations of the oligonucleotide transport literature: insights from nature. *Adv. Drug Deliv. Rev.* **44,** 59–70.
9. GarciaChaumont, C., Seksek, O., Grzybowska, J., Borowski, E., and Bolard, J. (2000) Delivery systems for antisense oligonucleotides. *Pharmacol. Ther.* **87,** 255–277.
10. Lukacs, G. L., Haggie, P., Seksek, O., Lechardeur, D., Freedman, N., and Verkman, A. S. (2000) Size-dependent DNA mobility in cytoplasm and nucleus. *J. Biol. Chem.* **275,** 1625–1629.
11. Stein, C. A. (1999) Two problems in antisense biotechnology: in vitro delivery and the design of antisense experiments. *Biochim. Biophys. Acta* **1489,** 45–52.
12. Kole, R. and Sazani, P. (2001) Antisense effects in the cell nucleus: modification of splicing. *Curr. Opin. Mol. Ther.* **3,** 229–234.
13. Helin, V., Gottikh, M., Mishal, Z., Subra, F., Malvy, C., and Lavignon, M. (1999) Cell cycle-dependent distribution and specific inhibitory effect of vectorized antisense oligonucleotides in cell culture. *Biochem. Pharmacol.* **58,** 95–107.
14. Pichon, C., Monsigny, M., and Roche, A. C. (1999) Intracellular localization of oligonucleotides: influence of fixative protocols. *Antisense Nucleic Acid Drug Dev.* **9,** 89–93.
15. White, P. J., Fogarty, R. D., Werther, G. A., and Wraight, C. J. (2000) Antisense inhibition of IGF receptor expression in HaCaT keratinocytes: a model for antisense strategies in keratinocytes. *Antisense Nucleic Acid Drug Dev.* **10,** 195–203.
16. White, P. J., Fogarty, R. D., McKean, S. C., Venables, D. J., Werther, G. A., and Wraight, C. J. (1999) Oligonucleotide uptake in cultured keratinocytes: influence of confluence, cationic liposomes, and keratinocyte cell type. *J. Invest. Dermatol.* **112,** 699–705.
17. Takagi, T., Hashiguchi, M., Hiramatsu, T., Yamashita, F., Takakura, Y., and Hashida, M. (2000) Effect of cationic liposomes on intracellular trafficking and

efficacy of antisense oligonucleotides in mouse peritoneal macrophages. *J. Drug Target* **7**, 363–371.

18. de Semir, D., Petriz, J., Avinyo, A., Larriba, S., Nunes, V., Casals, T., et al. (2002) Non-viral vector-mediated uptake, distribution, and stability of chimeraplasts in human airway epithelial cells. *J. Gene Med.* **4**, 308–322.

19. Hoque, A. T., Sedelnikova, O. A., Luu, A. N., Swaim, W. D., Panyutin, I. G., and Baum, B. J. (2000) Use of polyethylenimine-adenovirus complexes to examine triplex formation in intact cells. *Antisense Nucleic Acid Drug Dev.* **10**, 229–241.

20. Hughes, J., Astriab, A., Yoo, H., Alahari, S., Liang, S., Sergueev, D., et al. (2000) In vitro transport and delivery of antisense oligonucleotides. *Methods Enzymol.* **313**, 342–358.

21. Jaaskelainen, I., Honkakoski, P., and Urtti, A. (2002) In vitro delivery of antisense oligonucleotides. *Cell Mol. Biol. Lett.* **7**, 236–237.

22. Dass, C. R. (2002) Vehicles for oligonucleotide delivery to tumours. *J. Pharm. Pharmacol.* **54**, 3–27.

23. Maurer, N., Fenske, D. B., and Cullis, P. R. (2001) Developments in liposomal drug delivery systems. *Exp. Opin. Biol. Ther.* **1**, 923–947.

24. Audouy, S. and Hoekstra, D. (2001) Cationic lipid-mediated transfection in vitro and in vivo (review). *Mol. Membr. Biol.* **18**, 129–143.

25. Williams, S. A., and Buzby, J. S. (2000) Cell-specific optimization ofphosphorothioate antisense oligodeoxynucleotide delivery by cationic lipids. *Methods Enzymol.* **313**, 388–397.

26. Semple, S. C., Klimuk, S. K., Harasym, T. O., and Hope, M. J. (2000) Lipid-based formulations of antisense oligonucleotides for systemic delivery applications. *Methods Enzymol.* **313**, 322–341.

27. Matsuno, A., Nagashima, T., Katayama, H., and Tamura, A. (2000) In vitro and in vivo delivery of antisense oligodeoxynucleotides using lipofection: application of antisense technique to growth suppression of experimental glioma. *Methods Enzymol.* **313**, 359–372.

28. Tari, A. (2000) Preparation and application of liposome-incorporated oligodeoxynucleotides. *Methods Enzymol.* **313**, 372–387.

29. Bijsterbosch, M. K., Manoharan, M., Dorland, R., Waarlo, I. H., Biessen, E. A., and van Berkel, T. J. (2001) Delivery of cholesteryl-conjugated phosphorothioate oligodeoxynucleotides to Kupffer cells by lactosylated low-density lipoprotein. *Biochem. Pharmacol.* **62**, 627–633.

30. Dass, C. R. (2002) Immunostimulatory activity of cationic-lipid-nucleic-acid complexes against cancer. *J. Cancer. Res. Clin. Oncol.* **128**, 177–181.

31. Patil, S. D. and Rhodes, D. G. (2000) Conformation of oligodeoxynucleotides associated with anionic liposomes. *Nucleic Acid Res.* **28**, 4125–4129.

32. Lakkaraju, A., Dubinsky, J. M., Low, W. C., and Rahman, Y. E. (2001) Neurons are protected from excitotoxic death by p53 antisense oligonucleotides delivered in anionic liposomes. *J. Biol. Chem.* **276**, 32,000–32,007.

33. Lu, D. and Rhodes, D. G. (2002) Binding of phosphorothioate oligonucleotides to zwitterionic liposomes. *Biochim. Biophys. Acta* **1563**, 45–52.

34. Sasaki, M., Hayashi, J., Fujii, M., Koizumi, K., Fujita, H., Kobayashi, M., et al. (2001) Neutral liposome-mediated delivery process of fluorescein-modified oligonucleotides in cultured human keratinocytes. *J. Photochem. Photobiol. B* **60**, 120–128.

35. Hughes, M. D., Hussain, M., Nawaz, Q., Sayyed, P., and Akhtar, S. (2001) The cellular delivery of antisense oligonucleotides and ribozymes. *Drug Discov. Today* **6**, 303–315.

36. Hafez, I. M., Maurer, N., and Cullis, P. R. (2001) On the mechanism whereby cationic lipids promote intracellular delivery of polynucleic acids. *Gene Ther.* **8**, 1188–1196.

37. Cao, A., Briane, D., Coudert, R., Vassy, J., Lievre, N., Olsman, E., et al. (2000) Delivery and pathway in MCF7 cells of DNA vectorized by cationic liposomes derived from cholesterol. *Antisense Nucleic Acid Drug Dev.* **10**, 369–380.

38. Liang, E., Rosenblatt, M. N., Ajmani, P. S., and Hughes, J. A. (2000) Biodegradable pH-sensitive surfactants (BPS) in liposome-mediated nucleic acid cellular uptake and distribution. *Eur. J. Pharm. Sci.* **11**, 199–205.

39. de Oliveira, M. C., Rosilio, V., Lesieur, P., Bourgaux, C., Couvreur, P., Ollivon, M., and Dubernet, C. (2000) pH-sensitive liposomes as a carrier for oligonucleotides: a physico-chemical study of the interaction between DOPE and a 15-mer oligonucleotide in excess water. *Biophys. Chem.* **87**, 127–137.

40. Skalko-Basnet, N., Tohda, M., and Watanabe, H. (2000) Delivery of antisense oligonucleotides to neuroblastoma cells. *Neuroreport* **11**, 3117–3121.

41. Hafez, I. M. and Cullis, P. R. (2000) Cholesteryl hemisuccinate exhibits pH sensitive polymorphic phase behavior. *Biochem. Biophys. Acta* **1463**, 107–114.

42. Dokka, S. and Rojanasakul, Y. (2000) Novel non-endocytic delivery of antisense oligonucleotides. *Adv. Drug Deliv. Rev.* **44**, 35–49.

43. Stahn, R., Grittner, C., Zeisig, R., Karsten, U., Felix, S. B., and Wenzel, K. (2001) Sialyl Lewis(x)-liposomes as vehicles for site-directed, E-selectin-mediated drug transfer into activated endothelial cells. *Cell Mol. Life Sci.* **58**, 141–147.

44. Pagnan, G., Stuart, D. D., Pastorino, F., Raffaghello, L., Montaldo, P. G., Allen, T. M., et al. (2000) Delivery of c-myb antisense oligodeoxynucleotides to human neuroblastoma cells via disialoganglioside GD(2)-targeted immunoliposomes: antitumor effects. *J. Natl. Cancer Inst* **92**, 253–261.

45. Kaneda, Y. (2000) Virosomes: evolution of the liposome as a targeted drug delivery system. *Adv. Drug Deliv. Rev.* **43**, 197–205.

46. Kaneda, Y., Nakajima, T., Nishikawa, T., Yamamoto, S., Ikegami, H., Suzuki, N., et al. (2002) Hemagglutinating virus of Japan (HVJ) envelope vector as a versatile gene delivery system. *Mol. Ther.* **6**, 219–226.

47. Murray, K. D., Etheridge, C. J., Shah, S. I., Matthews, D. A., Russell, W., Gurling, H. M., and Miller, A. D. (2001) Enhanced cationic liposome-mediated transfection using the DNA-binding peptide mu (mu) from the adenovirus core. *Gene Ther.* **8**, 453–460.

48. Kondoh, M., Matsuyama, T., Suzuki, R., Mizuguchi, H., Nakanishi, T., Nakagawa, S., et al. (2000) Growth inhibition of human leukemia HL-60 cells by

an antisense phosphodiester oligonucleotide encapsulated into fusogenic liposomes. *Biol. Pharm. Bull.* **23,** 1011–1013.

49. Shi, F., Wasungu, L., Nomden, A., Stuart, M. C., Polushkin, E., Engberts, J. B., and Hoekstra, D. (2002) Interference of poly(ethylene glycol)-lipid analogues with cationic-lipid-mediated delivery of oligonucleotides; role of lipid exchangeability and non-lamellar transitions. *Biochem. J.* **366,** 333–341.

50. Hu, Q., Shew, C. R., Bally, M. B., and Madden, T. D. (2001) Programmable fusogenic vesicles for intracellular delivery of antisense oligodeoxynucleotides: enhanced cellular uptake and biological effects. *Biochim. Biophys. Acta* **1514,** 1–13.

51. Maurer, N., Wong, K. F., Stark, H., Louie, L., McIntosh, D., Wong, T., et al. (2001) Spontaneous entrapment of polynucleotides upon electrostatic interaction with ethanol-destabilized cationic liposomes. *Biophys. J.* **80,** 2310–2326.

52. Kwon, G. S. (1998) Diblock copolymer nanoparticles for drug delivery. *Rev. Ther. Drug Carrier Syst.* **15,** 481–512.

53. Lambert, G., Fattal, E., and Couvreur, P. (2001) Nanoparticulate systems for the delivery of antisense oligonucleotides. *Adv. Drug Deliv. Rev.* **47,** 99–112.

54. Texeira, H., Dubernet, C., Rosilio, V., Laigle, A., Deverre, J. R., Scherman, D., et al. (2001) Factors influencing the oligonucleotide release from O-W submicron caionic emulsions. *J. Control Rel.* **70,** 243–255.

55. Tenjarla, S. (1999) Microemulsions: an overview and pharmaceutical applications. *Crit. Rev. Ther. Drug Carrier Sys.* **16,** 461–521.

56. Zimmer, A. (1999) Antisense oligonucleotide delivery with polyhexylcyanoacrylate nanoparticles as carriers. *Methods* **18,** 286–295.

57. Berton, M., Turelli, P., Trono, D., Stein, C. A., Allemann, E., and Gurny, R. (2001) Inhibition of HIV-1 in cell culture by oligonucleotide-loaded nanoparticles. *Pharm. Res.* **18,** 1096–1101.

58. Fritz, H., Maier, M., and Bayer, E. (1997) Cationic polystyrene nanoparticles: preparation and characterization of a model drug carrier system for antisense oligonucleotides. *J. Colloid. Interface Sci.* **195,** 272–288.

59. Aynie, I., Vauthier, C., Chacun, H., Fattal, E., and Couvreur, P. (1999) Spongelike alginate nanoparticles as a new potential system for the delivery of antisense oligonucleotides. *Antisense Nucleic Acid Drug Dev.* **9,** 301–312.

60. Delie, F., Berton, M., Allémann, E., and Gurny, R. (2001) Comparison of two methods of encapsulation of an oligonucleotide into poly(D,L-lactic acid) particles. *Int. J. Pharm.* **214,** 25–30.

61. Freytag, T., Dashevsky, A., Tillman, L., Hardee, G. E., and Bodmeier, R. (2000) Improvement of the encapsulation efficiency of oligonucleotide-containing biodegradable microspheres. *J. Control Release* **69,** 197–207.

62. Lambert, G., Fattal, E., Pinto-Alphandary, H., Gulik, A., and Couvreur, P. (2001) Polyisobutylcyanoacrylate nanocapsules containing an aqueous core for the delivery of oligonucleotides. *Intl. J. Pharm.* **214,** 13–16.

63. Kakizawa, Y. and Kataoka, K. (2002) Block copolymer micelles for delivery of gene and related compounds. *Adv. Drug Deliv. Rev.* **54,** 203–222.

64. Chirila, T. V., Rakoczy, P. E., Garrett, K. L., Lou, X., and Constable, I. J. (2002) The use of synthetic polymers for delivery of therapeutic antisense oligodeoxynucleotides. *Biomaterials* **23,** 321–342.

65. Gill, J. S., Zhu, X., Moore, M. J., Lu, L., Yaszemski, M. J., and Windebank, A. J. (2002) Effects of NFKB decoy oligonucleotides released from biodegradable polymer microparticles on a glioblastoma cell line. *Biomaterials* **23,** 2773–2781.

66. Jeong, J. H. and Park, T. G. (2001) Novel polymer-DNA hybrid polymeric micelles composed of hydrophobic poly(D,L-lactic-co-glycolic acid) and hydrophilic oligonucleotides. *Bioconjug. Chem.* **12,** 917–923.

67. De Rosa, G., Quaglia, F., La Rotonda, M. I., Besnard, M., and Fattal, E. (2002) Biodegradable microparticles for the controlled delivery of oligonucleotides. *Int. J. Pharm.* **242,** 225–228.

68. Deshpande, M. C., Garnett, M. C., Vamvakaki, M., Bailey, L., Armes, S. P., and Stolnik, S. (2002) Influence of polymer architecture on the structure of complexes formed by PEG-tertiary amine methacrylate copolymers and phosphorothioate oligonucleotide. *J. Control. Rel.* **81,** 185–199.

69. Harada, A., Togawa, H., and Kataoka, K. (2001) Physicochemical properties and nuclease resistance of antisense-oligodeoxynucleotides entrapped in the core of polyion complex micelles composed of poly(ethylene glycol)-poly(L-lysine) block copolymers. *Eur. J. Pharm. Sci.* **13,** 35–42.

70. Kakizawa, Y., Harada, A., and Kataoka, K. (2001) Glutathione-sensitive stabilization of block copolymer micelles composed of antisense DNA and thiolated poly(ethylene glycol)-block-poly(L-lysine): a potential carrier for sytemic delivery of antisense DNA. *Biomacromolecules* **2,** 491–497.

71. Kabanov, A. V., Lemieux, P., Vinogradov, S., and Alakhov, V. (2002) Pluronic block copolymers: novel functional molecules for gene therapy. *Adv. Drug Deliv. Rev.* **54,** 223–233.

72. Edelman, E., Simons, M., Sirois, M., and Rosenberg, R. (1995) c-myc in vasculoproliferative disease. *Circ. Res.* **76,** 176–182.

73. Zobel, H. P., Junghans, M., Maienschein, V., Werner, D., Gilbert, M., Zimmermann, H., et al. (2000) Enhanced antisense efficacy of oligonucleotide adsorbed to monomethylmaminoethylmethacrylate methacrylate copolyopean Biophysics Journal with Biophysics Lettersmer nanoparticles. *Eur. J. Pharm. Biopharm.* **49,** 203–210.

74. Vinogradov, S., Bronich, T. K., and Kabanov, A. V. (2002) Nanosized cationic hydrogels for drug delivery: preparation, properties and interactions with cells. *Adv. Drug Deliv. Rev.* **54,** 135–147.

75. McGregor C., Perrin, C., Monck, M., Camilleri, P., and Kirby, A. J. (2001) Rational approaches to the design of cationic gemini surfactants for gene delivery. *JACS* **123,** 6215–6220.

76. Lebedeva, I., Benimetskaya, L., Stein, C. A., and Vilenchik, M. (2000) Cellular delivery of antisense oligonucleotides. *Eur. J. Pharm. Biopharm.* **50,** 101–119.

77. Pichon, C., Gonçalvez, C., and Midoux, P. (2001) Histidine-rich peptides and polymers for nucleic acids delivery. *Adv. Drug Deliv. Rev.* **53,** 75–94.

78. Junghans, M., Kreuter, J., and Zimmer, A. (2001) Phosphodiester and phosphorothioate oligonucleotide condensation and preparation of antisense nanoparticles. *Biochim. Biophys. Acta* **1544,** 177–188.
79. Welz, C., Neuhuber, W., Schreier, H., Metzler, M., Repp, R., Rascher, W., and Fahr, A. (2000) Nuclear transport of oligonucleotides in HepG2-cells mediated by protamine sulfate and negatively charged liposomes. *Pharm. Res.* **17,** 1206–1211.
80. Benimetskaya, L., Guzzo-Pernell, N., Liu, S., Lai, J. C. H., Miller, P., and Stein, C. A. (2002) Protamine-fragment peptides fused to an SV40 nuclear localization signal deliver oligonucleotides that produice antisense effects in prostate and bladder carcinoma cells. *Bioconj. Chem.* **13,** 177–187.
81. Astriab-Fisher, A., Sergueev, D. S., Fisher, M., Shaw, B. R., and Juliano, R. L. (2000) Antisense inhibition of P-glycoprotein expression using peptide-oligonucleotide conjugates. *Biochem. Pharmacol.* **60,** 83–90.
82. Pichon, C., Roufai, M. B., Monsigny, M., and Midoux, P. (2000) Histidylated oligolysines increase the transmembrane passage and the biological activity of antisense oligonucleotides. *Nucleic Acids Res.* **28,** 504–512.
83. Esfand, R. and Tomalia, D. A. (2001) Poly(amidoamione) (PAMAM) dendrimes: from biomimicry to drug delivery and biomedical applications. *Drug Dev. Today* **6,** 427–436.
84. Yoo, H., Sazani, P., and Juliano, R. L. (1999) PAMAM dendrimers as delivery agents for antisense oligonucleotides. *Pharm. Res.* **16,** 1799–1804.
85. Godbey, W. T., Wu, K. K., and Mikos, A. G. (2001) Poly(ethylenimine)-mediated gene delivery affects endothelial cell function and viability. *Biomaterials* **22,** 471–480.
86. Dheur, S. and SaisonBehmoaras, T. E. (2000) Polyethyleneimine-mediated transfection to improve antisense activity of 3'-capped phosphodiester oligonucleotides, in *Antisense Technology* vol. 313, (Phillips, M. I., ed.), Academic Press, San Diego, CA, pp. 56–73.
87. Junghans, M., Kreuter, J., and Zimmer, A. (2001) Phosphodiester and phosphorothioate oligonucleotide condensation and preparation of antisense nanoparticles. *Biochem. Biophys. Acta* **1544,** 177–188.
88. DeLong, R. K., Yoo, H., Alahari, S. K., Fisher, M., Short, S. M., Kang, S. H., et al. (1999) Novel cationic amphiphiles as delivery agents for antisense oligonucleotides. *Nucleic Acids Res.* **27,** 3334–3341.
89. Tung, C. H. and Stein, S. (2000) Preparation and applications of peptide-oligonucleotide conjugates. *Bioconjugate Chem.* **11,** 605–618.
90. Plank, C., Zauner, W., and Wagner, E. (1998) Application of membrane-active peptides for drug and gene delivery across cellular membranes. *Adv. Drug Deliv. Rev.* **34,** 21–35.
91. Freulon, I., Roche, A. C., Monsigny, M., and Mayer, R. (2001) Delivery of oligonucleotides into mammalian cells by anionic peptides: comparison between monomeric and dimeric peptides. *Biochem. J.* **354,** 671–679.
92. Rittner, K., Benavente, A., Bompard-Sorlet, A., Heitz, F., Divita, G., Brasseur, R., and Jacobs, E. (2002) New basic membrane-destabilizing peptides for plasmid-based gene delivery in vitro and in vivo. *Mol. Ther.* **5,** 104–114.

93. Pichon, C. K. A., Stewart, A. J., Duc Dodon, M., Gazzolo, L., Courtoy, P. J., Mayer, R., Monsigny, M., and Roche, A. C. (1997) Intracellular routing and inhibitory activity of oligonucleopeptides containing a KDEL motif. *Mol. Pharm.* **51**, 431–438.
94. Morris, M., Chalouin, L., Heitz, F., and Divita, G. (2000) Translocating peptides and proteins and their use for gene delivery. *Curr. Opin. Biotechnol.* **11**, 461–466.
95. Antopolsky, M., Azhayeva, E., Tengvall, U., Auriola, S., Jääskeläinen, I., Rönkö, S., et al. (1999) Peptide-oligonucleotide phosporothioate conjugates with membrane translocation and nuclear localization propoerties. *Bioconjugate Chem.* **10**, 598–606.
96. Kunisawa, J., Nakagawa, S., and Mayumi, T. (2001) Pharmacotherapy by intracellular delivery of drugs using fusogenic liposomes: application to vaccine development. *Adv. Drug Deliv. Rev.* **52**, 177–186.
97. Derossi, D., Calvet, S., Trembleau, A., Brunissen, A., Chassaing, G., and Prochiantz, A. (1996) Cell internalization of the third helix of hte Antennapedia homeodomain is receptor-independant. *J. Biol. Chem.* **271**, 18,188–18,193.
98. Lindgren, M., Hällbrink, M., Prochiantz, A., and Langel, U. (2000) Cell-penetrating peptides. *TiPS* **21**, 99–103.
99. Oehlke, J., Birth, P., Klauschenz, E., Wiesner, B., Beyermann, M., Oksche, A., and Bienert, M. (2002) Cellular uptake of antisense oligonucleotides after complexing or conjugation with cell-penetrating model peptides. *Eur. J. Biochem.* **269**, 4025–4032.
100. Garcia-Chaumont, C., Seksek, O., Jolles, B., and Bolard, J. (2000) A cationic derivative of amphotericin B as a novel delivery system for antisense oligonucleotides. *Antisense Nucleic Acid Drug Dev.* **10**, 177–184.
101. Tschoep, K., Hartmann, G., Jox, R., Thompson, S., Eigler, A., Krug, A., et al. (2001) Shock waves: a novel method for cytoplasmic delivery of antisense oligonucleotides. *J. Mol. Med.* **79**, 306–313.
102. Mann, M. J., Gibbons, G. H., Hutchinson, H., Poston, R. S., Hoyt, E. G., Robbins, R. C., and Dzau, V. J. (1999) Pressure-mediated oligonucleotide transfection of rat and human cardiovascular tissues. *Proc. Natl. Acad. Sci. USA* **96**, 6411–6416.
103. Liu, Y. and Bergan, R. (2001) Improved intracellular delivery of oligonucleotides by square wave electroporation. *Antisense Nucleic Acid Drug Dev.* **11**, 7–14.
104. Derenne, S., Monia, B., Dean, N. M., Taylor, J. K., Rapp, M. J., Harousseau, J. L., et al. (2002) Antisense strategy shows that Mcl-1 rather than Bcl-2 or Bcl-x(L) is an essential survival protein of human myeloma cells. *Blood* **100**, 194–199.

# 44

## Selection of Peptides for Specific Delivery of Oligonucleotides Into Cancer Cells

### Mohsen Shadidi and Mouldy Sioud

#### Summary

In this chapter, phage-display peptide technology has been used to select peptides that internalize into breast-cancer cells. The used biopanning procedure provides information on the best peptide to be used. When one of the selected peptides was conjugated to an antisense oligonucleotide against the ErbB2 receptor, specific delivery to breast-cancer cells was demonstrated. The established biopanning procedure should help in the rational selection of cancer-targeting peptides for specific delivery of DNA and RNA oligonucleotides into cancer cells.

**Key Words:** Phage display; peptide library; therapeutic peptides; gene therapy.

## 1. Introduction

Notably, the lack of selective delivery of therapeutics to tumor cells represents a major limitation in cancer therapy *(1)*. Thus, there is a great need to develop cell-selective delivery agents. Specific delivery of antisense, ribozymes, and small interfering RNAs (siRNA) to target cells would improve their efficacy and minimize potential adverse side effects *(2)*. During the last years, various targeting agents have been developed *(3–10)*. Because of their small size, peptides may become the most important agents for specific targeting of oligonucleotides to cancer cells. To select cell-binding peptides, the phage-display library is one of the biological approaches that have been used *(10,11–33; see* **Fig. 1A** and **Note 1**). **Table 1** summarizes the results obtained with peptide phage libraries. This chapter describes basic experiments that are useful for the selection of internalizing peptides along with additional experiments that can help to specifically target gene expression *(32)*.

From: *Methods in Molecular Biology, vol. 252: Ribozymes and siRNA Protocols, Second Edition*
Edited by: M. Sioud © Humana Press Inc., Totowa, NJ

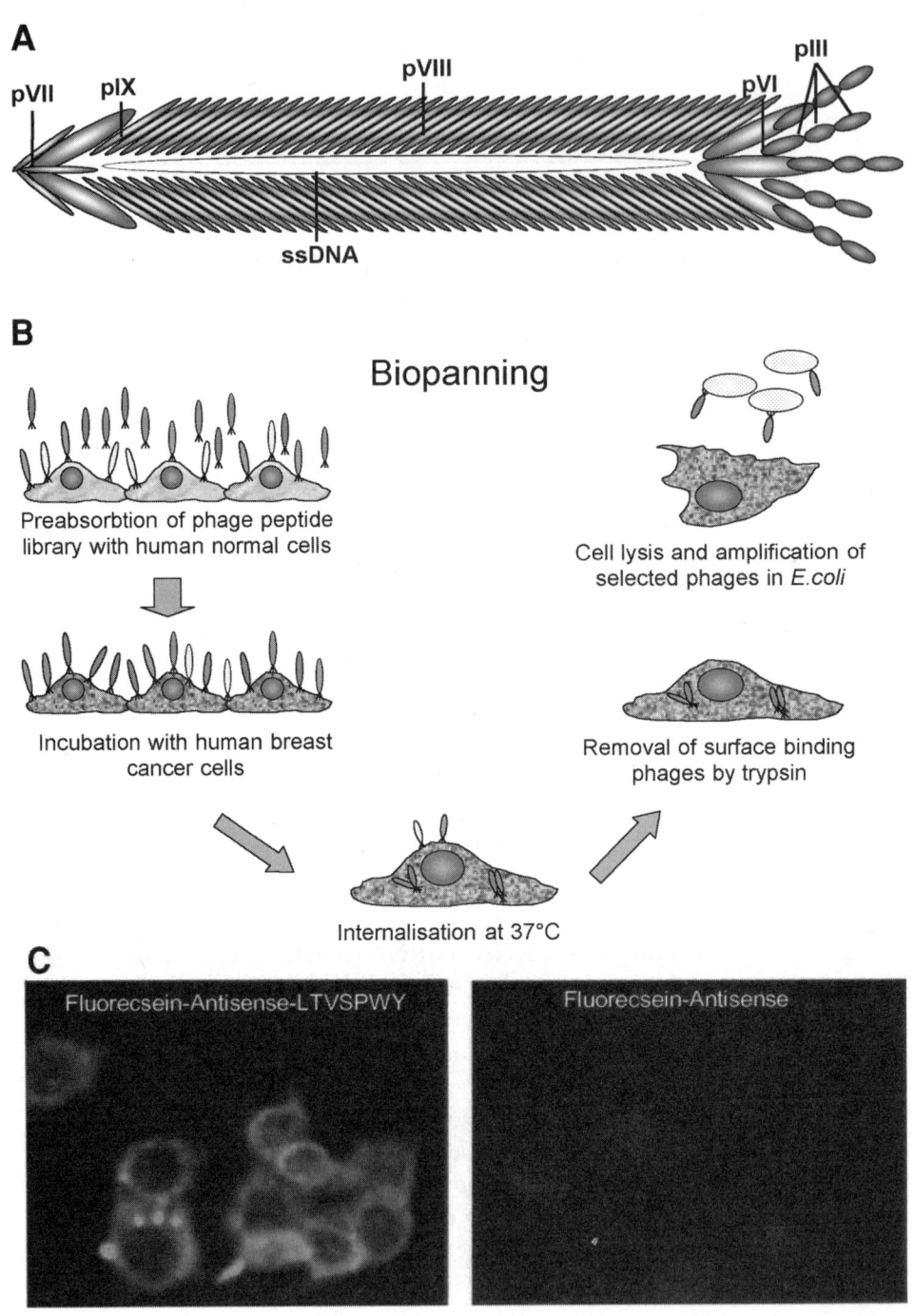

Fig. 1. **(A)** Schematic representation of the M13 phage. Coat proteins are indicated. **(B)** Experimental design. **(C)** Uptake of the fluorescein antisense-peptide conjugates by SKBR3 cells.

**Table 1**
**Target-Specific Peptides Identified by Phage-Display Library**

| Peptide | Cellular targets |
|---|---|
| Vasculature of various tumors | |
| CD**RGD**CFC, ACDC**RGD**CFCG *(11,12)* | $\alpha_v\beta_3$, $\alpha_v\beta_5$ integrins |
| C**NGR**CVSGCAGRC, CVC**NGR**MEC, **NGR**AHA *(13)* | Aminopeptidase N |
| CPGPEGAGC *(14)* | Aminopeptidase P |
| TAASGVRSMH, LTLRWVGLMS *(15)* | NG2 proteoglycan |
| C**GSL**VRC, C**GLS**DSC, NRSLKRISNKRIRRK *(16)* | Tumor vasculature |
| LRIKRKRRKRKKTRK, NRSTHI *(17)* | IC-12 rat trachea |
| SMSIARL, VSFLEYR *(18)* | Mice prostate |
| ATWLPPR, RRKRRR, ASSSYPLIHWRPWAR *(19–22)* | Vascular endothelial growth factor (VEGF) |
| CTTHWGFTLC *(23)* | Gelatinase |
| Cell surface of various tumors | |
| KNGPWYAYTGRO, NWAVWXKR, YXXEDLRRR, XXPVDHGL *(24)* | Surface idiotype of SUP-88 human B-cell lymphoma |
| LVRSTGQFV, LVSPSGSWT, ALRPSGEWL, AIMASGQWL, QILASGRWL, RRPSHAMAR, DNNRPANSM, LQDRLRFAT, PLSGDKSST *(25)* | Surface idiotype of human chronic lymphocytic lymphoma (CLL) |
| IELLQAR *(26)* | HL 60 human lymphoma and B-16 mouse melanoma |
| CVFXXXYXXC, CXFXXXYXYLMC, CVXYCXXXXCYVC, CVXYCXXXXCWXC *(27)* | Prostate-specific antigen (PSA) |
| DPRATPGS *(28)* | LNCaP prostate cancer |
| HLQLQPWYPQIS *(29)* | WAC-2 human neuroblastoma |
| VPWMEPAYQRFL *(29)* | MDA-MB435 breast cancer |
| TSPLNIHNGQKL *(30)* | Head and neck cancer cell lines |
| **SPL**W/F,R/K,N/H,**S**, V/H,**L** *(31)* | ECV304 endothelial-cell line |
| RLTGGKGVG *(31)* | HEp-2 human laryngeal carcinoma |
| **LTV**XPW**X** *(32)* | Human breast tumor |
| KCCYSL *(33)* | ErbB2 Tyrosine kinase type 2 receptor |

## 2. Materials

### 2.1. Buffers

1. Phosphate-buffered saline (PBS): 137 m$M$ NaCl, 3 m$M$ KCl, 8 m$M$ Na$_2$HPO$_4$, 1.5 m$M$ KH$_2$PO$_4$, pH 7.5.
2. PBST: PBS with 0.1% Tween-20.

### 2.2. Cells

1. Human mammary epithelial cell (HMEC) (*see* **Note 2**).
2. Human peripheral-blood mononuclear cells (PBMC).
3. Human breast-cancer cell line SKBR3 (American Type Culture Collection [ATCC]).

### 2.3. Media

1. Mammary epithelial-cell growth medium with bullet (MEGM) supplemented with10% fetal bovine serum (FBS).
2. RPMI-1640 medium supplemented with 10% FBS.

### 2.4. Biopanning

1. Random 7-mer peptide-displaying phage library (New England Biolabs) (*see* **Note 3**).
2. Luria-Bertani medium (LB): 10 g bacto-tryptone, 5 g bacto-yeast extract, 10 g NaCl. Adjust pH to 7.0.
3. Tetracycline stock solution: 20 mg/mL in ethanol. Store at –20°C in the dark and vortex before use.
4. LB-Tet plates. Keep at 4°C in the dark.
5. *E. coli* strain ER2738.
6. PEG/NaCl: 20% (w/v) polyethylene glycol (PEG) 8000 and 2.5 $M$ NaCl.
7. Phage elution buffer: 0.2 $M$ glycine-HCl, pH 2.2 with 1 mg/mL bovine serum albumin (BSA).
8. Neutralization solution: 2 $M$ Tris-Base.
9. Top agarose: 10 g bacto-tryptone, 5 g yeast extract, 5 g NaCl, 1 g MgCl$_2$. Dissolve in 900 mL of H$_2$O, add 7 g agarose, adjust the volume to 1 L, and autoclave.
10. IPTG/Xgal stock solution: 1.25 g isopropyl β-D-thiogalactoside (IPTG) and 1 g X-gal (5-bromo-4-chloro-3-indolyl-β-D-galactoside) in 25 mL dimethylformamide. Store at –20°C in the dark.
11. LB/IPTG/X-gal plates: LB medium with 15 g/L agar and 1 mL of IPTG/X-gal stock solution. Store at 4°C in the dark.

### 2.5. Flow Cytometry

1. 96-well microtiter plates with conical bottom.
2. 10 m$M$ ethylenediaminetetraacetic acide (EDTA) in PBS, pH 8.0.
3. Flow cytometry buffer: 1% FBS in PBS containing 0.005% azide (FC buffer).
4. Anti M13 mouse monoclonal antibody (MAb).

5. Goat phycoerythrin (PtdEtn) conjugated anti-mouse IgG.
6. Fluorescence-activated cell sorter (FACS) caliber.

## 2.6. DNA Sequencing

1. BigDye sequencing kit (ABI prism).
2. pIII oligonucleotide hybridizing to the 96 position of the insert.
3. ABI prism 310 genetic analyzer sequencing machine.

## 2.7. Antisense- and Fluorescein-Peptide Conjugates

1. Synthetic peptides: LTVSPWY**K** and LTVSPWY**C**.
2. 6-iodoacetamidofluorescein (6-IAF)
3. A synthetic antisense oligonucleotide against ErbB2: 5'-CTCCATGGT GCTCACSSpy-3'.
4. A scrambled oligonucleotide: 5'-CGCCTTATCCCGTAGC-SSpy-3' (*see* **Note 4**). Both oligonucleotides contain dithiodipyridine (SSpy) groups at their 3' end for peptide conjugation.

## 3. Methods

## 3.1. Biopanning

### 3.1.1. Biopanning on Whole Cells

The breast-cancer cell line SKBR3 was used as an affinity matrix to select phages from a 7-mer random peptide library as described in the following steps (*see* **Note 5**).

1. Plate SKBR3 and HMEC in T-25 tissue-culture flasks, so they are 70–80% confluent on the day of the biopanning.
2. Remove the medium, wash the cells once with 1X PBS, and add prewarmed complete RPMI medium containing10% FCS (*see* Note 6).
3. Add 1 mL of $1.0 \times 10^6$/mL freshly isolated PBMC in RPMI to the T-25 flask containing HMEC.
4. Add $1 \times 10^9$ pfu of the phage library to the HMEC/PBMC cells, and incubate for 1 h at 4°C.
5. Wash SKBR3 cells growing in T-25 tissue-culture flask with PBS, add prewarmed RPMI medium containing 10% FCS, and incubate at 37°C for 30 min.
6. Remove the medium from HMEC/PBMC cells and centrifuge for 10 min at 4000*g* at 4°C.
7. Transfer the supernatant, which contains unbound phages, to a new Falcon tube.
8. Remove the medium from the flask containing SKBR3 cells.
9. Add the supernatant containing the unbound phages to SKBR3 cells, and incubate for 1 h at 4°C with gentle shaking
10. Remove the medium and wash SKBR3 cells 10× (5 min each) with 1X PBS 0.2% Tween (PBST), pH 5.0, twice with PBST, pH 7.5, and twice with PBS, pH 7.5 (*see* **Note 7**).

11. Add 0.5 mL of trypsin/EDTA to the cells and incubate at 37°C for 5 min.
12. Add 3 mL of ice-cold RPMI with 10% FCS to the detached cells and centrifuge at 1200$g$ for 5 min at 4°C.
13. Wash the cells twice with PBS and resuspend in 50 μL of water. For complete lysis, vortex vigorously.
14. Add 450 μL of elution buffer to the cell lysates and incubate for 10 min at room temperature (*see* **Note 8**).
15. Neutralize the cell lysates with 1 mL of Tris-base (2 *M*).
16. Infect 5 mL of exponentially growing ER2537 (OD$_{600}$ ~0.5) with 100 μL of the phage eluate and incubate for 30 min at 37°C.
17. Dilute the culture to 1/50 with LB medium and incubate overnight at 37°C with shaking at 200 g/min.
18. In parallel, determine the titter of the phage eluate by infecting 250 μL of ER2537 with 1 mL of phage eluate. Incubate for 5 min at room temperature, add 3 mL of melted top agar, and plate on an IPTG/X-gal plate.
19. Invert the plate and incubate at 37°C overnight.

### 3.1.2. Phage Preparation

Spin down the overnight culture (**Subheading 3.1.1., step 17**) for 10 min at 8000$g$.

1. Transfer 80% of the cleared supernatant to a new tube. Add 1/5 vol of PEG/NaCl to the supernatant, mix well, and leave at 4°C for 2 h or more.
2. Centrifuge the precipitated phages for 15 min at 12,000$g$ at 4°C, aspirate, and discard the supernatant.
3. Resuspend the pellet in 1 mL of PBS.
4. Transfer the phage solution to a new microcentrifuge tube and spin for 5 min at 4000$g$ at 4°C to remove cellular debris.
5. Transfer the supernatant to a new microcentrifuge tube, and if necessary, repeat the PEG precipitation and the subsequent centrifugation steps.
6. Dry the phage pellet and redissolve it in 250 μL of PBS.
7. Store phages at 4°C.

We have described how to perform a single round of selection. However, in an in vitro selection scheme, it is strongly recommended that more than two rounds be performed in order to enrich for positive clones (*see* **Fig. 1B**). Use approx $10^{10}$ pfu for additional rounds of selection.

### 3.1.3. Phage Titration

After each round of selection determine the phage titers.

1. Grow a single colony of ER2738 in 5 mL LB medium until exponential phase (OD$_{600}$ ~0.5 ).
2. Prewarm LB/IPTG/X-gal plates at 37°C.

3. Melt top agarose in the microwave oven, pour 3 mL per Falcon tube, and incubate the tubes at 60°C until use.
4. Prepare 10-fold dilutions of each phage in LB medium.
5. Add 10 μL of each dilution to 3 mL melted top agarose, mix, and pour onto an LB/IPTG/X-gal plate.
6. Invert the plates and incubate at 37°C overnight.
7. Count the blue plaques of plates with $10^2$ plaques, and multiply each number by the dilution factor.

### 3.1.4. Single-Plaque Phage Amplification

After the desired round of selection, random phage clones should be tested for binding to cells.

1. Prepare a dilution series, up to $10^{12}$-fold, of the amplified phages (enriched library) from the last round of biopanning.
2. Plate the $10^8$–$10^{12}$ phage dilutions in agarose top onto LB/IPTG/X-gal plates as described previously.
3. Invert the plates and incubate overnight at 37°C.
4. Inoculate 2 mL of LB with a single colony of ER2738 and grow overnight.
5. Dilute the overnight culture 1/100 in LB medium.
6. Pick single blue plaques (**step 2**) from the plates, inoculate 2 mL culture of ER2738 cells (**step 5**), and incubate for 4–5 h at 37°C.
7. Spin down the bacteria cells and transfer 1 mL of the supernatant medium to new microcentrifuge tubes.
8. Add 200 μL of PEG/NaCl and leave at 4°C overnight.
9. Centrifuge the precipitated phages at 12,000*g* for 15 min at 4°C. Remove the supernatant and dry the pellets.
10. Dissolve the phage pellets in 100 μL of PBS.
11. To remove the rest of bacteria, spin down briefly at 4000*g* for 5 min.
12. Transfer 80 μL of the phage solution to new microcentrifuge tubes without touching the bacteria pellet.
13. Keep the amplified phages at 4°C.

### 3.2. Flow Cytometry

To evaluate the binding of the selected phages to human cancer cells, single phage clones from the enriched library should be purified (*see* **Subheading 3.1.4.**) and tested for binding to cells by flow cytometry.

1. Grow SKBR3 and HMEC cells in T-75 tissue-culture flasks.
2. Wash the cells once with PBS.
3. De-attach the cells from the culture flasks by treatment with 500 μL of 10 m*M* EDTA in PBS, pH 8.0. Incubate for 5 min at 37°C.
4. Add 5 mL of complete medium (RPMI + 10% FCS) to the flasks and transfer the de-attached cells to Falcon tubes.

5. Centrifuge at 1200$g$ for 5 min.
6. Remove the medium and resuspend the cell pellet in 5 mL of fresh complete medium.
7. Incubate the cells for 2 h at 37°C.
8. Divide of $10^4$ of each cell type into a conical 96-well microtiter plate and wash with FC buffer.
9. Centrifuge the plate at 1200$g$ for 5 min and discard the supernatant.
10. To each well, add 50 μL of phage particle ($10^8$ pfu) in FC and incubate for 30 min on ice. Different phage concentrations should be tested.
11. Wash the cells three times with FC buffer.
12. Add anti-M13 monoclonal IgG diluted 1/250 in FC buffer, and incubate for 30 min on ice.
13. Wash three times with the FC buffer.
14. Add PtdEtn conjugated polyclonal anti-mouse IgG diluted 1/250 in FC buffer, and incubate for 30 min on ice.
15. Wash three times with FC buffer.
16. Resuspend the cell pellet in 200 μL of PBS and analyze with FACS flow cytometry.

## 3.3. Analysis of the Peptide Sequences Displayed by the Positive Phages

Single-strand DNAs were prepared from positive phages *(34)*, and sequencing of the DNA inserts was carried out by automatic sequencing (ABIPRISM 310 Genetic Analyzer). The peptide sequences displayed by the positive phage clones shared a major core motif (**LTVxPWY**) that was not found in negative phages *(32)*. The phage that displayed the LTVSPWY exhibited the strongest binding and internalization. Therefore, this peptide was used for the delivery of antisense oligonucleotides into breast-cancer cell lines.

## 3.4. Synthetic Peptides and Antisense-Peptide Conjugates

To determine whether the SKBR3 -binding pepide LTVSPWY could be used to enhance the delivery of antisense oligonucleotides, a fluorescein-comjugated antisense oligonucleotide against the ErbB2 was designed and coupled via a disulfide bridge to the peptide (*see* **Note 9**). SKBR3 cells incubated with the antisense-peptide conjugates for 4 h at 37°C exhibited intracellular binding, yet no significant staining was obtained with the antisense alone (**Fig. 1C**). In contrast to the antisense alone, the antisense-peptide conjugates inhibited ErbB2 gene expression *(32)*.

## 4. Notes

1. Phage-display libraries are constructed both in lytic and non-lytic bacteriohages. The non-lytic M13 filamentous phage is the most commonly used vector. This phage is from the fd phage family, which infects Gram-negative bacteria by rec-

ognizing the F-pilus of their host strain. The coat proteins of M13, which are often used for fusion of library protein and peptides, are pIII, pVI, and pVIII (**Fig. 1A**). The pIII coat protein is often used for constructions of short peptide libraries. The M13 phage contains five copies of the pIII coat protein, but only one copy of this protein is required for infection. A classic M13 short peptide library constructed on the pIII coat protein contains 3–5 copies of the displayed foreign peptide, each fused to the N-terminus of each pIII protein.

2. The HMEC cells were used as a normal counterpart of SKBR3 cancer cells. Both the HMEC and SKBR3 cells originate form breast tissue and have epithelial morphology. PBMC was used to eliminate peptide binding to normal blood cells.

3. The Ph.D.7 phage library was purchased from New England Biolabs. The biopanning conditions are modified for selection of peptides that are taken up by the cells via receptor-mediated internalization. Notably, optimized panning conditions are important for successful ligand selection.

4. The phosphorothioate oligonucleotide against ErbB2 was described by Vaughn et al. *(36)*. The antisense and the control-scrambled oligonucleotide were fused to the synthetic peptides via the 3' SSpy group. The S-S bridge between the oligonucleotide and the synthetic peptide is expected to break up in the reducing conditions of the cytoplasm.

5. Various phage-display libraries were designed and found to be useful tools to study receptor-ligand interactions for a variety of target molecules. Successive selection of ligands from a phage-display library is called "Biopanning," which involves screening of a phage-display library against a target molecule. The high-affinity binders are then captured and amplified in 3-4 rounds of panning. Individual plaques of phage are analyzed for their specific binding and are sequenced. The target molecules for a biopanning assay can be known immobilized proteins or unknown like-surface receptors on certain cell lines (**Fig. 1B**).

6. We used RPMI medium containing the 10% FCS in the panning procedure to remove phages binding to serum proteins. The serum also functions as a blocker of nonspecific binders. It is crucial that the number of input phages do not exceed $1 \times 10^{10}$ pfu, since this will result in much higher background and increased nonspecific binding.

7. The acidic washing conditions were used to minimize the background and the nonspecific binding. Notably, many cell types do not endure these harsh conditions, and require a much more gentle treatment. The cells were treated with trypsin to remove the surface-binding phages and to select only the phages that are internalized.

8. M13 phage is extremely stable in an acidic environment, but elution in pH 2.2 must not exceed 10 min because it can destroy the phage. The eluted phages must be neutralized immediately with 2 *M* Tris-base.

9. A lysine residue was added to the C-terminus for antisense conjugation to the peptide. Peptides with cysteine residue at the C-terminus were directly conjugated to 6-1AF, as described by the manufacturer (Molecular Probes).

## References

1. Dachs, G. U., Dougherty, G. J., Stratford, I. J., and Chaplin, D. J. (1997) Targeting gene therapy to cancer: a review. *Oncol. Res.* **9,** 313–325.
2. Gewirtz, A. M., Sokol, D. L., and Ratajczak, M. Z. (1998) Nucleis acid therapeutics: State of the art and future prospects. *Blood.* **92,** 712–736.
3. Hurford. R. K. Jr., Dranoff, G., Mulligan, R. C., and Tepper, R. I. (1995) Gene therapy of metastatic cancer by in vivo retroviral gene targeting. *Nat Genet.* **10,** 430–435.
4. Robbins, P. D., Tahara, H., and Ghivizzani, S. C. (1998) Viral vectors for gene therapy. *Trends Biotechnol.* **16,** 35–40.
5. Carter, B. J. (2000) Adeno-associated viruses as gene transfer vehicles. *Contrib Microbiol.* **4,** 85–106.
6. Kasahara, N., Dozy, A. M., and Kan, Y. W. (1994) Tissue-specific targeting of retroviral vectors through ligand-receptor interactions. *Science.* **266,** 1373–1376.
7. Lee, R. J. and Huang, L. (1997) Lipidic vector systems for gene transfer. *Crit. Rev. Ther. Drug Carrier Syst.* **14,** 173–206.
8. Mahato, R. I. (1999) Non-viral peptide-based approaches to gene delivery. *J. Drug Target.* **7,** 249–268.
9. Morris, M. C., Chaloin, L. Heitz, F., and Divita, G. (2000) Translocating peptides and proteins and their use for gene delivery. *Curr. Opin. Biotechnol.* **11,** 461–466.
10. Aina, O. H., Sroka, T. C., Chen, M. L., and Lam, K. S. (2002) Therapeutic cancer targeting peptides, *Biopolymers* **66,** 184–199.
11. Pasqualini, R., Koivunen, E., and Ruoslahti, E. (1995) A peptide isolated from phage display libraries is a structural and functional mimic of an RGD-binding site on integrins. *J. Cell. Biol.* **130,** 1189–1196
12. Assa-Munt, N., Jia, X., Laakkonen, P., Ruoslahti, E. (2001) Structures and integrin binding activities of an RGD peptide with two isomers, *Biochemistry* **40,** 2373–2378.
13. Pasqualini, R., Koivunen, E., Kain, R., Lahdenranta, J., Sakamoto, M., Stryhn, A., et al. (2000) Aminopeptidase N is a receptor for tumor-homing peptides and a target for inhibiting angiogenesis. *Cancer Res.* **60,** 722–727.
14. Essler, M. and Ruoslahti, E. (2002) Molecular specialization of breast vasculature: a breast-homing phage-displayed peptide binds to aminopeptidase P in breast vasculature. *Proc. Natl. Acad. Sci. USA* **99,** 2252–2257.
15. Burg, M. A., Pasqualini, R., Arap, W., Ruoslahti, E., and Stallcup, W. B. (1999) NG2 proteoglycan-binding peptides target tumor neovasculature. *Cancer Res.* **59,** 2869–2874.
16. Arap, W., Pasqualini, R., and Ruoslahti, E. (1998) Cancer treatment by targeted drug delivery to tumor vasculature in a mouse model. *Science* **279,** 377–380.
17. Kennel, S. J., Mirzadeh, S., Hurst, G. B., Foote, L. J., Lankford, T. K., Glowienka, K. A., et al (2000) Labeling and distribution of linear peptides identified using in vivo phage display selection for tumors. *Nucl. Med. Biol.* **27,** 815–825.

18. Arap, W., Haedicke, W., Bernasconi, M., Kain, R., Rajotte, D., Krajewski, S., et al. (2002) Targeting the prostate for destruction through a vascular address. *Proc. Natl. Acad. Sci. USA* **99,** 1527–1531.

19. Binetruy-Tournaire, R., Demangel, C., Malavaud, B., Vassy, R., Rouyre, S., Kraemer, M., et al. (2000) Identification of a peptide blocking vascular endothelial growth factor (VEGF)-mediated angiogenesis. *EMBO J.* **19,** 1525–1533.

20. Bae, D. G., Gho, Y. S., Yoon, W. H., and Chae, C. B. (2000) Arginine-rich antivascular endothelial growth factor peptides inhibit tumor growth and metastasis by blocking angiogenesis. *J. Biol. Chem.* **275,** 13,588–13,596.

21. Oku, N., Asai, T., Watanabe, K., Kuromi, K., Nagatsuka, M., Kurohane, K., et al. (2002) Anti-neovascular therapy using novel peptides homing to angiogenic vessels. *Oncogene* **21,** 2662–2669.

22. Asai, T., Nagatsuka, M., Kuromi, K., Yamakawa, S., Kurohane, K., Ogino, K., et al. (2002) Suppression of tumor growth by novel peptides homing to tumor-derived new blood vessels. *FEBS Lett.* **510,** 206–210.

23. Koivunen, E., Arap, W., Valtanen, H., Rainisalo, A., Medina, O. P., Heikkila, P., et al. (1999) Tumor targeting with a selective gelatinase inhibitor. *Nat. Biotechnol.* **17,** 768–774.

24. Renschler, M. F., Bhatt, R. R., Dower, W. J., and Levy, R. (1994) Synthetic peptide ligands of the antigen binding receptor induce programmed cell death in a human B-cell lymphoma. *Proc. Natl. Acad. Sci. USA* **91,** 3623–3627.

25. Buhl, L., Szecsi, P. B., Gisselo, G. G., and Schafer-Nielsen, C. (2002) Surface immunoglobulin on B lymphocytes as a potential target for specific peptide ligands in chronic lymphocytic leukaemia. *Br. J. Haematol.* **116,** 549–554.

26. Fukuda, M. N., Ohyama, C., Lowitz, K., Matsuo, O., Pasqualini, R., Ruoslahti, E., and Fukuda, M. (2000) A peptide mimic of E-selectin ligand inhibits sialyl Lewis X-dependent lung colonization of tumor cells *Cancer Res.* **60,** 450–456.

27. Wu, P., Leinonen, J., Koivunen, E., Lankinen, H., and Stenman, U. H. (2000) Identification of novel prostate-specific antigen-binding peptides modulating its enzyme activity. *Eur. J. Biochem.* **267,** 6212–6220.

28. Romanov, V. I., Durand, D. B., and Petrenko, V. A. ( 2001) Phage display selection of peptides that affect prostate carcinoma cells attachment and invasion. *Prostate* **47,** 239–251.

29. Zhang, J., Spring, H., and Schwab, M. (2001) Neuroblastoma tumor cell-binding peptides identified through random peptide phage display. *Cancer Lett.* **171,** 153–164.

30. Hong, F. D. and Clayman, G. L. (2000) Isolation of a peptide for targeted drug delivery into human head and neck solid tumors. *Cancer Res.* **60,** 6551–6556.

31. Ivanenkov, V. V, Felici, F., and Menon, A. G. (1999) Targeted delivery of multivalent phage display vectors into mammalian cells. *Biochim. Biophys. Acta* **1448,** 463–472.

32. Shadidi, M. and Sioud, M. (2003) Identification of novel carrier peptides for the specific delivery of therapeutics into cancer cells. *FASEB J.* **17,** 256–258.

33. Karasseva, N. G., Glinsky, V. V., Chen, N. X., Komatireddy, R., and Quinn, T. P. (2002) Identification and characterization of peptides that bind human ErbB-2 selected from a bacteriophage display library. *J. Protein Chem.* **21**, 287–296.
34. Sambrook, J., Fritsch, E. F., and Maniatis T. (1989) *A Laboratory Manual, 2nd ed.* E3–E4.
35. Scott, J. K. and Smith G. P. (1990) Searching for peptide ligands with an epitope library. *Science.* **249,** 386–390.
36. Vaughn, J. P., Iglehart, J. D., Demirdji, S., Davis, P., Babiss, L. E., Caruthers, M. H., and Marks, J. R. (1995) Antisense DNA downregulation of the ERBB2 oncogene measured by a flow cytometric assay. *Proc. Natl. Acad. Sci. USA* **92,** 8338–8342.

# 45

## Clinical Gene Therapy Research Utilizing Ribozymes

*Application to the Treatment of HIV/AIDS*

**Frances K. Ngok, Ronald T. Mitsuyasu, Janet L. Macpherson, Maureen P. Boyd, Geoff P. Symonds, and Rafael G. Amado**

### Summary

Antiretroviral drug therapy can effectively reduce the viral load, and is associated with a degree of immune reconstitution in human immunodeficiency virus (HIV)-infected patients. However, the presence of a latent viral reservoir, the development of drug resistance, drug toxicity, and compliance problems are obstacles that impede full eradication of HIV through drug therapy. The cellular introduction of genetic elements that are capable of inhibiting HIV replication is conceptually appealing as a potential new treatment paradigm for aquired immunodeficiency syndrome (AIDS). In theory, this approach can lead to the development of regenerated hematopoiesis with cells that inhibit viral replication and are protected from the pathogenic effects of HIV. Ribozymes are catalytic RNA molecules that can efficiently and selectively cleave target RNA. By *ex vivo* retroviral transduction, we have introduced a HIV-1 *tat* gene-targeted ribozyme (RRz2) and a control construct (LNL6) into granulocyte-colony-stimulating factor (G-CSF) mobilized $CD34^+$ hematopoietic progenitor cells (HPC). Transduced autologous $CD34^+$ cells (an approximately equal mix of RRz2 and LNL6) were infused in 10 patients in this Phase I study. After a median follow-up of 2.5 yr, gene presence and expression were detected by a sensitive polymerase chain reaction (PCR) assay in a transduced-$CD34^+$ cell dose-dependent manner. In this chapter, we describe general considerations related to HIV hematopoietic progenitor-cell gene therapy trial design, implementation, and safety, with an emphasis on the critical steps of this process, namely vector production and characterization, target-cell selection, transduction, final product release testing, and evaluation of vector presence.

**Key Words:** Ribozyme; HIV; AIDS; gene therapy; hematopoietic stem cell; hematopoietic progenitor cell; ribozyme; hammerhead ribozyme; CD34 selection; CD34 transduction.

From: *Methods in Molecular Biology, vol. 252: Ribozymes and siRNA Protocols, Second Edition*
Edited by: M. Sioud © Humana Press Inc., Totowa, NJ

## 1. Introduction

Gene therapy can be conceptualized as a form of "intracellular immunization" *(1)*. Several novel gene-transfer approaches have been designed for the treatment of HIV-1 infection, and a number of reports have described the efficacy of these strategies in tissue-culture systems. Examples of these strategies include intracellular expression of transdominant proteins *(2–5)*, intracellular antibodies *(6,7)*, antisense ribonucleic acid (RNA) *(8–10)*, viral sequence decoys *(11–14)*, and catalytic ribozymes *(15–20)*. Clinical trials designed to test these gene therapy strategies using human hematopoietic cells as targets have been described *(14,21–24)*.

Ribozymes are small RNA moieties that can cleave specific RNA target molecules. For intracellular immunization against HIV-1, ribozymes offer several advantages over alternative anti-viral genes. Ribozymes can target several steps in the HIV-1 life cycle including i) genomic viral RNA in recently infected cells (prior to reverse transcription) and ii) viral RNA transcribed from the provirus before translation or prior to genomic RNA packaging *(18,25)*. RNA molecules are less likely to induce an immune response against cells that express the transduced gene than protein-based gene therapy strategies such as transdominant Rev or intracellular antibodies. Although antisense sequences may offer this same advantage, they are theoretically less effective than ribozymes because the latter are catalytic molecules that can bind to and cleave multiple RNA substrate molecules. The requirements for cleavage by a ribozyme are an accessible region of RNA, and in the case of the hammerhead ribozyme, a GUX target motif, (in certain cases, NUX may suffice in which N is any nucleotide and X is A, C, or U).

## 2. Preclinical Experience With Anti-HIV-1 Ribozymes

A number of laboratories (including our own) have demonstrated ribozyme cleavage activity in vitro, and protective effects in tissue-culture systems against laboratory and clinical isolates of HIV-1 *(15–20,26–31)*. These studies used either hammerhead or hairpin ribozymes. The ribozyme that was most effective in our studies was a hammerhead ribozyme moiety directed against a highly conserved region of the *tat* gene. The *tat* gene is essential for HIV-1 replication; it produces the Tat protein that is a transcriptional activator of an integrated HIV-1 provirus. This ribozyme is designated as Rz2 (**Fig. 1**). The Rz2 complementary hybridizing and target sequences comprise nucleotides 5865–5882 (GGAGCCA GUA GAUCCUA) of HIV-IIIB reference strain (Genbank accession number X01762).

For clinical experimentation, the Rz2 ribozyme sequence 5'-TTA GGA TCC TGA TGA GTC CGT GAG GAC GAA ACT GGC TC-3' was inserted as DNA

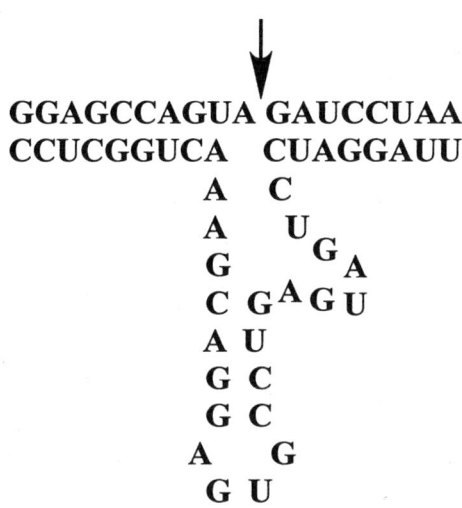

Fig. 1. Sequence of the hammerhead ribozyme Rz2, and representation of its complementary target and hybridizing sequence within the HIV-1 *tat* gene. The site of cleavage is indicated by an arrow.

into the 3' untranslated region (UTR) of the *neo*[R] gene within the replication-incompetent retroviral vector LNL6 (Genbank accession number M63653). This vector was known as RRz2. The ribozyme sequence is expressed as a *neo*[R]-ribozyme fusion transcript from the Moloney murine leukemia virus (MoMLV) long terminal repeat (LTR) in RRz2. In experiments with a number of laboratory strains (HIV-IIIB, HIV-SF2, and HIV-SF33) and clinical isolates (82H, P22, ASC3, PK2) of HIV-1, RRz2 has been shown to reduce HIV-1 replication and enhance cell viability in T-cell lines (CEMT4, and SupT1) and peripheral-blood lymphocytes (PBL) *(17,25,27)*. Compared to antisense or ribozymes directed against other HIV-1 regions, RRz2 showed superior inhibition of HIV-1 replication in T-cell lines, as well as in PBLs. This inhibition was consistently about 80% in multiple experiments. Against clinical isolates, the degree of inhibition of HIV-1 replication was greater than 2 logs in T-cell lines. In addition, in a resistance assay customarily used for anti-retroviral drugs, no resistance mutations in the Rz2 target region of HIV-1 developed during a 6-mo sequential passage assay *(27)*. In all of these studies, the expression of the ribozyme construct had no effect on cell viability, proliferation, phenotype, and—in an assay of monocytes derived from transduced CD34[+] cells—cellular differentiation.

In summary, Rz2 showed substantial anti-HIV-1 activity against both laboratory and clinical isolates in T-cell lines, and in PBLs, indicating that the expression of Rz2 in these cell types in vivo may lead to an anti-HIV effect. These results provided the foundation for clinical trials. A phase I study using this vector has been conducted, and preliminary results are described here.

## 3. Clinical Experience Using Ribozyme Gene Transfer Into Hematopoietic Progenitor Cells (HPC)

### 3.1. Use of CD34+ HPC as Targets for Gene Transfer

In adults, pluripotent HPCs are present in human bone marrow and peripheral blood (*32–35*). These cells can be collected from the bone marrow by aspiration or from peripheral blood by apheresis following mobilization from the bone-marrow compartment into the peripheral blood (*32*). Enrichment of early progenitor cells from the bone marrow or from the mobilized peripheral-blood progenitor-cell fraction can be achieved by selection of cells that express the CD34+ antigen. This membrane-bound 115-$K$d molecule is present on cells that are capable of giving rise to multilineage colony forming cells, and it is absent on mature hematopoietic cells (*36*). Results of transplant studies in primates and humans indicate that CD34+ enriched cells are capable of rapidly reconstituting lymphoid and myeloid hematopoiesis.

The introduction of a gene therapeutic into CD34+ cells *ex vivo* is an attractive approach for the treatment of genetic or infectious diseases that affect multiple hematopoietic lineages. In the case of the Rz2 ribozyme, such a strategy should result in ribozyme-construct presence and expression in both mature lymphoid and myeloid cells, rendering these cells at least partially protected from HIV-1 replication. Since it is known that CD34+ cells have reconstitution and repopulation potential, and given the evidence that most CD34+ cells are not directly infected by HIV-1 (*37,38*), targeting CD34+ cells should theoretically provide an ongoing source of protected cells within the patient.

The ability to transduce CD34+ cells with murine retroviruses depends in part on the cell-cycle status of the cells (*39*). We have demonstrated the introduction of RRz2 into mobilized CD34+ cells that are capable of differentiating into T-lymphocytes (*40*). High levels of transduction have been consistently observed when the recombinant fibronectin molecule CH-296 (or retronectin) has been used (*41–45*). Retronectin is a recombinant fragment of human fibronectin composed of the cell-binding domain (C-domain), heparin-binding domain II (H-domain), and CS1 site. Fibronectin enhances retrovirus-mediated gene transduction of CD34+ cells by co-localization. The RGD amino acid sequence in the C-domain is recognized through the integrin VLA-5, and the LDV amino acid sequence in the CS1-site is recognized by the integrin VLA-4. In addition, retrovirus particles bind to the H-domain of the protein. Consequently, retronectin can enhance transduction through binding to both retrovirus particles and target cells that express integrins VLA-5 and/or VLA-4.

## 3.2. Results Using Transduced CD34⁺ HPC in Clinical Trials

Outgrowth of a T-lymphocyte-cell population containing genes that produce a therapeutic protein has been demonstrated in gene-therapy studies of severe combined immunodeficiency (SCID) patients, using CD34⁺ HPC transduced with the adenosine deaminase (ADA) gene for ADA-SCID (46) or the gamma common chain-receptor gene for SCID-X1 (47,48). Although the ADA study did not result in a therapeutic benefit, mature T-lymphocytes carrying the ADA transgene accumulated in the peripheral blood, suggesting that preferential survival was conferred by the ADA gene product. The SCID-X1 study demonstrated that the genetically modified T-lymphocytes have a strong selective advantage. The transduced CD34⁺ cells differentiated and produced T-lymphocytes that proliferated and functioned in an apparently normal way, thereby correcting the immune deficiency. Reconstitution with a relatively normal number of T-lymphocytes and natural killer cells occurred within a period of 6 mo *(47,48)*. In the setting of HIV infection, one can envisage a survival advantage in the T-lymphocytes and monocytes (populations susceptible to HIV) that arise from the genetically modified CD34⁺ HP cells, without preferential survival occurring at the level of the CD34⁺ HPC. A potential strategy to enhance the number of gene-containing T-lymphocytes and monocytes is to increase the number of the transduced CD34⁺ cell dose until chimeric engraftment results in enough ribozyme-containing mature lymphoid (CD4⁺ T-lymphocytes) and myeloid (monocyte/macrophages) cells to impact on HIV disease progression.

## 3.3. Engraftment in a Non-Myeloablated Setting

Following myeloablation, CD34⁺ cells have been shown to reconstitute the bone marrow compartment in animals and humans *(49–51)*. In the absence of myeloablation, or in the setting of partial myeloablation, CD34⁺ infusion has been shown to produce a chimeric hematopoietic system *(52–54)*. Reconstitution of hematopoiesis was demonstrated in non-myeloablated dogs for up to 2 yr, indicating that myeloablation is not an absolute requirement to create "marrow space" for engraftment of infused cells *(55)*. In the rhesus macaque model, CD34⁺ cells transduced with retroviral vectors were infused into four animals that had undergone minimal myeloablative conditioning. In these animals, neutrophil recovery was found within 0–3 d, and there were no significant periods of severe thrombocytopenia. Marking of circulating granulocytes and mononuclear cells was extensive and durable in all animals (exceeding 12% in the mononuclear cells of one animal), and persisted beyond the final sampling time in all animals (up to 33 wk) *(53)*. More recently, the introduction of "mini-allogeneic" transplants has confirmed the rapid contribution of infused donor cells to hematopoietic chimerism, indicating that partial conditioning with immunosuppressive chemotherapy is sufficient to allow engraftment of allogenic progenitor cells *(56)*.

### 3.4. Results of a Phase I Clinical Trial Using an Anti-HIV Ribozyme

We conducted a Phase I study in ten patients using autologous CD34+ HPC collected from HIV-1 infected patients. The treatment schema of this study is outlined in **Fig. 2**. Following G-CSF mobilization and apheresis, CD34+ HPC were purified using a CD34-specific antibody. CD34+ cells were then cultured with the growth factors stem cell factor (50 ng/mL SCF) and megakaryocyte growth and development factor (100 ng/mL MGDF), and half the cells separately transduced on d 2 of culture (two separate transductions) with LNL6, and with RRz2 (an LNL6-based vector containing a hammerhead ribozyme directed against the *tat* region of HIV-1) *(22)*. At study entry, each patient had CD4+ counts between 300 and 700/mm3 and viral loads of <10,000 copies/ mL. Median follow-up was 2.5 yr. Gene presence was detected by a sensitive PCR assay. Relative amounts of vector were analyzed using mixed-effect linear regression.

No gene transfer-related serious adverse events were reported for any of the treated patients. No replication competent retrovirus (RCR) was detected by PCR. Long-term follow up and archival samples are collected according to the current Food and Drug Administration (FDA) recommendations. The degree and persistence of gene marking in progeny cells was dependent on the cell dose infused with a minimum dose of $0.5 \times 10^6$ marked CD34+ cells/kg required for long-term gene persistence. Expression of both constructs was detected in peripheral-blood mononuclear cells (PBMC) obtained more than 1 yr after CD34+ infusion. Marking persisted in multiple lineages and in mature and naïve T-cells, up to 3 yr post-infusion. Low levels of gene presence in naïve thymocytes were observed, even in patients with high levels of viremia. These findings demonstrate that *de novo* T-cell development can occur from genetically engineered hematopoietic progenitors in adult HIV-infected patients, supporting the concept of gene therapy as a modality to effect immune reconstitution.

### 4. Main Steps in the Implementation of HPC Gene Therapy Clinical Study

This section is intended to highlight the most critical steps and laboratory procedures required to support clinical gene therapy research. Methodological details of these procedures are provided in Chaper 46 of this volume. The following is a brief description of the steps involved.

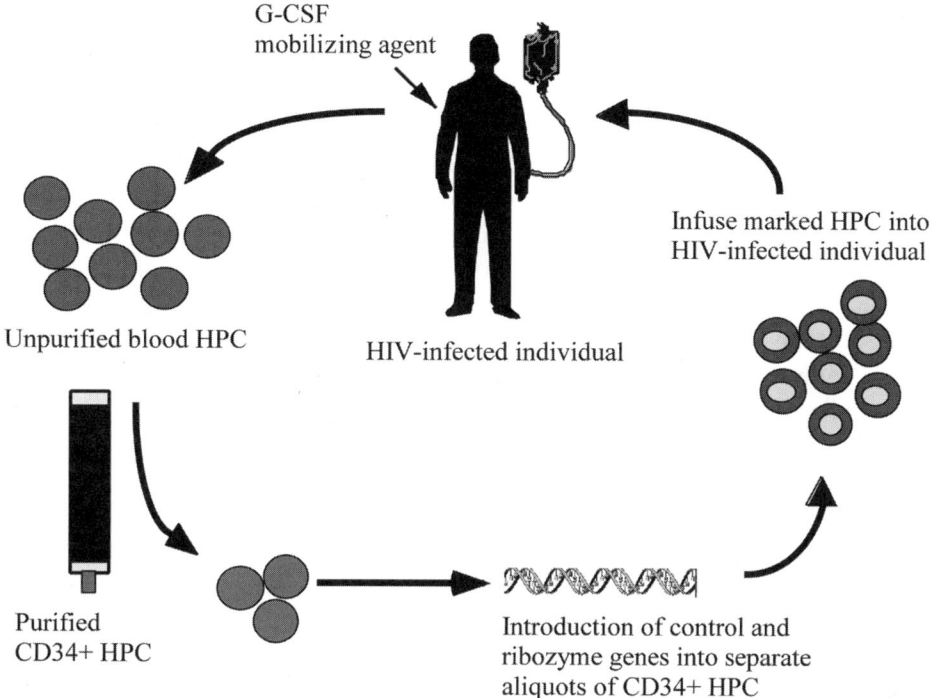

Fig. 2. Schematic representation of a HPC clinical trial using a ribozyme containing anti-HIV vector.

After obtaining informed consent and establishing patient eligibility, a date is set for protocol initiation. In our Phase I trial, patients receive G-CSF from d 1–5 to promote mobilization of the HPCs from the bone marrow. On d 5, patients continue to receive G-CSF, and undergo the first apheresis. The apheresis product containing HPC-enriched mononuclear cells is washed and stored in a good manufacturing process (GMP) facility. This procedure is repeated on d 6. The second apheresis product is then combined with the apheresis product collected on d 5 and the CD34⁺ HPC are isolated using the Isolex 300i, a system for immunomagnetic, large-scale purification of CD34⁺ cells (Baxter). The CD34⁺ cell-enriched product is cultured overnight in culture medium containing SCF and MGDF. On d 7 and 8, the cultured cells are divided in two aliquots of equal numbers, and are separately transduced with culture supernatant that contains one or other of the two retroviral vectors (LNL6 and RRz2). On d 9, the transduced progenitor cells are harvested by washing, and quality-control release testing is performed before the final product is infused into the patient. During the post-infusion period, patients are evaluated at predetermined time-points to assess study end points, including safety and gene marking.

## 4.1. GMP Production of Virus

For clinical applications, the virus production process begins with the manufacture of the producer-cell lines. Clonal amphotropic packaging-cell lines are generated using an ecotropic packaging-cell line and an amphotropic packaging-cell line. These are mammalian-cell lines selected for high retrovirus expression, following transfection with one or two plasmids, collectively containing the *gag/pol/env* genes. In a split-genome packaging-cell line, these plasmids are introduced independently. For safety, the plasmids contain mutations, such as deletion of the retroviral packaging signal, the 3' LTR, part of the 5' LTR, and the site for second-strand DNA synthesis. The location of the replicative genes on different plasmids and the mutations introduced in these plasmids reduce the likelihood of recombination events leading to RCR. A total of three independent recombination events would be required to permit the formation of RCR in a split-genome packaging line *(57)*.

For the production of retroviral vectors, the engineered producer-cell line is cloned, and high-titer producer clones are screened for provirus integrity and safety. These cells are engineered by inserting the MMLV-based vector containing the gene of interest using transfection methods. The initial cell line produces ecotropic vectors that infect the amphotropic-cell line. Such cell lines produce vectors that are capable of infecting cells from multiple mammalian species, including human cells. By single-cell cloning, a high-producer amphotropic clone is selected. The integrity of the inserted provirus in the selected clone, and the number of copies per cell, are verified by Southern blot using specific restriction enzymes. The growth characteristics of the cell lines are monitored, and the viral titer is usually determined by transfer of a marker gene such as neomycin (G418) resistance using NIH-3T3 cells as targets. Absence of RCR is evaluated by a marker rescue assay. Mycoplasma testing is conducted using biological assays, and sequencing of retroviral RNA is performed to verify the integrity of vector RNA. A Master Cell Bank (MCB) is prepared under GMP conditions. Vials of cells from the MCB are used for production of clinical-grade vector under GMP conditions. All manufacture and testing must be carried out using standardized and validated methods. The testing steps commonly employed prior to virus certification are outlined in Chapter 46 of this volume.

## 4.2. CD34+ Cell Mobilization

Mobilization of CD34+ cells from the bone marrow to the peripheral circulation is performed by the use of hematopoietic growth factors, such as G-CSF. G-CSF is generally administered subcutaneously at a dose of 10 µg/kg/d, for six consecutive days. Bone pain is a common side effect of this treatment.

Complete blood counts (CBCs) and differential and platelet counts are performed daily from d 2–6 of G-CSF administration to assess the extent of the leukocytosis and the effects of G-CSF on platelet counts. Blood samples for CD34$^+$ cell count (performed by an standardized validated flow cytometry assay) are generally drawn on d 4, d 5 (first apheresis procedure), and d 6 (second apheresis procedure) of G-CSF administration to evaluate the efficacy of CD34$^+$ cell mobilization. G-CSF is discontinued before d 6 if bone pain is severe (only responds partially to opioid analgesia), or if the patient develops leukocytosis greater than 75,000 cells/mm$^3$.

## 4.3. Aphaeresis

Before study entry, patients are evaluated for antecubital-vein quality at screening. Those who have inadequate peripheral-vein access—defined as access that is unlikely to tolerate the required inlet-flow rate—are asked to consider insertion of a large-bore catheter into a central vein. This procedure is performed under a separate consent. Non-gene therapy studies have shown that the time to, and degree of engraftment is dependent of the number of CD34$^+$ cells infused. The number of apheresis procedures varies with every protocol, but in general, studies should aim to collect at least a number of CD34$^+$ cells known to allow reconstitution of a myeloablated host ($2 \times 10^6$ CD34$^+$ cells/kg of body wt). In our study, cells are collected from an apheresis filtration volume of 10 L during each procedure. We have used a target inlet-flow rate of 55–85 mL/min and a product volume collection rate of 1 mL/min to achieve maximal cell numbers over a run time of less than 3 h. The system that we have used for progenitor-cell harvesting is the COBE Spectra Apheresis System (Gambro BCT, Inc. Lakewood, CO). If the inlet flow rate is not achievable during apheresis, the rate is reduced and the duration of the apheresis is extended. The volume of the final apheresis product is generally between 120 and 220 mL using these conditions. The cells obtained from each day's apheresis are transferred to a laboratory where the first days' apheresis product is stored overnight. Universal precautions are followed at all times by appropriately trained personnel in Good Laboratory Practice and GMP standards.

## 4.4. CD34$^+$ HPC Selection

In some protocols, the apheresis product from the first day's collection is washed and resuspended in autologous plasma for overnight storage. A preliminary debulking step can be introduced prior to storage to remove excess granulocytes and platelets. This procedure generally involves gradient centrifugation, and results in a several-fold CD34$^+$ cell enrichment. Cells from the second day's apheresis are washed separately (and de-bulked if applicable), and then pooled with product from the first day prior to CD34$^+$ cell enrichment. Other protocols call for individual separations of the apheresis product, particularly when each collection is cell-rich, surpassing the target cell-loading dose of the selection device.

The procedure for CD34 HPC enrichment generally utilizes large-scale CD34 purification devices. Of these, the device that is most commonly used and the one we have employed is the Isolex 300i (Baxter). This system has a maximum loading capacity of $8 \times 10^{10}$ mononuclear cells. The key principles of the positive cell selection process, as described in the Operator's Manual, are: sensitization, capture/rosette, separation, and release.

- Sensitization: The apheresis product is incubated with the primary antibody (anti-CD34).
- Capture/Rosette: After washing to remove the unbound antibody, Dynabeads® M-450 sheep anti-mouse IgG are mixed with the cell suspension. Dynabeads® M-450 are magnetic beads coated with a secondary antibody (sheep anti-mouse IgG).
- Separation: The rosetted cells are separated form the unbound cells by applying a magnetic field to the suspension chamber.
- Release: After removing non-target cells by washing, a releasing agent is added to separate the antibodies/beads from CD34+ cells. The beads and associated antibodies are retained within the disposable chamber by the magnetic field. The separated CD34+ cells are then washed to remove residual reagents, such as mouse and sheep antibodies, and collected for incubation.

### 4.5. Gene Transfer Into CD34+ Cells

Prior to transduction, the enriched CD34+ cells are cultured overnight under sterile, conditions. To enhance survival and to induce cell-cycle required for MMLV transduction, the CD34+ cells are cultured in the following growth factors: 50 ng/mL of SCF and 100 ng/mL of MGDF *(40)*. The transduction facilitating agent retronectin (3.5 µg/mL) is used to coat the tissue-culture vessels.

After 1 d of incubation, the culture medium is removed and GMP-certified viral supernatant supplemented with aforementioned growth factors is added to the culture. This transduction step takes place in retronectin-coated tissue-culture flasks or gas-permeable bags. The volume of viral supernatant added to the cells is calculated according to viral titer, cell number, and target multiplicity of infection (MOI = number of viral particles per cell) planned for the specific protocol. In general, MOI in excess of five are desirable. To optimize gene transfer, the transduction procedure is repeated in most protocols at least once. The interval between transduction is variable, but in general, greater than 4 h is allowed to ensure vector entry. Each transduction is performed with careful attention to minimize cell loss.

### 4.6. Testing and Infusion of Final Cell Product in HIV Gene Therapy

On the day of cell infusion, the cells are harvested, washed, and resuspended into RPMI medium (without phenol red) containing 5% human serum albumin for a final infusion volume of 50–100 mL. The following tests are generally performed on the cultured CD34$^+$ cell product, prior to release of the final product for patient infusion:

1. Endotoxin testing
2. Gram stain
3. Mycoplasma PCR assay

The following tests may be completed after infusion, although some may be required by regulatory agencies prior to cell infusion:

1. HIV p24 antigen enzyme-linked immunosorbent assay (ELISA) in pre- and post-CD34$^+$ cell-transduction culture supernatants. This assay is performed to ensure that HIV replication has not occurred during the culture period.
2. Sterility culture for both aerobic and anaerobic bacteria, or fungal contamination and Mycoplasma.
3. Biological RCR testing in supernatant and in transduced cells using *Mus dunni* amplification and co-cultivation assay, respectively, followed by subsequent S$^+$ L- end point assay.
4. Colony-forming unit (CFU-GM) assay for evaluation of HPCs.
5. Archival samples are stored for future analyses as needed, per FDA Center for Biologics Evaluation and Research (CBER) requirements.
6. Flow cytometry to evaluate CD34 purity.

After the product is certified to meet release criteria, it is delivered to the investigator or the study nurse for infusion. Patients are generally premedicated with an oral dose of acetaminophen and intravenous (iv) doses of diphenhydramine and hydrocortisone. After verification of patient identity by two operators, a test dose of 1% of the volume is administered intravenously. After 1-h observation period, and in the absence of untoward effects, the remaining volume is infused over approx 10 min. The infusion bag can then be rinsed with 50 mL normal saline, and this volume is infused, again over a period of 5–10 min.

Vital signs are monitored every 15 min during the test dose and for 1 h after cell infusion. The patient is monitored for a variable period of time thereafter for possible evidence of allergic reactions. Any adverse events are graded according to standardized charts, and are recorded. In most protocols, infusions are performed in an ambulatory setting.

### 4.7. Patient Monitoring for Assessment of Gene Construct(s)

Post-study safety/efficacy followup occurs at specified intervals during the Study period, usually at least quarterly in the first 12 mo post-infusion. As per FDA/CBER requirements for subjects who participate in gene-transfer studies, patients are requested to submit a blood sample on at least an annual basis after study completion. These samples are obtained for RCR testing, archival storage of serum and PBMCs, and investigation of integration events. At the time of their visit, patients are asked about the general status of their health and any adverse events they may have experienced since their previous visit. Patients who withdraw from the study at any point enter long-term annual post-study followup, as per the FDA/CBER guidelines.

### 4.8. Detection of Cells Containing and Expressing the Gene Construct(s)

We have used a sensitive PCR method to evaluate cells that contain and express the gene constructs. This PCR method involves nested reactions using primers located within the $neo^R$ gene, and a radioactive primer in the second reaction. Assay for gene presence and expression is generally performed in total PBMC, bone-marrow mononuclear cells, and at some time-points, in hematopoietic population subsets. Phenotypic separation of hematopoietic cells, such as T-cell subsets ($CD4^+$, $CD8^+$, naïve and memory phenotypes), monocytes and granulocytes is performed by gradient centrifugation (granulocytes), or by immune methods such as flow cytometry sorting, or magnetic bead immunoselection. The latter technique generally requires the use of two or more selection passages to ensure purity percentages greater than 90% (MACS, Miltenyi; Biotech Inc., Auburn, CA).

### 5. Safety Considerations in Retrovirus-Mediated HPC Clinical Research

Just as the promise of gene therapy is realized in the SCID-X1 study mentioned above, severe toxicity has also begun to emerge *(58)*. Patients in the SCID XI study developed a functional immune system in which 100% of T-lymphocytes carried the retroviral vector (a MMLV-derived vector) containing the gamma-common chain gene *(47,48)*. In contrast, only about 0.1% of cells in the myeloid lineage were found to have the vector, indicating that cells that require the transgene to confer a survival signal had a significant proliferative advantage. T-cells were found to be polyclonal, as evidenced by integration analysis and T-cell-receptor repertoire studies, and generally contained one copy of the provirus per cell. Tempering the apparent therapeutic success in this disease indication, 2 of the 11 children treated under this protocol developed a form of T-cell acute lymphocytic leukemia (ALL) approx 3 yr postinfusion *(59,60)*. For the two children, T-cell-receptor analysis revealed a monoclonal gamma delta and an oligoclonal alpha beta populations that contained intact integrated provirus in, or near the LMO-2 gene, respectively. LMO-2 is a proto-oncogene rearranged in translocations associated with childhood T-cell leukemia. LMO-2 transcripts were detected in leukemic blasts in these children. At the time the leukemic events had become clinically apparent, karyotypic abnormalities were also present.

These reports represent the first cases of malignant transformation caused by vector insertion in a human gene-therapy study. Before this, insertional mutagenesis had been described in a study in which rhesus macaques received a cell preparation containing RCR *(61)*. In the above human SCID XI study, RCR has not been demonstrated to date. An obvious question that is currently being considered is why this complication related to insertional mutagenesis had not been observed in any of about 217 previous studies using similar retroviral vectors involving more than 1700 patients worldwide. Indeed, some of these studies utilized HPC. The reason(s) that no vector safety issues have previously arisen may include: i) decreased integration events resulting from lower transduced-cell numbers, thereby reducing the potential for transforming mutations, and ii) the combination of SCIDXI phenotype, plus the use of the gamma common chain gene that provides a strong proliferative advantage that drives expansion of the cells, including those that contain an insertional event. At the time of writing, the children who developed leukemia in this study appear to have responded well to chemotherapy. However, their long-term prognosis is presently unknown. Regulatory authorities together with the wider scientific community are currently evaluating the impact of these events on ongoing and future hematopoietic stem/progenitor-cell retrovirus-mediated gene-transfer studies. As part of this process of evaluation, the American Society of Gene Therapy conducted a survey to review serious adverse events (SAE) related to retroviral vector use. To date, no SAE related to retroviral vectors have been reported in any other clinical study. This review included patients that received similar doses of transduced CD34$^+$ cells as the patients in the SCID-XI study *(63)*.

Other potential safety considerations for genetic manipulation of progenitor cells include the potential development of immune responses against cells expressing transgene-encoded proteins. Such a response could result in the elimination of the genetically corrected HPC and their progeny. Cell-mediated immunity against a transferred gene has been reported to occur in patients in a CD8$^+$ T-cell HIV-specific cytotoxic T-lymphocyte study in which the cells had been engineered to contain hygromycin/thymidine kinase genes. The elimination of transduced cells that was observed was associated with the emergence of specific CTLs against this marker *(62)*. This has not been reported to date for the *neo*$^R$ gene. A minimal cell dose, protein presentation, and immunocompetence are probable requisites for the development of this complication. Another potential risk associated with HPC gene transfer techniques using retroviral vectors is the transfer of genetic material to the germline. In the case of *ex vivo* cell manipulation, such an event would require the generation of RCR, and vector preparations are screened for RCR negativity. Aside from these potential toxic effects, there are unforeseen risks that may occur as this technology matures. Investigators must proceed cautiously and work in collaboration with experts in this field and regulatory bodies in order to maximize safety and therapeutic potential.

## References

1. Baltimore, D. (1988) Gene therapy. Intracellular immunization. *Nature* **335,** 395,396.
2. Bahner, I., Zhou, C., Yu, X. J., Hao, Q. L., Guatelli, J. C., and Kohn, D. B. (1993) Comparison of trans-dominant inhibitory mutant human immunodeficiency virus type 1 genes expressed by retroviral vectors in human T lymphocytes. *J Virol* **67,** 3199–3207.
3. Buchschacher, G. L., Jr., Freed, E. O., and Panganiban, A. T. (1995) Effects of second-site mutations on dominant interference by a human immunodeficiency virus type 1 envelope glycoprotein mutant. *J Virol* **69,** 1344–1348.
4. Fox, B. A., Woffendin, C., Yang, Z. Y., San, H., Ranga, U., Gordon, D., et al. (1995) Genetic modification of human peripheral blood lymphocytes with a transdominant negative form of Rev: safety and toxicity. *Hum Gene Ther* **6,** 997–1004.
5. Smythe, J. A., Sun, D., Thomson, M., Markham, P. D., Reitz, M. S., Jr., Gallo, R. C., and Lisziewicz, J. (1994) A Rev-inducible mutant gag gene stably transferred into T lymphocytes: an approach to gene therapy against human immunodeficiency virus type 1 infection. *Proc. Natl. Acad. Sci. USA* **91,** 3657–3661.
6. Marasco, W. A., Haseltine, W. A., and Chen, S. Y. (1993) Design, intracellular expression, and activity of a human anti-human immunodeficiency virus type 1 gp120 single-chain antibody. *Proc. Natl. Acad. Sci. USA* **90,** 7889–7893.
7. Mhashilkar, A. M., Bagley, J., Chen, S. Y., Szilvay, A. M., Helland, D. G., and Marasco, W. A. (1995) Inhibition of HIV-1 Tat-mediated LTR transactivation and HIV-1 infection by anti-Tat single chain intrabodies. *EMBO J.* **14,** 1542–1551.
8. Sczakiel, G. and Pawlita, M. (1991) Inhibition of human immunodeficiency virus type 1 replication in human T cells stably expressing antisense RNA. *J. Virol.* **65,** 468–472.
9. Rhodes, A. and James, W. (1991) Inhibition of heterologous strains of HIV by antisense RNA. *AIDS* **5,** 145–151.
10. Joshi, S., Van Brunschot, A., Asad, S., van der Elst, I., Read, S. E., and Bernstein, A. (1991) Inhibition of human immunodeficiency virus type 1 multiplication by antisense and sense RNA expression. *J. Virol.* **65,** 5524–5530.
11. Sullenger, B. A., Gallardo, H. F., Ungers, G. E., and Gilboa, E. (1990) Overexpression of TAR sequences renders cells resistant to human immunodeficiency virus replication. *Cell* **63,** 601–608.
12. Lisziewicz, J., Sun, D., Smythe, J., Lusso, P., Lori, F., Louie, A., et al. (1993) Inhibition of human immunodeficiency virus type 1 replication by regulated expression of a polymeric Tat activation response RNA decoy as a strategy for gene therapy in AIDS. *Proc. Natl. Acad. Sci. USA* **90,** 8000–8004.
13. Gervaix, A., Li, X., Kraus, G., and Wong-Staal, F. (1997) Multigene antiviral vectors inhibit diverse human immunodeficiency virus type 1 clades. *J. Virol.* **71,** 3048–3053.

14. Kohn, D. B., Bauer, G., Rice, C. R., Rothschild, J. C., Carbonaro, D. A., Valdez, P., et al. (1999) A clinical trial of retroviral-mediated transfer of a rev-responsive element decoy gene into CD34(+) cells from the bone marrow of human immunodeficiency virus-1-infected children. *Blood* **94,** 368–371.

15. Sun, L. Q., Warrilow, D., Wang, L., Witherington, C., Macpherson, J., and Symonds, G. (1994) Ribozyme-mediated suppression of Moloney murine leukemia virus and human immunodeficiency virus type I replication in permissive cell lines. *Proc. Natl. Acad. Sci. USA* **91,** 9715–9719.

16. Sun, L. Q., Wang, L., Gerlach, W. L., and Symonds, G. (1995) Target sequence-specific inhibition of HIV-1 replication by ribozymes directed to tat RNA. *Nucleic Acids Res* **23,** 2909–2913.

17. Sun, L. Q., Pyati, J., Smythe, J., Wang, L., Macpherson, J., Gerlach, W., and Symonds, G. (1995) Resistance to human immunodeficiency virus type 1 infection conferred by transduction of human peripheral blood lymphocytes with ribozyme, antisense, or polymeric trans-activation response element constructs. *Proc. Natl. Acad. Sci. USA* **92,** 7272–7276.

18. Sarver, N., Cantin, E. M., Chang, P. S., Zaia, J. A., Ladne, P. A., Stephens, D. A., and Rossi, J. J. (1990) Ribozymes as potential anti-HIV-1 therapeutic agents. *Science* **247,** 1222–1225.

19. Weerasinghe, M., Liem, S. E., Asad, S., Read, S. E., and Joshi, S. (1991) Resistance to human immunodeficiency virus type 1 (HIV-1) infection in human CD4+ lymphocyte-derived cell lines conferred by using retroviral vectors expressing an HIV-1 RNA-specific ribozyme. *J. Virol.* **65,** 5531–5534.

20. Yu, M., Ojwang, J., Yamada, O., Hampel, A., Rapapport, J., Looney, D., and Wong-Staal, F. (1993) A hairpin ribozyme inhibits expression of diverse strains of human immunodeficiency virus type 1. *Proc. Natl. Acad. Sci. USA* **90,** 6340–6344.

21. Morgan, R. A. and Walker, R. (1996) Gene therapy for AIDS using retroviral mediated gene transfer to deliver HIV-1 antisense TAR and transdominant Rev protein genes to syngeneic lymphocytes in HIV-1 infected identical twins. *Hum. Gene Ther.* **7,** 1281–1306.

22. Amado, R. G., Mitsuyasu, R. T., Symonds, G., Rosenblatt, J. D., Zack, J., Sun, L. Q., et al. (1999) A phase I trial of autologous CD34+ hematopoietic progenitor cells transduced with an anti-HIV ribozyme. *Hum. Gene Ther.* **10,** 2255–2270.

23. Cooper, D., Penny, R., Symonds, G., Carr, A., Gerlach, W., Sun, L. Q., and Ely, J. (1999) A marker study of therapeutically transduced CD4+ peripheral blood lymphocytes in HIV discordant identical twins. *Hum. Gene Ther.* **10,** 1401–1421.

24. Ranga, U., Woffendin, C., Verma, S., Xu, L., June, C. H., Bishop, D. K., and Nabel, G. J. (1998) Enhanced T cell engraftment after retroviral delivery of an antiviral gene in HIV-infected individuals. *Proc. Natl. Acad. Sci. USA* **95,** 1201–1206.

25. Sun, L. Q., Gerlach, W. L., and Symonds, G. (1998) The design, production and validation of an anti-HIV type1 ribozyme, in *Therapeutic Application of Ribozymes*, volume 11, Humana Press, Totowa, NJ, pp. 51–64.

26. Klebba, C., Ottmann, O. G., Scherr, M., Pape, M., Engels, J. W., Grez, M., et al. (2000) Retrovirally expressed anti-HIV ribozymes confer a selective survival advantage on CD4+ T cells in vitro. *Gene Ther.* **7,** 408–416.

27. Wang, L., Witherington, C., King, A., Gerlach, W. L., Carr, A., Penny, R., et al. (1998) Preclinical characterization of an anti-tat ribozyme for therapeutic application. *Hum. Gene Ther.* **9,** 1283–1291.
28. Jackson, W. H., Jr., Moscoso, H., Nechtman, J. F., Galileo, D. S., Garver, F. A., and Lanclos, K. D. (1998) Inhibition of HIV-1 replication by an anti-tat hammerhead ribozyme. *Biochem Biophys Res Commun* **245,** 81–84.
29. Koizumi, M., Ozawa, Y., Yagi, R., Nishigaki, T., Kaneko, M., Oka, S., et al. (1995) Design and anti-HIV-1 activity of ribozymes that cleave HIV-1 LTR. *Nucleic Acids Symp. Ser.* **34,** 125,126.
30. Chen, C. J., Banerjea, A. C., Harmison, G. G., Haglund, K., and Schubert, M. (1992) Multitarget-ribozyme directed to cleave at up to nine highly conserved HIV-1 env RNA regions inhibits HIV-1 replication—potential effectiveness against most presently sequenced HIV-1 isolates. *Nucleic Acids Res.* **20,** 4581–4589.
31. Yu, M., Poeschla, E., Yamada, O., Degrandis, P., Leavitt, M. C., Heusch, M., Yees, J. K., Wong-Staal, F., and Hampel, A. (1995) In vitro and in vivo characterization of a second functional hairpin ribozyme against HIV-1. *Virology* **206,** 381–386.
32. Hohaus, S., Goldschmidt, H., Ehrhardt, R., and Haas, R. (1993) Successful autografting following myeloablative conditioning therapy with blood stem cells mobilized by chemotherapy plus rhG-CSF. *Exp. Hematol.* **21,** 508–514.
33. Quesenberry, P. and Levitt, L. (1979) Hematopoietic stem cells. *N. Engl. J. Med.* **301,** 755–761.
34. Srour, E. F., Brandt, J. E., Briddell, R. A., Leemhuis, T., van Besien, K., and Hoffman, R. (1991) Human CD34+ HLA-DR- bone marrow cells contain progenitor cells capable of self-renewal, multilineage differentiation, and long-term in vitro hematopoiesis. *Blood Cells* **17,** 287–295.
35. Lowry, P. A. and Tabbara, I. A. (1992) Peripheral hematopoietic stem cell transplantation: current concepts. *Exp. Hematol.* **20,** 937–942.
36. Andrews, R. G., Singer, J. W., and Bernstein, I. D. (1986) Monoclonal antibody 12-8 recognizes a 115-kd molecule present on both unipotent and multipotent hematopoietic colony-forming cells and their precursors. *Blood* **67,** 842–845.
37. Slobod, K. S., Bennett, T. A., Freiden, P. J., Kechli, A. M., Howlett, N., Flynn, P. M., et al. (1996) Mobilization of CD34+ progenitor cells by granulocyte colony-stimulating factor in human immunodeficiency virus type 1-infected adults. *Blood* **88,** 3329–3335.
38. De Luca, A., Teofili, L., Antinori, A., Iovino, M. S., Mencarini, P., Visconti, E., et al. (1993) Haemopoietic CD34+ progenitor cells are not infected by HIV-1 in vivo but show impaired clonogenesis. *Br. J. Haematol.* **85,** 20–24.
39. Miller, D. G., Adam, M. A., and Miller, A. D. (1990) Gene transfer by retrovirus vectors occurs only in cells that are actively replicating at the time of infection. *Mol. Cell Biol.* **10,** 4239–4242.
40. Amado, R. G., Symonds, G., Jamieson, B. D., Zhao, G., Rosenblatt, J. D., and Zack, J. A. (1998) Effects of megakaryocyte growth and development factor on survival and retroviral transduction of T lymphoid progenitor cells. *Hum. Gene Ther.* **9,** 173–183.

41. Williams, D. A. and Moritz, T. (1994) Umbilical cord blood stem cells as targets for genetic modification: new therapeutic approaches to somatic gene therapy. *Blood Cells* **20,** 504–515.

42. Hanenberg, H., Hashino, K., Konishi, H., Hock, R. A., Kato, I., and Williams, D. A. (1997) Optimization of fibronectin-assisted retroviral gene transfer into human CD34+ hematopoietic cells. *Hum. Gene Ther.* **8,** 2193–2206.

43. Malech, H. L., Maples, P. B., Whiting-Theobald, N., Linton, G. F., Sekhsaria, S., Vowells, S. J., Li, F., Miller, J. A., et al. (1997) Prolonged production of NADPH oxidase-corrected granulocytes after gene therapy of chronic granulomatous disease. *Proc. Natl. Acad. Sci. USA* **94,** 12,133–12,138.

44. Bauer, T. R., Schwartz, B. R., Liles, W. C., Ochs, H. D., and Hickstein, D. D. (1998) Retroviral-mediated gene transfer of the leukocyte integrin CD18 into peripheral blood CD34+ cells derived from a patient with leukocyte adhesion deficiency type 1. *Blood* **91,** 1520–1526.

45. Rosenzweig, M., MacVittie, T. J., Harper, D., Hempel, D., Glickman, R. L., Johnson, R. P., et al. (1999) Efficient and durable gene marking of hematopoietic progenitor cells in nonhuman primates after nonablative conditioning. *Blood* **94,** 2271–2286.

46. Kohn, D. B., Hershfield, M. S., Carbonaro, D., Shigeoka, A., Brooks, J., Smogorzewska, E. M., et al. (1998) T lymphocytes with a normal ADA gene accumulate after transplantation of transduced autologous umbilical cord blood CD34+ cells in ADA- deficient SCID neonates. *Nat. Med.* **4,** 775–780.

47. Cavazzana-Calvo, M., Hacein-Bey, S., de Saint Basile, G., Gross, F., Yvon, E., Nusbaum, P., et al. (2000) Gene therapy of human severe combined immunodeficiency (SCID)-X1 disease. *Science* **288,** 669–672.

48. Hacein-Bey-Abina, S., Le Deist, F., Carlier, F., Bouneaud, C., Hue, C., De Villartay, J. P., et al. (2002) Sustained correction of X-linked severe combined immunodeficiency by ex vivo gene therapy. *N. Engl. J. Med.* **346,** 1185–1193.

49. Bomberger, C., Singh-Jairam, M., Rodey, G., Guerriero, A., Yeager, A. M., Fleming, W. H., et al. (1998) Lymphoid reconstitution after autologous PBSC transplantation with FACS- sorted CD34+ hematopoietic progenitors. *Blood* **91,** 2588–2600.

50. Dunbar, C. E., Seidel, N. E., Doren, S., Sellers, S., Cline, A. P., Metzger, M. E., Agricola, B. A., et al. (1996) Improved retroviral gene transfer into murine and Rhesus peripheral blood or bone marrow repopulating cells primed in vivo with stem cell factor and granulocyte colony-stimulating factor. *Proc. Natl. Acad. Sci. USA* **93,** 11,871–11,876.

51. Nachbaur, D., Fink, F. M., Nussbaumer, W., Gachter, A., Kropshofer, G., Ludescher, C., and Niederwieser, D. (1997) CD34+-selected autologous peripheral blood stem cell transplantation (PBSCT) in patients with poor-risk hematological malignancies and solid tumors. A single-centre experience. *Bone Marrow Transplant.* **20,** 827–834.

52. Stewart, F. M., Crittenden, R. B., Lowry, P. A., Pearson-White, S., and Quesenberry, P. J. (1993) Long-term engraftment of normal and post-5-fluorouracil murine marrow into normal nonmyeloablated mice. *Blood* **81,** 2566–2571.

53. Huhn, R. D., Tisdale, J. F., Agricola, B., Metzger, M. E., Donahue, R. E., and Dunbar, C. E. (1999) Retroviral marking and transplantation of rhesus hematopoietic cells by nonmyeloablative conditioning. *Hum Gene Ther* **10**, 1783–1790.

54. Barquinero, J., Kiem, H. P., von Kalle, C., Darovsky, B., Goehle, S., Graham, T., et al. (1995) Myelosuppressive conditioning improves autologous engraftment of genetically marked hematopoietic repopulating cells in dogs. *Blood* **85**, 1195–1201.

55. Bienzle, D., Abrams-Ogg, A. C., Kruth, S. A., Ackland-Snow, J., Carter, R. F., Dick, J. E., et al. (1994) Gene transfer into hematopoietic stem cells: long-term maintenance of in vitro activated progenitors without marrow ablation. *Proc. Natl. Acad. Sci. USA* **91**, 350–354.

56. Giralt, S., Khouri, I., and Champlin, R. (1999) Non myeloablative "mini transplants". *Cancer Treat. Res.* **101**, 97–108.

57. Markowitz, D., Goff, S., and Bank, A. (1988) A safe packaging line for gene transfer: separating viral genes on two different plasmids. *J. Virol.* **62**, 1120–1124.

58. Baum, C., Duellmann, J., Li, Z., Fehse, B., Meyer, J., Williams, D. A., and Von Kalle, C. (2003) Side effects of retroviral gene transfer into hematopoietic stem cells. *Blood* **101**, 2099–2114.

59. Hacein-Bey-Abina, S., von Kalle, C., Schmidt, M., Le Deist, F., Wulffraat, N., McIntyre, E., et al. (2003) A serious adverse event after successful gene therapy for X-linked severe combined immunodeficiency. *N. Engl. J. Med.* **348**, 255,256.

60. Marshall, E. (2003) Gene therapy. Second child in French trial is found to have leukemia. *Science* **299**, 320.

61. Vanin, E. F., Kaloss, M., Broscius, C., and Nienhuis, A. W. (1994) Characterization of replication-competent retroviruses from nonhuman primates with virus-induced T-cell lymphomas and observations regarding the mechanism of oncogenesis. *J Virol* **68**, 4241–4250.

62. Riddell, S. R., Elliott, M., Lewinsohn, D. A., Gilbert, M. J., Wilson, L., Manley, S. A., et al. (1996) T-cell mediated rejection of gene-modified HIV-specific cytotoxic T lymphocytes in HIV-infected patients. *Nat. Med.* **2**, 216–223.

63. Kohn, D., Cornetta, K., Brenner, M., Dunbar, C., and Malech, H. (2003) Report from the ASGT Ad Hoc Committee on Retroviral-Mediated Gene Transfer to hematopoietic Stem Cells, www.asgt.org.

# 46

## Critical Steps in the Implementation of Hematopoietic Progenitor-Cell Gene Therapy Using Ribozyme Vectors

### A Laboratory Protocol

**Maureen P. Boyd, Frances K. Ngok, Alison V. Todd, Geoff P. Symonds, Janet L. Macpherson, and Rafael G. Amado**

### Summary

The implementation of a hematopoietic progenitor-cell gene-therapy program involves the performance of laboratory procedures and compliance with the current code of Good Manufacturing Practices. This chapter explains the multiple laboratory steps used in our recent Phase I gene transfer study for HIV. This study employed a retroviral vector to deliver an anti-HIV ribozyme to CD34+ hematopoietic progenitor cells.

**Key Words:** Ribozyme; anti-HIV agent; gene therapy; CD34+ hematopoietic progenitor cells; hematopoietic stem-cell mobilization; leukapheresis; immunomagnetic separation; cell culture; retroviral vector transduction.

## 1. Introduction

Hematopoietic stem/progenitor cells (HPC) are primitive cells of the hematopoietic system that have the potential—through differentiation into a series of committed precursors—to give rise to all the mature cells of the hematopoietic lineages *(1)*. In humans, HPC can be obtained by bone-marrow collection *(2)*, mobilization into the peripheral blood followed by apheresis *(3)*, or cord blood collection *(4)*. CD34+ cells include a subset of cells that are capable of reconstituting fully myeloablated hosts. For this reason, CD34+ HPC have been used as targets for gene transfer in a variety of human marking studies and therapeutic trials, for indications such as congenital and acquired immunodeficiencies (e.g., HIV infection).

We have used CD34+ HPC in combination with ribozyme-containing retroviral vector transduction to conduct a Phase I clinical trial in HIV-infected

From: *Methods in Molecular Biology, vol. 252: Ribozymes and siRNA Protocols, Second Edition*
Edited by: M. Sioud © Humana Press Inc., Totowa, NJ

patients *(5)*. The trial involved the following steps, along with the relevant method number used within this chapter. Here, we describe the laboratory procedures that are most critical for the implementation of HPC-based gene-therapy studies.

1. Manufacture of retroviral vector. This step is required prior to initiation of any clinical study.
2. Mobilization of HPC from the bone marrow into the peripheral blood by the use of filgastrim (granulocye-colony-stimulating factor [G-CSF]).
3. Harvest of the mononuclear cell population by apheresis.
4. Purification of the HPC by CD34 antigen binding.
5. Culture of the CD34+ HPC.
6. Retroviral transduction with a ribozyme-containing vector.
7. Re-infusion of final product into the subject.
8. Determination of the percentage of cells that contain the retroviral vector.
9. Patient monitoring and peripheral-blood sample collection.
10. Analysis of peripheral-blood samples for gene-containing cells.

## 2. Materials
### 2.1. Laboratory Requirements and Equipments

All cell-culture procedures should be performed using an aseptic technique in a Biological Safety Cabinet. Procedures should follow Good Clinical Practice Guidelines and other applicable regulatory guidelines, including aspects of Good Tissue Practice and current Good Manufacturing Practice (GMP). All laboratories should establish Standard Operating Procedures, and should document procedures using Batch Records. In the case of human immunodeficiency virus (HIV) containing samples, they should be handled using Universal Precautions for blood-borne pathogens, and local regulations may require that they be manipulated within a suitable containment facility such as an authorized BSL3 (US) or PC3 Facility (Australia).

1. COBE Spectra Apheresis System (Gambro BCT, Lakewood, CO) or similar system.
2. Biological safety cabinet.
3. Centrifuge for spinning blood bags and tubes.
4. Standard $CO_2$ cell-culture incubator 37°C, 5%$CO_2$.
5. Phase-contrast microscope for viewing cell cultures.
6. Hemocytometer.
7. −80°C Ultrafreezer.
8. Refrigerator.
9. 37°C water bath.
10. Dry block heater.
11. Flow cytometer capable of two-color analysis, preferably three-color to allow phenotypic characterization of blood cells.

12. Isolex 300i (Baxter).
13. Serological pipets and Pipet Aid.
14. Automatic pipets.
15. Magnetic antibody cell separation (VarioMACS) system (Miltenyi Biotech Inc., Auburn, CA) or similar system.
16. Spectrophotometer.
17. Phosphorimager and Imagequant Software (Molecular Dynamics).
18. Thermocycler: PE2400 thermocycler (Perkin-Elmer).

## 2.2. Manufacturer of Retroviral Vectors

1. Dulbecco's modified Eagle's medium (DMEM), Iscove's modified Dulbecco's medium (IMDM), and FBC culture media.
2. 0.2-μm filter.
3. Bottles for storage of (VCM) (e.g., PTFE).

## 2.3. Isolation of CD34+ HPC

1. Phosphate-buffered saline (PBS), 3-L bag.
2. Sodium citrate anticoagulant, 360 mL.
3. Human serum albumin, 25% solution, 120 mL.
4. Isolex 300i magnetic cell separator disposable set, Baxter R4R9850.
5. Isolex stem cell reagent kit, Baxter R4R9752.
6. Magnetic particle concentrator (MPC-1), Dynal.
7. Intravenous (IV) immunoglobulin injection (Gammagard), such as Baxter 060384.
8. Syringes and needles or plastic cannulas.
9. Centrifuge tubes or Transfer packs.
10. Disposable serological pipets (various sizes).
11. Aerosol resistant pipet tips (various sizes).

## 2.4. Culture of the CD34+ HPC

1. T-175 tissue-culture flasks, or gas permeable bag (Baxter LifeCell).
2. IMDM (Gibco-BRL).
3. Fetal bovine serum (FBS) (Gibco-BRL).
4. Stem-cell factor (SCF) (Amgen, Thousand Oaks California).
5. Megakaryocyte growth and development factor (MGDF) (Amgen).
6. Nevirapine (Boehringer Ingelheim).
7. Syringes and needles or plastic cannulas.
8. Sterile centrifuge tubes or transfer packs.
9. Disposable serological pipets (various sizes).
10. Aerosol resistant pipet tips (various sizes).

## 2.5. Retroviral Transduction With a Ribozyme-Containing Vector

1. T-175 flasks (non-tissue-culture-treated), or gas-permeable bag (Baxter LifeCell).
2. IMDM (Gibco-BRL.)

3. FBS (Gibco-BRL).
4. SCF (Amgen).
5. MGDF (Amgen).
6. Nevirapine (Boehringer Ingelheim).
7. Syringes and needles or plastic cannulas.
8. 0.2-μm low-protein-binding syringe filter.
9. Centrifuge tubes or Transfer packs.
10. Disposable serological pipets (various sizes).
11. Aerosol resistant pipet tips (various sizes).
12. RetroNectin (Takara, Japan).
13. Sterile water for injection.
14. Retroviral supernatant (VCM), Manufactured for each study, 100–200 mL for each transduction.

## 2.6. Preparation of Cells for Re-Infusion Into Subject

1. PBS.
2. Centrifuge tubes or bags.
3. RPMI (Gibco-BRL).
4. Human serum albumin, 25%, 60 mL.
5. Mycoplasma polymerase chain reaction (PCR) kit (Roche).
6. Endotoxin (LAL) chromogenic assay (Biowhittaker QCL-1000).
7. Gram stain kit (Difco-BBL).
8. Gram stain QC slides.
9. Methylcellulose (Stem Cell Technologies).
10. G418 (geneticin) (Gibco-BRL.)
11. Cryogenic vials (1.8 mL).
12. Test tubes (various sizes for collection of samples for product testing and storage).
13. Anti-CD34-PerCP antibody.
14. Anti-CD45-FITC antibody.
15. Anti-CD38-PE antibody.
16. Isotype control antibodies.
17. Aerobic and anaerobic bacterial-culture bottles.

## 2.7. Determination of Transduction of CD34+ HPC

1. PBS.
2. Oligonucleotide primers.
- 5L1A (5'-GAG TTC TAC CGG CAG TGC AAA-3')
- 3L2A (5'-CAC TCA TGA GAT GCC TGC AAG-3')
- 5Nes1 (5'-GAT CCC CTC GCG AGT TGG TTC A-3')
3. PreTaq (Boehringer Mannheim).
4. Lysis buffer: 10 m$M$ Tris-HCl, 0.4% Triton-X, 50 m$M$ KCl, 3 m$M$ CaCl$_2$. (Prepared by adding 50 mL 100 m$M$ Tris-HCl, pH 8.0, 20 mL 10% (v/v) Triton

X-100, 50 mL 500 m$M$ KCl, 30 mL 50 m$M$ CaCl$_2$, and distilled water to final volume of 500 mL).

5. ACES buffer, pH 6.8 (prepared by adding 2.2 g Aces buffer [Sigma], 12.5 mL 0.5 $M$ NaOH, 12.5 mL Tween 20, concentrated HCl to adjust pH and make up to 50 mL).
6. 20 m$M$ NaOH.
7. 25 m$M$ MgCl$_2$ (Perkin Elmer.)
8. Ultrapure deoxynucleotide 5' triphosphate (dNTP) set (Pharmacia Biotech).
9. RTS T4 kinase labeling system (Gibco-BRL).
10. Buffer II (Perkin-Elmer, CA).
11. *AmpliTaq* DNA polymerase (Perkin-Elmer).
12. Wax beads (Ampliwax PCR Gem 100; Perkin Elmer).
13. 200 μL MicroAmp reaction tube with cap (Perkin Elmer), or microwell plates for PCR.
14. RNA-loading dye.
15. 5% denaturing polyacrylamide gel.

## 2.8. Isolation of Patient Peripheral-Blood-Cell Populations for Evaluation of Gene Marking

1. Ficol-Hypaque (Amersham).
2. One-step polymorphs (Accurate Chemical and Scientific Corporation, Westbury, NY).
3. Giemsa stain.
4. PBS with 10% FBS (MACS wash buffer).
5. MACS Microbeads for CD3, CD4, CD8, CD14 (Miltenyi Biotech Inc., Auburn, CA).
6. MACS Multisort Microbeads for CD4, CD8 (Miltenyi).
7. MACS MS and LS Columns (depending on cell numbers to isolate) (Miltenyi).
8. CD45RA fluorescein isothiocyanate (FITC) (Becton Dickinson, CA).
9. Anti-FITC Multisort kit (Miltenyi).
10. Murine IgG$_{2a}$ anti-CD62L antibody (Becton Dickinson, CA).
11. Rat anti-mouse IgG$_{2a+b}$ Microbeads (Miltenyi).
12. Sterile tubes (50-mL, 15-mL, 12 × 75 mm, 1.5-mL microcentrifuge).
13. Antibodies: Anti-CD4, -CD8, -CD45, -CD3, -CD14, -CD45RA and -CD62L conjugated to FITC, PE, or PerCP for flow cytometery analysis.

## 2.9. Analysis of Gene-Containing Cells

1. 5Nes1: 5'-GATCCCCTCGCGAGTTGGTTCA-3'.
2. RNeasy Blood Mini Kit (Qiagen).
3. Reverse transcriptase (Superscript, Gibco-BRL #18053-017).
4. RNasin (Promega).
5. RQ-1 DNase (Promega, cat. no.M6101).
6. Multicore buffer (Promega).
7. 25 m$M$ MgCl$_2$ ·

## 3. Methods

### 3.1. Manufacture of Retroviral Vectors

Retroviral vectors are made as plasmid DNA by cloning. The production of these vectors is not the subject of this chapter, and numerous studies have explored this topic *(6,7)*. The current method is summarized briefly, as follows: For the production of RRz2, the ribozyme Rz2 was cloned into the into a *Sal*-I site in the untranslated region (UTR) of the neomycin phosphotransferase *neo^R* gene of the Moloney murine leukemia virus (MMLV) retroviral vector LNL6. The plasmid is termed pRRz2 and Rz2 is expressed as a *neo^R* -ribozyme transcript from the MMLV long terminal repeat (LTR) *(8)*. The control retroviral vector LNL6 (termed pLNL6 in the plasmid form) is also used to control for ribozyme-specific effects on HPC engraftment and T-lymphoid development. The retroviral producer-cell lines were prepared in a two-stage process by transfecting the cDNA constructs, pLNL6, or pRRz2, into the ecotropic packaging-cell line to produce populations of ecotropic replication-incompetent virus. This population is then used to infect the PA317 or other amphotropic packaging-cell line, which becomes the vector-producer-cell line.

### 3.2. Manufacture of GMP Retroviral Vectors

1. Producer-cell lines are expanded in T225 tissue-culture flasks or roller bottles (Corning or Costar) at $36°C/5\%$ $CO_2$ using DMEM (Gibco-BRL) containing 10% FBS. Supernatant containing retroviral vector (VCM) is collected at defined periods—e.g., overnight.
2. Viral titers can be determined by infecting the NIH-3T3 cell line using serial dilutions of retroviral culture supernatant (VCM) and scoring G418 resistance. Generally, titers should be above $1 \times 10^6$ infectious viral particles/mL. Titers for LNL6 and RRz2 GMP/clinical-grade retroviral vectors were $1.4 \times 10^7$ and $0.8 \times 10^7$ infectious viral particles/mL, respectively. The clinical-grade virus-production material should be tested and certified following a plan determined with the relevant regulatory agencies. An example plan is shown here, and details of the individual tests can be obtained from an FDA-approved testing laboratory.
3. Preliminary screening of candidate seed bank to ensure integrity of construct and freedom from bacterial and Mycoplasma contaminants. In addition, the optimal cell-seeding density and timing for collection of viral supernatant should be analyzed for the following:
   a. Sterility (direct inoculation).
   b. Mycoplasma culture.
4. Expansion of cells, preparation of GMP Master Cell Bank (MCB), and certification testing involves the following:
   a. Sterility (direct inoculation).
   b. Mycoplasma culture.

    c. Isoenzyme analysis.

    d. In vivo assay for adventitious viral contaminants.

    e. In vitro assay for adventitious viral contaminants.

    f. RCR performed by co-cultivation of cells.

    g. RCR performed by amplification of culture supernatant.

    h. XC plaque assay for Ecotropic virus.

    i. Adventitious bovine viral agents.

    j. Adventitious porcine viral agents.

5. Virus production pilot run to validate production methods, check viral titer and key safety parameters, particularly replication-competent retrovirus (RCR).

    a. Viral titer.

    b. RCR (co-cultivation or amplification).

6. GMP production of clinical lot virus. This involves processing, packaging in labeled vials, and certification testing.

    a. Membrane filtration.

    b. General safety test.

    c. Bacterial endotoxin (Limulus Amebocyte Lysate).

    d. Sterility, direct inoculation.

    e. Mycoplasma culture.

    f. Isoenzyme analysis.

    g. In vivo assay for adventitious viral contaminants.

    h. In vitro assay for adventitious viral contaminants.

    i. RCR by co-cultivation of the End of Production cells.

    j. RCR performed by amplification of the culture supernatant.

    k. Integrity of product.

    l. Viral titer (potency).

## 3.3. Mobilization of CD34+ HPC

HPC are mobilized from the bone marrow into the peripheral blood by the use of a cytokine growth factor known as G-CSF *(9,10)*. A course of injections is given subcutaneously over a period of 5 or 6 d. G-CSF is available as Filgrastim (Neupogen, Amgen) injection single-dose vials, and should be stored at 2°C to 8°C. The dose is calculated based on body wt (10 μg/kg), and can either be self-administered or given in an ambulatory setting.

## 3.4. Harvest of the Mononuclear-Cell Population by Apheresis

The mononuclear cells, including the HSC, are collected by means of an apheresis procedure using standard apheresis instrumentation and techniques *(11)*. Technical details regarding collection settings and procedures are beyond the scope of this chapter, but the optimal implementation of this step, resulting in a mononuclear-cell-rich collection is critical for downstream cell processing. Acid citrate dextrose is the anticoagulant of choice.

### *3.5. Isolation of CD34+ HPC*

The leukapheresis product is transferred to a cell processing laboratory and washed with PBS. CD34+ selection/purification is then performed using a system such as the automated Isolex 300i magnetic cell-selection system (*see* **Note 1**).

#### *3.5.1. Purification of the HPC by CD34 Antigen Binding*

1. The Isolex 300i system combines a disposable set (incorporating a spinning membrane for cell washing) and a series of pumps, clamps, and weight scales as well as a rotating separation chamber and variable magnet position. Positive selection with the Isolex 300i relies on antigen/antibody interactions to specifically capture and release CD34+ cells. The entire selection process is controlled by software supplied with the instrument, and takes about 4 h.
2. Add 360 mL sodium citrate and 120 mL of 25% human serum albumin solution to a 3-L PBS buffer bag to prepare the Isolex working buffer.
3. Install the Isolex disposable set, and add buffer and reagents supplied in the reagent kit to the bags when prompted by the Isolex Software. Allow set to prime with the working buffer.
4. Pre-incubate apheresis product with the IV immunoglobulin solution. The immunoglobulin binds to any Fc receptors on the surface of cells, and thus reduces the likelihood of the murine CD34+ antibody binding nonspecifically to CD34 negative cells via the Fc receptor.
5. Initially, the cells are washed through the spinning membrane and filtered wash bag to remove platelets.
6. Cells are sensitized with a murine anti-human monoclonal antibody (MAb) directed against the CD34 antigen.
7. Any remaining unbound murine antibody is removed by washing.
8. The cells are incubated with immunomagnetic beads coated with polyclonal sheep anti-mouse antibody. Rosettes form between the beads and the CD34+ cells based on the strong antibody bridge.
9. The bead/CD34+ cell rosettes are then separated from the unbound non-CD34+ cells in the separation chamber using the array of magnets. The bead/CD34+ rosettes are held within the magnetic field as non-target cells are washed away.
10. The CD34+ cells remaining in the separation chamber are released using a stem cell-releasing agent (an octapeptide with high affinity for the CD34 antibody). The PR34 peptide competes for the antibody-binding site and effectively displaces the CD34+ cells from the beads.
11. The resulting bead/antibody/peptide complexes are magnetically held within the separation container while the released CD34+ cells are washed and pumped into a separate collection container.
12. Purity of the CD34+ cells is established by flow cytometry, and should be >50%.
13. Typical cell numbers should be of the order of $2-5 \times 10^6$ CD34+ HPC/kg.

### 3.6. Culture of the CD34+ HPC

CD34+ HPC are cultured at 37°C, 5% CO2 in humidified air at a maximum density of $5 \times 10^5$ in T-175-cm flasks from completion of CD34+ cell selection on d 6 to the beginning of d 8 when they are transduced with retroviral vector. The CD34+ culture medium consists of IMDM, supplemented with 10–20% FBS, 50 ng/mL SCF, 100 ng/mL MGDF (*see* **Note 2**) and Nevirapine at 500 n$M$ (to inhibit potential HIV replication).

1. To each 500 mL of IMDM add the following:
   a. 50–100 mL of FBS.
   b. 0.5 mL of 50 μg/mL 1000X stock solution of SCF (diluted in medium and stored at –70°C).
   c. 1 mL of 100 μg/mL 1000x stock solution MDGF (diluted in medium and stored at –70°C).
   d. 13.3 μL of 18.7 m$M$ (5 mg/mL) stock solution of Nevirapine (diluted in ethanol and stored in single-use aliquots at –20°C).
2. The CD34+ HPC are gently resuspended in the complete IMDM at a density of less than $5 \times 10^5$ cells/mL and transferred to cell-culture flasks or gas permeable culture bags.
3. The cells are cultured at 37°C in a humidified atmosphere of 5% $CO_2$ for 1–3 d to allow the cells to enter cell cycle and become susceptible to infection by the retroviral vector.

### 3.7. Retroviral Transduction With a Ribozyme-Containing Vector

Murine-based retroviral vectors can only infect cells in cell cycle. CD34+ HPC are quiescent or resting cells, and they must be stimulated to effect integration by retroviral vectors. The use of SCF and MGDF has been demonstrated to allow cells to enter cell cycle while maintaining long-term proliferative potential *(12)*. In order to maximize transduction, culture vessels should be coated with the CH296 fragment of human fibronectin, RetroNectin *(13)*. The addition of RetroNectin to the transduction protocol greatly enhances the efficiency with which the VCM can infect the target CD34+ HPC. It acts to co-localize the viral particles and cells on the surface of the culture vessel (**Fig. 1**). VCM is usually added twice on d 8 approx 6 h apart. Additional transductions can be performed on subsequent days at intervals of up to 24 h. VCM must be supplemented with cytokine growth factors before adding to the CD34+ cell cultures. Although it is best if the VCM is thawed immediately before use, VCM that has had cytokines added may be stored at 2–8°C overnight for use the following day if required. Avoid freeze-thaw cycles, as this will lead to a loss of viral infectivity.

1 RetroNectin is prepared as a 25 μg/mL solution in PBS, and is stored frozen.
   a. Reconstitute RetroNectin powder in sterile water at 1 mg/mL by the addition of 2.5 mL water for injection to 2.5 mg powder in a vial.

Without retronectin

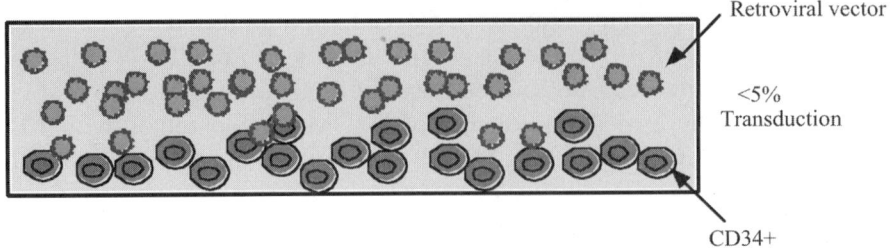

With retronectin

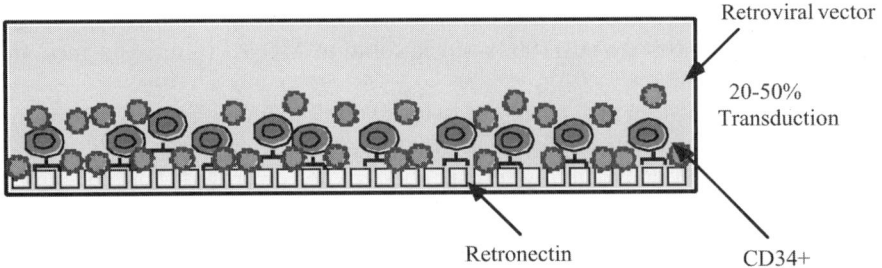

Fig 1. Effect of RetroNectin on transduction. Schematic representation of how the fibronectin protein CH296 (RetroNectin) facilitates retroviral transduction by bringing the CD34+ cells into close proximity with the retroviral vector. The percentages indicated are those reported commonly in the literature and demonstrated in our own laboratory.

      b. Dilute to 25 µg/mL by adding 100 mL PBS to the RetroNectin in a sterile container. Filter through a 0.2-µm low-protein-binding syringe filter prior to use. RetroNectin solution may be stored frozen for several months at –20°C, and thawed prior to use.

2. Add 25 mL of RetroNectin solution to T175 non-tissue-cultured-treated flasks. Allow to coat with RetroNectin at 3.5 µg/cm$^2$ by incubating at room temperature for 2 h. Alternatively, flasks may be coated overnight at 2–8°C. For each VCM, 2X T175 flasks are required.

3. RetroNectin solution is removed by aspiration, and the flasks are blocked with a 2% human serum albumin (HSA) or FBS in PBS solution for 30 min, then rinsed with IMDM. Flasks may be used immediately or stored at 2–8°C for 2–3 d.

4. Thaw 200-mL GMP-grade VCM (LNL6 and Rz2) by warming at 37°C in a water bath with gentle agitation.

5. Prepare VCM by the addition of culture supplements. Filter through 0.2-µm low-protein-binding syringe filter prior to use.

      a. FBS to final concentration of 20%; for 200 mL VCM, add 20 mL FBS (VCM already contains 10% FBS.)

  b.  0.2 mL of 50 µg/mL 1000X stock solution of SCF.
  c.  0.4 mL of 100 µg/mL 1000X stock solution MDGF.
  d.  5.3 µL of 5 mg/mL (18.7 m*M*) 37.4X stock solution of Nevirapine.
 6.  Collect pre-stimulated CD34+ cells from culture flask by gently mixing the cell suspension using a 25-mL pipet, then transfer to a tube that is suitable for centrifugation. Centrifuge tubes at 200*g* for 10 min.
 7.  Aspirate supernatant from cell pellets, and resuspend in supplemented and filtered VCM. The first transduction is with 50 mL VCM per flask. Culture cells at 37°C in 5%CO2 in a humidified atmosphere for approx 6 h. Culture can proceed for a period of up to 24 h.
 8.  At the completion of the first transduction, collect cells from culture flask by gently mixing the cell suspension, using a 25-mL pipet, then transfer to a tube that is suitable for centrifugation. Centrifuge tubes at 200*g* for 10 min.
 9.  Aspirate supernatant from cell pellets (retain), and resuspend cells in supplemented and filtered VCM. The second transduction is with 50 mL VCM per flask. Culture cells at 37°C in 5% CO2 in a humidified atmosphere for approx 16–18 h. Culture can proceed for a period of up to 24 h.
 10.  Samples of the retained culture supernatant may be submitted to overnight sterility culture, which can be used as an indicator on the day of harvest that the product is free of contamination (*see* **Note 3**).
 11.  A small sample of the supplemented VCM should be retained for pre-culture evaluation of HIV p24 content.

## 3.8. Preparation of Cells for Re-Infusion Into the Subject

Following transduction, cells are left in culture until the morning of d 9, and at this time, they are harvested and subjected to a series of release tests for evaluation of the purity and potency of the infused cell dose. The pre-infusion safety testing comprises a gram stain for bacteria as well as endotoxin (LAL) assay. A Mycoplasma test can also be performed. These tests must be negative in order to proceed to infusion, and results should be checked and signed off by a Quality Assurance individual. Quality Control measures should be built into the Standard Operating Procedures for each of these tests. The following is merely a summary of the type of tests required and how they can be performed. It is not intended to substitute for well-written Standard Operating Procedures and Quality Control. Sterility testing for both aerobic and anaerobic bacteria, and fungal contamination and for Mycoplasma culture, are set up at the time of preparation of the infusion product, but results will not be available until after the infusion has been completed.

 1.  Prepare CD34+ HPC wash buffer by adding 40 mL of 25% HSA to 1 L PBS.
 2.  Prepare infusion buffer by aseptically adding 80 mL RPMI without phenol red and 20 mL 25% HSA solution to give a final concentration of 5% HSA.
 3.  On the morning of d 9, remove cultures from incubator and collect cells from culture flask by gently mixing the cell suspension using a 25-mL pipet, then

transfer to a tube suitable for centrifugation. Cells may be difficult to remove from flasks coated with RetroNectin. If more vigorous pipetting still fails to dislodge the cells, a non-enzymatic cell-dissociation solution may be used. All cells should be pooled together for washing.

4. Collect samples for release testing. At this point, samples are collected for Mycoplasma testing and CD34+ estimation.

5. A rapid test for the presence of mycoplasma RNA can be performed using a commercial kit such as the Roche PCR kit. If the test is completed rapidly, it can be used as a release test (*see* **Note 4**).

6. Centrifuge tubes at 200$g$ for 10 min. Aspirate supernatant spent culture medium and retain for collection of additional samples for testing.

7. 5% of the culture supernatant should be retained for biological RCR bioassay using amplification *Mus dunni* cells if the transduced cells have been cultured for more than 4 d. The sample is stored frozen at –80°C until tested by a FDA-approved testing laboratory, and it is not a release criterion.

8. Prepare a smear of culture supernatant and air-dry and fix by passing through a flame. Stain using Gram stain with appropriate controls. This is a release test, and no organisms should be seen.

9. LAL is assessed using the chromogenic assay. This is a release test. A limit of 5 EU/kg can be infused within a 1-h period.

10. HIV *p24* antigen enzyme-linked immunosorbent assay (ELISA) in pre- and post-CD34+ cell transduction supernatants, to ensure that HIV replication has not occurred during the culture period. Although this test is not required for release of the infusion product, it is very useful as an indicator of HIV activation and replication during the period of cell culture and transduction.

11. The transduced CD34+ cells are washed three times in 200 mL of CD34+ HPC wash buffer (centrifuged for 15 min at 200$g$, aspirate supernatant, resuspend in 200 mL CD34+ HPC wash buffer) to ensure that VCM and other culture additives are removed from the cell infusate.

12. Resuspend in 100 mL RPMI without phenol red supplemented with 5% human serum albumin (prepared by adding 20 mL 25% HSA to 80 mL RPMI) and store at room temperature until ready for infusion.

13. A small sample is removed from the cell-infusion product for estimation of the cell count. The total cell dose is calculated, and the dose of CD34+/kg is determined.

14. 1% of the cell infusion product is cryopreserved for RCR assay by co-cultivation with *Mus dunni* cells followed by S+L- end point assay. This assay is required if cells have been cultured for more than 4 d.

15. A small number of CD34+ cells are cultured in the methylcellulose colony-forming unit (CFU-GM) assay for evaluation of HPCs in the presence or absence of 1 mg/mL G418. $10^5$ cells are plated in 35-mm gridded culture dishes with and without G418 at 1 mg/mL final concentration, and cultured at 37°C/5% $CO_2$ in a humidified atmosphere for 14–18 d. Colonies are counted and scored on the basis of morphology. The proportion of progenitor cells that have been transduced can be determined by comparison of the number of colonies in methylcellulose in the

presence of G418 (transduced cells) and absence of G418 (transduced and non-transduced cells).

16. Following safety assessments, cells can be infused into autologous recipient patients without myelosuppression. It is critical that the intended recipient is carefully identified.

## 3.9. Determination of Transduction of CD34+ HPC

Transduction efficiencies can be determined by performing quantitative competitive PCR on an aliquot of CD34+ HP cells *(14)*.

### 3.9.1. Quantitative Competitive PCR for Evaluation of Transduction Efficiency

1. Primers 5L1A (5'-GAG TTC TAC CGG CAG TGC AAA-3') and 3L2A (5'-CAC TCA TGA GAT GCC TGC AAG-3') amplify a region of the *neo*$^R$ gene within LNL6 in which the therapeutic gene, *Rz2*, is cloned.

2. Amplification of the control vector (LNL6) produces a 174-bp amplicon, and amplification of the anti-HIV ribozyme (RRz2) produces a 216-bp amplicon. A competitor template has been constructed from Rz2 so that the primer binding sequences are identical, but the amplicon produced is smaller (142 bp) than both the LNL6 and RRz2 amplicons *(14)*.

3. Standards (controls with known percentages of LNL6- and RRz2-containing cells) are produced by spiking unmarked CEMT4 cells (a human T-lymphoblastoid cell line from NIH AIDS Research and Reference Program) with known quantities of CEM T4 cells that were transduced with PA317/LNL6 and CEM T4 cells that were transduced with PA317/RRz2. The transduced cells have been selected and grown in the presence of G418, and generally contain one copy of the transgene per cell.

4. DNA from standards and CD34+ HPC aliquots is isolated from $5 \times 10^6$ cells using the Acest Polymer DNA extraction method *(15)* or Qiagen DNA extraction kit. A range of standards can be included in each PCR run to allow direct comparison of unknown CD34+HPC with the standard curve.

5. DNA (150 ng) is amplified in the presence of a series of known competitor concentrations under the following conditions:
   a. 0.2 m$M$ each dNTP (Pharmacia Biotech),
   b. 0.4 m$M$ each primer (5L1A radioactively end-labelled with $\gamma^{32}$P-deoxyadenosine 5' triphosphate (dATP) using the RTS T4 kinase labeling system [Gibco-BRL]),
   c. 2 m$M$ MgCl$_2$, 10 m$M$ Tris-HCl, pH 8.3, 50 m$M$ KCl, and
   d. 0.25 U *AmpliTaq* DNA polymerase (Perkin-Elmer).
   e. Reactions are 50 μL and include a wax bead (Ampliwax PCR Gem 100; Perkin-Elmer) to facilitate a "hot start."
   f. Samples are amplified for one round at 94°C for 3 min followed by 35 cycles of annealing at 68°C for 30 s and denaturation at 94°C for 30 s in a PE2400 thermocycler (Perkin-Elmer).

6. 10 μL of each PCR reaction is added to 5 μL of RNA loading dye, denatured (5 min, 95–100°C) and separated on a 5% denaturing polyacrylamide gel.
7. The gel is then fixed, dried, and exposed on a phosphorimaging screen (Molecular Dynamics, Sunnyvale, CA).
8. Diagnostic amplicons can be analyzed using ImageQuant software (Molecular Dynamics). Following subtraction of background intensity, the diagnostic/competitor ratio for the standard curve is calculated by dividing the sum of the intensity of the two diagnostic bands by the intensity of the competitor band, and this ratio plotted against the known percentages of the standards.
9. The transduction efficiency of unknown samples is calculated by linear regression using Cricket Graph software. In addition, the ratio of LNL6- vs RRz2-containing cells can be determined by dividing the intensity of the LNL6 amplicon by the intensity of the RRz2 amplicon.
10. Alternatively percent transduction can be determined by performing PCR for vector sequences in single colonies grown from the final transduced CD34+ cell product in the absence of G418. In addition, colony numbers in methylcellulose in the presence and absence of 1 mg/mL G418 are a measure of the number of transduced and non-transduced cells. Transduction efficiency is estimated by obtaining the ratio of surviving colonies in the presence of G418 over the number of colonies appearing in the absence of G418.
11. Both methods give similar results.

### 3.10. Isolation of Patient Peripheral-Blood-Cell Populations for Assessment of Gene Marking

Purification of hematopoietic cells from peripheral blood for analysis of vector presence and expression.

1. Granulocytes are isolated from peripheral blood by density-gradient centrifugation using 1-Step Polymorphs. A smear is prepared on a glass slide, and granulocyte purity is evaluated by Giemsa stain and microscopic examination.
2. Peripheral-blood mononuclear cells (PBMC) are isolated from peripheral blood by density-gradient centrifugation on Ficol-Hypaque. Blood is diluted with an equal volume of PBS and layered over Ficol-Hypaque at the ratio of 20 mL Ficol and 30 mL diluted blood. Centrifuge at 1200g for 30 min without brake. Recover the mononuclear cell layer with a pipet and transfer to a 50-mL tube. Wash with an excess of PBS, centrifugation 200g for 10 minutes.
3. Monocytes and T-lymphocytes are selected from PBMC using the magnetic antibody cell separation (MACS) system (Miltenyi Biotech Inc., Auburn CA) in which cells are labeled with a magnetic antibody (CD14 or CD3 direct MicroBeads, respectively) and passed through a magnetic-positive selection column on the VarioMACS, resulting in a negative fraction. The positive fraction can then be eluted. Details of this method are provided in the manufacturer's product insert.
4. T-cells can be separated into subpopulations as follows:
   a. CD4 cells can be separated by positive selection from PBMC using the MACS CD4 multisort bead kit.

b. The negative fraction can then be labeled with the MACS CD8 multisort bead kit and CD8+ cells can be selected.

c. The Multisort bead is removed, and the CD4- and CD8-positive fractions can both be labeled with a CD45RA antibody conjugated to FITC. Cells are then separated using the MACS anti-FITC multisort kit.

d. CD62L selection can also be performed on the CD4/CD45RA- and CD8/CD45RA-positive fractions using a murine $IgG_{2a}$ anti-CD62L antibody, followed by selection with rat anti-mouse $IgG_{2a+b}$ microbeads.

5. Purity is evaluated by flow cytometry. The exact methodology is beyond the scope of this chapter. Use a CD3 antibody to define the T-lymphocytes, CD14 to define monocytes, CD4/CD45RA/CD62L to differentiate naive and memory helper T-cells, and CD8/CD45RA/CD62L differentiate naive and memory cytotoxic T-cells. A total of at least 20,000 events should be acquired, and the percentage of positive cells can be determined using a gating strategy based on scatter properties or CD45 staining *(16)*. Purities for all sub-populations should be greater than 90%, as determined by flow cytometry.

6. Fine-needle aspirates of inguinal or axillary lymph nodes can be performed by a conventional technique using a 22-gage needle. The material is expelled from the syringe into a 15-mL vial containing 10 mL of RPMI plus 1%FCS. The cell suspension is counted and $5 \times 10^6$ cells used for flow cytometry analysis. The remainder can be used to prepare a cytospin, and following flash freezing, DNA for PCR analysis.

## *3.11. Analysis of Gene-Containing Cells*

Determination of vector copy number in mature hematopoietic cells can be performed in PBMC or selected cell populations using a nested competitive PCR method. This method requires DNA extraction from $5 \times 10^6$ cells. Evaluation of expression of vector requires RNA extraction from $10^7$ cells, and can thus only be used if the sample has sufficient cells. The two assays are very similar, involving multiple first-round reactions, pooling samples, and amplification in a second PCR. In the case of RNA, the cDNA is first prepared, pooled, and then subjected to the two rounds of PCR, as for the detection assay. The DNA assay utilizes standards prepared from DNA, and the RNA assay utilizes standards prepared from RNA.

1. PCR analysis of vector presence (DNA) in hematopoietic cells. The range of the standard can be set from 0.01–0.005%.

2. The efficiency of nested PCR should be tested on ratios of 1:1, 1:3.5, 1:5, and 1:10 marked cells (LNL6:RRz2) and the inverse ratio 10:1 (LNL6:RRz2) in a background of normal CEMT4 cells.

3. Normal or non-gene-containing PBMC can be used as a negative control in all PCR assays to screen for any vector contamination during DNA extractions and subsequent PCR reactions.

4. A DNA (or RNA) ratio control is constructed by diluting DNA (or RNA) from CEM T4 transduced with LNL6 and DNA (or RNA) from CEMT4 transduced

with RRz2 at ratios of 1:3.5 or 1:5 (where LNL6 = 1) in a background of PBL (negative) DNA (or RNA) to a concentration of 0.005% marked cells.

5. Refer to **step 10** for the RNA method.

6. Nested (hot start) DNA PCR is performed using primers 5Nes1 and 3L2A in the first round (10 replicates of 1 mg DNA, 17 cycles, annealing 68°C 30 s and denaturation 94°C 30 s). Other conditions are as described in **Method 8**.

7. The second-round PCR is with 5L1A and $P^{32}$ end-labeled 3L2A primer, (pooled replicates, 35 cycles), and other cycling conditions as previously noted.

8. This nested PCR assay has a lower limit of detection of 0.005%, and results are accepted if the ratio control falls within accepted limits set during validation experiments (e.g., for a ratio of 1:3.5 LNL6:RRz2 in a 0.005% vector containing sample, accepted range is 1:1.1 to 1: 6.9). Other acceptance criteria are five or more replicates per PCR reaction, DNA recovery of greater than 70% and 1 μg DNA per replicate.

9. Only PCR results in which the control fell within this range were accepted. We estimated copy number by comparing radioactivity volumes to these controls consisting of 1/25,000, 1/100,000 (for 1:5 dilution controls) and negative control.

10. PCR analysis of vector expression (RNA) in hematopoietic cells is performed using the assay described here, but preceded by a cDNA synthesis step from the cellular RNA.

11. Total cellular RNA is extracted from PBMC using the Qiagen RNeasy kit, following the manufacturer's instructions.

12. Residual DNA is removed by digestion with 5 μL DNase for 15 min at 37°C, and heat inactivation at 75°C for 5 min. 700–1000 ng RNA is reverse-transcribed using 3L2A and Gibco Superscript Reverse Transcriptase, with seven replicates of each sample.

13. Samples are then pooled, and cDNA amplification performed on 10 replicates as described for the DNA detection assay.

14. Expression levels in patient samples should be observed at levels that are within the range of the control dilutions.

## 4. Notes

1. The CliniMACS device from Miltenyi is approved for CD34+ cell selection outside the United States, but the Isolex 300i is the only device approved for use within the United States.

2. Other cytokine combinations can be used, and the ideal combination for your application should be determined empirically. One alternative combination is 300 ng/mL SCF, 300 ng/mL Flt3 ligand, 60 ng/mL IL-3, and 100 ng/mL MGDF.

3. Culture sterility can be performed in several ways. The current requirement for a pharmaceutical product is for the testing to comply with the USP method. However, many cell-processing laboratories prefer to use aerobic and anaerobic blood cultures performed in their local accredited microbiology/pathology testing laboratory.

4. There are several commercially available tests for the detection of Mycoplasma species commonly found in cell culture. When selecting which test you will use, consider the sensitivity of the method, the number of species detected, and the length of time required to obtain a result.

## References

1. Lowry, P. A. and Tabbara, I. A. (1992) Peripheral hematopoietic stem cell transplantation: current concepts. *Exp. Hematol.* **20,** 937–942.
2. Srour, E. F., Brandt, J. E., Briddell, R. A., Leemhuis, T., van Besien, K., and Hoffman, R. (1991) Human CD34+ HLA-DR- bone marrow cells contain progenitor cells capable of self-renewal, multilineage differentiation, and long-term in vitro hematopoiesis. *Blood Cells* **17,** 287–295.
3. Hohaus, S., Goldschmidt, H., Ehrhardt, R., and Haas, R. (1993) Successful autografting following myeloablative conditioning therapy with blood stem cells mobilized by chemotherapy plus rhG-CSF. *Exp. Hematol.* **21,** 508–514.
4. Wagner, J. E., Broxmeyer, H. E., Byrd, R. L., Zehnbauer, B., Schmeckpeper, B., Shah, N., et al. (1992) Transplantation of umbilical cord blood after myeloablative therapy: analysis of engraftment. *Blood* **79,** 1874–1881.
5. Amado, R. G., Mitsuyasu, R. T., Symonds, G., Rosenblatt, J. D., Zack, J., Sun, L. Q., et al. (1999) A phase I trial of autologous CD34+ hematopoietic progenitor cells transduced with an anti-HIV ribozyme. *Hum. Gene Ther.* **10,** 2255–2270.
6. Miller, D. G., Adam, M. A., and Miller, A. D. (1990) Gene transfer by retrovirus vectors occurs only in cells that are actively replicating at the time of infection. *Mol. Cell. Biol.* **10,** 4239–4242.
7. Miller, A. D. and Buttimore, C. (1986) Redesign of retrovirus packaging cell lines to avoid recombination leading to helper virus production. *Mol. Cell. Biol.* **6,** 2895–2902.
8. Sun, L. Q., Pyati, J., Smythe, J., Wang, L., Macpherson, J., Gerlach, W., and Symonds, G. (1995) Resistance to human immunodeficiency virus type 1 infection conferred by transduction of human peripheral blood lymphocytes with ribozyme, antisense, or polymeric trans-activation response element constructs. *Proc. Natl. Acad. Sci. USA* **92,** 7272–7276.
9. Schwartzberg, L. S., Birch, R., Hazelton, B., Tauer, K. W., Lee, P., Jr., Altemose, R., et al. (1992) Peripheral blood stem cell mobilization by chemotherapy with and without recombinant human granulocyte colony-stimulating factor. *J. Hematother.* **1,** 317–327.
10. Slobod, K. S., Bennett, T. A., Freiden, P. J., Kechli, A. M., Howlett, N., Flynn, P. M., et al. (1996) Mobilization of CD34+ progenitor cells by granulocyte colony-stimulating factor in human immunodeficiency virus type 1-infected adults. *Blood* **88,** 3329–3335.
11. Alcorn, M. J., Farrell, E., Barr, J., Pearson, C., Green, R., and Holyoake, T. (2000) The number of CD34+ cells mobilized into the peripheral blood can predict the quality of subsequent collections. *J. Hematother. Stem Cell Res.* **9,** 89–93.
12. Amado, R. G., Symonds, G., Jamieson, B. D., Zhao, G., Rosenblatt, J. D., and Zack, J. A. (1998) Effects of megakaryocyte growth and development factor on survival and retroviral transduction of T lymphoid progenitor cells. *Hum. Gene Ther.* **9,** 173–183.
13. Williams, D. A. and Moritz, T. (1994) Umbilical cord blood stem cells as targets for genetic modification: new therapeutic approaches to somatic gene therapy. *Blood Cells* **20,** 504–515.

14. Knop, A. E., Arndt, A. J., Raponi, M., Boyd, M. P., Ely, J. A., and Symonds, G. (1999) Artificial capillary culture: Expansion and retroviral transduction of CD4+ T lymphocytes for clinical application. *Gene Ther.* **6,** 373–384.
15. Ward, R., Hawkins, N., O'Grady, R., Sheehan, C., O'Connor, T., Impey, H., et al. (1998) Restriction endonuclease-mediated selective polymerase chain reaction: a novel assay for the detection of K-ras mutations in clinical samples. *Am. J. Pathol.* **153,** 373–379.
16. Hokland, M., Jorgensen, H., and Hokland, P. (1994) Isolation of peripheral blood mononuclear cells and identification of human lymphocyte subpopulations by multiparameter flow cytometry, in: *Cell Biology: A Laboratory Handbook, vol 1* (Celis, J. E., ed.), Academic Press, San Diego, CA, pp. 179–184.

# Index